# Mixed-Species Forests

Hans Pretzsch • David I. Forrester •
Jürgen Bauhus
Editors

# Mixed-Species Forests

## Ecology and Management

Springer

*Editors*
Hans Pretzsch
Faculty of Forest Science and Resource Management, Chair for Forest Growth and Yield Science
Technical University of Munich
Freising, Germany

David I. Forrester
Swiss Federal Institute for Forest, Snow and Landscape Research WSL
Birmensdorf, Switzerland

Jürgen Bauhus
Institute of Forest Sciences, Chair of Silviculture
University of Freiburg
Freiburg, Germany

ISBN 978-3-662-54551-5 ISBN 978-3-662-54553-9 (eBook)
DOI 10.1007/978-3-662-54553-9

Library of Congress Control Number: 2017942726



Cover photo: Mixed mountain forest near Kreuth in the Bavarian Alps. Photo Leonhard Steinacker

Printed on acid-free paper

This Springer imprint is published by Springer Nature
The registered company is Springer-Verlag GmbH Germany
The registered company address is: Heidelberger Platz 3, 14197 Berlin, Germany

# Preface

Mixed-species stands are widely thought to provide many forest functions and services to a higher level than monocultures. However, the scientific foundations of forest management, from analyses and measurements to modelling and knowledge application, have been largely derived from monocultures. More recently, there has been a growing body of evidence showing that the effects of mixing different tree species on ecosystem functions are not simply a combination of the effects of each participating species in monoculture. It is precisely this question, how the performance of polycultures compares with that of monocultures or more simple forests, that determines whether forest owners decide for or against establishing and maintaining mixed-species stands. However, the available knowledge was so far rather fragmented and could not provide a holistic picture of mixed-species forests. Until now, a textbook has been missing to synthesise the existing knowledge and provide answers to questions such as:

Which concepts of the dynamics and management of mixed-species stands are available, how generalisable are they, and what kind of experiments are necessary to develop them further?

How do mixed-species stands compare with monocultures in relation to productivity, ecological stability, and wood quality?

How do mixing effects on ecosystem functioning depend on the species composition, site conditions, and stand structure?

How does any over- or underyielding at the level of forest stands emerge from the tree and organ level, and what are the main mechanisms behind the mixing effects?

How can any facts, findings, and models in mixed-species stands be efficiently integrated in silviculture, management, and planning in forest practice?

Do ecological characteristics of mixed-species stands also translate into economic differences compared with monocultures?

These and other questions are in the focus of this book. In addition, it will address experimental designs and analytical approaches required to study mixed-species forests and provide empirical facts and findings, general relationships, models, and management tools. These differ in many ways from established concepts developed for monocultures. However, our work for this book has also identified many open questions. To better understand, model, and manage mixed-species forests, many additional measurements of resources and environmental factors, tree and stand structures, and growth and allocation are required. The ambition is to measure and quantify as many relevant system attributes as possible. Those variables which are not yet sufficiently measured or accessible are often replaced by nonspecific proxies such as site index, growing space, real niche, or crown volume. We hope that these gaps can be filled in the future to develop a more process-based understanding of mixed-species forests.

The content of this book is meant for students, scientists, lecturers, forest planners, forest managers, forest experts, and consultants.

The editors and main authors of the book Hans Pretzsch (HP), David Forrester (DF), and Jürgen Bauhus (JB) could not have completed the book without their co-authors Thomas Knoke (TK) and Thomas Rötzer (TR) from the Technical University of Munich, Germany. Several others have provided valuable contributions to individual chapters, where this is indicated.

The first seven chapters of the book synthesise the existing empirical knowledge about mixed-species forest and its theoretical underpinning development. Chapter 1 (Mixed-Species Forests: The Development of a Forest Management Paradigm. JB, DF, and HP) summarises the background and approaches of research in mixed-species stands. Chapter 2 (From Observations to Evidence About Effects of Mixed-Species Stands. JB, DF, and HP) presents theoretical concepts and hypotheses related to mixed-species and diverse forests and discusses approaches to study their ecological functioning and dynamics. Chapter 3 (Ecological and Physiological Processes in Mixed Versus Monospecific Stands. DF) introduces the basic processes behind any emergent properties of mixed-species stands compared with monocultures. Chapter 4 (Stand Dynamics of Mixed-Species Stands Compared with Monocultures. HP and DF), Chap. 5 (Size-Structure Dynamics in Mixed Versus Monospecific Stands. HP), and Chap. 6 (Individual Tree Structure and Growth in Mixed Compared with Monospecific Stands. HP) trace mixing effects from the stand level to community, population, tree, and organ level. The overview of mixed-species stand development is completed by Chap. 7 (Ecological Stability of Mixed-Species Forests. JB, DF, HP, and others) which analyses whether mixtures are more resistant and resilient in relation to abiotic and biotic stress and disturbance.

Chapter 8 (Modelling Mixed-Species Forest Stands. HP, TR, and DF) describes common models for monospecific stands and discusses the environmental conditions, processes, and structures that need to be included in forest growth models to be applied to mixed-species forests. The focus is on how these different processes are incorporated into models and the strengths and weaknesses of tree-level and stand-level approaches.

The next three chapters are dedicated to the integration of biophysical knowledge into silvicultural practice, forest management planning, and economic analyses. Chapter 9 (Silvicultural Options for Mixed-Species Stands. JB, DF, HP, and others) discusses approaches and guidelines for natural and artificial establishment, tending, thinning, and regeneration of mixed-species stands. Whereas the focus of the first chapters is at the stand level, application of the results in forest planning and management, in contrast, requires extension to the enterprise or landscape level, which is discussed in Chaps. 10 and 11 (Forest Management Planning in Mixed-Species Forests. HP and TK. Economics of Mixed Forests. TK). The book ends with Chap. 12 (Perspectives for Future Research Directions on Mixed-Species Systems. HP, DF, JB, TK, and Carola Paul) which presents identified knowledge gaps and perspectives for future research. A glossary including the key terms complements this first textbook on mixed-species forests.

When writing this book, Hans Pretzsch, Thomas Knoke, and Thomas Rötzer could build on their research and lectures at the Ludwig Maximilian University and Technical University of Munich. Jürgen Bauhus and David Forrester based their contributions on their research and teaching at the Albert-Ludwigs-University of Freiburg. This book is dedicated to all students, researchers, and colleagues at the above-mentioned institutions who have contributed to the realisation of this book. In addition, we are grateful to Hervé Jactel, Ramón Vallejo, Barry Gardiner, Adam Felton, Patrick Pyttel, Adam Benneter, and Carola Paul who made specific contributions to Chaps. 7, 9, and 12.

For the review of substantial parts and individual chapters of the manuscript, we are deeply indebted to Dan Binkley of Colorado State University, USA; Jette Bredahl Jacobsen of the University of Copenhagen, Denmark; Hubert Sterba and Manfred Josef Lexer both of the University of Natural Resources and Life Sciences, Vienna, Austria; Jerry Vanclay of Southern Cross University, Lismore, Australia; and Charles Nock of the University of Freiburg.

We further thank many students for prereviewing the manuscripts, Gerhard Schütze for data analyses, Ulrich Kern for the graphical artwork, Emily Kilham and Astrid Moser for language revision of some chapters, and Germar Csapek and Beate Felsl for proofreading and assistance with the glossary.

Finally, we also extend our thanks to the editor at Springer Publishing, Christine Eckey, for her reliable and congenial assistance and great patience.

Freising-Weihenstephan and Freiburg
Germany
Spring 2017

Hans Pretzsch
David I. Forrester
Jürgen Bauhus

# Contents


1 **Mixed-Species Forests: The Development of a Forest Management Paradigm** . . . . . 1
Jürgen Bauhus, David I. Forrester, and Hans Pretzsch

2 **From Observations to Evidence About Effects of Mixed-Species Stands** . . . . . 27
Jürgen Bauhus, David I. Forrester, and Hans Pretzsch

3 **Ecological and Physiological Processes in Mixed Versus Monospecific Stands** . . . . . 73
David I. Forrester

4 **Stand Dynamics of Mixed-Species Stands Compared with Monocultures** . . . . . 117
Hans Pretzsch and David I. Forrester

5 **Size-Structure Dynamics in Mixed Versus Monospecific Stands** . . . 211
Hans Pretzsch

6 **Individual Tree Structure and Growth in Mixed Compared with Monospecific Stands** . . . . . 271
Hans Pretzsch

7 **Ecological Stability of Mixed-Species Forests** . . . . . 337
Jürgen Bauhus, David I. Forrester, Barry Gardiner, Hervé Jactel, Ramon Vallejo, and Hans Pretzsch

8 **Modelling Mixed-Species Forest Stands** . . . . . 383
Hans Pretzsch, Thomas Rötzer, and David I. Forrester

9 **Silvicultural Options for Mixed-Species Stands** . . . . . 433
Jürgen Bauhus, David I. Forrester, Hans Pretzsch, Adam Felton, Patrick Pyttel, and Adam Benneter

10 **Forest Management Planning in Mixed-Species Forests** . . . . . 503
Hans Pretzsch and Thomas Knoke

**11 Economics of Mixed Forests** . . . . . . . . . . 545
Thomas Knoke

**12 Perspectives for Future Research on Mixed-Species Systems** . . . . . 579
Hans Pretzsch, Thomas Knoke, Carola Paul,
Jürgen Bauhus, and David I. Forrester

**Glossary** . . . . . . . . . . 607

**Index** . . . . . . . . . . 639

# Mixed-Species Forests: The Development of a Forest Management Paradigm

# 1

Jürgen Bauhus, David I. Forrester, and Hans Pretzsch

**Abstract**

In this chapter we show that most forests are naturally mixed and species diverse. However, the degree of species richness varies considerably and declines from tropical to temperate and boreal regions. The compositional and structural diversity of man-made or secondary forests is in most cases substantially lower than in the original, unmanaged native forest. The greater simplicity of managed, secondary forests has a number of causes including historical factors, current management goals focussing on commodity production, and the desire to reduce management complexity. The greater simplicity of these even-aged and monospecific managed forests was seen as the root cause for problems related to their ecological stability. This has triggered the development of alternative silvicultural approaches, initially in Central Europe, which promoted more natural, mixed-species, and structurally diverse forests. However, the initial movements towards a more natural, alternative forest management were not well supported by scientific evidence until the end of the twentieth century. Today, compositionally and structurally diverse forests represent an important element of approaches to deliver a wide range of ecosystem goods and services

J. Bauhus (✉)
Chair of Silviculture, Faculty of Environment and Natural Resources, University of Freiburg, Tennenbacherstr. 4, 79106, Freiburg, Germany
e-mail: juergen.bauhus@waldbau.uni-freiburg.de

D.I. Forrester
Swiss Federal Institute for Forest, Snow and Landscape Research WSL, Zürcherstrasse 111, 8903, Birmensdorf, Switzerland
e-mail: david.forrester@wsl.ch

H. Pretzsch
Chair for Forest Growth and Yield Science, Faculty of Forest Science and Resource Management, Technical University of Munich, Hans-Carl-von-Carlowitz-Platz 2, 85354, Freising, Germany
e-mail: Hans.Pretzsch@lrz.tum.de



H. Pretzsch et al. (eds.), *Mixed-Species Forests*, DOI 10.1007/978-3-662-54553-9_1

in the context of sustainable forest management. In addition, diverse forests are being regarded as more resistant, resilient, and adaptable in the face of global change.

## Contents

1.1 Natural Forests and Tree Species Diversity ... 2
1.2 Man-Made Forests and Tree Species Diversity ... 9
1.3 The Inception of Mixed and Heterogeneous Forests as Models for Management ... 13
1.4 Expectations Towards Mixed-Species Forests ... 15
Take-Home Messages ... 20
References ... 20

## 1.1 Natural Forests and Tree Species Diversity

Forests are among the least modified terrestrial ecosystems and harbour a large proportion of global biodiversity (Myers et al. 2000). The vast majority of vegetation diversity in most forest ecosystems occurs in the understorey (Ellenberg and Leuschner 2010; Halpern and Spiess 1995); however, this book concentrates on trees and forest stands, because it is the tree layer that shapes structure and processes in forest ecosystems and the other vegetation strata (Barbier et al. 2008). In most forests, trees store most of the biomass and have the strongest influence on energy and matter fluxes and hence ecosystem functioning, when compared to other vegetation layers (e.g. Whittaker and Woodwell 1969). Because of their size, firm position, and longevity, trees are the foundation species in ecosystems (Whitham et al. 2006) and determine the living conditions for many other organism groups and the extent of ecosystem functions and services (Hector and Bagchi 2007). In addition, trees are typically the economically most valuable component of forests, and therefore the tree layer is often the compartment of forest ecosystems that is most strongly modified by humans and their management activity, as we will see in the remainder of this book.

Most natural forests consist of a mixture of tree species that can reach high levels of tree species diversity (see Box 1.1). Tree species richness and diversity of forests is, from a global perspective, not evenly distributed (Currie and Paquin 1987). Tree species richness typically increases from cold, boreal to warm, tropical regions, and a range of hypotheses have been put forward to explain this global biogeographical pattern (Hansen and Rotella 1999). Species richness, including that of trees, may be controlled:

(a) By the available energy that can be partitioned among species
(b) By habitat complexity that provides a diversity of niches for the different species
(c) By disturbance regimes, which may favour both early and late successional species, if the disturbances are intermediate and not too intense, frequent, or expansive (or the opposite)

(d) By environmental stress, where species numbers may decrease, when the abiotic conditions become so harsh that they can be tolerated only by few species
(e) By environmental stability, where fewer species are adapted to tolerate highly fluctuating environmental conditions
(f) By area, where the species richness represents a sampling effect
(g) By biotic interactions among species which affect the niche partitioning (Hansen and Rotella 1999)

In addition, current tree species richness may not only be controlled by modern climate but present a legacy of past climatic changes that have caused transformations in species ranges and regional extinctions. Hence, modern tree species ranges may for example be still constrained by post-glacial recolonisation (Svenning and Skov 2004).

Empirical evidence can be found for each of the above hypotheses. Obviously, these hypotheses cannot be completely separated, and none of these hypotheses alone can be used to explain local or regional differences in tree species richness. For example, habitat complexity and disturbance regimes will be closely interlinked. Also, the influence of these factors cannot be directly translated into tree species diversity at the patch or stand level, where many other factors including history and site conditions need to be considered (e.g. Latham and Ricklefs 1993; Gaston 2000). For example, some species-rich forests may be found in temperate and even boreal regions, and there may also be forests with very few species in the tropics (Hart et al. 1989; Kazmierczak et al. 2016). At the same time, these hypotheses contain possible explanations why we do find mixtures of tree species in natural forests and why some forests are richer in tree species than others.

Natural stands in boreal forests of Canada, Fennoscandia, or Russia consist on average typically of one to three species, although they may be more in individual stands (e.g. Paquette and Messier 2011). In the boreal forests, many tree species are functionally different because most functional groups, if they equate to taxa, are represented by only one species. In many boreal forests, there is typically only one tree species representing each of the following genera *Picea*, *Abies*, *Pinus*, *Larix*, *Betula*, and *Populus*, and there are rarely more genera. These functional groups respond to and affect resource supply, disturbance regimes, and other important ecosystem processes and functions (Pastor et al. 1996). For example, the periodic, large-scale biotic disturbance through the eastern spruce budworm (*Choristoneura fumiferana*) in North America is largely confined to *Abies balsamea* and *Picea glauca* (Blais 1985). In the boreal forests of Fennoscandia, aspen (*Populus tremula*) has been identified as a keystone species that provides crucial habitat for biodiversity conservation, in particular for invertebrates (Niemelä 1997).

The functional contrast between tree species is less pronounced in temperate forests, where most genera are represented by more than one species and tree species richness at the stand level may range from 1 to 50 species (e.g. Ovington 1983; Masaki et al. 1992; Busing and White 1997; Wang et al. 2011) and even more so in subtropical and tropical forests. For example, in temperate forests of Japan, more than 32 species from 19 genera may be found, and 10 of these belong to just 2 genera (*Acer* and *Carpinus*) (Masaki et al. 1992). Tree species-rich temperate forests of Central Europe, which can be found on soils with a high base saturation, may contain between 9 and 16 tree species of which very few share the same genus (e.g. *Acer*, *Quercus*, *Ulmus*) (Ellenberg and Leuschner 2010). Some temperate evergreen forests appear to be less species-rich (Ovington 1983), but they can differ greatly in phylogenetic and hence functional diversity. For example, temperate evergreen forests of New Zealand may harbour 6–15 tree species belonging to 5–13 genera (Wardle et al. 1983). In contrast, the dominant stratum of temperate evergreen forests of Tasmania may comprise up to seven species, which may all belong to the same genus *Eucalyptus* (Wells and Hickey 1999).

In temperate forests, stand-level tree species richness appears to be higher in the regions of East Asia and North America than in Europe, where more species may have become extinct as a result of glaciation of large parts of the continent during ice ages (e.g. Latham and Ricklefs 1993; Hewitt 1999). In contrast to the other northern hemisphere continents, the east-west direction of major European mountain ranges such as the Alps and Pyrenees has likely increased species loss through restricting the north-south migration of species during ice ages to warmer refuge areas and has also hampered remigration. Hence, it is not surprising that more than three quarters of the area currently covered by forests in Europe comprises stands with not more than three tree species (Fig. 1.1), although there are also other reasons such as man-made modifications of forests, e.g. large-scale afforestations with single-species stands, and also the conventions used to quantify and record species richness (see Box 1.1). However, the dominance of few tree species and taxa in European mixed forests applies also to other forests in the world. For example, in North America, East Asia, and Europe, 55, 58, and 98% of the total growing stock in forests are attributable to the ten most common species (FAO 2010). Regarding these figures it has to be kept in mind that the information on tree species

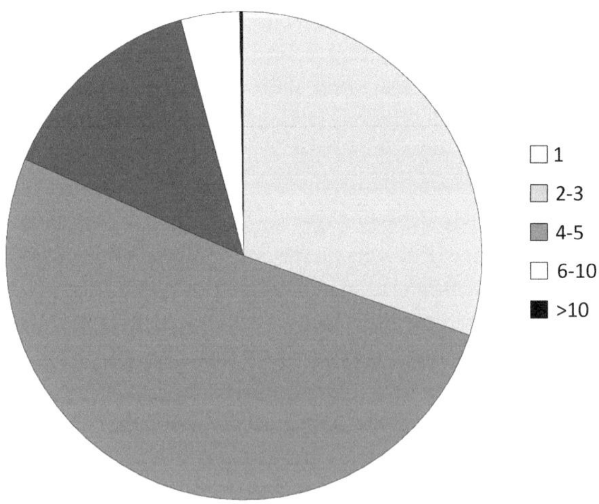


**Fig. 1.1** Proportion of forests with different levels of species richness in Europe (State of Europe's Forests 2011 Report). Country reports cover 75% of the total forest area of Europe (excluding the Russian Federation); data from 2005

represented in growing stock is poor. The information on the ten most common tree species covers only 61% of total forest area (FAO 2010). Globally, the seven most commonly reported tree genera reported in forest inventories comprise, in the sequence of importance, *Pinus*, *Quercus*, *Picea*, *Abies*, *Betula*, *Fagus*, and *Populus* (FAO 2006), further indicating a relatively low level of tree species diversity, in particular in boreal and temperate forests, where these species dominate. Even in tropical forests with extremely high levels of stand-level species richness and α-diversity, where up to 300 tree species (> 10 cm DBH) may be found per ha (Gentry 1988), tree communities are being dominated by a much smaller number of species. Among the ca. 16,000 tree species occurring in lowland Amazonia, only 1.4% of so-called hyperdominant species account for half of all tree individuals (ter Steege et al. 2013).

**Box 1.1 Mixed, Diverse, and Species-Rich Forests: How Are They Defined?**

A forest stand consisting of only one tree species, as is the case with many commercial plantations, is pure or monospecific. At the other extreme, a tropical lowland rainforest with 300 tree species per ha (e.g. Gentry 1988), is clearly species-rich and inevitably diverse. But what about forest stands that are dominated by one tree species and contain few other trees of one or more other tree species? In terms of tree species richness, which is simply the number of different tree species within a forest community or stand, these stands have a value for tree species richness equivalent to the number of tree species found within. Species richness, however, does not consider the abundances and thus the ecological significance of the different species or their importance from a management perspective. Therefore, in forest inventories, it is common practice to classify forest that have a dominant species and less than 10–20% canopy cover or stand density in other species still as monospecific stands (Schuck et al. 2002). This practice, however, may reflect a perspective that has its tradition in conceptual and management models derived from monospecific stands and which aims to simplify structures and management (c.f. Puettmann et al. 2009). Also, from an inventory point of view, trees and therefore species will only count towards species richness or diversity, if they are larger than a set diameter threshold, which is often around 7–10 cm DBH. Thus, there may be forests that have the appearance of mixed-species forests or are obviously developing towards a mixed composition, which are still classified as monospecific, because the trees of other species contribute less than a defined threshold of stand density or because they are yet too small to be counted.

In a measure of tree species diversity of forest stands, the abundance or evenness of individual species is being considered. One of the most commonly used diversity indices is the so-called Shannon index (Shannon 1948):

(continued)

**Box 1.1** (continued)

$$H' = -\sum_{i=1}^{S} p_i \ln p_i$$

where $p_i$ is proportion of the entire population made up of species $i$. This may be calculated as the number of individuals of species $i$ relative to the total number of individuals or in the case of forest stands also as the proportion of basal area occupied by species $i$. The latter may be more appropriate in highly irregular forests with trees of very different sizes and where the basal area is therefore a more meaningful measure of the relative importance in terms of site occupancy than only the number of individuals. $S$ is the number of species encountered within the forest stand (or the sample), and $\sum$ is the sum from species 1 to species $S$. There are a number of other indices such as Simpson's, the Berger-Parker, or Hill's index, which are used in the ecological literature to describe the species composition of communities (c.f. Magurran 2004). Each of these has its strengths and shortcomings and the indices are therefore often used in combination.

From the above we can conclude: not every forest stand that is species-rich is also species diverse, for example, if the stand is dominated by one or few species. Forests that are being classified as pure or monospecific may actually contain more than one species, which, however, do not contribute significantly to stand density. However, for that reason, forests are likely more species-rich than they appear in the statistics based on conventional forest inventories. In particular rare tree species (e.g. Hemery et al. 2010), those that contribute, for example, less than 1% to forest composition, are typically not adequately represented in these inventories, which are not designed to capture rare occurrences.

With regard to the functional aspects of biodiversity, i.e. the role that biodiversity plays for ecosystem functioning such as the rate and reliability of ecosystem processes, the functional diversity of the tree layer may be more relevant than its taxonomic diversity. As for taxonomic diversity, there are a number of different indices to calculate functional diversity: e.g. functional richness, functional evenness, and functional divergence (Mason et al. 2005). In a simple and easily applicable form, which has been often used in community ecology, species are assigned to subjectively chosen functional groups. In the case of trees, these may be related to morphological and physiological attributes such as shade tolerance, drought tolerance, root systems (deep or shallow), N-fixation capacity, wood density, maximum height, evergreen or deciduous, etc., to successional status or other attributes relevant to ecosystem functioning. Richness and diversity of these groups may then be calculated as for taxonomic diversity. Since niche complementarity may be the underlying mechanism to link species diversity to ecosystem processes

(continued)

**Box 1.1** (continued)

(e.g. Tilman et al. 2001), a categorical classification of functional groups may be inferior to the measurement of functional diversity on a continuous scale (Fonseca and Ganade 2001), which better captures the distribution and overlap of niches in functional space. Hence, a number of indices have been developed to capture functional diversity on a continuous scale (e.g. Mason et al. 2005; Petchey and Gaston 2002).

Among these different measures of biodiversity at the stand scale, a trade-off exists in relation to information content and ease of use/calculation. Measures such as species richness carry little or no information about the function of individual species, but they are relatively easy to obtain. In contrast, measures that contain detailed information about individual species, such as measures of functional diversity, are more difficult to obtain or to calculate (Petchey et al. 2004). This applies in particular to highly diverse forests such as lowland tropical rainforests that contain many species for which we have very little information on morphological, physiological, or ecological attributes. Given the large range of diversity indices that may be calculated for communities, it is important to know and decide from the outset which index to use for which purpose.

In addition to these stand or patch scale measurements of α-biodiversity, which are most relevant for the content of this book, there are measures to describe spatial forest heterogeneity at the landscape scale. Whittaker (1960) was one of the first to describe β-diversity and used it in a broad sense as a measure of compositional heterogeneity among places. Beta diversity can be calculated through partitioning of the total number of species found in a landscape (γ-diversity) into the number of compositional landscape units and the effective number of species per compositionally distinct landscape unit or habitat type (Tuomisto 2010), which may be represented by ecosystem types or forest stands as discrete sampling units. Or in other words and in a hierarchical sense, α-diversity describes within-habitat diversity, beta diversity represents among-habitat differentiation in a landscape, and γ-diversity refers to the total landscape diversity (Whittaker 1977). Hence β-diversity may, for example, be calculated as:

$$\beta = \frac{\gamma}{\alpha}$$

where $\gamma =$ total number of species found in a landscape and $\alpha =$ average number of species per sampling unit. Alternatively, β-diversity may be calculated in an additive way:

$$\beta = (S_1 - c) + (S_2 - c) + ...(S_n - c)$$

(continued)

**Box 1.1** (continued)

where $S$ = number of species in community 1, 2, . . .$n$ and $c$ = number of common species across the sampled communities).

In this book, the major focus of analyses of tree species interactions and the ecological function of tree species or functional diversity is at the level of forest stands and thus refers to α-diversity. However, with regard to forest management and economic analyses of mixed-species forests, the larger spatial scales are also considered.

As we can see from the above, most natural or seminatural forests are rich and diverse in tree species, regardless of the biogeographical region. Yet, there are also some native forests that are naturally monospecific or strongly dominated by one species. Examples of pure (or almost pure) forests comprise those that are dominated by a tree species that is a superior competitor, e.g. *Fagus sylvatica* L. in Europe (Peters 1997) and some of the tallest-growing species in the world such as *Sequoia sempervirens* (Busing and Fujimori 2002) along the west coast of North America and *Eucalyptus regnans* in south-eastern Australia (Ashton 1976). Naturally monospecific stands may also be comprised of tree species adapted to very harsh environments, e.g. *Picea mariana* on nutrient-poor bogs in North America (Groot and Horton 1994). Pure stands occur where one species is, under the prevailing disturbance regime, such a strong competitor that all other species that could potentially grow there are being outcompeted, or where the abiotic factors or the disturbance regime creates an environment that is only tolerated by one species. However, in the wider landscape, forest stands comprising only a single tree species are rather an exception than the rule.

According to Noss (1990), biodiversity of an ecosystem is characterised by its composition, structure, and functioning. Compositional characteristics are the number of species, genotypes, communities, and landscape types. Functional aspects include among others biomass production, competition, mortality and regeneration, decomposition, and nutrient and hydrological cycles. Structural attributes are related to plant morphology and habitus, stand structure, gap patterns after disturbances, spatial distribution of tree and stand development phases, and ecosystem fragmentation. All three components would be relevant for a comprehensive assessment of biodiversity, indicating how difficult it is to quantify biodiversity.

In addition to their characteristic tree species diversity, natural forests are typically also distinguished by a high level of structural diversity and compositional heterogeneity (see β-diversity). Their structural diversity can be attributed, for example, to the presence of different vegetation and canopy layers and variation and canopy density, to the variation in tree heights and diameters and their irregular spatial distribution, to the presence of standing and downed deadwood in different dimensions, distributions, and decay stages (McElhinny et al. 2005). Their compositional heterogeneity is a result of different habitats and successional dynamics.

Different habitats are occupied by different vegetation communities that are adapted to the niches provided by that habitat, e.g. to particular soil water or nutrient availability (Ellenberg and Leuschner 2010). Patches representing different successional stages following disturbances, for example in the form of gaps, may be similarly occupied by different vegetation communities (Watt 1947; Remmert 1991). For example, more light-demanding, early- or mid-successional species find opportunities to regenerate in forest gaps with increased light availability, which would be insufficient for them beneath the canopy of closed forest (Poulson and Platt 1989). Hence, there is an interaction between tree species diversity and ecosystem disturbances (e.g. Molino and Sabatier 2001). While disturbances are required to maintain species diversity, compositional heterogeneity will only result from disturbances such as gaps, when there are in fact different species available that can occupy these temporal niches. Similarly, structural diversity is likely higher in species-diverse forests, where, for example, tree species with different final heights and light requirements can form more canopy layers than would be the case in monospecific forests of less shade-tolerant species. Tree species with different growth rates and final heights will likely develop more structurally diverse forests than those composed of only one or few species. Also, different tree species with different wood decay resistances (Cornwell et al. 2009), dimensions, and mortality rates will likely provide for greater diversity in the amount and quality of deadwood (Jonsson et al. 2005). Therefore, structural diversity and compositional heterogeneity can, to a large extent, be attributed to species diversity (Noss 1990).

Both structural diversity and compositional heterogeneity can be dramatically changed in man-made, managed forests (Bauhus and Pyttel 2015), where successional processes are limited to early- and mid-successional phases, stands are managed in a way to avoid the occurrence of natural disturbances (e.g. Gardiner and Quine 2000), and species composition is strongly controlled, in the extreme so that just one species is cultivated across the landscape regardless of changes in soil substrate and original habitat differences. After all, successful tree species for plantations are those with a wide physiological amplitude and the capacity to grow well over a wide range of soil and site conditions (e.g. Burdon 2001).

## 1.2 Man-Made Forests and Tree Species Diversity

Historically, land-use changes and the management of forest have often led to a simplification in forest structure and composition (Puettmann et al. 2009). This is particularly pronounced where primary and natural forests have been replaced, often following an interim period of agricultural land use, by secondary forests (Ellenberg and Leuschner 2010). These landscape dynamics have shaped many traditional, cultural landscapes. Secondary forests have often been established and maintained as monocultures or with reduced tree species diversity when compared to the original, natural forest. The reasons for this development are many:

(a) Many secondary forests have developed from degraded forests, following clear-felling or following previous agricultural or agroforestry land use. These situations pose harsh conditions to tree regeneration. Often these were the poorest soils, most degraded sites, with frequent frost events or high insolation, continued browsing by domestic livestock or game animals, and other adverse factors which permitted the cultivation of only few species. Owing to these rather unfavourable conditions, the large reforestation and afforestation efforts were regarded as a huge cultural achievement (Pflug 1892). These factors have certainly contributed to the dominance of the conifers *Picea abies* (Norway spruce) and *Pinus sylvestris* (Scots pine) in secondary forests of Central Europe (Mosandl 2009) and many pine and eucalypt plantations in the tropics (Evans and Turnbull 2004). Afforestation following large-scale forest degradation and clearing in previous periods was often modelled on the agricultural approach to cultivate crops, namely, in monocultures (Mantel 1961).

(b) Often secondary forests have been established or are being managed with a focus on timber production for commodity products. For these commodity products, the market typically demands homogenous timber dimensions and properties, which may be most easily provided in monospecific plantations of trees with controlled genetic make-up and well-known growth and yield performance (e.g. Burdon 2001; Zobel and Van Buijtenen 1989).

(c) Also, in managed forests with a focus on timber production, the emphasis is typically placed on the most profitable tree species and production systems. In natural, mixed-species forests, there are typically a number of species that are much more valuable than others and some that have little commercial value. In addition, tree species differ with regard to their ease of management. This is strongly epitomised in tropical rainforests, where only a fraction of the tree species are actually used commercially (Wyatt-Smith 1987). Also in structurally diverse, managed temperate forests, species that were less valuable or more difficult to manage have been strongly reduced in abundance, for example, *Fagus sylvatica* in conifer-dominated Plenter forests (Schütz 2002). Hence, the focus on profitability of cultivating certain tree species may have led to a simplification in species composition in many forests.

(d) In addition, in forests with a production focus, there is a preference for species that come with a large research and knowledge base with regard to their physiology, nutrition, associated pests and pathogens, propagation, genetics, growth and yield, response to silvicultural treatments, and last but not least wood properties (e.g. Burdon 2001). For many native species, which might add to the diversity of plantations, in particular in the tropics, the knowledge base still needs to be developed (e.g. Haggar et al. 1998). Currently, the genome of only a few tree species has been analysed (e.g. *Populus*, *Picea abies*). Ideally, the selected species also have wide physiological amplitudes that facilitate their use over wide environmental gradients, which means that they can be grown on many different sites. It also means that individual stands of such species can be expansive because there is no need to change the species owing to variations in site conditions that may result from topography or changes in soil types. In contrast, when using species with a narrow physiological niche, it may be necessary to employ different species for different situations such as the bottom of the slope, mid-slope, and ridge tops. Often, the knowledge base on individual species with wide physiological amplitudes has been larger for exotic than native species, because the former have already been cultivated in many different places. Selecting such a species, if suitable to the site and environmental conditions, promises large savings on investments in research. These exotic species are typically cultivated in monocultures, since nothing or very little is known about their interactions with other species. Importantly, if we consider mixed-species forests, the knowledge base does not only need to be increased for every additional species but also for the interactions among species. This is one important purpose of this book.

(e) Monocultures may also result from a lack of available propagules and planting stock. Situations, in which forest management relies on seeds and seedlings from a limited number of tree species, which are typically easy to store as seed and propagate, may arise following large-scale disturbances, when there is a sudden large demand to regenerate forests. In the past, for example, in Central Europe, similar situations were faced after the wars, when large areas needed to be reafforested. In many tropical and subtropical regions, there is still very little practical knowledge on the propagation and establishment of most native species (e.g. Yang et al. 2013), which is a contributing factor to the use of a limited number of species that can be easily multiplied and established. In addition, the incentives for nurseries to cultivate rare species may be low unless a strong market for them has developed.

(f) Most importantly, increasing species diversity increases the complexity of silvicultural systems and forest management. As mentioned above, the complexity increases not only with every additional species but also through the interactions among species, because trees of the same species may grow differently in monocultures and mixtures. This will be explored in the remainder of this book. To exemplify this increasing complexity, let us consider a simple example. If the aim of forest management is to produce high-quality clear wood for sawn timber or veneer with tree species that require artificial pruning, then the optimal timing of pruning can be relatively easily determined by the growth dynamics of that species for a particular site (Alcorn et al. 2013). In practical terms, the prescriptions may stipulate the tree height, diameter, or age at which the pruning shall be carried out, perhaps in two lifts or more, to reach a certain branch-free height on the stem. In a mixed forest stand comprising two species of different growth dynamics and branching characteristics, it may be necessary to double the pruning interventions, to prune each species at the appropriate phase of stem and crown development. If the mixing pattern is not regular, with a variety of situations ranging from monospecific patches to intrinsically mixed arrangements of trees, which lead to large variations in crown competition and thus branch development, even more pruning interventions may be required. And the experience gained from one mixed stand is hard to apply to another, as the interactions between species may differ strongly among sites. In addition, different species require different propagation techniques, sowing or planting times, different thinning intensities, a different nutrient management, different production cycles, and so on. Also, the marketing and sale of a larger variety of products becomes more complex.

(g) Most forestry approaches and tools for inventory, planning, control, and management of forest stands, for example, yield tables, have been developed for monospecific forests, and their theoretical underpinning is still largely restricted to these simpler forest ecosystems. These approaches can be relatively simple, have a solid theoretical foundation, and are relatively inexpensive to apply. For mixed-species and diverse forest stands, however, a consistent analogue theoretical framework for the development of appropriate forestry tools is still largely lacking.

As we can see from the above, most man-made and managed forests are less tree species diverse than their natural and unmanaged counterparts. The reasons for this have been mostly economical. However, historical developments and traditions in forest management also play an important role (Puettmann et al. 2009). Interestingly, foresters were also the first to identify and criticise inherent problems associated with the management of monospecific forests (e.g. Gayer 1886; Möller 1922). Their critique focussed mostly on the low resistance of monocultures, mostly of secondary conifer forests, against abiotic and biotic stress and disturbances such as windthrow and bark beetles. It appears almost like a peculiar irony of forest history that these early critics of monoculture forestry laid the foundation for a movement towards more natural, species-rich, and structurally diverse forests in just the same country (Germany), where this type of forest management with single-species plantations underpinned by simple management models such as the normal forest (Hundeshagen 1826) was invented. Yet, it may be that this critique and change in mindset developed first there, where the problems became first apparent owing to widespread monocultures that were also often established on unsuitable sites and were not silviculturally treated in ways that increase their resistance to stress and disturbance (e.g. Wiedemann 1923; Powers 1999). A further factor that may have added to negative experiences with plantations in previous centuries is that these stands were often managed on long rotations, which increases any height- and age-related risks and in general the time for which one crop is exposed to unpredictable stress and disturbance events. Today, the critique of species-poor and structurally simple forests still focusses on low ecological stability, i.e. higher susceptibility to disturbance, and also on reduced biodiversity in these types of systems (e.g. Hartley 2002).

## 1.3 The Inception of Mixed and Heterogeneous Forests as Models for Management

Monospecific stands or plantations may be a successful model for the provision of a narrow suite of products and/or ecosystem services, and their management can be sustainable with regard to the 'sustainable yield' concept (e.g. Powers 1999). In addition, this type of forest management system can play an important role in the sustainable management of forest landscapes in concert with conservation reserves and less intensively managed and more natural forest areas (e.g. Seymour and Hunter 1999). In the so-called triad approach, a partial segregation of the landscape for intensive production systems permits also the segregation of strict conservation reserves while maintaining overall landscape productivity and functioning. These areas with contrasting management foci are then embedded in a matrix of multi-functional forests that serve as buffer and corridor between these systems and fulfil other important landscape functions.

Today, it has been recognised that there is an increasing demand for a larger variety of goods and ecosystem services, also from plantations or cultivated forests (Bauhus et al. 2010; Biber et al. 2015). This calls for a diversification of forest stands in terms of composition and structure (Bauhus and Schmerbeck 2010). However, this is not a new realisation, although it has been differently articulated. Earlier critics such as Karl Gayer (1886) or Alfred Möller (1922) had already promoted mixed, uneven-aged forests as a presumably more stable alternative to even-aged monocultures, but they still lacked scientific insights into specific advantages of mixed-species forest. In fact, earlier forest scientists had even promoted monospecific forests because of the difficulty in controlling the competition in mixed-species stands, where one species might overgrow the other (Hartig 1791). The first scientific investigations into tree species mixtures began much later (e.g. Wiedemann 1942) and were initially focussed mostly on growth and yield in stands comprising two or only a few species (e.g. Assmann 1961; Kelty 1992). And in this aspect, important early contributions were being made. Despite the incomplete understanding of the processes and dynamics governing mixtures, statistical models for growth and dynamics of mixed-species forests were developed in which the lack of knowledge was bridged by statistical relationships (e.g. Monserud 1975; Pretzsch et al. 2002, Pretzsch et al. 2015).

Research on mixed-species stands then excelled in the beginning of this century, although it still remained, in the forest sciences, focussed on specific mixtures with two or few tree species (e.g. Bauhus et al. 2000; Forrester et al. 2006; Liang et al. 2016; Pretzsch et al. 2010, 2013). Also at the beginning of this century, broader ecological research into the functioning of tree species diversity beyond specific mixtures used in forestry commenced (e.g. Scherer-Lorenzen et al. 2005; Bruelheide et al. 2013). Today there is a global network of experiments to investigate the role of tree species diversity in ecosystem functioning (Verheyen and Scherer-Lorenzen 2011), and our mechanistic understanding in this regard has greatly increased (e.g. Forrester and Bauhus 2016), as we will see later in this book.

Despite the lack of scientific underpinning, mixed-species stands formed a corner stone of forest management concepts such as close-to-nature forest management, which developed much earlier and may be traced back to the Dauerwald concept (perpetual, continuous cover forest) of Möller (1922). Möller's writings, in which forests are viewed as an organism (Lemmel 1939), represent an early attempt to describe what is now known as the ecosystem concept (Thomasius 1996). He adopted this perspective long before ecosystem ecology became of major interest to the scientific community. For that reason his writings reflect, from todays perspective, an incomplete understanding of ecosystems. This view expressed itself in the idea of bringing together productivity, diversity, and especially continuity and stability of forest conditions. Möller's view of ecosystems as an organism suggests close linkages with Clements' view of ecosystems as meta-organism dynamics (Clements 1936) and with the Gaia hypothesis (Levin 1998). Möller and subsequently his followers in the close-to-nature forestry movement (in German: Arbeitsgemeinschaft Naturgemäße Waldwirtschaft) believed that the maintenance of continuous cover (through abandoning clear-felling), the harvesting of single trees, creating mixed-species stands, the long tending of growing stock, natural regeneration, and the careful management of forest soils would provide forests with the capacity for self-regeneration and self-regulation. Close-to-nature forest management (CTNFM) was not designed as an approach to the conservation of forest biodiversity but may be regarded as a production system that developed from concepts of multifunctional forest management (Bauhus et al. 2013). Despite the increasing following of these ideas through practitioners (Münker 1958), there was no substantial change in species composition in the actual forests. Today, forest management according to these principles is carried out across Europe, and the national organisations promoting it are assembled in a federation of foresters called PROSILVA (PROSILVA 2013). Maintaining and promoting structurally diverse mixed-species forests can be regarded as a main element of CTNFM, and hence these have been strongly advocated. The principles of CTNFM, which has also been given a number of different labels, have been adopted by many private and public forest owner's and have become a paradigm of forest management in Central Europe, which is partially in response to the political promise to deliver sustainable forest management (Bauhus et al. 2013; Gamborg and Larsen 2003). However, the principles of CTNFM are principally tools that are aimed at managing forests for certain values. Despite its strong following and political support, many of the principles and ideas still lack scientific underpinning, and in many situations it is not clear to what extent they achieve the desired outcomes or may lead also to undesired outcomes (see also Brang et al. 2014). Therefore, this book will contribute to increasing our understanding regarding one core principle of CTNFM, the preference of mixed-species forests.

It is also interesting to note that despite the ever-growing number of supporters of these ideas in Central Europe, very little changed on the ground and in the forest, where the proportion of mixed-species forests remained relatively low. There is a wide range of factors that have been identified to impede the uptake of alternative forms of silviculture (Puettmann et al. 2015). It was not until the prevailing forest management paradigms were challenged by the realities of air pollution and the concomitant acid rain problem in the 1970–1980s, catastrophic wind storms in the 1990s, and an extreme drought in 2003 that mixed-species stands, which were regarded as less vulnerable and more resistant and resilient in relation to these disturbances, gained dramatically in importance. For example, in the German state of Baden-Württemberg, the proportion of monospecific stands declined from 53% in 1987 to 39% in 2012.

## 1.4 Expectations Towards Mixed-Species Forests

In democratic societies, societal expectations towards forest management should be reflected in forest policies and governance instruments. In most countries, policy-makers have recognised that sustainable management of forests needs to encompass the provision of a wide range of ecosystem goods and services to meet society's needs. At an international level, there have been a number of forest policy processes to implement and support sustainable forest management in different regions of the world. These include the Helsinki process for the European Union, the Montreal Process for temperate and boreal forests outside Europe, and the ITTO process for member countries of the International Tropical Timber Organisation (McDonald and Lane 2004). In these processes, the broad values for which forests should be managed have been defined as criteria. These comprise the conservation of biological diversity, the maintenance of the productive capacity of forest ecosystems, the maintenance of ecosystem health, conservation and maintenance of soil and water resources, maintenance of forest contribution to global carbon cycles, the maintenance and enhancement of long-term multiple social and economic benefits, and legal, institutional, and economic frameworks for forest management (McDonald and Lane 2004). The performance of forest management in relation to these criteria is then reported in the form of quantifiable indicators. Countries that are signatory to the above processes have adopted these criteria and indicator schemes for their national reporting of sustainable forest management. Voluntary certification schemes such as the Forest Stewardship Council (FSC) or the Programme for the Endorsement of Forest Certification Schemes (PEFC) have adopted certification schemes that fit under these international processes. The importance of mixed-species forests becomes apparent through the indicators that have been defined in these international processes. In the European context, for example, there are a number of indicators that relate directly to tree species composition (Rametsteiner and Mayer 2004). The criterion 'maintenance, conservation, and appropriate enhancement of biological diversity in forest ecosystems' is indicated by 'tree species composition' (area of forest, classified by number of tree species occurring), 'naturalness' (area of forests and other wooded land, classified as 'undisturbed by man', as 'seminatural', or as 'plantation'), and 'introduced tree species' (area of forest dominated by introduced tree species). In addition to these indicators directly related to tree species composition, there are a number of other indicators that are influenced by species composition. These include, for example, 'carbon stocks', 'soil condition', 'forest damage', 'increment and fellings', or 'cultural and spiritual values'. The way in which tree species composition influences these indicators will be discussed in the remainder of this book. The perceived function of species-diverse forests for provision of forest ecosystem goods and services may however not be mirrored by perceptions of forest practitioners or the general public. In the context of a case study from Belgium, the positive perception of forestry practitioners of the supporting, regulating, and cultural services in mixed-species forest stood in contrast to less conclusive results from the literature, where also some negative and neutral effects of mixed-species forests were reported, i.e. in relation to associated biodiversity (Carnol et al. 2014). In addition, provisioning services such as productivity were perceived as equal in mixed-species stands and monocultures, although there is much evidence now pointing to higher productivity in mixtures (Pretzsch et al. 2013; Forrester and Bauhus 2016). This discrepancy between the perceptions of the effects and functions of mixed-species stands on the one side and the scientific evidence for this on the other side underpins the need for this book.

There are two perceptions/hypotheses about biodiversity and ecosystem functioning,[1] which are central to the discussion of ecosystem services from mixed-species forests and also central to this book (Hector and Bagchi 2007; Hooper et al. 2005; Loreau et al. 2002; Baumgärtner 2007):

1. The level of ecosystem functioning increases with the level of tree species diversity (Fig. 1.2).
2. Tree species diversity may enhance ecosystem stability. An increase in the level of biodiversity decreases the temporal variability of the level at which ecosystem goods and services are provided under changing environmental conditions (Fig. 1.2).

However, it is not assumed that ecosystem functioning such as productivity and ecosystem stability increase linearly with tree species diversity but that the magnitude of these effects declines with the level of biodiversity, so that these functions follow the law of diminishing marginal benefit. These two perceptions will be dealt with briefly below and in more detail in the remainder of this book, when we review the evidence for the relationships between tree species diversity and productivity, other ecosystem functions, and ecological and economic stability.

The most important mechanisms behind the perceived functions of tree species diversity are niche complementarity and ecological insurance against stress and disturbances. Niche complementarity can lead to greater supply, capture, or use efficiency of resources through more diverse communities, in which a mixture of different growth forms of plants can fill the available limited space to a greater

[1] In this context, ecosystem functioning refers to ecosystem properties (e.g. compartments or pools and processes or fluxes) that facilitate the provision of ecosystem goods and services.

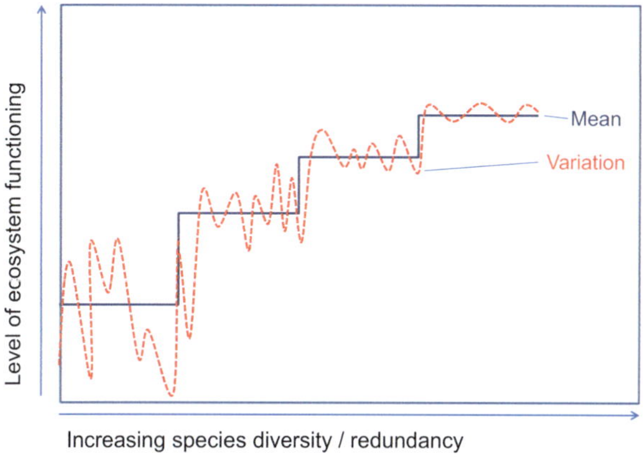


**Fig. 1.2** Hypothetical diagram depicting the relationship between species diversity and ecosystem functioning. With decreasing species diversity, e.g. through loss of species from the ecosystem, the level of ecosystem functioning, e.g. productivity, declines. In addition, the temporal variation in ecosystem functioning and subsequently in the provision of ecosystem goods and services in the face of stress and disturbances increases

extent and thus is able to exploit the available resources better than a monoculture under the same site conditions (Loreau and Hector 2001; Naeem 2002). Complementarity exists, when different species use different resources or the same resources but at different times or different points in space, for example, different soil layers. In this case, interspecific competition in the mixture is less than intraspecific competition in the monocultures. The concept of niche complementarity also comprises facilitation, the positive interactions among species, where certain species improve environmental conditions for others or supply a critical resource such as nitrogen through fixation. Both of these mechanisms, complementarity and facilitation, often lead to an 'overyielding effect', in which biomass production in species mixtures exceeds the productivity expected on the basis of the yields of the contributing species when grown in monoculture (see Chap. 4).

The ecological insurance concept postulates that more diverse communities are more likely to cope with new conditions when subject to unpredictable stress or disturbance (Yachi and Loreau 1999). With an increasing number of functionally different species, the probability increases that some of these species can respond in a differentiated manner to the external perturbations or changing environmental conditions. Also, the probability increases that one species can take over the role of another, redundant species that does not survive the disturbance or new conditions (Walker et al. 1999; Yachi and Loreau 1999, see Chap. 7). At a higher organisational level, plants in ecosystems act through organismic associations rather than as individuals. Hardly any plant persists without multiple biotic interactions (Zilber Rosenberg and Rosenberg 2008). These are typically expressed through competitive, facilitative, mutualistic, parasitic, or seemingly neutral relationships. For example, mycorrhizal fungi might reduce drought stress in host plants (e.g. Osonubi et al. 1991). Hence, individual trees and species can be viewed as elements of complex adaptive systems (Messier et al. 2013), where emergent properties resulting from ecological interactions at various scales influence adaptability and self-organisation of these systems. It can be assumed that with increasing tree species richness, the number of interactions and feedbacks also increases and this might further buffer mixed-species forests against stressors and disturbances. Following this logic, processes that are carried out by a small number of tree species and their interacting organisms are likely to be more sensitive to changes in diversity than those that are carried out by a large number of species.

Following from the above, it appears that it may be safer to manage forests sustainably with tree species-diverse forests rather than with monospecific or species-poor forests in particular if we consider the uncertainties of future developments (Fig. 1.3). Owing to a possibly higher ecological stability of mixed-species forests in relation to stress and disturbances, the influence of environmental changes on the capacity of forest ecosystems to provide goods and services may be lower when compared to monospecific or tree species-poor forests. These are, however, hypotheses that remain to be tested, which will be done in the following chapters.

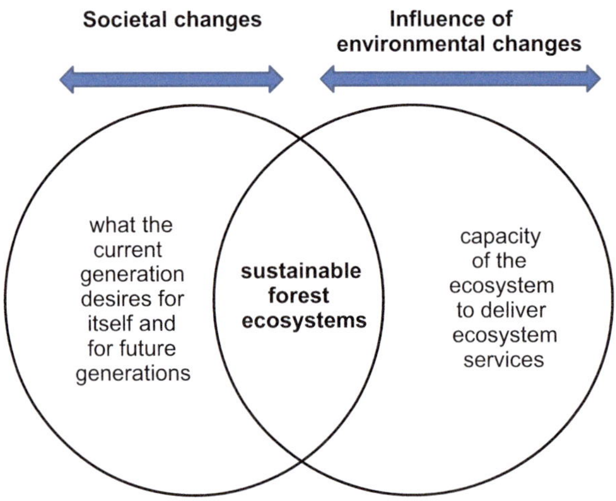


**Fig. 1.3** Sustainable forest ecosystems exist only where the aspirations of society for ecosystem goods and services from forests are actually met by the capacity of forest ecosystems to deliver these (adapted from Maser 1994). Owing to a possibly higher ecological stability of mixed-species forests in relation to stress and disturbances, the influence of environmental changes on the capacity of forest ecosystems to provide goods and services may be less when compared to monospecific or tree species-poor forests. In addition, mixed-species forests may have an inherently higher number of options for future forest functions that are desired by society (Wagner 2004). For example, it may be easier to convert the focus from a production orientation towards recreation in a species-diverse forest than in a monoculture of an exotic tree species

However, that does not mean that species-diverse and structurally complex forests are more sustainable than more simple forests. This depends to a large extent on the level of uncertainty faced in management (Fig. 1.4). In situations, where forests can be quickly adapted to changes in environmental conditions or management intent, e.g. through cultivation of new species or genotypes, there may be little need to maintain many options for the future through a high degree of species and structural diversity. However, in situations, where the species composition cannot be changed in the short term and where both management intent and future environmental conditions faced by current stands are rather uncertain, it may be prudent and advisable to manage forests with a high degree of species diversity. Thus, it is not surprising that promoting and maintaining tree species diversity is one of the most prominent elements of adaptation strategies to climate change in forest management (e.g. Reif et al. 2010).

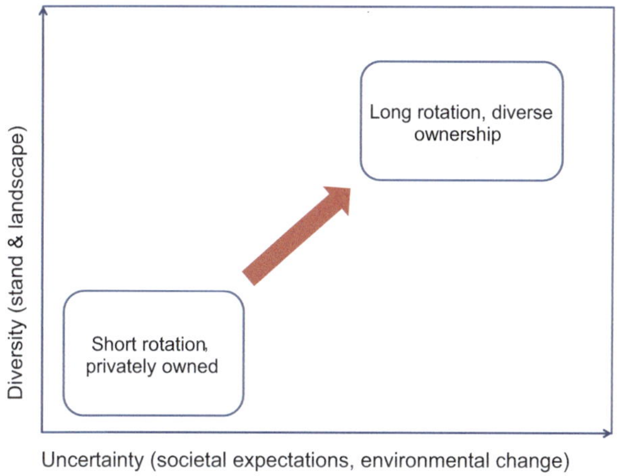


**Fig. 1.4** The demand to manage forest stands and landscapes with an increasing degree of tree species diversity increases with increasing uncertainty. If forests are managed on short rotations and the management objectives are undisputed and clearly defined (e.g. through a private landowner), there may be little need to employ the insurance function of tree species-diverse forests. If forests are no longer suited to new environmental conditions or management intent changes, the system can be quickly adapted to the new situation (e.g. through species change). However, managing forests on long rotations typically also increases risks as the forests age and grow taller they will have to face the medium- to long-term uncertain developments of global change. There may then be a greater need to promote or maintain tree species diversity to cope with the loss of individual tree species, as has been or is currently occurring in different parts of the world, for example, through Chestnut blight in the north-eastern USA (Anagnostakis 1987) or the ash dieback in Europe caused by *Chalara fraxinea* (Kowalski 2006). The strategy to manage for forest composition and structure that ensures many future options is also advisable, where the future goals of forest management are less predictable, e.g. in publicly owned forests, where the goals may change through elections

**Take-Home Messages**

1. Most forests are naturally mixed, but the degree of species richness varies considerably. Tree species richness declines from tropical to temperate and boreal regions. Regardless of the region, the tree species diversity in forests is often strongly influenced by few dominant species.
2. In addition to tree species richness and diversity at the stand level (α-diversity), biodiversity of forests is also characterised by structural diversity and compositional heterogeneity (β-diversity) at the landscape level, where the latter refers to variation in habitats with different vegetation communities and variation in successional dynamics.
3. The compositional and structural diversity of man-made or secondary forests is in most cases substantially lower than in the original, unmanaged native forest. The greater simplicity of managed forests is partially explained by historical factors, by management goals focussing on commodity production, and by the goal of reducing management complexity.
4. Problems related to the ecological stability of simple, monospecific forests led to the emergence of alternative approaches to forest management as early as the late nineteenth century and the first half of the twentieth century. However, the initial movements towards a more natural, alternative forest management were not well supported by scientific evidence. Research into mixed-species and uneven-aged forests took off only towards the end of the twentieth century.
5. Today, the establishment, restoration, and maintenance of compositionally and structurally diverse forests represent important approaches to deliver a wide range of ecosystem goods and services in the context of sustainable forest management. In addition, and based on the ecological insurance hypothesis, diverse forests are being perceived as more resistant, resilient, and adaptable in the face of global change.

## References

Alcorn PJ, Forrester DI, Smith RGB, Thomas DS, James R, Nicotra A, Bauhus J (2013) The influence of site quality on timing of pruning in *Eucalyptus pilularis* and *Eucalyptus cloeziana* plantations. Aust For 76:25–36

Anagnostakis SL (1987) Chestnut blight: the classical problem of an introduced pathogen. Mycologia 79:23–37

Ashton DH (1976) The development of even-aged stands of *Eucalyptus regnans* F. Muell. in central Victoria. Aust J Bot 24:397–414

Assmann EA (1961) Waldertragskunde. BLV, München Bonn Wien

Barbier S, Gosselin F, Balandier P (2008) Influence of tree species on understory vegetation diversity and mechanisms involved—a critical review for temperate and boreal forests. For Ecol Manage 254:1–15

Bauhus J, Pyttel P (2015) Managed forests. In: Peh KSH, Corlett RT, Bergeron Y (eds) Routledge handbook of forest ecology. Routledge, Oxon, pp 75–90

Bauhus J, Schmerbeck J (2010) Silvicultural options to enhance and use forest plantation biodiversity. In: Bauhus J, van der Meer P, Kanninen M (eds) Ecosystem goods and services from plantation forests. Earthscan, London, pp 96–139

Bauhus J, Khanna PK, Menden N (2000) Aboveground and belowground interactions in mixed plantations of *Eucalyptus globulus* and *Acacia mearnsii*. Can J For Res 30:1886–1894

Bauhus J, van der Meer P, Kanninen M (2010) Ecosystem goods and services from plantation forests. Earthscan, London, 254 p

Bauhus J, Puettmann KJ, Kühne C (2013) Close-to-nature forest management in Europe: does it support complexity and adaptability of forest ecosystems? In: Messier C, Puettmann KJ, Coates KD (eds) Managing forests as complex adaptive systems: building resilience to the challenge of global change. Routledge, The Earthscan Forest Library, pp 187–213

Baumgärtner S (2007) The insurance value of biodiversity in the provision of ecosystem services. Nat Resour Model 20:87–127

Biber P, Borges JG, Moshammer R, Barreiro S, Botequim B, Brodrechtová Y, Brukas V, Chirici G, Cordero-Debets R, Corrigan E, Eriksson LO, Favero M, Galev E, Garcia-Gonzalo J, Hengeveld G, Kavaliauskas M, Marchetti M, Marques S, Mozgeris G, Navrátil R, Nieuwenhuis M, Orazio C, Paligorov I, Pettenella D, Sedmák R, Smrecek R, Stanislovaitis A, Tomé M, Trubins R, Tucek J, Vizzarri M, Wallin I, Pretzsch H, Sallnäs O (2015) How sensitive are ecosystem services in European forest landscapes to silvicultural treatment? Forests 6(5):1666–1695

Blais JR (1985) The ecology of the eastern spruce budworm: a review and discussion. In: Sanders CJ, Stark RW, Mullins EJ, Murphy J (eds) Recent advances in spruce budworms research. CANUSA Spruce Budworms Research Sympsoium Proceedings. Bangor ME, Sept 1984. Canada Forest Services/USDA Forest Services, Ottawa, ON, pp 49–59

Brang P, Spathelf P, Larsen JB et al (2014) Suitability of close-to-nature silviculture for adapting temperate European forests to climate change. Forestry 87:492–503

Bruelheide H, Nadrowski K, Assmann T et al (2013) Designing forest biodiversity experiments: general considerations illustrated by a new large experiment in subtropical China. Methods Ecol Evol 5:74–89

Burdon RD (2001) Chapter 5. *Pinus radiata*. In: Last FD (ed) Ecosystems of the world 19 – Tree crop ecosystems. pp 99–161

Busing RT, Fujimori T (2002) Dynamics of composition and structure in an old Sequoia sempervirens forest. J Veg Sci 13:785–792

Busing RT, White PS (1997) Species diversity and small-scale disturbance in an old-growth temperate forest: a consideration of gap partitioning concepts. Oikos 78:562–568

Carnol M, Baeten L, Branquart E et al (2014) Ecosystem services of mixed species forest stands and monocultures: comparing practitioners' and scientists' perceptions with formal scientific knowledge. Forestry 87:639–653

Clements FE (1936) Nature and structure of the climax. J Ecol 24:252–284

Cornwell WK, Cornelissen JHC, Allison SD et al (2009) Plant traits and wood fates across the globe—rotted, burned, or consumed? Glob Chang Biol 15:2431–2449

Currie DJ, Paquin V (1987) Large-scale biogeographical patterns of species richness of trees. Nature 329:326–327

Ellenberg H, Leuschner C (2010) Vegetation Mitteleuropas mit den Alpen in ökologischer, dynamischer und historischer Sicht. Ulmer Verlag, Stuttgart

Evans J, Turnbull M (2004) Plantation forestry in the tropics, 3rd edn. Oxford University Press, Oxford

FAO (Food and Agriculture Organization of the United Nations) (2006) Global Forest Resources Assessment 2005 – Progress towards sustainable forest management. FAO Forestry Paper 147

FAO (Food and Agriculture Organization of the United Nations) (2010) Global Forest Resources Assessment 2010 – Main report. FAO Forestry Paper 163

Fonseca CR, Ganade G (2001) Species functional redundancy, random extinctions and the stability of ecosystems. J Ecol 89:118–125

Forrester DI, Bauhus J (2016) A review of processes behind diversity – productivity relationships in forests. Curr For Rep 2:45–61
Forrester DI, Bauhus J, Cowie AL, Vanclay JK (2006) Mixed-species plantations of Eucalyptus with nitrogen-fixing trees: a review. For Ecol Manage 233:211–230
Gamborg C, Larsen JB (2003) 'Back to nature'—a sustainable future for forestry? For Ecol Manage 179:559–571
Gardiner BA, Quine CP (2000) Management of forests to reduce the risk of abiotic damage – a review with particular reference to the effects of strong winds. For Ecol Manage 135:261–277
Gaston KJ (2000) Global patterns in biodiversity. Nature 405:220–227
Gayer K (1886) Der gemischte Wald – Seine Begründung und Pflege, insbesondere durch Horst- und Gruppenwirtschaft. Paul Parey Verlag, Berlin
Gentry AH (1988) Tree species richness of upper Amazonian forests. Proc Natl Acad Sci U S A 85:156–159
Groot A, Horton BJ (1994) Age and size structure of natural and second-growth peatland Picea mariana stands. Can J For Res 24:225–233
Haggar JP, Briscoe CB, Butterfield RP (1998) Native species: a resource for the diversification of forestry production in the lowland humid tropics. For Ecol Manage 106:195–203
Halpern CB, Spiess TA (1995) Plant species diversity in natural and managed forests of the Pacific Northwest. Ecol Appl 5:913–934
Hansen A, Rotella J (1999) Abiotic factors. In: Hunter ML (ed) Maintaining biodiversity in forest ecosystems. Cambridge University Press, Cambridge, pp 161–209
Hart TB, Hart JA, Murphy PG (1989) Monodominant and species-rich forests of the humid tropics: causes for their co-occurrence. Am Nat 133:613–633
Hartig GL (1791) Anweisung zur Holzzucht für Förster. Neue Akademische Buchhandlung, Marburg
Hartley MJ (2002) Rationale and methods for conserving biodiversity in plantation forests. For Ecol Manage 155:81–95
Hector A, Bagchi R (2007) Biodiversity and ecosystem multifunctionality. Nature 448:188–190
Hemery GE, Clark JR, Aldinger E et al (2010) Growing scattered broadleaved tree species in Europe in a changing climate: a review of risks and opportunities. Forestry 83:65–81
Hewitt GM (1999) Post-glacial re-colonization of European biota. Biol J Linn Soc 68:87–112
Hooper DU, Chapin FSI, Ewel JJ et al (2005) Effects of biodiversity on ecosystem functioning: a consensus of current knowledge and needs for future research. Ecol Monogr 75:3–36
Hundeshagen JC (1826) Die Forstabschätzung auf neuen wissenschaftlichen Grundlagen. Laupp, Tübingen
Jonsson BG, Kruys N, Ranius T (2005) Ecology of species living on dead wood–lessons for dead wood management. Silva Fennica 39:289–309
Kazmierczak M, Backmann P, Fedriani JM et al (2016) Monodominance in tropical forests: modelling reveals emerging clusters and phase transitions. J R Soc Interface 13(117):20160123
Kelty MJ (1992) Comparative productivity of monocultures and mixed-species stands. In: Kelty MJ, Larson BC, Oliver CD (eds) The ecology and silviculture of mixed-species forests: a Festschrift for David M. Smith. Springer, Netherlands, pp 125–141
Kowalski T (2006) *Chalara fraxinea* sp. nov. associated with dieback of ash (*Fraxinus excelsior*) in Poland. For Pathol 36:264–270
Latham RL, Ricklefs E (1993) Global patterns of tree species richness in mosit forests: energy-diversity theory does not account for variation in species richness. Oikos 67:325–333
Lemmel H (1939) Die Organismusidee in Möllers Dauerwaldgedanken. Springer, Berlin
Levin SA (1998) Ecosystems and the biosphere as complex adaptive systems. Ecosystems 1:431–436
Liang J, Crowther TW, Picard N, Wiser S, Zhou M, Alberti G, Schulze ED, McGuire AD, Bozzato F, Pretzsch H, de-Miguel S, Paquette A, Hérault B, Scherer-Lorenzen M, Barrett CB, Glick HB, Hengeveld GM, Nabuurs GJ, Pfautsch S, Viana H, Vibrans AC, Ammer C, Schall P, Verbyla D, Tchebakova N, Fischer M, Watson JV, Chen HYH, Lei X, Schelhaas MJ, Lu H,

Gianelle D, Parfenova EI, Salas C, Lee E, Lee B, Kim HS, Bruelheide H, Coomes DA, Piotto D, Sunderland T, Schmid B, Gourlet-Fleury S, Sonké B, Tavani R, Zhu J, Brandl S, Vayreda J, Kitahara F, Searle EB, Neldner VJ, Ngugi MR, Baraloto B, Frizzera L, Balazy R, Oleksyn J, Zawila-Niedzwiecki T, Bouriaud O, Bussotti F, Finér L, Jaroszewicz B, Jucker T, Valladares V, Jagodzinski AM, Peri PL, Gonmadje C, Marthy W, O'Brien T, Martin EH, Marshall AR, Rovero F, Bitariho R, Niklaus PA, Alvarez-Loayza P, Chamuya N, Valencia R, Mortier F, Wortel V, Engone-Obiang NL, Ferreira LV, Odeke DE, Vasquez RM, Lewis SL, Reich PB (2016) Positive biodiversity-productivity relationship predominant in global forests. Science 354(6309):aaf8957

Loreau M, Hector A (2001) Partitioning selection and complementarity in biodiversity experiments. Nature 412:72–76

Loreau M, Naeem S, Inchausti P (eds) (2002) Biodiversity and ecosystem functioning. Oxford University Press, New York

Magurran AE (2004) Measuring biological diversity. Blackwell, Malden

Mantel W (1961) Wald und Forst. Wechselbeziehungen zwischen Natur und Wirtschaft, Rowohlt, Reinbek bei Hamburg

Masaki T, Suzuki W, Niiyama K et al (1992) Community structure of a species-rich temperate forest, Ogawa Forest Reserve, central Japan. Vegetatio 98:97–111

Maser C (1994) Sustainable forestry: philosophy, science, and economics. St. Lucie Press

Mason NW, Mouillot D, Lee WG, Wilson JB (2005) Functional richness, functional evenness and functional divergence: the primary components of functional diversity. Oikos 111:112–118

McDonald GT, Lane MB (2004) Converging global indicators for sustainable forest management. For Policy Econ 6:63–70

McElhinny C, Gibbons P, Brack C, Bauhus J (2005) Forest and woodland stand structural complexity: Its definition and measurement. For Ecol Manage 218:1–24

Messier C, Puettmann KJ, Coates KD (eds) (2013) Managing forests as complex adaptive systems: building resilience to the challenge of global change. Routledge, London

Molino JF, Sabatier D (2001) Tree diversity in tropical rain forests: a validation of the intermediate disturbance hypothesis. Science 294:1702–1704

Möller A (1922) Der Dauerwaldgedanke – Sein Sinn und seine Bedeutung. Springer, Berlin

Monserud RA (1975) Methodology for simulating Wisconsin northern hardwood stand dynamics. Univ Wisconsin-Madison, PhD thesis Abstracts 36, No 11

Mosandl R (2009) Geschichte der Wälder in Mitteleuropa im letzten Jahrtausend. In: Herrmann B (ed) Beiträge zum Göttinger Umwelthistorischen Kolloquium 2008 – 2009. Universitätsverlag Göttingen, Göttingen

Münker W (1958) Dem Mischwald gehört die Zukunft. Ausschuß zur Rettung des Laubwaldes im Deutschen Heimatbund. Deutscher Heimatverlag, Bielefeld

Myers N, Mittermeier RA, Mittermeier CG et al (2000) Biodiversity hotspots for conservation priorities. Nature 403:853–858

Naeem S (2002) Ecosystem consequences of biodiversity loss: the evolution of a paradigm. Ecology 83:1537–1155

Niemelä J (1997) Invertebrates and boreal forest management. Conserv Biol 11:601–610

Noss RF (1990) Indicators for monitoring biodiversity: a hierarchical approach. Conserv Biol 4:355–364

Osonubi O, Mulongoy K, Awotoye OO et al (1991) Effects of ectomycorrhizal and vesicular-arbuscular mycorrhizal fungi on drought tolerance of four leguminous woody seedlings. Plant Soil 136:131–143

Ovington JD (1983) Temperate broad-leaved evergreen forests. Ecosystems of the World 10. Elsevier, Amsterdam

Paquette A, Messier C (2011) The effect of biodiversity on tree productivity: from temperate to boreal forests. Glob Ecol Biogeogr 20:170–180

Pastor J, Mladenoff DJ, Haila Y et al (1996) Biodiversity and ecosystem processes in boreal regions. Scope – Scientific Committee on Problems of the Environment International Council of Scientific Unions 55:33–69
Petchey OL, Gaston KJ (2002) Functional diversity (FD), species richness and community composition. Ecol Lett 5:402–411
Petchey OL, Hector A, Gaston KJ (2004) How do different measures of functional diversity perform? Ecology 85:847–857
Peters R (1997) Beech forests. Geobotany, Vol 24. Kluwer Academic, Dordrecht
Pflug A (1892) Die wirtschaftliche erschliessung öder und geringwertiger liegenschaften durch künstliche aufforstungen. Zeitschrift für die gesamte Staatswissenschaft/J Inst Theor Econ:79–122
Poulson TL, Platt WJ (1989) Gap light regimes influence canopy tree diversity. Ecology 70:553–555
Powers RF (1999) On the sustainable productivity of planted forests. New For 17:263–306
Pretzsch H, Biber P, Ďurský J (2002) The single tree based stand simulator SILVA. Construction, application and evaluation. For Ecol Manage 162:3–21
Pretzsch H, Block J, Dieler J et al (2010) Comparison between the productivity of pure and mixed stands of Norway spruce and European beech along an ecological gradient. Ann For Sci 67:1–12
Pretzsch H, Bielak K, Block J et al (2013) Productivity of mixed versus pure stands of oak (*Quercus petraea* (Matt.) Liebl. and *Quercus robur* L.) and European beech (*Fagus sylvatica* L.) along an ecological gradient. Eur J For Res 132:263–280
Pretzsch H, Forrester DI, Rötzer T (2015) Representation of species mixing in forest growth models. Ecol Model 313:276–292
PROSILVA (2013) PORSILVA Europe – Integrated forest management for resilience and sustainability across 25 countries. http://prosilvaeurope.wordpress.com/. Accessed 10.08.2014
Puettmann KJ, Coates KD, Messier C (2009) A critique of silviculture: managing for complexity. Island Press, Washington, DC
Puettmann KJ, Wilson SMG, Baker S et al (2015) Silvicultural alternatives to conventional even-aged forest management – what limits global adoption? For Ecosyst 2:8
Rametsteiner E, Mayer P (2004) Sustainable forest management and pan: European forest policy. Ecol Bull 51:51–57
Reif A, Brucker U, Kratzer R et al (2010) Waldbewirtschaftung in Zeiten des Klimawandels – Synergien und Konfliktpotenziale zwischen Forstwirtschaft und Naturschutz. Naturschutz und Landschaftsplanung 42:261–266
Remmert H (1991) The mosaic-cycle concept of ecosystems—an overview. In: Remmert H (ed) The mosaic-cycle concept of ecosystems. Ecol Stud 85:1–21
Scherer-Lorenzen M, Körner C, Schulze ED (eds) (2005) Forest diversity and function: temperate and boreal systems. Ecological studies, Vol 176. Springer, Berlin
Schuck A, Päivinen R, Hytönen T, Pajari B (2002) Compilation of forestry terms and definitions. European Forest Institute, Internal Report No. 6, p 48
Schütz JP (2002) Silvicultural tools to develop irregular and diverse forest structures. Forestry 75:329–337
Seymour R, Hunter ML Jr (1999) Principles of ecological forestry. In: Hunter ML Jr (ed) Maintaining biodiversity in forest ecosystems. Cambridge University Press, Cambridge
Shannon C E (1948) A mathematical theory of communication. Bell Syst Tech J 27:379–423 and 623–656
Svenning J-C, Skov F (2004) Limited filling of the potential range in European tree species. Ecol Lett 7:565–573
ter Steege H, Pitman NC, Sabatier D, Baraloto C, Salomão RP, Guevara JE, Fine PV (2013) Hyperdominance in the Amazonian tree flora. Science 342:1243092-1–1243092-9
Thomasius H (1996) Geschichte, Theorie und Praxis des Dauerwaldes. Landesforstverein Sachsen-Anhalt e. V, Straßfurt

Tilman D, Reich PB, Knops J et al (2001) Diversity and productivity in a long-term grassland experiment. Science 294:843–845
Tuomisto H (2010) A diversity of beta diversities: straightening up a concept gone awry. Part 1. Defining beta diversity as a function of alpha and gamma diversity. Ecography 33:2–22
Verheyen K, Scherer-Lorenzen M (2011) TreeDivNet – a unique platform for ecosystem research in tree diversity experiments worldwide. http://www.treedivnet.ugent.be/index.html. Accessed 10.08.2014
Wagner S (2004) Möglichkeiten und Beschränkungen eines funktionsorientierten Waldbaus. Forst und Holz 59:105–111
Walker B, Kinzig A, Langridge J (1999) Plant attribute diversity, resilience, and ecosystem function: the nature and significance of dominant and minor species. Ecosystems 2:95–113
Wang X, Wiegand T, Wolf A et al (2011) Spatial patterns of tree species richness in two temperate forests. J Ecol 99:1382–1393
Wardle P, Bulfin MJA, Dugdale J (1983) Temperate broad-leaved evergreen forests of New Zealand. In: Ovington JD (ed) Ecosystems of the World 10. Elsevier, Amsterdam
Watt AS (1947) Pattern and process in the plant community. J Ecol 35:1–22
Wells P, Hickey J (1999) Wet sclerophyll, mixed and swamp forest. In: Reid JB, Hill RS, Brown MJ, Hovenden MJ (eds) Vegetation of Tasmania. Flora of Australia Supplementary Series 8:224–243
Whitham TG, Bailey JK, Schweitzer JA, Schuster SM (2006) A framework for community and ecosystem genetics: from genes to ecosystems. Nat Genet 7:510–523
Whittaker RH (1960) Vegetation of the Siskiyou Mountains, Oregon and California. Ecol Monogr 30:279–338
Whittaker RH (1977) Evolution of species diversity in land communities. Evol Biol 10:1–67
Whittaker RH, Woodwell GM (1969) Structure, production and diversity of the Oak-Pine Forest at Brookhaven, New York. J Ecol 57:155–174
Wiedemann E (1923) Zuwachsrückgang und Wuchsstockungen der Fichte in den mittleren und unteren Höhenlagen der sächsischen Staatsforsten. Kommissionsverlag W Laux, Tharandt
Wiedemann E (1942) Der gleichaltrige Fichten-Buchen-Mischbestand. Mitteilungen aus Forstwirtschaft und Forstwissenschaft 13
Wyatt-Smith J (1987) The management of tropical moist forest for the sustained production of timber: some issues. International Union for Conservation of Nature and Natural Resources (IUCN)
Yachi S, Loreau M (1999) Biodiversity and ecosystem productivity in a fluctuating environment: the insurance hypothesis. Proc Natl Acad Sci U S A 96:57–64
Yang X, Bauhus J, Both S et al (2013) Establishment success in a forest biodiversity and ecosystem functioning experiment in subtropical China (BEF-China). Eur J For Res 132:593–606
Zilber-Rosenberg I, Rosenberg E (2008) Role of microorganisms in the evolution of animals and plants: the hologenome theory of evolution. FEMS Microbiol Rev 32:723–735
Zobel BJ, Van Buijtenen JP (1989) Wood variation: its causes and control. Springer, Berlin

# From Observations to Evidence About Effects of Mixed-Species Stands

# 2

Jürgen Bauhus, David I. Forrester, and Hans Pretzsch

**Abstract**

A critical understanding of the scientific evidence that we have about the effects of tree species diversity on ecosystem properties and processes is required to guide practical forest management as well as future research. However, current understanding is limited by the lack of an appropriate framework for evaluating the reported evidence. In this chapter we outline how research on mixed-species forests may fit into concepts of ecosystem hierarchy and how previous studies may be ranked regarding their level of evidence. We introduce the most important hypotheses and theories underpinning research on the relationship between tree diversity and ecosystem functioning and illustrate how these may be tested by analyses of forest inventories, experiments, and exploratory research platforms or a combination of these.

## Contents

2.1 Introduction .......... 28
2.2 Hierarchical Levels of Patterns and Processes for Analysing Mixed-Species Stands ... 30
2.3 Some Basic Ecological Concepts Related to Mixed-Species Stands .......... 33
2.3.1 Biodiversity and Ecosystem Functioning .......... 33

J. Bauhus (✉)
Chair of Silviculture, Faculty of Environment and Natural Resources, University of Freiburg, Tennenbacherstr. 4, 79106 Freiburg, Germany
e-mail: juergen.bauhus@waldbau.uni-freiburg.de

D.I. Forrester
Swiss Federal Institute for Forest, Snow and Landscape Research WSL, Zürcherstrasse 111, 8903 Birmensdorf, Switzerland
e-mail: david.forrester@wsl.ch

H. Pretzsch
Chair for Forest Growth and Yield Science, Faculty of Forest Science and Resource Management, Technical University of Munich, Hans-Carl-von-Carlowitz-Platz 2, 85354 Freising, Germany
e-mail: Hans.Pretzsch@lrz.tum.de

 
H. Pretzsch et al. (eds.), *Mixed-Species Forests*, DOI 10.1007/978-3-662-54553-9_2

2.3.2 Fundamental and Ecological Niches .... 36
2.3.3 Species Traits and Functional Diversity .... 42
2.3.4 Important Ecological Interactions .... 45
2.3.5 Tree Diversity and Diversity at Other Trophic Levels .... 49
2.4 Analysing Ecosystem Properties and Processes in Tree Species Mixtures Based on Experiments and Inventories .... 51
2.4.1 General Considerations Regarding Tree Diversity Experiments .... 51
2.4.2 Forest Inventories .... 53
2.4.3 Mixed-Species and Tree Species Diversity Experiments .... 55
2.4.4 Exploratories .... 63
Take-Home Messages .... 64
References .... 66

## 2.1 Introduction

It is important to have a framework for assessing the quality and reliability of our knowledge about mixed-species forests. Not all knowledge that we find documented in textbooks, journal articles, or other media is of equal value. The scientific evidence underpinning this information can vary greatly and so does the confidence that we can place in this information (Forrester and Pretzsch 2015). For example, already in 1828 the early forest scientist Heinrich von Cotta claimed 'Since not all tree species utilise resources in the same manner, growth is more lively in mixed stands and neither insects nor storms can do as much damage; also, a wider range of timber will be available everywhere to satisfy different demands. The quest to establish pure stands everywhere is based on an old and most damaging prejudice . . .' (Cotta 1828, p. 155, translated by authors). At that time, this expert judgement of Cotta was simply a notion and was likely based on observations in the field. Furthermore, since the history of modern scientific experiments in forestry began only at the end of the nineteenth century, we can safely assume that Cotta's impressions regarding growth in mixtures did not benefit from the rigour of controlled experiments. The foundation of the International Union of Forest Research Organizations (IUFRO) in 1892, which initiated coordinated experiments and international research collaboration, can be regarded as a landmark event in this regard.

Nowadays there is a strong demand to embrace standards of evidence in most scientific fields. The ideas and concepts developed to support 'evidence-based medicine' are now central to fields such as 'evidence-based conservation' (Sutherland et al. 2004) and are also influencing the forest sciences (Binkley and Menyailo 2005; Petrokofsky et al. 2012). The demand for evidence-based forestry, similar to evidence-based conservation practice, reflects the current situation where most decisions are based upon expert judgement and anecdotal sources, as opposed to evidence, and where very little evidence is collected on the consequences of particular practices (Sutherland et al. 2004). For that reason, Binkley and Menyailo (2005) proposed a rating scheme for the confidence warranted in forestry studies (Fig. 2.1). This scheme can be easily applied to all knowledge on mixed-species

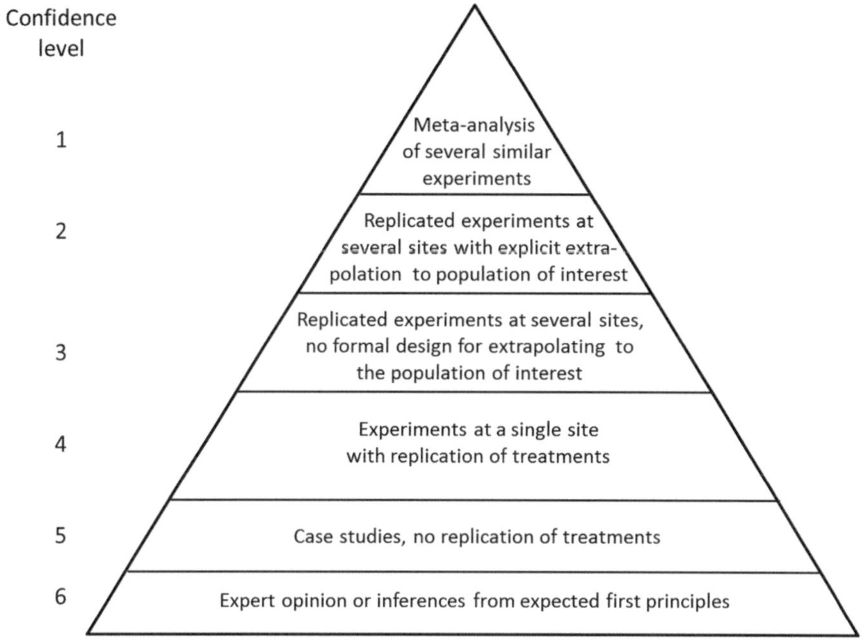


**Fig. 2.1** Hierarchy of confidence warranted in forestry studies and experiments ranging from expert opinions (6 = lowest level) to meta-analysis (1 = highest level); adopted from Binkley and Menyailo (2005). See text for examples of these different levels of evidence

forests. In this context, statements by early proponents for mixed-species silviculture can be regarded as expert opinions gained from observations (e.g. Möller 1922, and others in Chap. 1). Replicated experiments at single or multiple sites have greatly improved our understanding of processes and ecosystem functioning in mixed-species stands in tropical, subtropical, and temperate regions (e.g. Bauhus et al. 2000; Scherer-Lorenzen et al. 2007; Potvin and Gotelli 2008; Yang et al. 2013). Whereas, in the past, when studies were conducted at multiple sites, the variation in the response variable among sites was often treated as noise, recent studies explicitly analyse that variation to identify the influence of environmental factors, such as temperature or precipitation on performance of mixed-species communities (e.g. Forrester et al. 2013; Toïgo et al. 2015). A high level of evidence has been achieved in particular aspects of mixed-species forests. Meta-analyses have been carried out that confirmed consistent effects of mixtures on productivity (Piotto 2008; Liang et al. 2016), tree nutrition (Richards et al. 2010), or susceptibility to herbivory (Jactel and Brockerhoff 2007; Castagneyrol et al. 2014).

To facilitate a better appreciation of the information provided in this book, we will describe in this chapter the spatial and temporal scales and the different approaches that can be used to study ecological interactions between different tree species and the effects of mixed-species stands on ecosystem properties and processes. Beforehand, we will reflect on some important ecological hypotheses and theories related to mixed-species stands, since the exploration of these theories and testing of these hypotheses also determines the design of studies in mixed-species forests.

## 2.2 Hierarchical Levels of Patterns and Processes for Analysing Mixed-Species Stands

Forest ecosystems are enormously complex, self-organised systems that behave in non-linear ways (Messier et al. 2013). They contain vast numbers of species and individuals of a wide range of taxonomic groups that interact continuously with each other and with their abiotic environment. The complexity of such systems typically increases with the number of dominant tree species, which are again associated with specific species of different taxonomic groups (mycorrhiza, herbivores, etc.). In addition, processes in forest ecosystems occur at vastly different temporal and spatial scales, from seconds to thousands of years, and from the cellular to the continental level (Fig. 2.2). Scientists have therefore struggled to find appropriate approaches to analyse such systems (Looijen 1998). One such approach follows a reductionist paradigm. Here, the aim is to simplify complex systems by breaking them down into a number of subsystems that can be analysed separately one after the other. The results of these analyses of subsystems are then tried to be integrated and applied to the whole system. In ecology, this approach has been critiqued for a number of reasons including the possibility of failing to recognise important interactions between the subsystems and emergent properties of the whole system that result from these interactions.

| Scales | Observed vs. predicted | |
|---|---|---|
| Landscape | $L_{obs}$ | $L_{pred}$ |
| Stand | $S_{obs}$ | $S_{pred}$ |
| Plant | $P_{obs}$ | $P_{pred}$ |
| Organ | $O_{obs}$ | $O_{pred}$ |
| Cell | $C_{obs}$ | |

**Fig. 2.2** Mixed-species stands' research by scale-overlapping analysis of structure and processes reaching from cell to organ, plant, stand, and landscape level (denoted $C, O, P, S, L$). The observed system variables at a certain level ($C_{obs}$, $O_{obs}$, $P_{obs}$, $S_{obs}$, $L_{obs}$) can be used to predict the expected behaviour for the next higher level ($O_{pred}$, $P_{pred}$, $S_{pred}$, $L_{pred}$) through linear temporal or spatial upscaling (symbolised by the ladders). Deviations between the observed and predicted values at the next higher level indicate knowledge gaps and the limited relevance of lower-level processes for system behaviour at the next higher level (Pretzsch 2009, p 30)

A contrasting approach to deal with complexity of systems is holism, which regards the whole system as more than the sum of its parts of subsystems. This approach explicitly attempts to analyse the interactions between all parts of complex systems and considers all relevant spatial and temporal scales. It is obvious that holistic approaches to analyse complex ecosystems require an enormous amount of effort to study all the possible relationships and interactions between the entities of the system. This is in most cases not achievable.

Given the advantages and disadvantages of these two contrasting approaches, it is not surprising that a complementary approach combining the best aspects of reductionism and holism was developed. In hierarchy theory (Allan and Starr 1988), a reduction in complexity is accomplished on the one hand by concentrating on one single research question and relevant entities of the whole system. On the other hand, this approach considers different hierarchical levels of the system that operate at different spatial and temporal scales and explicitly analyses the processes and interactions between these levels.

The structures and processes at the different hierarchical levels of ecosystems are characteristically interwoven and obey their own principles: the upper levels exert pressure or constraints downwards through regulating processes (top-down control), whereas the lower levels influence the upper levels via other constraints and signals (bottom-up control) (Müller 1992; Ulrich 1994). For example, tree and stand structure regulate atmospheric deposition of elements into forest stands, as well the interception of radiation and precipitation, and determine the initial and limiting conditions for processes at the organ level (e.g. leaf). Processes between trees and the environment such as transpiration or processes at the individual scale such as assimilation and allocation of carbon are dominated by the temporal and spatial processes and structures at the levels above (Fig. 2.2).

Processes that operate on long time scales and moderate to large spatial scales possess a higher level of integration and are more than just the sum of the subordinate processes. The feedback mechanisms between processes at the same or different hierarchical levels shape the characteristic behaviour of biological systems, which cannot be deduced from an isolated view of the underlying elements and processes. Despite all the experimental accuracy, knowledge about soil chemical, biochemical, or physiological processes at a high spatial and temporal resolution (low level of hierarchy) cannot replace investigations at higher levels of integration such as forest growth research at long-term experimental plots. Hence, it is problematic to scale up from lower levels of ecosystem hierarchy to predict system responses at higher levels of hierarchy. For example, in the context of acid rain research, the predictions on growth and productivity of forests that were based on insights about the effects of acid deposition on tree organs, soil chemistry, input-output element balances, fine roots, and other plant organs did not match the actual observations that forest productivity increased over the same period (Spiecker et al. 1996; Pretzsch et al. 2014). Obviously, not all lower-level processes exert a strong enough signal to influence ecosystem behaviour at the higher level, and not all higher-level processes that buffer lower-level processes had been adequately considered.

Likewise, we can observe negative interactions between tree species in mixed stands that can be quantified as lower foliage nutrient concentrations (e.g. Richards et al. 2010) or reduced rooting depth of one species when compared to its performance in monocultures (Schmid and Kazda 2001). Yet, growth of this species, and others for that matter, may still be improved owing to compensatory mechanisms in other organs and at other spatial and temporal scales. Also the growth of individual trees for a given tree size may not differ between monocultures and mixtures. However, owing to niche complementarity, the packing of trees in mixed stands may be higher and thus a higher productivity can be achieved at the community level (e.g. Jucker et al. 2015, and see Chap. 4).

Currently the investigations of tree and stand growth are developing towards higher spatial and temporal resolution and refined causal explanations of structure and functioning. However, often scientific findings at the gene, cell, or organ level do not have any clear relevance at the tree or stand level. For example, stand-level growth across longer periods of time cannot be extrapolated effectively from short-term measurements of small parts of a few trees. Feedbacks between processes either within the same level or between different hierarchical levels influence the characteristic behaviour of ecosystems, which cannot be understood by considering the subprocesses in isolation. A solid understanding of mixed-species stand dynamics requires measurements at multiple hierarchical levels, e.g. stand and tree level, or tree and organ level. To understand the development of organs, trees, or stands, investigations should be conducted at multiple temporal and spatial scales as shown schematically in Fig. 2.2. When the results from observations at the plant level ($P_{\mathrm{obs}}$) (e.g. increment or mortality of individual trees in relation to stand density) are scaled up, they predict certain expected patterns at the stand level ($S_{\mathrm{pred}}$). The temporal or spatial upscaling to the higher level (symbolised through the ladders in Fig. 2.2) may be carried out through simple addition or multiplication or by modelling. If measurements at the stand level ($S_{\mathrm{obs}}$) (e.g. long-term records of stand development based on repeated inventories) do not match the expected pattern ($S_{\mathrm{obs}} \neq S_{\mathrm{pred}}$), this indicates some kind of buffering of the signal from the lower level and is probably attributable to reduced relevance of the processes at the individual plant or species level for explaining the system behaviour at the stand or community level. For example, through competitive reduction, mortality of species A in mixture may be lower than in monoculture. Obviously, the transition from tree to stand level introduces new effects, which cannot be deduced from the individual plant or species level owing to ecological interactions (e.g. facilitation, antagonism) and acclimation of individual species to the situation in mixed-species stands. The discrepancy between predictions and observations at the particular system level can be used as a basis to formulate new hypotheses. Hence, research approaches that integrate over different levels of system hierarchy facilitate assessment of the relevance of results at a certain level for the next higher hierarchical level and often help to identify knowledge gaps. In this book we therefore aim to study the effects of mixing tree species from the organ and tree level (Chaps. 4–6) to the cohort level (Chap. 5), community and stand level (Chap. 6), and even management block and forest company level (Chap. 10).

## 2.3 Some Basic Ecological Concepts Related to Mixed-Species Stands

### 2.3.1 Biodiversity and Ecosystem Functioning

Prior to reflecting on how knowledge about mixed-species forests has developed over time from observations, forest inventories, and experiments, it is important to reflect on some ecological hypotheses and theories related to mixed-species forests, because this will inform us about the adequacy of the different approaches to study mixed-species forests.

Most relevant to our considerations are the concepts that link biodiversity to ecosystem functioning. Here we will concentrate on the effects of tree species richness and diversity, which may also be expressed as the diversity of functional traits, on ecosystem processes. This is not a new question and has occupied ecologists and forest scientists for quite some time (Scherer-Lorenzen et al. 2005a). In ecology, this question has been mostly driven by the concern about the loss of biodiversity. A typical question that has been asked is: what happens to ecosystem functioning if we lose species from the ecosystem or community? This focus on the relationship between biodiversity and ecosystem functioning can be regarded as a paradigm shift in ecology, where previously a major objective was to understand the regulation and maintenance of biological diversity. Hence, biodiversity moved from a response variable to an explanatory variable (Hillebrand and Matthiessen 2009). The book by Schulze and Mooney (1993) on biodiversity and ecosystem functioning is often regarded as a landmark publication that represents this shift.

In forest sciences, the focus on mixtures started earlier, and this was initially not motivated by the question of what happens if we lose tree species from a forest community, but by the question of what can be achieved, mostly in relation to productivity and ecological stability, if we reintroduce species to forests that had previously been simplified, for example, through conversion to monocultures (see previous chapter). This has often resulted in the study of two-species mixtures (e.g. Wiedemann 1942; Pretzsch 1995, 1997; Forrester et al. 2006). Today, where forestry is concerned about providing the whole range of ecosystem goods and services, and where we are experiencing the loss of tree species from large regions, forest scientists have equally embraced the biodiversity and ecosystem functioning paradigm to carry out interdisciplinary studies jointly with ecologists to study the effects of tree species diversity on ecosystem functioning (e.g. Baeten et al. 2013; van der Plas et al. 2016; Verheyen et al. 2016). While a large range of response variables such as those related to nutrient and water cycles, herbivory, pathogens, decomposition, dependent taxonomic diversity, and many others are typically being analysed in these studies, they all include productivity as an important and integrative measure of ecosystem functioning. Productivity may be regarded as an important 'performance currency' of vegetation communities of forest stands (*sensu* McGill et al. 2006) and will therefore feature as an important dependent variable in this chapter and book.

Many different theories exist about how the magnitude of ecosystem processes or properties could respond to changes in species or functional diversity (e.g. Naeem 2002). The patterns of this relationship depend on a range of factors including the degree of dominance of the species lost/removed from, or added to a community, the type and strength of interactions with other species, the functional traits of species lost from the community, added to it, and from those remaining, and the relative strength of biotic versus abiotic control over the ecosystem process or property (Hooper et al. 2005).

Theories describing the relationship between species diversity and ecosystem functioning vary in attribution of importance to individual species effects on ecosystem functioning. For example, the **redundancy hypothesis** assumes that the magnitude of ecosystem processes or properties increases with the number of species in the ecosystem, but only up to a point beyond which additional species become redundant and do not have any further effect on the magnitude of the particular ecosystem process or property (Fig. 2.3). Or if we follow the curve from the right side, the loss of (redundant) species is compensated by other species remaining in the community that have similar functional traits to those that have been lost. The lower the total number of species in a community, the lower is the probability that species are functionally similar, for example, in a mixed-species boreal forest, where most genera are represented only by one or two species (Pastor et al. 1996). Hence, the redundancy hypothesis is closely linked to the insurance hypothesis, which assumes more stable ecosystem functioning at higher levels of species richness in the face of stress and disturbances (Yachi and Loreau 1999).

A similar response pattern may also be attributed to a **sampling effect** for dominant species, where the maximal magnitude of the response variable is also achievable with particular combinations of species at low levels of diversity or species richness. The increase in average magnitude of the response variable is attributable to the increasing probability of including the most effective species with increasing species richness (Wardle 1999; Hooper et al. 2005).

The **complementarity hypothesis** assumes that the magnitude of ecosystem processes and properties increases with the number of species through positive interactions such as complementarity or facilitation among species. Each species added to an ecosystem contributes to the magnitude of ecosystem processes, although the increase in the magnitude of a particular process increases decrementally as more species are added to the system, following the law of diminishing marginal benefit.

In addition, one might assume that there is no fixed relationship between species richness/diversity and ecosystem functions. This may be the case if the latter are dominated by abiotic factors, or when there is a strong overlap of resource use by different species, or if the ecosystem property or process is strongly dominated by one particular species that is present at all levels of diversity (Hooper et al. 2005). In this case, not the number but the identity of species present in a given community would be important.

Which of these and other theories receives most empirical support is not certain. This varies, for example, with the particular ecosystem process or property that is

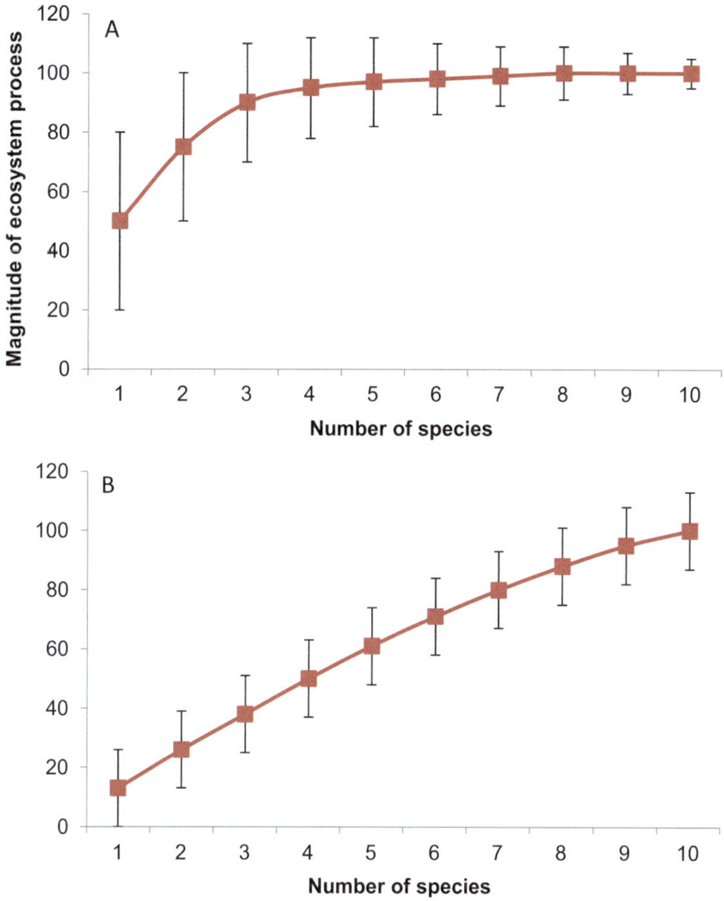


**Fig. 2.3** (**a** and **b**) Some hypotheses considering the relationship between species numbers and ecosystem functioning. (**a**) **Redundancy hypothesis**: magnitude of ecosystem processes increases with the number of species in the ecosystem, but only up to a point beyond which more species are redundant and do not have any additional effect on the average magnitude of the particular ecosystem process; however, the variability of that process may still decrease further with species richness. A similar response pattern may also be attributed to a **selection effect** for dominant species, where the maximal magnitude of the response variable is also achievable with particular combinations of species at low levels of diversity or species richness. The increase in average magnitude of the response variable is attributable to the increasing probability of including the most effective species with increasing species richness (Hooper et al. 2005). (**b**) **Complementarity hypothesis**: The magnitude of ecosystem processes increases with number of species through positive interactions such as complementarity or facilitation among species. Each species added to an ecosystem contributes to the magnitude of ecosystem processes, although the increase in the magnitude of a particular process increases decrementally as more species are added to the system. Once there is at least one representative species of each functional type, the effect of increasing species numbers should begin to saturate. At which point the relationship saturates depends on the degree of niche overlap among the different species (Hooper et al. 2005). Below this saturation point, any change in species numbers leads to a change in the magnitude of the ecosystem process

being regarded as a response variable and with the scale at which species richness or diversity is being analysed. In communities with few species, the left parts of Fig. 2.3a and b, it may be difficult to differentiate between the different response patterns and hence theories. This may be the case for many forest ecosystems in boreal and temperate regions (see Chap. 1). To identify which of these different hypotheses apply for certain ecosystems, it is necessary to compare the performances of individual species in mixtures and monocultures (e.g. Loreau and Hector 2001; Bruelheide et al. 2014).

Often relationships between tree diversity or species richness and ecosystem functioning are depicted for a measure of productivity, which is our 'performance currency' of tree communities *sensu* MCGill et al. (2006). On average, this relationship between tree diversity and forest productivity is positive across the major biomes (Forrester and Bauhus 2016; Liang et al. 2016). When other ecosystem functions are being regarded, the relationships between tree diversity/species richness and the particular ecosystem property or process are often less clear or follow different patterns; they may decline with high levels of species diversity (e.g. Gamfeldt et al. 2013), or there may be outright negative effects between species diversity and ecosystem functioning, for example, with regard to drought resistance or groundwater recharge in forest ecosystems (Grossiord et al. 2014).

The effects of tree species richness or diversity on ecosystem functioning depend largely on the type and intensity of ecological interactions that occur within particular species mixtures at a given site, and this is in turn driven by the particular attributes of species or their functional traits (Forrester and Bauhus 2016). For example, the magnitude of a range of ecosystem pools and processes in a mixed-species forest consisting only of evergreen coniferous species is likely to be less than in a forest consisting of the same number of species but covering a wider range of functional types, e.g. including pioneer small-leaved hardwood species, shade-tolerant and deep-rooting deciduous species, etc. Functional traits such as rooting depth, shade tolerance, wood density, specific leaf area, etc. determine the niches that species can occupy (fundamental niche), and the ecological interactions such as competition determine the actual niche of a species within a community at a given site (realised niche).

In the next sections we will therefore cover the concepts of niches, functional traits, and ecological interactions.

### 2.3.2 Fundamental and Ecological Niches

The concept of fundamental and realised niches as n-dimensional spaces, where every factor that influences the survival and fitness of a species such as temperature, moisture, or soil chemical reaction represents one dimension of this space, is commonly traced back to Hutchinson (1957). The fundamental niche describes the multidimensional space in the environment, where a species could potentially grow and reproduce. It is largely determined by the physiology of a given species, which reflects its evolutionary adaption to environmental conditions such as temperature, and the availability of water, light, and nutrients.

Fundamental niches are typically quantified by measuring growth or surrogates for it in relation to the variation of environmental variables, for example, to develop physiological response curves such a growth in relation to light availability (McGill et al. 2006). To describe the full physiological niche of a species, growth response curves would have to be determined in relation to all other factors that influence growth. The fundamental niche of a species is not constant and may change with the ontogeny of individuals (Poorter 2007). For example in trees, shade tolerance is often higher in seedlings than in mature trees of the same species. Hence, analyses of species differences commonly focus on a particular attribute, e.g. photosynthesis in relation to light availability or transpiration in relation to soil water potential, but do not attempt to quantify complete niches, which would be extremely difficult if not impossible.

The realised (or ecological) niche describes the multidimensional, not physical, space in the environment that is actually occupied by a species. Hence, it is a functional concept that describes the role that a species plays in an ecosystem and the factors that are required for its growth and survival. In most cases, the realised niche is a subset or proportion of the fundamental niche, where the actual occurrence of a species is limited by dispersal, disturbances, and biotic interactions. However, biotic interactions such as facilitation (see below) that benefit a species may also expand its realised niche, even if only in one or few dimensions of the environmental space. Current approaches to depict realised niches involve, for example, habitat modelling, where the occurrences of species are related to geographically referenced environmental variables. These concepts will be explained below using a two-species model.

The overlap in fundamental niches of species is a necessary, though insufficient prerequisite for determining whether species can co-occur in mixtures (Fig. 2.4a). A mixture of species may occur only at sites, where their fundamental niches overlap in terms of resource availability and environmental conditions (Fig. 2.4b, grey area). In man-made ecosystems such as plantations, these may be species that do not occur together in the natural environment, for example, when they originate from different forest types or even continents.

It is often assumed that, under steady-state conditions, the fitness of a species, measured as growth, reproductive rate, etc., is highest near the centre of the fundamental niche. If only one dimension of the niche space is depicted, this may be illustrated as an optimum curve in relation to the respective environmental variable (Fig. 2.5). Based on this concept we can assume that if a site lies near the centre of the fundamental niche of species 1 (Fig. 2.4b, site p), then intraspecific competition among individuals of this species is higher than interspecific competition with individuals from species 2, because species 2 occurs at the fringe of its niche at site p. However, under changing environmental conditions (Fig. 2.4c), site p represents the area of worst possible conditions or mortality for species 2.

Through the competitive influence of the two species on each other (e.g. shading, physical damage), the width of the fundamental niche of each species may be

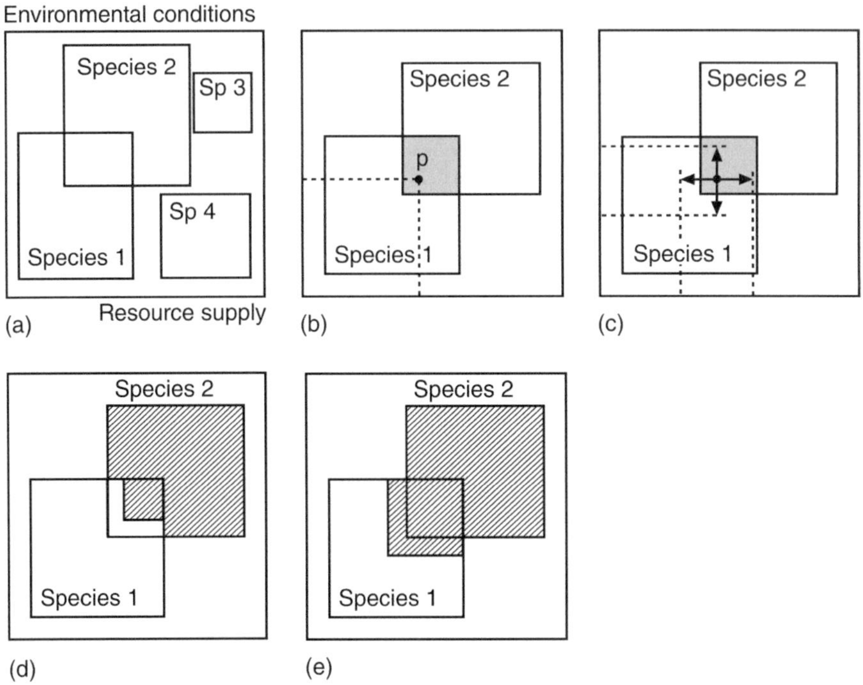


**Fig. 2.4** (**a–f**) Overlap of fundamental and realised niches as a prerequisite for a stable, productive species mixture: (**a**) fundamental niches of species defined by their potential occurrence in a matrix of resource supply and environmental conditions, (**b**) species 1 and 2 have a common niche (*grey area*) which includes site p, (**c**) resource supply or environmental conditions at site p can fluctuate (*arrows*) so that they extend beyond the fundamental niches of species 1 and 2, (**d**) the realised niche of species 2 can be restricted (*hatched area*) due to competition by species 1, or (**e**) the realised niche of species 2 can be extended (*hatched area*) with the admixture of species 1

restricted to a realised niche that is narrower than the fundamental niche. The area of possible coexistence of two species, therefore, may be smaller than the intersection of their fundamental niches. For example, species 1 causes a reduction in the fundamental niche of species 2, so that species 1 and 2 can coexist only in the hatched area (Fig. 2.4d). This reduction of the realised niche, or even the competitive exclusion of a species, occurs frequently when species differ greatly in competitive strength. In the case of trees, this may be related to differences in height growth development (species-specific ontogenesis), final height, and shade tolerance. The limited access of the inferior species to light can lead to a permanent decline and, ultimately, in the absence of disturbance, to the elimination of a species from a forest stand.

The realised niche of a species can also be extended through facilitative interactions with other species. For example, N-fixing species may facilitate establishment and growth of a species at nutrient poor sites where it could not otherwise

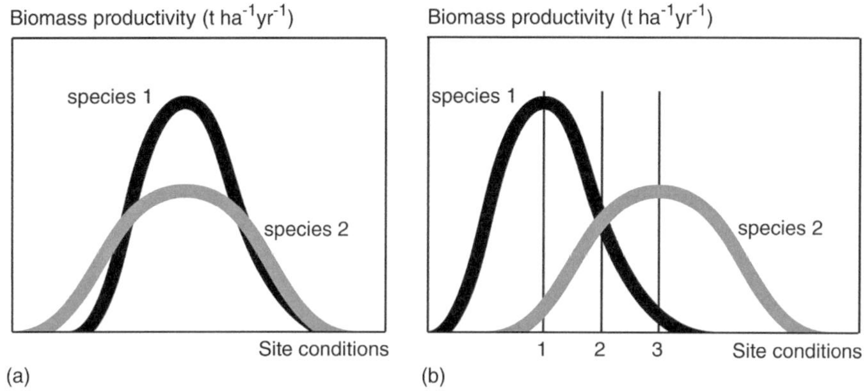


**Fig. 2.5** (**a**, **b**) Dry biomass productivity of different tree species in relation to site conditions: (**a**) growth of two tree species with similar fundamental niches but different levels of productivity; in this situation, species 1 would likely dominate the mixture at most site conditions, except at lower site quality; (**b**) growth of two tree species with different fundamental niches. Numbers 1–3 below the abscissa represent different site conditions with different growth responses by species 1 and 2; here species 2 would become increasingly dominant in mixture when moving from site condition 1 to 3 (Pretzsch 2005)

thrive. Figure 2.4e depicts this arrangement. Here, the area of possible coexistence of species 1 and 2 extends beyond the fundamental niche of species 2.

Coexistence of two or more species in the same niche does not mean that they use the same strategies to access the available resources, as has been mentioned above under the complementarity hypothesis. For a given site condition (site p in Fig. 2.4 a and b), species 1 and 2 may access resources in spatially and temporally different ways.

In contrast, in pure stands, genetically similar individuals with similar phenology, morphology, and physiology compete in space and time for the same pool of resources. The greater the niche differentiation achieved by mixing species, and hence the greater the diversity of functional traits, the higher the probability for complementary resource use and relaxed interspecific resource competition (Fig. 2.6). Considering the niche space of an ecosystem, depicted as two-dimensional space using two resource gradients, it is obvious that this niche space is more easily filled to a larger degree by many species rather than one or few (Fig. 2.6), even though some species might be very dominant and occupy a large share of the niche space; see also chapter one on natural forests that are almost monospecific. This filling of the niche space with additional species represents a complementary use of available resources. Clearly, the picture becomes much more complex, when we have mixtures of many species and consider the whole range of environmental factors.

Niche differentiation is also an important factor contributing to the **insurance hypothesis** in relation to biodiversity. On the one hand, a high level of niche differentiation in a community with few species reduces redundancy. This means

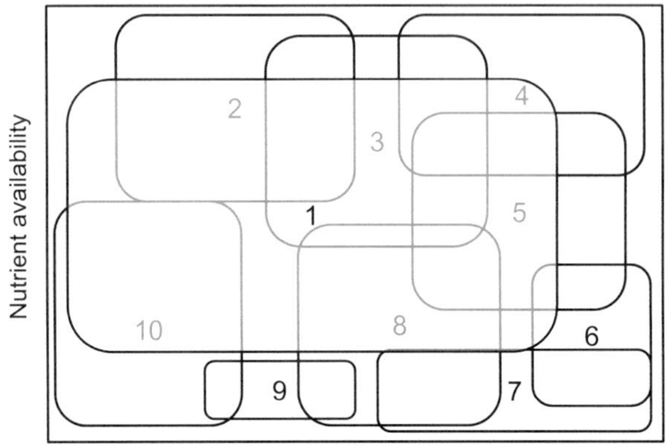


**Fig. 2.6** Two-dimensional niche space of an ecosystem with the fundamental niches of ten different species. Within an ecosystem, water and nutrient availability are not constant, and thus site conditions are variable. Under most situations one species such as species 1 in this example may be dominant (depicted by the shading of the niche space of other species). However, with changes in space or time and thus in resource availability, other species may gain dominance. Within this heterogeneous system, all species may coexist and the niche space provided within the ecosystem is filled to a large extent, leading to complementary resource use. The resources are used more fully by the whole community than by any single species or subset of species. Removing or losing one species from this system can be partially compensated by other species that share part of the fundamental niche

that the loss of one species with certain functional attributes cannot be fully compensated by another species with different functional attributes. On the other hand, in the case of changes in the type, frequency, or intensity of stress or disturbance factors, there is a greater likelihood that some species within the community can cope with the new conditions, when compared to a community with a low level of niche differentiation or diversity in functional traits. We can envisage this as a shift in the combination of environmental factors such as those depicted in Fig. 2.6 from one part of the niche space to another part; it is likely that there will be one or few species that can thrive under the new conditions. The greater the niche variation between the constituent species of a mixed stand, the more elastic its response will be to disturbances (Yachi and Loreau 1999). Returning to our example of a two-species mixture, if one species is more sensitive to a particular stress, the more robust species may then, on account of its better adaptation, profit from the weakening or mortality of the inferior species. Assuming a shift in environmental factors and resource availability reduces the fitness of species 1 (Fig. 2.7, arrow 1→3), this would lead to a significant decrease in productivity of pure stands of this species. However, if the same shift in environmental factors affected a stand where species 1 was mixed with a second species that was better adapted to the new growth conditions, stand growth would be much less affected. Species 2 would make better use of the available resources, and,

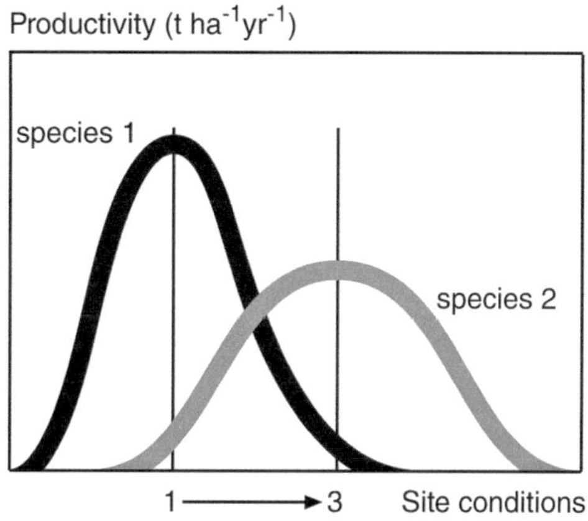


**Fig. 2.7** Dry biomass productivity for species 1 and 2 (*black* and *grey lines*, respectively) in relation to site conditions. The shift in site conditions from 1 to 3 (*arrows*) results in considerable growth loss for species 1. If species 1 and 2 are mixed, species 2 is able to compensate for a loss in biomass productivity

consequently, species 2 can improve its productivity and space sequestration. The same would happen if one species disappeared completely after a disturbance or through natural mortality (Fig. 2.8). In this case, the species remaining would contribute to ecosystem recovery and maintenance of ecosystem functioning through accelerated growth. This buffering or compensatory effect of the remaining species would be positively related to the density and regularity of its spatial distribution within the stand; it would be higher if the remaining species occurred in evenly spaced single tree or small group mixture rather than in irregular large patches. Managing forests in the form of mixed stands represents therefore a risk distribution through silvicultural diversification (Brang et al. 2014).

While fundamental and realised niches may provide a useful ecological concept, their use in quantitative analyses of the relationship between diversity and ecosystem functioning is limited. The concept has been widely applied to understand the coexistence of species within communities or to predict species distributions along environmental gradients or range shifts under future climatic conditions (Morin and Lechowicz 2008). It has had some success in explaining interactions in two or few-species systems, but has not been successful in providing general principles about interactions in and performance of communities with many species (McGill et al. 2006). One problem is that fundamental niches are rarely measured owing to the difficulty to quantify the performance of a species over a sequence of life stages and a large number of continuous abiotic gradients under controlled environmental conditions. This is one reason why different approaches based on functional traits have been proposed to study diverse communities (e.g. Violle and Jiang 2009). In addition, for many communities, interactions are diffuse, and considering each pairwise interaction between species in a highly diverse community as a separate process is difficult or impossible (McGill et al. 2006).

**Fig. 2.8** Bark beetles have killed all Norway spruce (*Picea abies*) trees in a mixed spruce-beech (*Fagus sylvatica*) forest. This biotic disturbance agent is specific to the conifer species and does not affect the deciduous hardwood species. The surviving beech trees maintain forest condition and ecosystem functioning (Photo: J. Bauhus)

### 2.3.3 Species Traits and Functional Diversity

A trait is a well-defined, measurable property of organisms, which is ideally measured as a continuous variable and in well-defined units at the individual level. To be useful for the purpose of analyses of diversity effects, traits should vary more between than within species. Functional traits are those that strongly influence organismal performance, i.e. plant attributes that enhance establishment, growth, survival, reproduction, etc. Functional traits are, for example, important drivers of the distribution of species along environmental gradients (McGill et al. 2006). The underlying concept of trait-based approaches is that traits, not species identity, are fundamental units of species sorting from a regional pool of species and that therefore traits should be a relevant tool to quantify niches. This may be applied by tracking plant functional traits related to the performance of individuals (Violle and Jiang 2009). There has been increasing support for the idea that many ecological questions can be addressed effectively by understanding how and why functional traits are related, and how these traits affect ecological outcomes. In the context of this book, relevant questions to which this concept has been applied comprise the vegetation community change along environmental gradients, effects of plants and communities on ecosystem functioning, and ecological interactions between species. An important question is which of the many possible traits have the strongest influence on the fundamental and realised niche of species. Fortunately, there is already some agreement about important plant traits that should form the basis of functional plant classifications, either because they have strong predictive power in relation to important ecosystem responses to environmental change or these traits themselves have a strong influence on ecosystem processes (Cornelissen et al. 2003). Based on the identification of important functional traits, protocols and databases have been developed to measure and document these traits and to make them available for trait-based analyses (Kattge et al. 2011). In concrete analyses of niche similarities and competitive abilities of species within a community using functional traits, however, the potentially long list is often reduced to a few key functional traits such as leaf mass per unit area (LMA), wood density (WD), seed mass (SM), and tree maximum height (Kunstler et al. 2012).

In the previous section, we have seen that a complementary effect in diverse communities would be maximised if all species contributed, with increasing species richness, proportionately to the increase of an ecosystem response variable such as productivity (Fig. 2.3b). This may be achieved by effective niche differentiation between the species. Applying a trait-based approach to this concept, maximum complementarity would be expected with maximum divergence of traits in the community. Hence, it may be expected more generally that strong relationships between response variables of ecosystem functioning and species richness are more likely in communities with trait divergence and hence high functional diversity, i.e. with great variation in specific leaf area, shade tolerance, maximum height, etc. In contrast, in communities with a high degree of trait convergence across participating species, ecosystem functioning may not be or only weakly related to species richness. Thus, instead of species richness or diversity, measures of functional diversity may be used as independent variables in approaches to quantify ecosystem functioning responses (e.g. Tilman et al. 1997). Accordingly, experimental designs have been developed that allow the separation between species richness and functional diversity (e.g. Tobner et al. 2014, Fig. 2.13).

It may be due to this difference between species and functional diversity that we see also different influences on the magnitude of ecosystem properties and processes between biodiversity experiments and culturally influenced production ecosystems on the one side and between natural communities on the other side. In natural communities, a convergence of traits, for example, those related to leaf morphology and function, are often observed (Reich et al. 1999) where plants share a similar life history, morphology, and physiology (Grime 2006). In contrast, experiments that have tested biodiversity ecosystem functioning relationships have often selected specifically different functional groups to include a high degree functional diversity (Scherer-Lorenzen et al. 2005b). Likewise, in cultivated forests, managers have used insights from empirical evidence to combine functionally different species, such as light-demanding and fast growing with shade-tolerant and slow growing species (e.g. Scots pine and European beech; Pretzsch et al. 2015) or nitrogen-fixing with non-nitrogen fixing species (e.g. eucalypts with acacias; Forrester et al. 2006). Although some of the combinations of tree species or genera also occur in nature (e.g. eucalypts with acacias), it has to be kept in mind when evaluating the majority of these experiments that their results should not be directly extrapolated to natural plant communities (Wardle 2016).

In the past, plant functional groups have often been used to depict different niches, and there have been many studies demonstrating the strong influence of functional group richness on ecosystem processes such as primary productivity (e.g. Hooper et al. 2005). Such plant functional groups can be defined as sets of species which perform similarly in an ecosystem based on a set of common traits. They can be defined either in relation to the role of species for certain ecosystem processes (such as C assimilation or nutrient cycling) or to the response of species to biotic and abiotic environmental factors such as temperature, pathogens, etc. (Lavorel et al. 1997). The number of functional groups represented by the different species within a community has been commonly used as a measure of functional diversity (e.g. Hector et al. 1999). However, owing to the large number of traits that may characterise the role of a species within an ecosystem, unambiguously assigning species to certain groups is not trivial and has been criticised as arbitrary. Therefore, other measures of functional diversity have been developed, which are based on the measured traits of species and their distance and provide a continuous measure of the natural variation among species (Petchey and Gaston 2002; Laliberté and Legendre 2010). However, measuring all relevant traits of all species in a community can be an onerous task, if these data are not already available. To overcome this, phylogenetic diversity has been used as a proxy of functional diversity. The underlying assumption of this approach is that plant function is phylogenetically conserved (Swenson and Enquist 2009), i.e. that species that are closely related have similar expression of functional traits because these traits are the results of the evolutionary history and inherited from ancestors (Cadotte et al. 2009). However, trait variation may be only partially related to phylogenetic distances between species, and hence values for functional diversity of plant communities may be only partially correlated with phylogenetic diversity (Flynn et al. 2011). Importantly, the use of phylogenetic diversity alone does not provide a mechanistic understanding of the relationships between species and their environment or the ecological interactions among species.

Functional traits determine the fundamental niche of species, and they shape the ecological interactions in mixed-species forests and thus also influence the ecological niche. In the following we will discuss some important ecological interactions in mixed-species communities.

### 2.3.4 Important Ecological Interactions

There is a range of positive and negative ecological interactions that may occur between different species in a community. In the context of interactions between tree species, the most important interactions comprise competition, facilitation, and tolerance. Competition can be defined as the ecological interaction in which one organism or species consumes a resource that would have been available to and could have potentially been consumed by the other, and hence the fitness of one is lowered by the presence of another (Begon et al. 1996). The limited resource for which conspecific or heterospecific individuals compete, comprise, in the case of plants, light, water and nutrients. Competition among individuals of the same species is defined as intraspecific competition, whereas competition between individuals of different species is interspecific competition. Negative net effects may occur between tree species in mixtures, when the **competition** for resources between species is stronger than any competitive reduction or facilitation (Fig. 2.9c) (Callaway and Walker 1997). In this case, the interspecific competition is stronger than the intraspecific competition. In other words, for a given subject tree, a tree neighbour from the other species exerts more competition for resources than a conspecific neighbour.

Positive, **complementary** effects that result from mixing trees species can be expected, when competitive reduction or facilitation between the species participating in a mixture outweigh any negative effects of competition (Kelty 1992). In the case of **competitive reduction**, interspecific competition in the mixture is lower than intraspecific competition in the monocultures (Fig. 2.9d). Competitive reduction may occur through a range of differences in patterns and processes between monocultures and mixed-species stands both above ground as well as below ground (e.g. Bauhus et al. 2000; Lei et al. 2012). These differences in patterns and processes may be the result of niche separation, where different species use either different resources or the same resources but at different points in space or time. This enables more of the total available resources to be used by the mixed-species stand when compared to monocultures of the participating species (Vandermeer 1989). If competitive reduction relates to resources that limit growth, then this mechanism should lead to greater total productivity and possibly also a decreased loss of resources from the ecosystem (e.g. through nutrient leaching) (Hooper et al. 2005). An example of above-ground competitive reduction may be when a tall light-demanding species that uses a relatively small proportion of the available light is mixed with a shorter shade-tolerant species that can use a large proportion of the light that is transmitted through the canopy of the light-demanding species (e.g. Bauhus et al. 2004). The competition between light-demanding and shade-tolerant species may be further reduced, if they have different growth dynamics such that the light-demanding species grows much faster initially and thus can achieve a large proportion of its final height, before the shade-tolerant species catches up (Fig. 2.10). A typical example for this pattern can be seen in mixtures of *Betula* or *Populus* with more shade-tolerant conifers such as *Picea* and *Abies* (e.g. Man and Lieffers 1999). In Sweden, the productivity of the *Picea abies*

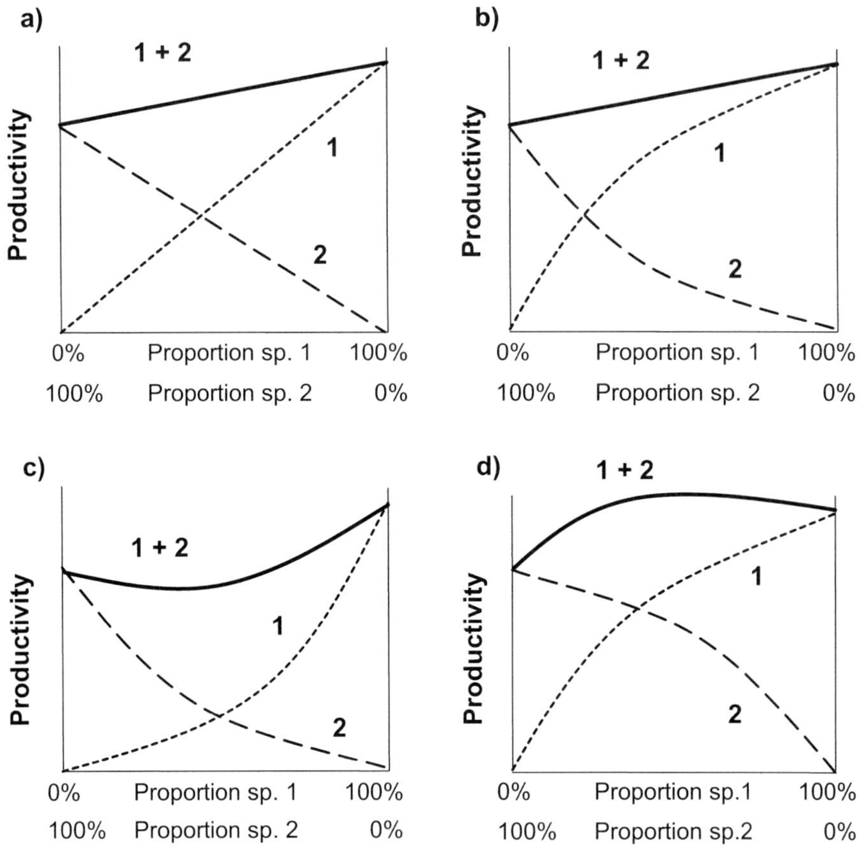


**Fig. 2.9** Different effects of mixing two tree species on the total productivity of the mixture (*fat line*) and the individual productivities of the participating species 1 and 2 (*dotted lines*) (after Harper 1977). The graphs can be thought of as depicting species replacement series, where the density of the mixture is the same as the monocultures and always 100% and where one species is gradually replaced by the other. In monoculture, species 1 is more productive than species 2. We can distinguish four different situations. (**a**) The two species do not interact, they tolerate each other. Interspecific competition equals intraspecific competition. Total productivity is the additive effect of the productivities of the individual species. (**b**) Species 1 benefits from the interaction, e.g. through facilitation or competitive reduction. The intraspecific competition for species 1 is higher than the interspecific competition with species 2; the reverse applies to species 2. However, these effects compensate each other so that the net effect for the mixture appears to be purely additive. (**c**) Interactions between the two species are antagonistic. Interspecific competition is higher than intraspecific competition for both species. At the level of the mixture, this leads to under-yielding. (**d**) Interactions between the two species are synergistic or mutualistic. Intraspecific competition is higher than interspecific competition for both species. This may result from facilitation or competitive reduction. This leads to overyielding at the level of the mixture

component in mixtures with a nurse canopy of 600 *Betula* trees ha$^{-1}$ was, up to the stand age of 25 years, not different from the productivity of pure *P. abies* stands (Tham 1994). Below-ground, competitive reduction may occur when trees with

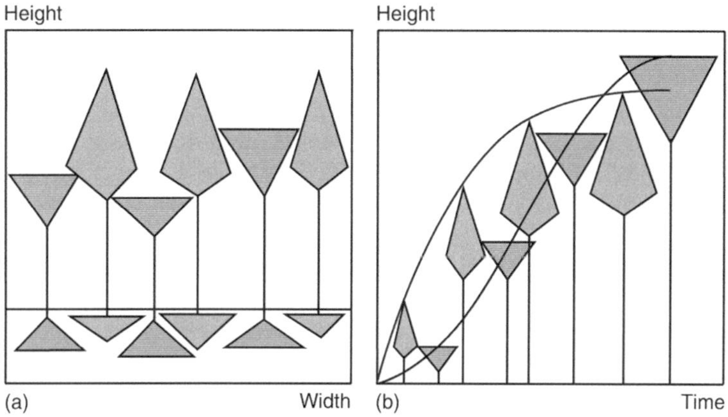


**Fig. 2.10** (**a**, **b**) Coexistence of species 1 and 2 can be facilitated and productivity increased through competitive reduction in the form of spatial or temporal niche differentiation by (**a**) stratification in the crown or root space or (**b**) asynchronous growth patterns of the species. Asynchronous growth may also take place below ground through different rates of the expansion of root systems, but this has so far not been documented

different root architectures and rooting depths (Fig. 2.10) take up water (e.g. Schwendenmann et al. 2014) or nutrients (Lang et al. 2014) from different soil horizons or take up different forms of the same nutrient (see Richards et al. 2010). In both cases, competitive reduction may also occur through differences in phenology, when evergreen and deciduous species are mixed or the seasonal growth of fine roots differs between species. In both cases, species will differ with regard to the time at which they can take up resources such as light, water, or nutrients. For example, evergreen species will benefit from mixtures with species that are deciduous during a seasonal drought period (Schwendenmann et al. 2014). Similarly evergreen species will likely benefit from adjacent leafless deciduous trees during periods in spring and fall, when they photosynthesise at higher levels of light availability when compared to situations with other evergreen trees as neighbours.

The concept of niche complementarity also comprises **facilitation**, which can lead to an increase in the magnitude of ecosystem pools or processes as species numbers or functional richness increase. Facilitation can be defined as a form of interaction among species that benefit at least one of the participants and cause harm to neither of the participating species. It can be classified as mutualism, in which both species benefit (Fig. 2.9d), or commensalism, in which one species benefits and the other remains unaffected (Stachowicz 2001). Facilitative interactions among species occur when certain species improve environmental conditions for others, for example, through accelerating litter decomposition (Gartner and Cardon 2004), through reducing damage to foliage or other organs through abiotic factors, such as lowering the probability of frost events through sheltering (e.g. Örlander 1993; Scowcroft et al. 2000), or biotic agents, such as disguising plants from insect herbivores (e.g. Castagneyrol et al. 2013), or where one species supplies a critical resource such as nitrogen through fixation to others (e.g. Forrester et al. 2006). Although mixing of N-fixing and non-N-fixing species represents differences in resource acquisition and benefits initially primarily the individuals of the N-fixing species, in the longer term, the N-enrichment of the system increases the turnover of plant material (nutrients in transported leaves and fine roots) and thereby the amount of resources introduced into the nutrient cycle (not just N), which become available to individuals of other species (Rothe and Binkley 2001; Piotto 2008). This does not mean, however, that the net effect of interactions between the N-fixing and non-N-fixing species is always positive for the latter (see examples in Callaway and Walker 1997).

**The Stress-Gradient Hypothesis**

The ecological interactions discussed above are not constant in space and time (Figs. 2.5 and 2.10; Chap. 3). That is one reason why it is so important to consider the extrapolation to the population of interest in the design of experiments on mixed species (Fig. 2.1). For example, synergistic effects between two species at an unfavourable site may not be representative for the same species combination at a site more favourable for plant growth, where the overall effect between the same species may be antagonistic.

Questions relating to how and why these interactions change, spatially or temporally, have received a considerable amount of attention, particularly in non-forest ecosystems (Bertness and Callaway 1994; Holmgren et al. 1997; Brooker 2006; Holmgren and Scheffer 2010). These interactions are often described using the stress-gradient hypothesis, which suggests that facilitation will increase and competition will decrease with increasing abiotic and biotic stress (Bertness and Callaway 1994). Hundreds of experiments have tested the stress-gradient hypothesis. While its generality has been questioned (Maestre et al. 2009), a recent meta-analysis that included 727 tests of shifts in the magnitude and type of ecological interactions between plant species concluded that there is a global generality of the stress-gradient hypothesis. It showed that plant interactions change with increasing stress through an outright shift to facilitation, when facilitation is quantified in terms of survival, or a reduction in competition, when facilitation is quantified in terms of growth and reproduction (He et al. 2013). A small number of studies found no change in ecological interactions to increasing stress but no studies found a shift towards increased competition.

It is important to note that there is usually a fundamental difference in the quantification of facilitation in forests, when compared with studies outside forests that were used to develop the stress-gradient hypothesis (Forrester 2014; Forrester and Bauhus 2016). In the latter, facilitation has been usually quantified by comparing the survival, growth, or fitness of a plant of a certain species with neighbours to a plant of the same species without neighbours, and these studies were often carried out in environments that are too harsh for forests to exist, either too cold or too dry. Under such harsh growing conditions, plants growing in a situation 'without neighbours' may not be unusual. However, this situation 'without neighbours' is rarely found in environments that support forests, which have higher stand densities. Therefore, facilitation or competitive reduction is often assumed to have occurred in forests, when growth (or survival) is greater with inter- than with intraspecific competition. As a result of this difference in quantification, the meta-analysis by He et al. (2013) did not include any studies of tree-tree interactions unless the trees were in open woodlands (not forests) or seedlings of tree species were examined.

It is often assumed that as stress increases, productivity will decline, and so facilitation will be more common in ecosystems with lower productivity, and this has been suggested for forest ecosystems (Toïgo et al. 2015; Pretzsch et al. 2013). However, while this may often be the case, there are also many exceptions such that complementarity increases with resource availability or site quality and often the interactions that occur in forests are competitive reduction and not facilitation, including all interactions that influence light absorption (see Table 1 in Forrester and Bauhus 2016). It is therefore problematic to apply the stress-gradient hypothesis, which is based on facilitation as opposed to competitive reduction, to forests unless most of the complementarity effect resulted from facilitation and not competitive reduction, and this is rarely known. Therefore, Chap. 3 (Sect. 3.6, Fig. 3.4) uses an alternative framework to the stress-gradient hypothesis to describe the spatial and temporal dynamics of complementarity in forests. This framework, which includes the stress-gradient hypothesis as a special case, is consistent with all mixed-forest studies in the literature and can be used to explain when and why complementarity can increase or decrease along spatial or temporal gradients in resource availability or climatic conditions.

In the context of this chapter, it is important to consider that, if we want to study the ecological interactions between certain combinations of species, this should ideally be done for the entire range of site conditions at which these species combinations occur (Confidence level 2 in Fig. 2.1). Otherwise we cannot extrapolate the results to the whole population of mixed stands of this species combination.

### 2.3.5 Tree Diversity and Diversity at Other Trophic Levels

Foundation species have been characterised as those individual species that define much of the structure of a community by creating locally stable conditions for other species and by modulating and stabilising fundamental ecosystem processes (Dayton 1972). Trees are often foundation species in forested ecosystems, because they shape forest structure, create a forest microclimate, and through their functional and physiological characteristics have a large influence on material cycling and energy flow (Ellison et al. 2005). Through their species-specific traits, they control ecosystem dynamics. Owing to specific interactions with symbionts, pathogens, herbivores, etc., a large number of other taxa typically depend on specific trees species. Hence, the loss of such foundation species might have dramatic consequences for the stability and functioning of forest ecosystems (Ellison et al. 2005). However, the number of other taxa that depend on individual tree species is difficult if not impossible to ascertain. For example, some early estimates of the global number of tropical forest beetles and other arthropods were based on the fogging with insecticide of the crowns of individual tropical tree species (Erwin 1982). Based on the count of arthropod species that were knocked down, which amounted to many hundreds and in some cases more than 1000 species per tree, estimates were derived for the richness of beetles and other arthropods based on an assumption of the proportion of host-specific species. In the early study of Erwin (1982), that proportion was estimated at 13.5%. The resulting number of species was then multiplied with the number of known tropical tree species to yield an estimate of 30 million arthropod species for tropical forests. However, owing to the unknown range of plant host-specificities of tropical insects, particularly of non-herbivore species, this and subsequent similar estimates are associated with very high uncertainty (Stork 1988). Although this figure has since been corrected downwards, recent studies, which have found close associations between floristic diversity and both herbivore and non-herbivore species richness in tropical forests, confirmed that the general modelling approach to scale up from plant species richness appears to be valid (Basset et al. 2012). The estimates of species richness in insect communities in the crowns of temperate forests are lower than those of tropical forests, but still in the order of hundreds of species (e.g. Southwood et al. 2004). If we consider host specificity or preference of additional taxonomic and functional groups such as fungal pathogens, wood inhabiting fungi (e.g. Unterseher et al. 2005; Purahong et al. 2014), saproxylic beetles (e.g. Floren et al. 2014), decomposers (e.g. Lodge 1997), mycorrhizal fungi (Ishida et al. 2007), root herbivores, and so forth, we can see that each tree species may provide habitat and resources for hundreds of other species at different trophic levels within the ecosystem. If we add or remove these tree species, these other species may be also added (if they can colonise the tree species) or lost from the system. Even non-native, introduced tree species can contribute substantially to species richness and diversity in forest ecosystems, when they are admixed in small proportions, for example, admixed *Pseudotsuga menziesii* in stands of native *Fagus sylvatica* (e.g. Goßner and Ammer 2006). Hence, many tree species can be viewed as foundation species that add much complexity to the ecological communities in forest ecosystems. Therefore, with increasing tree species richness, the number and frequency of interactions and feedbacks also increases, and this might buffer mixed-species forests against stressors and disturbances (Messier et al. 2013, see also Chap. 7). Increased stability of communities is of course not just linked to species-specific, dependent diversity. Generalist species add also to the complexity and stability of ecological networks, which typically increases with number of interactions between participating species. The strength of interactions typically decreases with the number of links between species in networks (Vázquez et al. 2007). The increased stability with increasing species richness is most likely to occur at the level of community and ecosystem processes and less at the level of populations of individual species, as has been observed for grasslands (Tilman 1996). However, increased stability with increasing complexity of ecological networks may be effective only in case of random removal of species. Complex networks may be very fragile in case of selective removal of the most connected species (Solé and Montoya 2001).

In terms of ecosystem hierarchy (Fig. 2.2), increasing species and functional diversity at the different levels of ecosystem organisation may dampen the signal from one level to the other.

## 2.4 Analysing Ecosystem Properties and Processes in Tree Species Mixtures Based on Experiments and Inventories

### 2.4.1 General Considerations Regarding Tree Diversity Experiments

Quantitative information about the effects of tree species diversity on ecosystem properties and processes stems mostly from the analyses of large-scale, permanent forest inventories that capture gradients in tree species richness (e.g. Caspersen and Pacala 2001; Gamfeldt et al. 2013; Vilà et al. 2013) and from controlled experiments or long-term forest plots comparing monocultures and tree species mixtures, mostly two-species mixtures (e.g. Forrester et al. 2006; Piotto 2008; Pretzsch et al. 2015). Whereas the first source of data stems mostly from all forms of forests (native, seminatural, and plantations), the latter stems mostly from plantations or seminatural forests. Only relatively recently have experiments been established that permit the analyses of tree species diversity and ecosystem functioning relationships for a wider range of species richness levels and functional diversity (e.g. Scherer-Lorenzen et al. 2007; Potvin and Gotelli 2008; Tobner et al. 2014; Bruelheide et al. 2014). In relation to the longevity of trees, these experiments can so far only provide information on the initial phase of stand development and species interactions.

Typical hypotheses to be addressed in such studies are (Baeten et al. 2013):

1. Ecosystem functioning is higher in tree species mixtures than in monocultures. This may be related to complementarity and selection effects (see above).
2. Species interactions and hence the mixing or diversity effects are not consistent along environmental gradients (Forrester et al. 2013) and with development of mixtures over time (Pretzsch and Schütze 2009).
3. Tree species mixtures have a higher resistance and resilience than monospecific stands to abiotic stress and disturbances as well as to pests and diseases.
4. Fewer tree species or a lower level of diversity is required to maximise a single ecosystem function such as productivity when compared to maximising multiple ecosystem functions simultaneously (e.g. van der Plas et al. 2016).

To be effective and make efficient use of resources for research, tree diversity studies should ideally be designed to permit addressing all of the above questions. Such an ideal design should consider and optimise three basic criteria, which have been termed orthogonality, comprehensiveness, and representativeness (Fig. 2.11) (Baeten et al. 2013).

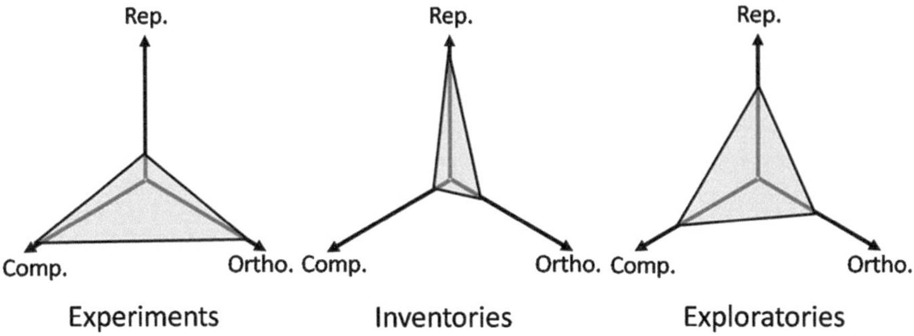


**Fig. 2.11** The relative differences in representativeness (Rep.), comprehensiveness (Comp.), and orthogonality (Ortho.) of three types of study designs used to quantify the effects of biodiversity: experiments, inventories, and exploratory plots in forests. This figure is reproduced with permission from the FunDivEUROPE project consortium (http://www.fundiveurope.eu/?page_id=30)

Orthogonality refers to the statistical independence of species diversity effects from other factors. Orthogonal designs allow the effects of species diversity to be completely uncorrelated with those of other variables and covariates (e.g., species identity, environmental gradients, management influences) (Nadrowski et al. 2010; Baeten et al. 2013). However, it is extremely difficult to design studies in which tree species diversity is completely independent (orthogonal) from other variables due to unavoidable biological correlations. For example, orthogonality would be reduced in an inventory-based analysis of tree diversity effects, if species diversity was not independent of successional stage or stand age, which in turn influences productivity. Orthogonality would be also reduced in a design, where the more diverse species mixtures can be found at the most productive sites (Vilà et al. 2005).

Comprehensiveness refers to the spectrum of ecosystem properties and processes and hence ecosystem functions quantified in a study (Nadrowski et al. 2010). Since forests are managed for many different functions and related ecosystem services, we would typically like to know how tree species diversity affects these functions. Therefore, it is desirable to design tree diversity studies in a way that as many as possible of these ecosystem functions (e.g. biomass production, carbon sequestration, nutrient cycling, provision of non-wood products, etc.) can potentially be analysed simultaneously (e.g. Gamfeldt et al. 2013). This typically requires large interdisciplinary teams of researchers to quantify these different functions. Inventory-based, observational studies have typically a low comprehensiveness because only a limited set of variables, from which ecosystem properties and processes can be derived, is quantified on inventory plots. However, these inventories are currently evolving towards multipurpose resource surveys to include increasingly more variables that are relevant for biodiversity (Corona et al. 2011). Low comprehensiveness is not a characteristic of any particular design per se; however, it is much easier to measure many functions in planted experiments, which are relatively small in area and have fewer trees, compared with inventory datasets, which can cover very large areas and include very high numbers of trees.

Representativeness refers to the proportion of the variation in the population of interest that is captured by the design. It indicates to which extent the results from the study may be extrapolated to the population of interest. For example, if we wanted to study tree diversity effects in temperate deciduous forests of Europe, then we would require a design that reflects the different existing forest types, soil types, stand density, and age structure to allow a transfer of the results to real-world conditions. This could be considered confidence level 2 in Fig. 2.1. A representative design relates also to the relevance of the results for policy and decision makers. Representativeness is typically high in large-scale, permanent inventories, which are designed to capture the entire variation of the population of interest, whereas experiments are likely to be far less representative because they tend to focus on a small range of environmental conditions, species compositions, stand densities, ages etc.

### 2.4.2 Forest Inventories

Forest inventories typically adopt sampling schemes in accordance with a spatial design in which sampling points are randomly selected, for example, as grid points, from the population of interest about we wish to infer certain properties (e.g. Corona et al. 2011). In many National Forest Inventories or forest enterprise-level inventories, sample plots with a sufficiently large radius or angle counts with a predefined basal area factor are established with their centres at the selected points, and a range of attributes are recorded for the trees included in the plots or in the angle counts (e.g. Schreuder et al. 1993). At the same sample points, smaller plots or angle counts with a low basal area factor may be established to capture regeneration and smaller tree individuals. A traditional focus of these inventories has been on the assessment of growing stock and its changes through periodical growth, recruitment, harvesting, and mortality and on tree species composition. More recently, forest inventories have broadened their scope to include variables related to forest carbon pools and carbon sequestration, forest health, and biodiversity (Corona et al. 2011). Typical forest inventories are theoretically designed to monitor forest changes through periodic remeasurements at 5–10 year periods. It has to be kept in mind, however, that the statistical design of these inventories is related to the sampling probability of important variables related to the original purpose of the inventory. Some variables that might be of interest with regard to additional ecosystem functions might not be sampled with sufficient intensity to provide reliable estimates for the population of interest. It may be very difficult, for example, to estimate the growth of rare tree species that contribute less than 0.1% to the total growing stock from nationwide forest inventories. One other problem that may be encountered in some regions when analysing forest inventory data for relationships between tree species richness or diversity and ecosystem functions is that the gradient in tree species richness can be rather short, because plots with high diversity are uncommon, either because the original sampling design did not take this into account or because the average stand condition on the landscape is relatively species poor. Therefore, most inventory plots cover only the lower end of the tree diversity gradient (e.g. Vilà et al. 2007). For example, in Europe, the probability of sampling plots with more than three species is less than 25% and to sample plots with more than five species is less than 10% (see Fig. 1.1 in Chap. 1). Another shortcoming of inventories is that 5–10 year measurement periods may be too long to examine relationships between growth and short-term climatic variability. For example, the effect of particular drought years on forest of different composition may not be captured through such inventories. However, the same problem applies to experiments or other plots, where processes are not recorded at a higher temporal resolution, for example, through measurement of tree growth through permanent dendrometer bands.

The strength of inventory datasets is their representativeness (Fig. 2.11). They typically cover a very wide range of environmental conditions, forest ages, species compositions and forest types, management approaches, and stand densities. Because they represent real-world ecosystems, they can help to put results obtained from artificial communities, such as created in experiments, into perspective. For example, in real-world ecosystems, species richness, if it is primarily related to the occurrence of rare species that contribute little to function, may not be a strong determinant of ecosystem functioning and ecosystem services provision (Winfree et al. 2015). However, inventories often only include data about a small number of ecosystem properties and processes, such as growth and standing wood volumes, biomass, and carbon stocks. They can also lack orthogonality, which can require careful consideration when interpreting the results (Vilà et al. 2005). Several studies have shown that variables such as edaphic and climatic conditions, stand age, and density can influence relationships between diversity and productivity and need to be considered when analysing inventory data (Firn et al. 2007; Vilà et al. 2007, 2013; Belote et al. 2011; Paquette and Messier 2011). For example, Caspersen and Pacala (2001) found an asymptotical increase of growth with increasing tree species richness when plotting the number of canopy tree species against stand growth using national forest inventory data from the USA. From their study, one might conclude that tree species richness enhances productivity owing to functional differences between species in accordance with the niche complementarity hypothesis outlined above (Sect. 2.3.1). However, one might also conclude that more productive sites that support more productive stands simply permit the coexistence of more species. Thus, cause and effect cannot be disentangled from observational and comparative studies in the absence of additional information (Scherer-Lorenzen et al. 2005b). In another inventory-based study by Vilà et al. (2013), no effect of species richness was observed in *Pinus sylvestris*-dominated forests, but a positive effect was detected in *Pinus halepensis* stands of Catalonia. In the latter case, however, tree species richness was no longer a significant independent variable, when climate, bedrock types, radiation, and successional stage of inventory plots were considered.

Also thinning may be heavier in stands, where tree species diversity effects accelerate the growth rate, to harvest the additional increment and to adjust species composition. This is often done to release the inferior and rare tree species from competition to keep them in the forest community. When inventory plots represent mainly managed stands, their stand density may be considerably below the maximum density which may best reveal any mixing effects. Hence, mixing effects may be eliminated or reduced by thinning (Chaps. 4 and 9).

Approaches such as structured equation modelling may be used to account for these confounding factors when analysing inventory data, for example, that basal area had a greater effect on productivity than tree species diversity (Paquette and Messier 2011; Vilà et al. 2013). However, extracting the signal of tree species diversity from forest inventory data requires a large amount of additional information.

When accounting for possible confounding factors, most analyses of large-scale forest inventories have found positive relationships between tree species richness and forest productivity, from Mediterranean, to temperate and boreal forests. Productivity is also the variable that has been most commonly addressed in these studies. However, additional variables that served as proxies for biodiversity and ecosystem services such as soil carbon storage, bilberry production, game production potential, understorey plant species richness, and dead wood have also been related to tree species richness on the basis of data from the National Forest Inventory of Sweden and the Swedish Survey of Forest Soils and Vegetation (Gamfeldt et al. 2013). In addition to the analyses of relationships between species richness and ecosystem properties and processes, inventory data might also be used to address other hypotheses, for example, regarding the interaction between site quality and complementarity. For example, Toïgo et al. (2015) were able to show with data from the French National Forest Inventory that for a number of two-species mixtures exhibiting overyielding, this effect was higher at sites of lower productivity.

### 2.4.3 Mixed-Species and Tree Species Diversity Experiments

Here we distinguish between experiments to study the mixing effects of two species, which we call mixed-species experiments, which have a long tradition in forest sciences (Pretzsch 2005), and tree diversity experiments, which comprise more than two tree species and permit the separation of diversity effects from species identity effects (Scherer-Lorenzen et al. 2005b). The latter type of experiment aims at identifying and quantifying relationships between biodiversity and ecosystem functioning, so-called BEF experiments (Bruelheide et al. 2014). Large-scale experiments with higher species richness levels have much less tradition than the former and have only been installed since the beginning of this century, which is currently still a weakness of this approach. The experimental designs can be broadly divided into (1) planted experiments and (2) exploratory plots in forests (Baeten et al. 2013; Bruelheide et al. 2014; Verheyen et al. 2016). Each of these has advantages and disadvantages, also when compared to inventories, and these can be summarised as depicted in Fig. 2.11.

Experiments are typically designed to analyse the influence of one or a few independent variables, here chiefly tree species richness, tree diversity, and functional diversity, and to exclude or control the influence of other factors and hence to maximise orthogonality (Fig. 2.11). Thus the particular strength of experiments is to detect effects of diversity per se and to identify and quantify the underlying mechanisms. The greater the extent to which they consider the population of inference, e.g. by replicating the diversity gradients along environmental gradients such as different site or soil types, the higher the level of confidence we can place in the results from these experiments (see Fig. 2.2). The compact nature of experiments with shared infrastructures typically permits the quantification of a wide range of ecosystem properties and processes, hence they provide for high comprehensiveness. In the following, we will discuss some commonly used designs for such experiments.

#### 2.4.3.1 Planted Experiments

By using planted experiments, it is possible to carefully control species compositions and proportions, functional diversity, stand density, and edaphic and climatic factors. This is a clear advantage over inventories and so-called exploratories (see below), where these factors can be confounded with diversity. Over the last two decades, a rapidly increasing number of planted biodiversity experiments have been established, and many of these have joined to form extensive networks (e.g. Verheyen et al. 2016). While all of these experiments are primarily interested in the influence of biodiversity on ecosystem functions, they also have their own unique design features. Some consider genetic diversity or functional diversity instead of, or in addition to, species diversity (e.g. Scherer-Lorenzen et al. 2007; Tobner et al. 2014). In addition to tree species diversity, some include understorey diversity treatments (Bruelheide et al. 2014).

A common and major focus of most of these BEF experiments is the question whether random species loss can affect ecosystem functioning (Hooper et al. 2005). This can be examined using extinction scenarios, whereby certain species from the total species pool are left out of the less diverse treatments (Fig. 2.13) while maintaining the same total initial tree density in experimental plots. In this regard, these BEF experiments usually follow a substitutive (also called replacement series) design (see below) rather than additive series, where diversity is confounded with density.

The extinction scenarios used in the experiments may follow a random pattern, if there is inadequate information about the drivers of extinction or when those drivers do not lead to a biased extinction of species (Bruelheide et al. 2014). Alternatively, non-random scenarios may be employed to reflect the preferential loss of species with certain traits and thus a particular contribution to the ecosystem and its functional diversity (Bruelheide et al. 2014). For example, in the BEF-China experiment, two types of non-random extinction scenarios were implemented. One removes species in the order of rarity starting with the rarest species, which may resemble what might happen in case of forest loss and fragmentation in the landscape. The other removes species in relation to the expression of the functional trait specific leaf area (SLA) starting with the highest SLA, assuming that they will be the first to decline in vitality as the climate becomes drier and warmer (Bruelheide et al. 2014). It is also important to have several extinction scenarios to separate diversity effects from those of a particular community composition, and in experiments such as BEF-China (Fig. 2.12), there are three overlapping random extinction scenarios at each of two sites.

The International Diversity Experiment Network with Trees (IDENT) focusses on the separation of tree species richness and functional diversity effects (Tobner et al. 2014) (Fig. 2.13).

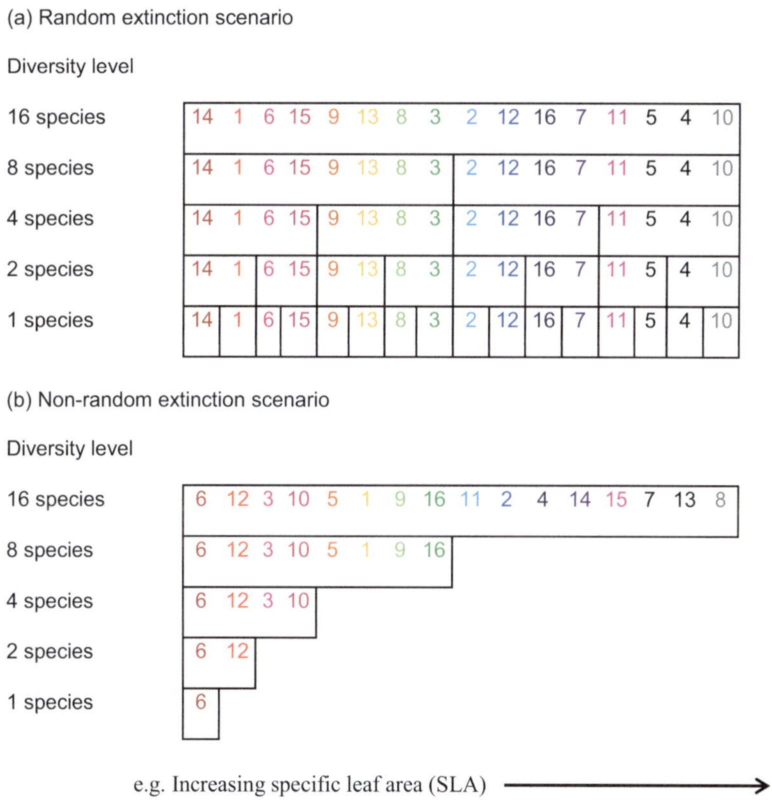


**Fig. 2.12** (**a**) The random extinction design used in the BEF-China experiment. Each number represents a different species. A pool of 16 species is randomly partitioned into two mixtures that each contains eight species. This partitioning is continued down to monocultures. Each species is included at each diversity level. (**b**) A non-random extinction scenario based on increasing specific leaf area. Most studies contain several extinction scenarios, and when they contain only one they are referred to as a dilution diversity gradient. Modified from Bruelheide et al. (2014)

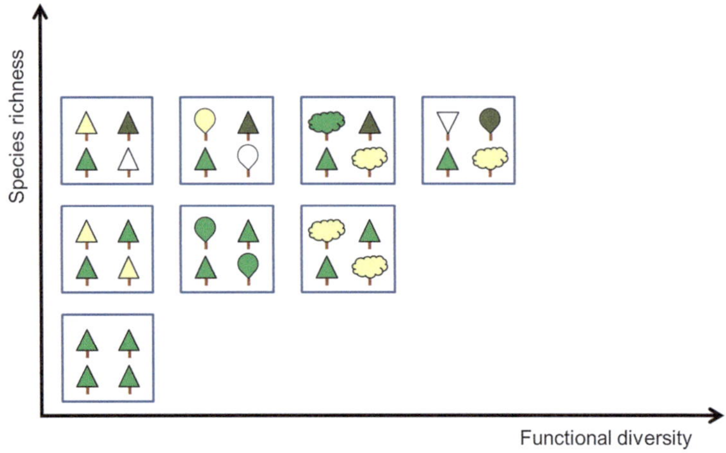


**Fig. 2.13** Outline of an experimental design used in the International Diversity Experiment Network with Trees (IDENT) comprising gradients of species richness (SR) and functional diversity (FD) (Tobner et al. 2014). The design includes monocultures of all species and mixtures of increasing species richness that have different degrees of functional diversity. For example, a two-species combination with low functional diversity may consist of two species of the same family with similar traits (e.g. *Picea abies* and *Pinus sylvestris*), whereas a higher level of functional diversity would be represented by a combination of a broadleaved deciduous species with an evergreen conifer, e.g. *A. pseudoplatanus* with *Picea abies*. A four species mixture representing a high level of functional diversity may comprise an evergreen (*P. sylvestris*) and deciduous conifer (*Larix europaea*) as well as a shade-tolerant broadleaved (*A. pseudoplatanus*) and a shade-intolerant deciduous species (*Quercus robur*) (see Wein et al. (2016) for the whole range of species combinations). Levels of functional diversity are repeated using different species combinations. For analytical purposes, these levels of functional diversity can be pooled in groups of similar value

In contrast to other experiments, the basic design of IDENT experiments comprises gradients in both, species richness and functional diversity, and it focusses on the early stages of stand development through employing a high planting density. The high-density design reduces the space required to implement the experiment and time needed for trees to start interacting. The variation of functional diversity within levels of constant species richness also allows the testing for underlying mechanisms such as complementarity or selection effects on ecosystem properties and processes (Tobner et al. 2014). Functional traits used to compute indices of functional diversity in some of these experiments included wood density, seed mass, and foliage N content; however, one could think of many other traits to be used for that purpose. For two-species combinations, the calculated functional diversity, measured as the extent of functional differences (distance) among species in the community (Laliberté and Legendre 2010), ranged from 0.3 in congeneric mixtures of *Picea glauca* and *P. abies* to 2.23 in mixtures of *Quercus robur* with *P. glauca*. The latter value was higher than in the six species mixtures of this experiment, indicating how species richness and functional diversity might diverge.

The IDENT network with a number of experiments in North America and Europe using the same design allows for later meta-analysis and synthesis that can provide a high level of evidence for possible findings (see Fig. 2.1).

#### 2.4.3.2 Planting Designs

The most common experimental designs to study species diversity effects are 'substitutive' or 'additive'. In the case of short-lived plants, which are often already planted with the final spacing between individuals and species, such designs can be kept through the whole rotation until the final harvest (if that is intended). This is typically not possible with long-lived plants such as trees, which continue to increase in size and typically at different rates, so that the initial composition of plots with regard to species proportions and density will change as the experiment progresses.

Establishment of a substitutive design (de Wit 1960) comprises the planting of monocultures of each species participating in mixed plots. This is a prerequisite to answer the question whether the average performance across all species is higher at higher levels of species diversity (the complementarity effect) or whether diverse species mixtures have a higher probability to contain particular species with a very high contribution to the performance of the community with regard to certain ecosystem processes or properties (sampling or selection effect). When monoculture plots of all species are included in the experiment, the additive partitioning method of Loreau and Hector (2001) can be employed to separate complementarity and selection effects.

In its most simple form with two species, the mixed stand can be established by planting half of the tree positions (or any other defined proportion) with one species and the remaining positions with the other species (Fig. 2.14), thus keeping the density constant across monocultures and mixtures of the replacement series. Ideally, the replacement series are also established at different densities to allow the separation of mixing and density effects. However, in practice such substitutive designs are mostly established at one density only and also at one site only, which limits their representativeness.

In experiments with a substitutive design, three main types of planting patterns may be employed to establish plots with an identical initial number of trees. For a two-species mixture, these are tree-by-tree (ABABABAB) (Fig. 2.15 top right), patches (AAABBBAAABBB), or random (ABBABAABA). Tree-by-tree or row-by-row mixing patterns, in which each individual has at least two heterospecific neighbours, maximise interactions between species during the early stages of development. This also has the advantage that for certain types of measurements, e.g. soil properties, fine roots, or light, a measurement point can be selected that represents the influence of the different species in that particular neighbourhood situation based on the distance to the surrounding trees (e.g. Bauhus et al. 2000). However, owing to differences in growth dynamics between species, it is possible that some species may outcompete others. To avoid this, the patch-planting design is sometimes used so that some individuals of a given species are surrounded by individuals of the same species (even in the diverse treatments) to ensure that some

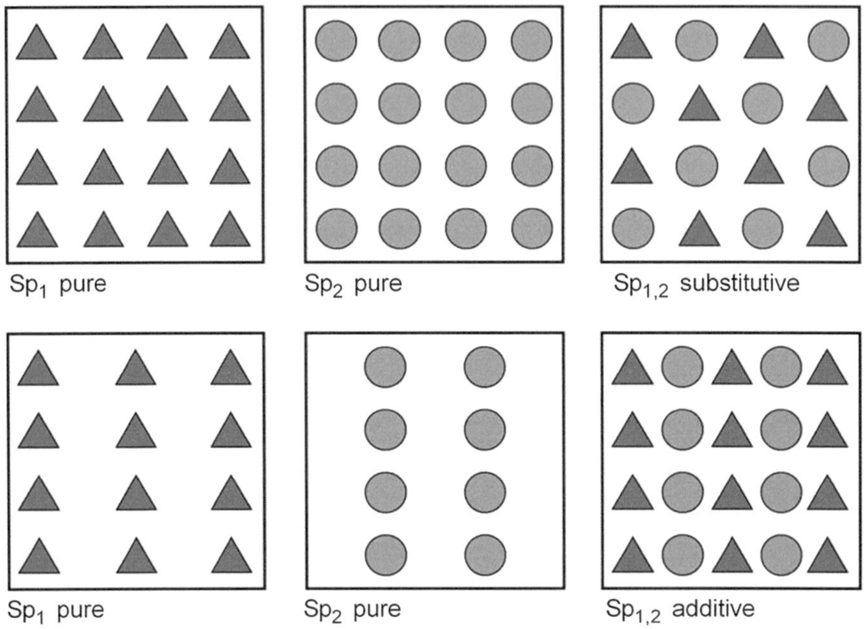


**Fig. 2.14** Illustration of substitutive design (above) and additive experimental design (below) for a simple two-species mixture. In case of the substitutive design, the monocultures are established by planting two species in a defined pattern (density, distribution), and the mixed stand is established by planting half of the positions (or any other defined proportion) with one species and the remaining positions with the other. Here we have depicted a 50:50 mixture; often mixtures of 25:75 percent have also been included in replacement series. The additive design also bases on the distribution pattern in the pure stands. But for creating the mixture, the plants are not substituted by the other species but both patterns are added

individuals initially experience only conspecific competition. This is done with the expectation that as the stand develops, the diversity treatment will be retained for longer without losing the less competitive species completely (Scherer-Lorenzen et al. 2007). The disadvantage of such a patch-planting design is that most of the early interactions among trees within the diverse plots will actually be intraspecific. Some studies try to avoid both of these disadvantages by using random patterns (e.g. BEF-China; Bruelheide et al. 2014). This also creates more variability at the tree level in terms of the neighbourhood compositions of individual trees.

In tree diversity experiments with more than two species, the population sizes per species decrease inversely with increasing species richness, while the community density within plots remains unchanged. The substitutive design avoids confounding community density and diversity effects, a problem that is encountered with additive designs.

The additive design also starts with the distribution pattern of trees in pure stands. However, in mixed-species plots the planting positions of one species are not partially substituted by the other species, but both patterns are added (Fig. 2.14).

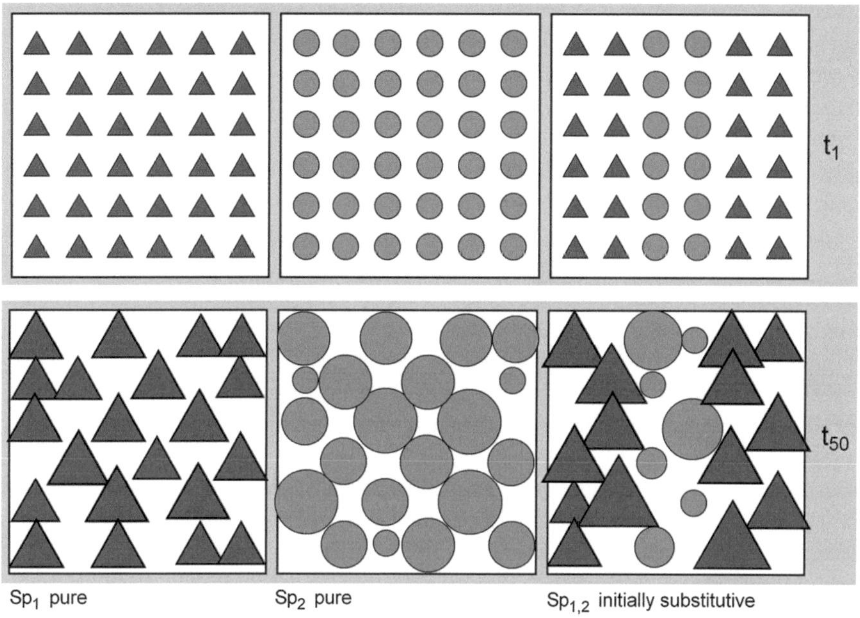


**Fig. 2.15** Possible development of an 'initially substitutive design' with two tree species, where the experiment starts with an unimpaired substitutive design with double rows. The spatial pattern progressively deviates from the initial design owing to species-specific size growth, density-dependent mortality, disturbances, and silvicultural interference in the course of stand development (below). In the analogue case of an additive design (not shown), this would be called 'initially additive design'

The substitutive design is common in even-aged forests, where the participating species with similar growth dynamics are planted at the same time (e.g. Forrester et al. 2004). In contrast, the additive design is more suitable in two-layered stands, where, for example, a shade-tolerant species may be planted beneath the canopy of a shade-intolerant species, e.g. *F. sylvatica* under *P. sylvestris*, and both species are also represented in monoculture as reference.

In planted experiments it is also possible to quantify the influence of site heterogeneity before any of the species have influenced the edaphic and climatic factors (Bruelheide et al. 2014). These details can be included as covariates later, and it also enables the examination of how the diversity treatments modify these environmental factors (Bruelheide et al. 2014).

Pot trials are also planted experiments that offer many of the same characteristics of planted experiments. An advantage of pot trials, where different species are placed in the same pot, is that it is easier to carefully control water and nutrient availability as well as climatic conditions. However, the pots restrict growth, in particular of the root systems, and the interactions between small plants growing in pots may often not resemble those of larger plants.

#### 2.4.3.3 Development of Initially Substitutive or Additive Designs in Forest Stands

Experiments in long-lived forest stands are often established either with a substitutive or an additive design but with stand development they progressively deviate from it. Since the original design in terms of stand density, mixing pattern, or proportion of the species commonly changes due to species-specific size growth, mortality, and thinning, the respective designs might be more appropriately called 'initially substitutive design' or 'initially additive design' (Fig. 2.15). This deviation from the original design can be accounted for by frequent surveying of pure and mixed plots to record their productivity, current species proportions, stand density, mixing patterns, and other variables as confounding factors for later analyses of the temporal variation of mixing effects during stand development (see Sect. 6.5). Another alternative to deal with this development is to move away from plot-based measurements and focus on diversity effects in tree neighbourhoods, where the tree species diversity may then be defined by the species identity, dimension, and distance of all neighbours of a subject tree (e.g. Forrester et al. 2013; Forrester 2015). However, when moving from the plot level to the neighbourhood level, only those properties and process in relation to tree species diversity can be analysed that occur at that spatial scale. These may be interactions regarding the competition for resources such as light, water, and nutrients. The neighbourhood scale is likely not relevant for questions regarding the interactions between tree diversity and ecological disturbances. In addition, the maximum tree species richness that may be analysed is likely limited to the number of immediate neighbours, which would typically be not more than eight.

An advantage of the longevity of experiments is that the tree species diversity effects can evolve and be studied much longer. When closed canopies can be maintained by only permitting self-thinning or light silvicultural thinnings, the effect of tree diversity on productivity and site carrying capacity in terms of maximum stand density can be quantified. In this sense, the longevity of experiments increases their representativeness since the results obtained over the course of the experiment can then be extrapolated to all ages or developmental stages covered in the study. Exceptions, where the initial planting design can be maintained during a complete rotation of the experiment, are short-rotation plantations.

Another interesting question is how much forest stands can change their maximum stand density and carrying capacity through species interactions and modifications of site conditions (see Chap. 4). To answer this question, it is necessary that mixed pure and mixed-species plots remain unthinned or fully stocked. However, as shown above leaving plots unthinned means that they will gradually drift away from the initially substitutive design with equal density, mixing proportions, and defined mixing pattern.

### 2.4.4 Exploratories

The third approach of **exploratories** represents a hybrid strategy, where existing mature stands are explicitly selected for diversity gradients with otherwise comparable environmental conditions and management regimes. This approach considers all three criteria of orthogonality, comprehensiveness, and representativeness, but without fully maximising them.

Exploratories combine some of the attributes of planted experiments with those of inventories and complement these approaches. Like inventories, they are done in existing forests so that tree size distribution, stand structures, and species compositions may be more representative for real-world systems than in planted experiments, and this representativeness may be improved further if the plots are replicated at different sites and comprise different stand ages. However, unlike inventories, and consistent with planted experiments, there is usually some control to ensure a gradient in species diversity at a given site, thereby making them more orthogonal than inventories. This diversity gradient may use existing stand variability or be created manually.

One older example using this approach is situated in a central *F. sylvatica* forest with a small-scale mosaic of tree diversity and otherwise similar soils and climatic conditions (Leuschner et al. 2009). A weakness of that research platform, apart from no replication, was related to the lack of monocultures for all tree species in the species pool, except for *F. sylvatica*, and the presence of one species, *F. sylvatica*, in all mixtures. The design might therefore be regarded as dilution design (Nadrowski et al. 2010) in which the effects of tree species richness or diversity could not be separated from the effects of *F. sylvatica* abundance.

Another possible approach is the adjustment of different levels of species richness in removal experiments, where different degrees of tree diversity at the stand level are created through removing one or more individual species from an initially species-diverse forest ecosystem. These types of experiments, which may emulate the loss of species as has happened as a result of specific pests or pathogens in the past, could potentially yield very interesting results. They may also be implemented in older stands, and they would permit the study of before and after effects in relation to a reduction in species diversity. However, there a number of strong limitations and conceptual problems with removal experiments (see also Díaz et al. 2003):

- It will be extremely difficult to find large enough natural stands with a homogenous distribution of a diverse range of species of similar abundance. These would have to be reasonably young stands capable of rapid canopy closure following species removal. Otherwise the removal of species would create gaps, and thus the removal/diversity effect could not be separated from a disturbance effect.
- The distribution of species in natural stands is in most situations not homogenous or random but clumped or clustered, and many species in species-rich and

diverse stands are rare, making it very difficult to establish different tree species diversities at a scale (neighbourhood), where trees do interact.

- This type of experiment would have to be based on the assumption that the distribution of species within a stand is random and not affected by site factors or other (unknown) co-variables.
- In order to analyse the effect of different diversities after removing a certain number of species, a similar biomass must be removed or a similar stand density retained in all treatments. Otherwise the diversity effect cannot be separated from a stand density effect. This would be extremely difficult to achieve in practice.
- The removal of a species would leave legacies, such as litter or the root system, which may not be desirable for analysis of species diversity effects following the removal.
- Such an experiment would have to be established as a long-term endeavour, since its value would be in the analysis of before and after removal effects. However, this requires the description of the baseline conditions through extensive measurements, which may need several years to capture temporal variation, before removal of one or more species.

When an adequate range of diversity or density already exists, it is not necessary to thin or plant in order to alter the diversity and density of the stands. An example of this is the exploratory platform within the FunDivEUROPE project (Baeten et al. 2013). This design includes study areas in six forest ecosystem types in different European regions, from boreal forests (Finland) to mediterranean (Spain). Here, many plots were established to explore the relationships between tree species diversity and ecosystem functioning. For that purpose, three to five co-occurring, important tree species were selected for each region, and then plots were identified in existing forest representing monocultures of each species and species mixtures with two to five species. Each species is usually represented at each level of diversity (to avoid dilution designs), and each of these species combinations is replicated in two or more plots. In practice, it is of course difficult to translate this design on the ground, since some combinations of species are very rare or clearly violate the attempt to have *ceteris paribus* conditions with regard to other environmental variables.

**Take-Home Messages**

1. Much of our knowledge about the effects of mixed-species forests stems from experiments and exploratory research plots with replications of species richness levels at the same site. In recent years, this knowledge has been greatly expanded by analyses of forest inventories using tree species richness or diversity as an independent variable. However, these latter analyses are restricted to those ecosystem properties and processes that are recorded in forest inventories; often they are limited to measures of productivity. Our

knowledge will gain substantially from a considerable number of experiments on the function of tree species diversity that have been established in the last one or two decades. However, compared to the longevity of trees, these experiments are still relatively young, and the results so far restricted to the initial phase of the growth of even-aged stands. In future, these experiments will facilitate meta-analyses of tree species diversity effects for a range of ecosystem functions and hence provide a high level of evidence for these findings. Owing to the different strengths and weaknesses of inventories, exploratories, and experiments, an approach based on a combination of these research platforms is likely to yield the most comprehensive information in the short term. The level of evidence for tree diversity effects will increase with the temporal and spatial variation that is captured in experiments or networks of experiments.

2. Tree diversity studies should optimise three basic criteria: orthogonality, comprehensiveness, and representativeness. An optimal combination of these criteria is likely not achievable at the level of individual studies but more so at the level of networks of coordinated experiments with a wide geographical coverage. Orthogonality refers to the statistical independence of species diversity effects from other factors, comprehensiveness refers to the spectrum of ecosystem properties and processes and hence ecosystem functions quantified in a study, and representativeness refers to the proportion of the variation in the population of interest that is captured by the design. Ideally, any experiment on the effects of tree diversity on ecosystem functioning should be established in a way that permits the comparison of structures and processes between mixtures (of different diversity) and monocultures at all levels of organisation from cells to the stand (or possibly landscape) scale to facilitate a mechanistic understanding of the underlying processes. To analyse the effects of tree diversity on ecosystem functioning, mixtures with different degrees of tree species richness or functional diversity are typically established. For tree diversity experiments, a number of different planting designs are available; most commonly substitutive designs are being used.
3. A number of ecological theories exist to explain the changes in ecosystem functioning (or properties and processes) with increasing tree species richness or diversity. The most relevant of these is the complementarity hypothesis which postulates that the magnitude of ecosystem processes increases decrementally with the number of species. Each species added to a community, which is functionally different from the species already present, contributes to the magnitude of ecosystem processes. However, this increase follows a saturating curve as more species are added to the system, because the overlap between species in functionality increases.
4. The most important ecological interactions between species in mixture are competition, competitive reduction, and facilitation; the latter two lead to complementary effects. The degree of competition or complementarity is likely dependent on the functional traits of the participating species and the

functional diversity of the community. Functional traits are an expression of the fundamental and realised niche of species. We assume that the lower the functional diversity between species the higher the competition and the lower the complementarity.

5. Since many tree species are foundation species in forest ecosystems, they have a strong influence on the diversity of other taxonomic groups and hence through the increasing complexity of ecological networks will influence ecosystem properties and processes at many levels of ecosystem hierarchy.
6. Diversity effects in mixed tree communities are not constant in space and time. They may be more pronounced on infertile sites, where species interactions improve nutrition of individual species, or they may be more pronounced on sites with good nutrient and water availability, if interactions improve the light use of participating species. The relative importance of the different mechanisms of species interactions further change with stand development. This spatial and temporal variation in interactions even within the same mixture or community makes it difficult to generalise and predict the outcomes of specific tree species mixtures.

## References


Allan TFH, Starr TB (1988) Hierarchy: perspectives for ecological complexity. The University of Chicago Press, Chicago

Baeten L, Verheyen K, Wirth C et al (2013) A novel comparative research platform designed to determine the functional significance of tree species diversity in European forests. Perspect Plant Ecol Evol Syst 15:281–291

Basset Y, Cizek L, Cuénoud P et al (2012) Arthropod diversity in a tropical forest. Science 338:1481–1484

Bauhus J, Khanna PK, Menden N (2000) Aboveground and belowground interactions in mixed plantations of *Eucalyptus globulus* and *Acacia mearnsii*. Can J For Res 30:1886–1894

Bauhus J, van Winden AP, Nicotra AB (2004) Above-ground interactions and productivity in mixed-species plantations of *Acacia mearnsii* and *Eucalyptus globulus*. Can J For Res 34:686–694

Begon M, Harper JL, Townsend CR (1996) Ecology: individuals, populations and communities. Blackwell Science, Oxford

Belote RT, Prisley S, Jones RH et al (2011) Forest productivity and tree diversity relationships depend on ecological context within mid-Atlantic and Appalachian forests (USA). For Ecol Manage 261:1315–1324

Bertness MD, Callaway RM (1994) Positive interactions in communities. Trends Ecol Evol 9:191–193

Binkley D, Menyailo O (2005) Gaining insights on the effects of tree species on soils. In: Binkley D, Menyailo O (eds) Tree species effects on soils: implications for global change. Kluwer Academic, Dordrecht, pp 1–16

Brang P, Spathelf P, Larsen JB et al (2014) Suitability of close-to-nature silviculture for adapting temperate European forests to climate change. Forestry 87:492–503

Brooker RW (2006) Plant-plant interactions and environmental change. New Phytol 171:271–284

Bruelheide H, Nadrowski K, Assmann T et al (2014) Designing forest biodiversity experiments: general considerations illustrated by a new large experiment in subtropical China. Methods Ecol Evol 5:74–89

Cadotte MW, Cavender-Bares J, Tilman D, Oakley TH (2009) Using phylogenetic, functional and trait diversity to understand patterns of plant community productivity. PLoS One 4:e5695
Callaway RM, Walker LR (1997) Competition and facilitation: a synthetic approach to interactions in plant communities. Ecology 78:1958–1965
Caspersen JP, Pacala SW (2001) Successional diversity and forest ecosystem function. Ecol Res 16:895–903
Castagneyrol B, Giffard B, Péré C, Jactel H (2013) Plant apparency, an overlooked driver of associational resistance to insect herbivory. J Ecol 101:418–442
Castagneyrol B, Jactel H, Vacher C et al (2014) Effects of plant phylogenetic diversity on herbivory depend on herbivore specialization. J Appl Ecol 51:134–141
Cornelissen JHC, Lavorel S, Garnier E (2003) Handbook of protocols for standardised and easy measurement of plant functional traits worldwide. Aust J Bot 51:335–380
Corona P, Chirici G, McRoberts RE (2011) Contribution of large-scale forest inventories to biodiversity assessment and monitoring. For Ecol Manage 262:2061–2069
Cotta H (1828) Anweisung zum Waldbau, 4th edn. Arnoldische Buchhandlung, Dresden Leipzig, 413 p
Dayton PK (1972) Toward an understanding of community resilience and the potential effects of enrichments to the benthos at McMurdo Sound, Antarctica. In: Proceedings of the colloquium on conservation problems in Antarctica. Allen Press, Lawrence, KS, pp 81–95
Díaz S, Symstad AJ, Chapin IS et al (2003) Functional diversity revealed by removal experiments. Trends Ecol Evol 18:140–146
Ellison AM, Bank MS, Clinton BD (2005) Loss of foundation species: consequences for the structure and dynamics of forested ecosystems. Front Ecol Environ 3:479–486
Erwin TL (1982) Tropical forests: their richness in Coleoptera and other arthropod species. Coleopt Bull 36:74–75
Firn J, Erskine PD, Lamb D (2007) Woody species diversity influences productivity and soil nutrient availability in tropical plantations. Oecologia 154:521–533
Floren A, Müller T, Dittrich M et al (2014) The influence of tree species, stratum and forest management on beetle assemblages responding to deadwood enrichment. For Ecol Manage 323:57–64
Flynn DF, Mirotchnick N, Jain M et al (2011) Functional and phylogenetic diversity as predictors of biodiversity-ecosystem-function relationships. Ecology 92:1573–1581
Forrester DI (2014) The spatial and temporal dynamics of species interactions in mixed-species forests: from pattern to process. For Ecol Manage 312:282–292
Forrester DI (2015) Transpiration and water-use efficiency in mixed-species forests versus monocultures: effects of tree size, stand density and season. Tree Physiol 35:289–304
Forrester DI, Pretzsch H (2015) Tamm Review: on the strength of evidence when comparing ecosystem functions of mixtures with monocultures. For Ecol Manage 356:41–53
Forrester DI, Bauhus J (2016) A review of processes behind diversity – productivity relationships in forests. Curr For Rep 2:45–61
Forrester DI, Bauhus J, Khanna PK (2004) Growth dynamics in a mixed species plantation of *Eucalyptus globulus* and *Acacia mearnsii*. For Ecol Manage 193:81–95
Forrester DI, Bauhus J, Cowie AL, Vanclay JK (2006) Mixed-species plantations of *Eucalyptus* with nitrogen-fixing trees: a review. For Ecol Manage 233:211–230
Forrester DI, Kohnle U, Albrecht AT, Bauhus J (2013) Complementarity in mixed-species stands of *Abies alba* and *Picea abies* varies with climate, site quality and stand density. For Ecol Manage 304:233–242
Gamfeldt L, Snäll T, Bagchi R et al (2013) Higher levels of multiple ecosystem services are found in forests with more tree species. Nat Commun 4:340
Gartner TB, Cardon ZG (2004) Decomposition dynamics in mixed-species leaf litter. Oikos 104:230–246
Goßner M, Ammer U (2006) The effects of Douglas-fir on tree-specific arthropod communities in mixed species stands with European beech and Norway spruce. Eur J For Res 125:221–235

Grime JP (2006) Trait convergence and trait divergence in herbaceous plant communities: mechanisms and consequences. J Veg Sci 17:225–260
Grossiord C, Granier A, Ratcliffe S et al (2014) Tree diversity does not always improve resistance of forest ecosystems to drought. Proc Natl Acad Sci U S A 111:14812–14815
Harper JL (1977) Population biology of plants. Academic Press, London. 892 p
He Q, Bertness MD, Altieri AH (2013) Global shifts towards positive species interactions with increasing environmental stress. Ecol Lett 16:695–706
Hector A, Schmid B, Beierkuhnlein C et al (1999) Plant diversity and productivity experiments in European grasslands. Science 286:1123–1127
Hillebrand H, Matthiessen B (2009) Biodiversity in a complex world: consolidation and progress in functional biodiversity research. Ecol Lett 12:1405–1419
Holmgren M, Scheffer M (2010) Strong facilitation in mild environments: the stress gradient hypothesis revisited. J Ecol 98:1269–1275
Holmgren M, Scheffer M, Huston MA (1997) The interplay of facilitation and competition in plant communities. Ecology 78:1966–1975
Hooper DU, Chapin FSI, Ewel JJ et al (2005) Effects of biodiversity on ecosystem functioning: a consensus of current knowledge and needs for future research. Ecol Monogr 75:3–36
Hutchinson MF (1957) Concluding remarks. In: Cold Spring Harbour symposia on quantitative biology, vol. 22, Population studies: animal ecology and demography. Cold Spring Harbor, New York, NY, pp 415–427
Ishida TA, Nara K, Hogetsu T (2007) Host effects on ectomycorrhizal fungal communities: insight from eight host species in mixed conifer–broadleaf forests. New Phytol 174:430–440
Jactel H, Brockerhoff EG (2007) Tree diversity reduces herbivory by forest insects. Ecol Lett 10:835–848
Jucker T, Bouriaud O, Coomes DA (2015) Crown plasticity enables trees to optimize canopy packing in mixed-species forests. Funct Ecol 29:1078–1086
Kattge J, Diaz S, Lavorel S et al (2011) TRY–a global database of plant traits. Glob Chang Biol 17:2905–2935
Kelty MJ (1992) Comparative productivity of monocultures and mixed-species stands. In: Kelty MJ, Larson BC, Oliver CD (eds) The ecology and silviculture of mixed-species forests. Kluwer Academic, Dordrecht, pp 125–141
Kunstler G, Lavergne S, Courbaud B et al (2012) Competitive interactions between forest trees are driven by species' trait hierarchy, not phylogenetic or functional similarity: implications for forest community assembly. Ecol Lett 15:831–840
Laliberté E, Legendre P (2010) A distance-based framework for measuring functional diversity from multiple traits. Ecology 91:299–305
Lang AC, Oheimb G, Scherer-Lorenzen M et al (2014) Mixed afforestation of young subtropical trees promotes nitrogen acquisition and retention. J Appl Ecol 51:224–233
Lavorel S, McIntyre S, Landsberg J, Forbes TDA (1997) Plant functional classifications: from general groups to specific groups based on response to disturbance. Trends Ecol Evol 12:474–478
Lei P, Scherer-Lorenzen M, Bauhus J (2012) Belowground facilitation and competition in young tree species mixtures. For Ecol Manage 265:191–200
Leuschner C, Jungkunst HF, Fleck S (2009) Functional role of forest diversity: pros and cons of synthetic stands and across-site comparisons in established forests. Basic Appl Ecol 10:1–9
Liang J, Crowther TW, Picard N et al (2016) Positive biodiversity–productivity relationship predominant in global forests. Science 354. doi:10.1126/science.aaf8957
Lodge DJ (1997) Factors related to diversity of decomposer fungi in tropical forests. Biodivers Conserv 6:681–688
Looijen RC (1998) Holism and reductionism in biology and ecology: the mutual dependence of higher and lower level research programmes. PhD thesis, University of Groningen. http://www.rug.nl/research/portal/files/10660346/thesis.pdf!null

Loreau M, Hector A (2001) Partitioning selection and complementarity in biodiversity experiments. Nature 412:72–76
Maestre FT, Callaway RM, Valladares F, Lortie CJ (2009) Refining the stress-gradient hypothesis for competition and facilitation in plant communities. J Ecol 97:199–205
Man R, Lieffers VJ (1999) Are mixtures of aspen and white spruce more productive than single species stands? For Chron 75:505–513
McGill BJ, Enquist BJ, Weiher E, Westoby M (2006) Rebuilding community ecology from functional traits. Trends Ecol Evol 21:178–185
Messier C, Puettmann KJ, Coates KD (eds) (2013) Managing forests as complex adaptive systems: building resilience to the challenge of global change. Routledge, The Earthscan Forest Library
Möller A (1922) Der Dauerwaldgedanke – Sein Sinn und seine Bedeutung. Springer, Berlin, 136 p
Morin X, Lechowicz MJ (2008) Contemporary perspectives on the niche that can improve models of species range shifts under climate change. Biol Lett 4:573–576
Müller F (1992) Hierarchical approaches to ecosystem theory. Ecol Model 63:215–242
Nadrowski K, Wirth C, Scherer-Lorenzen M (2010) Is forest diversity driving ecosystem function and service? Curr Opin Environ Sustain 2:75–79
Naeem S (2002) Ecosystem consequences of biodiversity loss: the evolution of a paradigm. Ecology 83:1537–1552
Örlander G (1993) Shading reduces both visible and invisible frost damage to Norway spruce seedlings in the field. Forestry 66:27–36
Paquette A, Messier C (2011) The effect of biodiversity on tree productivity: from temperate to boreal forests. Glob Ecol Biogeogr 20:170–180
Pastor J, Mladenoff DJ, Haila Y, et al (1996) Biodiversity and ecosystem processes in boreal regions. Scope – Scientific Committee on Problems of the Environment International Council of Scientific Unions 55:33–69
Petchey OL, Gaston KJ (2002) Functional diversity (FD), species richness and community composition. Ecol Lett 5:402–411
Petrokofsky G, Kanamaru H, Achard F et al (2012) Comparison of methods for measuring and assessing carbon stocks and carbon stock changes in terrestrial carbon pools. How do the accuracy and precision of current methods compare? A systematic review protocol. Environ Evid 1:6
Piotto D (2008) A meta-analysis comparing tree growth in monocultures and mixed plantations. For Ecol Manage 255:781–786
Poorter L (2007) Are species adapted to their regeneration niche, adult niche, or both? Am Nat 169:433–442
Potvin C, Gotelli NJ (2008) Biodiversity enhances individual performance but does not affect survivorship in tropical trees. Ecol Lett 11:217–223
Pretzsch H (1995) Analyse und Reproduktion räumlicher Betandesstrukturen – Methodische Überlegungen am Beispiel niedersächsischer Buchen-Lärchen-Mischbestände. Centralbl Gesamte Forstwesen 112:91–117
Pretzsch H (1997) Analysis and modeling of spatial stand structures. Methodological considerations based on mixed beech-larch stands in Lower Saxony. For Ecol Manage 97 (3):237–253
Pretzsch H (2005) Diversity and productivity in forests: evidence from long-term experimental plots. In: Scherer-Lorenzen M, Körner C, Schulze ED (eds) Forest diversity and function: temperate and boreal systems. Springer, Berlin, pp 41–64
Pretzsch H (2009) Forest dynamics, growth and yield. Springer, Berlin, 664 p
Pretzsch H, Schütze G (2009) Transgressive overyielding in mixed compared with pure stands of Norway spruce and European beech in Central Europe: evidence on stand level and explanation on individual tree level. Eur J For Res 128:183–204
Pretzsch H, Bielak K, Block J et al (2013) Productivity of pure versus mixed stands of oak (*Quercus petraea* (MATT.) LIEBL. and *Quercus robur* L.) and European beech (*Fagus sylvatica* L.) along an ecological gradient. Eur J For Res 132:263–280

Pretzsch H, Biber P, Schütze G et al (2014) Forest stand growth dynamics in Central Europe have accelerated since 1870. Nat Commun 5:4967

Pretzsch H, Del Río M, Ammer C et al (2015) Growth and yield of mixed versus pure stands of Scots pine (*Pinus sylvestris L.*) and European beech (*Fagus sylvatica* L.) analysed along a productivity gradient through Europe. Eur J For Res 134:927–947

Purahong W, Hoppe B, Kahl T et al (2014) Changes within a single land-use category alter microbial diversity and community structure: molecular evidence from wood-inhabiting fungi in forest ecosystems. J Environ Manage 139:109–119

Reich PB, Ellsworth DS, Walters MB et al (1999) Generality of leaf trait relationships: a test across six biomes. Ecology 80:1955–1969

Richards AE, Forrester DI, Bauhus J, Scherer-Lorenzen M (2010) The influence of mixed species tree plantations on the nutrition of individual species: a review. Tree Physiol 30:1192–1208

Rothe A, Binkley D (2001) Nutritional interactions in mixed species forests: a synthesis. Can J For Res 31:1855–1870

Scherer-Lorenzen M, Körner C, Schulze ED (eds) (2005a) Forest diversity and function: temperate and boreal systems. Springer, Berlin

Scherer-Lorenzen M, Potvin C, Koricheva J et al (2005b) The design of experimental tree plantations for functional biodiversity research. In: Scherer-Lorenzen M, Körner C, Schulze ED (eds) Forest diversity and function: temperate and boreal systems. Springer, Berlin, pp 347–376

Scherer-Lorenzen M, Schulze ED, Don A et al (2007) Exploring the functional significance of forest diversity: a new long-term experiment with temperate tree species (BIOTREE). Perspect Plant Ecol Evol Syst 9:53–70

Schmid I, Kazda M (2001) Vertical distribution and radial growth of coarse roots in pure and mixed stands of *Fagus sylvatica* and *Picea abies*. Can J For Res 31:539–548

Schreuder HT, Gregoire TG, Wood GB (1993) Sampling methods for multiresource forest inventory. Wiley, New York, 464 p

Schulze ED, Mooney HA (1993) Biodiversity and ecosystem function. Springer, Berlin, Heidelberg, 525 p

Schwendenmann L, Pendall E, Sánchez BR et al (2014) Tree water uptake in a tropical plantation varying in tree diversity: interspecific differences, seasonal shifts and complementarity. Ecohydrology 8. doi:10.1002/eco.1479

Scowcroft PG, Meinzer FC, Goldstein G et al (2000) Moderating night radiative cooling reduces frost damage to *Metrosideros polymorpha* seedlings used for forest restoration in Hawaii. Restor Ecol 8:161–169

Sole RV, Montoya M (2001) Complexity and fragility in ecological networks. Proc R Soc Lond B Biol Sci 268(1480):2039–2045

Southwood TRE, Wint GW, Kennedy CE, Greenwood SR (2004) Seasonality abundance, species richness and specificity of the phytophagous guild of insects on oak (Quercus) canopies. Eur J Entomol 101:43–50

Spiecker H, Mielikäinen K, Köhl M, Skovsgaard JP (1996) Growth trends in European forests: studies from 12 countries. Springer, Heidelberg, 372 p

Stachowicz JJ (2001) Mutualism, facilitation, and the structure of ecological communities. Bioscience 51:235–246

Stork NE (1988) Insect diversity: facts, fiction and speculation. Biol J Linn Soc 35:321–337

Sutherland WJ, Pullin AS, Dolman PM, Knight TM (2004) The need for evidence-based conservation. Trends Ecol Evol 19:305–308

Swenson NG, Enquist BJ (2009) Opposing assembly mechanisms in a Neotropical dry forest: implications for phylogenetic and functional community ecology. Ecology 90:2161–2170

Tham Å (1994) Crop plans and yield predictions for Norway spruce (*Picea abies* (L.) Karst.) and birch (*Betula pendula* Roth & *Betula pubescens* Ehrh.) mixtures. Studia Forestalia Suecica 195

Tilman D (1996) Biodiversity: population versus ecosystem stability. Ecology 77:350–363

Tilman D, Knops J, Wedin D et al (1997) The influence of functional diversity and composition on ecosystem processes. Science 277:1300–1302
Tobner CM, Paquette A, Reich PB et al (2014) Advancing biodiversity–ecosystem functioning science using high-density tree-based experiments over functional diversity gradients. Oecologia 174:609–621
Toïgo M, Vallet P, Perot T et al (2015) Overyielding in mixed forests decreases with site productivity. J Ecol 103:502–512
Ulrich B (1994) Process hierarchy in forest ecosystems: an integrative ecosystem theory. In: Godbold DL, Hüttermann A (eds) Effects of acid rain on forest processes. Wiley-Liss, New York, pp 353–398
Unterseher M, Otto P, Morawetz W (2005) Species richness and substrate specificity of lignicolous fungi in the canopy of a temperate, mixed deciduous forest. Mycol Prog 4:117–132
Vandermeer JH (1989) The ecology of intercropping. Cambridge University Press, Cambridge
van der Plas F, Manning P, Allan E et al (2016) 'Jack-of-all-trades' effects drive biodiversity-ecosystem multifunctionality relationships in European forests. Nat Commun 7:11109
Vázquez DP, Melián CJ, Williams NM et al (2007) Species abundance and asymmetric interaction strength in ecological networks. Oikos 116:1120–1127
Verheyen K, Vanhellemont M, Auge H et al (2016) Contributions of a global network of tree diversity experiments to sustainable forest plantations. Ambio 45:29–41
Vilà M, Inchausti P, Vayreda J et al (2005) Confounding factors in the observational productivity – diversity relationship in forests. In: Scherer-Lorenzen M, Körner C, Schulze E-D (eds) Forest diversity and function: temperate and boreal systems. Springer, Berlin, pp 65–86
Vilà M, Vayreda J, Comas L et al (2007) Species richness and wood production: a positive association in Mediterranean forests. Ecol Lett 10:241–250
Vilà M, Carrillo-Gavilá A, Vayreda J et al (2013) Disentangling biodiversity and climatic determinants of wood production. PLoS One 8:e53530
Violle C, Jiang L (2009) Towards a trait-based quantification of species niche. J Plant Ecol 2(2):87–93
Wardle DA (1999) Is "Sampling Effect" a problem for experiments investigating biodiversity-ecosystem function relationships? Oikos 87:403–407
Wardle DA (2016) Do experiments exploring plant diversity–ecosystem functioning relationships inform how biodiversity loss impacts natural ecosystems? J Veg Sci 27:646–653
Wein A, Bauhus J, Bilodeau-Gauthier S et al (2016) Tree species richness promotes invertebrate herbivory on congeneric native and exotic tree saplings in a young diversity experiment. PLoS One 11(12):e0168751
Wiedemann E (1942) Der gleichaltrige Fichten-Buchen-Mischbestand. Schaper Verlag, Hannover, 88 p
de Wit CT (1960) On competition. Verslagen Landbouwkundige Onderzoekingen 66:1–82
Winfree R, Fox JW, Williams NM et al (2015) Abundance of common species, not species richness, drives delivery of a real-world ecosystem service. Ecol Lett 18:626–635
Yachi S, Loreau M (1999) Biodiversity and ecosystem productivity in a fluctuating environment: the insurance hypothesis. Proc Natl Acad Sci U S A 96:57–64
Yang X, Bauhus J, Both S et al (2013) Establishment success in a forest biodiversity and ecosystem functioning experiment in subtropical China (BEF-China). Eur J For Res 132:593–606

# Ecological and Physiological Processes in Mixed Versus Monospecific Stands

# 3

David I. Forrester

**Abstract**

Tree growth depends on the resource availability, the proportion of resources acquired, and the efficiency with which those resources are used. Each of these variables can be influenced by species interactions. These interactions are dynamic and change spatially and temporally as resource availability and climatic conditions change. It is important to understand these processes when designing and managing mixed-species stands and also when modelling these processes. These interactions and their dynamics are the focus of this chapter. To begin with, the production ecology equation is described because it provides a useful framework to quantify the types of processes that influence the growth of forests and how these are influenced by species interactions. This equation describes growth as a function of resource availability, resource acquisition, and resource-use efficiency. Then, while referring to this equation, some of the main types of processes are described in terms of how they influence these variables and hence the productivity of mixtures. This is done for nutrients, then light, and then water. The influence of a given type of interaction on growth is not static. Instead, it changes with spatial and temporal variability in resource availability and climatic conditions and as a stand develops. Therefore, the next section describes a framework that explains these spatial and temporal dynamics and indicates when different types of interactions are important. Finally, stand density can influence the effect of these interactions. As stand density increases, interactions may become more favourable or less favourable, depending on how, and which, resources are influenced by the change in density. The final section therefore shows why stand density needs to be taken into account when examining how species interact.

D.I. Forrester (✉)
Swiss Federal Institute for Forest, Snow and Landscape Research WSL, Zürcherstrasse 111, 8903 Birmensdorf, Switzerland
e-mail: david.forrester@wsl.ch



H. Pretzsch et al. (eds.), *Mixed-Species Forests*, DOI 10.1007/978-3-662-54553-9_3

## Contents


3.1 The Production Ecology Equation ... 74
3.2 Types of Mixture Comparisons and Levels of Analyses ... 76
3.3 Nutrient Availability, Acquisition, and Use Efficiency ... 77
3.3.1 Nutrient Availability ... 77
3.3.2 Nutrient Uptake ... 80
3.3.3 Nutrient-Use Efficiency ... 81
3.4 Light Availability, Absorption, and Use Efficiency ... 82
3.4.1 Changes in Stand Structure and APAR ... 84
3.4.2 Changes in Tree Allometry and APAR ... 85
3.4.3 Changes in Tree Physiology ... 85
3.5 Water Availability, Use, and Use Efficiency ... 88
3.5.1 General Patterns and the Consideration of Spatial Scale ... 88
3.5.2 Processes That Influence Transpiration or WUE ... 92
3.5.3 The Influence of Stand Density and Water Stress ... 94
3.6 Spatial and Temporal Dynamics of Species Interactions ... 95
3.6.1 A Framework for Understanding the Dynamics of Species Interactions ... 95
3.6.2 Spatial Effects of Interactions That Influence Nutrient Availability ... 98
3.6.3 Spatial Effects of Interactions That Influence Water Availability ... 101
3.6.4 Spatial Effects of Interactions That Influence Light Absorption and Use ... 102
3.6.5 Temporal Dynamics of Species Interactions ... 102
3.6.6 Simultaneously Occurring Species Interactions ... 104
3.7 Stand Density ... 104
3.8 Synopsis ... 105
Take-Home Messages ... 107
References ... 108


## 3.1 The Production Ecology Equation

The relationship between growth, resource availability, resource acquisition, and the efficiency with which resources are used to fix carbon can be described using the production ecology equation (Monteith 1977):

$$\begin{aligned}\text{Gross primary production} = {} & \text{resource supply} \\ & \times \text{fraction of resource acquired} \\ & \times \text{resource use efficiency}\end{aligned} \tag{3.1}$$

When examining forest growth, the focus might be on above-ground biomass or wood production (Mg ha$^{-1}$ year$^{-1}$), which can be described using Eq. (3.1) as a function of the supply (Resource units ha$^{-1}$ year$^{-1}$), acquisition (a fraction), and use efficiency of light, water, or nutrients (Mg$_{\text{biomass}}$ Resource units$^{-1}$), minus allocation to respiration and nonwoody tissues (Binkley et al. 2004).

Interactions in mixtures are often described using terms such as facilitation, competition reduction, and competition (Chap. 2). However, it can be very difficult to separate the contribution of each of these on the growth dynamics of forests, and this is rarely attempted. The production ecology equation provides a framework that is useful for quantifying the processes driving these effects in order to understand how and why growth can change in mixtures (Richards et al. 2010). It has also been

**Table 3.1** Many of the processes or species interactions that could influence the growth of mixtures compared with monocultures and an indication of whether they are facilitative or competitive reduction processes

| Name of process or pattern | Facilitation (F) or competitive reduction (Cr) | Section |
|---|---|---|
| *Nutrient related* | | |
| Symbiotic nitrogen fixation | F | 3.3.1 |
| Nutrient cycling | F | 3.3.1 |
| Chemical, spatial, or temporal stratification | Cr | 3.3.2 |
| Nutrient mineralisation | F | 3.3.1 |
| Canopy nitrogen retention | F | 3.3.1 |
| *Light related* | | |
| Space occupation—canopy stratification | Cr | 3.4.1 |
| Space occupation—complementary crown shapes and architectures | Cr | 3.4.1 |
| Space occupation—intraspecific variability in crown architecture and size | Cr | 3.4.2 |
| Physiological differences | Cr | 3.4.3 |
| Phenology and interspecific effects on these | Cr | 3.4.1 |
| *Water related* | | |
| Hydraulic redistribution | F | 3.5.2 |
| Shared mycorrhizal networks | F | 3.5.2 |
| Other spatial stratification | Cr | 3.5.2 |
| Canopy interception | Cr | 3.5.2 |
| Transpiration and WUE | Cr | 3.5.1, 3.5.2 |
| Litter layer as a sponge or barrier | F or Cr | 3.5.2 |
| Isohydric vs. anisohydric | Cr | 3.5.2 |
| Inter-and intraspecific differences in phenology | Cr | 3.5.2 |
| Modified within-stand environmental conditions | F | 3.5.2 |
| *Biotic* | | |
| Insect herbivory and leaf pathogens | F | |

The section column indicates where in the text their effect on the production ecology is described. See also Sect. 3.6 for the spatial and temporal implications of these interactions. Modified from Forrester and Bauhus (2016)

used in several reviews about the mechanisms driving growth responses to fertiliser application, irrigation, pruning, thinning, spacing, genotypes, species, stand age and geographic gradients (Binkley 2012; Forrester 2013) and even to examine the influence of species interactions on plant nutrition (Richards et al. 2010), transpiration (Forrester 2015), and light (Forrester and Albrecht 2014; Forrester et al. in press). Table 3.1 lists many of the processes and species interactions that have been found to influence mixtures. These are discussed in the text in relation to their influence on the production ecology of mixtures. The production ecology equation will provide the foundation for this chapter.

## 3.2 Types of Mixture Comparisons and Levels of Analyses

There are many ways experiments have been designed to make comparisons between mixtures and monocultures. Each type of comparison is useful for answering different questions, so it is important to describe the types of comparison that are often made and that will be referred to in this chapter.

*Type 1* Comparing different species growing in a mixture in terms of their growth, morphology, physiology, and phenology (interspecific comparisons). This might be done to determine whether different species have traits that could complement each other and may not actually involve any measurements in monocultures at all.

*Type 2* Comparisons of mixtures and monocultures in terms of total stand variables, such as growth, light absorption, transpiration, litterfall, etc. However, this total stand information does not show how each species contributes to the mixing effect. The individual species contributions are important because sometimes only one of the species contributes to the mixing effect. It is possible that an absence of any mixture effect at the total stand level is not due to an absence of a mixing (complementarity) effect, but rather results from opposing responses by different species. This Type 2 comparison does not provide information about individual species contributions to mixing effects.

*Type 3* Compare the growth, morphology, physiology, or phenology of a given species in mixture with that of the same species in monoculture (intraspecific comparisons). This enables us to determine whether the species interactions actually changed the performance of a given species. This type of comparison, and the production ecology equation, has been used to examine mixture effects on nutrients (Richards et al. 2010), light absorption (Forrester and Albrecht 2014; Forrester et al. in press), and water pools or fluxes (Forrester 2015) and is frequently considered in this chapter.

A closely related issue is the level of analysis in terms of the scale of the variable being measured (Forrester and Pretzsch 2015). The main levels considered in this and the following chapters are the tree, neighbourhood, species, community, and landscape or estate. *Tree-level* analyses examine individual trees, such as when regression is used to examine how the relationship between tree diameter and tree biomass varies between treatments. *Neighbourhood-level* analyses are a form of tree-level analysis that take account of the characteristics of the trees' neighbourhood in terms of factors such as basal area and species composition (e.g. in terms of basal area, species composition; Vanclay 2006b; Boyden et al. 2005). The more typical tree-level analyses contrast with neighbourhood-level analyses because they either ignore the characteristics of the trees' neighbourhood or consider those characteristics only in terms of the stand-level treatment (where all trees within the plot have the same (mean plot) neighbourhood) (Forrester and Pretzsch 2015).

The next level up, *stand-level* analysis, considers totals and means of all trees within the plot, such as total above-ground biomass ($B_T$) or mean tree height. Stand-level analyses include *species-level* and *total stand-level* analyses. Species-level analyses divide the total stand by species to provide the totals and means for each species within the stand. For example, in a three-species mixture, $B_T = B_{species1} + B_{species2} + B_{species3}$. A species-level analysis could consider the $B_{species1}$, $B_{species2}$, and $B_{species3}$, whereas the total stand-level analysis could compare the $B_T$ of different stands. The total stand-level analyses are commonly referred to as *community-level* analyses because they consider the totals or means of the whole community. The next level up includes the landscape level or the estate level. These are commonly used in forest planning and are referred to in Chaps. 10 and 11. While many other levels exist such as the leaf level and organ level (e.g. branches, roots) or coarser scales such as the region or continental levels, they are rarely considered in this book.

## 3.3 Nutrient Availability, Acquisition, and Use Efficiency

### 3.3.1 Nutrient Availability

Nutrient availability can increase when the size of the nutrient pool is increased, such as by symbiotic N fixation and atmospheric deposition or by increasing the proportion of the pool that is actually available to plants and the speed at which it is cycled (Forrester and Bauhus 2016). The available portion of the nutrient pool can be increased by increasing mineralisation rates, or when rates of nutrient cycling are accelerated by the production of more readily decomposable litter, by providing environments more favourable to litter decomposition and when nutrients that were inaccessible for one species are acquired and then cycled by another species. Roots require water to absorb and transport nutrients, so even when there are sufficient nutrient pool sizes, the availably can be low if the soils are dry or frozen. Therefore, processes that increase soil water availability (see Sect. 3.5) could also increase the available portion of nutrients. The proportion of nutrients available to a given species may also be increased when uptake by one species is lower than that of another, thereby leaving more nutrients for the latter, e.g. when there are interspecific differences in growth, in nutrient-use efficiency, or in the source of the given nutrient. Availability may also be increased by reducing nutrient losses that occur via leaching and erosion, or by combining species or altering stand structures to increase atmospheric deposition, but due to the more limited information on this aspect in mixtures, it will not be discussed (for Type 1 or 2 effects on soil fertility, see Augusto et al. 2002, 2015; Berger et al. 2009a, b). More detailed descriptions of processes that influence nutrition and nutrient cycling in forests are available in several reviews (Richards et al. 2010; Hinsinger et al. 2011; Binkley and Giardina 1998; Rothe and Binkley 2001; Knops et al. 2002; Gartner and Cardon 2004).

#### 3.3.1.1 Nutrient Mineralisation

There can be large differences in the rates of mineralisation under different species (Binkley and Giardina 1998), and mixing species can result in increased mineralisation rates compared with monocultures of the species with slower rates of mineralisation. For example, N mineralisation rates and N uptake were 54% and 41% higher, respectively, in mixtures of *Picea sitchensis* and *Pinus sylvestris* than in *P. sitchensis* monocultures (Williams 1992). Mixtures of *Eucalyptus globulus* and *Acacia mearnsii* had N mineralisation rates that were about twice as high as those in *E. globulus* monocultures (Khanna 1997). N availability also increased with increasing proportions of *Albizia falcataria* in mixtures with *Eucalyptus saligna*, while P availability followed the opposite trend (Kaye et al. 2000). These effects clearly result from interspecific differences (Type 1 effects). It is also possible that mineralisation rates under a given species change when it is growing in a mixture compared with a monoculture (Type 3 effects). For example, Binkley and Valentine (1991) found that mineralisation rates under *Fraxinus pennsylvanica*, *Pinus strobus*, and *Picea abies* trees were modified by the identity of their neighbours. In contrast, antagonistic effects can reduce nutrient availability in mixtures compared with monocultures, and this occurred in mixtures of *Larix laricina* with either *Picea mariana* or *Pinus strobus*. In these mixtures, N mineralisation was lower than expected based on monocultures; however, aboveground biomass growth was greater than expected in *L. laricina–P. mariana* mixtures (Dijkstra et al. 2009). The mixing of lignin-rich litter from the conifers to the N-rich *L. laricina* litter was suggested to have suppressed the formation of lignolytic enzymes or formed complexes that were highly resistant to microbial degradation (Dijkstra et al. 2009).

#### 3.3.1.2 Symbiotic N Fixation

Rates of symbiotic N fixation can be anything from 1 to 200 kg ha$^{-1}$ year$^{-1}$, and this can be 10% to >70% of the total N used by the N-fixing plant (Binkley and Giardina 1997; Khanna 1998; Fisher and Binkley 2000). This N can be transferred to non-N-fixing species after the plant and microbial tissues decompose (Van Kessel et al. 1994; Fisher and Binkley 2000). Smaller quantities may also be transferred via root exudation and common mycorrhizal connections (He et al. 2003). Soil N availability may also be increased for non-N-fixing species when the N-fixing species strongly rely on fixed N rather than soil N. Rates of N fixation and the proportion of N that is derived from the atmosphere ($N_{dfa}$) can increase with increasing soil P and decreasing with soil N and tend to vary with the same factors that influence growth. These effects have been the focus of several reviews (Crews and Peoples 2005; Peoples et al. 1995). However, the temporal dynamics of N fixation and the effects of competition from non-N-fixing species have received far less attention. These temporal dynamics clearly influence the balance between competition and facilitation and need to be understood in order to apply appropriate silvicultural treatments, such as thinning the N-fixing species.

It might be expected that as stands develop, rates of N fixation and $N_{dfa}$ will increase to a peak and then decline as N availability increases, and internal N cycling within the trees provides some of the N demands of the trees. However, while this has been observed in some stands, it has not in others, and there has not been enough research on N fixation dynamics in forests or plantations to confidently describe and generalise about these trends. In *Elaeagnus angustifolia*, rates of N fixation peaked at around age 2 or 3 years before declining at ages 4 and 5 years, but $N_{dfa}$ continued to increase slowly (Khamzina et al. 2009). Rates of N fixation by *Acacia mearnsii* and *Acacia dealbata* in mixtures with *Eucalyptus* were about 40–90 kg ha$^{-1}$ year$^{-1}$ in at ages 5–10 years (May and Attiwill 2003; Forrester et al. 2007b) but were lower in older mixtures of *A. dealbata* or *A. melanoxylon* with *Eucalyptus* (Pfautsch et al. 2009a, b). In contrast, high rates of N fixation (75–85 kg ha$^{-1}$ year$^{-1}$) were still occurring at age 55 years in two *Alnus rubra-Pseudotsuga menziesii* mixtures (Binkley et al. 1992b). Rates of N fixation remained high between ages 1 and 3.5 years for *Leucaena leucocephala* and *Casuarina equisetifolia* (73–74 kg ha$^{-1}$ year$^{-1}$), and while $N_{dfa}$ for *L. leucocephala* declined from 98% and 38%, there was no decline in that of *C. equisetifolia*, which fluctuated between 43% and 62% (Parrotta et al. 1996).

Competition from non-N-fixing species appears to have little influence on rates of N fixation. Rates of N fixation per tree and linear relationships between N accretion and the proportion of N-fixing species in mixtures have been found in several mixtures (ordered as N-fixing species with non-N-fixing species) including *Falcataria moluccana* with *E. saligna*, *A. mearnsii* with *E. globulus*, and *Alnus rubra* with *Populus trichocarpa* (DeBell and Radwan 1979; Kaye et al. 2000; Forrester et al. 2007b). In most of these mixtures, the growth of each species increased compared with monocultures. However, when competition from the non-N-fixing species is intense enough to significantly reduce the growth of the N-fixing species, it may also reduce rates of N fixation. This occurred in mixtures of *E. grandis* (1111 trees ha$^{-1}$) and a N-fixing tree, *A. mangium* (556 trees ha$^{-1}$), where growth of *A. mangium* was reduced and N fixation at age 30 months was 7–31 kg ha$^{-1}$ compared with 66 kg ha$^{-1}$ in *A. mangium* monocultures (1111 trees ha$^{-1}$) (Bouillet et al. 2008).

#### 3.3.1.3 Accelerating Rates of Nutrient Cycling

As plant material decomposes, the nutrients it contains return to a mineral form in the soil, which plants can use. Nutrient inputs into forests are often low, and growth is often limited by nutrient availability, so the rate at which litter decomposition and nutrient cycling occurs is important. Litter decomposition depends on the microclimate and quality (e.g. nutrient concentrations, lignin contents) of the litter and surrounding soil, which influence decomposer composition and activity, and these factors differ between mixtures and monocultures (Rothe and Binkley 2001; Gartner and Cardon 2004; Hättenschwiler et al. 2005). The differences between mixtures and monocultures can be a combination of interspecific differences (Type 1 effects) as well as intraspecific differences, such as when the quality of litter from a given species varies between mixtures and monocultures (Type 3 effects) (Hättenschwiler et al. 2005; Richards et al. 2010). Several reviews of litter decomposition found that, in about half of the studies, litter decomposition was faster in mixtures than expected based on monocultures (mean of 17% faster), but in about

30% of studies, there was no effect, and in about 20% of studies, there was an antagonistic effect (Hättenschwiler et al. 2005; Gartner and Cardon 2004). These reviews contained tree mixtures but also grass mixtures. A review that focussed on tree mixtures found that the most common result was no effect but that synergistic and antagonistic effects all occurred (Rothe and Binkley 2001).

### 3.3.2 Nutrient Uptake

The N and P uptake of mixed-species stands was reviewed by Richards et al. (2010). In >50% of studies, the N or P uptake, by a given species, increased by at least 10% in mixtures compared with monocultures, and this was not dependent on whether the mixtures contained N-fixing species. It is important to note that nutrient uptake was quantified as the nutrient content of above-ground biomass (kg nutrient $ha^{-1}$), so this indicates the net effect of changes in availability as well as the proportion of nutrient taken up. Some of the processes described in Sect. 3.3.1 that influence nutrient availability will also increase the proportion of nutrients that are taken up. In addition to these processes, the proportion of nutrients taken up by plants might increase when there are differences (Type 1 or 3) in nutrient preferences and symbiotic associations, temporal differences in nutrient uptake, contrasting spatial distributions of fine roots or more fine roots, and hence a greater potential nutrient uptake.

Spatial fine-root distributions vary between species (Type 1 differences), which may result in the use of a higher proportion of soil nutrients when the species are mixed. The spatial distribution of fine roots in mixtures can also be influenced by the species interactions, such that the fine-root distributions for a given species in mixture are significantly different to its monocultures (Type 3) (Rothe and Binkley 2001). Rothe and Binkley (2001) suggested that differences in fine-root distributions could be exploited if deeper rooting species are used to 'pump' nutrients from the subsoil to upper layers where shallow-rooted species can make use of them.

Several studies have found that fine-root biomass or fine-root production is higher in mixtures than at least one of the monocultures (sometimes with transgressive below-ground overyielding) and that fine-root biomass or production of a given species is greater than expected based on monocultures (Fredericksen and Zedaker 1995; Wang et al. 2002; Brassard et al. 2011; Schmid and Kazda 2002; Laclau et al. 2013). The mixing effect appears to be greater when the species are mixed more homogeneously within the forest enabling a greater proportion of the soil space to be filled due to interspecific differences in fine-root positioning and proliferation (Brassard et al. 2013), compared with mixed-species forests with a less uniform distribution of species where there is a higher probability that individual trees will be growing in neighbourhoods dominated by other individuals of the same species. In some cases, the fine-root biomass of one species increases at the expense of another species, whose fine-root biomass production is suppressed (Laclau et al. 2013; Bolte and Villanueva 2006).

These studies rarely link the changes in fine-root distributions, biomass, or growth with actual nutrient (or water) uptake, which would indicate whether the below-ground complementarity effects were associated with increased nutrient uptake. This is important because greater growth or nutrient uptake in mixtures might result from, but does not require, any differentiation in fine-root distributions between species or that mixtures have greater fine-root biomass or fine-root growth. Several studies have found no spatial differences in fine-root distributions and/or no differences in fine-root biomass in mixtures and monocultures, even though above-ground biomass (shown in Fig. 4.16) was much greater in mixtures (Bauhus et al. 2000). One explanation is that nutrient uptake was not a major process influencing this complementarity effect; complementarity may have been driven more by increased nutrient availability (which may even reduce fine-root biomass) or light or water resources. Another explanation is that nutrient uptake does not directly relate to fine-root distributions, biomass, or growth. Nutrient uptake also depends on the fine-root architecture, mycorrhizal associations, and the root physiology and phenology (Richards et al. 2010).

Chemical stratification is another process whereby competition for nutrients may be reduced and the proportion of the total nutrient pool taken up is increased in mixtures compared with monocultures (McKane et al. 2002; Richards et al. 2010). Chemical stratification occurs when different species take up different forms, or sources, of a given nutrient (e.g. organic or inorganic forms) due to contrasting mycorrhizal associations and the ability to influence nutrient acquisition by secreting enzymes and organic acids (Ewel 1986; Turner 2008; Richards et al. 2010; Hinsinger et al. 2011). Interspecific differences in N preferences (Type 1), such as symbiotically fixed N, nitrate, ammonium, and amino acids, have been found in mixtures (McKane et al. 2002; Pfautsch et al. 2009b; Kranabetter and MacKenzie 2010), and there may also be differences in P preferences (Turner 2008; Hinsinger et al. 2011).

### 3.3.3 Nutrient-Use Efficiency

When comparing the N- or P-use efficiency of trees growing in mixtures with that of the same species growing in monocultures, Richards et al. (2010) found that decreases, increases, and no changes in N- and P-use efficiency all occurred and that declines were more common than increases. In some studies, N- and P-use efficiency declined even when growth increased in mixtures. It is important to note that Richards et al. quantified nutrient-use efficiency as above-ground biomass growth per unit of nutrient in litterfall (following Vitousek 1982) or in foliage mass (following Harrington et al. 1995). These nutrient contents are rough approximations of actual nutrient uptake and require an assumption that litterfall mass or foliage mass is a fixed proportion of growth, which is often not the case. Nitrogen-use efficiency calculated from actual measurements of nutrient uptake is not common in mixtures. One example is mixtures that contained *Pseudotsuga menziesii* with a N-fixing species, *Alnus rubra* (Binkley et al. 1992b). The growth of

*P. menziesii* was significantly faster in mixtures but only on low N sites, and this was related to increases in nutrient uptake (N, Mg, and K); N-use efficiency in mixtures was only 25%, and 16% of those in *P. menziesii* stands on low- and high-N sites, respectively (Binkley et al. 1992b). However, N-use efficiency can also increase as nutrient uptake increases. For example, in *Eucalyptus* monocultures, growth and N-use efficiency increased with increasing nitrogen use and increasing precipitation (Stape et al. 2004). These studies indicate that there may not be a general trend for changes in nutrient-use efficiency in mixtures, unlike those for water- and light-use efficiency (see below).

## 3.4 Light Availability, Absorption, and Use Efficiency

The absorption of photosynthetically active radiation (APAR) by trees and stands is the basis for their growth. Increases in tree growth rates are often accompanied by increases in APAR, light-use efficiency (LUE), or both, according to a literature review by Binkley (2012) that included a wide range of treatments, such as productivity gradients, fertiliser application, irrigation, mixed-species systems, stand structures, and stand age. When growth increased, APAR also increased by a median of 40% and a mean of 85%. In about 90% of studies, when growth rates increased, LUE also increased by a median of 50% and a mean of 70%. In about half of the studies, there were increases in APAR and LUE, while in the other half, either one or the other increased, not both.

So what does this mean for mixtures? It suggests that for a given species, when its growth increases in mixture, due to some sort of complementarity effect, this may be associated with increased APAR and/or LUE (Type 3 difference). In addition, Type 1 differences will probably exist because the species in mixture are likely to vary in terms of their APAR and LUE due to interspecific differences in physiology, morphology, and phenology. For example, if a fast-growing species with high LUE overtops a slower-growing and more shade-tolerant species that is capable of high LAI and APAR, then mixtures could have a greater LUE than monocultures of the more shade-tolerant species and could have a greater APAR than monocultures of the less shade-tolerant species in the upper canopy.

Only a few studies have measured and compared APAR and LUE in mixtures with monocultures (Fig. 3.1), and Type 1 and Type 3 differences have both been observed (Forrester and Bauhus 2016). For example, mixtures of *Eucalyptus saligna* with *Falcataria moluccana* in Hawaii and *Eucalyptus globulus* with *Acacia mearnsii* in Australia had greater APAR and LUE than monocultures (Binkley et al. 1992a; Forrester et al. 2012b). The *Eucalyptus* species had higher LUE than the other species (Type 1 difference), but their growth was limited by N. *F. moluccana* and *A. mearnsii* are both N-fixing species. In mixtures, the higher LUE of the eucalypts (Type 1 difference) was combined with the higher N availability under the N fixers, which increases the growth of the eucalypts (Type 3). Since the LUE of the mixtures was greater than each monoculture, rather than a weighted average, the LUE of at least one of the species must have increased in each mixture compared

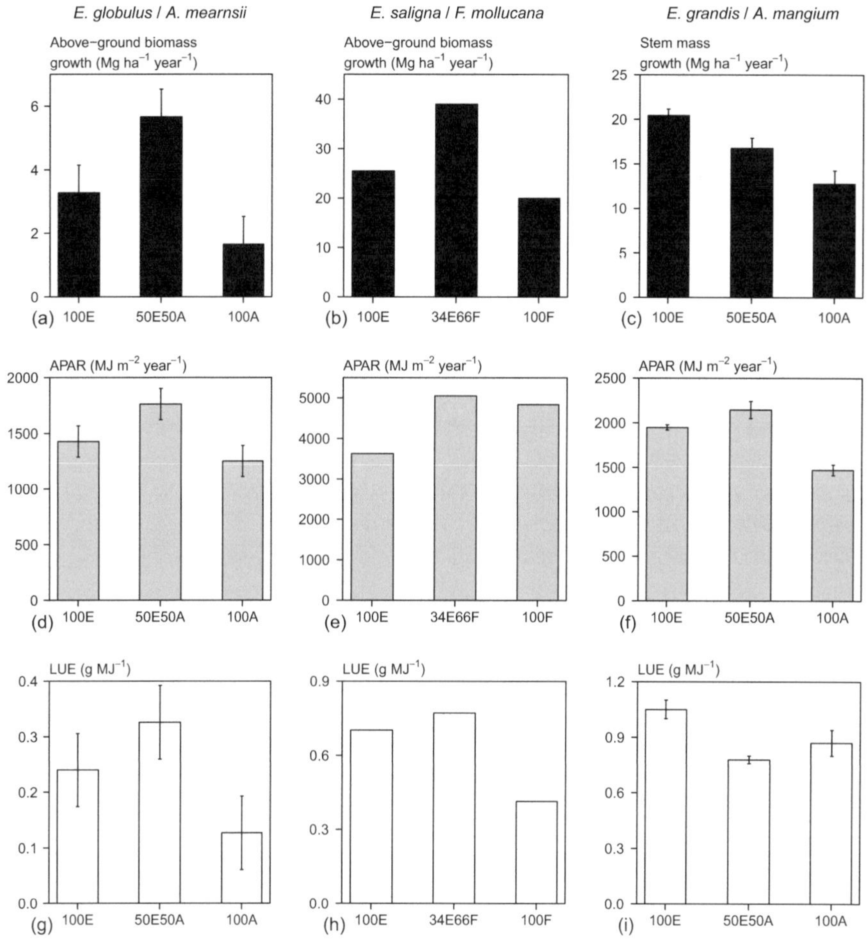


**Fig. 3.1** Biomass growth, absorption of photosynthetically active radiation (APAR), and light-use efficiency (LUE) in monocultures and mixtures of (**a**, **d**, **g**) *Eucalyptus globulus* with *Acacia mearnsii* between ages 14 and 15 years (Forrester et al. 2012b), (**b**, **e**, **h**) *Eucalyptus saligna* with *Falcataria moluccana* between ages 5 and 6 years (Binkley et al. 1992a), and (**c**, **f**, **i**) *Eucalyptus grandis* and *Acacia mangium* during a 6-year rotation (le Maire et al. 2013). Error bars are standard errors of difference for *E. globulus*/*A. mearnsii*, or standard deviations for *E. grandis*/*A. mangium*. The numbers on the *x*-axes are the percentage of all planted trees belonging to a given species (e.g. 100E is 100% *Eucalyptus*; 34E66F is 34% *Eucalyptus* and 66% *Falcataria*)

with its monoculture (Type 3). The increases in APAR resulted, at least in part, from greater leaf areas associated with increases in growth due to a higher availability of, or reduced competition for, other resources. Several processes that could increase resource availability are described in Sect. 3.3 (nutrients) and Sect. 3.5 (water).

### 3.4.1 Changes in Stand Structure and APAR

Contrasting stand structures between mixtures and monocultures can result in higher APAR for some of the species in mixtures, often the taller species (Forrester and Bauhus 2016). The total planting densities (trees per ha) of all mixtures and monocultures were kept constant in each experiment in Fig. 3.1, and the eucalypts overtopped the N-fixing species. Therefore, before the eucalypt canopy closes above the N-fixing species, there should be more space between the eucalypt crowns in mixtures than in monocultures. This could result in less shading of eucalypt crowns, at least from the sides, and higher APAR per eucalypt tree in mixtures compared with monocultures. In the mixtures of Fig. 3.1a, b, the trees of both species were larger in mixtures than in monocultures, except for *F. moluccana*, which was a similar size in both treatments. In extreme cases, such as Fig. 3.1c, the growth and APAR per tree of one species can increase because it outcompetes and reduces the growth of the other species. This was the case in 1:1 mixtures of *E. grandis* and *A. mangium*. In these mixtures, stand APAR was higher than that of monocultures of either species (Fig. 3.1f) (le Maire et al. 2013), similar to the *E. saligna–F. moluccana* and *E. globulus–A. mearnsii* mixtures described in Fig. 3.1. However, the LUE of both *E. grandis* and *A. mangium* were lower in mixtures than monocultures, and as a result, the growth of mixtures was intermediate between faster-growing *E. grandis* monocultures and slower-growing *A. mangium* monocultures (Fig. 3.1c, f, i). That is, mean tree *E. grandis* growth and APAR increased compared with trees in monocultures, but this was at the expense of lower mean tree *A. mangium* growth and APAR in the mixtures.

In addition to changes in the vertical structure of the canopy, there can be temporal changes in structure when mixtures contain species that have leaves when others do not, in which case the APAR of the given species can increase while the other species are leafless. This interspecific difference in phenology (Type 1) may not be very useful in forests where the deciduous trees lose their leaves outside most of the main growing season of the other species, such as during cold temperate or boreal winters when the other species are not growing much anyway (Forrester et al. in press). However, this may be particularly useful in forest types where trees grow all year, such as in the tropics and subtropics (Sapijanskas et al. 2014).

Individual tree APAR (for a given crown leaf area) could also be increased when the different species have crowns with complementary shapes that fit together more efficiently in the mixtures than in the monocultures (Bauhus et al. 2004). For example, in mixtures of *A. alba* and *P. abies*, the APAR of *A. alba* crowns with a given leaf area, and at a given stand density, was greater when its neighbourhood was mostly composed of *P. abies* than when it was composed of other *A. alba* trees, even though the mean height of *A. alba* trees was usually lower than that of *P. abies* trees (Forrester and Albrecht 2014). More efficient 'packing' (see also Chap. 4) of crowns is one explanation for this increase in APAR. Another is changes in crown architecture (Type 3) for a given species when growing in mixture compared with monocultures (Forrester et al. in press).

### 3.4.2 Changes in Tree Allometry and APAR

The crown architecture of a given species often varies in response to competition and changes in resource availability (Ryan et al. 2004; Forrester et al. 2012a), such that trees in mixture might have wider or longer crowns compared with monocultures even when they have the same stem diameters (Bauhus et al. 2004; Getzin and Wiegand 2007; Bayer et al. 2013; Forrester et al. 2016). One possible reason for this change in allometry is to increase APAR and to shade competitors. In mixtures of *P. abies* and *A. alba*, a 10% increase in crown size, in terms of either width, length, or leaf area, resulted in predicted increases in APAR of up to about 15% (Forrester and Albrecht 2014). Interestingly, increasing leaf area by 10% resulted in the smallest increase (only about 2.5%) in APAR. Increasing leaf area without also increasing crown volume will result in a higher leaf area density and self-shading, and this appears to be a relatively inefficient way for trees to increase APAR. Leaf area density tends to change much less than crown lengths and widths in response to many different treatments, including thinning, fertiliser application, and pruning (Forrester et al. 2012a; Ligot et al. 2014; Guisasola et al. 2015). Inter- and intraspecific changes in crown architecture were also found to influence tree and stand APAR in *P. sylvestris* and *F. sylvatica* mixtures (Forrester et al. in press) and to increase total stand APAR in a planted biodiversity experiment (Sapijanskas et al. 2014).

### 3.4.3 Changes in Tree Physiology

So what are the processes that lead to increases in LUE? LUE could increase if photosynthesis increases and/or if there is a higher allocation of biomass above ground, to capture more light per unit of net primary production. Photosynthetic rates were higher for *E. globulus* in mixture (same experiment shown in Fig. 3.1a, d, g) with the N-fixing *A. mearnsii* (Forrester et al. 2012b), and many studies have found that photosynthesis can increase following fertiliser application (Forrester 2013). However, photosynthesis certainly does not always increase following fertiliser application, and many studies have found that increases in above-ground growth resulted from increases in carbon partitioning to above-ground growth (Box 3.1) (Forrester 2013). Two studies where carbon partitioning was measured in mixed-species stands found that when mixtures of *Eucalyptus* and N-fixing *Acacia* species were more productive than monocultures, they also allocated a higher proportion of carbon above ground than monocultures (Box Fig. 3.1-1 in Box 3.1).

**Box 3.1 Carbon Partitioning and Implications for Mixtures**

Carbon (C) partitioning is important to consider when comparing mixtures with monocultures because even when there are no differences in net primary production, the above-ground biomass or wood mass may still differ, simply due to differences in C partitioning (Box Fig. 3.1-1). Two important

(continued)

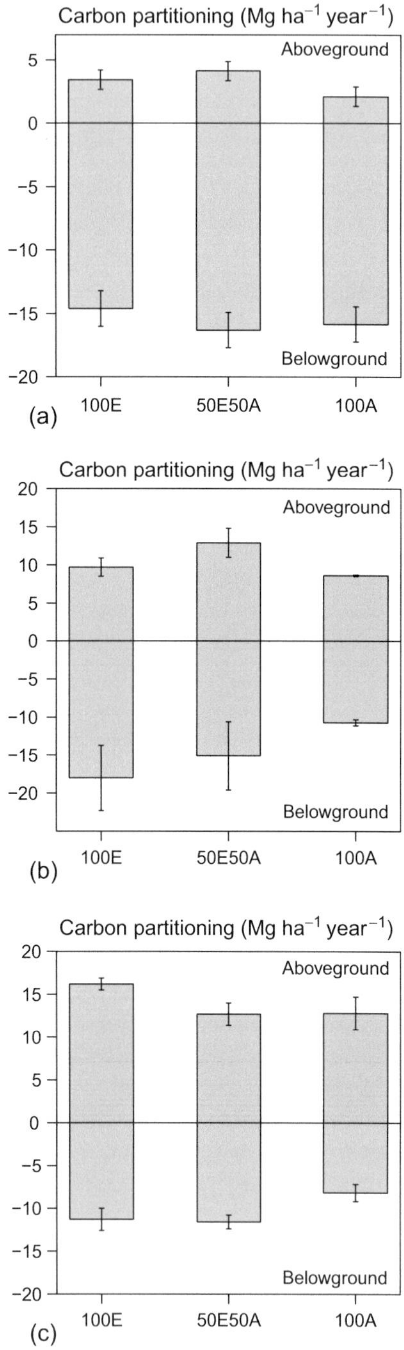


**Box Fig. 3.1-1** Carbon partitioning to above-ground net primary production (including above-ground biomass + litterfall – morality) and the total below-ground flux (including respiration), in (**a**) *Eucalyptus globulus* and *Acacia mearnsii* stands between ages 10.5 and 11.5 years, at Cann River, Australia (Forrester et al. 2006a); (**b**) *Eucalyptus grandis* and *Acacia mangium* between ages 4 and 6 years at Itatinga, Brazil; and (**c**) *E. urophylla* × *grandis* clone and *A. mangium* between ages 6 to 7 years at Kissoko, Congo (Epron et al. 2013). Error bars are standard errors of difference for *E. globulus*/ *A. mearnsii*, or standard deviations for *E. grandis*/ *A. mangium*. The numbers on the *x*-axes are the percentage of all planted trees belonging to a given species (100E is 100% *Eucalyptus*, 50E50A is 50% *Eucalyptus* and 50% *Acacia*, and 100A is 100% *Acacia*)

**Box 3.1** (continued)
factors that influence C partitioning are (1) ontogeny and (2) resource availability. Both of these are likely to differ between mixtures and monocultures. In terms of ontogeny, as stands age and trees get larger, there is often a decline in the partitioning of C below ground and to leaves and an increase in partitioning to stems (Litton et al. 2007; Poorter et al. 2012). When growth is increased in mixtures, there may therefore be changes in C partitioning due to changes in tree size, and the rate and magnitude of this change may vary between species.

Regarding the second factor, as resource availability increases, plants tend to partition less C to the plant tissues that forage for that resource. So as nutrient or water availability increase, partitioning to roots decreases, and plants allocate more resources above ground, to capture more light and fix more C (Litton et al. 2007; Poorter et al. 2012; Reich et al. 2014).

The photosynthetic capacity of the canopy might also be improved if mixtures have a closer-to-optimal distribution of nutrients in relation to light availability (Wacker et al. 2009). This 'optimality theory' (Field 1983; Hirose and Werger 1987; Anten et al. 1995; Sands 1995) predicts that canopy photosynthesis will be maximised when leaf mass and leaf N concentrations are high at the top of the canopy and decrease continuously towards the lower canopy in relation to the vertical gradient in light availability. The higher leaf N concentration at the top means that leaves can make use of the higher light availability and have high rates of photosynthesis, while the lower leaf N at the bottom of the canopy prevents N from being wasted where there is not enough light to maintain high rates of photosynthesis. Studies with grasses, shrubs, and trees have shown that monocultures generally do not develop such an 'optimal' distribution of leaf N for reasons including the following:

- Vertical distributions in leaf-specific hydraulic conductance that are not optimal. Optimal N distribution is of no use if there is not enough water for photosynthesis (Peltoniemi et al. 2012).
- The availability of light within the entire canopy can be very variable. Even though there can be a general decline in light from the top to the bottom of the canopy, it is likely that some high leaves will be more shaded than some lower leaves, but leaves on branches of the upper canopy may still have higher leaf N concentrations than those on branches in the lower canopy, thereby preventing an optimal relationship between light availability and foliar nutrition (Osada et al. 2014). That is, Osada et al. (2014) found that for a given range in light availability, there was a smaller range in leaf N concentrations for leaves within a single branch than there was for leaves in different branches throughout the canopy.

- Leaf N may be more dependent on the light availability that existed when the leaf was produced than current conditions. The light availability for a given leaf may therefore change more rapidly than the leaves can re-acclimate (Niinemets 2012).
- The optimality theory assumes that there is no competition between individual plants and that they distribute the N within their crowns in a way that is optimal for the whole stand rather than the individual. However, there is typically strong competition between individuals, and stand leaf area is often higher than optimal, perhaps in an attempt to outcompete neighbours (Anten 2005). This strategy maximises the carbon gain of individuals but not the entire canopy.
- There is an upper bound constraint on specific-leaf area (leaf area per unit leaf mass) that leaves cannot go above, which is determined by a minimal physical strength that prevents structural damage from wind, herbivory, etc., and this constrains the morphology and nutrition of the lower canopy leaves. This can prevent an optimal distribution of leaf N concentrations such that N concentrations decline less rapidly than light availability from the top to the bottom of the canopy (Dewar et al. 2012).

In contrast, interspecific differences in leaf nutrition and leaf morphology, as well as differences in biomass allocation in terms the trade-off between height growth and leaf area, may enable mixtures to achieve a closer-to-'optimal' distribution than monocultures (Anten 2005; Wacker et al. 2009). The interspecific differences may be exaggerated further by intraspecific variability in mixtures resulting from interspecific interactions. This has not been tested in tree mixtures, but in grass mixtures, there was a shift in leaf biomass to greater heights and a greater APAR than in monocultures, although only a few of the mixtures examined had an optimal vertical distribution of leaf N (Wacker et al. 2009).

## 3.5 Water Availability, Use, and Use Efficiency

### 3.5.1 General Patterns and the Consideration of Spatial Scale

A major argument that is often used in favour of mixtures is their greater productivity; however, the faster trees grow, the more water they are likely to use (Law et al. 2002), which may in turn make them more susceptible to drought periods, as well as reducing water supplies for downstream users. Despite this, few studies have compared the transpiration ($E_T$), water-use efficiency (WUE, tree, or stand growth per unit $E_T$), and their seasonality in mixtures and monocultures (Forrester 2015). In contrast, there is far more information about the production ecology of monocultures with regard to tree or stand $E_T$ and WUE, and these studies indicate that generally when tree or stand growth increases, there is also an increase in $E_T$ and/or WUE (Binkley 2012; Binkley et al. 2004), and there are usually no reductions in $E_T$ or WUE when growth increases. This pattern has been found for a wide range of treatments including genetics, tree age, irrigation/drought, fertiliser application, pruning, thinning, species comparisons, and geographic gradients (Binkley 2012; Forrester 2013). Specifically, when (monospecific) stands were more productive than control treatments, in 90% of cases, they also used more water (median of 25%) and used it more efficiently (median of 70%). There were also increases in $E_T$ and WUE in about half of the studies, while in the other half, there were increases in one or the other, and not both.

Based on these patterns, it could be expected that if the growth of a given species is greater in mixture, there were probably also increases in $E_T$ and/or WUE. Furthermore, the interspecific interactions in mixtures, which improve growth, could be used and manipulated to improve tree or stand WUE. The few studies that have compared rates of water use in mixtures with monocultures have shown that, when mixtures were more productive, they also used more water and they were also more water-use efficient (Fig. 3.2) (see review by Forrester 2015). For example, by age 15 years, the above-ground biomass of mixed-species stands of *Eucalyptus globulus* and *Acacia mearnsii* was 85% and 58% greater than *E. globulus* or *A. mearnsii* monocultures, respectively. Between ages 14 and 15 years, the mixtures used 17% and 93% more water than the monocultures, respectively, but the mixtures were also more water-use efficient than the monocultures because the water-use efficiency of *E. globulus* trees was 80% greater in the mixtures (Fig. 3.2). The processes that increased the growth of these particular *Eucalyptus-Acacia* mixtures are described in detail as a case study in Sect. 4.2.2.

In contrast, when there are no complementarity effects on growth in mixtures, there are also generally no complementarity effects on $E_T$ or WUE (Forrester 2015). In such stands, the growth, $E_T$, or WUE of the mixtures is a function of the properties of the monocultures of each individual species and the proportion of stand basal area, sapwood area, or stand crown projection area that each species contributed to the mixture (Moore et al. 2011; Gebauer et al. 2012).

All of the patterns described so far in this section are at the species stand level or total stand level. However, stand-level patterns are determined by individual trees and the interactions between them. Stand-level responses are simply the net effect of a much wider range of individual tree-level responses. This is because there is considerable spatial and temporal variability in soil and canopy conditions within mixed-species forests (Schume et al. 2004; Boyden et al. 2012; Canham et al. 1999; He et al. 2014), and this, together with inter-tree variability in genetics, pest damage, etc., is reflected in tree-level relationships. Therefore, the tree-level patterns include fundamental information about the processes underlying the species stand-level and total stand-level responses.

Tree-level patterns of growth, $E_T$, or WUE were therefore examined in the same stands shown in Fig. 3.2. Even within the same mixtures, growth, $E_T$, or WUE only increased for some of the trees (compared with trees in monocultures) and not for others (Fig. 3.3). The response of the individual trees depended on their size and the species composition or basal area of their neighbourhoods (Fig. 3.3), and the stand-level patterns simply reflected the mean tree-level response (Forrester 2015).

In contrast to these species-specific patterns (at the tree or stand levels), it is far more difficult to make total stand-level generalisations about how the net effect of

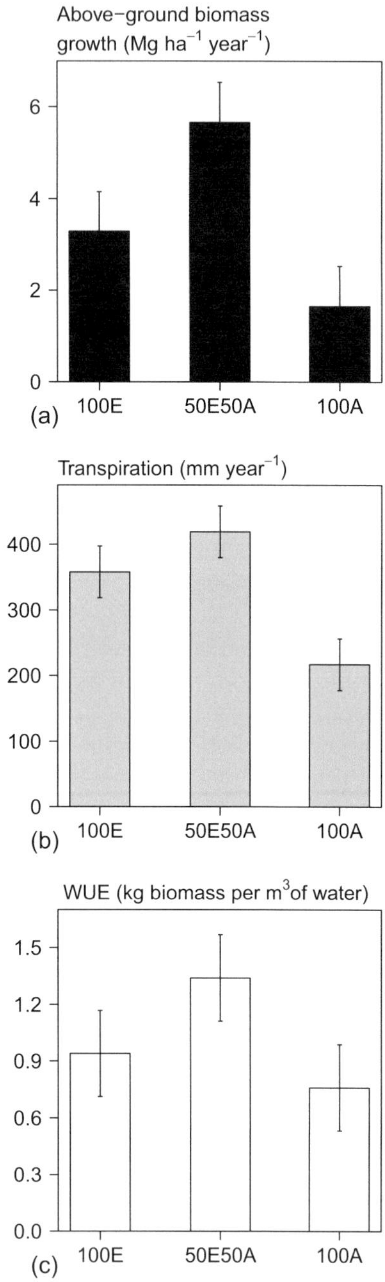


**Fig. 3.2** Above-ground biomass growth (**a**), transpiration (**b**), and water-use efficiency (**c**; WUE) in monocultures and mixtures of *Eucalyptus globulus* with *Acacia mearnsii* between ages 14 and 15 years (Forrester et al. 2010b). Error bars are standard errors of difference. (100E = 100% *Eucalyptus*, 50E50A = 50% *Eucalyptus* and 50% *Acacia*, and 100A = 100% *Acacia*)

all species in the mixture will influence stand water availability, $E_T$, and WUE. This is because regardless of the complementarity effects (significant Type 3 differences), there may be significant interspecific differences in $E_T$ and WUE, and this Type 1 difference will also contribute to total stand-level differences

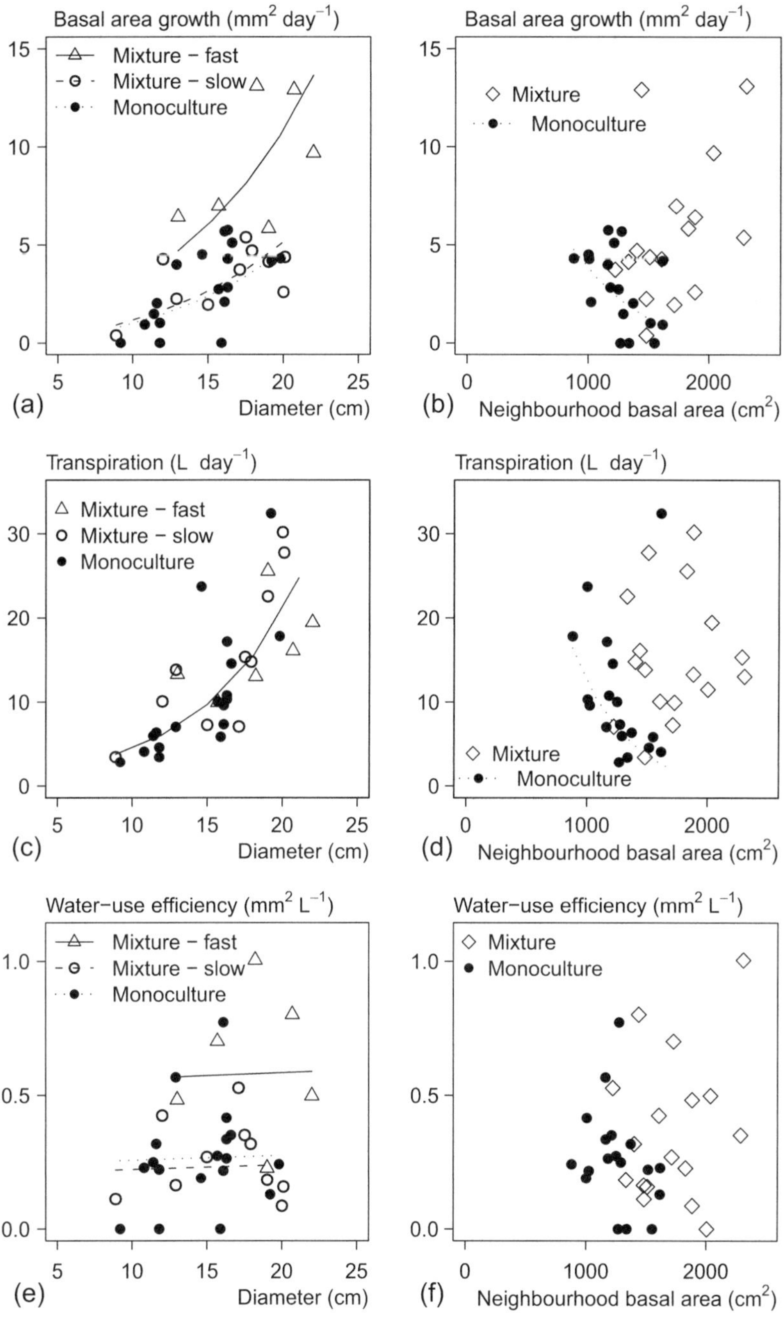


**Fig. 3.3** The relationship between diameter and basal area growth (**a**), transpiration ($E_T$; **c**) and water-use efficiency (**e**), or between neighbourhood basal area and basal area growth (**b**), transpiration ($E_T$; **d**), and water-use efficiency (WUE; **f**) of *E. globulus* trees growing in monocultures or in 1:1 mixtures with *A. mearnsii* trees. For (**d**) the monoculture line is fitted without the outlier with a transpiration of 32 L day$^{-1}$. Modified from Forrester (2015)

between mixtures and monocultures. If a management aim is to increase growth and WUE of mixtures, the best suggestion appears to be to combine very water-use-efficient species with species that will increase the growth of that water-use-efficient species, which could be via any process appropriate for the site and species. Another useful management tool is the observation that for a given species, if there is no increase in its growth in the mixtures, then there is unlikely to be any increase in its $E_T$ or WUE (Forrester 2015). Therefore, growth measurements, which are much easier and cheaper than $E_T$ measurements, could provide a good initial indication about whether mixtures are likely to be using more water (Forrester 2015).

### 3.5.2 Processes That Influence Transpiration or WUE

There are many processes that could influence the water availability or water stress of trees in mixtures compared with monocultures (Forrester and Bauhus 2016). However, few have actually been quantified and compared between mixtures and monocultures. Therefore, even if they occur in mixtures, it is unclear whether they often have a significant effect on growth, water availability, $E_T$, or WUE. Some of these processes are described below.

- *Interception losses*: Part of the precipitation received by a forest never actually reaches the soil because it is caught by the canopy and evaporates back into the atmosphere. The proportion of precipitation that is 'intercepted' by the canopy depends on variables such as canopy leaf area (larger leaf areas have higher interception), the roughness of the bark and funnel-like architecture (rough-barked and porous-barked trees have less stem flow and more interception loss), and the size of the rainfall events; following small rainfall events, a higher proportion of precipitation will probably be intercepted by, and evaporated from, the canopy (Gash et al. 1995; Augusto et al. 2002, 2015; Levia and Frost 2003; Schume et al. 2004). The funnel-like crown architectures might also help to distribute the water around the base of a given tree where it is harder to reach by neighbours (Gerrits et al. 2010). These traits clearly vary a lot between species and may also be influenced by the species interactions, such as when growth and leaf areas are increased in mixtures. For example, even with similar leaf areas, *P. abies* monoculture canopies intercepted more rain than *F. sylvatica* monocultures or mixtures of these species (Schume et al. 2004). Therefore, the mixtures might have more available water than *P. abies* monocultures.
- *Increased storage*: Water availability could be increased if the mixtures develop a thicker O horizon that is capable of storing more water than the O horizons in one, or all, of the monocultures (Ilek et al. 2015). In contrast, water availability could be lower if infiltration of precipitation into the O horizon is reduced resulting in higher evaporation and runoff (Schume et al. 2004).

- *Contrasting water requirements*: If one species transpires (or intercepts) less water per unit area than another, then water availability will increase for the other species, while the less water demanding species may have to deal with more intense interspecific competition for water compared with the intraspecific competition in its monocultures. High water-using species could be those with high growth rates or low water-use efficiency.
- *Contrasting sources*: If one species obtains water from deeper depths than another, it could have less competition in the mixture than in its monoculture. Several studies have shown interspecific differences in water sources by species growing in mixtures. For example, *Q. petraea* was found to use deeper soil water than *F. sylvatica* (Zapater et al. 2011). Similarly, *E. globulus* dried the soil out more at deeper layers than *A. mearnsii* and may therefore have experienced less competition in the mixtures (Forrester et al. 2010b). Contrasting fine-root distributions (Sect. 3.3.2) could also differ between species or change as a result of species interactions, thereby influencing competition for water.
- *Differences in phenology* might leave one species free of competition or with reduced competition for water during parts of its growing season (e.g. Vandermeer 1989; Moore et al. 2011; Roupsard et al. 1999; Schwendenmann et al. 2015). For example, many forests are composed of deciduous and evergreen species, and an extreme case is the deciduous N fixer *Faidherbia albida* that has a 'reverse phenology' where it loses its leaves during the wet season (Roupsard et al. 1999). Most studies that have compared the transpiration of mixtures and monocultures were done over complete growing seasons (Forrester et al. 2010b; Kunert et al. 2012; Moore et al. 2011; Gebauer et al. 2012) rather than for short periods (several days or weeks) to account for the seasonality in the ranking of $E_T$ and WUE between mixtures and monocultures that results from interspecific differences in phenology (e.g. time of peak $E_T$) and the effects of intra-annual climatic conditions on $E_T$ or WUE.
- *Hydraulic redistribution*: This is a process where roots take up water from moist soil and release it into drier soil (Prieto et al. 2012; Neumann and Cardon 2012). Reported mean magnitudes of hydraulic redistribution vary between about 0.04 to 1.3 mm of water per day, and these have come from a wide range of forest types and from the tropics to temperate and Mediterranean climates (Neumann and Cardon 2012). The significance of this process will clearly depend on its magnitude and timing, and the potential benefits can include increasing dry-season transpiration and photosynthesis, lifting water so shallow-rooted plants can use it, increasing plant nutrient uptake (which requires soil moisture), extending the life span of roots, and moving water into deeper layers where it does not evaporate (Neumann and Cardon 2012).
- *Anisohydric or isohydric*: Different species tend to show one of two types of stomatal behaviour in response to the soil drying out. Isohydric species close their stomata during earlier stages of drought. This conserves water and reduces the risk of embolisms in their water transport systems. However, closing their stomata also prevents them from fixing carbon, and if the drought goes for a long time, they may use up all their carbon reserves and die. This conservative

strategy may result in higher water availability for more anisohydric species during the early stages of drought. Anisohydric species open their stomata for longer into drought periods. They invest more carbon into their transport systems to reduce their vulnerability to embolisms. But if the drought is severe enough, then eventually tension within the water transport system builds up, the water columns break, and the trees suffer an embolism. If this occurs at a faster rate than the trees can repair themselves, some branches will begin to die, and eventually the tree may die. If they do not die, then after the drought they will need to rebuild their canopies, and the isohydric species may then have less competition.
- *Changes in canopy microclimate*: Transpiration from one species (in the overstorey) could reduce the vapour pressure deficit within the canopy and hence facilitate an understorey plant (Saccone et al. 2009). This potential process has received very little attention with respect to tree–tree interactions in forests (as opposed to tree-seedling or shrub-seedling interactions).
- *Processes that improve light or nutrient availability and use*: Growth, $E_T$, or WUE can increase in response to processes that improve light and nutrient availability or uptake. When these processes increase growth, they are also likely to increase $E_T$. They can also increase WUE by shifting C partitioning from below ground to above ground or by increasing the availability or uptake of nutrients or light enabling the plants to increase photosynthesis and make more efficient use of their water.

The processes listed here will often occur simultaneously and in opposite directions. Therefore, information about one process may not give a good indication of the total or average effects of all water-related interactions. For example, in monocultures and mixtures of *F. sylvatica* and *P. abies*, the *F. sylvatica* (a deciduous deeper rooter) used more water per crown projection area, but this was compensated for by higher interception rates of *P. abies* (an evergreen shallower rooter) (Schume et al. 2004).

### 3.5.3 The Influence of Stand Density and Water Stress

Mixed-species forests are often recommended for their potential to provide higher levels of ecosystem services than monocultures, including a reduced susceptibility to droughts (Grossiord et al. 2014). At first glance, this might appear to contradict the case studies described above where $E_T$ was higher in mixtures than monocultures and could therefore result in reduced water availability and increased drought stress in mixtures. However, while this may sometimes be true, it will certainly not always be the case and may be related to stand density. This is because $E_T$ is only one of the processes listed above that could influence the water availability and drought stress of mixtures. The other water-related processes could be beneficial during periods of water stress.

There is likely to be a trade-off between increasing productivity (and hence $E_T$), but not increasing it so much that there is a large enough reduction in water availability that outweighs the complementarity effect on water stress during a drought period (Forrester 2015). This trade-off may require minimal differences in stand density between the mixtures and monocultures. If a mixture is growing faster than a monoculture, it probably has a higher stand density in terms of basal area, sapwood area, leaf area, and biomass and is therefore likely to transpire more water. If stand density and productivity are significantly greater in mixtures, then mixtures may use more water than monocultures, but if stand density is similar, the water-related processes listed above could reduce the water stress of trees in mixtures. Controlling stand density is a common and important silvicultural treatment used in forests to manage water availability (Hawthorne et al. 2013), and density can be similar in mixtures and monocultures when it is managed by thinning or when each species has similar growth rates. An example of the interaction between stand density and growth, $E_T$, and WUE is shown in Fig. 3.3 (see also Sect. 3.7), where *E. globulus* growth and $E_T$ are negatively correlated with neighbourhood density (basal area) in the monocultures but not in the mixtures.

## 3.6 Spatial and Temporal Dynamics of Species Interactions

Interactions between a given pair of species are dynamic and change as resource availability or climatic conditions change (see reviews by Forrester 2014; Forrester and Bauhus 2016). Net complementarity interactions between a given pair of species can transform into net competitive interactions and vice versa (Forrester et al. 2011; Bouillet et al. 2013; Binkley 2003; Pretzsch et al. 2010; Boyden et al. 2005). It is important to understand how these interactions might change when managing mixed-species stands and designing mixed-species plantations. Furthermore, mixed-species forests are an important component of climate adaptation and risk-reduction strategies (Reif et al. 2010), and to ensure that mixtures are used appropriately, it is critical to understand how climatic variability influences species interactions. These spatial and temporal dynamics were reviewed by Forrester (2014), and this forms the basis for the following sections.

### 3.6.1 A Framework for Understanding the Dynamics of Species Interactions

The spatial and temporal dynamics of species interactions can be summarised in Fig. 3.4. This framework is based on the assumption that complementarity effects are related to (1) the types of species interactions (e.g. N fixation) and (2) how resource availability changes along the spatial or temporal gradient (Forrester 2014). Complementarity increases as the availability of resource 'X' declines (or climatic condition 'X' becomes harsher) if the species interactions improve

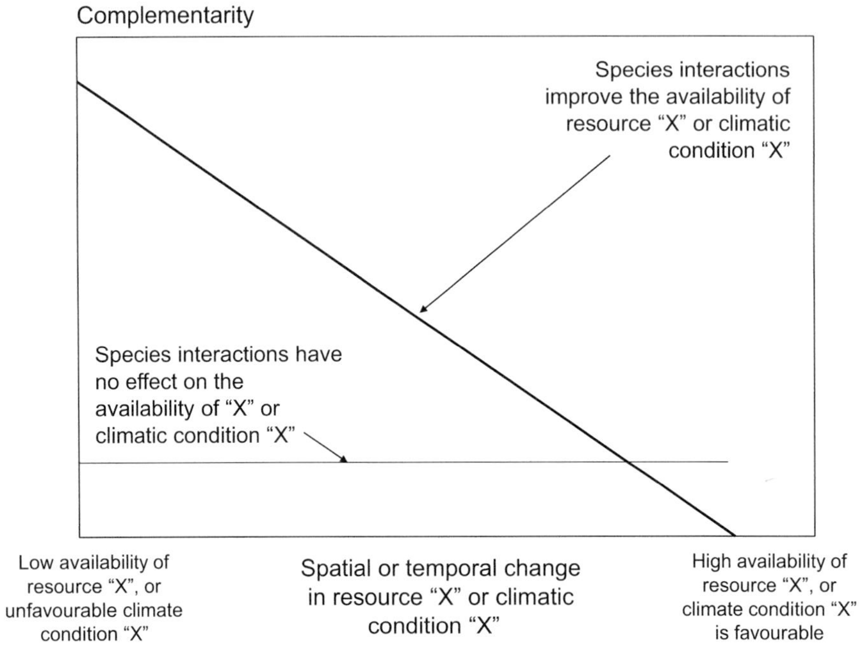


**Fig. 3.4** A framework to describe the relative complementarity response of a species growing in a mixture in relation to gradients in growing conditions and the types of species interaction that occur between the species in that mixture. The thick diagonal line shows a pattern where complementarity increases as the availability of resource "X" declines or climatic condition "X" becomes harsher. This occurs when the species interactions improve the availability, uptake or use efficiency of resource "X" or interactions improve climatic condition "X". For example, complementarity could increase as competition for light becomes more intense (and light availability per tree declines) when interactions increase light interception or light-use efficiency. This type of interaction would be less useful if nutrients or water limit growth, but its usefulness should increase as soil resource availability increases (or climatic conditions become more favourable). The thin horizontal line shows what could result when the species interactions do not lead to any change in complementarity along the gradient because complementarity does not result from interactions that influence "X". This figure ignores the possible interactions between multiple X factors on the $x$-axis (see Sect. 3.6.6). Modified from Forrester (2014) and Forrester and Bauhus (2016)

the availability, uptake, or use efficiency of resource 'X' (or interactions improve climatic condition 'X') (Forrester and Bauhus 2016).

The spatial and temporal shifts in the way species interact can occur on small to large scales. There is spatial variability in light and soil resource availability in forests at the scale of a few square metres (Boyden et al. 2012) and climatic conditions within a stand vary due to canopy gaps, shading, or wind protection from neighbours (Rao et al. 1998). At larger spatial scales, abiotic factors vary along slopes and with aspect and of course from one stand or region to another. Temporally, climatic conditions vary from 1 year to another and may even change in the long term. Species interactions can also change as stands develop because the

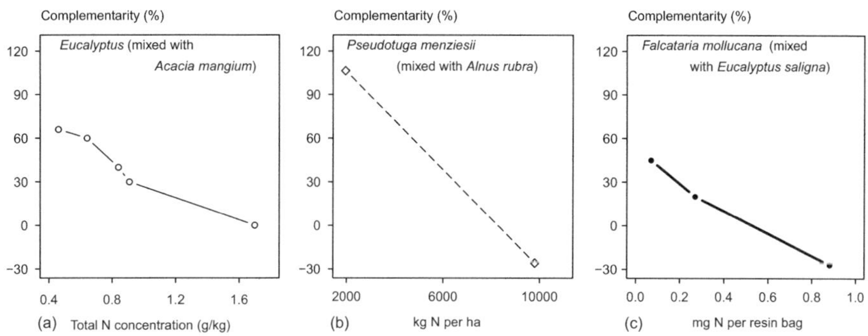


**Fig. 3.5** Relationships between complementarity (%), calculated using Eqs. (3.2) and (3.3), and soil N availability, which were quantified differently in each study. The stands contained (**a**) *E. grandis* or *E. urophylla* × *grandis* clones with *A. mangium* (Bouillet et al. 2013), (**b**) *P. menziesii* with *A. rubra* (Binkley 2003), and (**c**) *F. moluccana* with *E. saligna* (Boyden et al. 2005)

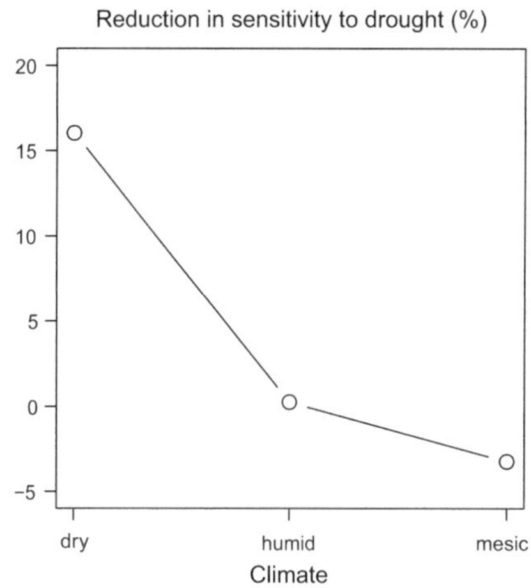


**Fig. 3.6** The relative reduction in drought sensitivity (%) of *A. alba* (in mixtures with *P. abies*) in relation to climate (Lebourgeois et al. 2013). The reduction in sensitivity is the mean sensitivity of trees in mixtures compared with monocultures. Figure modified from Forrester (2014)

demand and availability of light, water, and nutrients will often change as growth rates and stand biomass changes.

Using studies from the literature, Figs. 3.5, 3.6, and 3.7 summarise the relationships between complementarity and gradients in resource availability or climatic conditions. The complementarity effects that are shown in these figures were calculated using tree or stand species-level information. Using tree-level information, complementarity was calculated for mean tree sizes and medium density stands based on the relative productivity (Eq. 3.2) (Forrester and Pretzsch 2015).

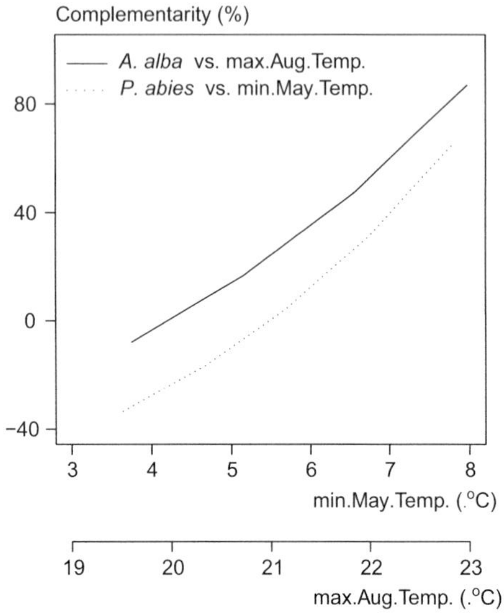


**Fig. 3.7** Relationships between complementarity (%), calculated using Eq. (3.2), and the mean maximum temperature in August or the mean minimum temperature in May, both of which varied between six sites at which there were *A. alba* and *P. abies* forests (Forrester et al. 2013)

$$\text{Relative Productivity } (\%) = \frac{\text{growth in mixture} - \text{growth in monoculture}}{\text{growth in monoculture}} \times 100 \tag{3.2}$$

Using stand-level information, complementarity was calculated using Eq. (3.3)

$$\text{Relative Productivity } (\%) = \left(\frac{\text{growth or yield in mixture}}{\text{growth or yield in monoculture} \times \text{species proportion}} - 1\right) \times 100 \tag{3.3}$$

where the species proportion is the stand density of the given species in mixture divided by the total stand density, and density can be expressed as the initial planting density (trees $ha^{-1}$) or stand biomass.

### 3.6.2 Spatial Effects of Interactions That Influence Nutrient Availability

Species interactions that improve nutrient availability, uptake, and use efficiency often result in greater facilitative effects on sites where those nutrients are limited (Forrester 2014, Fig. 3.5). The most well-known example is probably the increasing facilitative effect of N-fixing species on the growth of non-N-fixing species as soil N becomes more limiting (Binkley 2003; Forrester et al. 2006b, c; Bouillet et al.

2013). For example, the N-fixing *Alnus rubra* significantly increased the growth of *Pseudotsuga menziesii* on a low-N site, but not on a high-N site (Binkley 2003). The complementarity effect by age 70 years was as high as 100% (Fig. 3.5b) and was related to greater nutrient uptake rather than changes in availability or efficiency. The rates of N, Mg, and K uptake were greater in mixtures than in *P. menziesii* stands at both sites, but the relative increases were much greater at the low-N site (Binkley et al. 1992b). Rates of N fixation by *A. rubra* were high at both sites, and the N-use efficiency of mixtures was <30% of *P. menziesii* stands on both sites (Binkley et al. 1992b).

The studies in Fig. 3.5 imply that facilitative (complementarity) effects that result from increases in the availability of a given resource (e.g. N) can be outweighed by competition when other resources are strongly limiting growth. There are many examples where mixture productivity was not improved by N-fixing species because N was not limiting enough or the N-fixing species were outcompeted by the non-N-fixing species (Forrester et al. 2007a; Bouillet et al. 2013; Epron et al. 2013). Managers could make use of these dynamics to increase facilitative effects. That is, the availability of resources that are not greatly influenced by species interactions, but that are limiting growth, could be increased, thereby increasing the likelihood that the resources influenced by the species (e.g. N) are the major limiting resources. For example, as phosphorus availability increases, the facilitative effects of N-fixing species on eucalypts also increase (Boyden et al. 2005; Forrester et al. 2006c), so the application of phosphorus fertiliser might increase the facilitative effect of the N-fixing species in these stands.

N fixation is an example of how N availability can be increased by increasing the size of the N pool. Nutrient availability can also be increased by increasing the proportion of the total soil nutrient pool that is actually available to plants. Section 3.3 discussed some of the mechanisms, including accelerating rates of nutrient cycling, contrasting mycorrhizal associations, and contrasting fine-root distributions. When any of these do actually improve nutrient availability, the facilitative effect on growth should be more useful on sites where the given nutrient is limiting growth. No studies could be found that measured complementarity in growth and nutrient cycling along gradients in nutrient availability. However, traits or interactions that reduced competition for, or improved, the availability of nutrients, including faster nutrient cycling, were suggested to have contributed to the increasing complementarity effects of *F. sylvatica* on the growth of *P. abies* (Pretzsch et al. 2010), *Quercus petraea*, or *Q. robur* (Pretzsch et al. 2013a) as site quality increased (Box Fig. 3.2-1 in Box 3.2). This was also suggested for *Picea mariana* and *Populus tremuloides* mixtures (Cavard et al. 2011).

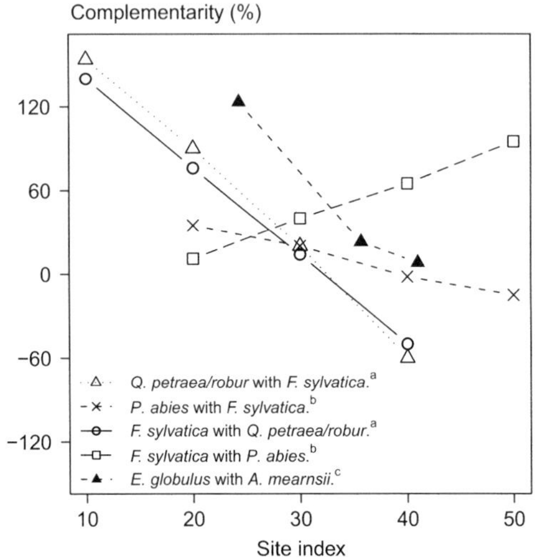


**Box Fig. 3.2-1** Relationship between complementarity (C%) and site indices. C% was calculated using Eq. (3.3). The site index is the mean or maximum height at age 100 years (**a**, **b**) or the stand volume (of monocultures) at age 15 years ($m^3$ $ha^{-1}$/3).[a] Pretzsch et al. (2013a),[b] Pretzsch et al. (2010),[c] Forrester et al. (2011). Figure modified from Forrester (2014)

**Box 3.2 Examining the Spatial Dynamics of Species Interactions Using Site Indices**

Site quality or site indices are often used to quantify differences in growing conditions. Site quality or site indices are the net effect of many different factors, including climatic variables such as temperature, vapour pressure deficit, precipitation and climatic extremes, and soil properties relating to the availability of water and many different nutrients. Site quality and site indices are used a lot in forestry because they 'summarise' the net effects of all these factors. However, this strength becomes a weakness and is problematic when examining the spatial dynamics of species interactions, which are related directly to actual gradients in specific resources (Figs. 3.5, 3.6, and 3.7), and not the productivity (site index) of the site per se (Box Fig. 3.2-1). For example, some of the largest facilitative effects of N-fixing trees on the growth of eucalypts have been found in relatively productive plantations (Binkley et al. 2003; DeBell et al. 1987), and a review of these effects found that there was no significant relationship between the mean annual biomass or volume growth of eucalypt monocultures (a measure of site quality) and the mixing effect in 1:1 mixtures (Forrester et al. 2006b). That is, very productive sites had plenty of water, phosphorus, and favourable climatic conditions but were N limited, whereas sites that were not so N limited had low productivities because other resources, such as water availability, limited growth.

Similarly, complementarity effects were found to increase with increasing site quality for *F. sylvatica* mixed with *P. abies* (Pretzsch et al. 2010) and for

(continued)

**Box 3.2** (continued)
*A. alba* mixed with *P. abies* (Forrester et al. 2013). However, in different studies, but for the same species combinations, complementarity effects for *F. sylvatica* that occurred during low-growth years were absent during more favourable growth periods (del Río et al. 2014), or complementarity effects for *A. alba* that were found on dry sites were not significant on mesic or humid sites (Lebourgeois et al. 2013). Thus, in some studies, complementarity effects increased as growing conditions improved, and in other studies, complementarity for the same species in the same mixture combinations decreased as growing conditions improved. As explained in Sect. 3.6.6, this is probably the result of measurements of complementarity during average conditions in one study (when interactions influencing light were important) but during slow-growth periods in another study (when interactions influencing water were important). This shows the risk of relying on site quality or site indices when studying how plant interactions change (perhaps with the exception of light-related interactions).

### 3.6.3 Spatial Effects of Interactions That Influence Water Availability

As water availability declines, any interactions or mechanisms that influence water availability and uptake should have a greater complementarity effect (Fig. 3.4). For example, the sensitivity of *Abies alba* to summer droughts declined in mixtures with *Picea abies* or *Fagus sylvatica* compared with monospecific neighbourhoods, but this complementarity effect was only found at dry sites and not at mesic or humid sites (Fig. 3.6). The mechanisms responsible for this effect were not measured, but it was suggested that competition for water was reduced for *Abies alba* because the other species have shallower roots systems, and the canopies of *F. sylvatica* may intercept less rain, allowing more to reach the soil. Increasing complementarity effects of *Quercus petraea* or *Q. robur* on *F. sylvatica* with declining site quality were suggested to result from mechanisms that improved water availability, including hydraulic redistribution (see Sect. 3.5.2) (Pretzsch et al. 2013a). However, water availability or use was not measured in that study, and another study using the same species found that although *Q. petraea* redistributed water from deeper to more shallow regions of the soil, there was no evidence that *F. sylvatica* actually acquired that water (Zapater et al. 2011). The contribution of processes that could improve water relations and complementarity does not appear to have been quantified along gradients in water availability in forests. All examples of complementarity effects in this paragraph were observed at drier ends of spatial gradients; however complementarity involving improved water relations may also occur at sites that are not considered to be dry but that experience seasonal water deficits or periodic droughts (see Sect. 3.6.5).

### 3.6.4 Spatial Effects of Interactions That Influence Light Absorption and Use

Mixtures that contain a fast-growing and light-use efficient species that overtops a slower-growing but more shade-tolerant species could have a greater LUE than monocultures of the shade-tolerant species and greater APAR than monocultures of the less shade-tolerant upper canopy species. Complementarity effects resulting from interactions that improve APAR or LUE should increase as light availability declines (Fig. 3.4). As climatic conditions improve or the availability of water and nutrients increase, stand leaf area and APAR should also increase. As this happens, light availability will decline, and trees tend to respond by allocating a higher proportion of their growth above ground, which improves light absorption (Box 3.1) (Litton et al. 2007; Poorter et al. 2012). Several studies have found increases in complementarity with improvements in growing conditions, such as better climatic conditions and site quality (Box Fig. 3.2-1 in Box 3.2 and Fig. 3.7). Increasing relative and absolute complementarity effects were found with improved growing conditions in *Picea abies* and *Abies alba* stands in south-western Germany (Forrester et al. 2013). The complementarity appeared to result from greater APAR and LUE due to contrasting crown physiologies, crown architectures, and canopy structure (Forrester and Albrecht 2014). Similarly, the complementarity effect for *Picea abies* growing with *Abies alba* in Switzerland also increased with increasing site quality (Huber et al. 2014). Complementarity effects for *F. sylvatica* mixed with *P. abies* increased with increasing site quality in Central European forests (Pretzsch et al. 2010). *F. sylvatica* trees of a given diameter had larger crown projection areas in mixtures than in monocultures, and this effect increased with site quality (Dieler and Pretzsch 2013). These greater crown sizes may have improved light interception by trees in mixtures. In Canadian forests, relative growth reductions of *Picea glauca* × *engelmannii* growth (compared with hypothetical free-growing trees) due to competition from *Pinus contorta* were lower at higher-quality sites (Coates et al. 2013), which may indicate that species interactions resulted in improved APAR or LUE of *Picea glauca* × *engelmannii*.

### 3.6.5 Temporal Dynamics of Species Interactions

As stands develop, there are often changes in complementarity effects or the relative dominance of each species (Fig. 3.8) (Binkley 2003; Filipescu and Comeau 2007; Cavard et al. 2011; Forrester et al. 2011; Bouillet et al. 2013; Forrester 2014). These temporal changes can result from abiotic factors, such as climatic conditions and stand disturbances, or stands developing and influencing the availability of light and soil resources. The temporal dynamics in Fig. 3.8 illustrate the value of long-term measurements and the likelihood that measurements at a single point in time may not reflect actual long-term complementarity effects (see also Figs. 4.17 and 4.18). As an example, several studies have shown that complementarity can change with temporal changes in water availability. As even-aged stands develop, leaf

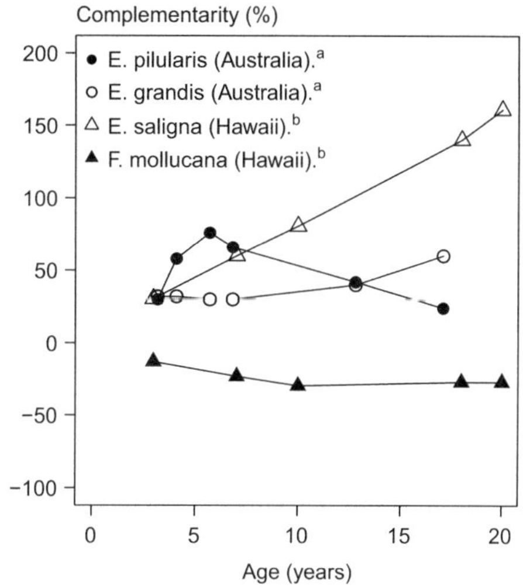


**Fig. 3.8** Temporal dynamics of complementarity in 1:1 mixtures of *E. pilularis* with *E. grandis* or *E. saligna* with *F. moluccana*. [a]Forrester and Smith (2012), [b]Binkley et al. (2003). Figure modified from Forrester (2014)

areas, growth rates, and transpiration initially increase before peaking and then declining (Dunn and Connor 1993; Vertessy et al. 1996; Ryan et al. 1997; Forrester et al. 2010a). As transpiration increases, soil water availability may decline, leaving the stands more water limited and susceptible to dry or hot conditions. When mixtures are more productive than monocultures, they can also use more water (Forrester et al. 2010b; Kunert et al. 2012), and this may increase their susceptibility to water-limiting conditions. Several studies have reported early complementarity effects in mixtures that were lost later due to excessive competition for water (Snowdon et al. 2003; Forrester et al. 2007a; Bouillet et al. 2013). In these mixtures, the main complementarity effects resulted from higher N availability due to N-fixing species. In contrast, when interactions or contrasting species traits reduce competition for water, the mixtures may be less sensitive to hot and dry periods. For example, complementarity effects in mixtures of *Q. petraea*, *P. abies*, and *F. sylvatica* were observed during low-growth years but not during high-growth years (del Río et al. 2014). The *F. sylvatica* trees were also more resistant and resilient to droughts when growing in mixtures (Pretzsch et al. 2013b), and it was speculated that the complementarity effects resulted from improved water relations.

The magnitude of complementarity effects could also vary with the intensity of light competition (Sect. 3.6.4). As stands develop and leaf areas increase, so does competition for light, and trees can allocate a higher proportion of growth above ground, which improves light interception (Litton et al. 2007). As competition for light increases, it becomes more important for shade-intolerant species to overtop more shade-tolerant species (Kelty 1992). When shade-intolerant species, such as eucalypts, are unable to overtop the admixed species, complementarity effects are more likely to be smaller or absent (Forrester et al. 2006b, 2011).

### 3.6.6 Simultaneously Occurring Species Interactions

In any one mixture, there are almost certainly several different processes that influence the net complementarity effects. For example, it has been well documented that complementarity effects of mixtures containing N-fixing species do not only result from improved N availability, but there can also be accelerated rates of P cycling, reductions in competition for light and greater APAR or LUE, and greater transpiration or WUE (Binkley et al. 1992a; Forrester et al. 2005; Richards et al. 2010; Hinsinger et al. 2011). Furthermore, as discussed above, the contribution that each process makes to the net complementarity effects changes with spatial and temporal variability in resource availability and climatic conditions (Forrester 2014). This also implies that for a given species combination, there are likely to be several complementarity—resource availability relationships depending on the resource gradients that are being examined and when they are being examined (see Box 3.2). Many studies focus, understandably, on a single resource gradient and may therefore underestimate the total complementarity effects. Similarly, productivity-biodiversity relationships will be the net result of many interactions between each specific species combination and their spatial and temporal dynamics (Forrester and Bauhus 2016). The asymptotic productivity-biodiversity relationships that are often found (Forrester and Bauhus 2016) might suggest that some species are redundant in more diverse stands. However, grassland mixture studies have shown that when more functions (growth, nutrient cycling, etc.) and more growing conditions are considered (ages, resources, climatic conditions), the number of species required to reach an asymptote is higher because there is a greater chance that each will be useful under specific conditions (Hector and Bagchi 2007; Isbell et al. 2011). The absence of productivity-biodiversity relationships in some forests (Vilà et al. 2013) may result because the species in those forests do not interact in ways that improve the availability or use of the limiting resources during the measurement period (Forrester and Bauhus 2016). For example, when water limits growth, having a wide range of nutrient uptake strategies may be of little benefit.

## 3.7 Stand Density

The interactions between species can also be influenced by stand density, which is important for two reasons. The first is that stand density can be positively correlated with species diversity (e.g. Vilà et al. 2013; Chisholm et al. 2013), and so greater productivity in mixtures may sometimes, at least partly, result from higher densities or higher structural diversity, rather than from direct effects of species diversity or interactions per se (Forrester and Bauhus 2016). Therefore, if stand density is not taken into account, complementarity effects may be over- or underestimated. These density effects are well known, and various approaches for dealing with them have been developed (Chap. 4). At the stand level, replacement and additive series designs have been used for planted experiments (Kelty and Cameron 1995) (see

Chap. 2), while in forests, plots that are close to their maximum density are used (see Chap. 4). At the tree level, competition indices are sometimes used to separate the effects of inter- and intraspecific interactions and stand density (Boyden et al. 2005; Vanclay 2006a; Forrester et al. 2013; Forrester 2015).

The second reason why density is important is that all of the complementarity effects described in this chapter could potentially be influenced by stand density (e.g. light-related effects in Forrester and Albrecht 2014, or water-related effects in Forrester 2015; Fig. 3.3). At low densities (and following thinning), species interactions will be weak whether they are complementary or competitive because the trees need to be close enough to interact in the first place. At the other extreme, when stand density is very high, competition will probably be high enough to outweigh any complementarity effects. This has been shown in *Eucalyptus* and *Acacia* mixtures, for which there are many examples of strong complementarity effects (Forrester et al. 2006b). However, at very high densities, when up to 20,000 *Acacia* weeds invade monospecific *Eucalyptus* plantations, there are net competitive effects on *Eucalyptus* growth, even though soil nitrogen increases with *Acacia* density (Hunt et al. 1999). Between these extremes of very low and very high densities, the relationship between density and complementarity effects depend on the resources that are limiting and that are influenced by changes in density, as well as the types of species interactions (following Fig. 3.4). Depending on the study, increasing stand density has increased or reduced complementarity effects in mixed-species forests or plantations (Fig. 3.9) (Garber and Maguire 2004). For example, complementarity effects increased with stand density on sites where mixtures of *P. abies* and *A. alba* were more productive than monocultures but increased competition on sites where mixtures were less productive (Fig. 3.9, Forrester et al. 2013). Complementarity effects were higher at lower densities for each species in *P. sylvestris* and *Q. pyrenaica* mixtures (Fig. 3.9, del Río and Sterba 2009). Complementarity effects increased with increasing density for *F. sylvatica*, which is a relatively strong competitor, but declined for associated *P. sylvestris* (Condés et al. 2013). Complementarity effects increased or decreased with stand density for *Falcataria moluccana* and *Eucalyptus saligna*, depending on how soil N and P availability changed (Boyden et al. 2005).

These effects of density on complementarity indicate that thinning is likely to significantly influence complementarity effects and may therefore be a useful management tool in mixtures. However, it also has important implications for studies that use plots in forests where previous thinning intensities are unknown, in which case it may be difficult to separate complementarity effects from density effects that are an artefact of previous management.

## 3.8 Synopsis

This chapter describes how species interactions can influence some of the main physiological and ecological processes that occur in mixed-species forests and plantations. Firstly, the production ecology equation is introduced as a framework

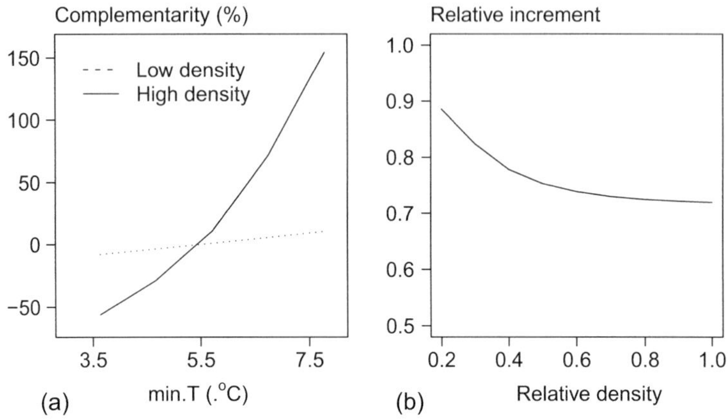


**Fig. 3.9** Increasing stand density (**a**) increased complementarity effects for *P. abies* in mixtures with *A. alba* along a 6-site gradient in mean minimum temperature in May (min.T) (Forrester et al. 2013). Increasing stand density (**b**) decreased complementarity effects (in terms of relative increment) for *Pinus sylvestris* in mixture with *Quercus pyrenaica* (modified from del Río and Sterba 2009). Relative increment is the increment ($m^3$ $ha^{-1}$ $year^{-1}$) of *P. sylvestris* in mixture divided by its increment in monoculture, and relative density is the basal area per ha of *P. sylvestris* in monoculture divided by the maximum observed basal area per ha of *P. sylvestris* stands

for quantifying the types of processes that influence tree growth and how these are influenced by species interactions. This equation describes growth as a function of resource availability, resource acquisition, and resource-use efficiency. Nutrient availability can increase when the size of the nutrient pool is increased, such as by symbiotic N fixation or by increasing the proportion of the pool that is actually available to plants and the speed at which it is cycled. Nutrient-use efficiency can increase, decrease, or remain unchanged when productivity increases in mixtures. This contrasts with light- and water-use efficiency (tree growth per unit of light absorption or transpiration), which increase or remain unchanged, but generally do not decline, as productivity increases. With regard to light, if a fast-growing species with high light-use efficiency overtops a slower-growing and more shade-tolerant species that is capable of high leaf areas and high light absorption, then mixtures could have a greater light-use efficiency than monocultures of the more shade-tolerant species and could absorb more light than monocultures of the less shade-tolerant species of the upper canopy. The few studies that have compared the water availability, transpiration, and water-use efficiency of a given species growing in a mixture compared with a monoculture have indicated that when growth increases (at the tree or stand levels), there are also likely to have been increases in transpiration or water-use efficiency, and several different processes can contribute to this response.

The influence of a given type of interaction on growth is not static; it changes with spatial and temporal variability in resource availability and climatic conditions and as a stand develops. A recently developed framework is used to explain these spatial and temporal dynamics and to indicate when different types of interactions are important. This shows that complementarity effects depend on the types of species interactions and how resource availability changes along the spatial or temporal gradients. When interactions improve N availability (e.g. via symbiotic N fixation), then relative complementarity effects are highest when N availability is low, and they decline as N availability increases because there is then less need for any additional N. Similarly, interactions that improve water availability are likely to lead to greater relative complementarity effects when water availability is low, and these effects are likely to decline as water availability increases. In contrast, some studies have found that complementarity effects increase with increasing site qualities. These patterns appear to occur because as growing conditions improve, competition for light increases, and complementarity effects will result from interactions that improve light absorption or light-use efficiency. It is also important to consider that multiple types of interactions are likely to occur simultaneously in mixtures, such as interactions that influence nutrients, others that influence water, and others that influence light. Therefore, different resource availability-complementarity patterns can occur for a given pair of species, depending on the resource being examined. Finally, stand density can influence these interactions. As stand density increases, interactions may become more favourable or less favourable, depending on how, and which, resources are influenced by the change in density.

### Take-Home Messages

1. The production ecology equation provides a useful framework to quantify the processes driving species interactions. This equation describes growth as a function of resource availability multiplied by the proportion of resource acquired by the trees, multiplied by the resource-use efficiency.
2. Nutrient availability can increase when the size of the nutrient pools increase (e.g. via N fixation, atmospheric deposition) and when a higher proportion of the nutrient pool is available to the trees (e.g. faster mineralisation or nutrient cycling). The proportion of nutrients used (nutrient uptake) can be increased when there are also spatial, chemical, or temporal differences in nutrient uptake. Nutrient-use efficiency can increase, decrease, or not change as growth increases.
3. Combining a shade-tolerant species that is capable of high light interception, with a very light-use efficient species, could result in mixtures that absorb more light than monocultures of the less shade-tolerant species and that are more light-use efficient than the more shade-tolerant species. In addition, when interactions increase growth, they are also likely to increase leaf area and light absorption, as well as light-use efficiency.
4. The few studies that have examined the transpiration and WUE of mixtures have found that when mixtures are more productive than monocultures, they

often also use more water and have a higher WUE. There are several processes that could increase the water availability in mixtures, but their influence on the complementarity effects that are sometimes observed in mixtures has rarely been quantified.

5. Species interactions are dynamic and vary spatially and temporally with resource availability and climatic conditions. The direction depends on the types of species interactions and how the resources change along the gradients. For example, interactions that increase water availability, transpiration, or WUE will become more useful as water availability declines along spatial gradients, or during periods of water stress, or as stands develop and competition for water changes.
6. For a given species combination, there are probably several different types of interactions occurring simultaneously (e.g. N fixation, reduced competition for light or water). Therefore, there can be different complementarity-resource availability relationships for a given species combination depending on the resource gradients that are being examined. For this reason, site quality or site indices can be problematic because they do not differentiate between the different resources that limit growth. Complementarity-site index relationships are the net effect of different types of interactions, and opposing relationships have been found for the same combinations of species in different studies that use site indices.
7. Stand density can modify the way species interact because it changes the balance between competitive and complementarity interactions. Furthermore, whether or not complementarity increases or decreases with increasing stand density depends on the types of species interactions that occur in the given mixture.

## References


Anten NPR (2005) Optimal photosynthetic characteristics of individual plants in vegetation stands and implications for species coexistence. Ann Bot 95:495–506

Anten NPR, Shieving F, Werger MJA (1995) Patterns of light and nitrogen distribution in relation to whole canopy carbon gain in $C_3$ and $C_4$ mono- and dicotyledonous species. Oecologia 101:504–513

Augusto L, Ranger J, Binkley D, Rothe A (2002) Impact of several common tree species of European temperate forests on soil fertility. Ann For Sci 59:233–253

Augusto L, Schrijver AD, Vesterdal L, Smolander A, Prescott C, Ranger J (2015) Influences of evergreen gymnosperm and deciduous angiosperm tree species on the functioning of temperate and boreal forests. Biol Rev 90:444–466

Bauhus J, Khanna PK, Menden N (2000) Aboveground and belowground interactions in mixed plantations of *Eucalyptus globulus* and *Acacia mearnsii*. Can J For Res 30:1886–1894

Bauhus J, van Winden AP, Nicotra AB (2004) Above-ground interactions and productivity in mixed-species plantations of *Acacia mearnsii* and *Eucalyptus globulus*. Can J For Res 34:686–694

Bayer D, Seifert S, Pretzsch H (2013) Structural crown properties of Norway spruce (*Picea abies* [L.] Karst.) and European beech (*Fagus sylvatica* [L.]) in mixed versus pure stands revealed by terrestrial laser scanning. Trees Struct Funct 27:1035–1047

Berger TW, Inselsbacher E, Mutsch F, Pfeffer M (2009a) Nutrient cycling and soil leaching in eighteen pure and mixed stands of beech (*Fagus sylvatica*) and spruce (*Picea abies*). For Ecol Manag 258:2578–2592

Berger TW, Untersteiner H, Toplitzer M, Neubauer C (2009b) Nutrient fluxes in pure and mixed stands of spruce (*Picea abies*) and beech (*Fagus sylvatica*). Plant Soil 322:317–342

Binkley D (2003) Seven decades of stand development in mixed and pure stands of conifers and nitrogen-fixing red alder. Can J For Res 33:2274–2279

Binkley D (2012) Understanding the role resource use efficiency in determining the growth of trees and forests. In: Schlichter T, Montes L (eds) Forests in development: A vital balance. Springer, The Hague, pp 13–26

Binkley D, Giardina C (1997) Nitrogen fixation in tropical forest plantations. In: Nambiar EKS, Brown AG (eds) Management of soil, nutrients and water in tropical plantation forests. Australian Centre for International Agricultural Research, Canberra, pp 297–337

Binkley D, Giardina C (1998) Why do tree species affect soils? The Warp and Woof of tree-soil interactions. Biogeochemistry 42:89–106

Binkley D, Valentine D (1991) Fifty-year biogeochemical effects of green ash, white pine, and Norway spruce in a replicated experiment. For Ecol Manag 40(1–2):13–25

Binkley D, Dunkin KA, DeBell D, Ryan MG (1992a) Production and nutrient cycling in mixed plantations of *Eucalyptus* and *Albizia* in Hawaii. For Sci 38(2):393–408

Binkley D, Sollins P, Bell R, Sachs D, Myrold D (1992b) Biogeochemistry of adjacent conifer and alder-conifer stands. Ecology 73:2022–2033

Binkley D, Senock R, Bird S, Cole TG (2003) Twenty years of stand development in pure and mixed stands of *Eucalyptus saligna* and N-fixing *Facaltaria moluccana*. For Ecol Manag 182:93–102

Binkley D, Stape JL, Ryan MG (2004) Thinking about efficiency of resource use in forests. For Ecol Manag 193:5–16

Bolte A, Villanueva I (2006) Interspecific competition impacts on the morphology and distribution of fine roots in European beech (*Fagus sylvatica* L.) and Norway spruce (*Picea abies* (L.) Karst.) Eur J For Res 125:15–26

Bouillet JP, Laclau JP, Goncalves JLM, Moreira MZ, Trivelin PCO, Jourdan C, Silva EV, Piccolo MC, Tsai SM, Galiana A (2008) Mixed-species plantations of *Acacia mangium* and *Eucalyptus grandis* in Brazil 2: nitrogen accumulation in the stands and biological $N_2$ fixation. For Ecol Manag 255:3918–3930

Bouillet J-P, Laclau J-P, Gonçalves JLM, Voigtlaender M, Gava JL, Leite FP, Hakamada R, Mareschal L, Mabiala A, Tardy F, Levillain J, Deleporte P, Epron D, Nouvellon Y (2013) Eucalyptus and Acacia tree growth over entire rotation in single- and mixed-species plantations across five sites in Brazil and Congo. For Ecol Manag 301:89–101

Boyden S, Binkley D, Senock R (2005) Competition and facilitation between *Eucalyptus* and nitrogen-fixing *Falcataria* in relation to soil fertility. Ecology 86(4):992–1001

Boyden S, Montgomery R, Reich PB, Palik B (2012) Seeing the forest for the heterogeneous trees: stand-scale resource distributions emerge from tree-scale structure. Ecol Appl 22 (5):1578–1588

Brassard BW, Chen HYH, Bergeron Y, Paré D (2011) Differences in fine root productivity between mixed- and single-species stands. Funct Ecol 25:238–246

Brassard BW, Chen HYH, Cavard X, Laganière J, Reich PB, Bergeron Y, Paré D, Yuan Z (2013) Tree species diversity increases fine root productivity through increased soil volume filling. J Ecol 101:210–219

Canham C, Coates KD, Bartemucci P, Quaglia S (1999) Measurement and modeling of spatially explicit variation in light transmission through interior cedar-hemlock forests of British Columbia. Can J For Res 29:1775–1783

Cavard X, Bergeron Y, Chen HYH, Paré D, Laganière J, Brassard B (2011) Competition and facilitation between tree species change with stand development. Oikos 120:1683–1695

Chisholm RA, Muller-Landau HC, Rahman KA, Bebber DP, Bin Y, Bohlman SA, Bourg NA, Brinks J, Bunyavejchewin S, Butt N, Cao H, Cao M, Cardenas D, Chang L-W, Chiang J-M, Chuyong G, Condit R, Dattaraja HS, Davies S, Duque A, Fletcher C, Gunatilleke N, Gunatilleke S, Hao Z, Harrison RD, Howe R, Hsieh C-F, Hubbell SP, Itoh A, Kenfack D, Kiratiprayoon S, Larson AJ, Lian J, Lin D, Liu H, Lutz JA, Ma K, Malhi Y, McMahon S, McShea W, Meegaskumbura M, Razman SM, Morecroft MD, Nytch CJ, Oliveira A, Parker GG, Pulla S, Punchi-Manage R, Romero-Saltos H, Sang W, Schurman J, Su S-H, Sukumar R, Sun I-F, Suresh HS, Tan S, Thomas D, Thomas S, Thompson J, Valencia R, Wolf A, Yap S, Ye W, Yuan Z, Zimmerman JK (2013) Scale-dependent relationships between tree species richness and ecosystem function in forests. J Ecol 101:1214–1224

Coates KD, Lilles EB, Astrup R (2013) Competitive interactions across a soil fertility gradient in a multispecies forest. J Ecol 101(3):806–818

Condés S, Rio MD, Sterba H (2013) Mixing effect on volume growth of *Fagus sylvatica* and *Pinus sylvestris* is modulated by stand density. For Ecol Manag 292:86–95

Crews TE, Peoples MB (2005) Can the synchrony of nitrogen supply and crop demand be improved in legume and fertilizer-based agroecosystems? A review. Nutr Cycl Agroecosyst 72:101–120

DeBell DS, Radwan MA (1979) Growth and nitrogen relations of coppiced black cottonwood and red alder in pure and mixed plantations. Bot Gaz 140(Suppl.):S97–S101

DeBell DS, Whitesell CD, Crabb TB (1987) Benefits of *Eucalyptus-Albizia* mixtures vary by site on Hawaii Island. USDA For. Serv. Res. Paper PSW-187

del Río M, Sterba H (2009) Comparing volume growth in pure and mixed stands of *Pinus sylvestris* and *Quercus pyrenaica*. Ann For Sci 66:502

del Río M, Schütze G, Pretzsch H (2014) Temporal variation of competition and facilitation in mixed species forests in Central Europe. Plant Biol 16:166–176

Dewar RC, Tarvainen L, Parker K, Wallin G, McMurtrie RE (2012) Why does leaf nitrogen decline within tree canopies less rapidly than light? An explanation from optimization subject to a lower bound on leaf mass per area. Tree Physiol 32:520–534

Dieler J, Pretzsch H (2013) Morphological plasticity of European beech (*Fagus sylvatica* L.) in pure and mixed-species stands. For Ecol Manag 295:97–108

Dijkstra FA, West JB, Hobbie SE, Reich PB (2009) Antagonistic effects of species on C respiration and net N mineralization in soils from mixed coniferous plantations. For Ecol Manag 257:1112–1118

Dunn GM, Connor DJ (1993) An analysis of sap flow in Mountain Ash (*Eucalyptus regnans*) forests of different age. Tree Physiol 13:321–336

Epron D, Nouvellon Y, Mareschal L, MoreiraeMoreira R, Koutika L-S, Geneste B, Delgado-Rojas JS, Laclau J-P, Sola G, Gonçalves JLM, Bouillet J-P (2013) Partitioning of net primary production in *Eucalyptus* and *Acacia* stands and in mixed-species plantations: two case-studies in contrasting tropical environments. For Ecol Manag 301:102–111

Ewel JJ (1986) Designing agricultural ecosystems for the humid tropics. Annu Rev Ecol Syst 17:245–271

Field C (1983) Allocating leaf nitrogen for the maximization of carbon gain: leaf age as a control on the allocation program. Oecologia 56:341–347

Filipescu CN, Comeau PG (2007) Competitive interactions between aspen and white spruce vary with stand age in boreal mixedwoods. For Ecol Manag 247:175–184

Fisher RF, Binkley D (2000) Ecology and management of forest soils, 3rd edn. Wiley, New York, Chichester

Forrester DI (2013) Growth responses to thinning, pruning and fertiliser application in *Eucalyptus* plantations: a review of their production ecology and interactions. For Ecol Manag 310:336–347

Forrester DI (2014) The spatial and temporal dynamics of species interactions in mixed-species forests: from pattern to process. For Ecol Manag 312:282–292
Forrester DI (2015) Transpiration and water-use efficiency in mixed-species forests versus monocultures: effects of tree size, stand density and season. Tree Physiol 35:289–304
Forrester DI, Albrecht AT (2014) Light absorption and light-use efficiency in mixtures of *Abies alba* and *Picea abies* along a productivity gradient. For Ecol Manag 328:94–102
Forrester DI, Bauhus J (2016) A review of processes behind diversity – productivity relationships in forests. Curr Forest Rep 2:45–61
Forrester DI, Pretzsch H (2015) Tamm Review: on the strength of evidence when comparing ecosystem functions of mixtures with monocultures. For Ecol Manag 356:41–53
Forrester DI, Smith RGB (2012) Faster growth of *Eucalyptus grandis* and *Eucalyptus pilularis* in mixed-species stands than monocultures. For Ecol Manag 286:81–86
Forrester DI, Bauhus J, Cowie AL (2005) Nutrient cycling in a mixed-species plantation of *Eucalyptus globulus* and *Acacia mearnsii*. Can J For Res 35(12):2942–2950. doi:10.1139/x05-214
Forrester DI, Bauhus J, Cowie AL (2006a) Carbon allocation in a mixed-species plantation of *Eucalyptus globulus* and *Acacia mearnsii*. For Ecol Manag 233:275–284. doi:10.1016/j.foreco.2006.05.018
Forrester DI, Bauhus J, Cowie AL, Vanclay JK (2006b) Mixed-species plantations of *Eucalyptus* with nitrogen fixing trees: a review. For Ecol Manag 233:211–230. doi:10.1016/j.foreco.2006.05.012
Forrester DI, Cowie AL, Bauhus J, Wood J, Forrester RI (2006c) Effects of changing the supply of nitrogen and phosphorus on growth and interactions between *Eucalyptus globulus* and *Acacia mearnsii* in a pot trial. Plant Soil 280(1-2):267–277. doi:10.1007/s11104-005-3228-x
Forrester DI, Bauhus J, Cowie AL, Mitchell PA, Brockwell J (2007a) Productivity of three young mixed-species plantations containing $N_2$-fixing *Acacia* and non-$N_2$-fixing *Eucalyptus* and *Pinus* trees in Southeastern Australia. For Sci 53(3):426–434
Forrester DI, Schortemeyer M, Stock WD, Bauhus J, Khanna PK, Cowie AL (2007b) Assessing nitrogen fixation in mixed- and single-species plantations of *Eucalyptus globulus* and *Acacia mearnsii*. Tree Physiol 27:1319–1328. doi:10.1093/treephys/27.9.1319
Forrester DI, Collopy JJ, Morris JD (2010a) Transpiration along an age series of *Eucalyptus globulus* plantations in southeastern Australia. For Ecol Manag 259:1754–1760. doi:10.1016/j.foreco.2009.04.023
Forrester DI, Theiveyanathan S, Collopy JJ, Marcar NE (2010b) Enhanced water use efficiency in a mixed *Eucalyptus globulus* and *Acacia mearnsii* plantation. For Ecol Manag 259:1761–1770. doi:10.1016/j.foreco.2009.07.036
Forrester DI, Vanclay JK, Forrester RI (2011) The balance between facilitation and competition in mixtures of *Eucalyptus* and *Acacia* changes as stands develop. Oecologia 166(1):265–272
Forrester DI, Collopy JJ, Beadle CL, Baker TG (2012a) Interactive effects of simultaneously applied thinning, pruning and fertiliser application treatments on growth, biomass production and crown architecture in a young *Eucalyptus nitens* plantation. For Ecol Manag 267:104–116. doi:10.1016/j.foreco.2011.11.039
Forrester DI, Lancaster K, Collopy JJ, Warren CR, Tausz M (2012b) Photosynthetic capacity of *Eucalyptus globulus* is higher when grown in mixture with *Acacia mearnsii*. Trees Struct Funct 26:1203–1213
Forrester DI, Kohnle U, Albrecht AT, Bauhus J (2013) Complementarity in mixed-species stands of *Abies alba* and *Picea abies* varies with climate, site quality and stand density. For Ecol Manag 304:233–242
Forrester DI, Benneter A, Bouriaud O, Bauhus J (2016) Diversity and competition influence tree allometry – developing allometric functions for mixed-species forests. J Ecol. doi:10.1111/1365-2745.12704
Forrester DI, Ammer C, Annighöfer PJ, Barbeito I, Bielak K, Bravo-Oviedo A, Coll L, del Río M, Drössler L, Heym M, Hurt V, Löf M, den Ouden J, Pach M, Pereira MG, Plaga B, Ponette Q,

Skrzyszewski J, Sterba H, Svoboda M, Zlatanov T, Pretzsch H (in press) Effects of crown architecture and stand structure on light absorption in mixed and monospecific *Fagus sylvatica* and *Pinus sylvestris* forests along a productivity and climate gradient through Europe. J Ecol

Fredericksen TS, Zedaker SM (1995) Fine root biomass, distribution, and production in young pine-hardwood stands. New For 10:99–110

Garber SM, Maguire DA (2004) Stand productivity and development in two mixed-species spacing trials in the Central Oregon Cascades. For Sci 50:92–105

Gartner TB, Cardon ZG (2004) Decomposition dynamics in mixed-species leaf litter. Oikos 104:230–246

Gash JHC, Lloyd CR, Lachaud G (1995) Estimating sparse forest rainfall interception with an analytical model. J Hydrol 170:79–86

Gebauer T, Horna V, Leuschner C (2012) Canopy transpiration of pure and mixed forest stands with variable abundance of European beech. J Hydrol 442–443:2–14

Gerrits AMJ, Pfister L, Savenije HHG (2010) Spatial and temporal variability of canopy and forest floor interception in a beech forest. Hydrol Process 24:3011–3025

Getzin S, Wiegand K (2007) Asymmetric tree growth at the stand level: random crown patterns and the response to slope. For Ecol Manag 242:165–174

Grossiord C, Granier A, Ratcliffe S, Bouriaud O, Bruelheide H, Chećko E, Forrester DI, Dawud SM, Finér L, Pollastrini M, Scherer-Lorenzen M, Valladares F, Bonal D, Gessler A (2014) Tree diversity does not always improve resistance of forest ecosystems to drought. Proc Natl Acad Sci U S A 111(41):14812–14815

Guisasola R, Tang X, Bauhus J, Forrester DI (2015) Intra- and inter-specific differences in crown architecture in Chinese subtropical mixed-species forests. For Ecol Manag 353:164–172

Harrington RA, Fownes JH, Meinzer FC, Scowcroft PG (1995) Forest growth along a rainfall gradient in Hawaii: *Acacia koa* stand structure, productivity, foliar nutrients, and water- and nutrient-use efficiencies. Oecologia 102:277–284

Hättenschwiler S, Tiunov AV, Scheu S (2005) Biodiversity and litter decomposition in terrestrial ecosystems. Annu Rev Ecol Evol Syst 36:191–218

Hawthorne SND, Lane PNJ, Bren LJ, Sims NC (2013) The long term effects of thinning treatments on vegetation structure and water yield. For Ecol Manag 310:983–993

He X-H, Critchley C, Bledsoe C (2003) Nitrogen transfer within and between plants through common mycorrhizal networks (CMNs). Crit Rev Plant Sci 22(6):531–567

He Z-B, Yang J-J, Du J, Zhao W-Z, Liu H, Chang X-X (2014) Spatial variability of canopy interception in a spruce forest of the semiarid mountain regions of China. Agric For Meteorol 188:58–63

Hector A, Bagchi R (2007) Biodiversity and ecosystem multifunctionality. Nat Lett 448:188–191

Hinsinger P, Betencourt E, Bernard L, Brauman A, Plassard C, Shen J, Tang X, Zhang F (2011) P for two, sharing a scarce resource: soil phosphorus acquisition in the rhizosphere of intercropped species. Plant Physiol 156:1078–1086

Hirose T, Werger MJA (1987) Maximizing daily canopy photosynthesis with respect to the leaf nitrogen allocation pattern in the canopy. Oecologia 72(4):520–526

Huber MO, Sterba H, Bernhard L (2014) Site conditions and definition of compositional proportion modify mixture effects in *Picea abies* – *Abies alba* stands. Can J For Res 44:1281–1291

Hunt MA, Unwin GL, Beadle CL (1999) Effects of naturally regenerated *Acacia dealbata* on the productivity of a *Eucalyptus nitens* plantation in Tasmania, Australia. For Ecol Manag 117:75–85

Ilek A, Kucza J, Szostek M (2015) The effect of stand species composition on water storage capacity of the organic layers of forest soils. Eur J For Res 134:187–197

Isbell F, Calcagno V, Hector A, Connolly J, Harpole WS, Reich PB, Scherer-Lorenzen M, Schmid B, Tilman D, Ruijven J, Weigelt A, Wilsey BJ, Zavaleta ES, Loreau M (2011) High plant diversity is needed to maintain ecosystem services. Nature 477:199–203

Kaye JP, Resh SC, Kaye MW, Chimmer RA (2000) Nutrient and carbon dynamics in a replacement series of *Eucalyptus* and *Albizia* trees. Ecology 81(12):3267–3273

Kelty MJ (1992) Comparative productivity of monocultures and mixed-species stands. In: Kelty MJ, Larson BC, Oliver CD (eds) The ecology and silviculture of mixed-species forests. Kluwer Academic Publishers, Dordrecht, pp 125–141
Kelty MJ, Cameron IR (1995) Plot designs for the analysis of species interactions in mixed stands. Commonwealth For Rev 74(4):322–332
Khamzina A, Lamers JPA, Vlek PLG (2009) Nitrogen fixation by *Elaeagnus angustifolia* in the reclamation of degraded croplands of Central Asia. Tree Physiol 29(6):799–808
Khanna PK (1997) Comparison of growth and nutrition of young monocultures and mixed stands of *Eucalyptus globulus* and *Acacia mearnsii*. For Ecol Manag 94:105–113
Khanna PK (1998) Nutrient cycling under mixed-species tree systems in southeast Asia. Agrofor Syst 38:99–120
Knops JMH, Bradley KL, Wedin DA (2002) Mechanisms of plant species impacts on ecosystem nitrogen cycling. Ecol Lett 5:454–466
Kranabetter JM, MacKenzie WH (2010) Contrasts among mycorrhizal plant guilds in foliar nitrogen concentration and $\delta^{15}N$ along productivity gradients of a boreal forest. Ecosystems 13:108–117
Kunert N, Schwendenmann L, Potvin C, Hölscher D (2012) Tree diversity enhances tree transpiration in a Panamanian forest plantation. J Appl Ecol 49:135–144
Laclau J-P, Nouvellon Y, Reine C, Gonçalves JLM, Krushe AV, Jourdan C, le Maire G, Bouillet J-P (2013) Mixing *Eucalyptus* and *Acacia* trees leads to fine root over-yielding and vertical segregation between species. Oecologia 172:903–913
Law BE, Falge E, Guc L, Baldocchi DD, Bakwind P, Berbigier P, Davis K, Dolmang AJ, Falk M, Fuentes JD, Goldstein A, Granier A, Grelle A, Hollinger D, Janssensm IA, Jarvis P, Jensen NO, Katul G, Mahli Y, Matteucci G, Meyers T, Monsont R, Munger W, Oechel W, Olson R, Pilegaard K, Paw KT, Thorgeirsson H, Valentini R, Verma S, Vesala T, Wilson K, Wofsy S (2002) Environmental controls over carbon dioxide and water vapor exchange of terrestrial vegetation. Agric For Meteorol 113:97–120
le Maire G, Nouvellon Y, Christina M, Ponzoni FJ, Gonçalves JLM, Bouillet J-P, Laclau J-P (2013) Tree and stand light use efficiencies over a full rotation of single- and mixed-species *Eucalyptus grandis* and *Acacia mangium* plantations. For Ecol Manag 288:31–42
Lebourgeois F, Gomez N, Pinto P, Mérian P (2013) Mixed stands reduce *Abies alba* tree-ring sensitivity to summer drought in the Vosges mountains, western Europe. For Ecol Manag 303:61–71
Levia DF, Frost EE (2003) A review and evaluation of stemflow literature in the hydrologic and biogeochemical cycles of forested and agricultural ecosystems. J Hydrol 274(1–4):1–29
Ligot G, Balandier P, Courbaud B, Claessens H (2014) Forest radiative transfer models: which approach for which application? Can J For Res 44(5):385–397
Litton CM, Raich JW, Ryan MG (2007) Carbon allocation in forest ecosystems. Glob Chang Biol 13:2089–2109
May BM, Attiwill P, M. (2003) Nitrogen-fixation by *Acacia dealbata* and changes in soil properties 5 years after mechanical disturbance or slash-burning following timber harvest. For Ecol Manag 181:339-355
McKane RB, Johnson LC, Shaver GR, Nadelhoffer KJ, Rastetter EB, Fry B, Giblin AE, Kielland K, Kwiatkowski BL, Laundre JA, Murray G (2002) Resource-based niches provide a basis for plant species diversity and dominance in arctic tundra. Nature 415:68–71
Monteith JL (1977) Climate and the efficiency of crop production in Britain. Philos Trans R Soc Lond B 281:277–294
Moore GW, Bond BJ, Jones JA (2011) A comparison of annual transpiration and productivity in monoculture and mixed-species Douglas-fir and red alder stands. For Ecol Manag 262:2263–2270
Neumann RB, Cardon ZG (2012) The magnitude of hydraulic redistribution by plant roots: a review and synthesis of empirical and modeling studies. New Phytol 194:337–352

Niinemets Ü (2012) Optimization of foliage photosynthetic capacity in tree canopies: towards identifying missing constraints. Tree Physiol 32:505–509

Osada N, Yasumura Y, Ishida A (2014) Leaf nitrogen distribution in relation to crown architecture in the tall canopy species, *Fagus crenata*. Oecologia 175:1093–1106

Parrotta JA, Baker DD, Fried M (1996) Changes in dinitrogen fixation in maturing stands of *Casuarina equisetifolia* and *Leucaena leucocephala*. Can J For Res 26(9):1684–1691

Peltoniemi MS, Duursma RA, Medlyn BE (2012) Co-optimal distribution of leaf nitrogen and hydraulic conductance in plant canopies. Tree Physiol 32:510–519

Peoples MB, Herridge DF, Ladha JK (1995) Biological nitrogen fixation: an efficient source of nitrogen for sustainable agricultural production? Plant Soil 174(1-2):3–28

Pfautsch S, Gessler A, Adams MA, Rennenberg H (2009a) Using amino-nitrogen pools and fluxes to identify contributions of understory *Acacia* spp. to overstory *Eucalyptus* regnans and stand nitrogen uptake in temperate Australia. New Phytol 183:1097–1113

Pfautsch S, Rennenberg H, Bell TL, Adams MA (2009b) Nitrogen uptake by *Eucalyptus regnans* and *Acacia* spp. – preferences, resource overlap and energetic costs. Tree Physiol 29:389–399

Poorter H, Niklas KJ, Reich PB, Oleksyn J, Poot P, Mommer L (2012) Biomass allocation to leaves, stems and roots: meta-analyses of interspecific variation and environmental control. New Phytol 193:30–50

Pretzsch H, Block J, Dieler J, Dong PH, Kohnle U, Nagel J, Spellmann H, Zingg A (2010) Comparison between the productivity of pure and mixed stands of Norway spruce and European beech along an ecological gradient. Ann For Sci 76:712–723

Pretzsch H, Bielak K, Block J, Bruchwald A, Dieler J, Ehrhart H-P, Kohnle U, Nagel J, Spellmann H, Zasada M, Zingg A (2013a) Productivity of mixed versus pure stands of oak (Quercus petraea (MATT.) LIEBL. and *Quercus robur* L.) and European beech (*Fagus sylvatica* L.) along an ecological gradient. Eur J For Res 132:263–280

Pretzsch H, Schütze G, Uhl E (2013b) Resistance of European tree species to drought stress in mixed versus pure forests: evidence of stress release by inter-specific facilitation. Plant Biol 15:483–495

Prieto I, Armas C, Pugnaire FI (2012) Water release through plant roots: new insights into its consequences at the plant and ecosystem level. New Phytol 193:830–841

Rao MR, Nair PKR, Ong CK (1998) Biophysical interactions in tropical agroforestry systems. Agrofor Syst 38:3–50

Reich PB, Luo Y, Bradford JB, Poorter H, Perry CH, Oleksyn J (2014) Temperature drives global patterns in forest biomass distribution in leaves, stems, and roots. Proc Natl Acad Sci U S A 111:13721–13726

Reif A, Brucker U, Kratzer R, Schmiedinger A, Bauhus J (2010) Waldbewirtschaftung in Zeiten des Klimawandels – Synergien und Konfliktpotenziale zwischen Forstwirtschaft und Naturschutz. Naturschutz und Landschaftsplanung 42(9):261–266

Richards AE, Forrester DI, Bauhus J, Scherer-Lorenzen M (2010) The influence of mixed tree plantations on the nutrition of individual species: a review. Tree Physiol 30:1192–1208. doi:10.1093/treephys/tpq035

Rothe A, Binkley D (2001) Nutritional interactions in mixed species forests: a synthesis. Can J For Res 31(11):1855–1870

Roupsard O, Ferhi A, Granier A, Pallo F, Depommier D, Mallet B, Joly HI, Dreyer E (1999) Reverse phenology and dry-season water uptake by *Faidherbia albida* (Del.) A. Chev. in an agroforestry parkland of Sudanese west Africa. Funct Ecol 13:460–472

Ryan MG, Binkley D, Fownes JH (1997) Age-related decline in forest productivity: pattern and process. Adv Ecol Res 27:213–262

Ryan MG, Binkley D, Fownes JH, Giardina CP, Senock RS (2004) An experimental test of the causes of forest growth decline with stand age. Ecol Monogr 74:393–414

Saccone P, Delzon S, Pagès J-P, Brun J-J, Michalet R (2009) The role of biotic interactions in altering tree seedling responses to an extreme climatic event. J Veg Sci 20:403–414

Sands PJ (1995) Modelling canopy production. I Optimal distribution of photosynthetic resources. Aust J Plant Physiol 22:593–601

Sapijanskas J, Paquette A, Potvin C, Kunert N, Loreau M (2014) Tropical tree diversity enhances light capture through crown plasticity and spatial and temporal niche differences. Ecology 95:2479–2492

Schmid I, Kazda M (2002) Root distribution of Norway spruce in monospecific and mixed stands on different soils. For Ecol Manag 159(1-2):37–47

Schume H, Jost G, Hager H (2004) Soil water depletion and recharge patterns in mixed and pure forest stands of European beech and Norway spruce. J Hydrol 289:258–274

Schwendenmann L, Pendall E, Sanchez-Bragado R, Kunert N, Hölscher D (2015) Tree water uptake in a tropical plantation varying in tree diversity: interspecific differences, seasonal shifts and complementarity. Ecohydrology 8:1–12

Snowdon P, Wichiennopparat W, Khanna PK (2003) Growth, above-ground biomass and nutrient content of eucalypts and acacias grown in mixture in a tropical environment – evaluation for one full rotation. In: International Conference on Eucalypt Productivity, Hobart, Australia, 10–15 November 2003. pp 57–58

Stape JL, Binkley D, Ryan MG (2004) *Eucalyptus* production and the supply, use and efficiency of use of water, light and nitrogen across a geographic gradient in Brazil. For Ecol Manag 193:17–31

Turner BL (2008) Resource partitioning for soil phosphorus: a hypothesis. J Ecol 96:698–702

Van Kessel C, Farrell RE, Roskoski JP, Keane KM (1994) Recycling of the naturally-occurring $^{15}$N in an established stand of *Leucaena leucocephala*. Soil Biol Biochem 26:757–762

Vanclay JK (2006a) Experiment designs to evaluate inter- and intra-specific interactions in mixed plantings of forest trees. For Ecol Manag 233:366–374

Vanclay JK (2006b) Spatially-explicit competition indices and the analysis of mixed-species plantings with the Simile modelling environment. For Ecol Manag 233:295–302

Vandermeer J (1989) The ecology of intercropping. Cambridge University Press, New York

Vertessy RA, Hatton TJ, Benyon RG, Dawes WR (1996) Long-term growth and water balance predictions for a mountain ash (*Eucalyptus regnans*) forest catchment subject to clear-felling and regeneration. Tree Physiol 16(1-2):221–232

Vilà M, Carrillo-Gavilán A, Vayreda J, Bugmann H, Fridman J, Grodzki W, Haase J, Kunstler G, Schelhaas M, Trasobares A (2013) Disentangling biodiversity and climatic determinants of wood production. PLoS One 8(2):e53530

Vitousek PM (1982) Nutrient cycling and nutrient efficiency. Am Nat 119:553–572

Wacker L, Baudois O, Eichenberger-Glinz S, Schmid B (2009) Effects of plant species richness on stand structure and productivity. J Plant Ecol 2(2):95–106

Wang XL, Klinka K, Chen HYH, de Montigny L (2002) Root structure of western hemlock and western redcedar in single- and mixed-species stands. Can J For Res 32:997–1004

Williams BL (1992) Nitrogen dynamics in humus and soil beneath Sitka spruce (Picea sitchensis (Bong.) Carr.) planted in pure stands and in mixture with Scots pine (Pinus sylvestris L.) Plant Soil 144:77–84

Zapater M, Hossann C, Bréda N, Bréchet C, Bonal D, Granier A (2011) Evidence of hydraulic lift in a young beech and oak mixed forest using $^{18}$O soil water labelling. Trees 25:885–894

# Stand Dynamics of Mixed-Species Stands Compared with Monocultures

4

Hans Pretzsch and David I. Forrester

**Abstract**

Based on well-established principles and using reference values for monocultures, this chapter examines how mixed-species stands differ from monocultures in terms of stand growth and yield, self-thinning, alien thinning, stand density, and density-growth relationships. This chapter also considers how the species interactions and the resulting mixing effects on growth can vary spatially and temporarily in response to the prevailing environmental conditions.

Beyond the basic principles of mixed stand behaviour, this chapter will provide mean and sum values for mixed versus monospecific stand growth and yield for temperate Central European ecosystems. Both the growth principles and dendrometrical measures for mixed compared with monospecific stands are required for systematically designing, and managing, more resource-use-efficient mixed-species production systems. The overyielding of mixed versus monospecific stands revealed in this chapter is clearly of a magnitude that is worthwhile exploring further and tracing from the stand to the tree and organ levels.

H. Pretzsch (✉)
Chair for Forest Growth and Yield Science, Faculty of Forest Science and Resource Management, Technical University of Munich, Hans-Carl-von-Carlowitz-Platz 2, 85354 Freising, Germany
e-mail: Hans.Pretzsch@lrz.tum.de

D.I. Forrester
Swiss Federal Institute for Forest, Snow and Landscape Research WSL, Zurcherstrasse 111, 8903 Birmensdorf, Switzerland
e-mail: david.forrester@wsl.ch



H. Pretzsch et al. (eds.), *Mixed-Species Forests*, DOI 10.1007/978-3-662-54553-9_4

## Contents

4.1 Species Diversification and Interspecific Interactions Can Modulate Short- and Long-Term Growth ... 119
4.1.1 Theoretical Consideration ... 119
4.1.2 Empirical Evidence ... 121
4.2 Effect of Mixing on the Volume Growth, Yield, and Standing Stock ... 125
4.2.1 Theoretical Considerations ... 125
4.2.2 Empirical Evidence of Overyielding and Underyielding in Mixed Compared with Monospecific Forest Stands ... 146
4.3 Self-Thinning, Alien-Thinning, and Stand Density ... 160
4.3.1 Theoretical Considerations ... 160
4.3.2 Evidence of Mixing Effects on Stand Density, Self-Thinning, and Alien Thinning ... 167
4.4 Effect of Tree Species Mixing on the Total Yield and the Yield Level ... 176
4.4.1 Analysing the Rule by Eichhorn (1902) and the General Yield Level by Gehrhardt (1923) and Assmann (1961) ... 176
4.4.2 Mixing Modified Standing Volume and Total Yield Rather Than Tree Height and Site Index ... 177
4.5 Spatial and Temporal Variation of Mixing Effects ... 181
4.5.1 Theoretical Considerations ... 181
4.5.2 Empirical Evidence of Spatial and Temporal Variation in Overyielding and Underyielding at the Stand Level ... 183
4.6 Spacing, Thinning, and Other Small-Scale Interference in Canopy Closure as Modifiers of Mixing Effects ... 190
4.6.1 Theoretical Considerations ... 190
4.6.2 Empirical Evidence of Modified Growth Responses in Relation to Density Reduction in Mixed Compared with Monospecific Forest Stands ... 192
4.7 Reference Values from Long-Term Experimental Plots ... 197
4.7.1 Reference Values for Growth and Yield of Monocultures and Even-Aged Mixed Stands ... 197
4.7.2 Reference Values for Standing Stock or Volume and Biomass Growth in Even-Aged Monocultures Compared with Uneven-Aged Mixed-Species Stands ... 199
4.8 Synopsis of Mixing Effects at the Forest Stand Level ... 201
Take-Home Messages ... 203
References ... 204

To better understand species mixing effects, it is useful to trace them from the stand to the size distribution, tree, and organ levels. This chapter starts with the stand level, and the subsequent chapters will proceed down to the lower levels. The main question is to what extent any differences between mixture and monoculture are just a weighted average of the monocultures, also referred to as an additive effect, or whether the mixture characteristics depart from the weighted average of the monocultures, sometimes referred to as non-linear or multiplicative effects (Forrester and Pretzsch 2015; Kelty 1992). The term 'additive effect' underlines that this kind of mixing effect results from nothing more than selecting the species and adding up the characteristics of the monocultures (Kelty 1992; Forrester 2014). Multiplicative mixing effects emerge from the species interactions and cannot be predicted by only studying the species in their monocultures. Therefore, multiplicative effects are of particular relevance for both forest practice and forest science.

For further explanation and evaluation of additive and multiplicative mixing effects, see Sect. 4.2.1 and Box 4.1. The majority of the mixing effects reported in this chapter are based on replacement experiments (Jolliffe et al. 1984; Jolliffe 2000), which are common in forest science (see Chap. 2).

Monocultures might be an ecologically questionable reference especially when they are completely artificial and unrepresentative of most stands where the species occur naturally. Nevertheless, they are useful as references because they often represent the silvicultural business as usual. Mixed stands, which are often more complex, should offer some advantages over the reference before being implemented as an alternative.

Beyond the basic principles of mixed stand behaviour, this chapter also provides mean and sum values for mixed versus monospecific stand growth and yield for temperate Central European ecosystems. Both the growth principles and dendrometrical measures for mixed compared with monospecific stands are required for systematically designing, and managing, more resource-use-efficient mixed-species production systems. The overyielding of mixed versus monospecific stands revealed in this chapter is clearly of a magnitude that is worthwhile exploring further and tracing from the stand to the tree and organ levels.

## 4.1 Species Diversification and Interspecific Interactions Can Modulate Short- and Long-Term Growth

The development of mixed-species forest stands depends on both the intraspecific and interspecific interactions between the associated tree species. The result of all competitive reduction or facilitation interactions depends on the species' ecological traits and the environmental conditions. The mixed stands do not usually perform as expected based on the weighted mean of the respective monocultures (Pretzsch and Schütze 2009). The species interactions can strongly modify the level and rhythm of the species-specific trajectories of mean or dominant height, tree number, and volume growth development. We start with basic principles which are well established for monocultures and use them to characterise the considerably differing dynamics of mixed stands at the species and whole-stand levels. We outline how mixing can modify the key characteristics of stand growth—height, density, and volume growth—and consider the mixing effects on both long- and short-term stand dynamics.

### 4.1.1 Theoretical Consideration

As the height development of a forest stand is linked to many other stand attributes, it is commonly used for characterising the species-specific dynamics of monocultures (Skovsgaard and Vanclay 2008). Figure 4.1a shows distinct differences in the sigmoid height growth curve of early-successional and light-demanding (e.g. *Betula*, *Larix*, *Pinus*), late-successional and shade-tolerant (*Abies*, *Fagus*, *Picea*), and intermediate species (*Acer*, *Tilia*, *Fraxinus*). Early-successional, intermediate, and late-

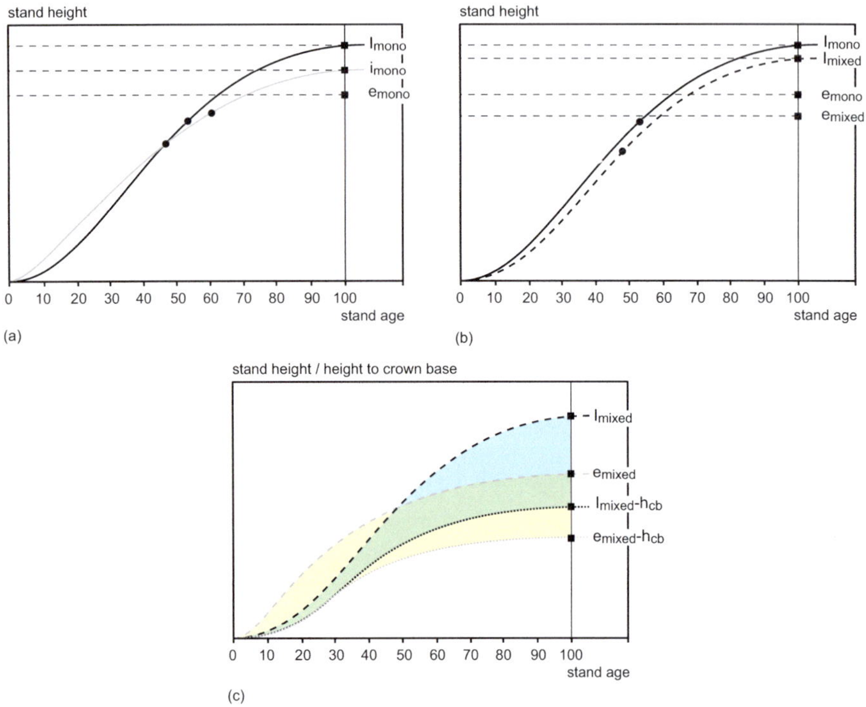


**Fig. 4.1** (**a**–**c**) Courses of species-specific height and height to crown base provide key information about the dynamics of mixed-species stands. Schematic representation of (**a**) sigmoid height growth in monospecific stands for early-successional ($e_p$), intermediate ($i_p$), and late-successional tree species ($l_p$), (**b**) slowing down of height growth by interspecific competition in mixed-species stands, and (**c**) ingrowth of the late-successional and shade-tolerant species in the mixed stand ($l_m$) into the crown layer of the early-successional fast grower ($e_m$)

successional species are indicated by e, i, and l in Fig. 4.1. In monospecific stands, intraspecific competition only has a minor effect on the mean height development and a negligible effect on top height (see $e_p$, $i_p$, and $l_p$ in Fig. 4.1a). That changes when the mixed species have height curves that differ in level or that intersect (Fig. 4.1b). The species-specific levels of the height curves and the age of intersection reveal the height and light competition to be expected when mixing those species. They suggest how competition can be released by temporal separation of the species, such as by staggering, planting, or using advanced natural regeneration, or spatial separation, such as via group-by-group mixing as opposed to tree-by-tree mixing in order to increase the probability of intra- in addition to interspecific competition.

Figure 4.1b shows a slowing down of the height growth of both associated species (Wiedemann 1951, pp 131–133). The species-specific height growth in mixture, and especially the comparison of this with that in monocultures (observed or taken from models), reflects the outcome of the interspecific competition. The taller species commonly reduce the growth of the shorter ones not only by competing for soil resources but also by pre-empting the light. Conversely, the shorter species may reduce the growth of the taller species by appropriating any soil resources that are limiting, such as water and mineral nutrients. The point of intersection of the height curves reveals at which stand age the previously shorter species begins to take the lead. Note that the shorter species already start competing for light and reducing the growth of their taller neighbours when they enter (from below) and shade the lower crowns of the taller species (double-hatched area in Fig. 4.1c).

### 4.1.2 Empirical Evidence

Figure 4.2 shows how the height development of Norway spruce and European beech on the mixed-species experimental plots in the Bavarian Forest can be slowed down compared with the development in the monospecific stands in the early stand phase. However, the mixed stand later catches up or even overtakes the monospecific stand in terms of height development.

Reductions of one or more species in the mixed stand compared with the monospecific stand were also reported by Wellhausen et al. (2016) for stands of Scots pine and Norway spruce and by Wiedemann (1951, pp 131–133) for mixtures of sessile oak and European beech, Scots pine and Norway spruce, Scots pine and European beech, and Norway spruce and European beech. The quicker growing species can also be further accelerated in height growth as reported by Pretzsch et al. (2015) for Scots pine when mixed with European beech. Growth of silver fir can be strongly reduced and nearly stopped when suppressed and kept in the

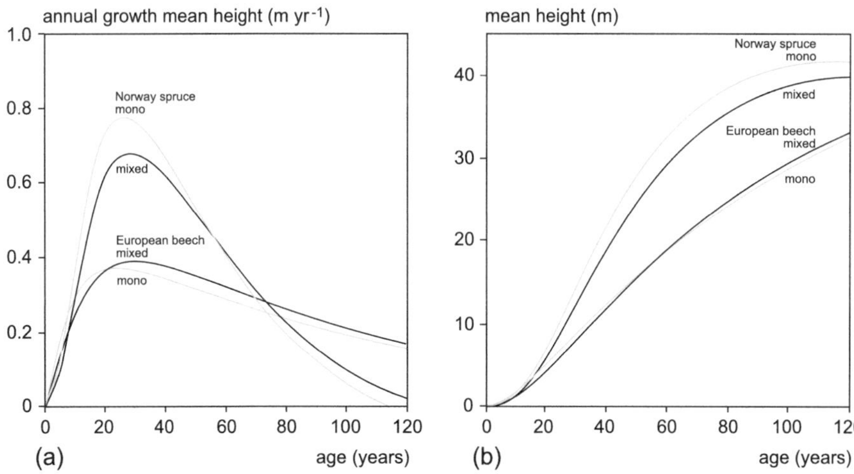


**Fig. 4.2** Tree species mixing can modify (**a**) the course of the annual stand height growth and (**b**) mean height development of Norway spruce and European beech in mixed-species stands compared with monospecific stands. The graph shows the results of a long-term survey of the mixed-species experiments Zwiesel 111, 134, and 135 (according to Pretzsch 1992, pp 193–198, see also Fig. 4.7)

understorey under a dense canopy of tall European beeches and Norway spruces (Kern 1966; Preuhsler 1979; Magin 1959).

Species can also differ considerably in terms of their intra-annual or even daily course of growth (see Box 4.2). The effects of differences in the species' behaviour within a year or day on their common resource use and growth in mixed-species stands are even less understood than the long-term effects during rotation. Measurements based on permanent girth tapes on trees in monospecific and mixed-species stands can reveal the modulation of the intra-annual and daily growth course through mixing (Box 4.2).

#### 4.1.2.1 Overview of Responses in Common Tree Species Mixtures in Central Europe

Figure 4.3 provides an overview of the mixing effects on the mean height of the tree as well as the quadratic mean diameter for various common species assemblages in Central Europe. More details are provided by Pretzsch et al. (2016b). The figure is based on 141 triplets of fully stocked and even-aged mixed-species stands and neighbouring monospecific stands of silver fir/Norway spruce, Norway spruce/Scots pine, Norway spruce/European larch, Norway spruce/European beech, Norway spruce/red alder, Scots pine/European beech, European larch/European beech, European beech/sessile oak, and European beech/Douglas-fir. On average, the mean height and diameter in the mixed-species stands are rather similar to the weighted mean tree height and mean tree diameter of the respective monospecific stands. On average, the mean heights were 2% lower in mixed compared with

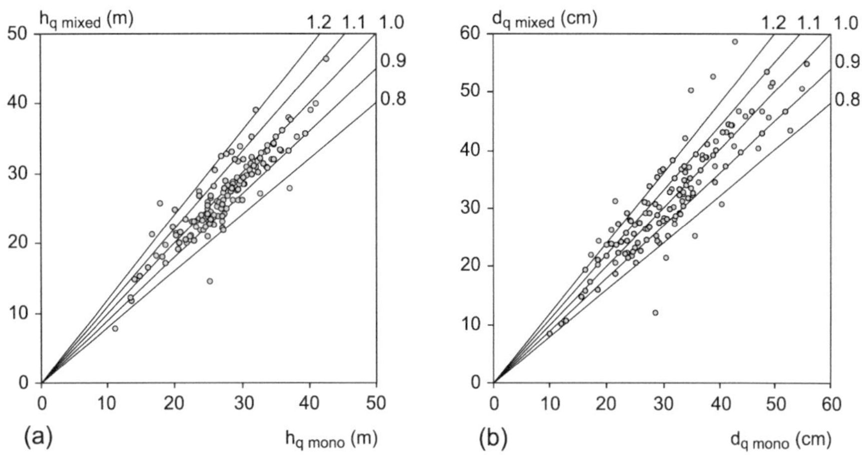


**Fig. 4.3** Mean tree height (**a**) and quadratic mean tree diameter (**b**) in mixed-species stands are, on average, rather similar to the respective mean tree dimensions in the neighbouring monocultures. Values near the bisector line (1.0 line) indicate similarity between the mean tree dimensions in the mixed-species stands and the weighted mean tree dimensions of the neighbouring monocultures

monospecific stands. The mean tree diameters and tree volumes were 1–5% higher in mixed than in monospecific stands; however, the differences were not significant.

For five very common mixtures, Table 4.1 shows the differences between mixed and monospecific stands at the species level. Analogous to the overarching analysis for all species, there were few significant differences between the mean tree dimensions of the mixed stand and the weighted mean of the monospecific stands. The weighted mean represents the behaviour of the mixed stand with additive effects only, i.e. when each species sticks to its monospecific behaviour (see Table 4.1, column total stand mixed/mono). The examination at the species level reveals that the lead of one species is mostly compensated by the delay of the other species (Table 4.1, columns species 1 mixed/mono and species 2 mixed/mono), so that on average, the height and diameter of mixed stands resemble the weighted mean of the respective monospecific stands. An exception is the mixture of sessile oak and European beech, where in the mixed stand, both species' size growths fall 5% behind the size development of the neighbouring monospecific stands.

Using a study of 32 triplets of monospecific and mixed stands of Scots pine and European beech along a productivity gradient through Europe, we can illustrate how the intraspecific behaviour of these species can be modified by decades of interspecific competition, competitive reduction, and facilitation (Pretzsch et al. 2015). Figure 4.4 shows the mean tree height and stem diameter as well as the ratio between height and diameter in mixed compared with neighbouring monocultures of the same age, about 40–60 years (Pretzsch et al. 2015).

In mixed stands, the mean height of pine was 6% higher, and the mean diameter was 20% higher than in monocultures, so the ratio between $h_q/d_q$ was 8% lower in mixed than in monospecific stands (Fig. 4.4a, c, e). In contrast, the mean height and

**Table 4.1** Mean tree height, $h_g$, and mean tree diameter, $d_g$, in mixed-species stands in relation to the neighbouring monocultures calculated separately for five selected species assemblages

| Variables | Species combination | $n$ | Species 1 mixed/ mono (± SE) | Species 2 mixed/ mono (± SE) | Total stand mixed/ mono (± SE) |
|---|---|---|---|---|---|
| *Mean stand height $h_q$ (m)* | | | | | |
| | N. spruce/S. pine | 7 | **0.86 (±0.06)** | 1.05 (±0.03) | 0.94 (±0.04) |
| | N. spruce/E. larch | 10 | 0.71 (±0.31) | 1.07 (±0.10) | 0.95 (±0.10) |
| | N. spruce/E. beech | 52 | 1.01 (±0.01) | 0.99 (±0.02) | 1.00 (±0.01) |
| | S. pine/E. beech | 17 | 1.04 (±0.03) | 0.97 (±0.05) | 1.01 (±0.04) |
| | s. oak/E. beech | 24 | 0.98 (±0.01) | **0.90 (±0.02)** | **0.95 (±0.01)** |
| *Mean diameter $d_q$ (cm)* | | | | | |
| | N. spruce/S. pine | 7 | 0.82 (±0.09) | 1.03 (±0.03) | 0.89 (±0.05) |
| | N. spruce/E. larch | 10 | 0.74 (±0.21) | 1.11 (±0.08) | 1.08 (±0.13) |
| | N. spruce/E. beech | 28 | **1.12 (±0.02)** | **0.95 (±0.02)** | 1.05 (±0.02) |
| | S. pine/E. beech | 9 | **1.13 (±0.06)** | 0.94 (±0.04) | 1.04 (±0.04) |
| | s. oak/E. beech | 12 | **0.94 (±0.02)** | 0.95 (±0.03) | **0.95 (±0.02)** |

Ratios for mixed/mono that are above/below 1.00 indicate a superiority/inferiority of the species' performance in mixed-species stands versus monocultures. Ratios in bold numbers indicate significant differences ($p < 0.05$) between mixed-species stands and monocultures.

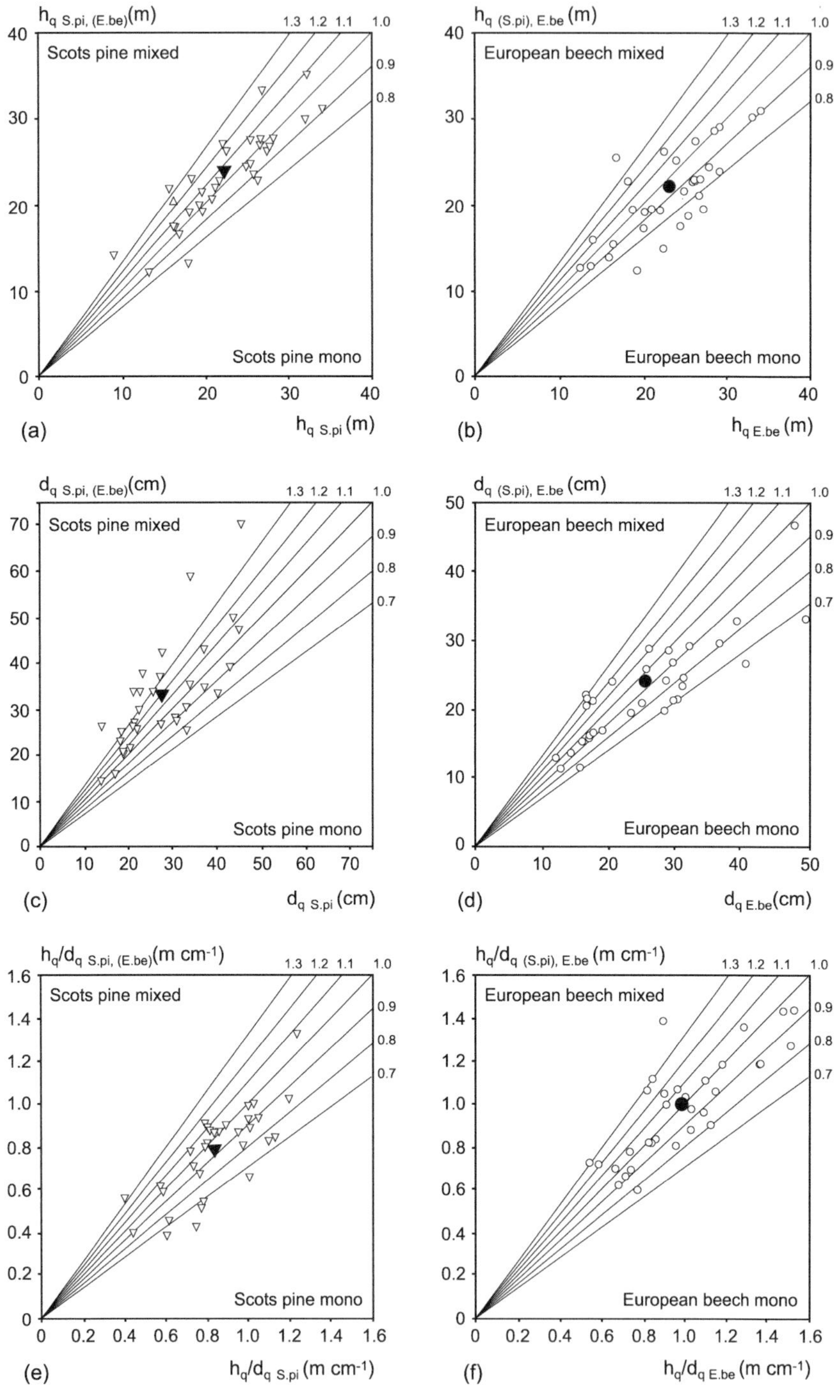


**Fig. 4.4** (**a–f**) Stand characteristics of Scots pine (*triangles*) and European beech (*circles*) in the monospecific stand (*x*-axis) compared with the neighbouring mixed stand (*y*-axis). Values

mean diameter of European beech were 8% lower in mixture, and the mean $h_q/d_q$ was 2% higher in mixed than in monospecific stands, i.e. due to the superior height of Scots pine; the growth of beech was reduced and modified in favour of height compared with diameter growth (Fig. 4.4b, d, f).

The height increase in pine and height decrease in beech in mixed compared with monospecific stands underline that height growth is superimposed by inter-specific competition and is less indicative of the site quality than height growth in monocultures. The finding that the current mean diameters of Scots pine and European beech are considerably higher and lower, respectively, in mixed than in monospecific stands is a result of interspecific versus intraspecific neighbourhoods. Any differences in the current growth of mixed versus monospecific stands may result partly from the size differences and partly from differences in the current environmental conditions within the mixed versus monospecific stands. Differences in the $h_q/d_q$ ratios indicate that mixing can affect stem shape and thereby the mechanical stability (von Lüpke and Spellmann 1999) and wood quality (Pretzsch and Rais 2016).

## 4.2 Effect of Mixing on the Volume Growth, Yield, and Standing Stock

### 4.2.1 Theoretical Considerations

#### 4.2.1.1 Modulation of the Course of Growth and Yield Through Mixing

The species-specific level and rhythm of the course of annual growth known from monocultures can be modulated through species mixture (Fig. 4.5a) (see Mitscherlich 1970, pp 112–123). In this example, the trajectories of the two species are different in the monospecific stands ($sp_1$ and $sp_2$). They differ even more when growing in mixture ($sp_{1,(2)}$ and $sp_{(1),2}$). The resulting total mixed stand growth curve ($sp_{1,2}$) has a temporarily lower level but a much broader plateau than the mono-specific stand curve for species 1 (Fig. 4.5b).

The total yield of the mixed stand can be higher than the weighted mean of the monocultures' yield (Fig. 4.5c, upper curves). The mixed stand may have a higher maximum standing stock and final yield according to Weiner and Freckleton (2010).

**Fig. 4.4** (continued) on the bisector line indicate equality of mixed with monospecific stand characteristics. (**a** and **b**) Show the height of the tree with the quadratic mean diameter, $h_q$ (m), of Scots pine and European beech, respectively. (**c** and **d**) Show the quadratic mean diameter, $d_q$ (cm), of Scots pine and European beech, respectively. (**e** and **f**) Show the ratios between mean height and mean diameter, $h_q/d_q$ (m cm$^{-1}$) for Scots pine and European beech, respectively. The small symbols represent the observed values, and the large symbols indicate the mean values of all 32 triplets

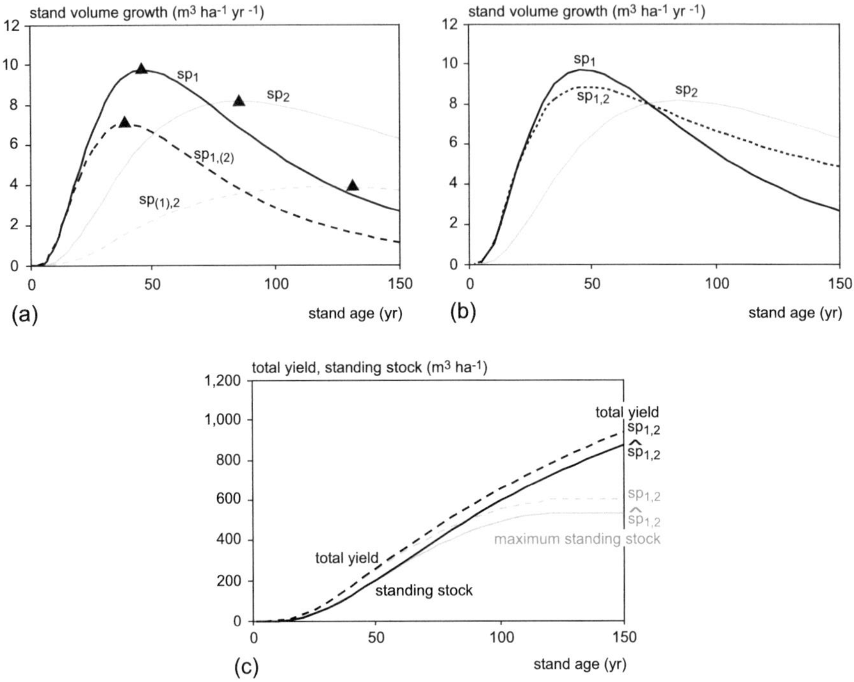


**Fig. 4.5** Mixing can modulate the course of stand growth, yield, and standing stock. (**a**) Compared with the course of annual volume growth in monospecific stands ($sp_1$ and $sp_2$); the level and the age of culmination (*filled triangle*) of both species in a mixed stand ($sp_{1,(2)}$ and $sp_{(1),2}$) can differ (**b**) due to different rhythms of the annual growth and positive species interactions; the course of total stand growth can be lower but have a broader plateau in mixed ($sp_{1,2}$) compared with monospecific stands ($sp_1$ and $sp_2$). (**c**) The course of the total yield (*upper lines*) may indicate overyielding, and the maximum standing volume (end of *lower line*) may be higher in mixed ($sp_{1,2}$) compared with the weighted mean of monocultures ($sp_2$ and $sp_2$). In this model example, total yield (*black*) and standing stock (*grey*) are compared. The courses for $\widehat{sp}_{1,2}$assume that mixed stands have the weighted mean of the monospecific stand dynamics, while $sp_{1,2}$ indicates an additional benefit through mixing. Note that this example shows positive mixing effects. However, negative mixing responses can also be quantified by the changes in terms of the level, rhythm, and final yield in the mixed-species stand

**Box 4.1 Basic Variables, Measures, and Cross Diagrams for Quantifying and Illustrating Mixing Effects on Stand Productivity**

*Basic Variables*

The standard setup for comparing the performance of mixed with monospecific stands comprises plots of both species in monoculture and in mixture. The productivities of species 1 and 2 in monoculture are named $p_1$ and $p_2$. The productivity of the mixed stand in total is $p_{1,2}$, the share of species 1 and 2 is

(continued)

**Box 4.1** (continued)

$pp_{1,(2)}$ and $pp_{(1),2}$ ($p_{1,2} = pp_{1,(2)} + pp_{(1),2}$), and their mixing portions are $m_1$ and $m_2$ ($m_1 + m_2 = 1$). The productivities per ha of species 1 and 2 in the mixed stand are called $p_{1,(2)}$ and $p_{(1),2}$ and result from upscaling by the species-specific mixing portions $p_{1,(2)} = pp_{1,(2)}/m_1$ and $p_{(1),2} = pp_{(1),2}/m_2$, respectively. Comparisons between mixed and monospecific stands with respect to other tree and stand variables follow the same nomenclature and algorithm, e.g. tree size, stand volume, and stand density yield.

*Measures of RPA for the Relationship Between Mixed and Monospecific Stand Productivity*

*R*atio of *P*roductivity based on *A*bsolute productivity (RPA) results from the observed productivity of the mixed stand $p_{1,2}$ divided by the expected stand productivity $\widehat{p}_{1,2}$

$$\mathrm{RPA}_{1,2} = p_{1,2}/\widehat{p}_{1,2}.$$

The expected productivity $\widehat{p}_{1,2}$ is derived from the productivity of both species in the neighbouring monospecific stands, $p_1$ and $p_2$, and their mixing portions $m_1$ and $m_2$($\widehat{p}_{1,2} = m_1 \times p_1 + m_2 \times p_2$). The mixing portions $m_1$ and $m_2$ are calculated, e.g. on the basis of the species' share of the stock of dry mass $M$ ($m_1 = M_1/(M_1 + M_2)$,$m_2 = M_2/(M_1 + M_2)$). For other calculations of species mixing proportions, see Box 4.2 or Keller (1995), Pretzsch (2009, pp 354–356), or Dirnberger and Sterba (2014).

A positive effect of mixing on stand growth, equivalent to overyielding, is indicated by a productivity $p_{1,2}$ greater than the productivity of the linear combination of two monocultures on the same areas ($p_{1,2} > \widehat{p}_{1,2}$). Transgressive overyielding means that the mixed stand productivity even exceeds the productivity of both monocultures ($p_{1,2} > \max(p_1, p_2)$). Analogously, underyielding means $p_{1,2} < \widehat{p}_{1,2}$ and degressive underyielding $p_{1,2} < \min(p_1, p_2)$.

The relative productivity of species 1 in mixed versus monospecific stands is

$$\mathrm{RPA}_{1,(2)} = pp_{1,(2)}/m_1/p_1 = p_{1,(2)}/p_1,$$

with the share of productivity of species 1 in the mixed stand, $pp_{1,(2)}$, mixing portion, $m_1$, and productivity of the monoculture, $p_1$. For species 2, $\mathrm{RPA}_{(1),2} = pp_{(1),2}/m_2/p_2$. Note that $pp_{1,(2)}$and $pp_{(1),2}$ are the productivity of species 1 and 2 in the mixed stand, which add up to $p_{1,2}$.

The difference $\mathrm{AP}_{1,2} = p_{1,2} - \widehat{p}_{1,2}$ reflects the absolute gain or loss caused by species mixing in terms of stand productivity. Analogously,

(continued)

**Box 4.1** (continued)

$AP_{1,(2)} = p_{1,(2)} - p_1$ and $AP_{(1),2} = p_{(1),2} - p_2$ indicate the absolute gains and losses at the species level.

*Visualisation of Mixing Effects by Cross Diagrams*

Cross diagrams on either absolute or relative scales (Box Fig. 4.1-1a and b) are widely used for graphical display of mixing effects (see Kelty 1992; Pretzsch et al. 2010; Vandermeer 1992). In the first form, such diagrams represent absolute productivity for the monospecific stand of species 1, $p_1$, scaled on the left-hand ordinate, and for species 2, $p_2$, scaled on the right-hand ordinate (Box Fig. 4.1-1a). The upper, connecting line (accentuated by diamonds) represents $p_{1,2}$ of the mixed stand with the mixing proportions scaled on the abscissa. In this example, we depict for clarity the mixing effect for just one mixed stand; for analogous cross diagrams with plots of differing mixing portions, see Figs. 4.12, 4.13, 4.14. In addition, the share of both species' productivity $pp_{1,(2)}$ and $pp_{(1),2}$ in the mixed stand is shown (triangles and circles, respectively). Their sum $p_{1,2} = pp_{1,(2)} + pp_{(1),2}$ is represented by the upper diamond symbol, which makes the gable of the roof-shaped upper line. $m_1$, $m_2$ are the species' mixing portions scaled on the abscissa (Box 4.2). Note that the mixing portion of species 2 increases from left to right and that $m_1 + m_2 = 1$; in our example, $m_1 = 0.4$, $m_2 = 0.6$.

(continued)

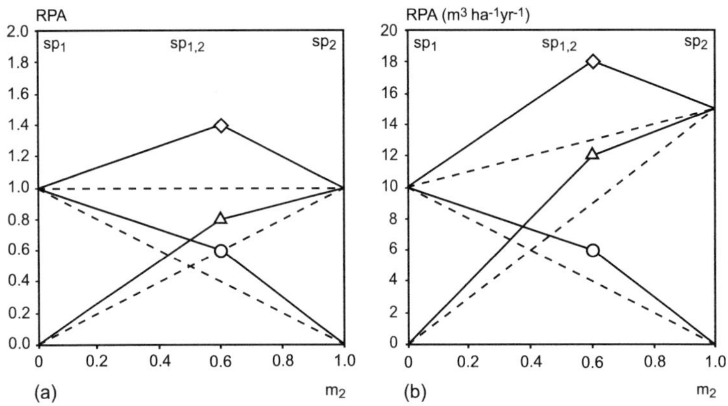


**Box Fig. 4.1-1** Cross diagrams for display of mixing effects on productivity in two-species mixtures by comparing (**a**) their relative productivity and (**b**) their absolute productivity. The *left* and *right* ordinates represent the productivity of the species in the monocultures. The range in between represents the relative productivity in the mixture depending on the mixing portion. Broken lines represent the productivity expected for neutral mixing effects on the stand level (horizontal 1.0 line) and the species level (*decreasing* or *increasing lines*). The upper connecting lines (accentuated by *diamonds*) represent the total production of the mixed stand, while the lower connecting lines (*triangles* and *circles*) show the productivity on the species level

**Box 4.1** (continued)

Assuming neutral mixing effects, the productivity of the mixed stand would lie on the straight line connecting the two monospecific stand records. Deviations in the observed values from the straight reference line reveal gains or losses at the whole-stand level. Concave (seen from below) upper lines indicate positive mixing effects, while convex lines indicate growth losses due to species mixing. In the same way, the comparison between the observed productivity of the individual species $pp_{1,(2)}$ and $pp_{(1),2}$ with the reference lines (forming a cross) indicates whether a species benefits or suffers from mixing. Again, concavity indicates a benefit, and convexity indicates a loss due to the mixing. Note that the ratio between observation (solid line) and prediction (broken line) reflects the over- or underyielding in terms of volume or biomass growth at the stand level ($RPA_{1,2} = p_{1,2}/\widehat{p}_{1,2}$) or at the species level ($RPA_{1,(2)} = p_{1,(2)}/p_1$ and $RPA_{(1),2} = p_{(1),2}/p_2$).

The vertical distance between the observed values and the reference represent the absolute gains and losses by mixing at the total stand level ($AP_{1,2} = p_{1,2} - \widehat{p}_{1,2}$) and the species level ($AP_{1,(2)} = p_{1,(2)} - p_1$ and $AP_{(1),2} = p_{(1),2} - p_2$).

The cross diagrams based on relative instead of absolute productivity enable an integrated view on mixing effects of stands with different productivity levels in one graph (Box Fig. 4.1-1a). Later, they will be used to examine RPA values of mixed-species stands located along an ecological gradient. As shown in Box Fig. 4.1-1a, they set the productivity of the neighbouring monoculture to 1.0 and show how the mixed stand and the species perform in relation to this reference. The broken straight lines represent the expected productivity of the mixed stand in total (horizontal line 1.0 line) as well as the share of species 1 and 2 on stand productivity (descending line connecting (0|1) with (1|0) and the rising line connecting (0|0) with (1|1), respectively) in relation to the mixing portion, which is scaled on the abscissa. The three solid lines represent the observed relative productivities: The upper curve reflects $RPA_{1,2} = p_{1,2}/\widehat{p}_{1,2}$ for the whole stand, and the two lower lines represent the species-specific ratios $pp_{1,(2)}/p_1$ and $pp_{(1),2}/p_2$, respectively. As long as not specially mentioned, all analyses of mixing effects in this chapter are based on RPA, i.e. *R*atio of *P*roductivity based on *A*bsolute productivity.

*Ratio RPR for Comparing Mixed with Monospecific Stand Productivity (Equivalent to the Land Equivalent Ratio (LER) Used in Agronomy)*

$RPR_{1,2} = pp_{1,(2)}/p_1 + pp_{(1),2}/p_2$ (*R*atio of *P*roductivity based on *R*elative productivity) also sets the observed productivity in the mixed stand in relation to the productivity of neighbouring monocultures. The RPR does not explicitly consider the mixing portions and is based on the two relative productivities $pp_{1,(2)}/p_1$ and $pp_{(1),2}/p_2$. RPR quantifies the required land area of the

(continued)

**Box 4.1** (continued)

monospecific stand in relation to the mixed stand for producing the same yield. We use Vandermeer's (1992, pp 20–21) example to explain the concept behind RPR.

Suppose 1 ha of species 1 in monoculture 1 produces $p_1 = 20\ \mathrm{t\ ha^{-1}}$, and 1 ha of species 2 produces $p_1 = 75\ \mathrm{t\ ha^{-1}}$, we further assume that the mixed stand produces $pp_{1,(2)} = 10\ \mathrm{t\ ha^{-1}}$ of species 1 and $pp_{(1),2} = 50\ \mathrm{t\ ha^{-1}}$ of species 2 (total productivity $60\ \mathrm{t\ ha^{-1}}$). We then calculate that to produce 10 t of species 1 as monoculture, an area of 0.5 ha would be needed ($pp_{1,(2)}/p_1 = 10/20 = 0.50$), whereas to produce 50 t of species 2 in monoculture, an area of 0.67 ha would be required ($pp_{(1),2}/p_2 = 50/75 = 0.67$), and in total 1.17 ha would be required ($RPR_{1,2} = pp_{1,(2)}/p_1 + pp_{(1),2}/p_2 = 0.50 + 0.67 = 1.17$). Thus, the total area of species 1 and 2 needed to produce the equivalent of a single hectare of species 1 and 2 in combination is 1.17, which is referred to as RPR in the following. $RPR_{1,2}$ is equal to the relative yield total (RYT) and Land Equivalent Ratio (LER) which are common in herbaceous plant biology and agronomy where mixing proportions in terms of required growing space are difficult to measure and neglected when measuring mixing effects [see Harper (1977); Vandermeer (1992)].

*Differences Between RPA and RPR*

By setting the productivity per unit area of the mixed stand in relation to the productivity per unit area in the monoculture, RPA indicates the mixing effect at the stand and species levels with respect to the efficiency of area and resource use. As it considers the mixing proportions of the species, RPA represents general scientific stand and species level information of mixing responses suitable for further analyses of, and dependent on, site conditions, stand density, or mixing pattern.

As RPR does not consider the species-specific space requirement in the respective mixed stand, its indication is case related, not general and rather ambiguous as shown by analysing the following model stands 1–8. In order to reveal the often neglected or confusing differences between RPA and RPR, we use model stands in Box Table 4.1-1 where the species differ in productivity level and mixing proportions. Triplets 1 and 2 (Box Table 4.1-1 upper two lines) represent two special cases, where $p_1$ and $p_2$ are equal or similar, so that $RPA_{1,2}$ and $RPR_{1,2}$ are also similar. In contrast, triplets 3 and 4 represent positive and negative mixing responses, respectively, in terms of RPA, but RPR is not sufficiently sensitive to reflect those differences. The same applies for triplets 5 and 6, as well as triplets 7 and 8.

Box Figure 4.1-2a–i reflects this important deficit of RPR using cross diagrams. According to RPA, triplet 3 indicates overyielding by 29% and triplet 4 underyielding by 5% (Box Fig. 4.1-2a and b); however, RPR indicates overyielding of 10% for both stands (Box Fig. 4.1-2c). While

(continued)

**Box Table 4.1-1** Eight model triplets for explaining the ambiguity of RPR in comparison with RPA values (see also Box Fig. 4.1-2)

| Model triplet | $p_1$ (t ha$^{-1}$ year$^{-1}$) | $p_2$ (t ha$^{-1}$ year$^{-1}$) | $m_1$ | $pp_{1,(2)}$ (t ha$^{-1}$ year$^{-1}$) | $m_2$ | $pp_{(1),2}$ (t ha$^{-1}$ year$^{-1}$) | $RPA_{1,2}$ | $RPR_{1,2}$ |
|---|---|---|---|---|---|---|---|---|
| 1 | 10.0 | 10.0 | 0.25 | 5.0 | 0.75 | 6.0 | 1.100 | 1.100 |
| 2 | 7.0 | 8.0 | 0.50 | 6.0 | 0.50 | 7.0 | 1.733 | 1.732 |
| 3 | 15.0 | 10.0 | 0.25 | 10.5 | 0.75 | 4.0 | 1.289 | 1.100 |
| 4 | 10.0 | 15.0 | 0.25 | 7.0 | 0.75 | 6.0 | 0.945 | 1.100 |
| 5 | 15.0 | 7.5 | 0.75 | 6.0 | 0.25 | 4.5 | 0.800 | 1.000 |
| 6 | 7.5 | 15.0 | 0.75 | 3.0 | 0.25 | 9.0 | 1.280 | 1.000 |
| 7 | 15.0 | 10.0 | 0.60 | 4.5 | 0.40 | 6.0 | 0.808 | 0.900 |
| 8 | 5.0 | 15.0 | 0.60 | 1.5 | 0.40 | 9.0 | 1.167 | 0.900 |

When the productivity level or mixing portions of the two combined species are similar (triplets 1 and 2), their RPA and RPR values are comparable. The larger the difference in the species-specific productivity level and mixing portion, the more misleading the RPR values in comparison to RPA. The mixed stand of triplet 3 considerably overyields the monocultures ($RPA_{1,2} = 1.289$) and the triplet 4 represents underyielding ($RPA_{1,2} = 0.945$), but for both stands result in $RPR = 1.100$. The misjudgement of RPR in relation to RPA is even worse for the triplets 5 and 6 or triplets 7 and 8

**Box 4.1** (continued)

triplets 5 and 6 represent considerable underyielding and overyielding, respectively, in terms of RPA (Box Fig. 4.1-2d and e), RPR incorrectly indicates neutral effects for both stands (Box Fig. 4.1-2f). Finally, triplets 7 and 8 show under- and overyielding, respectively (Box Fig. 4.1-2g and h); however, RPR indicates underyielding for both (Box Fig. 4.1-2i).

Distinctive positive deviations of RPA from RPR values occur when one species is much more productive than the other (e.g. $p_1 >> p_2$) but only represented with a small portion (e.g. $m_1 << m_2$) in the mixed stand (e.g. Box Table 4.1-1, triplets 3, 6, and 8). Negative deviations of RPA from RPR values emerge when one species is much more productive than the other (e.g. $p_1 >> p_2$) and also much higher in its mixing portion (e.g. $m_1 >> m_2$) (e.g. Box Table 4.1-1, triplets 4, 5, and 7).

Under average site conditions, Central European species are rather similar in their stem mass productivity in monocultures. In addition, in practice as well as on long-term experimental plots, the mixing portions are mostly rather balanced. In such stands, the differences between mixing analysis based on RPA differ only slightly from RPR. Analysis of 213 mixed stands of Norway spruce and European beech yielded on average RPA = 1.18 and RPR = 1.22, whereas for 217 stands of sessile oak and European beech, they yielded RPA = 1.17 and RPR = 1.20; for silver fir and Norway spruce, RPA = 1.14 and RPR = 1.15; and for Scots pine and European beech, RPA = 1.78 and

(continued)

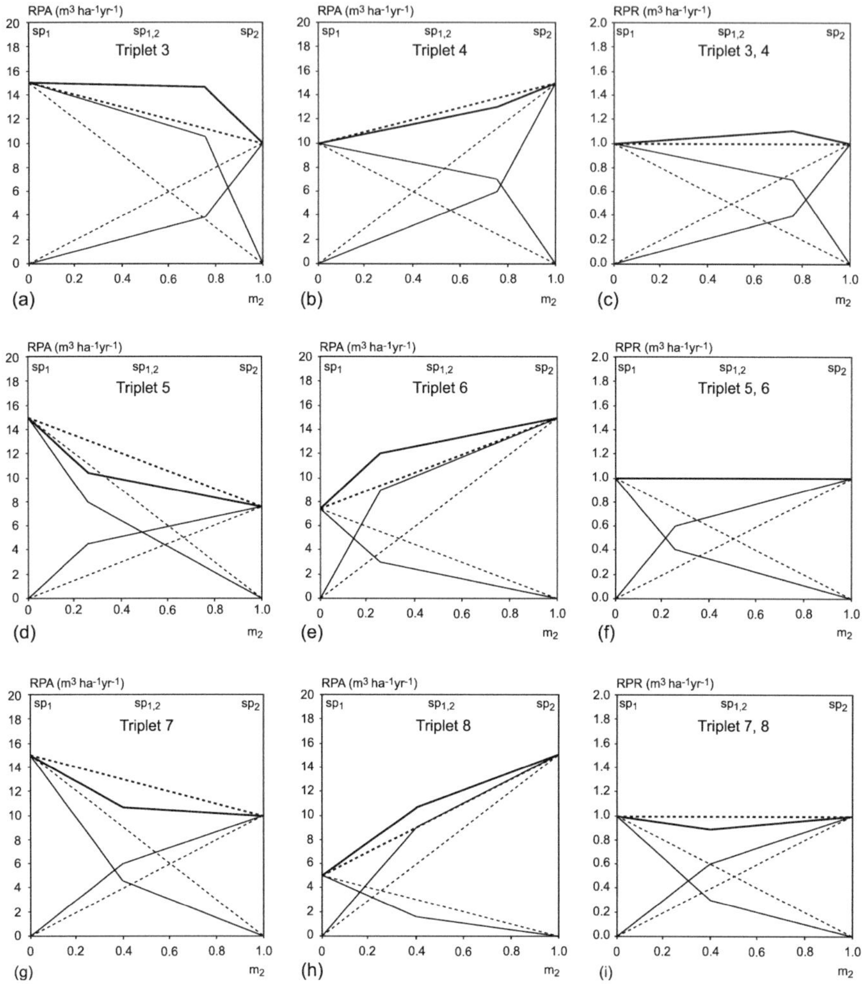


**Box Fig. 4.1-2** (**a**–**i**) Ambiguity of RPR values shown by cross diagrams for model triplets 3–8 introduced in Box Table 4.1-1. Different RPA findings (*left* and *central column*) can result in equal RPR values (*right column*)

**Box 4.1** (continued)

RPR = 1.80. In the following, we mostly apply the RPA values, which are sensitive to mixing portions, more conservative, and, on average, 1–3% lower than the RPR values, which are commonly used in agriculture (Vandermeer 1992).

#### 4.2.1.2 Evaluation of the Mixing Effects on Stand Growth and Total Yield

Findings regarding mixing effects at the stand level in this chapter and in general are mostly based on long-term experiments, on permanent forest inventories, or on a clever combination of both (Fig. 4.6).

Long-term mixing experiments with substitutive or additive designs (Chap. 2), especially when unmanaged reference variants are included, provide an excellent data source for comparing mixed versus monospecific stands under ceteris paribus conditions. However, due to their longevity, such experiments are costly and are therefore infrequently established or surveyed in temperate forest stands, although they may be more frequently conducted in shorter rotation planted forests in subtropical and tropical latitudes. Permanent inventories at the national and enterprise level are a potentially useful data source (Nagel et al. 2012) but have rarely been exploited for mixed stand research. When repeated, they provide growth and yield variables for a broad range of site and management conditions, mixing proportions, and patterns. However, they often lack fully stocked unmanaged stands as references, and mixing effects can be analysed statistically by correlation rather than causally. Many permanent inventories only contain rather short time series, but in combination with long-term experimental plots, they can provide new insights into mixed stand productivity (Fig. 4.6).

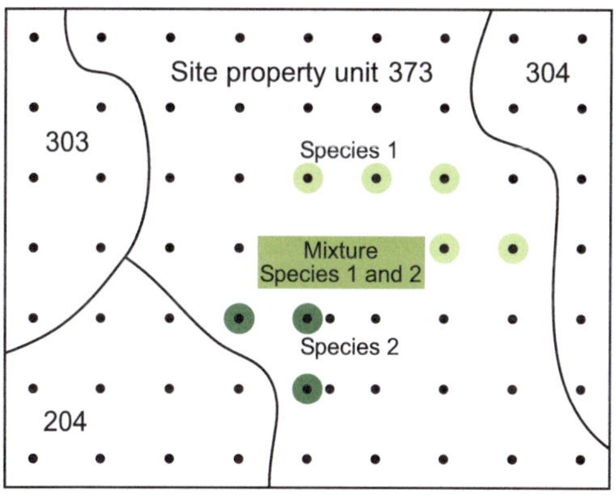


**Fig. 4.6** A schematic representation of the supplementation of long-term experimental data by forest enterprise inventory plot data. This figure shows two long-term experimental plots with a two-species mixture (*rectangles*) that lacks monospecific reference plots. The lacking long-term reference plots are compensated for by inventory plots of a similar age, density, and silvicultural treatment that are selected from the raster-based inventory plots (grid marked by points) with the same site classification unit 342, while other units are excluded. In this example, five inventory plots are pooled to represent the monospecific stand growth of species 1, and three plots for the monospecific stand growth of species 2. Permanent forest inventory plots which have a grid of 100 or 200 m are most suitable for this approach

**Box 4.2 Species-Specific Differences in the Intra-Annual Course of Size Growth**

*Temporal Complementarity by Species-Specific Intra-Annual Courses of Resource Capture and Growth*

Species can vary in their intra-annual course of growth and when mixed can use resources in a complementary way over time. Box Figure 4.2-1 shows the mean and standard error of stem diameter growth for 24 Norway spruce trees (64 ± 2 years in 2015) and 24 European beech trees (84 ± 4 years in 2015) on the mixed-species plots of the Kranzberg Forest experiment (degree of longitude 11°39′42″E, latitude 48°25′12″N, 490 m a. s. l) in the years 2014 and 2015 measured using electronic dendrometers (Pretzsch et al. 2016a). To show the species-specific behaviour, we chose the intra-annual courses for the year 2014 (Box Fig. 4.2-1a) with average weather conditions (mean temperature and sum of precipitation in the growing season from April to September of 15.2 °C and 423 mm, respectively) and for the drought year 2015 (Box Fig. 4.2-1b) (mean temperature and sum of precipitation in the growing season from April to September of 16.6 °C and 328 mm, respectively). The mean diameters of Norway spruce and European beech were 34.3 cm and 28.9 cm, respectively, and the mean heights were 29.0 and 26.1 m, respectively, indicating that Norway spruce was ahead of beech in terms of size growth.

This is substantiated by the superior stem diameter growth of Norway spruce in years with average weather conditions (Box Fig. 4.2-1a). Norway spruce started early, accelerated at a higher rate, and ended up with a larger diameter than European beech. By starting earlier, Norway spruce benefited from the mostly water-saturated soil in spring 2014 and also in 2015;

(continued)

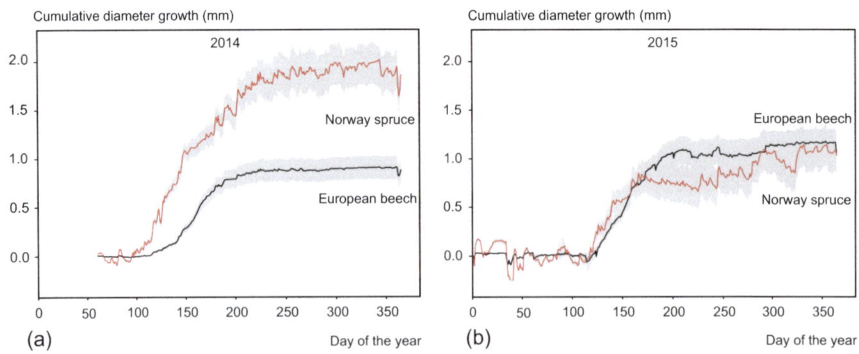


**Box Fig. 4.2-1** Asynchrony of the intra-annual courses of mean stem diameter growth (at height 1.30 m) of Norway spruce and European beech of the Kranzberg Forest experiment (Pretzsch et al. unpublished) in (**a**) the year 2014 with average weather conditions and (**b**) the drought year 2015 (see Sect. 12.1)

**Box 4.2** (continued)
however, the drought in 2015 resulted in significant growth reduction in June 2015 (Box Fig. 4.2-1b). European beech started later and slower in both years. While Norway spruce shows a strong reduction in growth due to drought, European beech shows no reduction at all. When growing in mixture, stand growth is higher in normal years due to the faster growth of Norway spruce. In contrast, in drought years, European beech can stabilise stand growth because it partly compensates the growth losses of Norway spruce (Pretzsch 2005b; del Río et al. 2014).

In addition to their species-specific courses in monocultures, the intra-annual courses of growth can be further modulated through intraspecific competition in such a way that the temporal complementarity even increases when growing in mixed compared to pure stands (Vospernik and Nothdurft 2016).

In the following, we mainly refer to species mixtures represented by a triplet design as shown in Fig. 4.7 for Norway spruce and European beech. This presents an initially substitutive design, as differences in the species' tree size development and maximum stand densities altered the originally equal density and spacing. Experiments like ZWI 111 (Fig. 4.7) are particularly informative when they include unmanaged or at least only slightly thinned and fully stocked plots, because they deliver information about mixing effects on stand density. Box 4.1 introduces the basic variables in detail and cross diagrams for quantifying and illustrating mixing effects on stand productivity for triplet designs.

It is often of interest to know whether the mixed-species stand with species proportions $m_1$ and $m_2$ produces more than monospecific stands with area portions $m_1$ and $m_2$ ($p_{1,2} > p_1 \times m_1 + p_2 \times m_2$) or whether the mixed stand produces even more than the monospecific stand that has the maximum productivity ($p_{1,2} > \max(p_1, p_2)$).

A mixed stand productivity equal to the weighted mean productivity of the monocultures would indicate a simple additive effect. A productivity above or below (over- or underyielding, respectively) the weighted mean would mean a multiplicative mixing effect, which is of special scientific interest because it indicates beneficial or detrimental interspecific interactions. A mixed stand productivity even greater than the productivity of the most productive monoculture (transgressive overyielding) is of special interest for decision making in forest practice. The next section considers both over- or underyielding and transgressive overyielding or degressive underyielding.

The comparison of mixed versus monospecific stands can refer to a given growth period (e.g. age 20), to the total yield at a given stand age (e.g. total yield until age 50), or to the total yield in the mature stand phase (e.g. rotation age). Such

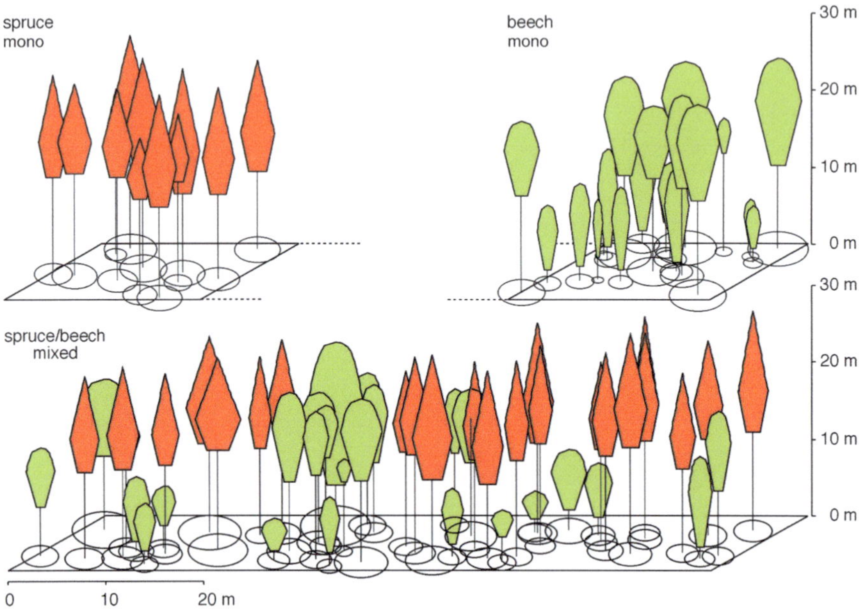


**Fig. 4.7** Sections of the Norway spruce and European beech mixing experiment Zwiesel 111, in the Bavarian Forest in southern Germany. The experiment ZWI 111 encompasses monospecific stands of spruce and beech as well as mixed plots with different mixing proportions of spruce and beech. In such triplets, the monospecific stands provide the productivities $p_1$ and $p_2$ as references, and the mixed plots deliver the total productivity of the mixed stand, $p_{1,2}$, the mixing portions, $m_{sp}$, $m_{be}$, and the productivity of both species in the mixture, $p_{1,(2)}$, $p_{(1),2}$, for detailed analysis (see Box 4.1)

comparisons are examined below for a two-species substitutive mixing experiment, but apply analogously for $n$-species mixtures.

In Fig. 4.8, a and b show the unimodal course of the current annual stand growth, $p_1, p_2$, and the S-shaped total yield curve (integral of $p_1$ resp. $p_2$ over time), $Y_1, Y_2$, for species 1 and 2 growing in monospecific stands. In Fig. 4.8c and d show the growth, $p_{1,2}$, and total yield, $Y_{1,2}$, of a neighbouring mixed stand compared with the expected productivity ($\widehat{p}_{1,2} = p_1 \times m_1 + p_2 \times m_2$) and yield ($\widehat{Y}_{1,2} = Y_1 \times m_1 + Y_2 \times m_2$) as based on the monocultures. In order to keep the graph simple, both species make up half of the mixed stand ($m_1 = m_2 = 0.5$). For calculation of the mixing proportions, see Box 4.3.

Most studies compare the productivity of a mixed stand, $p_{1,2}$, at a defined age or in a defined survey period with the expected performance as based on the monospecific stands $\widehat{p}_{1,2} = p_1 \times m_1 + p_2 \times m_2$. Depending on the chosen age ($t_1$, $t_2$, and $t_3$), the comparison can reflect underyielding ($t_1$, $\mathrm{RPA}_{1,2} = p_{1,2}/\widehat{p}_{1,2} = 0.91$), equality ($t_2$, $\mathrm{RPA}_{1,2} = 1.0$), and overyielding ($t_3$, $\mathrm{RPA}_{1,2} = 1.26$) ($t_4$, $\mathrm{RPA}_{1,2} = 1.50$) of

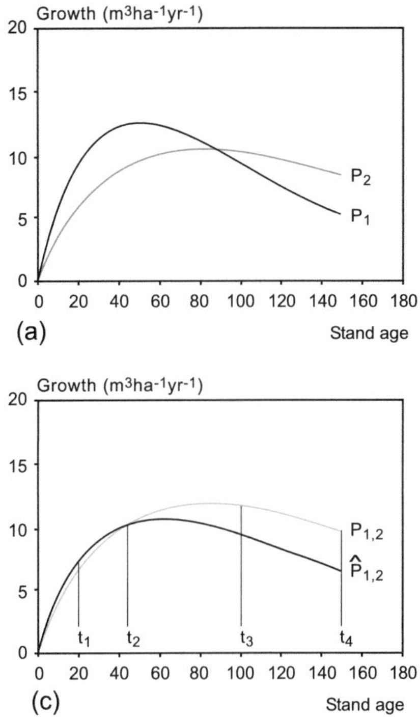


**Fig. 4.8** (**a–d**) Calculation of mixing effects on growth and yield by comparing mixed-species stands with neighbouring monocultures at different stand ages $t_1 \ldots t_4$. (**a** and **b**) The growth, $p_1$ and $p_2$, of monospecific stands and their yield, $Y_1$ and $Y_2$, are used as references. (**c** and **d**) Based on the monospecific stand performance and their proportion in the mixed stands, the expected growth and yield, $\widehat{p}_{1,2}$, $\widehat{Y}_{1,2}$, are calculated. The observed growth and yield, $p_{1,2}$, $Y_{1,2}$, are compared with the expected values. The comparison shows underyielding of mixed stands at time $t_1$, equality for $t_2$, and overyielding for $t_3$ and $t_4$

mixtures in relation to monospecific stand productivity (Fig. 4.8c). Analogous comparisons between observed and expected yield (Fig. 4.8d) indicate small differences up to the intermediate ($t_1$, $t_2$) ages and overyielding of $\text{RYA}_{1,2} = Y_{1,2}/\widehat{Y}_{1,2} = 1.05 - 1.15$ at the advanced ages ($t_3$ and $t_4$).

**Box 4.3 Mixing Proportions**

To identify any productivity benefit or loss due to species interactions, mixture proportions of species 1 and 2 (e.g. $m_{sp}$ and $m_{be}$) in the mixed stands need to be calculated and applied in the formulas for calculating the cross-species diagrams for illustrating mixing effects. However, the different

(continued)

**Box 4.3** (continued)
conventions for calculating mixture proportions can produce contrasting results (Dirnberger et al. 2016).

Simple approaches for deriving the mixture proportions use tree number $N$ (trees ha$^{-1}$), stand basal area BA (m$^2$ ha$^{-1}$) or standing stem volume $V$ (m$^3$ ha$^{-1}$) for the total stand and for each species. Suppose we analyse a mixed stand of Norway spruce and European beech, then the mixture proportion of Norway spruce based on tree number is determined from $m_{sp}(N) = N_{sp}/(N_{sp} + N_{be})$. The portions $m_{sp}$(BA) and $m_{sp}(V)$ are calculated analogously.

These methods do not adequately take into account species-specific mean tree size, growing space requirements, and tree packing densities. Norway spruce may comprise numerous small trees $N_{sp}$ and European beech only a few tall trees $N_{be}$. In this case, the mixture proportion $m_{sp}(N)$ would overestimate Norway spruce's share of the stand area and resources. Mixture proportions based on BA or $V$ take into account species-specific sizes. However, depending on the species-specific crown allometry and specific wood density $R$ (Pretzsch (2009), Table 2.8), very different leaf area, living tree biomass, and growing space requirements may underlie a given basal area (BA) or standing volume ($V$). Thus, it is important to consider species-specific density and growing space requirements when calculating the mixture proportions $m_1 \ldots m_n$.

To consider the species-specific growing space requirements, forest practice adjusts observed basal area proportions in mixed stands ($\text{BA}_{\text{obs sp}}$, $\text{BA}_{\text{obs be}}$) by the basal area predicted in the yield tables for monospecific stands ($\text{BA}_{\text{Y sp}}$, $\text{BA}_{\text{Y be}}$) on the given site. The mixture proportion of Norway spruce $m_{sp}(\text{BA}_\text{Y})$, e.g., is

$$m_{sp}(\text{BA}_\text{Y}) = \frac{\text{BA}_{\text{obs sp}}/\text{BA}_{\text{Y sp}}}{\text{BA}_{\text{obs sp}}/\text{BA}_{\text{Y sp}} + \text{BA}_{\text{obs be}} + \text{BA}_{\text{Y be}}}.$$

In other words, the relationship between the basal area of both species in the monoculture, which expresses the species-specific packing density, is applied to adjust observed basal areas of Norway spruce and European beech to their share of stand space and resources. As, in our case, neighbouring plots represent the species basal area in monocultures; we can use the measured basal areas on the nearby reference plots to adjust the observed basal area proportions in mixed stands giving,

$$m_{sp}(\text{BA}_\text{ref}) = \frac{\text{BA}_{\text{obs sp}}/\text{BA}_{\text{ref sp}}}{\text{BA}_{\text{obs sp}}/\text{BA}_{\text{ref sp}} + \text{BA}_{\text{obs be}} + \text{BA}_{\text{ref be}}}.$$

(continued)

**Box 4.3** (continued)

If total crown projection area $CPA_T$ ($m^2$ $ha^{-1}$) is available for both of the species occurring ($CPA_{T\ sp}$ and $CPA_{T\ be}$, respectively), Assmann (1961) and Kennel (1965) favour its application for calculating the mixture proportions: e.g. $m_{sp}(CPA_T) = CPA_{T\ sp}/(CPA_{T\ sp} + CPA_{T\ be})$. The same method can be based only on the projection area of the dominant crowns and yields, $m_{sp}(CPA_D)$ and $m_{be}(CPA_D)$, respectively. However, again, a given crown projection area may reflect very different leaf areas and living stem biomass and, hence, very different resource consumption.

The following three methods aim to provide an even better quantification of the mixing proportions by considering the species-specific biomass. Assmann (1961) and many others apply specific wood density for the species present (e.g. $R_{sp} = 0.39$ g $cm^{-3}$ and $R_{be} = 0.56$ g $cm^{-3}$ for Norway spruce and European beech, respectively) to calculate the mixture proportions on the basis of basal area proportions:

$$m_{sp}(BA_R) = \frac{BA_{obs\ sp}\, R_{sp}}{BA_{obs\ sp}\, R_{sp} + BA_{obs\ be}\, R_{be}}.$$

Keller (1995) criticised the adjustment with the untransformed specific wood density $R$. He argued that $R$ represents a density in relation to space but that the basal area should be corrected by relating density to area, i.e. $\sqrt[3]{R^2}$. Supposing a stem is a stereometric body with the volume $v$, basal area ba, and diameter $d$, then simple geometrical scaling yields allometry $v \propto d^3$ and ba $\propto d^2$, and thus ba $\propto v^{2/3}$. As volume is proportional to weight ($v \propto w$), ba $\propto w^{2/3}$. This means the relationship between two volumes corresponds to a relationship between the respective basal areas raised to the power of 2/3. Applying the ratio of the specific wood densities of Norway spruce to European beech, 0.39:0.56, the ratio of the basal areas becomes $0.39^{2/3}:0.56^{2/3} = 0.534:0.679 = 1:1.272$. Given this relationship, the basal area of Norway spruce should be weighted by $\sqrt[3]{R_{sp}{}^2}$ and that of European beech by $\sqrt[3]{R_{be}{}^2}$ in to the following equation,

$$m_{sp}(BA_R 2/3) = \frac{BA_{obs\ sp}\, \sqrt[3]{R_{sp}}}{BA_{obs\ sp}\, \sqrt[3]{R_{sp}} + BA_{obs\ be}\, \sqrt[3]{R_{be}}}.$$

A similar approach applies the aboveground biomass of the species present, calculated with biomass functions of the type $w = f$(tree diameter and/or tree height), to obtain $m_{sp}(W) = W_{sp}/(W_{sp} + W_{be})$.

In the following sections, we mostly apply $m\,(W)$. This approach assumes that the share of aboveground biomass of each species best expresses their competitive access to above and belowground resources, and thus their

(continued)

**Box 4.3** (continued)
proportion in the mixture. As the productivity and mixture proportions always relate to one hectare, a mixture proportion of $m = 0.8$ would assume that 0.8 ha or 80% of the growing space and resource are sequestrated by the respective species.

Another method, based on the stand density index (Box 4.5), takes into account both aspects of density, mean size, and tree number, in a standardised size development phase of $d_q = 25$ cm.

Based on the total stand density standardised to species 1 ( $\underline{SDI}_{1,2} = SDI_{1,(2)} + SDI_{(1),2} \times e_{2\Rightarrow 1}$ ) and the shares of species 1 and 2 ( $SDI_{1,(2)} / \underline{SDI}_{1,2}$, respectively, $\left(SDI_{(1),2} \times e_{2\Rightarrow 1}\right)/\underline{SDI}_{1,2}$ ), the mixing proportions $m_1$ and $m_2$ of species 1 and 2 are

$m_1 = SDI_{1,(2)}/(SDI_{1,(2)} + SDI_{(1),2} \times SDIMAX_1/SDIMAX_2)$

$m_2 = (SDI_{(1),2} \times SDIMAX_1/SDIMAX_2)/(SDI_{1,(2)} + SDI_{(1),2} \times SDIMAX_1/SDIMAX_2)$.

The results of such comparisons can be visualised using the cross diagrams shown in Fig. 4.9. In each pane, the $x$-axes start (far left) at the monoculture of species 1 before progressing through the mixtures with increasing proportions of species 2, before ending (far right) with the monocultures of species 2. This shows how the productivity of the mixtures depends on the mixing portion. Broken lines represent the productivity expected for additive mixing effects on the stand level (upper horizontal line) and the species level (lower crossing lines). The upper solid lines represent the observed total production of the mixed stand, while the lower solid lines show the observed productivity at the species level. The cross diagrams in Fig. 4.9a–d compare mixed with monospecific stands concerning the absolute productivity, while Fig. 4.9e–h uses the stand productivity of both monocultures as a reference (1.0 line) and reflects the mixed productivity in relation to this reference. The cross diagrams in Fig. 4.9 indicate underyielding for the stand as a whole and the two species at ages $t_1$ and $t_2$, and increasing levels of overyielding of mixed in comparison to monospecific stands created by both species at ages $t_3$ and $t_4$. For a general explanation of cross diagrams, see Box 4.1.

Finally, we return to the question of whether the mixed stand performs even better than the best growing of the two monocultures. Figure 4.10 shows that the mixed stand productivity, $p_{1,2}$, exceeds the productivity of the monospecific stand of species 2 from the beginning. However, it does not exceed the productivity of the monospecific stand of species 1 (transgressive overyielding) until age 60. After time $t_5 = 60$, transgressive overyielding starts. Total yield of the mixed stand, $Y_{1,2}$, exceeds both monocultures from age 120.

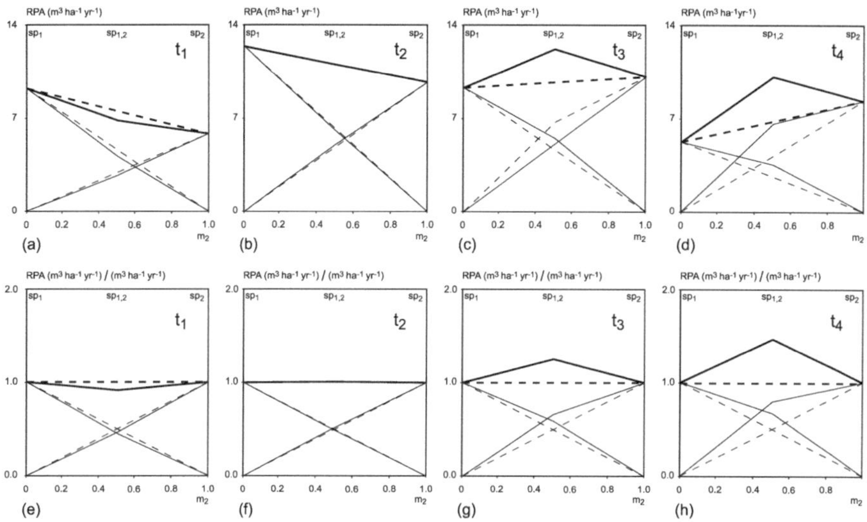


**Fig. 4.9** (**a**–**h**) Cross diagrams reflecting the mixing effects in terms of absolute productivity (**a**–**d**) and relative productivity (**e**–**h**) for the model stands in Fig. 4.8 at times $t_1 \ldots t_4$. The mixing effects can vary considerably during stand development depending on the growth rhythm of the species in mixed and monospecific stands

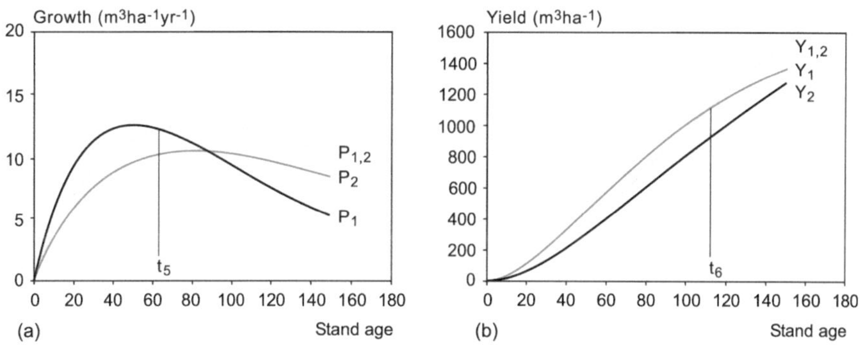


**Fig. 4.10** Schematic representation of transgressive overyielding of mixed versus monospecific stands shown for the model stands in Figs. 4.8 and 4.9. (**a**) Starting at time $t_5 = 60$ years, the growth of the mixed-species stand, $p_{1,2}$, exceeds the productivity of both monocultures, $p_1$ and $p_2$. (**b**) Transgressive overyielding of $Y_{1,2}$ with respect to $Y_1$ and $Y_2$ does not start before time $t_6 = 110$ years

#### 4.2.1.3 Elimination of the Size Effect Through Allometric Adjustment

Usually, neighbouring monocultures are used as references for the quantification of species' mixing effects on tree and stand growth. This comparison can be rather misleading when the size development of one species is quicker than that of the other. Most tree and stand characteristics such as stem volume growth or stand density change continuously with tree size. So, a difference between tree and stand characteristics in mixed and monospecific stands might come either from the fact that one of the stands is simply ahead in size growth (size effect which might result from a mixing effect in the past), from a superior performance even when the trees in the stands are a similar size (current mixing effect), or from a combination of size and current mixing effects.

If trees in mixed and monospecific stands perform differently, but only because of differences in size, this may be called an indirect mixing effect or size effect. Behind the size difference is an historical acceleration or slowdown in size growth which is of interest on its own, but which should not be conflated with a current mixing effect or direct mixing effect. Box 4.4 introduces how the size effect can be eliminated in order to reveal a direct mixing effect.

**Box 4.4 Elimination of the Size Effect Through Allometric Adjustment**

*Separation of Indirect from Direct Mixing Effects*

Usually, the performance of neighbouring monocultures serves as a reference for the detection of positive or negative mixing effects on tree growth or stand density. However, this comparison can be misleading when one species has a quicker size development than the other, as shown in Box Fig. 4.4-1a. In this example, mixed mean tree volume $\bar{v}_{1,(2)}$ of species 1 is inferior to the mean volume of this species $\bar{v}_1$ in the monoculture at time $t_1$, equal at time $t_2$, and superior at time $t_3$. Many tree and stand variables, e.g. mean tree growth, $\bar{p}$, and tree number, $N$, change continuously with stand development depending mainly on mean size, e.g. represented by mean tree volume (Box Fig. 4.4-1b and c). So, a difference in $\bar{p}$ or $N$ between mixed and monospecific stands might either come mainly from the fact that one of the stands is ahead in size growth as in this example where $\bar{v}_{1,(2)} > \bar{v}_1$, or from a superiority even when the trees in the stands have similar sizes ($\bar{v}_{1,(2)} = \bar{v}_1$).

If trees in mixed and monospecific stands perform differently, but simply because of their different size, we call this an indirect mixing effect. Behind the size difference is a historical acceleration or slowdown in size growth which is of interest on its own, but which should not be conflated with a direct mixing effect. If trees of the same size differ in their performance in mixed versus monospecific stands, we consider it a direct mixing effect.

*Allometric Adjustment Through Elimination of Size Effects*

In order to eliminate indirect mixing effects, which are simply caused by different sizes at the beginning of a period, from those of direct effects, which come from species interactions (competitive reduction, facilitation) within the period, the allometric adjustment, illustrated on Box Fig. 4.4-2, can be applied. As an indicator for the mean tree size and development phase, we use the volume of the mean tree $\bar{v}$ of the mixed stand $\bar{v}_{1,(2)}$ and the monoculture $\bar{v}_1$. If both are equal, adjustment is not necessary because both stands are in the

(continued)

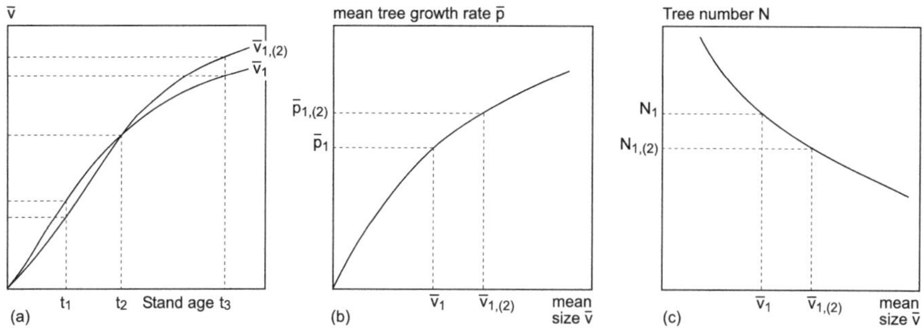


**Box Fig. 4.4-1** (**a–c**) Asynchronous development of a species' mean tree volume in the mixed-species stand, $\bar{v}_{1,(2)}$, and the neighbouring monoculture, $\bar{v}_1$, (**a**) and the relationship between mean tree volume, $\bar{v}$, and mean productivity, $\bar{p}$, and stocking density, $N$, (**b** and **c**)

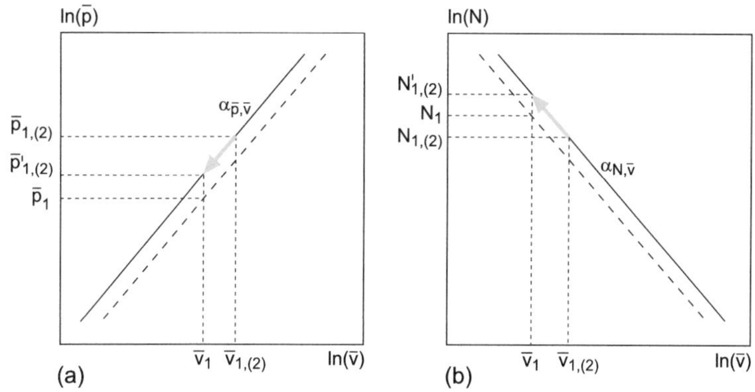


**Box Fig. 4.4-2** Elimination of allometric drift when comparing mixed versus monospecific stand attributes. Development of (**a**) volume growth of mean tree and (**b**) tree number per unit area in relation to mean tree volume, $\bar{v}$. The *grey arrows* display the elimination of size effects on (**a**) mean periodic tree growth through allometric scaling with exponent $\alpha_{\bar{p},\bar{v}}$ and (**b**) stand density by exponents $\alpha_{N,\bar{v}}$. Further explanation and formulas in the text

**Box 4.4** (continued)

same phase. If they differ, this indicates a phase difference. This means that apart from any mixing effects on productivity, the stands differ in productivity because they are in different development phases. As the size-related change in productivity of forest stands follows general allometric relationships, the respective allometric exponents, e.g. $\alpha_{\bar{p},\bar{v}}$ in $\bar{p} \propto \bar{v}^{\alpha_{\bar{p},\bar{v}}}$, are used to extrapolate from the observed productivity $\bar{p}_{1,(2)}$ in the mixed stand with $\bar{v}_{1,(2)}$ to the expected productivity of this stand, $\bar{p}'_{1,(2)}$, supposing it has the mean dimension of the monoculture, $\bar{v}_1$, ($\bar{p}'_{1,(2)} = \bar{p}_{1,(2)} \times \left(\bar{v}_1/\bar{v}_{1,(2)}\right)^{\alpha_{\bar{p},\bar{v}}}$).

(continued)

**Box 4.4** (continued)

Box Figure 4.4-2a shows this allometric adjustment for mean tree productivity. The broken and solid lines represent the $\ln(\bar{p}) - \ln(\bar{v})$-allometry with general slopes but different levels in monospecific and mixed stands. Unadjusted analysis of mean productivity would compare $\bar{p}_{1,(2)}$ with $\bar{p}_1$; however, in this case, initial size differences would be neglected. For elimination of initial size differences, $\bar{p}_{1,(2)}$ is transformed to $\bar{p}'_{1,(2)}$ by scaling along the slope $\alpha_{\bar{p},\bar{v}}$ (arrow along the slope); $\bar{p}'_{1,(2)}$ is also higher than $\bar{p}_1$, but the ratio $\bar{p}'_{1,(2)}/\bar{p}_1$ is much smaller than $\bar{p}_{1,(2)}/\bar{p}_1$, i. e. the latter ratio reflects to a greater extent the indirect mixing effect. Box Figure 4.4-2a illustrates analogously the allometric adjustment for the tree number. In this example, the unadjusted comparison between mixed and monospecific stands, ratio $N_{1,(2)}/N_1$, predicts an inferiority of mixed versus monospecific stand density; however, if we take into consideration that the trees in the mixed stands are already taller than in the monospecific stand and eliminate this difference (arrow along the slope), the ratio $N'_{1,(2)}/N_1$ indicates a higher density in mixed versus monospecific stands ($N'_{1,(2)}/N_1 > 1$). Through application of the stand density index for comparing mixed with monospecific stands, any differences in the development state are also eliminated (see Sect. 4.3.1).

According to the theory of allometry (West et al. 1997, 1999; Yoda et al. 1963), the scaling of mean tree productivity versus tree volume should be $\alpha_{\bar{p},\bar{v}} = 3/4$, and tree number versus tree volume should be $\alpha_{N,\bar{v}} = -3/4$. For further application, we used yield tables to derive species-specific $\alpha_{\bar{p},\bar{v}}$-values of 0.417, 0.605, 0.437, 0.341, 0.592, 0.577, and 0.521 for Norway spruce, silver fir, Scots pine, European larch, European beech, sessile/common oak, and Douglas-fir, respectively. According to Pretzsch (2006), $\alpha_{N,\bar{v}}$-values amount to −0.618, −0.678, −0.685, −0.535 for Norway spruce, Scots pine, European beech, and sessile/common oak, respectively. If such empirically derived values are not available, the above theoretically based values can be used.

*Example for Size Differences and Allometric Adjustment*

Box Figure 4.4-3 shows the mean tree size effect for two-species mixed stands in south Germany on mixed versus monospecific plots in terms of stem volume. Norway spruce (a) generally maintains similar average tree volumes when growing in mixed and monospecific stands. European beech (b) achieves higher tree volumes when growing in monospecific compared with mixed stands. Its stem growth is probably reduced by upperstorey oaks and spruces. Oak behaves in the opposite way; its mean tree volume is ahead in mixtures compared with monocultures. Except for equality of mean tree size, a sound comparison between the direct effects of mixing would require an allometric elimination of any size difference before comparison, or at least

(continued)

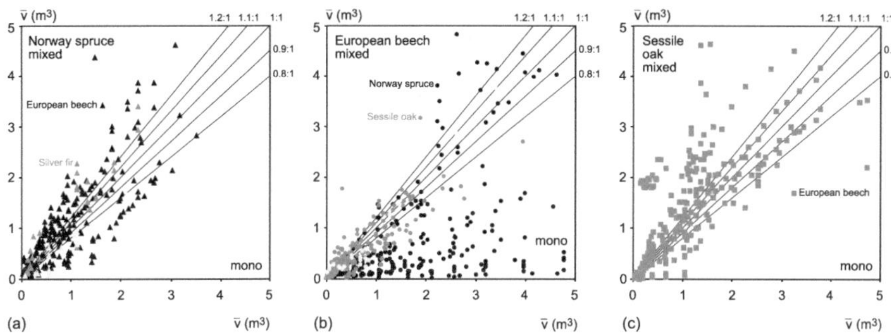


**Box Fig. 4.4-3** Differences between mean tree volume of mixed versus monospecific stands for (**a**) Norway spruce in mixed stands with European beech, silver fir, or Scots pine, (**b**) European beech in mixed stands with Norway spruce, sessile oak, or Scots pine, and (**c**) oak in mixture with European beech. The closer the observed values lie to the bisection line, the smaller the effect of the mixture on the mean tree size in terms of mean tree volume, $\bar{v}$

**Box 4.4** (continued)

a selection and restriction on those mixed and monospecific stand plots with comparable size development.

Suppose mean tree growth of Norway spruce (empirically derived $\alpha_{\bar{p},\bar{v}} = 0.42$) in both mixed and monospecific stands amounts to $\bar{p}_{1,(2)} = \bar{p}_1 = 0.10\ \text{m}^3\ \text{year}^{-1}$. This is not necessarily evidence for neutral mixing effects on tree productivity. Suppose $\bar{v}_{1,(2)} = 1.0\ \text{m}^3$ and $\bar{v}_1 = 1.50\ \text{m}^3$, i.e. the tree size in the monoculture is considerably ahead and already in a more advanced development phase. When the productivity in the mixed stand is scaled to $\bar{v}_1$ by the formula $\bar{p}'_{1,(2)} = \bar{p}_{1,(2)} \times (\bar{v}_1/\bar{v}_{1,(2)})^{\alpha_{\bar{p},\bar{v}}}$, then $\bar{p}'_{1,(2)} = 0.10 \times (1.50/1.00)^{0.42} = 0.12$. While the ratio $\bar{p}_{1,(2)}/\bar{p}_1 = 1$ reflects no mixing effect, $\bar{p}'_{1,(2)}/\bar{p}_1 = 1.20$ reveals a direct benefit of mixing on mean tree productivity of 20%. On the other hand, mixing effects diagnosed on the basis of uncorrected productivities can vanish when scrutinised using this method: If a naively calculated mixing effect of 5% $\bar{p}_{1,(2)}/\bar{p}_1 = 1.05$ (= 0.105/0.10 m$^3$ year$^{-1}$) is coupled with mean tree volumes of $\bar{v}_1 = 0.89$ and $\bar{v}_{1,(2)} = 1.0\ \text{m}^3$, then adjustment may show that $\bar{p}'_{1,(2)} = 0.105 \times (0.90/1.0)^{0.42} = 0.10$, so that $\bar{p}'_{1,(2)}/\bar{p}_1 = 1$. In other words, there seems to be a 5% mixing effect; however, the mixed stand is simply in a more advanced development phase ($\bar{v}_{1,(2)} > \bar{v}_1$) with higher growth rates, and the mixing effect vanishes when this lead in development is eliminated.

Box Figure 4.4-3 underlines, especially for beech and oak, a considerable size difference between monospecific and mixed stands, which suggests the need to eliminate size differences in order to quantify direct mixing effects.

### 4.2.2 Empirical Evidence of Overyielding and Underyielding in Mixed Compared with Monospecific Forest Stands

#### 4.2.2.1 Early Findings in Temperate and Boreal Forests

Most older works on the growth and yield of mixtures were searching for overyielding of mixed versus monospecific stands. In Norway spruce/European beech mixtures in northern Germany, Wiedemann (1942, 1943, 1951) found approximately the same dry biomass in mixtures as in corresponding monocultures on sites of mediocre quality. However, on poorer sites with moderate growth of European beech, the mixed stand biomass production was about 19% less than in monocultures on neighbouring sites. Kennel (1965) studied mixtures of Norway spruce and European beech in the Bavarian alpine foothills, the Bavarian Forest, and the Harz in Lower Saxony as did Burger (1941) in Switzerland. Their results range from slight overyielding to underyielding of mixed versus monospecific stands. Zöhrer (1969) provides evidence that the biomass production of European larch/Norway spruce mixtures in the Salzburger Land is superior to that of neighbouring monocultures. The European larch/Norway spruce mixture surpassed the Norway spruce monoculture by 22–28% and the European larch monoculture by 2–13%. For mixed stands of sessile oak and European beech, Scots pine and Norway spruce, and Scots pine and European beech, Bonnemann (1939) and Wiedemann (1943, 1951) found similar beneficial effects from species interactions after 50 years of observation. For long-term Scots pine and European beech experimental areas in the Dübener Heide, Dittmar et al. (1986) report beneficial interaction effects, compared with a monospecific stand, of 7–25%, depending on the age and structure of the mixture. Burger (1941) and Wimmenauer (1941) note a similar superiority in European larch/European beech mixtures. The contrasting results of these older studies do not deliver a consistent insight into the extent and causes of mixing effects on productivity, which at least partly results from the very different approaches of evaluating mixing responses, e.g. on the basis of volume or biomass growth, growth or yield, temporary or long-term plots, neighbouring monocultures, or yield tables as references.

#### 4.2.2.2 Current Reviews and Meta-Analyses

The reason for the increasing attention given to mixed-species stands is that close-to-nature approaches, which often involve the use of mixed-species stands, are widely held to supply ecological, economic, and sociocultural forest goods and services at a similar or higher level than many monocultures (Hooper et al. 2005; Hector and Bagchi 2007). However, sound knowledge about mixing effects, even for the most common tree species combinations, is rather rare and scattered. Just in the last few years, after the failure of some monocultures and rethinking on risk distribution (Knoke et al. 2005), resource use efficiency (Richards et al. 2010), and the functional significance of species diversity (Scherer-Lorenzen et al. 2005), mixed stand dynamics have become a focus of forest science (see e.g., Scherer-Lorenzen et al. 2005; Pretzsch et al. 2010; Forrester et al. 2006).

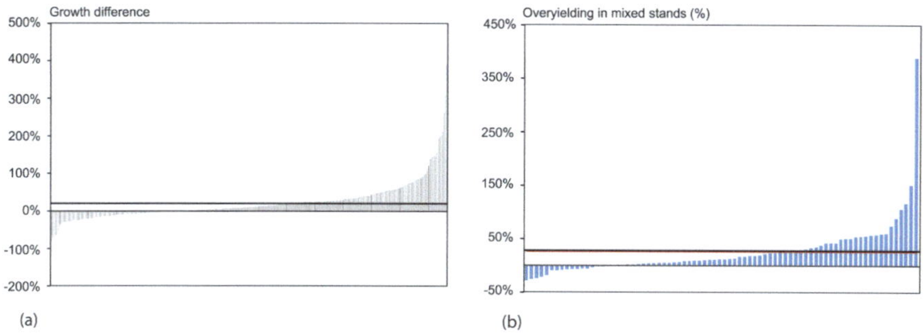


**Fig. 4.11** Results of a meta-analysis on over-/underyielding in terms of volume growth at (**a**) the species level and (**b**) the whole-stand level in boreal, temperate, and tropical forests (according to Gritti et al. unpublished). The individual studies are represented by bars ranked from the most negative to the most positive mixing effects on productivity (from *left* to *right*). (**a**) Evaluation of 51 publications referring to 319 case studies revealed an average overyielding at the species level of +14.5% (mean effect size). (**b**) Results at the whole-stand level are based on 29 publications referring to 78 case studies. They show that mixed stands produce an average of about +26.5% (mean effect size) more than the weighted mean of neighbouring monocultures

A crucial question for relating our knowledge about mixtures and their practical application is how their productivity compares with monocultures. Knowledge on the advantages or disadvantages of mixed versus monospecific stands, with respect to productivity, decisively influences the forest owners' decisions in favour or against mixed-species stands (Olsthoorn et al. 1999). Meta-analyses by Piotto (2007) and Zhang et al. (2012) which provide evidence that mixed stands have the potential to outyield monocultures are also of considerable relevance for forest management.

Figure 4.11 illustrates (a) a mean overyielding of +14.5% at the species level and (b) of +26.5% at the stand level. A recent meta-analysis by Gritti et al. (unpublished) broadens the scope of earlier studies by increasing the number of case studies, publications, the range of climate zones (e.g. boreal, temperate, Mediterranean, subtropical), and species combinations (e.g. conifers and broadleaved, early-successional and late-successional species, atmospheric-nitrogen-fixing species and non-nitrogen-fixing species). Analysis at the species and stand level (Fig. 4.11a and b) revealed a broad variation in mixing effects, including strong overyielding but also underyielding (bar reaching above and below the horizontal line of the mean effect, respectively).

Another meta-analysis was compiled specifically for this chapter. It is based on long-term experiments but is restricted to common species combinations in Germany. It revealed a mean overyielding of 19% for Norway spruce and European beech (Fig. 4.12a), 24% for sessile oak and European beech (Fig. 4.12b), and 20% for the species combination of Norway spruce, silver fir, and European beech (Fig. 4.12c).

The latter three-species mixture is very common in the mountainous areas of the European Alps. The meta-analysis was based on 46 plots. The mean relative

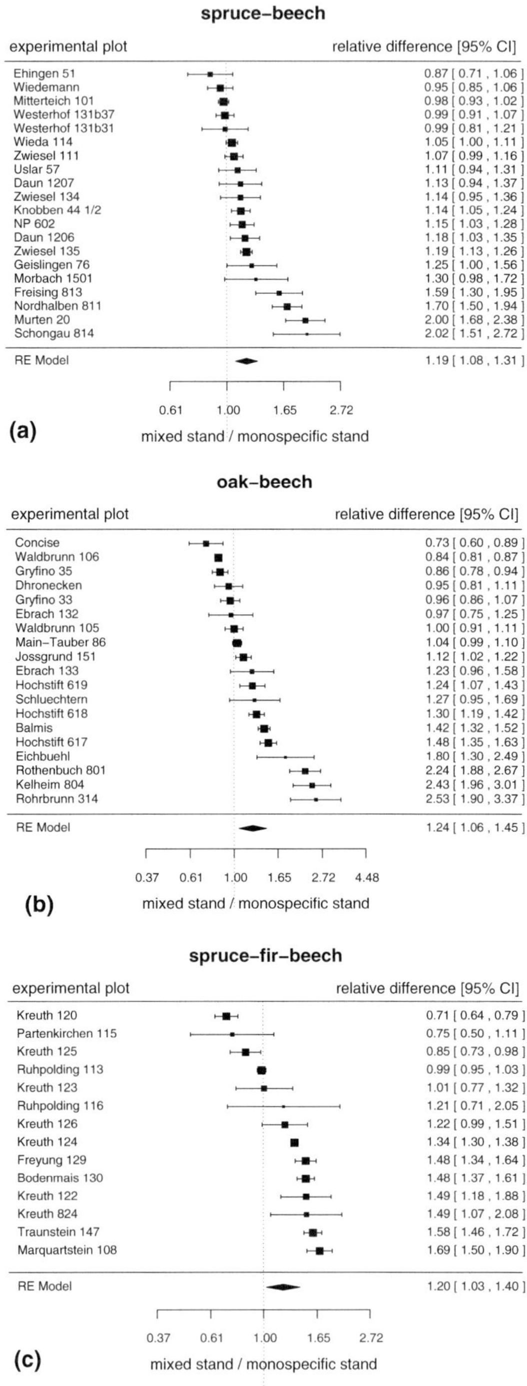


**Fig. 4.12** Comparison of the stand productivity of mixed versus monospecific forest stands based on long-term experimental plots in Central Europe (**a**) Norway spruce and European beech, (**b**) sessile oak and European beech, (**c**) mixed mountain forests of Norway spruce, silver fir, and European beech. On average, the mixed stands produced 119%, 124%, and 120%, respectively, of the weighted mean of the neighbouring monocultures. All three analyses revealed a broad variation in mixing effects including underyielding and strong overyielding. The entry 'Wiedemann' in Fig. 4.12b refers to the results of the study by Wiedemann (1942) about the growth of mixed versus monospecific stands of Norway spruce and European beech

productivity amounted to 120% of the neighbouring monocultures, while the absolute gain in productivity amounted, on average, to 1.60 t $ha^{-1}$ $year^{-1}$. Figure 4.12c shows considerable variation around the mean mixing effect. A similar evaluation for two experiments with six plots and three-species mixtures of sessile oak, European beech, and Scots pine in the hill country of Steigerwald and Spessart also yielded, on average, a higher productivity in mixtures than in monocultures (amounting to 143%) and an absolute productivity gain of 1.89 t $ha^{-1}$ $year^{-1}$. Although based on a rather small database, these results are indicative given the scarce information available to date on the effect of three-species mixtures. The size of the mixing responses (on average 120–143% in relation to the neighbouring monocultures) stresses that in three-species mixtures, the absolute gain as well as the variation can be even higher than in two-species mixtures (Fig. 4.12c), consistent with many studies that have examined the relationships between tree species richness and growth (Forrester and Bauhus 2016).

At the species level, the mean gain in productivity for a given species combination in terms of RPA ranges from +13% to +80%. With one exception (silver fir in the mixture N. spruce/s. fir/E. beech), the mixed-species stands and the species in the mixture increased significantly ($p < 0.05$) in terms of productivity compared with the neighbouring monocultures (Pretzsch et al. 2013b).

#### 4.2.2.3 Big Data Analysis

The still rather fragmentary knowledge of mixing effects on forest stand productivity has been complemented by analyses of forest inventory data (Gamfeldt et al. 2013; Vallet and Perot 2011; Vilà et al. 2013). Compared with the restricted number of long-term mixed-species experiments, forest inventories represent mixing effects for a much broader range of site conditions, species assemblages, stand ages, mixing proportions, mixing patterns, and stand densities (Condés et al. 2013; Forrester and Pretzsch 2015; Toïgo et al. 2015). But in contrast to the ceteris paribus conditions and causal evidence of mixing effects on experimental plots (Pretzsch et al. 2010, 2013a), inventory data analyses only provide the statistical relationship between species composition and productivity; e.g. an overyielding of mixed in comparison to monospecific stands may result from a predominance of mixed stands on slightly better sites compared with monocultures. A complete disentanglement of the species composition–productivity relationship from all confounding co-variables is rarely possible using inventory data because information about, e.g. stand history, site conditions, thinning, and mortality, is often simply lacking in the case of forest inventory data. Thus, inventory data analyses may substantiate but not replace mixed-species stand analyses based on experimental data (Nagel et al. 2012).

Using forest inventory data at the regional or national level, several studies found that overyielding of mixed versus pure stands was modified by site conditions (Toïgo et al. 2015), stand density (Condés et al. 2013), and various other stand characteristics (Vilà et al. 2013). In contrast to stand productivity, dominant stand height was hardly affected by tree species mixing (Vallet and Perot 2011). Liang et al. (2016) quantified the effect of tree species richness on the productivity of

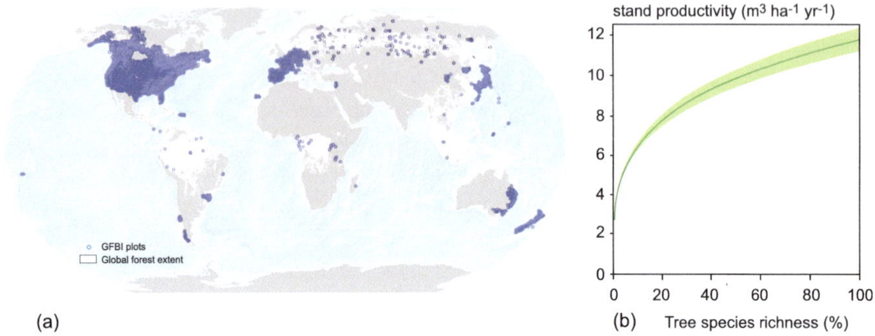


**Fig. 4.13** Global effect of tree species richness on forest productivity according to Liang et al. (2016). Analysis of data from 777,126 global permanent sample plots reflected by *blue dots* in (**a**) revealed a consistent positive and concave-down tree species richness-productivity relationship (**b**)

many of the world's forested ecosystems in the most comprehensive study so far. They used repeated forest inventories from 777,126 permanent sample plots that contain over 30 million trees representing 8737 tree species and span most of the global terrestrial biomes. They found a consistently positive concave-down relationship between stand productivity and tree species richness at landscape, country, ecoregion, and global levels (Fig. 4.13).

Globally, a 10% loss of tree species richness corresponded on average to a 6–7% decline in productivity, and the rate of this decline increased significantly with a further reduction of biodiversity. This relationship revealed that continued biodiversity loss at landscape scales would result in an accelerating decline in forest productivity worldwide. The relationship between tree species richness and stand productivity was strikingly similar across global climates and biomes despite some geographic variation. Liang et al. (2016) estimated that the economic value of biodiversity in maintaining commercial forest productivity alone is more than twice that of the total costs that would be necessary if we were to effectively conserve all terrestrial ecosystems at a global scale.

The study by Liang et al. (2016) highlighted both the negative effect of biodiversity loss on forest productivity and the potential benefits from the transition of monocultures to mixed-species stands in forestry practices. The relationship between tree species richness and productivity that they discovered across forest ecosystems worldwide corresponded well with experimental and observational studies on forest and non-forest ecosystems.

#### 4.2.2.4 Special Examples of Temperate Forest Species

In order to unify the somewhat scattered sources of information about mixing effects, Pretzsch et al. (2010) compiled and analysed existing growth and yield plots of monospecific and mixed stands of Norway spruce and European beech. The database contains information from 23 long-term plots, covering an ecological gradient from nutrient-poor and dry to nutrient-rich and moist sites throughout

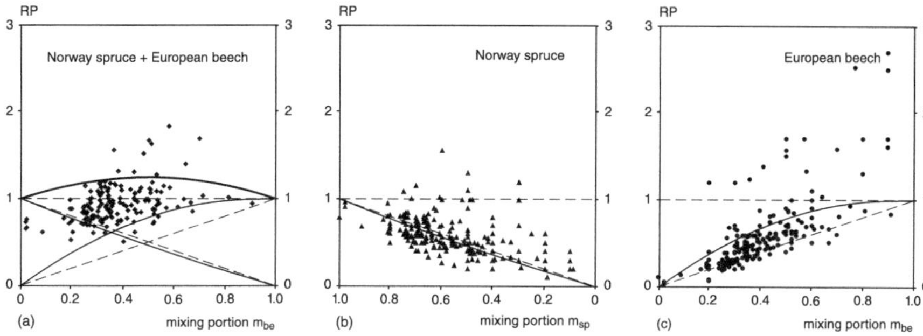


**Fig. 4.14** Relative productivity of volume growth of (**a**) the mixed stand in total and (**b**) spruce and (**c**) beech in relation to the neighbouring monocultures. The points represent the observed relative volume productivity of mixed versus monospecific stands. The curves represent the average mixing responses of spruce, beech, and total stand according to Pretzsch et al. (2010)

Central Europe. Depending on site condition, dry mass growth in mixed stands can range from −46% to +138% of the growth yielded by neighbouring monocultures. Figure 4.14 shows the observed relative volume productivity, RPA, of mixed versus monospecific stands. On average, the relative productivity of the mixed stands amounts to 120% of that expected on the basis of the neighbouring monocultures (Fig. 4.14a). The gain in absolute productivity in terms of biomass amounts to 1.5 t $ha^{-1}$ $year^{-1}$. The analysis at the species level (Fig. 4.14b and c) shows that Norway spruce can profit but also suffer from mixture, resulting in a neutral average response (Fig. 4.14b). European beech also shows broad variation in performance; however, on average, the outcome is positive. The latter result is indicated in Fig. 4.14c by the transgression of the concave solid model line above the level of the increasing broken straight reference line. The functions representing the mean observed relative productivity in relation to the mixing proportion result from model fitting by non-linear regression analysis.

The species combination of oak (*Quercus petraea* (Matt.) Liebl. and *Quercus robur* L.) and beech is already of considerable importance and will become even more relevant in forests as the climate changes. Data from 37 long-term mixing experimental plots in Poland, Germany, and Switzerland were pooled for analysis of mixing effects on stand productivity in relation to mixing proportions and site conditions (Fig. 4.15). On average, mixed stands of oak and beech produce 30% or 1.7 t $ha^{-1}$ $year^{-1}$ more than monocultures, as both species profit from the mixture (Pretzsch et al. 2013a).

Both species, oak and beech, as well as the total mixed stand, are more productive than the monospecific stands on poor sites, but this mixing effect is reversed on fertile sites. Mixed and monospecific stands only have similar productivity on mediocre sites. Facilitation, competitive reduction, and competition may occur simultaneously (Vandermeer 1989), and the net effect is indicated by productivity gains or losses on the stand level. The analysis of the productivity relationships suggests that on poor sites, facilitation, even mutual facilitation, or competitive reduction has the upper hand. The observations represented by the points above the

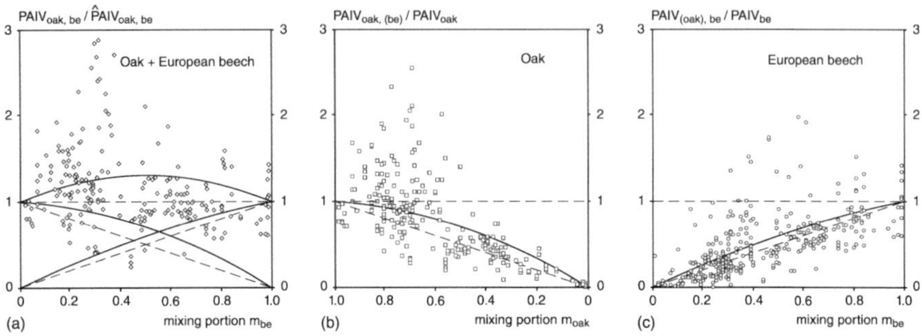


**Fig. 4.15** Relative productivity in terms of volume growth for (**a**) oak, (**b**) beech, and (**c**) the mixed stand in total in relation to the productivity of the neighbouring monocultures. The points represent the observed relative volume productivity of mixed versus monospecific stands. The curves represent the average mixing responses of oak, beech, and the whole stand. For explanation of the included reference lines (broken), see Box 4.1

**Table 4.2** Mixing effects on stand productivity of various tree species mixtures in Central European forests derived from long-term experiments

| Species combination | N. sp/ E. be | s. pi/ E. be | s. oak/ E. be | E. be/ D-fir | S. pi/ N. sp | E. la/ N. sp | N. sp/ s. fir | Mean |
|---|---|---|---|---|---|---|---|---|
| Overyielding (± SE) in % | 21 (± 3) | 30 (± 9) | 20 (± 3) | 11 (± 8) | 21 (± 11) | 25 (± 6) | 13 (± 6) | |
| Corr. factor | 1.10 | 1.20 | 1.10 | 1.10 | 1.20 | 1.20 | 1.10 | 1.10 |

The relative overyielding (%) refers to the productivity of the mixed-species stands in relation to the weighted mean of the neighbouring monospecific stands (RPA values, see Box 4.1). The correction factors may be used to conservatively adjust the stand productivity of monospecific stands to the expected stand productivity of the respective species assemblages (Pretzsch 2016). N. sp. Norway spruce, E. be. European beech, S. pi. Scots pine, s. oak sessile oak, D-fir Douglas-fir, E. la. European larch, s. fir silver fir

broken 1.0 lines in Fig. 4.15 indicate this mutual facilitation or competitive reduction. They reflect that the productivity of a fully stocked, closed oak stand can be increased when beeches are added, and vice versa. In other words, the competition effect produced by adding the other species is more than outweighed by facilitation or competitive reduction on the species and stand levels. On fertile sites, competition between the two species seems to gain the upper hand.

Table 4.2 summarises the overyielding of common two-species assemblages in Central Europe and underlines the fact that the mixing effects are not only scientifically evident but also practically relevant. Long-term experiments show that in relation to the weighted mean of the monospecific stands, mixed-species stands produce 11–30% more stem volume (Pretzsch 2016). Combinations of more similar species (e.g. Norway spruce and European beech, Norway spruce and silver fir) result in lower overyielding than mixtures of more contrasting species (e.g. Scots pine and European beech, European larch and European beech), although it is worth noting that even mixing very similar species (from the same genus) can result in significant overyielding (Forrester and Smith 2012). In addition to the mean overyielding and standard error, Table 4.2 shows conservative correction factors which may be used to estimate mixed stand productivity based on the productivity of neighbouring monocultures. The correction factors indicate that the productivity of monocultures should be multiplied by 1.10–1.20 to estimate the productivity of the respective mixed-species stands, and they apply for fully stocked mixed-species stands, individual to groupwise mixing patterns, and mixing proportions of about 50:50.

#### 4.2.2.5 Mixing Effects in Stands Containing $N_2$-Fixing Species

There are many cases where the growth of mixed-species plantations containing *Eucalyptus* and $N_2$-fixing trees, such as *Acacia*, is significantly more productive than monocultures, sometimes by as much as 50% and in extreme cases 100% (Forrester et al. 2006). For example, Fig. 4.16 shows overyielding in mixed versus monospecific stands of *Eucalyptus globulus* Labill (eucalypt) and *Acacia mearnsii* De Wild (black wattle). This example can be used to stress the difference between the absolute gain or loss in productivity of a species or the stand as a whole (Fig. 4.16a) and the relative productivity (Fig. 4.16b) in a mixed compared with a neighbouring monoculture. Forest practice is mainly interested how much mixtures can outyield monocultures in the absolute amount of biomass (t $ha^{-1}$ $year^{-1}$) or volume ($m^3$ $ha^{-1}$ $year^{-1}$). The absolute outcome determines management decisions (production economy). In this example, the mixed stands with equal mixing proportions of both species overyield the monospecific reference stands by a

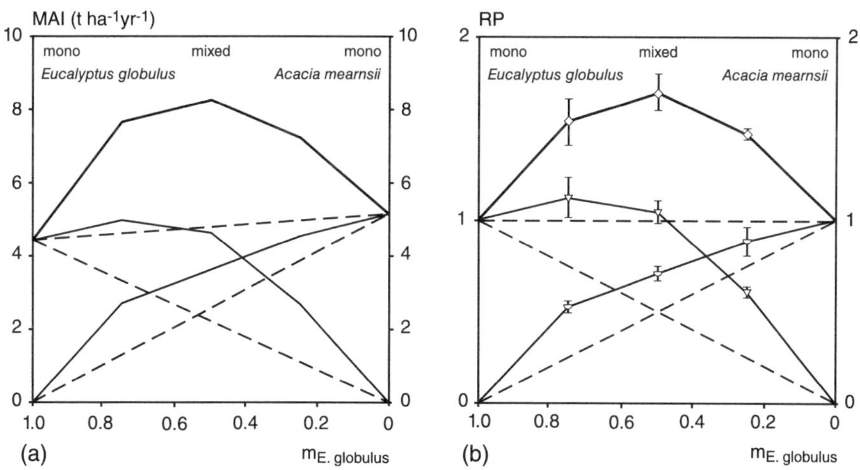


**Fig. 4.16** Overyielding in mixed versus monospecific stands of *Eucalyptus globulus* Labill and *Acacia mearnsii* De Wild in terms of (**a**) the *absolute* productivity of a species in mixture versus its performance in a monoculture and (**b**) the *relative* gain or loss of productivity of a species or the stand as a whole in mixed compared with a neighbouring monoculture (see also Forrester et al. 2006)

remarkable 3–4 t $ha^{-1}$ $year^{-1}$. Science is primarily interested in how productive a species is in mixture relative to the monospecific stand and how two or more species benefit or lose from the mixture. The relative outcome contributes to the understanding of species-specific competitiveness and fitness on a given site. The relative MAI of the mixed stand is 1.70 times that of the monoculture; in other words, the overyielding amounts to 70%. There are now over $20 \times 10^6$ ha of eucalypt plantations worldwide from the tropics to the temperate and Mediterranean zones. When these plantations are nitrogen (N) limited, $N_2$-fixing species may provide an alternative N source and provide it more continuously than synthetic N inputs, so that productivity and yield can be improved to a relevant extent.

The common perception that the benefits of $N_2$ fixation in mixtures will improve as site quality declines is only partly true. Eucalypts could potentially benefit from symbiotic $N_2$ fixation on any site where growth is limited by N, and the more limiting N is compared to all other resources, the bigger the relative facilitative effect might be. Some of the largest reported mixing effects are on very productive sites in Hawaii (Binkley et al. 2003). In these stands, growth was examined until age 20 years, and best-fit trends indicated that the most productive mixtures would be those composed (in trees $ha^{-1}$) of about 40% eucalypts and 60% *Falcataria moluccana*, which would be 64% more productive than eucalypt monocultures and 105% more productive than *F. moluccana* monocultures. Mixing effects are likely to increase as the relative availability of the resource that is improved by mixing (e.g. N) declines (see also Figs. 3.4 and 3.5 in Chap. 3).

While there are many examples of the facilitative effect that $N_2$-fixing species can have on the growth of mixtures, there are also examples where productivity was not improved. Inappropriate matching of species or sites can lead to the suppression or loss of some of the species from the mixture. For example, very fast growing eucalypt clones have been planted with *Acacia mangium* (brown salwood) in Brazil (Laclau et al. 2008). Few trees would be competitive with the eucalypt clones on these sites, and eucalypts grew faster in these mixtures because they suppressed the brown salwood trees. However, the faster growth of the eucalypt was not enough to compensate for the reduced growth of the brown salwood, so the mixtures were not as productive as the eucalypt monocultures.

Other examples of stands containing $N_2$-fixing species include *Alnus rubra* in Douglas-fir forests of North America (Binkley 2003), *Alnus glutinosa* [L.] Gaertn. in Norway spruce (Schwaiger 2013) forests in Central Europe, and *Lupinus arboreus* in *Pinus radiata* plantations in New Zealand (Beets and Madgwick 1988).

#### 4.2.2.6 Change in Mixing Effects Within the Rotation Period of Plantations Containing Eucalyptus and Nitrogen-Fixing Acacia

Figure 4.17 shows the periodic annual volume growth (a–c) and the mean annual volume growth (d–f) for black wattle and eucalypt in total and separately in a substitutive mixing experiment in southeastern Australia. This is a plantation, but these species sometimes coexist in natural forests. We start with such an experiment

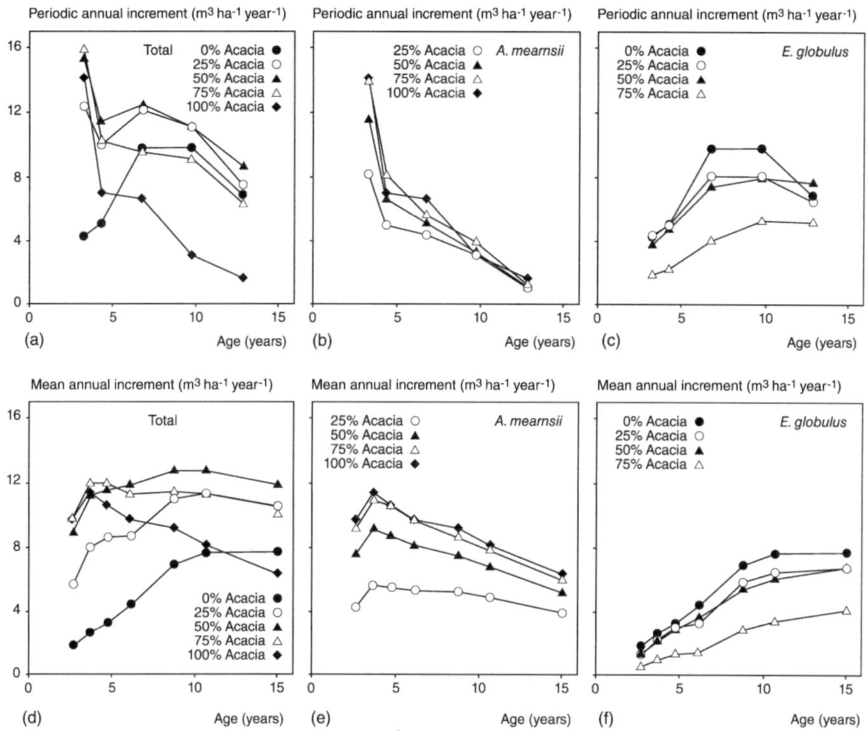


**Fig. 4.17** (**a**–**f**) Mean periodic annual volume increment (**a**–**c**) and mean annual volume increment (**d**–**f**) of *Acacia mearnsii*, *Eucalyptus globulus*, and the total of both species in mixtures and monocultures of plantations in southeastern Australia. The percentages in the legend indicate the percentage of the planting stocking (total of 1515 trees $ha^{-1}$ in all treatments) that were *Acacia mearnsii*. The graph is based on data from Forrester et al. (2004, 2010)

which displays the mixing effects during the complete rotation because it delivers facts that are of interest for temperate forests.

The modification of the growth courses by the different mixing proportions of *acacia* and *eucalyptus* (b and c) indicates that mixing changes not only the level but also the rhythm and course of the species growth curves. The higher the proportion of the second species, the lower the peak of *Acacia* growth and the later the peak of *Eucalyptus* growth (Fig. 4.17b and c). The curves of mean annual growth (Fig. 4.17d–f) represent the mean of the accumulated yield, and they therefore have a smoother course than the periodic annual growth which is more affected by the prevailing temporal environmental conditions.

Figure 4.17 clearly shows that mixtures were about 50% more productive than both monocultures, and at various ages, this mixing effect was even greater. The finding that growth courses of the mixed variants are mostly running at a higher level than both the monospecific *Acacia* and *Eucalyptus* stand rather than between them indicates prevailing overyielding of mixed versus monospecific stands for most of the rotation (Fig. 4.17a and d). It is also clear that the total growth of the mixtures was the net effect of two species that have contrasting growth dynamics. The periodic and mean annual volume increment curves show that the black wattle trees had faster early growth rates peaking at 3–4 years, or earlier, before declining, whereas the periodic annual increments of eucalypt trees did not peak until age 7.5 years, and mean annual increments increased until at least age 10 years. These contrasting growth dynamics worked well in these stands. The fast early growth by black wattle probably led to increases in N stocks and N availabilities. This fast early growth slowed in time for eucalypt, a very shade-intolerant species, to develop without being suppressed and at the same time may have improved the stem form of the eucalypt trees by shading their lower crowns.

When the volume growth curves of the monospecific and mixed variants cross each other, the ranking with respect to overyielding of the mixed versus monospecific variants must have changed within the survey period. Figure 4.18 illustrates these changes in the mixing effect using cross diagrams for the mean annual growth until stand ages 4, 9, and 15 (Fig. 4.18a–c). From the initial state until age 15, the cross diagrams indicate transgressive overyielding of the total stand and a contribution of both species to this superiority of mixed versus monospecific stands. The contrasting growth course of both species is reflected by the superiority of *Acacia* during the first half of the rotation which turns into a superiority of *Eucalyptus* during the second half. With progressing stand development, the 50:50 mixture of both species gains the upper hand in terms of total yield.

While intercropping in agriculture lasts about a year and, after that, mixing effects can be analysed rather easily by bulk harvesting (Vandermeer 1992), mixed-forest stands need decades or centuries to become mature and are more difficult to analyse. The variability in mixing responses means that results of mixing experiments can vary between survey periods. This insight is important for mixing analyses in European temperate forests with rotation periods of 100–200 years. When mixing analyses in such stands cover only a couple of decades or only one singular survey period, the results are relatively vague; however, they become sounder and more general as the recording period increases.

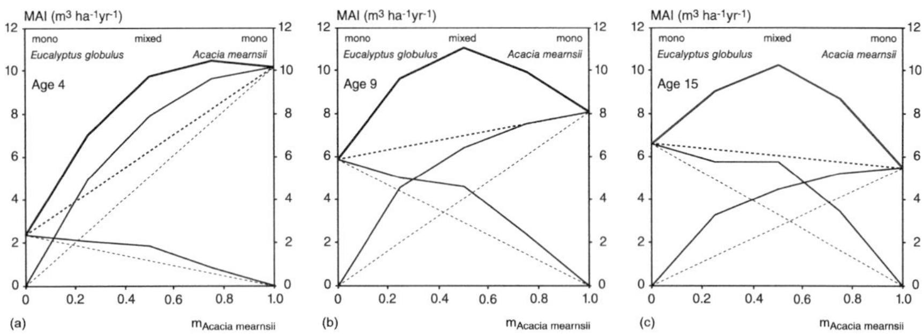


**Fig. 4.18** Cross diagrams reflecting the mixing effects concerning mean annual volume growth (MAI in $m^3$ $ha^{-1}$ $year^{-1}$) until age 4, 9, and 15 (from **a** to **c**) for the *Eucalyptus globulus*/*Acacia mearnsii* experiment (see Fig. 4.17)

#### 4.2.2.7 Change of Mixing Effects on Productivity During Stand Development in Long Rotation Temperate Forests

For this section, we selected mixing experiments in Europe where the variability in mixing responses has been examined for periods as long as 100 years. Differences between short- and long-term responses become obvious and show the risk of prematurely generalising mixing effects found in short-term experiments (e.g. in phytotrons, open-top chambers, repeated inventories after 10 years).

Figure 4.19 uses the long-term surveyed triplets Waldbrunn 105 and Wieda 114 to illustrate how mixing effects change during stand development. The performance of the two species and the mixed stand as a whole is compared with the monocultures (1.0 line) in each survey period. In addition to this period-wise evaluation, we calculated the total mixing effect until the last survey.

The strong mixing effects in favour of oak in the case of the sessile oak/-European beech experiment Waldbrunn 105 (Fig. 4.19a and c) subside as the stand develops. The analysis of periodic annual growth on the sessile oak/European beech triplet WAL 105 at age 49 (period 1935–1952) shows an overyielding of 20% caused by oak, which changes to a rather neutral mixing effect at age 72 (1966–1974). At age 91 (1990–1993), there is underyielding of about 10%, which is mainly caused by beech. Integration in terms of total yield until the present survey at age 105 shows a slight positive mixing response of oak ($RPA_{1,(2)} = 1.10$) and beech ($RPA_{(1),2} = 1.02$) which together result in an overyielding of 4% ($RPA_{1,2} = 1.04$). This clearly shows that mixing effects in terms of relative growth can vary considerably from period to period depending on the stand development phase, periodic environmental conditions, and disturbances like thinning or acid deposition.

The Norway spruce/European beech triplet Wieda 114 (Fig. 4.19b and d) shows minimal mixing responses during the first decades of the survey period. The depression of spruce between ages 80 and 100 in Wieda 114 can probably be attributed to acid deposition. However, beech compensates the losses of spruce so that total stand productivity has been higher in mixtures than in monocultures for most of the time since stand establishment. During the final third of the survey period, beech and Norway spruce overyield in the mixed stands. An integrated evaluation over the whole-stand development until age 103 results in a relative yield of 0.95–1.20 at the species level and 1.04 at the stand level (Fig. 4.19d).

Table 4.3 uses a set of long-term experiments to show the variability in mixing effects and corresponding differences in mixing analyses based on short-term productivity ratios and long-term survey yield records.

In the case of the sessile oak and European beech experiments at Waldbrunn, the mixing effects are at a maximum during early stand development phases and subside with progressing stand development. Mean productivity predicts underyielding of 0.95 and 0.92, respectively, in both cases. However, final yields indicate overyielding of 1.04 for WAB 105. Mixing analysis at age 49 indicated an overyielding for oak of 1.84, 1.20 for beech, and 1.24 for the total stand; a premature generalisation of this level across the whole rotation would involve a considerable overestimation of mixing effects. The difference between the

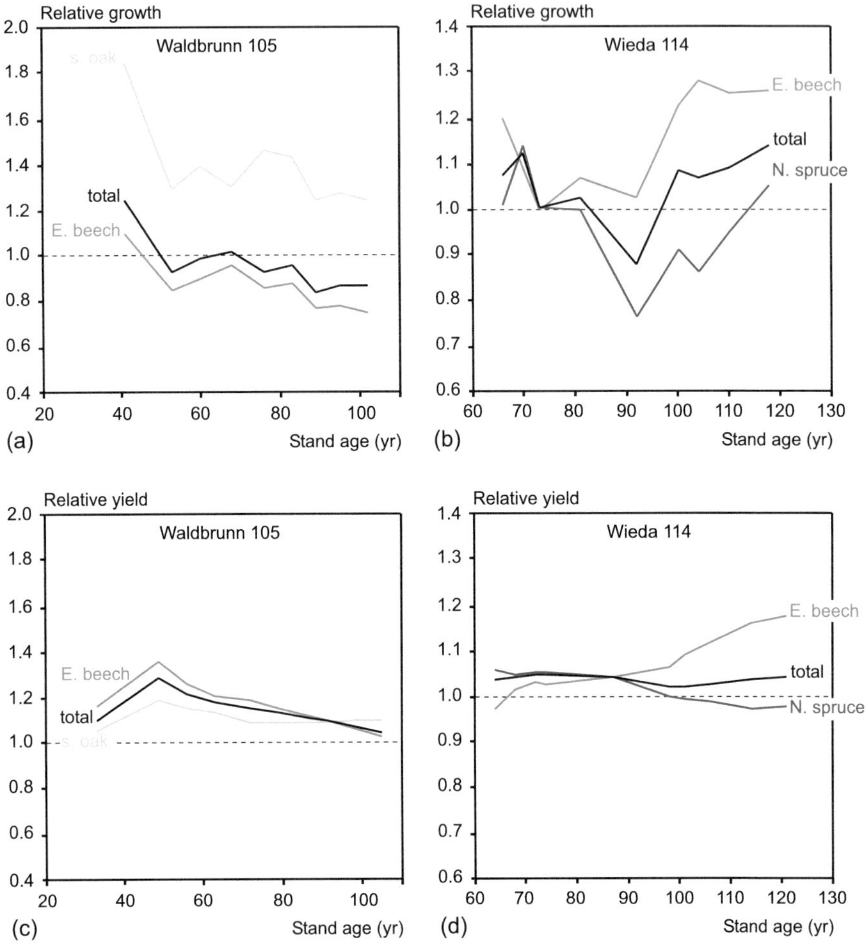


**Fig. 4.19** (**a**–**c**) Long-term course of mixing effects for (**a** and **c**) the sessile oak/European beech experiment Waldbrunn 105/1-3 since stand establishment in 1935 and (**b** and **d**) the Norway spruce/European beech experiment Wieda 114 since stand establishment in 1949. Periodic annual growth (above) and total yield (below) of stem mass of the species in the mixed stand and the mixed stand as a whole in relation to the monospecific stand performance (1.0 reference line). The relative periodic annual growth is presented over the period means, and the relative yield is presented at the ends of the respective periods

responses of Waldbrunn 105 on a poor site and Waldbrunn 106 (WAB 105, WAB 106) on a fertile site reflects that mixing effects are sometimes low or even negative on rich sites and stronger and positive on poor sites.

The analysis of mixing effects on the basis of relative periodic growth and yield shows rather similar results on experimental plot Wieda 114. The minimum and maximum values of growth differ from the yield characteristics, but on

**Table 4.3** The difference between periodic and long-term mixing effects on biomass productivity shown for selected long-term experiments

| Trial | Age first–last survey year | Surveys | Relative growth RPA | | | | | Relative total yield RYA | | | |
|---|---|---|---|---|---|---|---|---|---|---|---|
| | | *n* | Mean | Min | Age | Max | Age | Final | Age | Min | Max |
| WAB 105 | | | | | | | | | | | |
| s. oak | 33–105 | 10 | 1.38 | 1.24 | 105 | 1.84 | 49 | 1.10 | 105 | 1.05 | 1.18 |
| E. beech | 33–105 | 10 | 0.87 | 0.75 | 105 | 1.10 | 49 | 1.02 | 105 | 1.02 | 1.35 |
| Total | 33–105 | 10 | 0.95 | 0.84 | 91 | 1.24 | 49 | 1.04 | 105 | 1.04 | 1.28 |
| WAB 106 | | | | | | | | | | | |
| s. oak | 44–116 | 10 | 1.04 | 0.92 | 60 | 1.20 | 109 | 0.92 | 116 | 0.81 | 0.92 |
| E. beech | 44–116 | 10 | 0.85 | 0.76 | 102 | 0.95 | 60 | 0.96 | 116 | 0.96 | 1.14 |
| Total | 44–116 | 10 | 0.92 | 0.85 | 102 | 0.95 | 116 | 0.94 | 116 | 0.94 | 0.95 |
| WIE 114 | | | | | | | | | | | |
| N. spruce | 46–103 | 10 | 0.96 | 0.76 | 80 | 1.14 | 54 | 0.97 | 103 | 0.97 | 1.05 |
| E. beech | 64–121 | 10 | 1.16 | 1.00 | 74 | 1.28 | 106 | 1.17 | 121 | 0.97 | 1.17 |
| Total | | 10 | 1.05 | 0.88 | 98 | 1.14 | 121 | 1.04 | 121 | 1.02 | 1.05 |
| MIT 101 | | | | | | | | | | | |
| N. spruce | 50–123 | 7 | 1.03 | 0.86 | 57 | 1.60 | 123 | 0.92 | 123 | 0.92 | 1.05 |
| E. beech | 50–123 | 7 | 1.01 | 0.91 | 57 | 1.08 | 69 | 1.32 | 123 | 1.06 | 1.34 |
| Total | 50–123 | 7 | 0.98 | 0.88 | 57 | 1.07 | 117 | 1.14 | 123 | 1.05 | 1.14 |
| ZWI 111/1 | | | | | | | | | | | |
| N. spruce | 59–113 | 9 | 1.01 | 0.79 | 74 | 1.43 | 102 | 1.03 | 113 | 1.02 | 1.12 |
| E. beech | 78–132 | 9 | 1.14 | 0.97 | 93 | 1.42 | 111 | 1.11 | 132 | 1.07 | 1.12 |
| Total | | 9 | 1.05 | 0.85 | 74 | 1.35 | 102 | 1.06 | 132 | 1.03 | 1.12 |

The relevance of repeated or long-term surveys for reliable indications of mixing responses becomes obvious when considering the differences between the minimum, maximum, and mean of the species-specific and total stand ratios for individual periods (relative growth, RPA) and the integrated mixing effect over the whole observation period (relative yield, RYA) average both approaches show similar overyielding for beech (1.16 resp. 1.17), underyielding for spruce (0.96 resp. 0.97) and overyielding for the total stand (1.05 resp. 1.04). Beech also significantly overyields in the spruce-beech experiments Zwiesel 111 and Mitterteich. In those plots, the difference between growth-based and yield-based mixing analysis is distinct. Both experiments show a strong variation in the relative growth records including short periods with periods of severe underyielding (0.79 and 0.86, resp.) and overyielding (1.43 and 1.60, resp.). The yield-based evaluation reflects higher benefits from mixture (1.06 and 1.14, resp.) because short-term periods with a low performance in mixture reduce the mean of the relative growth, but as the low-growth periods occur during late stand development phases with low productivity, they have minor effects on the relative total yield. The latter is mainly driven by positive mixing effects during high-growth stand development phases, which might have low relative benefits, but due to the high productivity level, they can have strong positive effects on the total yield.

The ultimate data sources for assessing the performance of mixed versus monospecific stands at a given site are long-term records of the total yield covering the whole rotation period. However, with the exception of short-term rotation plantations, such records are extremely rare. Therefore, most mixed-species stand analyses are based on growth records from single or short-term surveys. The longer the growth period covered by the survey, the more the results from growth-based evaluations converge to total yield-based findings. As a result of mixing effects changing with age, insight into long-term mixing effects requires either long-term survey over longer periods or sampling of stands at different ages (real or artificial time series). However, sampling should also cover different time spans in terms of calendar years because mixing effects can also vary depending on periodic environmental conditions. Restriction of the analysis to a limited time span may produce biased results when environmental conditions during the measurement period have been above or below the average. Sound findings require sampling over a broad range of stand ages and calendar years and surveys over a broad range of site conditions.

## 4.3 Self-Thinning, Alien-Thinning, and Stand Density

### 4.3.1 Theoretical Considerations

#### 4.3.1.1 Lines of Self-Thinning and Alien Thinning

Figure 4.20 shows how mixed-species stands can differ from monocultures in terms of the level and the slope of their tree number-tree size trajectories. We start with monospecific stands as a reference (Fig. 4.20a). During their early open grown stage, monospecific stands remain rather constant in tree number ($\alpha \cong 0.0$). When they become fully stocked and competition intensifies, the tree number-tree size trajectory follows the self-thinning line ($\alpha \cong -1.6$). During this main stand phase, the stand volume can accumulate as the mean tree growing area and the mean tree

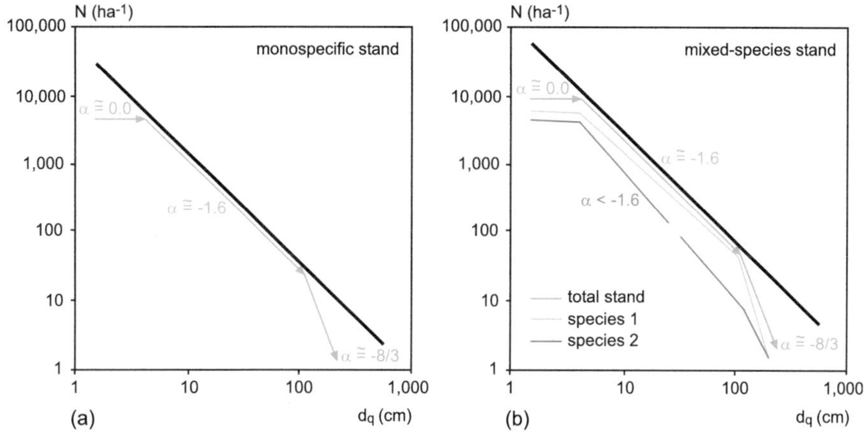


**Fig. 4.20** Decrease in tree number, $N$, per unit area over mean tree size, $d_q$, caused by (**a**) self-thinning in monospecific stands and (**b**) combined self-thinning and alien thinning in mixed-species stands. (**a**) The $\ln(N)$–$\ln(d_q)$ trajectory is shallow or even parallel to the $x$-axis during the early stage, approximately follows the $-1.6$ slope during the phase of stand differentiation, and changes to a slope of around $-8/3$ when the maximum standing stock is achieved in the mature and old stand phases. (b) Mixing can modify the species-specific $\ln(N)$–$\ln(d_q)$ slopes, raise the level of the whole-stand $\ln(N)$–$\ln(d_q)$ slope, and shift the arrival of the stage of maximum standing stock when the slope falls off to around $\alpha = -8/3$

leaf area increase under-proportionally in relation to the mean tree volume. The standing stock can increase until it reaches the maximum standing stock $V$ which is determined by the site conditions. From this stage on, any increase in mean tree diameter is coupled with a decrease in tree number according to $N \propto d^{-8/3}$, as allometric theory predicts $V = \text{const.} \propto N \times \bar{v}$, with $\bar{v} \propto \bar{d}^{8/3}$, so that $V = \text{const.} \propto N \times d^{8/3}$ and $N \propto d^{-8/3}$ (West et al. 2009).

Figure 4.20b shows the same principles for mixed-species stands. However, three aspects of the tree number-tree size trajectories can be changed by species mixing (Pretzsch et al. 2012). We consider these three aspects more closely because they may be used to characterise mixed-species stands. Firstly, the level of the stand self-thinning line may be shifted upwards (or downwards) as the packing density in an interspecific neighbourhood might be higher than under intraspecific conditions. Secondly, as the species interact, the slopes of their tree number-tree size trajectories may deviate distinctly from $\alpha \cong -1.605$. Figure 4.20b shows the rather common case where mixing results in a release of competition and a shallower slope for species 1 ($\alpha > -1.605$). This competition release is partly at the expense of species 2, as indicated by a steeper slope ($\alpha < -1.605$). At the whole-stand level, the slope stays close to $\alpha \cong -1.605$. Thirdly, the bending of the tree number-tree size trajectory from $\alpha \cong -1.605$ to $\alpha \cong -8/3$ may occur later or earlier as the density level can be higher, and the stands need more or less time to reach the maximum standing stock.

Suitable measures for describing the mixed stand dynamics in terms of tree number-tree size dynamics include the level of the total stand self-thinning line (e.g. via the stand density index by Reineke (1933) or the intercept of the self-thinning line of the total stand), the slopes of the self-thinning lines of the total stand and of each individual species ($\alpha_{1,2}$, $\alpha_{1,(2)}$, $\alpha_{(1),2}$), and the arrival time or mean tree size when maximum standing stock is achieved. For the level and arrival time of the maximum standing stock, see also the curve of the standing stock in Fig. 4.5c ($sp_{1,2}$ versus $\widehat{sp}_{1,2}$).

#### 4.3.1.2 Quantification and Standardisation of Stand Density

For quantifying stand density and mixing portions, stem density (N $ha^{-1}$) is a rather ambiguous measure as the mixed species may differ in terms of current mean tree size and growing area requirement. Stand basal area (BA $m^2$ $ha^{-1}$) is a better measure, but better still is the stand density index (Box 4.5) because it considers both tree number and size and is based on allometric theory (Reineke 1933; Pretzsch 2006).

**Box 4.5 Quantifying Stand Density Using the Stand Density Index**

The self-thinning line and stand density index, which characterises its level, are based on the pivotal assumption of stand dynamics that after their juvenile phase, fully stocked stands arrive at a maximum stand leaf area, LA, which represents their maximum stand density and is determined by the environmental conditions. In contrast, the mean tree leaf area, $\overline{la}$, increases at a decreasing rate with mean tree size (e.g. mean tree volume $\bar{v}$), following the allometric relationship $\overline{\mathrm{la}} \propto \bar{v}^{3/4}$. Note that an increase in tree size growth is coupled with an under-proportional increase in leaf area and growing area requirement. Stand leaf area is the product of tree number and mean tree leaf area, $\mathrm{LA} = N \times \overline{\mathrm{la}}$. So later equations can be written as $\mathrm{LA} \propto N \times \bar{v}^{3/4}$, and as LA is constant in the advanced stand development phase further as const $\propto N \times \bar{v}^{3/4}$ or $N \times \bar{v}^{-3/4}$ or $\bar{v} \times N^{-4/3}$. The analogue relationships based on the empirically found allometric exponent of $\alpha_{N,d} = -1.605$ are $N \times \bar{d}^{-1.605}$ or $\bar{d} \times N^{-0.623}$, respectively. This is the theoretical foundation of the self-thinning line and the explanation for why the mean size growth and reduction in tree number are related non-linearly and mass and volume accumulate in stands although the maximum leaf area is not exceeded.

*Quantifying Stand Density by the Level of the Self-Thinning Line and the Stand Density Index (SDI)*

The observed number of trees of a defined reference size (mean diameter or mean volume), the stand density index (SDI), is often used for standardised

(continued)

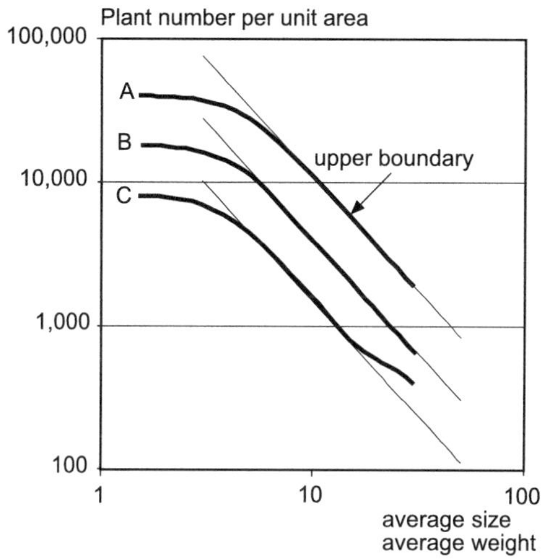


**Box Fig. 4.5-1** The common principle of Reineke's rule (1933) and the −3/2 power law by Yoda et al. (1963) for even-aged plant populations. The relationship between average plant size or average weight and plant number forms a straight line in a double-logarithmic scale. The upper, middle, and lower straight lines represent the self-thinning line of the plant populations A, B, and C and delineate the decrease in plant number under optimal, medium, and poor site conditions, respectively. Initially, the population density remains rather constant until the limited resources enhance competition and lead to self-thinning. A deviation from the self-thinning line, such as the last section of the trajectory C, occurs when environmental conditions and resource supply change

**Box 4.5** (continued)

characterisation of relative stand density, the maximum number of trees of a defined reference size for quantification of maximum stand density.

As plants grow in size, their demands on resources and growing space increase. If resources are no longer sufficient for all individuals, self-thinning commences, and the number of plants N per unit area decreases. Although the principle of allometry was derived for individual trees, its application to plant communities in which self-thinning occurs produces some important conclusions (Reineke 1933; Yoda et al. 1963). Box Figure 4.5-1 is a schematic representation of the relationship between average plant size and density on the double-logarithmic scale. The upper self-thinning, or limiting boundary line (solid line), marks the maximum possible density for a species at a given average plant size, or weight in even-aged monocultures under optimum site conditions. The lower self-thinning line (dashed line) marks the characteristic boundary relationship for any stand under suboptimum growing conditions. Given two stands A and B growing under optimum and suboptimum conditions, respectively, the size-density relationships of each stand

(continued)

**Box 4.5** (continued)

initially approximate their stand-specific self-thinning lines and, subsequently, deviate from this line, at different absolute levels, but with similar gradients.

The stand density rule by Reineke (1933) describes the relationship between mean diameter $d_q$ and stem number $N$ per hectare in a fully stocked, unmanaged, monospecific even-aged stand $N = a\ d_q{}^b$. This relationship can be represented by a straight line in the double-logarithmic coordinate system, $\ln(N) = \ln a + b \ln(d_q)$, with the intercept ln $a$ and slope $b$. Reineke (1933) postulated $N \propto d_q{}^b$, with $b = -1.605$; i.e. an increase in the mean diameter $d_q$ by 1% results in a decrease in tree number $N$ of $b = 1.605\%$ for all species. Reineke's exponent was analysed and differentiated for different tree species (Pretzsch and Biber 2005; Pretzsch 2005b), site conditions (von Gadow 1986), and stand histories (Montero et al. 2001). Reineke (1933) based his stand density index $\mathrm{SDI} = \mathrm{N} \times (25/d_q)^{-1.605}$ on this allometric relationship between stem number N and quadratic mean diameter $d_q$. Sterba (1987), Sterba and Monserud (1993), and Gül et al. (2005) studied the influence of the stand structure on Reineke's slope and the maximum stand density.

For a stand with an observed quadratic mean diameter $d_q$ and an observed tree number $N$ per hectare, the SDI indicates the expected number of trees in a stand with an index diameter of $d_q = 25$ cm. In other words, SDI reflects the number of trees a stand has when $d_q = 25$ cm. It is assumed that the number of trees decreases according to the stand density rule with gradient $b = -1.605$.

The extrapolation from an observed quadratic mean diameter $d_q$ and tree number $N$ to an index diameter of $d_q = 25$ is illustrated in Box Fig. 4.5-2. The tree numbers $N_1, \ldots, N_6$ are plotted against the mean diameter $d_{q\,1}, \ldots, d_{q\,6}$ in a double-logarithmic grid for a forest stand at the survey time points $t_1, t_2, t_3, \ldots, t_6$. To quantify the stand density at the survey time point $t_1$ or $t_6$ with the relative stand density index, a line with a slope $b = -1.605$ is drawn through the points $t_1$ and $t_6$ so that the expected stem number for a stand with a quadratic mean diameter $d_q = 25$ can be read off the graph. Stand density indices result, which, in our example, have the values $\mathrm{SDI}_{t_1} = 200$ and $\mathrm{SDI}_{t_6} = 2000$, and indicate the tree number per hectare this stand would have if the stand density remained constant in the time taken for the stand to attain a quadratic mean diameter $d_q = 25$.

Using the slope of the self-thinning line, the absolute stand densities in different stand development phases can be scaled to the reference diameter of 25 cm in order to eliminate the size effects and make the comparison sounder.

Results similar to this adjustment by mean diameter ratio using the Reineke exponent $\alpha_{N,d} = -1.605$ can be achieved by mean tree volume

(continued)

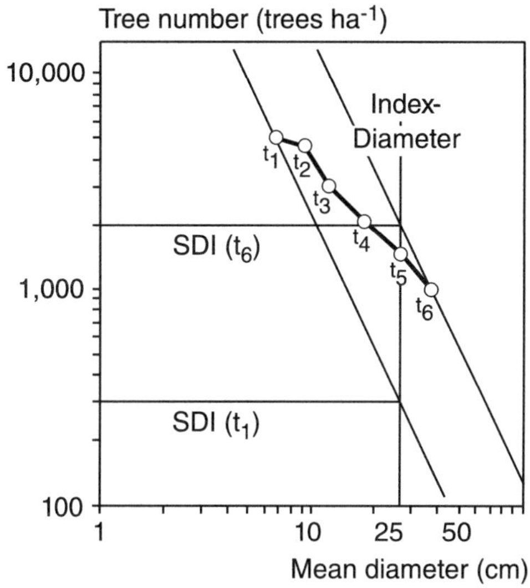


**Box Fig. 4.5-2** To quantify the stand density with the stand density index (SDI), stem number development can be plotted against mean diameter giving the trajectory $t_1 \rightarrow t_2 \rightarrow t_3, \ldots, t_6$ in the double-logarithmic net. To quantify the density in the development stage, $t_1$, or $t_6$, e.g. a line is drawn through these points with the gradient $b = -1.605$. One can then determine the stem number that a stand with an index diameter of $d_q = 25$ cm would have for the SDI $(t_1) = 200$ and SDI $(t_6) = 2000$, respectively

**Box 4.5** (continued)

ratio applying the Yoda exponent $\alpha_{N,\bar{v}} = -2/3$ or the exponent by West et al. (1997, 1999) $\alpha_{N,\bar{v}} = -3/4$ (see Box 4.4). Here we applied the empirically confirmed Reineke exponent, although discussion is still ongoing as to what extent scaling is species specific (Pretzsch and Biber 2005; Zeide 1987, 2004).

When comparing the SDI values of different species and calculating a combined SDI of species in a mixed stand, it is necessary to consider their species-specific growing area requirements. The species-specific growing area requirement of associated species can be indicated by the maximum stand densities ($SDIMAX_1$, $SDIMAX_2$) of the unmanaged, fully stocked monospecific stands of the triplets. The maximum stand densities can be used to derive equivalence coefficients $e_{2 \Rightarrow 1} = SDIMAX_1/SDIMAX_2$ and $e_{1 \Rightarrow 2} = SDIMAX_2/SDIMAX_1$ for converting the SDI from one species to the other.

When analysing common tree species in Central Europe, Pretzsch and Biber (2016) found the following mean equivalent coefficients (mean ± SE) for conversion between European beech (be), Norway spruce (sp), Scots pine (pi), and Douglas-fir (D-f): $e_{be \Rightarrow sp} = 1.32 \pm 0.09$, $e_{be \Rightarrow pi} = 1.10 \pm 0.06$, $e_{be \Rightarrow D\text{-}f} = 1.61 \pm 0.15$, and $e_{pi \Rightarrow sp} = 0.87 \pm 0.12$. The equivalence coefficient may be used to combine species 1 and 2 in the mixed stand ($\mathrm{SDI}_{\underline{1},2} = \mathrm{SDI}_{1,(2)} + \mathrm{SDI}_{(1),2} \times e_{2\Rightarrow 1}$) and to make the mixed stands' density comparable to the monoculture in order to obtain the relative density of the mixed stand in relation to the monoculture ($\mathrm{RD}_{1,2} = \mathrm{SDI}_{\underline{1},2}/\mathrm{SDI}_1$) and thus estimate any overstocking ($\mathrm{SDI}_{\underline{1},2} > \mathrm{SDI}_1$). Note that any differences between the species-specific slopes will distort the comparison more as the mean tree diameters differ from 25 cm.

#### 4.3.1.3 Quantification of the Slope of the Self-Thinning and Alien-Thinning Line

Allometric theory predicts that stands always develop on the same species-specific $\ln(N)-\ln(d_q)$ trajectories but run more quickly through these on fertile compared with poor sites (Long et al. 2004). However, light supply is widely held to modify the self-thinning line in terms of its level and slope (Harper 1977). Whether the slope of the $\ln(N)$–$\ln(d)$ trajectory is modified by mixing can easily be examined on the basis of the allometric exponent $\alpha_{N,d} = (\ln(N_{t_2}) - \ln(N_{t_1}))/(\ln(d_{t_2}) - \ln(d_{t_1}))$ with the tree number ($N_{t_2}, N_{t_1}$) and tree diameter ($d_{t_2}, d_{t_1}$) of a stand at the beginning and end of a survey period (Fig. 4.21). Based on this exponent for a species in the monospecific stand $\alpha_{N_1,d_1}$ and a neighbouring mixed stand $\alpha_{N_{1,(2)},d_{1,(2)}}$, the ratio $R\alpha_{N,d} = \alpha_{N_{1,(2)},d_{1,(2)}}/\alpha_{N_1,d_1}$

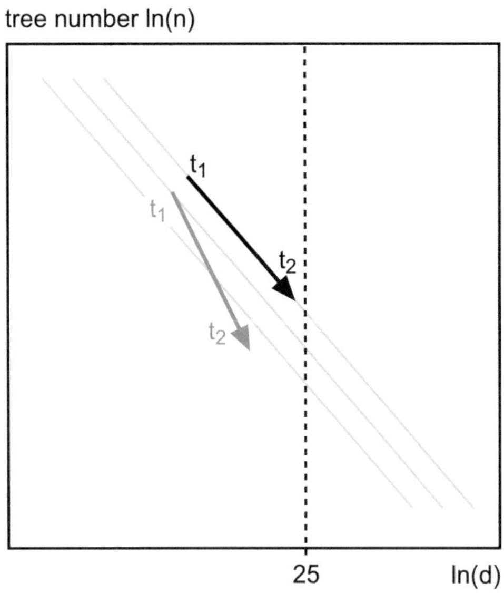


**Fig. 4.21** Analysing the effect of mixing on the slope of the self-thinning line. The upper self-thinning process (*black arrow* $t_1 \rightarrow t_2$) follows the common self-thinning trajectories, while the lower self-thinning process (*grey arrow* $t_1 \rightarrow t_2$) reflects a much stronger reduction in tree number with increasing mean size. Whether the slope of the $\ln(N)$–$\ln(d)$ trajectory is modified by mixing can easily be compared on the basis of the allometric exponent $\alpha_{N,d} = (\ln(N_{t_2}) - \ln(N_{t_1}))/(\ln(d_{t_2}) - \ln(d_{t_1}))$ with the tree number ($N_{t_2}, N_{t_1}$) and tree diameter ($d_{t_2}, d_{t_1}$) of a stand at the beginning and end of a survey period

reflects whether the thinning process is relaxed by mixing ($R\alpha_{N,d} < 1.0$), unaffected ($R\alpha_{N,d} = 1.0$), or aggravated ($R\alpha_{N,d} > 1.0$).

### 4.3.2 Evidence of Mixing Effects on Stand Density, Self-Thinning, and Alien Thinning

Silvicultural operations such as spacing, thinning, or fertiliser application often accelerate tree size development and might occasionally result in a temporary increase in the total leaf area beyond the inherent maximum stand density for a species and site (Long et al. 2004). In contrast, mixing can permanently increase the maximum density by improving the availability and uptake of resources such as water, nutrients, and light (Kelty 1992; Richards et al. 2010; Pretzsch et al. 2013a; Forrester et al. 2006). In fully stocked stands, the level of the self-thinning line or the stand density index (SDI) indicates the maximum stand density (see Box 4.5). Knowledge on the stand density of mixtures compared with monocultures, i.e. the maximum number of trees of a given size that the site can sustain, is among the most relevant for the assessment of site quality, development of silvicultural prescriptions, and stand modelling and prognosis (Pretzsch 2009).

#### 4.3.2.1 Meta-Analysis Based on Long-Term Experiments, Artificial Time Series, and a Study Along a Gradient

Based on 432 triplets of fully stocked monospecific and mixed stands of long-term experiments (L) and artificial time series (A) in Germany and similar triplets along a productivity gradient (G) through Europe, Fig. 4.22 shows how mixing can change the maximum stand density for the mixture of Norway spruce and European beech ($n = 167$), European beech and sessile oak ($n = 148$), silver fir and Norway spruce ($n = 15$), Scots pine and Norway spruce ($n = 18$), Scots pine and European beech ($n = 65$), and lastly European beech and Douglas-fir ($n = 19$).

The mean ratios between the stand densities in mixed versus monospecific stands are rather similar for the various species mixtures and in most cases reveal an increase in stand density through mixing. It is lowest for mixed stands of Norway spruce and European beech (1.00–1.13) and highest for the combination of Scots pine and European beech (1.03–1.39). On average, the mixed stands are 15% denser than the weighted mean of the neighbouring monocultures (mean ratio [95% CI] = 1.15 [1.07, 1.24]).

#### 4.3.2.2 Results of a Study Along a Gradient Through Europe

For the above-mentioned study about mixing effects between Scots pine and European beech along a productivity gradient through Europe, Fig. 4.23 illustrates the standardised stand density as well as further stand characteristics. The SDI and standing volume are, on average, about 15% higher in mixtures compared to the weighted mean of the neighbouring monocultures (Fig. 4.23a and b). In other words, fully stocked mixed stands with a mean diameter of 25 cm maintain about

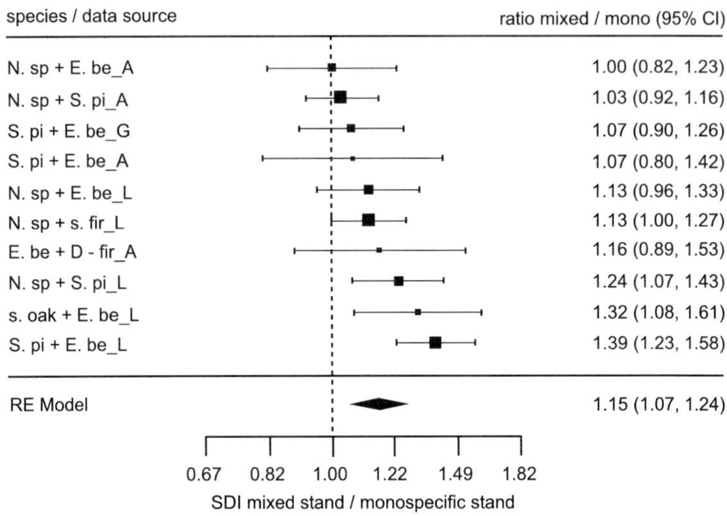


**Fig. 4.22** Comparison of stand density in mixed versus monospecific stands at the stand level for common tree species mixtures in Central Europe. The results are based on a total of 432 triplets from long-term experiments (L), artificial time series (A), and a study along a productivity gradient though Europe (G). The mixture types include Norway spruce and European beech (N. sp. + E. be.), Norway spruce and Scots pine (N. sp. + S. pi.), Scots pine and European beech (S. pi. + E. be.), Norway spruce and silver fir (N. sp. + s. fir), European beech and Douglas-fir (E. be. + D-fir), and sessile oak and European beech (s. oak + E. be.)

a 50–100 trees, or 50–100 $m^3$ $ha^{-1}$, more than monocultures. The mean periodic increment of stand basal area and stem volume in the last 5 years lies about 15% and 10%, respectively, above the weighted mean of the monospecific stands (Fig. 4.23c and d resp.).

If maximum density is higher in mixed than in monospecific stands, and this higher stand density is not reflected in silvicultural guidelines, e.g. in stand management diagrams, the stand might always be kept below the optimal density with respect to productivity. On the other hand, species mixing that reduces the maximum stand density is indicative of an antagonistic relationship. In the following section, we show how species mixing can modify the maximum stand density, the species-specific packing density, and the slope of the ln($N$)–ln($d$) trajectories, which are called self-thinning lines in monospecific stands and which we name alien-thinning lines in mixed-species stands.

#### 4.3.2.3 Effect of Species Mixing on the Slope of the Self-Thinning and Alien-Thinning Line

The Figs. 4.24a and b show that the reduction in spruce tree numbers occurs in mixed stands ($\alpha_{N,d} = -2.76$) more strongly than in monospecific stands ($\alpha_{N,d} = -1.97$), whereas beech is released by mixing ($\alpha_{N,d} = -1.59$ versus 1.669). Figure 4.24c shows this pattern for a broad set of mixed stands that contain beech. Self-thinning rates for beech are at a maximum in monospecific stands ($\alpha_{N,d} = -1.52$), and

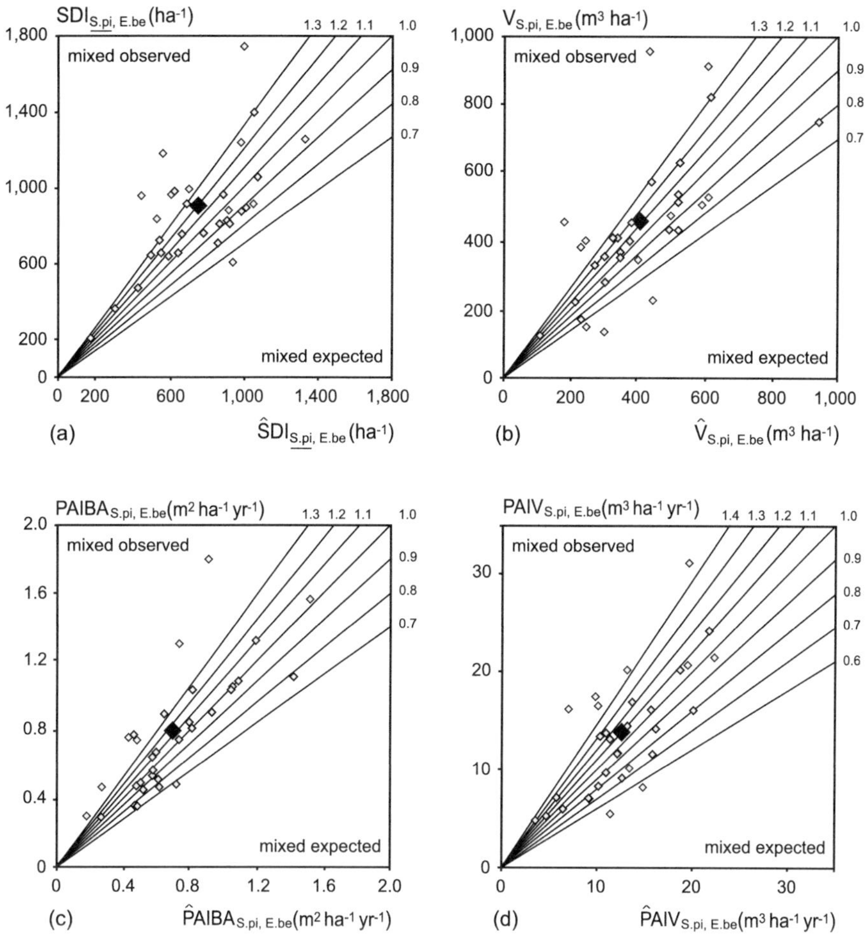


**Fig. 4.23** (**a**–**d**) Characteristics of the expected ($x$-axis, weighted mean of the neighbouring monospecific stands) and observed ($y$-axis) growth and yield of mixed stands of Scots pine/European beech. Values on the ends of the diagonal lines indicate that observed and expected mixed stand characteristics are equal. Values above the 1.0 (1:1) line indicate higher stand density, standing volume, and mean periodic stand growth in mixed compared with monospecific stands. (**a**) Expected stand density index, $\widehat{\mathrm{SDI}}_{\underline{\mathrm{S.pi}},\ \mathrm{E.be}}$, compared with the observed stand density index, $\mathrm{SDI}_{\underline{\mathrm{S.pi}},\ \mathrm{E.be}}$ (the *underline* in $\mathrm{SDI}_{\underline{\mathrm{S.pi}},\ \mathrm{E.be}}$ indicates that the SDI was standardised to species $1 =$ Scots pine) (**b**) expected standing stem volume, $\widehat{V}_{\mathrm{S.pi,\ E.be}}$ ($\mathrm{m}^3\ \mathrm{ha}^{-1}$, of merchantable stem volume $\geq$7 cm over bark) versus observed volume, $V_{\mathrm{S.pi,E.be}}$ (**c** and **d**) expected growth of stand basal area, $\widehat{\mathrm{IBA}}_{\mathrm{S.pi,\ E.be}}$ ($\mathrm{m}^2\ \mathrm{ha}^{-1}$), and tree volume, $\widehat{\mathrm{IV}}_{\mathrm{S.pi,\ E.be}}$ ($\mathrm{m}^3\ \mathrm{ha}^{-1}$), compared with the observed stand basal area, $\mathrm{IBA}_{\mathrm{S.pi,E.be}}$ ($\mathrm{m}^2\ \mathrm{ha}^{-1}$), and tree volume, $\mathrm{IV}_{\mathrm{S.pi,E.be}}$ ($\mathrm{m}^3\ \mathrm{ha}^{-1}$), during the last 5 years before tree-ring analysis and stand inventory

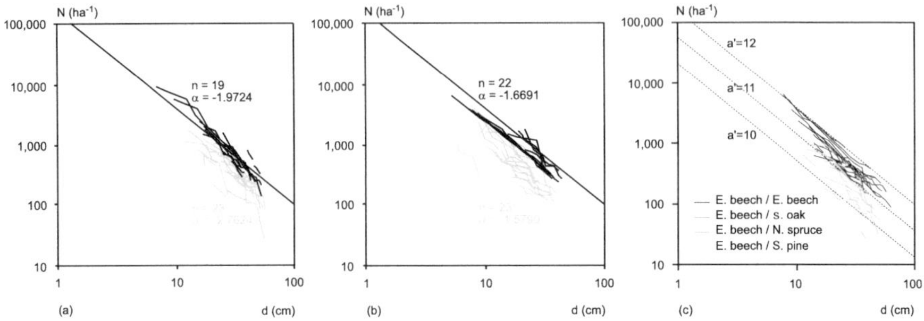


**Fig. 4.24** Modification of the ln($N$)–ln($d$) trajectories through species mixing. (**a**) On average, the alien-thinning slope of Norway spruce in mixture with beech ($\alpha_{N,d} = -2.76$) is steeper than the self-thinning slope in Norway spruce monocultures ($\alpha_{N,d} = -1.97$). (**b**) In the case of European beech, the self-thinning results in $\alpha_{N,d} = -1.669$, while mixing with spruce diminishes tree number reduction and steepness of the slope ($\alpha_{N,d} = -1.57$). (**c**) The mean slope and standard error of the thinning line $N \propto \bar{d}^{\alpha_{N,\bar{d}}}$ of a broad set of different species combinations with European beech are $\alpha_{N,\bar{d}} = -1.52$ ($\pm$0.07 SE) in the European beech monoculture and become more shallow in the admixture with Norway spruce ($-1.25 \pm 0.19$ SE), sessile oak ($-1.12 \pm 0.14$ SE), and Scots pine ($-0.51 \pm 0.32$ SE)

this species is continuously released when growing in mixture with oak ($\alpha_{N,d} = -1.12$), spruce ($\alpha_{N,d} = -1.25$), or pine ($\alpha_{N,d} = -0.51$). Whereas intra-specific competition is high and results in steep self-thinning slopes in monospecific beech stands, in mixture, competition turns against the admixed species and saves beeches from some mortality.

Based on the species-specific alien-thinning slopes in mixture ($\alpha_{1,(2)}$, $\alpha_{(1),2}$) and their species-specific level characterised, e.g. by the SDI values ($SDI_{1,(2)}$, $SDI_{(1),2}$), the slope of the mixed stand in total can be scaled up as $\alpha_{1,2} \cong (\alpha_{1,(2)} \times SDI_{1,(2)} + \alpha_{(1),2} \times SDI_{(1),2})/(SDI_{1,(2)} + SDI_{(1),2})$. This formula for two species can similarly be applied to mixtures with n species.

The slope $\alpha_{N,d}$ reveals the self-tolerance of a tree species growing in monospecific stands (Zeide 1985). The larger the $\alpha_{N,d}$, the lower the number of dying trees $\Delta N/N$ for a given diameter increment $\Delta\bar{d}/\bar{d}$ and the greater the self-tolerance of the species in monocultures. The ranking we revealed for the mean species-specific $\alpha_{N,d}$-values of European beech ($-1.789$) < Norway spruce ($-1.664$) < Scots pine ($-1.593$) < sessile oak ($-1.424$) indicates that in comparison to Norway spruce and European beech, sessile oak and Scots pine are more tolerant of trees of the same species (Pretzsch and Biber 2005). For instance, in European beech stands, a mean diameter increase of 1% causes a 1.79% decrease in the number of stems. Given the same diameter increment, the decrease in the number of stems is 1.66%, 1.59%, and 1.42% for Norway spruce, Scots pine, and sessile oak, respectively, or 7%, 11%, and 21% lower than the decrease in stem number for European beech. This underlines the low self-tolerance of European beech and its space-consuming investment strategy. The causes for this may be its wider and more dynamic lateral crown extension, as discussed in Sect. 6.5. For mixed stands on comparable sites, mean $\alpha_{N,d}$-values determined for European beech, Norway spruce, Scots pine, and sessile oak were $\alpha_{N,\bar{d}} = -0.40$, $-1.02$, $-1.06$, and $-2.01$, respectively, which indicates a reversal of the ranking in monospecific stands, e.g. European beech > Norway spruce > Scots pine > sessile oak (Pretzsch and Biber 2005).

The significant crown expansion and space occupation abilities under intraspecific conditions evidently conferred great assertive power in the mixed stands. While intraspecific competition is high and results in steep N-$\bar{d}$-self-thinning slopes in beech monocultures, in mixed-species stands, this competitive growth strategy turns against the admixed species and saves beeches from losses. Intraspecific competition turns into interspecific competition and self-thinning into alien thinning, and the thinning slope of beech becomes shallower (Fig. 4.24c versus b).

#### 4.3.2.4 Stand Density and Mean Tree Productivity at The Species Level

A change in resource supply can modify stand dynamics in terms of the number of trees (stand density), growth rate per tree (mean tree productivity), or both. Note that the stand productivity, $p$, is the product of stand density, $N$, and mean tree productivity, $\bar{p}$, ($p = N \times \bar{p}$).

For four common species combinations, Fig. 4.25a–d, illustrates how stand density (abscissa) and mean tree productivity (ordinate) in mixed stands deviate from the species performance in neighbouring monocultures (1.0 reference lines). Figure 4.25a, e.g. represents how Norway spruce (above) and European beech (below) come off in terms of stand density and mean tree productivity in the

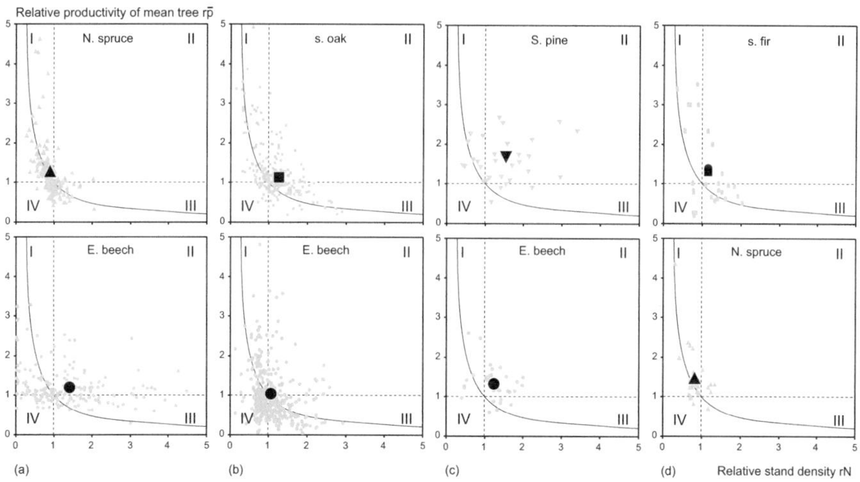


**Fig. 4.25** (**a**–**d**) Effect of species mixing on the relative stand density (abscissa) and relative tree productivity (ordinate) in relation to neighbouring monocultures (reference monospecific stand = 1.0). Results for mixed stands of (**a**) spruce/beech, (**b**) sessile oak/beech, (**c**) pine/beech, and (**d**) silver fir/spruce are shown. The small symbols represent all individual observations on long-term plots; the large symbols indicate average responses to admixture in all plots for the respective species. For a detailed explanation of the graph, the hyperbola, and the meaning of the quadrants I–IV, see Box 4.6

mixed stands of Norway spruce and European beech compared with neighbouring monocultures of the respective species. The iso-productivity hyperbola reflects all combinations of productivity and density effects which yield equality of mixed and monospecific stand productivity. All observations above the line indicate overyielding; all stands below indicate underyielding of the species in mixed versus monospecific stands (see Box 4.6 for further explanation of the graph).

For all species assemblages, the observations are scattered over the quadrants I–IV (Fig. 4.25a–d); however, on average (large symbols) the species achieve overyielding in mixed versus monospecific stands. The fact that in all four mixtures the means of both species lie above the iso-productivity hyperbola indicates that underyielding of one species does not necessarily occur at the expense of the other but that on average both species benefit from mixing. The extent of vertical and horizontal scattering of the observations indicates a very high plasticity for beech and rather low plasticity for the conifers.

The species behave rather differently in mixture in terms of their stand density and mean tree growth. In the mixed stand of spruce and beech (Fig. 4.25a), spruce increases mean tree productivity by about 25% ($R\bar{p}' = 1.25$) and decreases stand density by 10% ($RN' = 0.90$). For further explanation of $R\bar{p}'$ and $RN'$, see Box 4.6. European beech increases productivity by 22% ($R\bar{p}' = 1.22$) and stand density by 44% ($RN' = 1.45$). In other words, spruce benefits from mixture via higher growth efficiency (in terms of growth per growing area) and beech via higher stand density and growth efficiency. While beech mostly benefits from mixing via stand density (Fig. 4.25a–c, below), spruce and fir mainly profit via an increase in growth efficiency (Fig. 4.25a and d, above).

The mean mixing effect on tree growth efficiency ranges between $R\bar{p}' = 1.05 - 1.74$ and for stand density between $RN' = 0.84 - 1.74$. The conifers profited from mixing mainly via an increase in efficiency, while the broadleaved species profited mainly via an increase in stand density. Thus, thinning may be more likely to cancel mixing effects for broadleaves rather than for conifers.

**Box 4.6 Differentiation Between stand Density and Efficiency Effects of Species Mixing**

Here we introduce an approach to examine whether the responses of a species in terms of overyielding or underyielding at the stand level come from a change in stand density or a change in growing space efficiency. This distinction is relevant for both scientific analyses and practical profiting from mixing effects. Supposing mixing simply increased the packing density of trees in a stand and did not change the individual tree's size growth, mixing effects could not be revealed through common experiments which keep the stocking level constant and below the maximum density. In forest practice, mixing effects in terms of a rise in density might be 'thinned away', while accelerated size growth would not be cancelled through density reduction.

(continued)

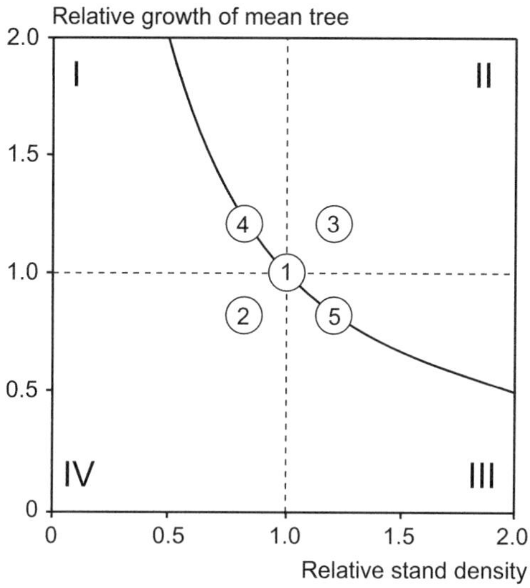


**Box Fig. 4.6-1** Graphical illustration of a species' response to mixing considering stand density and mean tree productivity in relation to its performance in a neighbouring monoculture. The stand density and mean tree productivity in the monoculture are represented by the reference values $RN = R\bar{p} = 1.0$ on the hyperbola. Observations above the hyperbolic isoline indicate overyielding and those below it indicate underyielding of the species in mixed versus the monospecific stand. Quadrants I–IV indicate whether the performance in multi-species conditions differs from that in monocultures, in terms of mean tree productivity and not in terms of density (quadrant I), superior in terms of both mean tree productivity and density (quadrant II), superior in terms of density but inferior in terms of productivity (quadrant III), and inferior compared to a monospecific stand in terms of both mean tree productivity and stand density (quadrant IV). The circles with numbers represent the outcome of the five generic mixing reaction patterns shown in Box Fig. 4.6-2

**Box 4.6** (continued)

Box Figure 4.6-1 shows a diagram for graphically analysing whether a change in stand density or in mean tree productivity is behind a species' mixing response compared with a neighbouring monoculture. In the graph, the species' stand density and mean tree productivity in the monoculture are represented by the reference values $RN = R\bar{p} = 1.0$ on the hyperbola. The points in the graph represent the performance of a species in a mixed stand in relation to a neighbouring monoculture. The abscissa scales the relative density, RN, which results as a ratio between the species density in the mixed and monospecific stand (e.g. $RN_{1,(2)} = N_{1,(2)}/m_1/N_1$). The ordinate scales the relative mean tree productivity $R\bar{p}_{1,(2)} = \bar{p}_{1,(2)}/\bar{p}_1$. Note that in the case of size differences between mixed and monospecific stands, the comparison should be based on $RN'$ and $R\bar{p}'$ as introduced in Box 4.5 in order to avoid confusion of mean tree size differences with stand density effects.

(continued)

**Box 4.6** (continued)

Points, i.e. RN–R$\bar{p}$ combinations, on the hyperbola indicate combinations of relative stand density and mean tree productivity which cancel each other out so that mixing changes both density and efficiency, but the productivity in total is equal to the monoculture ($RN = R\bar{p}^{-1}$ and $RN \times R\bar{p} = 1$). Points above the hyperbola reflect overyielding ($RN \times R\bar{p} > 1$), and points below this borderline indicate underyielding ($RN \times R\bar{p} > 1$). Their location in the quadrants I–IV indicates whether density or growth is behind the mixing response. Points above the iso-productivity hyperbola indicate overyielding: When they lie in quadrant I, this indicates overyielding caused by an increase in mean tree productivity only; location in quadrant II indicates an increase in both density and productivity; and points in quadrant III indicate that density, but not mean tree growth, is ahead of the monoculture. The same diagnosis can be applied analogously for the cases of underyielding, i.e. for points below the iso-productivity hyperbola lying in quadrants I, III, and IV.

For further understanding, we show schematically how mixing effects in terms of efficiency and density emerge on the tree level (Box Fig. 4.6-2a and b), scale to the stand level (Box Fig. 4.6-2c and d), and are reflected in the RN–R$\bar{p}$ diagrams (see circles with numbers in Box Fig. 4.6-1).

Line 1 in Box Fig. 4.6-2a illustrates schematically the typical decrease in mean tree growth with increasing stand density in a monospecific or mixed stand. Line 2 reflects a decrease in both growth and density, and line 3 is of special interest when mixing increases growth as well as stand density. The latter case may result from strong niche complementarity (e.g. combination of light-demanding species in the upperstorey with shade-tolerant species in the understorey) which allows foraging of more trees per hectare and an increased maximum stand density, combined with a complementarity effect, such as in terms of water supply due to the improvement of the water-holding capacity of the humus layer. Box Figure 4.6-2b again reflects Line 1 which represents a monospecific stand or the nonspectacular case that mixing leaves growth and density of the particular species unmodified. In comparison, reference line 4 reflects a stand where maximum density is lower but mean productivity at maximum density is higher, and line 5 represents the opposite case where the maximum density is higher but mean productivity at maximum density is lower.

Upscaling of the mean growth-density lines 1–5 to the stand level ($p = N \times \bar{p}$) produces the respective unimodal stand productivity-stand density curves in Box Fig. 4.6-2c and d. They are useful for discussing snares of mixed stand analysis:

Suppose line 1 represents the growth of a species in a monoculture (or a species with neutral mixing effects in mixture) and line 3 the performance of this species in the neighbouring mixed stand and both species are kept at a stand density of 0.6 (see Box Fig. 4.6-2c, vertical line). Then the relationship

(continued)

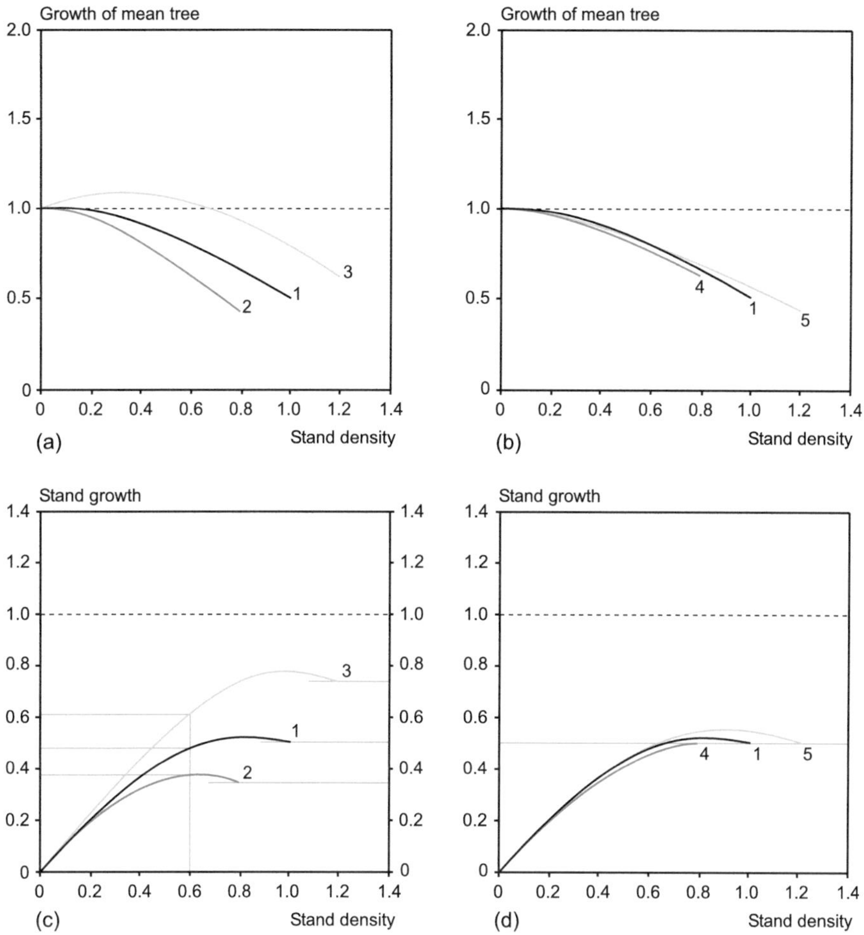


**Box Fig. 4.6-2** (**a**–**d**) Schematic representation of generic mixing responses (Lines 1–5) represented by the productivity of the mean tree in relation to stand density (**a** and **b**) and by stand growth in relation to stand density (**c** and **d**). Upscaling of curves at the tree level (above) results in the well-known unimodal growth-density relationship (below). Line 1: Growth-density response in a monoculture and in a mixed stand with a neutral mixing response, Line 2: decrease in growth and maximum density caused by mixtures on rich sites, Line 3: increase in relative growth rate and maximum density caused by mixtures on poor sites, Line 4: decrease in maximum density coupled with increase in productivity at maximum density, Line 5: increase in maximum density coupled with a decrease in productivity at the maximum stand density (see further explanation in text)

**Box 4.6** (continued)
between productivity of the mixed (0.62) versus monospecific stands (0.48) is 1.3, i.e. the overyielding amounts to 30%. However, if both stands are allowed to exploit all resources and approach maximum density, then the relationship between the mixed (0.74) and monospecific (0.51) stands would amount to 1.46, e.g. the overyielding is 17%. As most of the mixing effect has simply been thinned away in the first case (density 0.6), the mixing effect will be underestimated.

Suppose line 1 represents the monospecific stand and line 2 the neighbouring mixed stand and again both are kept at a stand density of 0.6. Then the mixed stands underyield the monospecific stands by 21%. When both stands are kept unthinned (maximum density), the underyielding would increase to 32%; in other words, in this case, thinning disguises part of the underyielding of mixed versus monospecific stands.

Only in the case of a full balance of stand density and mean tree productivity (see Box Fig. 4.6-2d, lines 1, 4, and 5) does the density control become irrelevant for the detection and quantification of mixing effects in terms of gains or losses of total stand productivity. Note that lines 1, 4, and 5 end at different stand densities but at the same level of stand productivity of $\bar{p} = 0.5$. In the RN– R$\bar{p}$ diagrams, this is reflected by the location of all three observations on the iso-productivity hyperbola (Box Fig. 4.6-1).

These theoretical considerations show that data for analysing mixing effects on stand growth should include records from unthinned or fully closed monospecific and mixed stands because these reveal a species' potential performance in terms of density and efficiency close to the maximum stand density, while lower densities may lead to misinterpretations.

## 4.4 Effect of Tree Species Mixing on the Total Yield and the Yield Level

### 4.4.1 Analysing the Rule by Eichhorn (1902) and the General Yield Level by Gehrhardt (1923) and Assmann (1961)

The rule by Eichhorn (1902) describes the relationship between stand volume and mean height ($V = f(h)$), whereas the general yield level by Gehrhardt (1923) and Assmann (1961, pp 158–160) describes the relationship between total yield and mean tree height ($\mathrm{TY} = f(h)$). The following analysis of how tree species mixing affects total yield and the yield level (Eichhorn 1902; Assmann 1961) was based on 141 combinations of mixed stand plots and neighbouring monocultures of silver fir/Norway spruce, Norway spruce/Scots pine, Norway spruce/European larch, Norway spruce/European beech, Norway spruce/red alder, Scots pine/European beech, European larch/European beech, European beech/sessile oak, and European beech/Douglas-fir that represent maximum stand density (Pretzsch et al. 2016b).

**Table 4.4** Species mixing can significantly increase the stand density and standing stock but leaves the mean tree values almost unchanged compared with neighbouring monocultures

| | | Sample sizes | Group means (± SE) | | Ratios |
|---|---|---|---|---|---|
| Variables | Units | $n$ | mixed | mono | mixed/mono (± SE) |
| Mean tree dimensions | | | | | |
| $h_q$ | m | 141 | 29.25 (± 0.52) | 29.85 (± 0.50) | **0.98* (± 0.008)** |
| $d_q$ | cm | 141 | 32.10 (± 0.88) | 32.18 (± 0.85) | 1.01 (± 0.100) |
| $v_q$ | $m^3$ | 141 | 1.36 (± 0.09) | 1.37 (± 0.09) | 1.05 (± 0.033) |
| Stand density and standing volume | | | | | |
| $N$ | Trees $ha^{-1}$ | 141 | 752 (± 54) | 635 (± 40) | **1.22*** (± 0.040)** |
| BA | $m^2\ ha^{-1}$ | 141 | 42.12 (± 1.43) | 38.09 (± 1.12) | **1.12** (± 0.024)** |
| SDI | Trees $ha^{-1}$ | 141 | 793 (± 27) | 717 (± 20) | **1.16*** (± 0.025)** |
| $V$ | $m^3\ ha^{-1}$ | 141 | 561.38 (± 21.66) | 525.59 (± 19.52) | **1.08** (± 0.026)** |
| Total stand volume yield | | | | | |
| TY | $m^3\ ha^{-1}$ | 79 | 979.85 (± 42.50) | 883.85 (± 37.61) | **1.12** (± 0.027)** |
| Eichhorn's rule and general yield level | | | | | |
| $V_{hq20}$ | $m^3\ ha^{-1}$ | 141 | 325.57 (± 13.06) | 282.06 (± 8.40) | **1.16** (± 0.028)** |
| $TY_{hq20}$ | $m^3\ ha^{-1}$ | 79 | 509.42 (± 21.59) | 419.78 (± 13.13) | **1.21*** (± 0.030)** |

The table displays the stand characteristics for mixed-species stands (group mean for mixtures) in relation to the weighted mean of neighbouring monocultures (group mean for monocultures) and ratios mixed/mono. Ratios above/below 1.00 indicate a superiority/inferiority of the species' performance in mixed-species stands versus monocultures. Ratios in bold numbers indicate significant differences ($p < 0.05$) between mixed-species stands and monocultures. The mean of the ratios (mixed/mono) is not necessarily equal to the ratio of the means (mean mixed/mean mono). So, we report both the groupwise arithmetic means (mean mixed and mean mono) as well as the mean ratios of the pairwise comparison (mixed/mono). The mean ratios of the pairwise comparison (mixed/mono) were used for testing group differences. Further explanation can be found in the text. Mean of the tree height, diameter, and individual tree volume, $d_q$, $h_q$, $v_q$; tree number per hectare, $N$; stand density index by Reineke (1933), SDI; standing stock of volume, $V$; total yield, TY; standing volume at index stand height 20 m, $V_{hq20}$; total yield at index stand height 20 m, $TY_{hq20}$

The mean stand height (−2%) and tree diameter (+1%) of mixed stands hardly differed from the weighted mean values of the neighbouring monocultures (Table 4.4). Minor inferiorities in the mean dimension of one species are mostly compensated by a minor superiority of the other species.

### 4.4.2 Mixing Modified Standing Volume and Total Yield Rather Than Tree Height and Site Index

The stand density, indicated by the tree number, the stand density index, the stand basal area, and the standing volume, in contrast, are 8–22% higher in mixed stands than in monocultures (Table 4.4). In most mixtures, both species increase their stand density compared with the neighbouring monocultures. The combination of

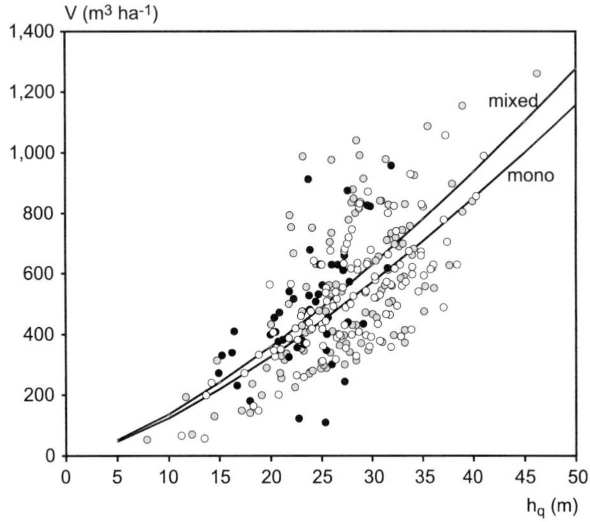


**Fig. 4.26** Standing volume, $V$, over stand mean height, $h_q$, for monoculture (*empty symbols*) compared with mixed-species stands (*filled symbols*). The regression lines are based on $\ln(V) = 1.13\ (\pm 0.282) + 1.526\ (\pm 0.085) \times \ln(h_q) + 0.086\ (\pm 0.041) \times$ mixed mono with mixed mono = 0 for monocultures and mixed mono = 1 for mixed-species stands ($n = 282$, $R^2 = 0.536$, $p < 0.001$)

invariable mean height and increases in stand density and total yield results in a significant increase in the level of the relationship by Eichhorn (+16%) and the general yield level by Assmann (1961) (+21%) in mixed versus pure stands (Fig. 4.26, Table 4.4).

**Box 4.7 Analysing the Tree Species Mixing Effect on the Rule by Eichhorn (1902) and the General Yield Level by Assmann (1961)**

Stand volume, $V$, and stand total volume yield, TY, increase progressively with advancing stand age and height (Box Fig. 4.7-1a). The rule by Eichhorn (1902) describes the relationship between stand volume and mean height ($V = f(h)$), and the general yield level by Gehrhardt (1923) and Assmann (1961, pp 158–160) describes the relationship between total yield and mean tree height ($\mathrm{TY} = f(h)$).

For the subsequent analysis, we derived the species-specific allometric exponents $\alpha_{V,h_q}$ and $\alpha_{\mathrm{TY},h_q}$ for the relationships between volume and mean height ($V \propto h_q^{\alpha_{V,h_q}}$) and total yield and mean height ($\alpha_{\mathrm{TY},h_q}$). Based on these relationships, we could project a standing volume of $V$ which was measured in a stand with mean tree height $h_q$ onto an index height of $h_q = 20$ m using the formula $V_{hq20} = V \times (20/h_q)^{\alpha_{V,h_{q20}}}$. Analogously, $\mathrm{TY}_{hq20} = \mathrm{TY} \times (20/h_q)^{\alpha_{\mathrm{TY},h_{q20}}}$ was applied to project the total yield onto the index height of $h_q = 20$. The values $V_{hq20}$ and $\mathrm{TY}_{hq20}$ reflect the standing stand volume and total volume yield which could be expected for a stand with mean height $h_q = 20$. Box Figure 4.7-1a and d

(continued)

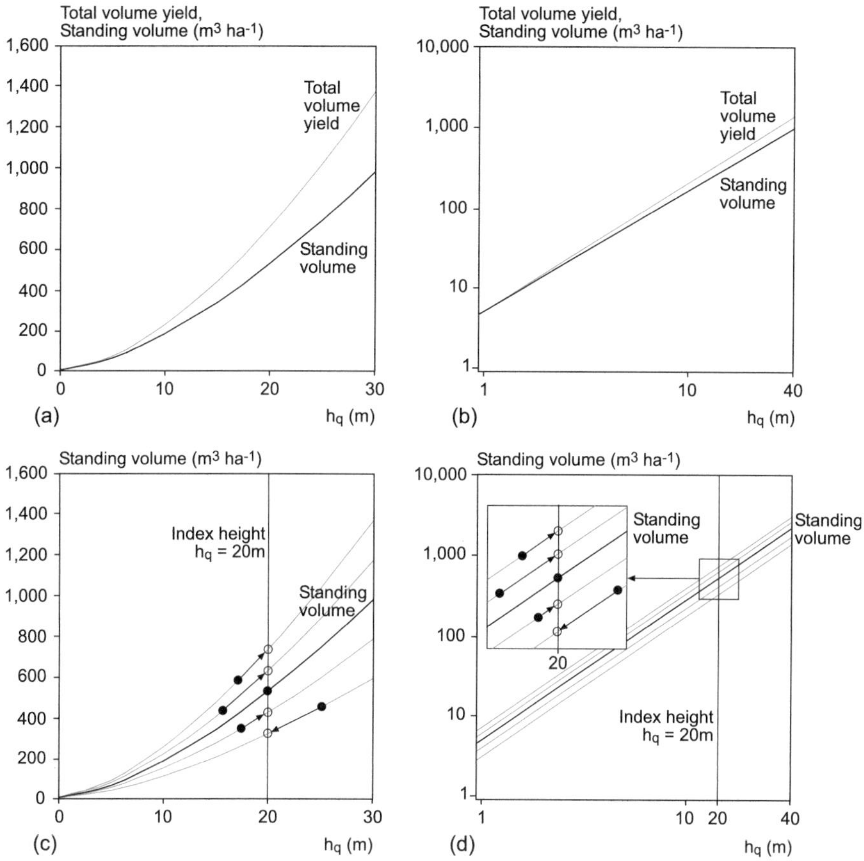


**Box Fig. 4.7-1** Relationship between standing volume and mean tree height and total yield and mean tree height in linear (**a** and **c**) and double-logarithmic (**b** and **d**) representation and derivation of the indices $V_{\mathrm{hq20}}$ and $\mathrm{TY}_{\mathrm{hq20}}$ for quantifying the mixing effect on the relationship by Eichhorn (1902) and on the general yield level by Gehrhardt (1923) (see Assmann (1961), pp 158–160). (**a**) The dependency of the standing volume, $V$, of the mean tree height, $h_g$, reflects Eichhorn's (1902) relationship, and the dependency of total yield, TY, of the mean tree height represents Gehrhardt's general yield level. (**b**) Eichhorn's rule and Gehrhardt's yield level in double-logarithmic representation. (**c**) Derivation of the index $V_{\mathrm{hq20}}$ for the quantification of the mixing effect on Eichhorn's rule in linear representation. (**d**) Derivation of the index $V_{\mathrm{hq20}}$ in the double-logarithmic scale; the inset shows graphically how observed standing volumes, $V$, and mean tree heights, $h_{\mathrm{q}}$, are projected to the index height $h_{\mathrm{q}} = 20$ m by the approach $V_{hq20} = V \times \left(20/h_q\right)^{\alpha_{V,\,h_{q20}}}$. The index $\mathrm{GWL}_{\mathrm{hq20}}$ for characterising the mixing effect on the general yield level (Gehrhardt 1923) was derived analogously to $V_{\mathrm{hg20}}$. For further explanations, see text

**Box 4.7** (continued)
illustrate this approach which is analogous to the evaluation of Reineke's stand density index (1933) for $V_{hq20} = V \times (20/h_q)^{\alpha_{V,hq20}}$ (see also Box 4.5).

The starting point is the allometric relationship between stand volume and mean stand height ($V \propto h_q^{\alpha_{V,hq}}$) shown in Box Fig. 4.7-1a and b in linear and double-logarithmic representation. Box Figures 4.7-1c and d show graphically how the standing volume $V_{hq20}$ at index height 20 m can be read off by extrapolation along the allometric relationships with slope $\alpha_{V,hq}$. Analogously, the $TY_{hq20}$ values can be read off based on extrapolation along the allometric relationship with slope $\alpha_{TY,h_q}$. The expected stand volumes and total yield at index height 20 m can be derived in this way for monocultures and mixed-species stands; in Sect. 4.4, this approach was applied for comparing Eichhorn's rule and Gehrhardt's and Assmann's general yield level in mixed stands with monocultures.

The allometric exponents $\alpha_{V,h_q}$ and $\alpha_{TY,h_q}$ for the above analyses were derived through regression analyses from the yield tables by Schober (1975) for moderately thinned stands of site index I, II, and III. This resulted in values for Norway spruce ($\alpha_{V,h_q} = 1.899$), Scots pine (1.398), European beech (2.165), sessile oak (2.017), Douglas-fir (1.491), and European larch (1.481). Analogously, $TY_{h20}$ values were derived for Norway spruce ($\alpha_{TY,h_q} = 2.199$), Scots pine (1.998), European beech (2.335), sessile oak (2.677), Douglas-fir (1.827), and European larch (1.882).

Comparison between the standing volume, $V$, and total yield, TY, of mixed and monospecific stands can be based on general allometric relationships which allow the extrapolation of $V$ and TY to an index stand height of 20 m (see Box 4.7). This allows the comparison of Eichhorn's rule and the general yield level in mixed-species stands versus monocultures even when the underlying long-term experiments differ in terms of the stand development phase.

While Table 4.4 reflects how Eichhorn's rule and the general yield level differ on average between mixed-species stands and monocultures, Table 4.5 shows the species-specific differences for the $V_{hq20}$ values. All the species assemblages examined have higher $V$–$h_q$ relationships (+5–43%) than expected for monocultures, and the superiority is significant in the cases of N. spruce/S. pine (+22%), S. pine/E. beech (+40%), and S. oak/E. beech (+11%).

Mixing modified stand density, standing volume, and total yield rather than tree height and site index (Pretzsch 2016; Vallet and Perot 2016). The increase in stand density, Eichhorn's rule, and general yield level should be taken into consideration when modelling stand growth (e.g. modification of self-thinning slopes and mortality modules developed for monocultures), inventory of standing volume by yield tables (e.g. correction of yield table estimates by the measured degree of stocking),

**Table 4.5** Standing stock at a mean tree height of 20 m of mixed-species stands in relation to the neighbouring monocultures calculated separately for five selected species assemblages

| Variable $V_{hq20}$ ($m^3\ ha^{-1}$) | Species combination | $n$ | Species 1 mixed/mono (± SE) | Species 2 mixed/mono (± SE) | Total stand mixed/mono (± SE) |
|---|---|---|---|---|---|
| | N. spruce/S. pine | 7 | **1.19 (±0.07)** | **1.38 (±0.04)** | **1.22 (±0.04)** |
| | N. spruce/E. larch | 10 | 1.84 (±0.94) | 1.32 (±0.19) | 1.43 (±0.31) |
| | N. spruce/E. beech | 52 | 1.07 (±0.04) | 1.05 (±0.04) | 1.05 (±0.03) |
| | S. pine/E. beech | 17 | **1.41 (±0.07)** | **1.40 (±0.09)** | **1.40 (±0.07)** |
| | s. oak/E. beech | 24 | 1.02 (±0.04) | **1.21 (±0.07)** | **1.11 (±0.04)** |

Ratios mixed/mono above/below 1.00 indicate a superiority/inferiority of Eichhorn's relationship in mixed-species stands versus monocultures. Ratios in bold numbers indicate significant differences ($p < 0.05$) between mixed-species stands and monocultures

or development and application of silvicultural prescriptions in mixed-species stands (e.g. increased normal density and number of selected future crop trees).

Whereas findings in mixed-species stands versus monocultures have been consistent for these Central European species (e.g. Vallet and Perot 2016), further empirical investigations would be necessary to provide evidence of these patterns for different species assemblages.

## 4.5 Spatial and Temporal Variation of Mixing Effects

### 4.5.1 Theoretical Considerations

One of the most commonly used ecological hypotheses is the stress-gradient hypothesis (Callaway and Walker 1997; Holmgren et al. 1997), which predicts that facilitation occurs under harsh conditions whereas competition is more likely under favourable conditions (Fig. 4.27a). This means that the relative benefit of species richness, e.g. the transition from monocultures to two-species or three-species mixture in terms of productivity gains, is higher on poor than on rich sites if facilitation is occurring within the stand.

This concept could help to better understand the controversially debated relationship between productivity and species richness (Fig. 4.27b and 4.13b). There has been debate as to whether this relationship is represented by an increasing straight line, a saturation curve, an optimum curve, or even a stepped, noncontinuous trajectory (Körner 2002, p 985). In Fig. 4.27b, we assumed a saturation curve according to Hector et al. (1999) and Liang et al. (2016) and drew in the trajectories observed on our long-term experimental plots, which will be reported in this section.

In the example shown in Fig. 4.27, the transition from monospecific to two-species stands results in a strong increase in productivity on poor sites (A), moderate increase on mediocre sites (B), and constant, or even slightly decreasing

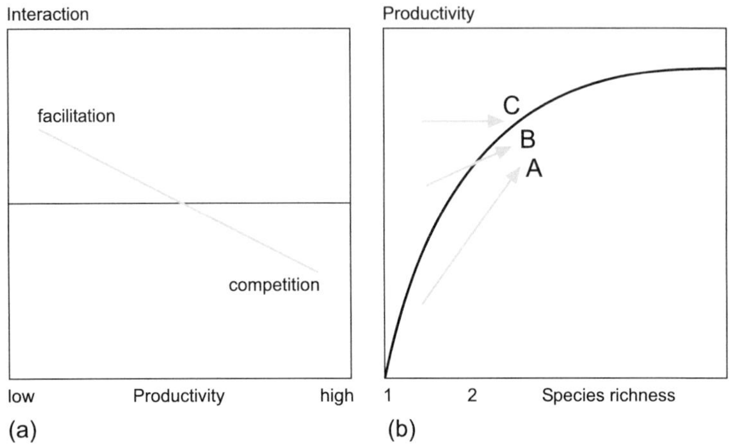


**Fig. 4.27** Dependency of species mixing effects on site conditions and species richness. (**a**) Change from facilitation dominated interactions to competition along the gradient from low to high productivity sites as predicted by the stress-gradient hypothesis (Holmgren et al. 1997; Callaway and Walker 1997). (**b**) Site-dependent relationships between productivity and species richness observed in this study (*grey lines* A, B, and C for poor, mediocre, and rich sites) and expected relationship (*solid saturation curve*) according to Körner (2002)

productivity on fertile sites (C). This illustrates how species richness-productivity experiments may bring different results depending on the initial site conditions and that the apparently contradictory findings might converge when the site conditions are taken into consideration as a modifier and third dimension. In this example, the more fertile the initial site conditions the shallower the slope of the observed trajectories (A > B > C) which could indicate a mainly positive but attenuating feedback between stand and local environment: The mixture has a positive effect and improves the site conditions, so that the additional benefit gradually becomes smaller. This attenuating feedback effect supports the hypothesis that the species richness-productivity curve follows a saturation curve. It is important to note that this interpretation, based on the stress-gradient hypothesis, requires that the mixing effects were caused by facilitation and not competitive reduction, because competitive reduction can lead to increasing levels of mixing effects with increasing site conditions.

Several other ecological hypotheses have been used to describe how interactions between a given combination of species changes from site to site and over time. Chapter 2 (2.3.3) and Chapter 3 (3.6.1) describe and apply a recent framework that was developed to explain these spatial and temporal dynamics. Chapter 2 also describes some of the ecological hypotheses that have been used to describe these dynamics. All of these hypotheses can be seen as special cases within the general framework described in Sect. 3.6.1.

### 4.5.2 Empirical Evidence of Spatial and Temporal Variation in Overyielding and Underyielding at the Stand Level

#### 4.5.2.1 Spatial Variability in Terms of Site Conditions

Only very few works address the dependency of mixing effects on site conditions (Forrester (2014) and Sect. 3.6). The study by Jensen (1983) along a west–east transect through Jutland, Denmark, gives a model example of site condition effects on the growth relationship between Norway spruce and silver fir. In the coastal dune belt, silver fir is superior to Norway spruce, whereas in the adjacent Riss-glacial landscape, the growth of silver fir and Norway spruce is similar and, on the old inland moraines of the Würm glacial period, silver fir is inferior to Norway spruce in terms of dry biomass production. A probable determining factor for the inland superiority of Norway spruce is its adaptability to lower water supplies on acidic sites. In contrast, silver fir profits from higher water availability and the more favourable nutrient supply in the coastal region. Frivold and Frank (2002) and Frivold and Kolström (1999) studied silver birch (*Betula pendula*), Scots pine, and Norway spruce growth in Finland, Sweden, and Norway. Like Jensen (1983), they emphasise that the over- or underyielding of mixed versus monospecific stands is related to site conditions. In Southern and Central Finland, Scots pine/silver birch mixtures surpass Scots pine/silver birch monocultures by 10% and 14%, respectively (Mielikäinen 1980). For Norway spruce/silver birch mixtures, a 10–15% increase in production compared to corresponding monocultures of these species may occur depending on the site (Mielikäinen 1985). In the oceanic regions of Norway and Sweden, silver birch loses some of its increment capacity compared with coniferous species. There, Scots pine/silver birch mixtures do not achieve greater yield than monospecific stands, whereas Norway spruce/silver birch mixtures show a beneficial effect from mixtures only during the juvenile stand development phase (Frivold and Frank 2002).

Pretzsch et al. (2013b) pooled 37 long-term mixing experimental plots of oak and beech in Poland, Germany, and Switzerland which cover a broad range of site conditions. To analyse any relationships between mixing effects and site quality, they used the height of the tree with the quadratic mean diameter at an age of 100 years, referred to as the site index $h_{qoak}$ and $h_{qbe}$ (see Pretzsch 2009, pp 200–203) for the definition and calculation of $h_q$ as a surrogate for site quality. As most of the survey series included species up to the age of 100 the $h_{qoak}$ and $h_{qbe}$, observed values from the monocultures were available. The height of the oak at age 100 ranged from 20 to 35 m; in the case of beech, the range was from 20 to 45 m or even wider. Based on the relative productivities (mixed versus monospecific stands), the mixing portion, and the site index, Pretzsch et al. (2013b) fitted the model shown in Fig. 4.28. This reveals how the volume growth in mixed stands changed from gains of 50% to losses of 10% depending on the site conditions and mixing portions. For both species, an increase in site index has a significantly negative effect on the relative productivity. Figure 4.28a and b show the model curves against the scattergrams for oak and beech in mixture. In order to demonstrate the site effect, we inserted height values of 10 . . . 40 into the model. On sites

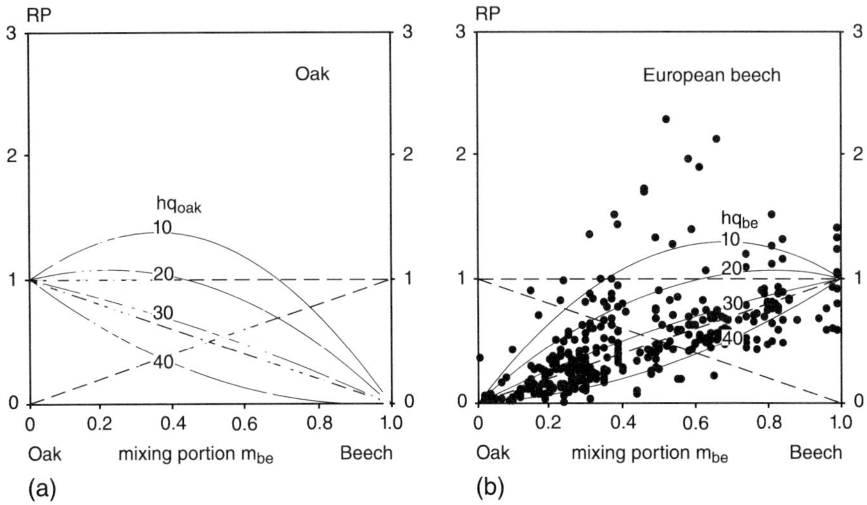


**Fig. 4.28** Species-specific relative productivity in mixed versus monospecific stands of (**a**) oak and (**b**) European beech depending on mixing portion, $m$, and height of the tree with the quadratic mean diameter at age 100, $h_q$, as an indicator for site fertility (Pretzsch et al. 2013b). (**a**) Observed relative productivity for oak plotted over the mixing portion of beech, $m_{be}$, (*filled rectangles*) and model prediction in relation to the admixture of beech and site fertility (curves with $h_{q\ oak} = 10 \ldots 40$ m). (**b**) Observed relative productivity for beech plotted over the mixing portion of beech, $m_{be}$, (*filled circles*) and model prediction in relation to the admixture of beech and site fertility (curves with $h_{q\ be} = 10 \ldots 40$ m)

with a low site quality ($h_q < 25$ m), both species profit considerably from the mixture. Their relative productivity even exceeds the 1.0 lines. On mediocre sites ($h_q = 25$–$30$ m), the species still show a positive mixing effect, but do not exceed the level of the neighbouring monocultures. On fertile sites ($h_q > 35$ m), mixing causes a loss of productivity in relation to neighbouring monocultures.

Figure 4.29 helps to illustrate even more clearly the relationship between site quality and the mixing responses of oak and beech (a) and the mixed stand total (b). It displays the relationship between the relative productivity (mixed versus monospecific stand) in relation to the height of the tree with the quadratic mean diameter at age 100 of oak $h_{qoak}$ and beech $h_{qbe}$, fitted by a simple linear equation. In the case of oak, as well as beech, the relative periodic annual volume increment decreases significantly with increasing site quality, indicated by significantly negative slopes. Insertion of $h_q = 20$ m in the linear model yields $RPA_{oak,be} = 1.49$ and $RPA_{(oak),be} = 1.38$, equivalent to an increase in productivity by mixing of 49% in the case of oak and 38% in the case of beech. Medium site conditions (indicated by $h_q = 30$ m) yield +11% and 15% for the productivity of oak and beech, respectively, in mixture versus monospecific stands. On very fertile sites ($h_q = 40$ m), the model predicts a loss of 27% for oak and 8% for beech. On the stand level, the gains and losses in total for $h_{qoak}$-values of 20, 30, and 40 m result in +32%, +7%, and −18%, respectively. The often positive total stand response shown in Fig. 4.29b can be interpreted as a mutualistic mixing interaction between oak and beech on poor sites,

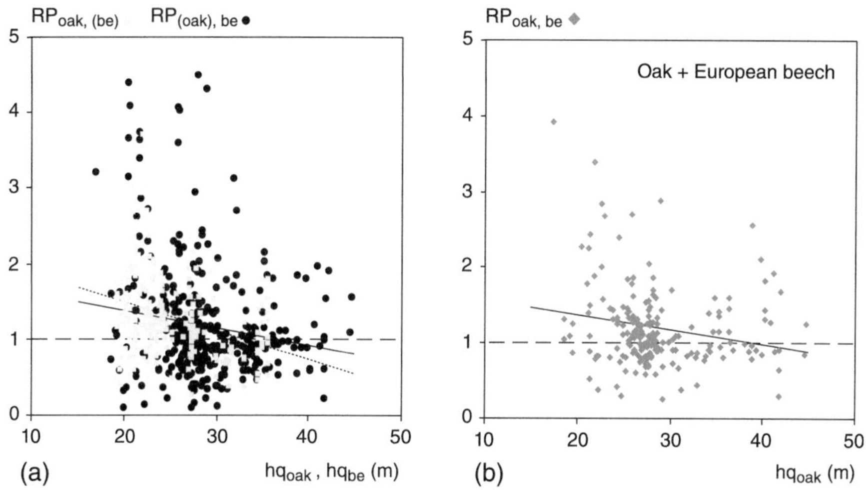


**Fig. 4.29** Relationship between (**a**) species-specific relative productivity and (**b**) whole-stand relative productivity of mixed versus monospecific stands in relation to site fertility, indicated by the height of the tree with the quadratic mean diameter at age 100, $h_q$ (Pretzsch et al. 2013b). The graph shows the relative productivity expected for neutral mixing responses (broken 1.0 line), the observed relative productivity of oak (*squares*), beech (*circles*), and the stand as a whole (*diamonds*), and the regressions lines for oak (Fig. 4.29a, *grey line*), beech (Fig. 4.29a, *black line*), and the total stand (Fig. 4.29b, *black line*)

a neutral interaction on medium sites, and an antagonistic interaction along with a reduction in productivity on fertile sites.

In order to stress the site-specific relative mixing effects, we display them in cross diagrams as introduced in Box 4.1. Using the statistical relationships between mixing effects and site conditions, it allows a prediction of how oak and beech interact on a poor site (represented by the height of the tree with the quadratic mean diameter at age 100 of $h_{qoak} = 21$ m, $h_{qbe} = 25$ m), on a mediocre site ($h_{qoak} = 26$ m, $h_{qbe} = 29$), and on a rich site ($h_{qoak} = 33$ m, $h_{qbe} = 36$). The results are shown in Fig. 4.30: The response at the stand level (bold curve in the upper part of the cross diagrams) is the sum of the species-specific curves. Depending on site conditions, mixing can trigger strong mutual facilitation or competitive reduction with a relative productivity at the stand level of about 1.66 (+66%) on poor sites, a moderate mixing effect of 1.35 (+35%) on medium sites, or a negative effect indicated by a relative productivity of 0.88 (−12%) on fertile sites (Fig. 4.30a–c). Whereas on the poor and mediocre sites, both species contribute to the productivity gain; on the rich site both respond slightly negatively and contribute to the overall loss. This enables the variability in observed mixing effects to be explained by the site conditions of the analysed stands.

Deeper insight into the modification of mixing responses by site conditions was provided by a statistical analysis of the relationship between the species-specific mixing effects (indicated by $RPA_{1,(2)}$ and $RPA_{(1),2}$) and site index along an ecological gradient (Pretzsch et al. 2013b). Site index was used as a general

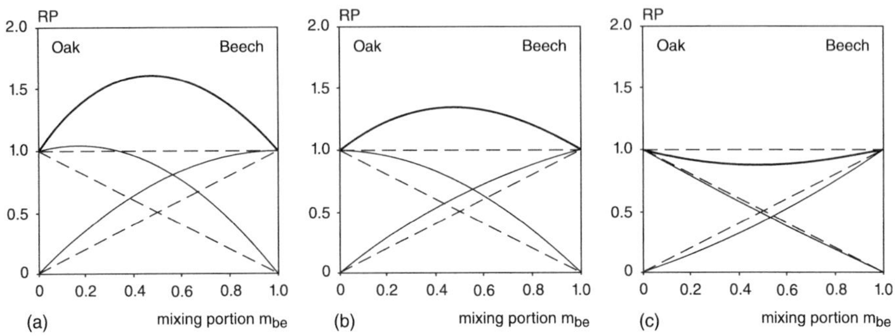


**Fig. 4.30** (**a–c**) Essential mixing response patterns of oak and beech observed along a gradient from poor to fertile sites. The site indices of oak and beech are 25 m and 21 m (**a**), 29 m and 26 m (**b**), and 36 m and 33 m (**c**), respectively. The cross diagrams indicate that the relative productivity was modified by site conditions

indicator for site quality because more detailed site characteristics were not available for the survey, which dated back to 1870. For site indexing, we used the yield tables by Assmann and Franz (1963) for Norway spruce, Schober (1972) for European beech, Wiedemann (1943) for Scots pine, Hausser (1956) for silver fir, and Jüttner (1955) for sessile/common oak. In addition, stand characteristics as a mixing portion, height relationship between the mixed species, and mean height as an indicator for the stand development phase were included in the analysis.

For all included species—Norway spruce, Scots pine, silver fir, European beech, and sessile/common oak—the gain achieved through mixing was at a maximum on poor sites and decreased significantly with site index (Fig. 4.31). Further analysis revealed that the mixing effect also depended on the admixed species, the mixing portion, and in some cases on the height relationship between the two species and the stand development phase. To illustrate these results, Fig. 4.31 (straight lines) shows the species-specific mixing responses in relation to the admixed species and site index. The other variables were held constant ($m_1 = m_2 = 0.5$; $h_{q1}/h_{q2} = 1.0$; $h_{q1} = h_{q2} = 20$ m). Thus, the lines represent mixing effects in a 50:50 mixture where both species have similar heights and are more or less in the middle phase of age development.

Studies based on other experiments and sites report the opposite effects, i.e. an increase of overyielding with increasing site fertility (Thurm and Pretzsch 2016; Pretzsch et al. 2010). The comprehensive study by Gritti et al. (unpublished) found that mixed-species stands in the second half of their rotation period had a significant increase in both overyielding and transgressive overyielding with increasing water availability (Fig. 4.32). The overyielding increased by as much as 50% and the transgressive overyielding by as much as 30% on moist sites in the subtropical regions. This study showed no differences in this response pattern for evergreen/evergreen and evergreen/deciduous mixtures. The underlying meta-analysis was based on 51 publications that included 319 cases studies (see also Sect. 4.2.2). They resulted in an average overyielding of 14.5% at the species level and 26.5% at the stand level for boreal, temperate, Mediterranean, and subtropical stands. The

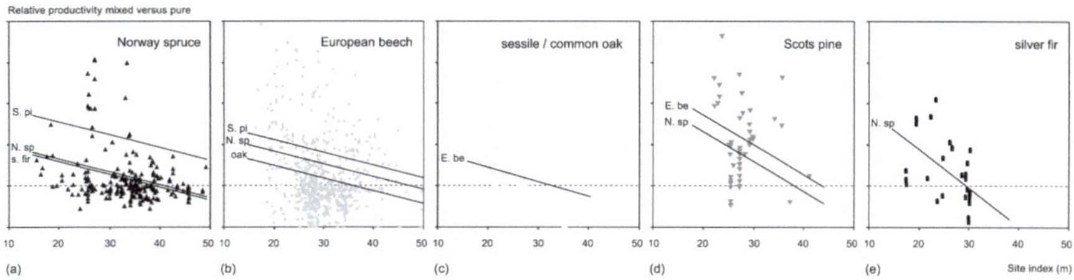


**Fig. 4.31** (**a**–**e**) Mixing responses of Norway spruce, European beech, sessile oak, Scots pine, and silver fir depending to site conditions indicated by the site index. The regression lines display the different species' responses to different admixtures (N. sp = Norway spruce, E. be = European beech, oak = sessile/common oak, S. pi = Scots pine, s. fir = silver fir). The 1.0 line represents neutral mixing effects on productivity. For further statistical results, see Pretzsch et al. (2013b)

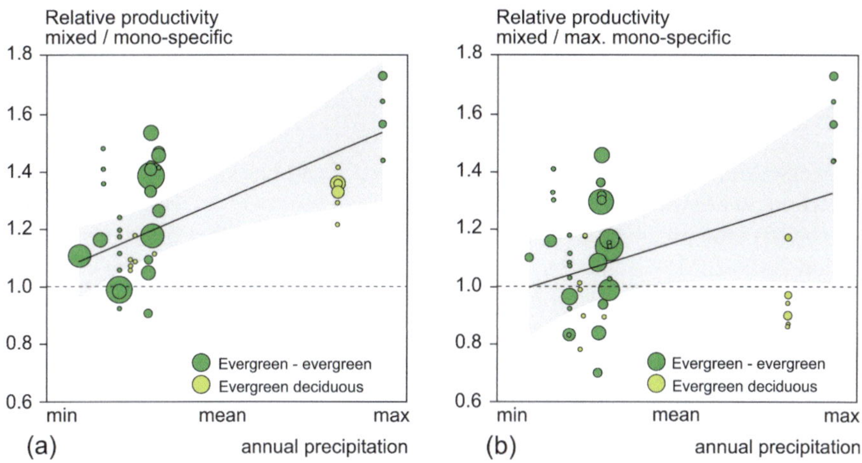


**Fig. 4.32** Significant increase in (**a**) overyielding and (**b**) transgressive overyielding of mixed compared to monospecific stands with increasing water availability according to a worldwide study by Gritti et al. (unpublished). The results refer to stands in the second half of the rotation. Min, mean, and max on the $x$-axis indicate minimum, mean, and maximum water supply for the given experiments. The circle sizes reflect how the different case studies were weighted when fitting the linear models

regression lines shown in Fig. 4.32 for older stands are based on 29 publications and 78 case studies.

#### 4.5.2.2 Temporal Variability in Relation to Periodic-Specific Environmental Conditions

In addition to changes relating to local site conditions, mixing effects can vary over time depending on the prevailing weather conditions. Tree level analyses revealed that low-growth years can trigger productivity gains, while high-growth years can enhance competition and cancel overyielding. Figure 4.33 illustrates a stimulation

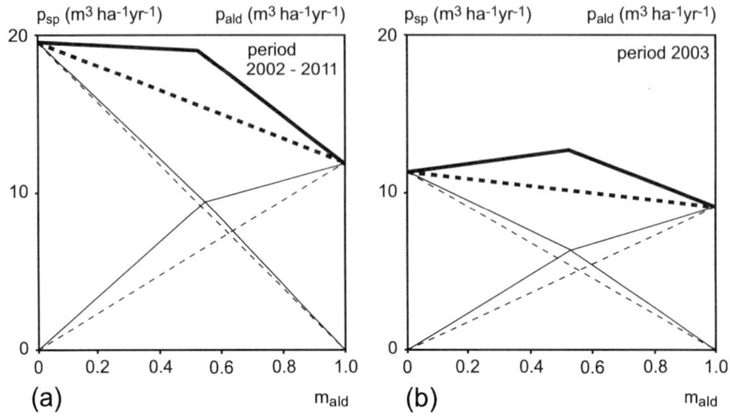


**Fig. 4.33** Stimulation of overyielding in mature mixed stands versus monospecific stands induced by drought. (**a**) Mean mixing effects in 80-year-old stands of Norway spruce and (**b**) black alder in the period 2001–2011. Mixing effect in drought year 2003 (according to Schwaiger (2013), pp 34–37). For explanation of cross diagrams and RPA, see Box 4.1

of overyielding in mature stands of Norway spruce and black alder during drought according to Schwaiger (2013, pp 34–37).

The left cross diagram (Fig. 4.33a) shows the long-term effect of mixing for the period 2001–2011. Calculation of the RPA value reveals an overyielding of mixed versus monospecific stands of about 20% (RPA = 1.2). The mixed stand exceeds the level expected for the linear combination of partial monospecific stands (upper broken line), but not the level of the neighbouring monocultures. Figure 4.33b shows the analogous evaluation for the mixed versus monospecific stand performance in the year 2003.

The summer heat of 2003 was among the severest in Europe's recent climate history (Schwaiger 2013). The temperatures from May to August were 5–6 °C above the long-term mean, and the precipitation in February to April and in summer from July to September was at least 50% below the mean level. In 2003, the drought started in spring and became more severe in late summer, but was eased temporarily by a rather moist period from May to June. To characterise the average water supply of the sampled stands, we calculated the index, Ma, of de Martonne (1926) on the basis of the annual precipitation (Pa in mm) and mean annual temperature (Ta in °C) $Ma = Pa/(Ta + 10)$. We used data from the 30-year period 1971–2000 in order to obtain stable mean values. Ma values of the sites in Bavaria lie in between (Ma = 37.7–56.5 mm $^{\circ}C^{-1}$). During the drought event of 2003, the temperature, precipitation, and consequently the Ma value were reduced to 60–70% of this long-term mean value.

Of course, the level of productivity in both monospecific and mixed stands was lower during the dry year of 2003 than the long-term average. However, in this year, the mixed stands produced 30% (RPA = 1.3) more than expected (broken line), and the mixed stand productivity exceeded the performance of both neighbouring monocultures, a phenomenon we introduced as transgressive overyielding in Sect. 4.2.1.

Bielak et al. (2014) analysed the extent to which the mixing effect was modified by drought stress. The study was conducted in north-east Poland and based on three long-term mixed-species experiments in Scots pine and Norway spruce, which have been under continuous survey for more than 100 years. In order to analyse the temporal variation of the mixing effects and any relationship to the temporal growing conditions, they calculated the periodic annual volume growth scaled up to 1 ha by mixing proportions for all available 4–5-year survey periods for the mixtures and the neighbouring monocultures. The ratio between the periodic annual volume growth in mixed and monospecific stands reflects any over- or underyielding of the mixed stands. To analyse the relationship between the mixing effects and the temporal growing conditions, the mean values of the index by de Martonne (1926) ($Ma = P/(T + 10)$ in mm °C$^{-1}$) were used for each survey period, on the basis of the precipitation ($P$, in mm) and mean temperature ($T$, in °C) of the summer months June, July, and August.

Figure 4.34 shows the decrease in overyielding of Scots pine in mixed stands with Norway spruce with increasing values of the de Martonne index. Remember that the higher the Ma index, the better the water supply for plant growth. The grey line results from linear model fitting to the relationship between the $RPAIV_{pi,(sp)}$ values and the de Martonne index using a linear mixed model. The analysis shows a significant decrease in the gains through mixing with increasing de Martonne index. Obviously, the benefit of Scots pine from mixing is higher in years with scarce water supply compared to moist years under our investigated site conditions.

The mixing effects are consistent with the general framework described in Sect. 3.6. They would be consistent with the stress-gradient hypothesis (Callaway and

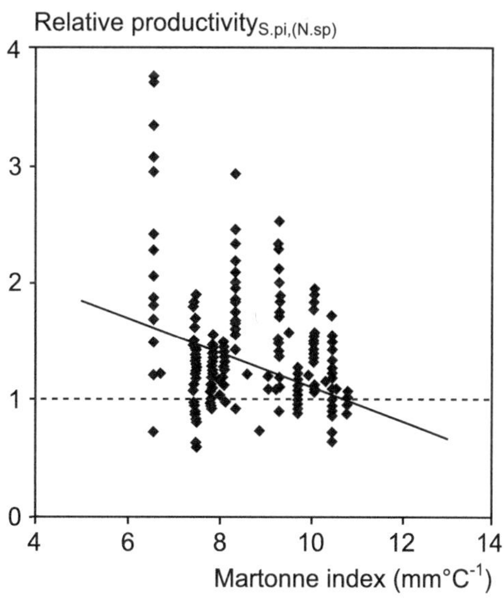


**Fig. 4.34** The relationship between the mixing effect and the de Martonne index for Scots pine. The mean relative productivity (RPAIV) of pine is shown in relation to the de Martonne index. Regression analysis revealed a significant decrease of overyielding with increasing de Martonne index. Broken line indicates the productivity in monospecific stands (1.0 line) (according to Bielak et al. 2014)

Walker 1997) if we assume that facilitative interactions caused most of the mixing effects. Other tree species mixtures react differently. For further testing or specification of the spatial and temporal dynamics of mixing effects for forest stands, we need many more analyses covering different species groups, environmental conditions, and resource supply.

## 4.6 Spacing, Thinning, and Other Small-Scale Interference in Canopy Closure as Modifiers of Mixing Effects

### 4.6.1 Theoretical Considerations

When mixed stands are very sparsely stocked, the trees of the different species will probably not directly affect each other and may not overyield neighbouring monocultures of the same density. However, if their stand density increases, the likelihood of interactions between crowns and roots increases, and beneficial interactions above and below ground are more likely to cause a net benefit and overyielding as shown by Amorosos and Turnblom (2006) and Condés et al. (2013). In addition to this, species mixing may increase the maximum stand density, above the level of neighbouring monocultures (see Fig. 4.23a and b). Without any positive or negative interactions between the trees of the standing stock, stand growth would increase proportionally with stand density. Stand growth would follow the course of a straight line to the origin when maximum density is reduced by thinning. However, because of compensation effects, it follows a concave or even unimodal curve. Low and medium density can yield overproportional growth (Assmann 1970, pp 227–235), and maximum stand density can cause suboptimal growth. The density with optimal growth can lay 10–20% below the maximum stand density (Box 4.8).

For further confirmation, we need clarification of the maximum natural stand density in mixed stands, the natural relative density, and the implications of species-specific growing area requirements and mixing proportions (see Sect. 12.3). Neglecting any increase in productivity and maximum stand density or application of monoculture thinning guidelines to mixed-species stands may result in severe growth losses and undesirable mixing portions.

**Box 4.8 Assmann's (1961) Concept of Maximum, Optimum, and Critical Density**

Assmann (1961) found that stand growth can increase with a reduction in stand density up to a certain optimum and then decreases with a further reduction in density. The relationship between stand density and stand growth can be described by the following three density levels (Box Fig. 4.8-1a). Maximum stand density (black circle) is achieved in unthinned stands.

(continued)

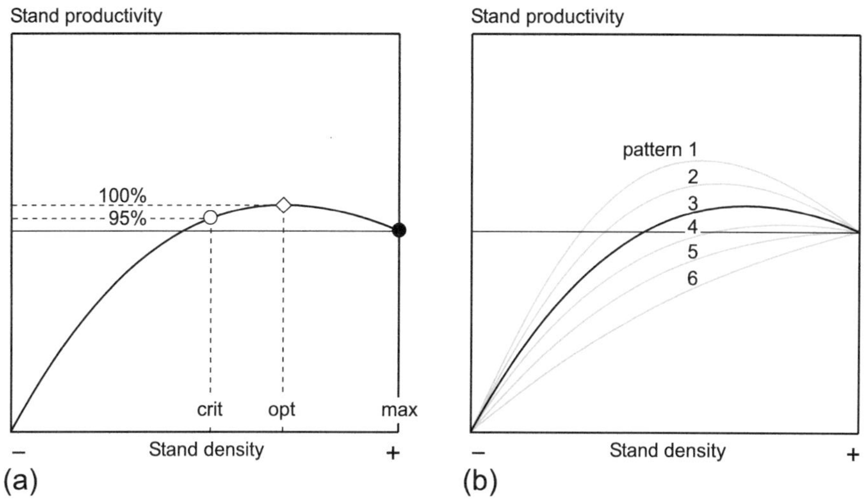


**Box Fig. 4.8-1** Schematic representation of the effect of stand density on the stand productivity of monocultures according to Assmann (1961, pp 222–228). (**a**) A reduction in maximum stand density can affect an increase in productivity. Maximum productivity can be achieved at medium stand densities (optimum stand density). A further reduction in stand density causes a decrease in stand productivity. Assmann (1961) called the density with maximum productivity 'optimal stand density' and the density still providing 95% of the maximum productivity 'critical stand density'. (**b**) The density-productivity trajectories are bulging (lines 1–2) for plastic shade-tolerant species and are rather shallow (lines 3–4) for nonplastic light-demanding species. They generally become shallower with increasing stand age

**Box 4.8** (continued)

Optimum stand density is the stand density at which the stand volume growth is at a maximum (diamond). Critical stand density is named the stand density where 95% of maximal productivity is reached (white circle). When stand density is reduced further, stand productivity declines almost linearly (see left branch of the dose–response curve). Stand density can be quantified, e.g. using the stand basal area (Avery and Burkhardt 1975), mean stand basal area (Assmann 1961), or the relative stand density, in terms of the stand density index by Reineke (1933) introduced in Box 4.5.

The shape of the density-growth relationship, especially the curve's bulge or plateau (Box Fig. 4.8-1b), reflects a stand's capacity to compensate or even overcompensate thinning removals through growth acceleration of the remaining stand through crown plasticity. The latter varies between tree species, stand development phases, and site conditions (Assmann 1970; Zeide 2001; Pretzsch 2005a). From young to mature stands, the productivity-density response pattern can change from unimodal curves in

(continued)

**Box 4.8** (continued)
the upper part of the curve bundle towards more shallow courses at the lower periphery of the curve spectrum (pattern 1 respective 2). Plastic tree species such as European beech follow upper productivity-density curves (pattern 1), and nonplastic species such as Scots pine develop along the lower border of the curve spectrum (pattern 6).

## 4.6.2 Empirical Evidence of Modified Growth Responses in Relation to Density Reduction in Mixed Compared with Monospecific Forest Stands

### 4.6.2.1 Effect of Spacing and Thinning

Mixed stands growing at low density are less likely to show mixing effects than more densely packed stands where there are stronger interactions between tree crowns and roots. Up to this point, this chapter has dealt with mixed stands at high or maximum density; however, management usually keeps density below the maximum. In order to show how density modifies mixing effects, we show the results of a spacing experiment.

For a 12-year-old 50:50 mixed plantation of Douglas-fir and western hemlock, Amorosos and Turnblom (2006) showed the importance of stand density for the relationship between the productivity of mixed versus monospecific stands. In terms of RPA, the fully stocked stands with a density level of 1729 N $ha^{-1}$ showed overyielding effects, while the advantage of mixing vanished when the stands were stocked with only 1111 or 494 trees per hectare (Fig. 4.35a). It was mainly the contribution of Douglas-fir that caused the overyielding. The graph for absolute yield shows that under high stand density, the mixed plots even transgressively overyield, as total yield at age 12 exceeds the total yield of neighbouring monocultures of western hemlock and Douglas-fir (Fig. 4.35b). This experiment indicates that high positive mixing responses and overyielding (even transgressive overyielding) may be more likely at medium and high density levels compared with stands where the stand density has been lowered, e.g. through thinning or other disturbances.

Stands with species of varying crown and root plasticity, stratification in different layers or cohorts in canopy and soil horizons, are widely assumed to exploit resources more extensively in closed stands and to have higher resilience after disturbances due to a more rapid closure of gaps and return to full exploitation of above- and below-ground resources (see Chap. 6, Sect. 6.8). Figure 4.36a–c, corroborates this assumption by showing the stem dry biomass yield for Norway spruce and European beech monocultures and mixed Norway spruce-common beech stands on plots subjected to various grades of thinning in comparison with the untreated A-grade (= 100%) at age 100. If we first consider the response pattern

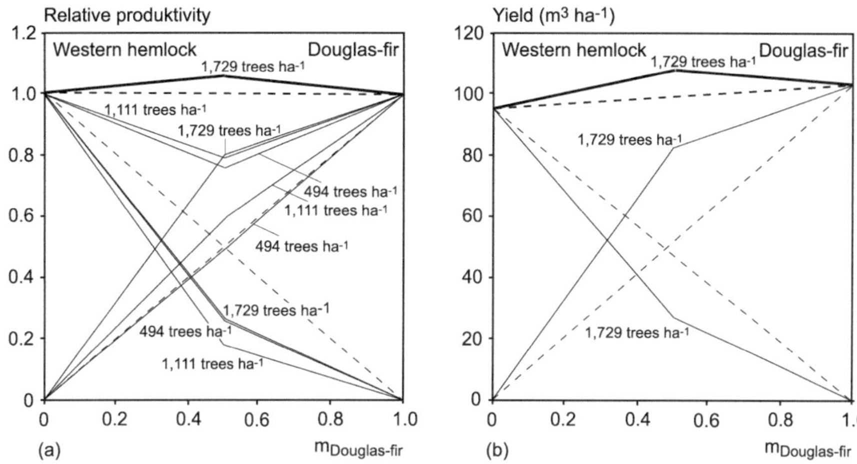


**Fig. 4.35** Effect of stand density on mixing effects. (**a**) Relative and (**b**) absolute yield for a 12-year-old 50:50 mixed plantation of Douglas-fir and western hemlock adapted from Amorosos and Turnblom (2006, Fig. 10). (**a**) Stands with high initial densities (1729 trees ha$^{-1}$) show the strongest positive mixing effect, while the benefit obtained through mixing decreases when stocking density is just 1111 or 494 N ha$^{-1}$. (**b**) In fully stocked stands with 1729 trees per hectare, mixed stands transgressively overyield both monocultures

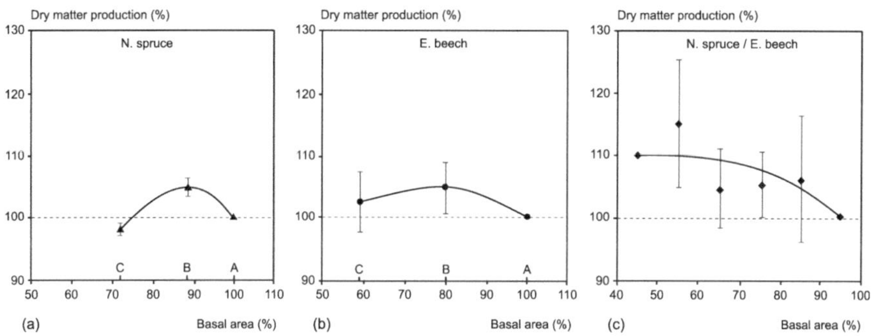


**Fig. 4.36** (**a–c**) Density-growth relationships in monospecific and mixed stands for Norway spruce and common beech at age 100. A-, B-, and C-grade refer to slight, moderate, and heavy thinning from below, respectively, in the monocultures. In the mixed stands, the experimental design included untreated plots, slight, moderate, and strong thinning (according to Pretzsch 2005a). (**a**) Results from nine Norway spruce thinning experiments with 26 plots, with consistent slight, moderate, and heavy thinning since 1870. (**b**) Results from ten common beech thinning experiments with 30 plots, consistently subjected to slight, moderate, and heavy thinning from below since 1870. (**c**) Results from 23 mixed stands with spruce and beech on 78 plots under observation since 1954

in monospecific stands (a and b), we can see that dry biomass production increases by 5–10% in both monocultures when transitioning from A-grade to B-grade. In the transition from B-grade to C-grade, we note a decrease in total growth. In European beech stands, total growth yield for C-grade is even higher than for A-grade. Means and standard error bars in Fig. 4.36 reveal differences between A-grade (= 100%-value) and B- and C-grades. After 130 years of experimental research, we are therefore able to state that for Norway spruce and European beech, there is a significant increase ($p < 0.05$) in growth from A-grade to B-grade and for Norway spruce, there is a significant decrease ($p < 0.05$) in growth in the transition from B-grade to C-grade. The relationship between density and growth represents an optimum curve. Any approach to the maximum stand density is concomitant with growth reductions of 5–10%.

In contrast, the biomass production in Norway spruce/European beech mixtures (Fig. 4.36c) achieves similar values over a wide range of densities. There is no significant reduction in biomass production even when stand density is reduced by 50%. In comparison with corresponding monocultures, Norway spruce/European beech mixtures are able to compensate for disturbances caused by thinnings through accelerated growth in the remaining stand.

The cause for this response pattern may be the complementary space sequestration by dominant Norway spruce and dominant but also subdominant European beech. In mixed stands, removals from or losses in the upper crown layer can be compensated for by stronger growth in the lower layer. This buffer effect becomes particularly effective with increasing age. In the mixed stand, the remaining trees are able to close any gaps that may form as a result of mortality, which slows down age-related breakup of the crown layer and has a stabilising effect on stand biomass production.

#### 4.6.2.2 Response to Partial Damage in Mixed Versus Monospecific Stands

The productivity relationship between monospecific and mixed stands under 'normal' conditions may shift considerably once risks are included. The decisive factor here, in essence, is the probability of the occurrence of disturbances and damage. The temporarily dazzling superiority in productivity of monospecific stands can be reduced if disturbances are included in the calculation. Mixtures can overcome interruptions of the crown space caused by windthrow, storm damage, or biotic calamities, which is equivalent to an abrupt reduction in stand density from medium to lower levels, without greater losses in growth. The higher elasticity of growth in mixed stands of spruce and beech is reflected by Fig. 4.37. It shows the relationship between the percentage of intermediate yield RIY (%) and the relative total dry matter production (%). The survey results from the mixed plots ZWI 111, 134, 135 (black dots) and long-term plots in spruce and beech monocultures (black line). The figure emphasises an important consequence for forest management.

Based on these data, total dry mass stem production is expected to occur if approximately 40% of the total yield is removed regularly from the stand by moderate thinning prior to final harvest or due to natural causes. Total production is expected to decrease when RIY is higher or lower than 40%. Indeed, RIY was found to be between 30 and 50%, which is close to the maximum.

In contrast, RIY of the mixed stands varied between 10 and 70% without any systematic effect on total dry mass stem production. This means that intensive

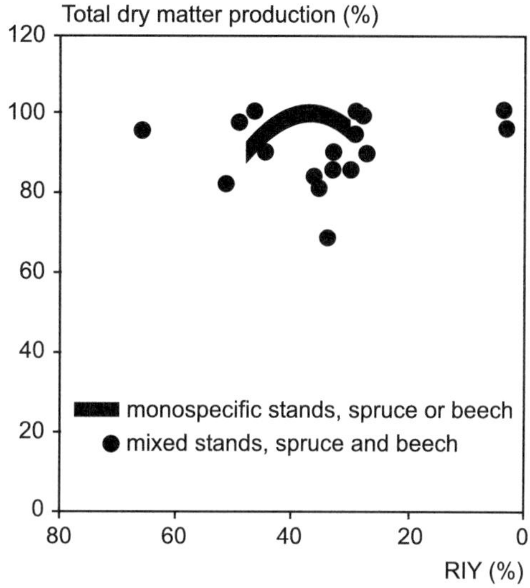


**Fig. 4.37** Relationship between the percentage of intermediate fellings (RIY) and the relative total stem dry mass production in monospecific stands (*black unimodal curve*) and mixed stands of Norway spruce and European beech black points (according to Pretzsch 2003)

thinning or partial disturbances (e.g. partial stand damage caused by storm, ice breakage, or bark beetle) that result in a percentage of intermediate yield RIY = 50–70% can be compensated for by increasing growth of the remaining individuals in mixed stands but not in monospecific stands. Also, under discontinued silvicultural treatments or self-thinning (RIY = 10–30%), mixed stands can be more stable in terms of total dry matter production, whereas monospecific stands show a significant decrease. This pattern represents an important stabilising advantage of mixed stands of spruce and beech compared to monocultures (see Chap. 11).

#### 4.6.2.3 Characteristics of Growth Resilience Against Disturbances

Figure 4.38 (upper panes) shows stand structures of monospecific and mixed stands. The deepness of the canopy increases from even-aged monospecific stands to uneven-aged mixed stands. How silvicultural interference (thinning, regeneration cuts) and other structural disturbances (windthrow, snow breakage, bark beetle damage) affect stand productivity depends mainly on the stand composition and stand structure. The deeper and more multilayered a canopy is, the higher the growth resilience to structural disturbances (see Chap. 7). Productivity losses due to tree removal through thinning or losses through damage can be better compensated for by the presence of a second or third layer of trees which might immediately use the resources that are no longer intercepted by the removed trees.

The two density growth relationships (Fig. 4.38, lower panes) are unimodal, i.e. with the lowest productivity level occurring at low densities, then increasing productivity when stand density increases and a slight decrease in productivity when approaching maximum stand density. However, while the patterns are similar in principle, the position and extent of the curves' peak are rather different.

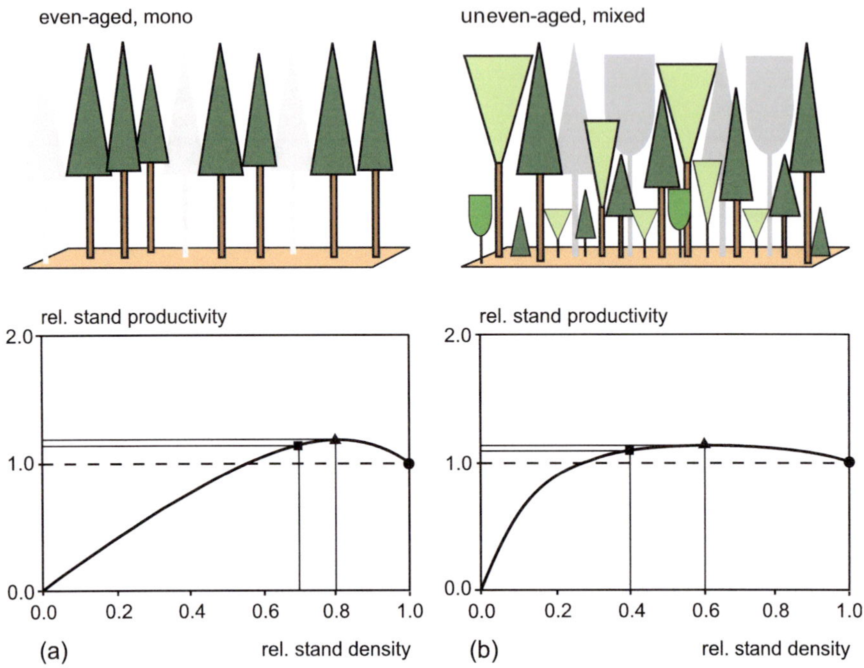


**Fig. 4.38** The growth resilience of stands to tree removal through thinning or losses through damage can differ between monospecific stands (**a**) and mixed-species stands (**b**). Schematic representation of the stand structure (*above*) and the stand density-productivity relationship (*below*). The course of the curve can be quantified by the origin (*filled circle*, $r_d = 1.0$, $r_p = 1.0$), the culmination point of the curve (*filled triangle*), and the point of critical density (*filled square*) where the productivity losses arrive at 5% of the maximum stand productivity

According to Assmann (1970, 227–233) (Assmann 1970), key characteristics of the unimodal relationship between stand density and productivity in monocultures (Fig. 4.38a) are the location of the maximum (black triangles) and the point where density reductions reduce stand productivity below 95% of the maximum productivity (black rectangles). The origin of the curve (black circle) is the maximum density (1.0) and productivity at maximum density 1.0. In order to stress the principle differences between growth resilience in monospecific and mixed stands (Fig. 4.38a and b), we show the course of relative productivity (1.0 = productivity in fully stocked and unmanaged stands) over relative stand density (1.0 = maximum stand density of fully stocked and unmanaged stands). In mixed-species stands, the relationship between stand density and productivity can have a broader plateau than in monocultures (see Fig. 4.38a versus b). The greater the height variation, the more growth losses due to the removal of trees (e.g. by thinning or natural disturbances) in one layer will be buffered by the remaining trees in another canopy layer.

In the schematic example (Fig. 4.38a), a maximum relative productivity of $rp_{max} = 1.25$ (i.e. 125% of the productivity in fully stocked stands) is achieved at a relative stand density of $rd_{max} = 0.80$ (i.e. 20% below the maximum stand density). The critical relative density where only 95% of the maximum productivity is achieved (i.e. density reduction causes a productivity loss of 5%) amounts to $rd_{crit} = 0.7$. So the value pair $rp_{max}$ and the corresponding $rd_{max}$ as well as the $rd_{crit}$ characterise the monocultures' growth resilience.

In contrast, in the mixed stand (Fig. 4.38b), a relative stand productivity of $rp_{max} = 1.20$ is achieved at a relative density of $rp_{max} = 0.60$. The critical stand density causing productivity losses of 5% amounts to $rd_{crit} = 0.4$. The relative density and productivity of both the maximum productivity and critical productivity characterise growth resilience which is relevant for system understanding and regulation.

## 4.7 Reference Values from Long-Term Experimental Plots

### 4.7.1 Reference Values for Growth and Yield of Monocultures and Even-Aged Mixed Stands

The following tables present some characteristic growth and yield values for even-aged monospecific stands (Table 4.6), even-aged mixed stands (Table 4.7), and uneven-aged mixed stands (Table 4.8) at experimental sites in Germany. The average growth and yield data collated for up to 130 years, provided for trees and stands, represent moderately thinned and otherwise largely undisturbed stands from experimental research plots with moderate to good site conditions. The volume data refer to merchantable volume (>7 cm at the smaller end) for broadleaved trees and to stem volume for conifers.

In the even-aged monospecific stands (Table 4.6), the quadratic mean diameter attains 43.3–55.6 cm, stand basal area of 28.2–88.4 $m^2\ ha^{-1}$, and standing wood volumes of 428–1480 $m^3\ ha^{-1}$ at age 93–178 years. Norway spruce and Douglas-fir

**Table 4.6** Stand characteristics of even-aged monospecific stands of various tree species in Central Europe after long-term inventory

| Tree species | Experiment | Age (years) | *d* (cm) | *h* (m) | st. basal area ($m^2\ ha^{-1}$) | st. volume ($m^3\ ha^{-1}$) | $PAI_{V\ max}$ ($m^3\ ha^{-1}\ year^{-1}$) | $MAI_V$ ($m^3\ ha^{-1}\ year^{-1}$) | $GY_V$ ($m^3\ ha^{-1}$) | $IY_V$ (%) |
|---|---|---|---|---|---|---|---|---|---|---|
| Norway spruce | DEN 05 | 143 | 54.4 | 40.3 | 88.4 | 1480 | 21.3 | 15.4 | 2199 | 32.7 |
| Scots pine | BAY 52 | 131 | 43.3 | 36.3 | 44.0 | 700 | 12.1 | 10.5 | 1370 | 48.9 |
| silver fir | WOL 97 | 134 | 53.9 | 34.0 | 45.1 | 637 | 12.0 | 9.5 | 1268 | 49.8 |
| European larch | MIS 47 | 178 | 55.6 | 36.7 | 28.2 | 428 | 10.4 | 5.4 | 953 | 55.1 |
| Douglas-fir | FSE 85 | 93 | 44.6 | 39.3 | 61.9 | 1012 | 33.4 | 17.2 | 1603 | 36.9 |
| seesile oak | LOH 59 | 168 | 50.3 | 33.0 | 37.4 | 605 | 10.6 | 7.7 | 1301 | 53.5 |
| European beech | FAB 15 | 178 | 51.5 | 38.0 | 48.0 | 950 | 15.9 | 8.7 | 1551 | 38.7 |

Several inventories date back to 1870 (Pretzsch 2009). *d*, *h* mean stand diameter and height; stand basal area; standing stem or merchantable volume, respectively; $PAI_{V\ max}$ maximum periodic annual volume increment since the beginning of the inventory; $MAI_V$ mean annual volume increment; $GY_V$ gross volume yield since stand establishment; $IY_V$ intermediate volume yield ($IY_V = 1 - $ [standing volume/gross volume yield] 100)

**Table 4.7** Stand characteristics of even-aged mixed stands in Bavaria (Pretzsch 2009)

| Tree species | Experiment | Age (years) | $d$ (cm) | $h$ (m) | $V$ ($m^3$ $ha^{-1}$) | $MAI_V$ ($m^3$ $ha^{-1}$ $year^{-1}$) | $GY_V$ ($m^3$ $ha^{-1}$) | $IY_V$ (%) |
|---|---|---|---|---|---|---|---|---|
| Norway spruce | SON 814 | 110 | 50.3 | 38.0 | 562 | 11.5 | 1267 | 55.6 |
| European beech | | 110 | 37.4 | 33.5 | 359 | 6.9 | 764 | 53.0 |
| Total | | 110 | | | 921 | 18.5 | 2031 | 54.8 |
| Scots pine | NEU 841 | 100 | 39.4 | 30.0 | 496 | 9.7 | 969 | 48.8 |
| Norway spruce | | 100 | 24.9 | 25.1 | 380 | 7.9 | 786 | 51.7 |
| Total | | 100 | | | 876 | 17.6 | 1755 | 50.1 |
| Scots pine | GEI 832 | 136 | 59.1 | 32.3 | 144 | 5.4 | 732 | 80.3 |
| European beech | | 136 | 23.0 | 26.1 | 250 | 10.1 | 1380 | 81.9 |
| Total | | 136 | | | 394 | 15.5 | 2112 | 81.0 |
| European larch | GEM 871 | 109 | 53.6 | 38.1 | 197 | 4.3 | 467 | 57.8 |
| European beech | | 109 | 45.9 | 35.3 | 383 | 7.5 | 814 | 52.9 |
| Total | | 109 | | | 580 | 11.8 | 1281 | 54.7 |
| sessile oak | KEH 804 | 146 | 48.8 | 33.6 | 548 | 8.8 | 1288 | 57.5 |
| European beech | | 146 | 27.6 | 26.5 | 140 | 3.6 | 519 | 73.0 |
| Total | | 146 | | | 688 | 12.4 | 1807 | 61.9 |

The evaluation is based on time series of mixed stands with up to 10 plots per time series and 2–3 successive inventories per plot. Mean annual volume increment ($MAI_V$), gross yield of volume ($GY_V$), and percentage of intermediate yield ($IY_V$) from stand establishment to the age of the oldest plot (ages 110, 100, 136, 109, and 146, respectively)

**Table 4.8** Stand characteristics of uneven-aged mixed stands of Norway spruce, Silver fir, and European beech in the temperate mountain forests of Bavaria

| Tree species | Experiment | Age (years) | $d_{max}$ (cm) | $h_{max}$ (m) | st. basal area ($m^2$ $ha^{-1}$) | $V$ ($m^2$ $ha^{-1}$) | $PAI_{Vmean}$ ($m^3$ $ha^{-1}$ $year^{-1}$) | $PAI_{Vmax}$ ($m^3$ $ha^{-1}$ $year^{-1}$) |
|---|---|---|---|---|---|---|---|---|
| N. spruce-silver fir-European beech | FRY 129/32 | 2–239 | 42.9 | 31.0 | 42.6 | 663 | 10.7 | 13.8 |
| N. spruce-silver fir-European beech | BOM 130/22 | 2–286 | 41.2 | 30.8 | 44.5 | 610 | 10.2 | 11.3 |
| N. spruce-silver fir-European beech | PAR 115/1 | 2–203 | 44.6 | 32.8 | 57.2 | 871 | 8.1 | 9.8 |
| N. spruce-silver fir-European beech | KRE 120/3 | 2–158 | 39.2 | 29.4 | 40.1 | 502 | 4.7 | 5.5 |
| N. spruce-silver fir-European beech | MAR 108/1 | 2–142 | 58.1 | 37.4 | 66.9 | 1028 | 15.9 | 19.6 |
| N. spruce-silver fir-European beech | RUH 110/2 | 25–165 | 55.3 | 36.1 | 29.7 | 477 | 7.9 | 10.4 |
| N. spruce-silver fir-European beech | RUH 116/1 | 2–183 | 45.3 | 33.7 | 52.6 | 784 | 9.8 | 11.4 |

Inventories date back to the year 1950 (Pretzsch 2009). $d_{max}$, $h_{max}$ mean stand diameter and height of the species with maximum size; st. basal area; $V$ total standing volume; $PAI_{V\ mean}$, $PAI_{V\ max}$ mean and maximum periodic annual volume growth since start of the survey rank highest, whereas European larch and sessile oak rank lowest. Maximum PAI values lie between 10.4 and 33.4 $m^3$ $ha^{-1}$ $year^{-1}$, MAI between 5.4 and 17.2 $m^3$ $ha^{-1}$ $year^{-1}$, and gross yield between 953 and 2199 $m^3$ $ha^{-1}$. Of this gross yield, 33–55% died or was removed during the stand lifetime ($IY_V$ (%) = intermediate volume yield in percent), giving net yield values of 428–1480 $m^3$ $ha^{-1}$ standing volume.

The values for the even-aged mixed stands (Table 4.7) are based on artificial time series comprising up to 10 adjacent plots of different ages, which cover the stand lifespan (Pretzsch 2005a; Pretzsch and Schütze 2009) (see Chap. 2). The stands were inventoried between 1992 and 2008 to derive growth and yield characteristics for the most relevant types of mixed forest in south Germany and parameterisation of the individual tree growth simulator SILVA (see Pretzsch et al. 2002, Chap. 8). The mean height and quadratic mean diameter for the two species present in the stand are listed separately. The first species mentioned generally exceeds the second in height and diameter, resulting in canopy stratification. At 100–146 years of age, the standing volume ranged from 580 to 921 $m^3$ $ha^{-1}$. Even-aged mixed stands, with MAI values of 11.8–18.5 $m^3$ $ha^{-1}$ $year^{-1}$ and gross yields of 1281–2112 $m^3$ $ha^{-1}$ at 100–146 years of age, do not fall behind the monocultures. However, the intermediate yields ($IY_V$ in %) in mixed stands of 50–81% are clearly higher than in monocultures, which means the final standing volume of 580–921 $m^3$ $ha^{-1}$, which is equivalent to the net yield, is lower than that of the monocultures.

The uneven-aged mixed stands (Table 4.8) represent selection forests and moderately thinned mountain forests, consisting mainly of Norway spruce, silver fir, and European beech. On the research plots, trees range from seedlings to almost 300-year-old trees. The quadratic mean diameter of the dominant species (mostly Norway spruce, sometimes silver fir) attains 39.2–58.1 cm, height of 29.4–37.4 m, and basal area of 29.7–66.9 $m^2$ $ha^{-1}$. The standing volume lies between 477 and 1028 $m^3$ $ha^{-1}$. In contrast to the previous, comparably unstratified monospecific and mixed stands, the standing volumes do not show the final net yield values reached near the end of the rotation cycle but give continuous averages that vary little due to the periodic removal of individual trees. The mean PAI and maximum PAI attain values of 4.7–15.9 $m^3$ $ha^{-1}$ $year^{-1}$ and 5.5–19.6 $m^3$ $ha^{-1}$ $year^{-1}$, respectively. In these uneven-aged forests, the long-term PAI averages are most comparable with the MAI values of even-aged rotation forests. A comparison shows that the PAI of uneven-aged stands are of the same order as the MAI values in even-aged stands (Table 4.7), and slightly lower than the MAI values in the even-aged mixed stands (Table 4.8), which are highest.

### 4.7.2 Reference Values for Standing Stock or Volume and Biomass Growth in Even-Aged Monocultures Compared with Uneven-Aged Mixed-Species Stands

In Table 4.9, the merchantable wood volume of the long-term experimental plots in Tables 4.6, 4.7, 4.8 is converted into biomass. The range in standing volume of the even-aged monospecific stands, 428–1480 $m^3$ $ha^{-1}$, is equivalent to a total biomass range of 378 t $ha^{-1}$ (European larch) and 1012 t $ha^{-1}$ (Norway spruce) (Table 4.9,

**Table 4.9** Upscaling from standing merchantable volume to total standing biomass, and from mean annual increment of merchantable volume to NPP on selected long-term experimental plots in Germany (see Tables 4.6, 4.7, 4.8)

| Tree species | Experimental plot | Standing volume ($m^3$ $ha^{-1}$) | Total st.biomass (t $ha^{-1}$) | MAI (PAI) ($m^3$ $ha^{-1}$ $year^{-1}$) | NPP (t $ha^{-1}$ $year^{-1}$) | NPP (kg $m^{-2}$ $year^{-1}$) |
|---|---|---|---|---|---|---|
| Pure even-aged | | | | | | |
| Norway spruce | DEN 05 | 1480 | 1012 | 15.4 | 13.7 | 1.4 |
| Scots pine | BAY 52 | 700 | 546 | 10.5 | 10.6 | 1.1 |
| silver fir | WOL 97 | 637 | 435 | 9.5 | 8.4 | 0.8 |
| European larch | MIS 47 | 428 | 378 | 5.4 | 6.1 | 0.6 |
| Douglas-fir | FSE 85 | 1012 | 756 | 17.2 | 16.7 | 1.7 |
| sessile oak | LOH 59 | 605 | 646 | 7.7 | 10.8 | 1.1 |
| European beech | FAB 15 | 950 | 983 | 8.7 | 11.7 | 1.2 |
| Total min | | 428 | 378 | 5.4 | 6.1 | 0.6 |
| Total max | | 1480 | 1012 | 17.2 | 16.7 | 1.7 |
| | | | | | | |
| Mixed even-aged | | | | | | |
| Norway spruce-European beech | SON 814 | 918 | 756 | 18.5 | 19.7 | 2.0 |
| Scots pine-Norway spruce | NEU 841 | 876 | 647 | 8.8 | 8.4 | 0.8 |
| Scots pine-European beech | GEI 832 | 394 | 371 | 15.5 | 19.0 | 1.9 |
| European larch-European beech | GEM 871 | 580 | 570 | 11.8 | 15.0 | 1.5 |
| sessile oak-European beech | KEH 804 | 688 | 730 | 12.4 | 17.1 | 1.7 |
| Total min | | 394 | 371 | 8.8 | 8.4 | 0.8 |
| Total max | | 918 | 756 | 18.5 | 19.7 | 2.0 |
| | | | | | | |
| Mixed uneven-aged | | | | PAI | | |
| N. spruce-s. fir-E. beech | FRY 129/32 | 663 | 500 | 10.7 | 10.5 | 1.0 |
| N. spruce-s. fir-E. beech | BOM 130/22 | 610 | 481 | 10.2 | 10.5 | 1.0 |
| N. spruce-s. fir-E. beech | PAR 115/1 | 871 | 626 | 8.1 | 7.6 | 0.8 |
| N. spruce-s. fir-E. beech | KRE 120/3 | 502 | 414 | 4.7 | 5.0 | 0.5 |
| N. spruce-s. fir-E. beech | MAR 108/1 | 1028 | 721 | 15.9 | 14.5 | 1.4 |

(continued)

**Table 4.9** (continued)

| Tree species | Experimental plot | Standing volume ($m^3$ $ha^{-1}$) | Total st.biomass (t $ha^{-1}$) | MAI (PAI) ($m^3$ $ha^{-1}$ $year^{-1}$) | NPP (t $ha^{-1}$ $year^{-1}$) | NPP (kg $m^{-2}$ $year^{-1}$) |
|---|---|---|---|---|---|---|
| N. spruce-s. fir-E. beech | RUH 110/2 | 477 | 351 | 7.9 | 7.6 | 0.8 |
| N. spruce-s. fir-E. beech | RUH 116/1 | 784 | 591 | 9.8 | 9.6 | 1.0 |
| Total min | | 477 | 351 | 4.7 | 5.0 | 0.5 |
| Total max | | 1028 | 721 | 15.9 | 14.5 | 1.4 |

Total standing biomass (t $ha^{-1}$) was estimated on the basis of standing merchantable volume using specific wood density $R$ and expansion factors $e_{br}$, $e_l$, $e_r$. NPP was estimated on the basis of MAI ($m^3$ $ha^{-1}$ $year^{-1}$) in the case of monospecific even-aged and mixed even-aged stands and on the basis of PAI ($m^3$ $ha^{-1}$ $year^{-1}$) in the case of mixed uneven-aged stands. For the upscaling, we applied specific wood density $R$, expansion factors $e_{br}$, $e_l$, $e_r$, and turnover factors $t_{org}$ [see Pretzsch (2009), pp 66–78]

above). For the even-aged mixed stands, the total biomass calculated from merchantable wood volumes of 394–918 $m^3$ $ha^{-1}$ ranges from 371 t $ha^{-1}$ (Scots pine/European beech) to 756 t $ha^{-1}$ (Norway spruce/European beech) (Table 4.9, middle). For the uneven-aged mixed stands, the merchantable wood volumes of 477–1028 $m^3$ $ha^{-1}$ correspond to a total biomass of 351–721 t $ha^{-1}$ (Table 4.9, below). In comparison with the wood volume values, the total biomass values of conifers and broadleaved species lie closer together. Often, the higher wood density of broadleaved species compensates for the lower packing density of the trees in the stand compared to conifer species.

## 4.8 Synopsis of Mixing Effects at the Forest Stand Level

This chapter examined mixing effects at the stand level and their relationship to growing conditions (e.g. spatial and temporal variation) and management (thinning, mixing pattern, mixing portion) and other disturbances (e.g. storm, plague of insects).

Many studies have found strong mixing responses in terms of overyielding on poor sites and in low-growth years. However, in Sect. 4.5, we also showed examples of opposite responses or with site invariant mixing responses. In line with Chap. 3, we hypothesise that the kind of mixing response and level of overyielding mainly depend on the potential of the species assemblage to remedy the respective growth-limiting factors on a given site.

The mixed stands shown in this chapter were, on average, 10–30% more productive than neighbouring monocultures (Fig. 4.39a). The benefit of mixing when mixing portion increases follows the diminishing marginal utility principle, i.e. small proportions cause progressive benefit but the additional benefit diminishes when mixing proportions rise (Fig. 4.39a, left and right branch of the concave course). Even small mixing portions may bring significant benefits in productivity. Part of the overyielding of mixed versus monospecific stands can result from the increase in stand density, which is shown in Fig. 4.39b and is represented by the

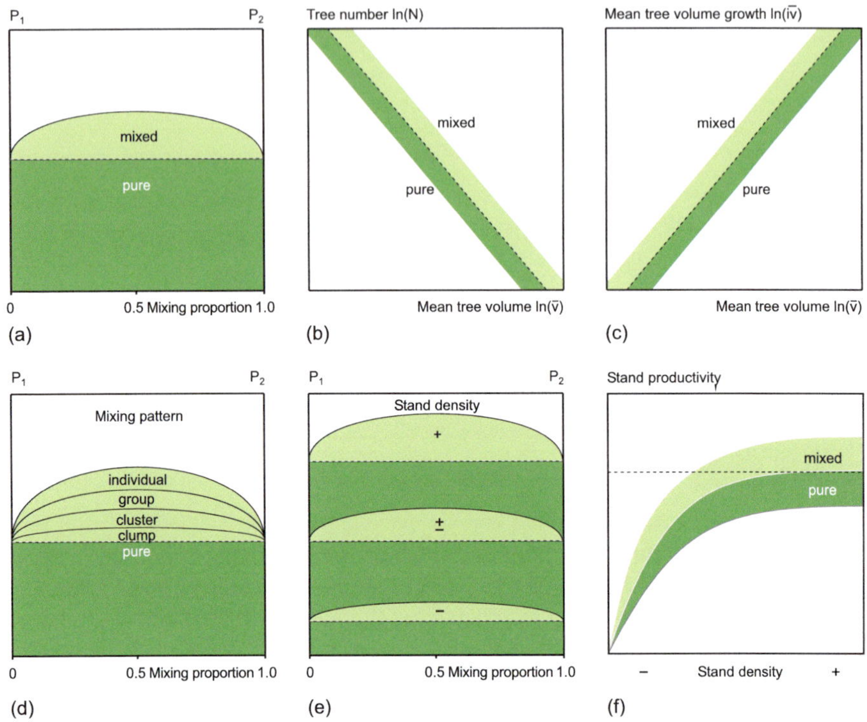


**Fig. 4.39** (**a–f**) Graphical abstract of the mixing responses in temperate forests reported in this chapter. Overyielding of mixed versus monospecific stands (**a**) can result from an increase in maximum stand density, indicated by an upward shift of the self-thinning line (**b**) and an increase in mean tree productivity, indicated by a rise in the mean tree growth-tree size relationship (**c**). Positive species interactions and overyielding are at a maximum in individual tree mixtures and decrease when species are mixed in larger units or simply border each other as monospecific stands (**d**). As species mixing can increase the maximum stand density, mixing effects are often higher in dense stands (**e**). Especially when multilayered, mixed stands are more resilient to density reductions compared with monocultures (**f**)

upward shift of the self-thinning line (relationship between $\ln(N) - \ln(\bar{v})$). Another part of the overyielding results from the increase in the relative growth rate of trees of a given size and (Fig. 4.39c) is represented by the upward shift in the relationship between the mean tree volume increment and mean tree volume ($\ln\left(\overline{iv}\right) - \ln(\bar{v})$).

Mixing effects normally increase with the intensity of intermingling (Fig. 4.39d). So, the effects decrease from individual tree mixing, mixing in small clusters, and groups to large clumps. The mixing effects continuously diminish (individual mixture > group > cluster > clump > monospecific stand), as more species are segregated. From this perspective, neighbouring monospecific stands are borderline cases of a mixture.

At low densities or even solitary growth, competitive reduction and facilitation are less effective because competition for resources and space is low. As the packing density increases, the benefit and competition release due to interspecific interactions are likely to increase (Fig. 4.39e). The highest mixing effects in many of the stands examined in this chapter occurred in unthinned or slightly thinned stands. However, as overyielding due to mixing can result from an increase in maximum stand density and growing space efficiency (Fig. 4.39b and c, respectively) in heavily thinned stands, only parts of the mixing effects may be 'thinned away'.

Behind the increasingly concave-shaped response pattern in the cross diagrams as the stand density rises (Fig. 4.39e) is the broad plateau of the growth-density relationship in mixed compared with monospecific stands (Fig. 4.39f). Whereas in monospecific stands, the growth-density relationship resembles an asymptotic or unimodal curve with a narrow maximum; the corresponding relationship in mixed stands can have a steeper left branch and a much broader plateau as well as a higher level. This means that the superiority of the mixed stand often increased with increasing stand density or moderation of thinning.

This chapter delivers facts on the performance of mixtures at the stand level relevant for decision making in practice. It is a preliminary synopsis, partially based on assumptions, because many knowledge gaps still remain. Most findings of over- and underyielding refer to two-species mixtures, but recent results (Forrester and Bauhus 2016), including global studies by Liang et al. (2016) suggest that they similarly apply for combinations of *n*-species mixture.

### Take-Home Messages

1. Forest management occurs at the stand or management block level. Therefore, essential performance elements of mixed-species stands versus monocultures as identified in this chapter may primarily serve to help improve establishment, design, and silvicultural treatment of mixed-species stands. In contrast, the tracing of mixing effects from the stand to the tree and organ levels in the subsequent chapters will contribute to understanding, modelling, and theory.
2. Species combinations with complementary ecological traits have the potential for higher resource supply and yield compared with monocultures. Combinations of shade-demanding and tolerant species, shallow- and deep-rooting species, early and late successional species, and mixtures with $N_2$-fixing tree species are likely to be useful.
3. The mixing responses and how they change along gradients of increasing resource availability can be positive, negative, or neutral. They depend on the respective limiting factors and the species' potential to remedy the limitation.
4. Common measures use the productivity of neighbouring monocultures as a reference and the species proportion in the mixed stand for quantifying how the mixed stand as a whole comes off in relation to the neighbouring monocultures and how each species contributes to any over- or underyielding.
5. The temperate mixed-forest stands examined in this chapter produce 10–30% more stem volume than monocultures; stands containing $N_2$-fixing tree species

can overyield monocultures by 50–100%. As mixing effects vary within a stand's life, analyses should cover a large proportion or the whole rotation.

6. Even small mixing proportions can trigger strong positive mixing effects. With an increasing mixing proportion, the effect increases at a diminishing rate. The intensity of species interactions in a mixed-species forest depends firstly on the spatial and temporal mixing pattern. The more the species are spatially and temporally separated, the weaker the interaction. Neighbouring monocultures can be understood as the borderline case of mixed stands.
7. Species mixing can increase the stand density and growth rate. Stand density, standing volume, and yield level were more affected by mixing than stand height or stand diameter in the stands examined in this chapter. As mixing effects are caused by both increased stand density and the growth velocity of individual trees, they get only partly eliminated by reducing density via thinning.
8. Due to their wider vertical structuring, mixed stands can potentially close gaps caused by thinning or other small-scale disturbances more quickly than monolayered monospecific stands. Growth losses through canopy opening are smaller, which is reflected by a broader plateau of the density-growth relationship of mixed versus monospecific stands.
9. Combining several species distributes and limits the risk to productivity caused by acute or chronic stress and damage. This applies for biological risks (e.g. insect gradation, fungi attacks, browsing by mammals), abiotic risks (storm, acidic emissions, drought), and technical and economic uncertainty (timber price, harvesting techniques).
10. The mixing effects under normal undisturbed stand development accumulate continuously and can increase the stand density and stand fertility. The effects of damage avoidance and reduction become effective and should be considered when disturbances affect mixtures less than monocultures.

## References


Amorosos MM, Turnblom EC (2006) Comparing productivity of pure and mixed Douglas-fir and western hemlock plantations in the Pacific Northwest. Can J For Res 36:1484–1496

Assmann E (1961) Waldertragskunde. BLV Verlagsgesellschaft, München, Bonn, Wien

Assmann E (1970) The principles of forest yield study. Pergamon Press, Oxford, New York

Assmann E, Franz F (1963) Vorläufige Fichten-Ertragstafel für Bayern. Institut für Ertragskunde der Forstl Forschungsanstalt München: 103

Avery TE, Burkhardt HE (1975) Forest measurements, 3rd edn. McGraw-Hill, New York

Beets PN, Madgwick HAI (1988) Aboveground dry matter and nutrient content of *Pinus radiata* as affected by lupin, fertilizer, thinning, and stand age. N Z J For Sci 18:43–64

Bielak K, Dudzinska M, Pretzsch H (2014) Mixed stands of Scots pine (*Pinus sylvestris* L.) and Norway spruce (*Picea abies* (L.) Karst) can be more productive than monocultures. Evidence from over 100 years of observation of long-term experiments. For Syst 23(3):573–589

Binkley D (2003) Seven decades of stand development in mixed and pure stands of conifers and nitrogen-fixing red alder. Can J For Res 33:2274–2279

Binkley D, Senock R, Bird S, Cole TG (2003) Twenty years of stand development in pure and mixed stands of *Eucalyptus saligna* and N-fixing *Falcataria moluccana*. Can J For Res 33:2274–2279

Bonnemann A (1939) Der gleichaltrige Misehbestand von Kiefer und Buche. Mitt aus Forstwirtschaft und Forstwissenschaft 10:439–483

Burger H (1941) Beitrag zur Frage der reinen oder gemischten Bestände. Mitt Schweiz Anst Forstl Versuchsw XXII:164–203

Callaway RM, Walker LR (1997) Competition and facilitation: a synthetic approach to interactions in plant communities. Ecology 78(7):1958–1965

Condés S, Del Rio M, Sterba H (2013) Mixing effect on volume growth of *Fagus sylvatica* and *Pinus sylvestris* is modulated by stand density. For Ecol Manag 292:86–95

de Martonne E (1926) Une novelle fonction climatologique: L'indice d'aridité. La Météorologie 21:449–458

del Río M, Schütze G, Pretzsch H (2014) Temporal variation of competition and facilitation in mixed species forests in Central Europe. Plant Biol 16(1):166–176

Dirnberger GF, Sterba H (2014) A comparison of different methods to estimate species proportions by area in mixed stands. For Syst 23(3):534–546

Dirnberger G, Sterba H, Condés S, Ammer C, Annighöfer P, Avdagic A, Bielak K, Brazaitis G, Coll L, Heym M, Hurt V, Kurylyak V, Motta R, Pach M, Ponette Q, Ruiz Peinado R, Skrzyszewski J, Šrámek V, Streel G, Svoboda M, Zlatanov T, Pretzsch H (2016) Species proportions by area in mixtures of Scots pine (*Pinus sylvestris* L.) and European beech (*Fagus sylvatica* L.) For Ecol Manag 373:149–166

Dittmar O, Knapp E, Zehler H (1986) Die langfristige Versuchsfläche Tornau im StFB Dübener Heide, ein Beispiel für den Weg vom Kiefernreinbestand zum Buchennaturverjüngungsbetrieb. Soz Forstw 36:344–348

Eichhorn F (1902) Ertragstafeln für die Weißtanne. Verlag Julius Springer, Berlin

Forrester DI (2014) The spatial and temporal dynamics of species interactions in mixed-species forests: From pattern to process. For Ecol Manag 312:282–292

Forrester DI, Bauhus J (2016) A review of processes behind diversity – productivity relationships in forests. Curr For Rep 2:45–61

Forrester DI, Pretzsch H (2015) Tamm review: on the strength of evidence when comparing ecosystem functions of mixtures with monocultures. For Ecol Manag 356:41–53

Forrester DI, Smith RGB (2012) Faster growth of *Eucalyptus grandis* and *Eucalyptus pilularis* in mixed-species stands than monocultures. For Ecol Manag 286:81–86

Forrester DI, Bauhus J, Cowie AL, Vanclay JK (2006) Mixed-species plantations of Eucalyptus with nitrogen-fixing trees: a review. For Ecol Manag 233:211–230

Forrester DI, Bauhus J, Khanna PK (2004) Growth dynamics in a mixed-species plantation of *Eucalyptus globulus* and *Acacia mearnsii*. For Ecol Manage 193:81–95

Forrester DI, Theiveyanathan S, Collopy JJ, Marcar NE (2010) Enhanced water use efficiency in a mixed *Eucalyptus globulus* and *Acacia mearnsii* plantation. For Ecol Manage 259:1761–1770

Frivold LH, Frank J (2002) Growth of mixed birch-coniferous stands in relation to pure coniferous stands at similar sites in South-eastern Norway. Scand J For Res 17:139–149

Frivold LH, Kolström T (1999) Yield and treatment of mixed stands of boreal tree species in Fennoscandia. In: Olsthoorn AFM, Bartelink HH, Gardiner JJ, Pretzsch H, Hekhuis HJ, Franc A (eds) Management of mixed-species forest: silviculture and economics, vol 15. IBN Scientific Contributions, pp 37–45

Gamfeldt L, Snäll T, Bagchi R, Jonsson M, Gustafsson L, Kjellander P, Ruiz-Jaen MC, Fröberg M, Stendahl J, Philipson CD, Mikusiński G, Andersson E, Westerlund B, Andrén H, Moberg F, Moen J, Bengtsson J (2013) Higher levels of multiple ecosystem services are found in forests with more tree species. Nat Commun 4:1340

Gehrhardt E (1923) Ertragstafeln für Eiche, Buche, Tanne, Fichte und Kiefer. Verlag Julius Springer, Berlin

Gül AU, Misir M, Misir N, Yavuz H (2005) Calculation of uneven-aged stand structures with the negative exponential diameter distribution and Sterba's modified competition density rule. For Ecol Manag 214(1):212–220
Harper JL (1977) Population biology of plants. Academic Press, London, New York
Hausser K (1956) Tannen-Ertragstafel. In: Schober R (ed) Ertragstafeln wichtiger Baumarten. JD Sauerländer's Verlag, Frankfurt am Main, p 154
Hector A, Bagchi R (2007) Biodiversity and ecosystem multifunctionality. Nature 448:188–190
Hector A, Schmid B, Beierkuhnlein C, Caldeira CM, Diemer M, Dimitrakopoulos PG, Finn JA, Freitas H, Giller PS, Good J, Harris R, Högberg P, Huss-Danell K, Joshi J, Jumpponen A, Körner C, Leadly PW, Loreau M, Minns A, Mulder CP, O'Donovan G, Otway SJ, Pereira JS, Prinz A, Read DJ, Scherer-Lorenzen M, Schulze ED, Siamantziouras ASD, Spehn EM, Terry AC, Troumbis AY, Woodward FI, Yachi S, Lawton JH (1999) Plant diversity and productivity experiments in European Grasslands. Science 286:1123–1127
Holmgren M, Scheffer M, Huston MA (1997) The interplay of facilitation and competition in plant communities. Ecology 78:1966–1975
Hooper DU, Chapin FS, Ewel JJ, Hector A, Inchausti P, Lavorel S, Lawton JH, Lodge DM, Loreau M, Naeem S, Schmid B, Setälä H, Symstad AJ, Vandermeer J, Wardle DA (2005) Effects of biodiversity on ecosystem functioning: a consensus of current knowledge. Ecol Monogr 75(1):3–35. doi:10.1890/04-0922
Jensen AM (1983) Growth of Silver Fir (*Abies alba* Mill.) compared with the growth of Norway Spruce (*Picea abies* (L) Karst.) in pure and mixed stands on sandy soils in the Western parts of Denmark. In: Reports from department of Forestry, vol 14. Royal Veterinary and Agricultural University, pp 1–498
Jolliffe PA (2000) The replacement series. J Ecol 88(3):371–385
Jolliffe PA, Minjas AN, Runecles VC (1984) A reinterpretation of yield relationships in replacement series experiments. J Appl Ecol 21:227–243
Jüttner O (1955) Ertragstafeln für Eichen. In: Etragstafeln wichtiger Baumarten bei verschiedener Durchforstung, neubearbeiet von R. Schober, vol 2. Auflage 1975. JD Saulerländer's Verlag, Frankfurt, p 154
Keller W (1995) Zur Oberhöhenberechnung in Mischbeständen aus standortkundlicher Sicht. In: Proc Dt Verb Forstl Forschungsanst, Joachimsthal, pp 52–60
Kelty MJ (1992) Comparative productivity of monocultures and mixed stands. In: Kelty MJ, Larson BC, Oliver CD (eds) The ecology and silviculture of mixed-species forests. Kluwer Academic Publishers, Dordrecht, pp 125–141
Kennel R (1965) Untersuchungen über die Leistung von Fichte und Buche im Rein- und Mischbestand. Allgemeine Forst- und Jagdzeitung 136:149–161, 173–189
Kern G (1966) Wachstum und Umweltfaktoren im Schlag- und Plenterwald. Bayerischer Landwirtschaftsverlag, München Basel Wien
Knoke T, Stimm B, Ammer C, Moog M (2005) Mixed forests reconsidered: a forest economics contribution on an ecological concept. For Ecol Manag 213(1–3):102–116. doi:10.1016/j.foreco.2005.03.043
Körner C (2002) Ökologie. In: Sitte P, Weiler EW, Kadereit JW, Bresinsky A, Körner C (eds) Strasburger Lehrbuch für Botanik, 35th edn. Spektrum Akademischer Verlag, Heidelberg, Berlin, pp 886–1043
Laclau J-P, Bouillet J-P, Gonçalves JLM, Silva EV, Jourdan C, Cunha MCS, Moreira MR, Saint-André L, Maquère V, Nouvellon Y, Ranger J (2008) Mixed-species plantations of *Acacia mangium* and *Eucalyptus grandis* in Brazil 1. Growth dynamics and aboveground net primary production. For Ecol Manag 225:3905–3917
Liang J, Crowther TW, Picard N, Wiser S, Zhou M, Alberti G, Schulze E-D, McGuire AD, Bozzato F, Pretzsch H, de-Miguel S, Paquette A, Hérault B, Scherer-Lorenzen M, Barrett CB, Glick HB, Hengeveld GM, Nabuurs G-J, Pfautsch S, Viana H, Vibrans AC, Ammer C, Schall P, Verbyla D, Tchebakova N, Fischer M, Watson JV, HYH C, Lei X, Schelhaas M-J, Lu H, Gianelle D, Parfenova EI, Salas C, Lee E, Lee B, Kim HS, Bruelheide H, Coomes DA,

Piotto D, Sunderland T, Schmid B, Gourlet-Fleury S, Sonké B, Tavani R, Zhu J, Brandl S, Vayreda J, Kitahara F, Searle EB, Neldner VJ, Ngugi MR, Baraloto C, Frizzera L, Bałazy R, Oleksyn J, Zawiła-Niedźwiecki T, Bouriaud O, Bussotti F, Finér L, Jaroszewicz B, Jucker T, Valladares F, Jagodzinski AM, Peri PL, Gonmadje C, Marthy W, O'Brien T, Martin EH, Marshall AR, Rovero F, Bitariho R, Niklaus PA, Alvarez-Loayza P, Chamuya N, Valencia R, Mortier F, Wortel V, Engone-Obiang NL, Ferreira LV, Odeke DE, Vasquez RM, Lewis SL, Reich PB (2016) Positive biodiversity-productivity relationship predominant in global forests. Science 354(6309). doi:10.1126/science.aaf8957

Long JN, Dean TJ, Roberts SD (2004) Linkages between silviculture and ecology: examination of several important conceptual models. For Ecol Manag 200:249–261

Magin R (1959) Struktur und Leistung mehrschichtiger Mischwälder in den bayerischen Alpen. Mitt Staatsforstverwaltung Bayerns 30:161

Mielikäinen K (1980) Mänty-koivusekametsiköiden rakenne ja kehitys. Summary: structure and development of mixed pine and birch stands. Commun Inst For Fenn 99:1–82

Mielikäinen K (1985) Koivusekoituksen Vaikutus Kuusikon Rakenteeseeen ja kehitykseen – Effect of an admixture of birch on the structure and development of Norway Spruce Stands. Commun Inst For Fenn 133:1–79

Mitscherlich G (1970) Wald, Wachstum und Umwelt, vol 1. Band. Form und Wachstum von Baum und Bestand. JD Sauerländer's Verlag, Frankfurt am Main

Montero G, Cañellas I, Ortega C, del Río M (2001) Results from a thinning experiment in a Scots pine (*Pinus sylvestris* L.) natural regeneration stand in the Sistema Ibérico Mountain Range (Spain). For Ecol Manag 145(1):151–161

Nagel J, Spellmann H, Pretzsch H (2012) Zum Informationspotenzial langfristiger forstlicher Versuchsflächen und periodischer Waldinventuren für die waldwachstumskundliche Forschung. Allgemeine Forst- und Jagdzeitung 183. Jg. (5/6):111–116

Olsthoorn AFM, Bartelink HH, Gardiner JJ, Pretzsch H, Hekhuis HJ, Franc A (1999) Management of mixed-species forest: silviculture and economics. IBN Scientific Contributions 15:389

Piotto D (2007) A meta-analysis comparing tree growth in monocultures and mixed plantations. For Ecol Manag 255:781–786

Pretzsch H (2003) The elasticity of growth in pure and mixed stands of Norway spruce (*Picea abies* [L.] Karst.) and common beech (*Fagus sylvatica* L.) J For Sci 49(11):491–501

Pretzsch H (2005a) Diversity and productivity in forests. In: Scherer-Lorenzen M, Körner C, Schulze E-D (eds) Forest diversity and function, Ecol Studies 176. Springer, Berlin, pp 41–64

Pretzsch H (2005b) Stand density and growth of Norway spruce (*Picea abies* (L.) Karst.) and European beech (*Fagus sylvatica* L.). Evidence from long-term experimental plots. Eur J For Res 124(3):193–205

Pretzsch H (2006) Species-specific allometric scaling under self-thinning: evidence from long-term plots in forest stands. Oecologia 146(4):572–583

Pretzsch H (2009) Forest dynamics, growth and yield. Springer, Berlin

Pretzsch H (2016) Ertragstafel-Korrekturfaktoren für Umwelt- und Mischungseffekte. AFZ Der Wald 14:47–50

Pretzsch H, Biber P (2005) A re-evaluation of reinekes rule and stand density index. For Sci 51 (4):304–320

Pretzsch H, Biber P (2016) Tree species mixing can increase maximum stand density. Can J For Res. 46, 1179–1193. doi:10.1139/cjfr-2015-0413

Pretzsch H, Rais A (2016) Wood quality in complex forests versus even-aged monocultures: review and perspectives. Wood Sci Technol 50:845–880

Pretzsch H, Schütze G (2009) Transgressive overyielding in mixed compared with pure stands of Norway spruce and European beech in Central Europe: evidence on stand level and explanation on individual tree level. Eur J For Res 128:183–204

Pretzsch H, Block J, Dieler J, Dong PH, Kohnle U, Nagel J, Spellmann H, Zingg A (2010) Comparison between the productivity of pure and mixed stands of Norway spruce and European beech along an ecological gradient. Ann For Sci 67:712

Pretzsch H, Matthew C, Dieler J (2012) Allometry of tree crown structure. relevance for space occupation at the individual plant level and for self-thinning at the stand level. In: Growth and defence in plants. Springer, Berlin, Heidelberg, pp 287–310

Pretzsch H, Bielak K, Block J, Bruchwald A, Dieler J, Ehrhart H-P, Kohnle U, Nagel J, Spellmann H, Zasada M, Zingg A (2013a) Productivity of mixed versus pure stands of oak (*Quercus petraea* (Matt.) Liebl. and *Quercus robur* L.) and European beech (*Fagus sylvatica* L.) along an ecological gradient. Eur J For Res 132(2):263–280

Pretzsch H, Bielak K, Bruchwald A, Dieler J, Dudzińska M, Ehrhart H-P, Jensen AM, Johannsen VK, Kohnle U, Nagel J, Spellmann H, Zasada M, Zingg A (2013b) Species mixing and productivity of forests. Results from long-term experiments. German title: Mischung und Produktivität von Waldbeständen. Ergebnisse langfristiger ertragskundlicher Versuche. Allgemeine Forst- und Jagdzeitung 184:177–196

Pretzsch H, del Río M, Ammer C, Avdagic A, Barbeito I, Bielak K, Brazaitis G, Coll L, Dirnberger G, Drössler L, Fabrika M, Forrester DI, Godvod K, Heym M, Hurt V, Kurylyak V, Löf M, Lombardi F, Matović B, Mohren F, Motta R, den Ouden J, Pach M, Ponette Q, Schütze G, Schweig J, Skrzyszewski J, Sramek V, Sterba H, Stojanović D, Svoboda M, Vanhellemont M, Verheyen K, Wellhausen K, Zlatanov T, Bravo-Oviedo A (2015) Growth and yield of mixed versus pure stands of Scots pine (Pinus sylvestris L.) and European beech (Fagus sylvatica L.) analysed along a productivity gradient through Europe. Eur J For Res 134(5):927–947. doi:10.1007/s10342-015-0900-4

Pretzsch H, Bauerle T, Häberle KH, Matyssek R, Schütze G, Rötzer T (2016a) Tree diameter growth after root trenching in a mature mixed stand of Norway spruce (*Picea abies* [L.] Karst) and European beech (*Fagus sylvatica* [L.]). Trees 30:1–13. doi:10.1007/s00468-016-1406-5

Pretzsch H, Schütze G, Biber P (2016b) Zum Einfluss der Baumartenmischung auf die Ertragskomponenten von Waldbeständen. Allgemeine Forst- und Jagdzeitung 187 (7/8):122–135

Preuhsler T (1979) Ertragskundliche Merkmale oberbayerischer Bergmischwald-Verjüngungsbestände auf kalkalpinen Standorten im Forstamt Kreuth. Forstl Forschungsber München 45:372 p

Reineke LH (1933) Perfecting a stand-density index for even-aged forest. J Agric Res 46:627–638

Richards AE, Forrester DI, Bauhus J, Scherer-Lorenzen M (2010) The influence of mixed tree plantations on the nutrition of individual species: a review. Tree Physiol 30(9):1192–1208

Scherer-Lorenzen M, Körner C, Schulze E-D (2005) Forest diversity and function, Ecol Studies, vol 176. Springer, Berlin, Heidelberg

Schober R (1972) Die Rotbuche. SchrReihe forstl Fak Univ Göttingen u Mitt Nieders VersAnst Bd. 43/44:333 p

Schober R (1975) Ertragstafeln wichtiger Baumarten. Saulerländers Verlag, Frankfurt a. M

Schwaiger S (2013) Wachstum von Fichte und Schwarz-Erle im Rein- und Mischbestand.

Skovsgaard JP, Vanclay JK (2008) Forest productivity: a review of the evolution of dendrometric concepts for even-aged stands. Forestry 81(1). doi:10.1093/forestry/cpm041

Sterba H (1987) Estimating potential density from thinning experiments and inventory data. For Sci 33(4):1022–1034

Sterba H, Monserud RA (1993) The maximum density concept applied to uneven-aged mixed-species stands. For Sci 39(3):432–452

Thurm EA, Pretzsch H (2016) Improved productivity and modified tree morphology of mixed versus pure stands of European beech (*Fagus sylvatica*) and Douglas-fir (*Pseudotsuga menziesii*) with increasing precipitation and age. Ann For Sci 73:1–15. doi:10.1007/s13595-016-0588-8

Toïgo M, Vallet P, Perot T, Bontemps JD, Piedallu C, Courbaud B (2015) Overyielding in mixed forests decreases with site productivity. J Ecol 103(2):502–512

Vallet P, Perot T (2011) Silver fir stand productivity is enhanced when mixed with Norway spruce: evidence based on large-scale inventory data and a generic modelling approach. J Veg Sci 22 (5):932–942

Vallet P, Perot T (2016) Tree diversity effect on dominant height in temperate forest. For Ecol Manag 38:106–114
Vandermeer JH (1989) The ecology of intercropping. Cambridge University Press, Cambridge
Vandermeer J (1992) The ecology of intercropping. Cambridge University Press
Vilà M, Carrillo-Gavilán A, Vayreda J, Bugmann H, Fridman J, Grodzki W, Haase J, Kunstler G, Schelhaas M, Trasobares A (2013) Disentangling biodiversity and climatic determinants of wood production. PLoS One 8(e53530)
von Gadow K (1986) Observations on self-thinning in pine plantations. S Afr J Sci 82(7):364–368
von Lüpke B, Spellmann H (1999) Aspects of stability, growth and natural regeneration in mixed Norway spruce-beech stands as a basis of silvicultural decisions. In: Olsthoorn AFM, Bartelink HH, Gardiner JJ, Pretzsch H, Hekhuis HJ, Franc A (eds) Management of mixed-species forest: silviculture and economics, vol 15. IBN Scientific Contributions, pp 245–267
Vospernik S, Nothdurft A (2016) Intra-annueller Zuwachs von Fichte, Buche und Zirbe. Jahrestagung der Sektion Ertragskunde im DVFFA 9. Lyss, Schweiz
Weiner J, Freckleton RP (2010) Constant final yield. Annu Rev Ecol Evol Syst 41:173–192
Wellhausen K, Heym M, Pretzsch H (2016) Mischbestände aus Kiefer und Buche. Ökologie, Ertrag und waldbauliche Behandlung. Allgemeine Forst- und Jagdzeitung 187
West GB, Brown JH, Enquist BJ (1997) A general model for the origin of allometric scaling laws in biology. Science 276:122–126
West GB, Brown JH, Enquist BJ (1999) A general model for the structure and allometry of plant vascular systems. Nature 400:664–667
West GB, Enquist BJ, Brown JH (2009) A general quantitative theory of forest structure and dynamics. PNAS 106(17):7040–7045
Wiedemann E (1942) Der gleichaltrige Fichten-Buchen-Mischbestand. Mitt Forstwirtsch u Forstwiss 13:1–88
Wiedemann E (1943) Der Vergleich der Massenleistung des Mischbestandes mit dem Reinbestand. Allgemeine Forst- und Jagdzeitung 119:123–132
Wiedemann E (1951) Ertragskundliche und waldbauliche Grundlagen der Forstwirtschaft. Frankfurt am Main
Wimmenauer K (1941) Zur Frage der Mischbestände. Allgemeine Forst- und Jagdzeitung 90:90–93
Yoda KT, Kira T, Ogawa H, Hozumi K (1963) Self-thinning in overcrowded pure stands under cultivated and natural conditions. J Inst Polytech, Osaka Univ D 14:107–129
Zeide B (1985) Tolerance and self-tolerance of trees. For Ecol Manag 13:149–166
Zeide B (1987) Analysis of the 3/2 power law of self-thinning. For Sci 33:517–537
Zeide B (2001) Thinning and growth: a full turnaround. J For 99:20–25
Zeide B (2004) How to measure density. Trees 19:1–14
Zhang Y, Chen HYH, Reich PB (2012) Forest productivity increases with eveness, species richness and trait variation: a global metaanalysis. J Ecol. doi:10.1111/j.1365-2745.2011.01944.x
Zöhrer F (1969) Bestandeszuwachs und Leistungsvergleich montan, subalpiner Lärchen-Fichten-Mischbestände. Forstwissenschaftliches Centralblatt 88:1–64

# Size-Structure Dynamics in Mixed Versus Monospecific Stands

# 5

Hans Pretzsch

**Abstract**

The community of mixed tree species is conceptualised in this chapter by the dynamics of its size distribution. We use experiments of mixed and neighbouring monocultures to show how mixture changes the size-structure dynamics and thereby the productivity of mixed versus monospecific stands. As the main cause of this modification, we identify the predominantly more size-asymmetric partitioning of the growth in favour of dominant trees in mixed stands. Furthermore, mixing can modify the growth allocation at the tree level in favour of the crown size. These differences in the inter- and intra-individual growth distribution may just slightly favour the size of tall trees at the expense of small ones and may slightly increase the canopy density in mixed compared to monospecific stands in a single growth period. However, via the feedback functioning→structure→environmental conditions→functioning, small initial and continuous advantages may result in a significantly different canopy structure, denser stocking, and higher productivity of mixed versus monospecific stands, especially in the advanced and mature phase. We characterise the size and growth distribution pattern emerging in mixed stands using various measures, indices, and relationships. The denser canopy space filling by trees with complementary light ecology may contribute to the general pattern of overyielding of mixed versus monospecific stands.

The community of mixed tree species is conceptualised in this chapter by the dynamics of its size distribution. Analysing the evolution of tree size distributions will provide essential insight into mixed-species population dynamics and provide a link between the stand level addressed in Chap. 4 and the tree level presented in Chap. 6. Mixing effects at the stand level are relevant

H. Pretzsch (✉)
Chair for Forest Growth and Yield Science, Faculty of Forest Science and Resource Management, Technical University of Munich, Hans-Carl-von-Carlowitz-Platz 2, 85354 Freising, Germany
e-mail: Hans.Pretzsch@lrz.tum.de



H. Pretzsch et al. (eds.), *Mixed-Species Forests*, DOI 10.1007/978-3-662-54553-9_5

for decision making and forest planning and indicate that advantageous species interactions are worth further exploration but are insufficient for revealing the underlying causes. Analyses at the tree level may reveal basic growth response patterns for better understanding competition and facilitation. They may also reveal how competition and facilitation result from a modification in the supply, capture, and use efficiency of resources. But the findings at individual tree level are barely sufficient for scaling up to stand-level behaviour because of system properties emerging through interindividual interactions. The size distribution dynamics provide a link between the stand and tree level and a transition between both levels of organisation.

## Contents

5.1 The Tree Size Distribution as Essential Stand Characteristic .......... 213
5.2 Abstracting Stand Dynamics as Evolving Tree Size Distribution .......... 216
5.2.1 Size-Structure Dynamics in Monocultures: Principle and Example .......... 217
5.2.2 Size-Structure Dynamics in Mixed-Species Stands: Principle and Example .... 218
5.3 Density, Shape, and Location of Size Distributions in Mixed Versus Monospecific Stands .......... 223
5.3.1 Measures for Characterising Density, Shape, and Location of Size Distributions .......... 223
5.3.2 Empirical Evidence of Mixing Effects .......... 224
5.3.3 Characteristic Modification of Tree Size Distribution Through Species Mixing .......... 228
5.4 Relationship Between Size Growth and Size of Trees in Mixed Versus Monospecific Stands .......... 229
5.4.1 Modes of Competition and Growth Distribution Between the Trees in a Forest Stand .......... 229
5.4.2 Symmetry and Asymmetry of Competition .......... 233
5.5 Modification of Tree Distribution by Removal Trees .......... 236
5.5.1 Ratio Between the Diameter of the Removal and Total Stand for Characterising the Mode of Mortality .......... 236
5.5.2 Mode of Tree Mortality in Mixed Versus Monospecific Stands .......... 237
5.6 Hierarchy of Tree Size, Tree Growth, and Growth Dominance .......... 239
5.6.1 Gini Coefficients, $GC_v$ and $GC_{iv}$, and Growth Dominance Coefficient, GDC ... 239
5.6.2 More Unequal Tree Sizes and Tree Growth Partitioning in Mixed Compared with Monospecific Stands but Similar Growth Dominance .......... 242
5.6.3 Remarkable Growth-Structure Elasticity of Mixed-Species Stands .......... 245
5.7 Modification of the Crown Allometry and thereby the Canopy Space Filling in Mixed versus Monospecific Stands .......... 246
5.7.1 From Stem Size Distribution to Crown Size Distribution .......... 246
5.7.2 Canopy Space Filling in Mixed versus Monospecific Stands .......... 248
5.7.3 Stand Density and Crown Density in Mixed and Monospecific Stands ....... 255
5.8 Effect of the Size-Structure Dynamics on Overyielding at the Stand and the Species Level .......... 258
5.8.1 Overyielding at the Stand and Species Level .......... 258
5.8.2 Relationship Between Size Structure and Over- and Underyielding at the Stand Level .......... 258
5.8.3 Increase in Canopy Space Filling and Light Interception .......... 259

5.9 Modification of Stand Structure in Monospecific and Mixed-Species Stands by Site Conditions .......... 260
5.10 Scientific Evidence and Practical Relevance .......... 261
5.10.1 Size Structure as Link Between Tree-Level and Stand-Level Research on Mixed Stands .......... 261
5.10.2 Consequences for Forest Management .......... 262
5.11 Synthesis .......... 263
Take-Home Messages .......... 265
References .......... 266

The familiar size distribution of monocultures can be modified with regard to location, shape, and scaling when tree species are mixed. In this chapter we use experiments of mixed and neighbouring monocultures to show how mixture changes the size-structure dynamics and thereby the productivity of mixed versus monospecific stands. As the main cause of this modification, we identify the predominantly more size-asymmetric partitioning of the growth in favour of dominant trees in mixed stands. Furthermore, mixing can modify the growth allocation at the tree level in favour of the crown size. These differences in the inter- and intra-individual growth distribution may just slightly favour the size of tall trees at the expense of small ones and may slightly increase the canopy density in mixed compared to monospecific stands in a single growth period. However, via the feedback functioning→structure→environmental conditions→functioning (see Chap. 6, Fig. 6.5), small initial and continuous advantages may result in a significantly different canopy structure, denser stocking, and higher productivity of mixed versus monospecific stands, especially in the advanced and mature phase. We characterise the size and growth distribution pattern emerging in mixed stands using various measures, indices, and relationships. The denser canopy space filling by trees with complementary light ecology may increase the light interception and light-use efficiency and contribute to the general pattern of overyielding of mixed versus monospecific stands that was shown in Chap. 4.

## 5.1 The Tree Size Distribution as Essential Stand Characteristic

The distribution of tree sizes for remaining and removal trees and the underlying growth partitioning between the trees goes far beyond a simple inventory and statistical overview of the state of a population. Rather, it reflects how successfully the involved social classes of one species or different associated species cope with crowding during stand development and their role in the population. As the relative size of trees in a population is strongly coupled with their access to resources, growth, and probability of survival, the size distribution also reflects the present social state and the potential further development of the represented trees of different sizes.

Figure 5.1a–c shows a schematic representation of three common stand structures (left) and how these translate into tree diameter distribution (middle). It

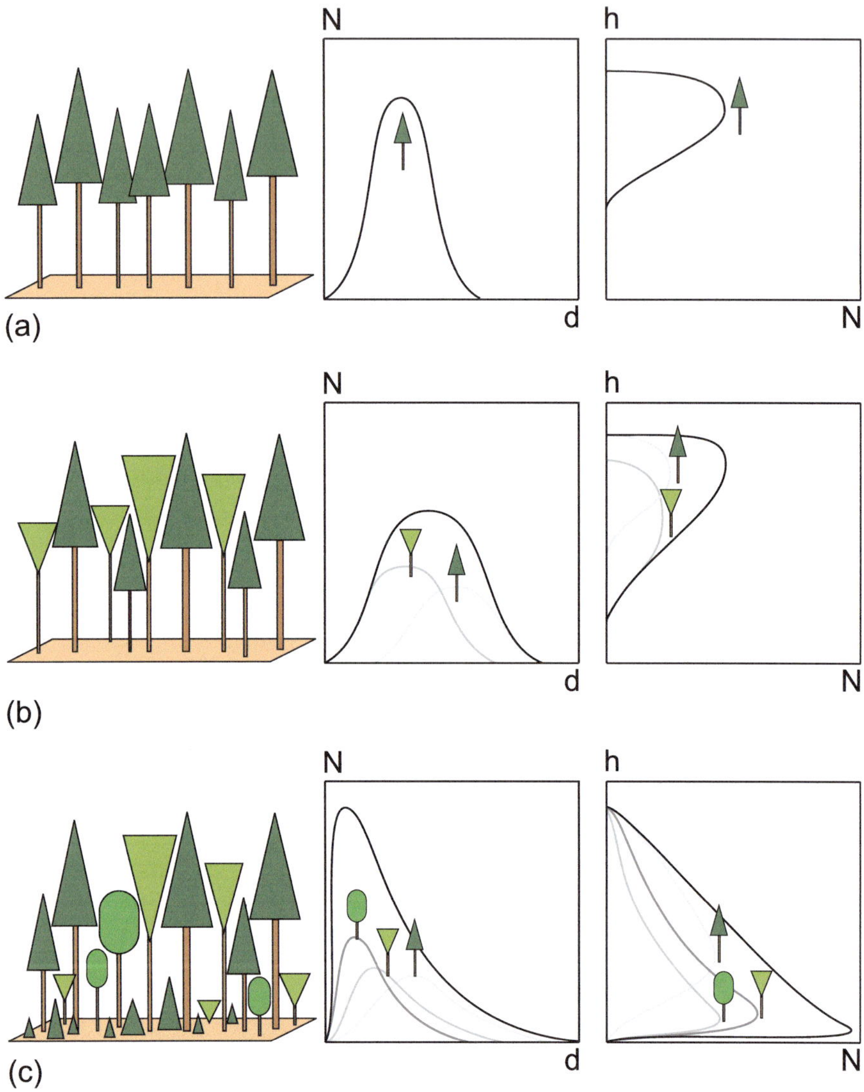


**Fig. 5.1** (**a**–**c**) The structure of forest stands (*left*) can be described by their specific tree diameter distributions (*centre*) and reflect the vertical layering of the canopy (*right*). From monolayered monospecific stands (**a**) to monolayered mixed-species stands (**b**) to multilayered mixed-species stands (**c**), the vertical heterogeneity increases, and the standing stock, light, and growth are partitioned among an increasingly wider range of tree sizes

also indicates the vertical height distribution and canopy space filling (right). Monocultures (Fig. 5.1a), especially when even-aged, tend towards a monolayered structure, Gaussian normal distribution of tree number over tree diameter, and a concentration of trees in mainly one height layer. In even-aged (Fig. 5.1b) or uneven-aged mixed-species stands (Fig. 5.1c), the size distribution may widen and indicate the species' interaction with regard to competition and their spatial niche occupation on the basis of complementarity in ecological traits. Thus, comparing the size-structure dynamics in mixed stands with those in monospecific stands of the same species may considerably contribute to understanding the mechanisms behind the mixing effects which were revealed at the stand level in the previous chapters. The widening of the tree diameter and height distribution of the stands shown in Fig. 5.1a–c indicates an increasing niche separation, inequality of size, and growth partitioning between small and tall trees.

Figure 5.1 illustrates that a variation in tree diameter distribution is coupled with a variation in tree height distribution which again indicates a richer structuring, indentation, and roughness of the crown surface area. Forests and trees provide structures on which, in which, or through which physical, biochemical, ecological, and socio-economic processes take place. Thus, size distribution characteristics can be useful indicators of the state and development of forest ecosystems and for applied management practice (MCPFE 1993). In comparison with direct quantitative measurements of biodiversity, stability, or sustainability (counting plant and animal species, assessment of matter balance, etc.), the use of structural parameters is advantageous as the data can be readily collected or is already included in forest inventory data.

According to Noss (1990), the structural aspects may positively affect compositional and functional aspects of biodiversity (Chap. 10, Fig. 10.8). They can positively influence the diversity of habitats as well as plant and animal species (Fig. 5.2). Stand structures determine the occurrence and population dynamics of owls and woodpeckers to such an extent that habitat suitability and the development of a population can be inferred directly on the basis of given structures (Letcher

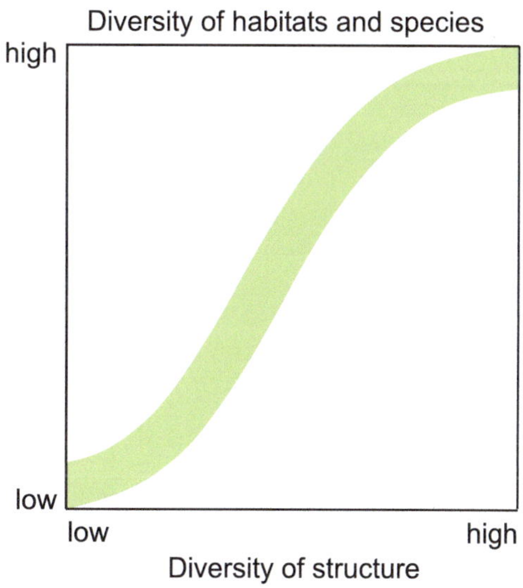


**Fig. 5.2** Diversity of structure analysed in this chapter has an impact on biodiversity and many other ecosystem functions and services. Schematic representation of the relationship between structure and diversity of habitats and species (adapted from Begon et al. (1998))

et al. 1998; McKelvey et al. 1993). Ammer and Schubert (1999) and Müller et al. (2007, 2012), among others, point to the close relationship between tree and stand structure and the presence of birds, beetles, spiders, lacewings, and soldier beetles. Consequently, size distributions may serve as indicators of the ecological diversity and stability of forest ecosystems and the type of management.

In heavily managed forests, size distributions may represent the silvicultural interference more than the species-specific ability to appropriate resources or adapt to competition. Particularly the lowering of stand density through wide spacing at stand establishment and thinning during further stand development may keep lower social classes or inferior species in subdominant and understorey positions, although these would be outcompeted in fully stocked stands (Fig. 5.1b and c). As the behaviour of trees of different social classes and species becomes most obvious under maximum density, we use mainly fully stocked stands to reveal basic size-growth dynamics in mixed versus monospecific stands in the following.

## 5.2 Abstracting Stand Dynamics as Evolving Tree Size Distribution

Given that evidence in the literature of mixing effects on the size-structure dynamics is sparse, this chapter is mainly based on our own evaluations of the following datasets: The thinning experiment in European beech Hain 27/1 (survey from 1871, stand age 38 years till present >180 years) was used to illustrate the characteristic size-structure dynamics of a monoculture (Kennel 1973). The species mixing experiment Waldbrunn 105 in sessile oak and European beech (survey from 1935 till present) was used to demonstrate the long-term effect of intra- versus interspecific competition on the size distribution, crown, and stem allometry (Freist-Dorr (1992), Sect. 4.2.2). A set of 42 triplets comprising 126 even-aged (mostly aged between 40 and 60 years), two-species mixed and monospecific stands of Scots pine (*Pinus sylvestris* L.), Norway spruce (*Picea abies* [L.] Karst.), Douglas-fir (*Pseudotsuga menziesii* (Mirb.) Franco), European beech (*Fagus sylvatica* L.), and sessile oak (*Quercus petraea* (Matt.) Liebl.) has been pooled and evaluated for this chapter (Pretzsch and Schütze 2014; Pretzsch et al. 2016). To analyse the mode of mortality in mixed versus monospecific stands, we used 11 long-term experiments in Germany in monospecific and mixed stands of European beech and sessile oak and European beech and Norway spruce (Pretzsch and Schütze 2014). To more closely analyse the canopy space filling, we used 253 crown maps in monospecific ($n = 87$), two-species ($n = 111$), and $\geq$ three-species ($n = 55$) mixed stands in Germany (Pretzsch 2014).

The data will be introduced briefly before the passages where it is applied in order to reveal mixing effects on the size-structure dynamics. All the data come from temperate forests where trees compete for nutrients and water and mainly for light through above-ground structural acclimation.

### 5.2.1 Size-Structure Dynamics in Monocultures: Principle and Example

The development of the trees in a monoculture or a species in a community can be characterised by tree size distribution, the growth distributions between the trees, and the mortality (Hara 1993). In single-cohort monocultures, the diameter distribution is narrow, asymmetric, and right skewed in the early stage and becomes increasingly symmetrical and Gaussian shaped with progressing stand development (Prodan 1965, pp 129–130). Silvicultural treatment cuts mainly the left branch by thinning from below, the right branch by thinning from above, or simply reduces the level of the size distribution through systematic thinning, such as elimination of every *n*th tree or tree row (Kramer 1988, pp 200–203). Shade-tolerant species tend towards wider size distributions than light-demanding species as a lower light compensation point allows for enhanced persistence of small trees in deep shade (Assmann 1970, pp 92–98).

The size distribution (tree number over tree diameter, basal area, height, or volume) of a forest stand at a given time (Fig. 5.3a) represents the composition and structure of the population. The height of the frequency distribution indicates the population density. The minimum, mean, maximum, coefficient of variation, and Gini coefficient indicate the heterogeneity of the structure, e.g. whether all trees are rather similar or significantly different in size (Peck et al. 2014).

Density and structuring determine the growth distribution between the trees in the stand. In homogeneous monospecific stands, the size growth-size relationship may be linear; in heterogeneous stands it tends to be convex from below as shown

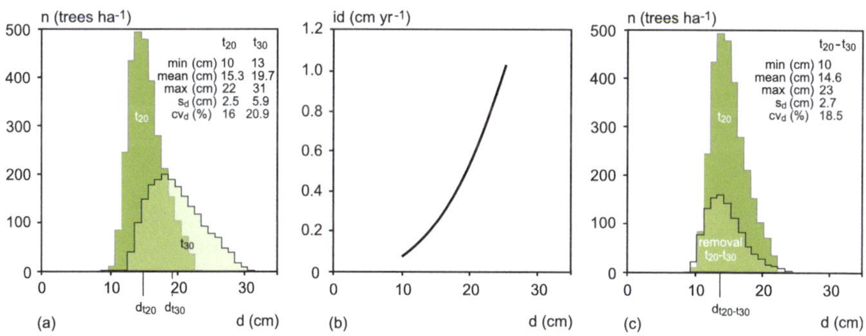


**Fig. 5.3** (**a**–**c**) Abstracting stand dynamics as evolving tree size distribution. (**a**) Stem diameter distribution of a stand at age $t_{20}$ and 10 years later $t_{30}$. (**b**) The change in size distribution from $t_{20}$ to $t_{30}$ is caused by tree growth (diameter increment in cm year$^{-1}$) depending on size in centimetre. (**c**) Removal stand in the period $t_{20}$–$t_{30}$ depending on size. The minimum, mean, maximum, and standard deviation and coefficient of variation of the distributions are 10, 15.3, 22, and 2.5 cm and 16% for $t_{20}$; 13, 19.7, 31, and 5.9 cm and 20.9% for $t_{30}$; and 10, 14.6, 23, and 2.7 cm and 18.5% for the removal stand between $t_{20}$ and $t_{30}$. The transformation function (b) is $id = 0.0001 \times d^{2.80}$, with id (cm year$^{-1}$) and $d$ (cm)

in Fig. 5.3b. This size growth-size relationship causes a change in the size distribution from $t_{20}$ to $t_{30}$, as it represents the growth rate of every tree depending on its initial size.

The evolution of the population from $t_{20}$ to $t_{30}$ is also shaped by the removal trees which leave the population due to mortality (e.g. self-thinning, diseases) or harvest. The mode of mortality can again be described by the size distribution of the removal tree number (Fig. 5.3c, frequency distribution for removal trees $t_{20}$–$t_{30}$). The drop out of trees can be quantified by number, minimum, mean, maximum size, or by the ratio between the mean tree size of removal and initial stand. For the removal stand shown in Fig. 5.3c applies $d_{out}/d_{in} = 14.6/15.3 = 0.95$. If $d_{out}/d_{in} = 1.0$, the size of both collectives is equal, whereas small ratios indicate prevailing drop out of small trees, and high ratios mean losses at the right branch of the diameter distribution, i.e. tree losses in the upper canopy.

In this chapter we will apply these three components of size-structure dynamics (initial distribution, transformation function, and distribution of the removal stand) and measures of their position and shape to analyse differences between mixed and monospecific stands at species and whole-stand level. A widening of the initial size distribution (Fig. 5.3a) in mixed versus monospecific stands, for example, would indicate a stronger vertical differentiation and deeper canopy. A shallower id–*d* relationship (Fig. 5.3b) would indicate a more even distribution of growth and resources between the trees in a stand. A shift in the removal stand towards the taller diameter classes would indicate that small trees have better conditions for growth and survival in mixed versus monospecific stands.

We use the development of the diameter distribution in the European beech experiment Hain 27/1 from age 51 to 172 in order to illustrate that as the mean diameter increases, the tree number and thereby the kurtosis and peak decrease, the range of the distribution ($d_{max}$–$d_{min}$) widens, and the number of small trees and thereby the skewness decrease because self-thinning eliminates mainly the small trees, i.e. it cuts mainly the left branch of the size distribution (Fig. 5.4).

## 5.2.2 Size-Structure Dynamics in Mixed-Species Stands: Principle and Example

### 5.2.2.1 Using Monocultures as a Reference for Distinguishing Between Multiplicative Mixing Effects and Additive Effects

The difference between the characteristic of mixed-species stands and monocultures may be due to a simple additive effect or a multiplicative effect (see introduction to Chap. 4). We use the tree size distributions in Fig. 5.5 to illustrate how to reveal both of these and to distinguish between them. Suppose the tree size distributions D (D stands for frequency distribution) of species 1 and 2 in the monoculture are $D_1$ and $D_2$ (Fig. 5.5a and c); then the weighted mean of both distributions in the case of a

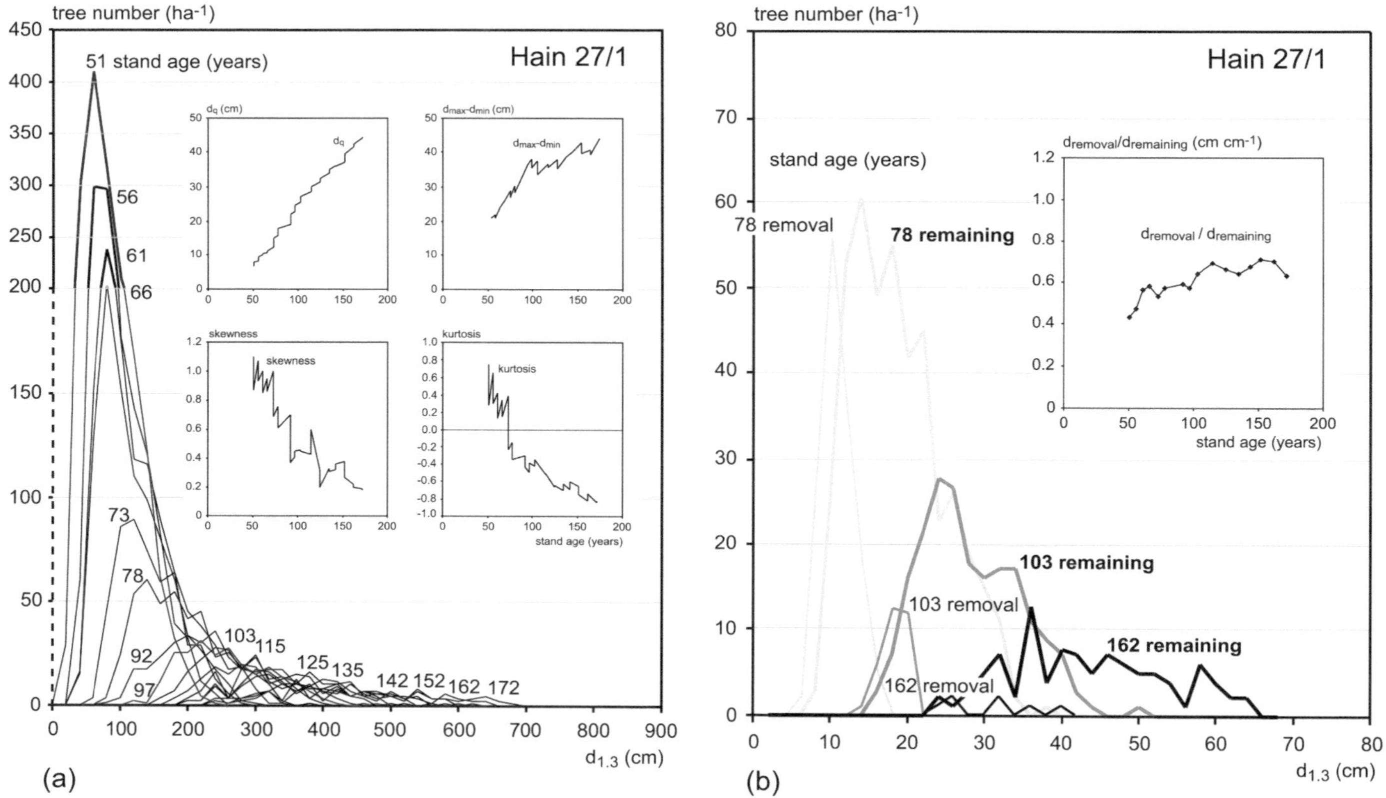


**Fig. 5.4** Long-term evolution of the diameter distribution in the European beech experiment Hain 27/1 (slightly thinned A-grade) from age 51 to 172. (**a**) Development of the remaining stand; note that the *y*-axis is shortened due to space restrictions. For development of mean diameter, $d_q$, $d_{max}$–$d_{min}$, skewness, and kurtosis see insets. (**b**) Remaining and removal stand for selected inventories at age 78, 103, and 162. For development of $d_{removal}/d_{remaining}$, see inset

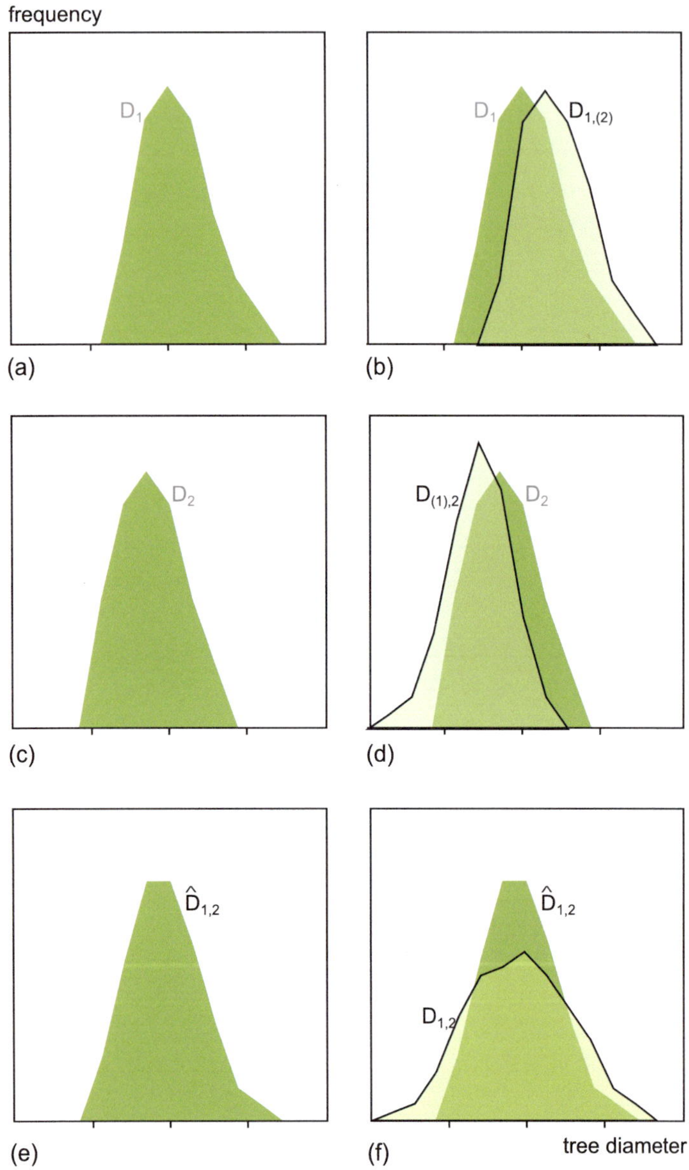


**Fig. 5.5** Schematic representation of the comparison between tree diameter distribution in monocultures and in mixed-species stands to quantify multiplicative effects (resulting from species interactions) as opposed to additive effects (resulting only from mixing species with different morphological or physiological traits). At the species level, size distributions $D_1$ and $D_2$ in monospecific stands can be compared with the respective distributions $D_{1,(2)}$ and $D_{(1),2}$ in neighbouring mixed stands (**a**–**d**). To quantify the mixing effect at the whole-stand level, the weighted mean of both monoculture distributions $\widehat{D}_{1,2}$ can be compared with the observed whole-stand distribution $D_{1,2}$ (**e** and **f**). Differences between the reference distributions (**a**, **c**, **e**) and the observed size distribution (**b**, **d**, **f**) can indicate interspecific interactions and multiplicative mixing effects

mixture with $m_1$ as the proportion of species 1 and thus $1-m_1$ for the proportion for species 2, $\widehat{D}_{1,2}$, represents the mean of $D_1$ and $D_2$, weighted by the proportions $m_1$ and $1-m_1$, respectively. $\widehat{D}_{1,2}$ is the weighted mean of both monocultures (Fig. 5.5e). It represents the expected distribution under the assumption that mixing simply causes an additive effect, i.e. retains the structural traits of the species as they are in the monoculture. In our example, $\widehat{D}_{1,2}$ ($\widehat{D}_{1,2} = D_1 \times m_1 + D_2 \times m_2$), where $m_1$ and $m_2$ are species proportions $m_1 = m_2 = 0.5$, differs clearly from the two monocultures $D_1$ and $D_2$ (compare the distributions shown in (e) with both (a) and (c)). The differences between $\widehat{D}_{1,2}$ and $D_1$ and $D_2$ are referred to as an additive effect because they are simply the effect of species identity. However $\widehat{D}_{1,2}$ also differs from the observed size distribution in the mixed-species stand, $D_{1,2}$ $(f)$. The latter difference between $\widehat{D}_{1,2}$ and $D_{1,2}$ indicates a multiplicative mixing effect, i.e. the size distribution of the mixed stand differs from the weighted mean of the neighbouring monospecific stands.

The comparisons between mixed stand and monoculture at the species level ($D_{1,(2)}$ versus $D_1$ and $D_{(1),2}$ versus $D_2$) shown in Fig. 5.5 (b and d) reveal how any differences at the stand level (f) emerge from the species level.

The approach for comparing mixed with monospecific stands (as introduced and applied above with reference to their tree diameter distribution) will also be applied for various tree attributes, e.g. for crown projection area, crown length, and individual tree growing area. One reason for using monocultures for this comparison is that mixed stands are often considered as alternatives to monocultures, and the frequency of tree attributes yielded by mixture compared with monocultures may be a basis for decision. Beyond this practical reason, using monocultures as a reference may be the best possible means of revealing the effect of inter- versus intraspecific competition on tree growth and structure.

#### 5.2.2.2 Mixing Can Modify the Size Distribution

Long-term observations of mixed species and general findings on the effect of mixing on the stand diameter distribution dynamics in mixed versus monospecific stands are still rare. In Fig. 5.6 we illustrate the development of the diameter frequency distribution of mixed (grey) versus monospecific (black) plots in the mixed-species experiment Waldbrunn 105 in sessile oak and European beech from 1935 to 2007 as a model example to point out some basic principles.

The diameter frequency distributions of the beech and oak monocultures (Fig. 5.6a and b) are used as references for analysing the mixed stands. Their development is typical for fully stocked even-aged monocultures, usually starting with a high-peaked, left steep, and right shallow tree diameter distribution, i.e. many rather equal-sized individuals with a majority of small ones and only a few tall ones. The initially greater size of tall trees increases through a compound interest effect, while initially small trees tend to drop out due to competition. Due to

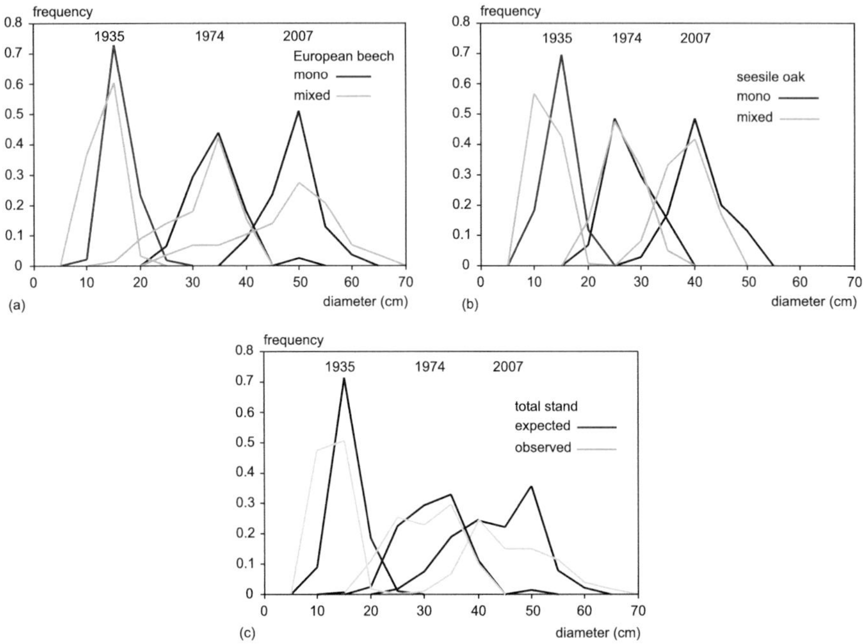


**Fig. 5.6** (**a**–**c**) Development of the diameter frequency distribution of mixed (*grey*) versus monospecific (*black*) plots in the mixed-species experiment Waldbrunn 105 in sessile oak and European beech from 1935 to 2007. (**a**) The diameter distribution of beech in mixture varies much more widely and shows a greater number of both smaller and taller trees than in the monoculture. (**b**) Diameter distribution of oak in mixture is narrower and behind the development of oak in monoculture. (**c**) Due to the opposite behaviour of beech and oak in the mixed stand, the observed total diameter distribution of both species (*grey*) is not significantly different from the expected (*black*). Notice that in (**a**) and (**b**) the distributions of the species in the mixed stands are scaled up to the unit area of 1 ha in order to make them comparable with the pure stands. In (**c**) the observed diameter distribution of the mixed stand as a whole is compared with the weighted mean distribution of the two pure stands (expected)

the diameter growth which is different for dominant and suppressed trees, the distribution moves rightwards and widens with progressing stand development. The combination of accelerated forward movement at the right side and self-thinning at the left side yields an ever widening, flattening, and symmetric diameter distribution. Thus, the right skewness and kurtosis are at maximum at the beginning and decrease as the stand ages.

The frequency distributions in the mixed stands at species level (Fig. 5.6a and b) and at whole-stand level (Fig. 5.6c) differ from the monocultures as follows: While at the beginning of the survey in 1935 (stand age 45), the diameter distributions of monospecific and mixed stands were rather similar, they increasingly deviated from each other during the following decades of stand development till 2007 (stand age 105). The diameter distribution of beech in mixture varies much more widely and shows a greater number of both smaller and taller trees than in the monoculture (Fig. 5.6a). The diameter distribution of oak in mixture is narrower and behind the development of oak in monoculture (Fig. 5.6b). Due to the opposite behaviour of beech and oak in the mixed stand, the observed total diameter distribution of both species (Fig. 5.6c, grey) is not significantly different from the expected (Fig. 5.6c, black).

This example underlines that although the whole-stand diameter distribution in the mixed stand does not differ much from the expected distribution, i.e. the weighted mean of the two monocultures' distributions, there might be considerable differences between mixed and monospecific stands at the species level. In our example, this similarity at whole-stand level results from the fact that beech is ahead and oak behind in mixed compared with monospecific stands, so that the first replaces the role of the latter in the mixture.

## 5.3 Density, Shape, and Location of Size Distributions in Mixed Versus Monospecific Stands

### 5.3.1 Measures for Characterising Density, Shape, and Location of Size Distributions

To compare the tree diameter distribution of mixed versus monospecific stands, we use measures such as tree number, arithmetic mean diameter, minimum and maximum diameter, diameter range (max-min), and standard deviation of tree diameter.

#### 5.3.1.1 Skewness, Kurtosis, and Other Measures for Characterising Diameter Distributions

To analyse any differences in the shape of the respective distributions, we use the skewness $\mathrm{skew} = \frac{\sum_{i=1}^{n} [(x_i - \bar{x})/s]^3}{n}$.

In the case of symmetric distribution, skew $= 0$ (Fig. 5.7a). Where an observed diameter distribution is equipped with many small trees and a low number of tall trees, it is left steep (right shallow) and yields skew $> 0$. If the distribution is equipped with many tall trees but small trees are rare, it is right steep (left shallow). Skew is useful for characterising the effect of any kind of thinnings (including self-thinning and alien thinning in unmanaged stands) on the shape of the distribution.

Furthermore, we calculated the kurtosis $\mathrm{kurt} = \frac{\sum_{i=1}^{n} [(x_i - \bar{x})/s]^4}{n} - 3$ which characterises the degree of concentration of tree sizes around the mean. If the concentration resembles the Gaussian normal distribution, kurt $= 0$. Stronger concentrations around the mean (peaked shapes) are indicated by kurt $> 0$, whereas lower concentrations

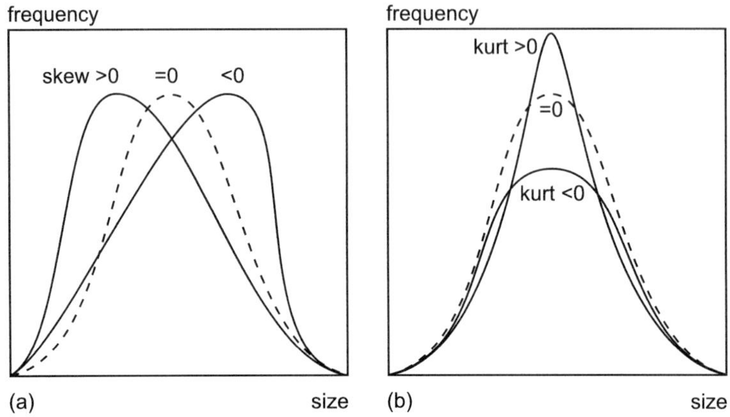


**Fig. 5.7** Illustration of skewness (**a**) and kurtosis (**b**) of tree diameter distributions

(shallow shapes) yield kurt $< 0$ (Fig. 5.7b). The kurtosis is appropriate for characterising the degree of restriction of a species caused by intra- and interspecific competition.

For the further evaluation, it is important to note that both skewness and kurtosis are invariant to linear transformation, i.e. if the tree diameter distribution of a species occupying a certain portion of the mixed stand is scaled up to 1 ha, the skewness and kurtosis remain unchanged.

## 5.3.2 Empirical Evidence of Mixing Effects

### 5.3.2.1 Dataset of 42 Triplets with 126 Plots in Monospecific and Mixed-Species Stands

As a dataset we used 42 triplets in monospecific and mixed stands of Norway spruce/European beech, Scots pine/European beech, Douglas-fir/European beech, and Norway spruce/Scots pine (Table 5.1). Each triplet consists of three plots with two in monospecific stands and one in a mixed stand of the respective species.

Only those species which play a major role in forestry in both monospecific and mixed-species stands (Brus et al. 2011) were included in the analysis. In order to cover a wide range of growing conditions, the triplets were selected in the lowlands of Poland, the uplands of south Germany, in the low and medium mountain range in south-west Germany, and in the foothills of the Bavarian Forest and Alps (Table 5.1). They lie between 60 and 785 m a.s.l., the mean temperature ranges from 5.5 to 9.5 °C, and the annual precipitation amounts to 556–1350 mm year$^{-1}$. Brown soils and para-brown soils from loess, loam, or sandy loam are dominant. All included plots represent more or less even-aged and monolayered forest stands.

The plots have not been recently thinned and represent high to maximum stand density. Mixed plots should represent individual tree mixture and a mixing proportion of about 50:50.

#### 5.3.2.2 Density, Shape, and Location of Size Distributions in Mixed-Species Stands Versus Monocultures

The first section of Table 5.1 shows that apart from the mixture of Norway spruce/European beech (mixed/mono = 0.96–1.00), all mixed stands have considerably higher tree numbers than monocultures. On average, there are 9–58% more trees (mixed/mono = 1.09–1.59) in mixed-species stands compared with monocultures (Table 5.1, first section). The minimum tree volumes are mostly lower in mixed compared with pure stands (Table 5.1, second section). The tallest trees, in contrast, can have significantly higher stem volumes in mixed-species stands than in neighbouring monocultures (Table 5.1, third section). However, the mean tree volumes are partly lower and partly significantly higher in mixed compared with pure stands (Table 5.1, fourth section). The differences in the shape of the size distribution were further substantiated by mostly higher right-skewed (skewness >> 0) and higher-peaked (kurt >> 0) size distributions in mixed-species stands compared with neighbouring monocultures (Pretzsch and Schütze 2015).

Table 5.2 shows the range of the frequency distribution for the tree diameter, tree height, and tree volume (from top to bottom). It illustrates that the widening of the tree volume distribution shown in Table 5.1 is coupled with an even stronger widening of the diameter and height distribution. In all cases, the ratios between the range in mixed and pure stands are above 1.0, i.e. the mixed stands are more heterogeneously structured than pure stands. In seven out of eight cases, the range of the diameter distribution is significantly wider in mixed versus pure stands. Regarding the range of the height distribution, mixed stands have a significantly stronger vertical structure in five out of eight cases. This tendency is corroborated by the mostly higher coefficients of variation of tree diameters, tree heights, and tree volumes in mixed compared with pure stands (Pretzsch and Schütze 2015).

In Tables 5.1, 5.2, 5.3, 5.4, 5.5, the columns ‘mean mixed’ and ‘mean mono’ display the arithmetic means of all *n* observations within the respective groups. The columns ‘mean mixed/mono’, in contrast, report the ratios resulting from the pairwise division of the characteristic of the mixed stand by the respective value of the neighbouring monoculture. The mean of these ratios (mixed/mono) is not necessarily equal to the ratio of the means (mean mixed/mean mono). So, we report both the groupwise arithmetic means (mean mixed and mean mono) as well as the mean ratios of the pairwise comparison (mixed/mono). The mean ratios of the pairwise comparison (mixed/mono) can be used for testing group differences. Our focus was on the relationships between neighbouring mixed-species stands and monocultures (reflected by their pairwise comparison) rather than on their differences in general (reflected by their overall means).

**Table 5.1** Tree number per hectare, minimum, maximum, and mean tree volume (mean ± standard deviation) on the 126 plots of the 42 triplets of Norway spruce/European beech, Scots pine/European beech, Douglas-fir/European beech, and Norway spruce/Scots pine

| Species | *n* | Tree number ($ha^{-1}$) | | | Tree number mixed/mono 1 | | | | Tree number mixed/mono 2 | | | |
|---|---|---|---|---|---|---|---|---|---|---|---|---|
| 1/2 | | Mean mixed | Mean mono 1 | Mean mono 2 | Min | Max | Mean | SE | Min | Max | Mean | SE |
| N.sp./E.be | 11 | 571 ± 72 | 629 ± 83 | 681 ± 112 | 0.39 | 1.68 | 1.00 | 0.11 | 0.44 | 2.33 | 0.96 | 0.15 |
| Sc.p./E.be | 7 | 1093 ± 268 | 945 ± 256 | 1170 ± 280 | 0.63 | 4.01 | 1.55 | 0.43 | 0.51 | 1.65 | 1.09 | 0.16 |
| D.-fir./E.be | 18 | 1051 ± 218 | 682 ± 127 | 1123 ± 233 | 0.58 | 3.83 | 1.58 | 0.19 | 0.32 | 2.61 | 1.19 | 0.16 |
| N.sp./Sc.p. | 6 | 1075 ± 78 | 916 ± 99 | 976 ± 170 | 0.65 | 1.84 | 1.25 | 0.16 | 0.67 | 1.84 | 1.27 | 0.21 |
| Species | *n* | Minimum tree volume | | | Min tree volume mixed/mono 1 | | | | Min tree volume mixed/mono 2 | | | |
| 1/2 | | Mean mixed | Mean mono 1 | Mean mono 2 | Min | Max | Mean | SE | Min | Max | Mean | SE |
| N.sp./E.be | 11 | 0.03 ± 0.01 | 0.21 ± 0.10 | 0.06 ± 0.02 | 0.03 | 1.21 | 0.41 | 0.12 | 0.17 | 2.10 | 0.93 | 0.20 |
| Sc.p./E.be | 7 | 0.06 ± 0.02 | 0.32 ± 0.11 | 0.07 ± 0.02 | 0.03 | 1.54 | 0.51 | 0.23 | 0.23 | 2.02 | 1.09 | 0.21 |
| D.-fir./E.be | 18 | 0.04 ± 0.01 | 0.60 ± 0.20 | 0.20 ± 0.14 | 0.00 | 3.29 | 0.51 | 0.22 | 0.06 | 2.29 | 0.99 | 0.15 |
| N.sp./Sc.p. | 6 | 0.03 ± 0.01 | 0.15 ± 0.09 | 0.23 ± 0.08 | 0.04 | 4.09 | 0.95 | 0.64 | 0.03 | 0.45 | 0.16*** | 0.06 |
| Species | *n* | Maximum tree volume | | | Max tree volume mixed/mono 1 | | | | Max tree volume mixed/mono 2 | | | |
| 1/2 | | Mean mixed | Mean mono 1 | Mean mono 2 | Min | Max | Mean | SE | Min | Max | Mean | SE |
| N.sp./E.be | 11 | 3.51 ± 0.55 | 2.83 ± 0.50 | 2.63 ± 0.30 | 0.68 | 3.02 | 1.39 | 0.19 | 0.78 | 1.96 | 1.35** | 0.12 |
| Sc.p./E.be | 7 | 1.91 ± 0.36 | 1.49 ± 0.29 | 1.68 ± 0.36 | 0.47 | 2.80 | 1.48 | 0.28 | 0.52 | 2.60 | 1.32 | 0.25 |
| D.-fir./E.be | 18 | 5.85 ± 1.15 | 5.68 ± 1.15 | 2.50 ± 0.51 | 0.37 | 2.28 | 1.01 | 0.11 | 0.79 | 5.77 | 2.49*** | 0.26 |
| N.sp./Sc.p. | 6 | 2.25 ± 0.29 | 2.10 ± 0.31 | 1.65 ± 0.29 | 0.66 | 2.07 | 1.15 | 0.20 | 1.11 | 1.92 | 1.43*** | 0.13 |
| Species | *n* | Mean tree volume | | | Mean tree volume mixed/mono 1 | | | | Mean tree number mixed/mono 2 | | | |
| 1/2 | | Mean mixed | Mean mono 1 | Mean mono 2 | Min | Max | Mean | SE | Min | Max | Mean | SE |
| N.sp./E.be | 11 | 1.25 ± 0.23 | 1.30 ± 0.30 | 0.88 ± 0.15 | 0.53 | 2.74 | 1.15 | 0.20 | 0.80 | 2.55 | 1.48** | 0.15 |
| Sc.p./E.be | 7 | 0.61 ± 0.12 | 0.76 ± 0.18 | 0.61 ± 0.17 | 0.33 | 1.49 | 0.92 | 0.16 | 0.55 | 2.50 | 1.26 | 0.26 |
| D.-fir./E.be | 18 | 1.34 ± 0.29 | 2.61 ± 0.60 | 1.34 ± 0.29 | 0.21 | 2.18 | 0.61 | 0.10 | 0.76 | 2.66 | 1.60*** | 0.13 |
| N.sp./Sc.p. | 6 | 0.61 ± 0.09 | 0.72 ± 0.12 | 0.79 ± 0.09 | 0.55 | 2.32 | 0.98 | 0.27 | 0.50 | 0.91 | 0.76 | 0.06 |

Note that in the columns 'mean mixed', 'mean mono 1', and 'mean mono 2', we report the arithmetic means of all n observations within the respective groups. In the columns 'mean mixed/mono 1' and 'mean mixed/mono 2' in contrast, we report the mean of the ratio resulting from the pairwise division of the characteristic of the mixed-species stands by the respective value of the neighbouring monocultures. '*', '**', and '***' indicate significant differences between mixed-species stands and monocultures at the level $p < 0.05$, 0.01, and 0.001

**Table 5.2** Range of stem diameter, $d$, tree height, $h$, and stem volume, $v$, (mean ± standard deviation) on the 126 plots of the 42 triplets of Norway spruce/European beech, Scots pine/European beech, Douglas-fir/European beech, and Norway spruce/Scots pine

| Species | | Range *d* | | | Range *d* mixed/mono 1 | | | | Range *d* mixed/mono 2 | | | |
|---|---|---|---|---|---|---|---|---|---|---|---|---|
| 1/2 | *n* | Mean mixed | Mean mono 1 | Mean mono 2 | Min | Max | Mean | SE | Min | Max | Mean | SE |
| N.sp./E.be | 11 | 45.26 ± 2.73 | 33.63 ± 3.17 | 36.72 ± 1.76 | 0.83 | 2.09 | 1.42*** | 0.12 | 0.99 | 1.61 | 1.24*** | 0.06 |
| Sc.p./E.be | 7 | 31.31 ± 3.05 | 19.41 ± 2.06 | 26.79 ± 2.67 | 0.98 | 2.82 | 1.74** | 0.27 | 0.67 | 1.70 | 1.21 | 0.12 |
| D.-fir./E.be | 18 | 54.38 ± 6.48 | 42.67 ± 4.42 | 29.06 ± 3.11 | 0.50 | 2.45 | 1.26* | 0.12 | 0.80 | 3.29 | 1.90*** | 0.15 |
| N.sp./Sc.p. | 6 | 35.77 ± 1.76 | 30.63 ± 1.95 | 22.73 ± 2.73 | 0.90 | 1.47 | 1.19* | 0.09 | 1.24 | 2.12 | 1.65*** | 0.15 |
| Species | | Range *h* | | | Range *h* mixed/mono 1 | | | | Range *h* mixed/mono 2 | | | |
| 1/2 | *n* | Mean mixed | Mean mono 1 | Mean mono 2 | Min | Max | Mean | SE | Min | Max | Mean | SE |
| N.sp./E.be | 11 | 19.56 ± 1.27 | 13.53 ± 1.73 | 12.98 ± 1.72 | 0.90 | 2.90 | 1.62** | 0.19 | 1.06 | 5.85 | 1.90* | 0.41 |
| Sc.p./E.be | 7 | 11.80 ± 1.65 | 5.53 ± 1.25 | 11.04 ± 1.06 | 0.85 | 6.24 | 2.85* | 0.72 | 0.72 | 1.67 | 1.07 | 0.11 |
| D.-fir./E.be | 18 | 22.54 ± 2.15 | 11.09 ± 2.12 | 13.17 ± 1.49 | 0.58 | 15.03 | 3.91** | 0.94 | 1.15 | 36.30 | 3.54 | 1.93 |
| N.sp./Sc.p. | 6 | 21.55 ± 1.13 | 15.52 ± 2.27 | 7.61 ± 2.05 | 1.01 | 3.74 | 1.70 | 0.44 | 1.45 | 9.22 | 4.13** | 1.19 |
| Species | | Range *v* | | | Range *v* mixed/mono 1 | | | | Range *v* mixed/mono 2 | | | |
| 1/2 | *n* | Mean mixed | Mean mono 1 | Mean mono 2 | Min | Max | Mean | SE | Min | Max | Mean | SE |
| N.sp./E.be | 11 | 3.48 ± 0.55 | 2.62 ± 0.48 | 2.57 ± 0.30 | 0.69 | 3.14 | 1.48* | 0.20 | 0.78 | 1.97 | 1.36** | 0.12 |
| Sc.p./E.be | 7 | 1.85 ± 0.37 | 1.17 ± 0.22 | 1.62 ± 0.35 | 0.52 | 3.68 | 1.82* | 0.37 | 0.48 | 2.63 | 1.33 | 0.25 |
| D.-fir./E.be | 18 | 5.81 ± 1.15 | 5.09 ± 1.00 | 2.30 ± 0.45 | 0.35 | 2.65 | 1.10 | 0.13 | 0.79 | 5.87 | 2.64*** | 0.27 |
| N.sp./Sc.p. | 6 | 2.23 ± 0.28 | 1.95 ± 0.23 | 1.42 ± 0.25 | 0.67 | 2.04 | 1.20 | 0.19 | 1.23 | 2.35 | 1.67 | 0.18 |

For further explanation see caption of Table 5.1

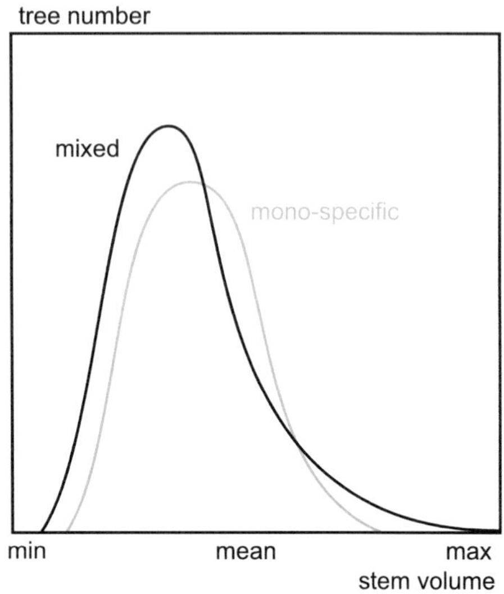


**Fig. 5.8** Characteristic size distribution of mixed stands compared with monocultures. In this analysis, the minimum tree sizes in mixed stands were lower, and the maximum tree sizes were higher, but the mean size was rather similar compared with monocultures. In mixed stands, the tree numbers were higher, and the size distributions more left steep (skewness >> 0) and more peaked (kurtosis >> 0). In contrast, the size distributions in monocultures were more symmetrical and flatter

### 5.3.3 Characteristic Modification of Tree Size Distribution Through Species Mixing

In the temperate forests of our study, the trees compete strongly above ground for light. As a result of this competition for light, tree size distribution, growth partitioning between the trees, and canopy density of mixed-species stands can change fundamentally and may contribute to an overyielding compared with neighbouring monocultures.

Figure 5.8 is a schematic representation of the effect of mixing on the tree size distribution. The total number of trees can be higher in mixed stands compared with monocultures. In addition, the minimum tree size can be smaller and the maximum size larger. The skewness and the kurtosis can also be higher. The higher tree number and lower minimum tree size result mainly from shade-tolerant species like beech which had reduced growth but can survive below more light-demanding and transparent species such as pine. The alien thinning can proceed less rigorously than the self-thinning, especially in the case of beech, which has the lowest self-tolerance of the species investigated (Pretzsch 2006; Zeide 1985). Trees of the species with superior size can accelerate growth and grow ahead. Due to its size superiority, this species can slow down the other species so that the latter lags behind and takes on a more subdominant position. This explains the higher maximum tree size and wide range of sizes in mixed stands compared with monocultures also found by Buongiorno et al. (1994) and Coomes and Allen (2007).

In most cases, one of the two even-aged species is ahead of the other regarding size development. Figure 5.9 is a schematic representation showing that species 1 is ahead of species 2 in the monoculture. The leading species is often more light demanding and grows more quickly, while the slower species is often more shade

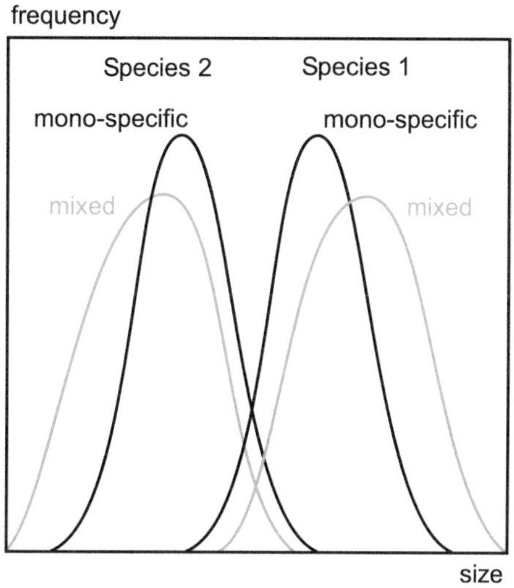


**Fig. 5.9** In the mixed stand, the accelerated forward shift in the size distribution of species 1 (*right*) can slow down and modify the shape of the size distribution of species 2 (*left*). Schematic representation of how the species' tree size frequency distribution in a two-species mixed stand (*grey lines*) can differ from the size distribution in neighbouring monocultures (*black lines*)

tolerant. Growing in mixture can trigger an additional forward shift in the size distribution of species 1 which is already ahead in the monoculture (e.g. Norway spruce or Douglas-fir in mixture with European beech). Species 2, which is behind the size growth of species 1, may cause a thinning-from-below effect on species 1 by outcompeting smaller population members of species 1. In contrast, species 1 may reduce species 2, similar to a thinning-from-above effect. As a consequence, the total tree size distribution in mixed stands is often broader, with higher skewness and kurtosis than in monospecific stands (Fig. 5.9).

## 5.4 Relationship Between Size Growth and Size of Trees in Mixed Versus Monospecific Stands

### 5.4.1 Modes of Competition and Growth Distribution Between the Trees in a Forest Stand

When growing solitarily, it is mainly size that determines a plant's growth-size trajectory (Box 5.1). In temperate forests with light as a limiting factor for individual tree growth, taller trees can pre-empt the light and shade and thus reduce growth of their smaller neighbours. Thus, in a cohort of even-aged trees, growth of smaller trees can fall much further below the potential growth curve and behind that of their fitter neighbours. Diameter growth plotted above diameter (or volume growth versus volume) reveals the interindividual competition and growth partitioning between the trees in a stand (Wichmann 2001, 2002).

**Box 5.1 From the Potential Size-Growth Curve of an Individual Tree to the InterIndividual Size-Growth Relationship at a Given Point in Time in a Stand**

*Potential Size Growth as a Function of Tree Size as Reference*

In the juvenile phase of tree development, anabolism has the upper hand and drives growth exponentially. However, simultaneously with size growth, maintenance costs increase and affect culmination of the growth rate which then finally tapers off (Zeide 1993). Open-grown trees or trees with constant inhibition follow a unimodal growth-size trajectory which comprises convex (seen from below) curve sections in the juvenile phase (1–3), concave sections in the middle age (4–6), and again convex sections in the mature phase (7–9) (see Avery and Burkhardt (1983, p 266); Schütz (1989, pp 4–5)). The higher the site fertility, the lower the growth limitation and the higher the level of the curve due to the greater resource supply per plant for a given size (Pretzsch and Biber 2010). Box Figure 5.1-1a illustrates this by curves expected for high, medium, and low resource supply. Such growth curves can be derived from long-term survey (real time series) or chronosequences (artificial time series) of trees grown under solitary conditions.

*Effect of Interindividual Competition*

When growing solitarily it is mainly size that determines the plants' unimodal growth trajectory shown in Box Fig. 5.1-1a. Within a stand, particularly those trees coping with crowding or stress can fall below this trajectory. For trees within a stand, size denotes access to resources, especially to light. In most cases, the taller a tree, the more privileged is its access to resources, space occupation, and repression effect on neighbours (Biging and Dobbertin 1995; Pretzsch 2009). But size is an ambiguous trait; it can also mean higher susceptibility to windthrow (Peltola 1996; Valinger et al. 1993), drought (Condit et al. 1995; Skov et al. 2004), or bark beetle attacks (Coggins et al. 2010). However, in temperate forests with light as a limiting factor for individual tree growth, taller trees shade and thus reduce the growth of their smaller neighbours. In a cohort of even-aged trees, growth of smaller trees falls behind the taller ones. Subsequently, it drops further below the potential than that of fitter neighbours.

This size-dependent growth reduction is behind the phenomenon which forest scientists refer to as diameter increment-diameter line (id-d line) and which is often used to describe and model the growth-size relationship of even-aged stands in a given developmental phase (e.g. Prodan (1965, pp 474–476)). While Box Fig. 5.1-1a shows the potential growth-size trajectories of individual trees with high, medium, and low resource supply, Box Fig. 5.1-1b exemplarily illustrates for trees with high resource supply how competition can transform the unimodal curve of individuals to the linear interindividual growth-size relationships. Diameter growth

(continued)

**Box 5.1** (continued)
plotted above diameter (or volume growth versus volume) yields a straight line with a steep positive slope in early stand phases and an increasingly flatter slope with progressive stand development (Box Fig. 5.1-1b, segments 1–9).

*Site Conditions Can Modify the Size Growth-Size Relationship*

Under ceteris paribus conditions, trees on fertile sites with a high supply of resources can make more use of their privileged position and exert a more negative effect on the growth of their smaller neighbours (Wichmann 2001, 2002). This should be reflected by steeper slopes of the interindividual size-growth relationships (see straight lines in Box Fig. 5.1-1c), while nutrient limitation should diminish their superiority and the slope of the growth-size relationship (Pretzsch and Dieler 2011).

The absolute growth rate in a defined period, such as 1 year, plotted against plant size at the beginning of the respective period can result in different patterns of growth allocation, representing different modes of competition between trees. Figure 5.10 displays a set of linear (lines 1, 3, 4, and 5) and non-linear (lines 2, 6) growth-size relationships (Weiner 1990). A steeper slope gradient indicates a stronger concentration of growth rates and resources on tall trees in the stand. The

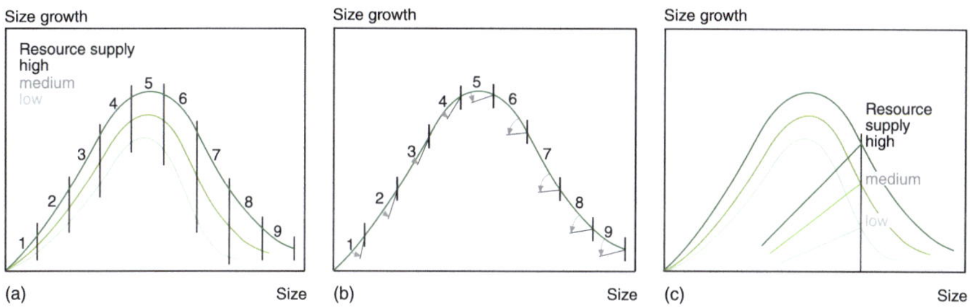


**Box Fig. 5.1-1** (**a**–**c**) Schematic representation of the relationship between size and growth of individual trees (**a**) growing without competition, (**b**) modified by competition, and (**c**) resulting size-growth relationships for different levels of resource limitation. (**a**) Potential size-growth trajectories of open-growing individual trees on sites with high, medium, and low resource supply. Without competition, the course of growth in dependence on size consists of convex (segments 1–3 and 7–9) and concave (segments 4–6) parts. (**b**) Unimodal individual growth curves and linear interindividual size-growth relationships caused by competition. Illustration of the relationship between size and growth without competition (unimodal curve) and the growth reduction due to competition (*arrows*) which results in more or less linear interindividual size-growth relationships (*short straight lines*). This growth reduction through competition results in linear size-growth relationships with steep positive slopes in the early stand phase and increasingly flatter slopes with progressive stand development (segments 1–9). (**c**) On sites with high resource supply, the interindividual growth reduction due to competition causes stronger asymmetry and steeper size-growth relationships than on sites with medium or low resource supply (decreasing slope of the size-growth straight lines with decreasing resource availability)

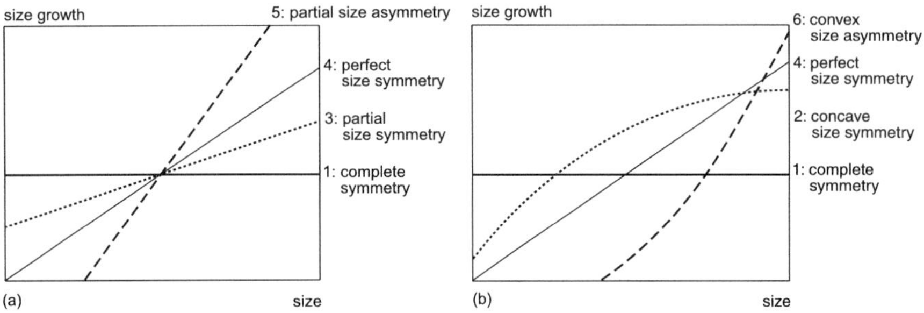


**Fig. 5.10** Hypotheses about the relationship between plant size and absolute growth rate. (**a**) Different linear relationships between size and growth and (**b**) non-linear relationships between plant size and growth. Line 1 represents the more theoretical case of a completely symmetrical size-growth relationship where all plants receive the same budget of growth irrespective of size. Line 2 displays non-linear concave size symmetry where growth increases less digressively with size. Line 3 reflects partial size symmetry where growth increases linearly with size. Line 4 represents perfect size symmetry and means that growth increases proportionally with size. Line 5 stands for partial size asymmetry where growth increases linearly with size. Line 6 represents non-linear convex size asymmetry where growth increases progressively with size

case of complete size asymmetry, indicated by a line parallel to the *y*-axis (slope $= \infty$; a sub-cohort of large plants receives all growth), is rarely observed and not integrated in Fig. 5.10. Note that all relationships in Fig. 5.10a are linear. However, only line 4 represents a linear and proportional increase in the absolute growth rate with increasing size, meaning that only in this case is the relative growth rate equal for all individuals.

Complete symmetry (Fig. 5.10a, line 1) would mean that growth and the resources that competitors receive are independent of their size. Tendency towards complete symmetry (line 1) or partial size symmetry (Fig. 5.10a, line 3; Fig. 5.10b, line 2) is assumed to prevail where below-ground resources (water and mineral nutrients) are a limiting factor, as these are mobile, diffuse quickly and are difficult for larger individuals to pre-empt (van Kuijk et al. 2008). Partial or strong size asymmetry (Fig. 5.10a, line 5; Fig. 5.10b, line 6) means that larger individuals obtain a disproportionately higher share of resources and growth (see, e.g. Fig. 5.3b). This mode of growth-size relationship can be expected on high-quality sites where light is the limiting factor and, as a vectorial resource, can be pre-empted by the larger individuals (Cannell and Grace 1993; Weiner and Thomas 1986).

In order to quantify the mode of competition in mixed versus monospecific stands, tree size growth can be plotted over size (stem volume increment, iv, over stem volume, $v$). Fitting a straight line ($\mathrm{iv} = a + b \times v$) through the iv–$v$ scattergram by OLS regression yields the intercept $a$ and slope $b$. This procedure can be applied for trees in monocultures and mixed-species stands for comparing them. Based on the resulting intercepts and slopes (Fig. 5.11), it can be analysed whether and how species mixing modifies the mode of competition (Hara 1992; Wichmann 2001, 2002).

The slope b of the relationship $\mathrm{iv} = a + b \times v$ indicates and integrates several aspects of the size-growth dynamics of a population. The steeper the slope b, the stronger the pre-emption of light by the highly efficient tall trees, the higher the size

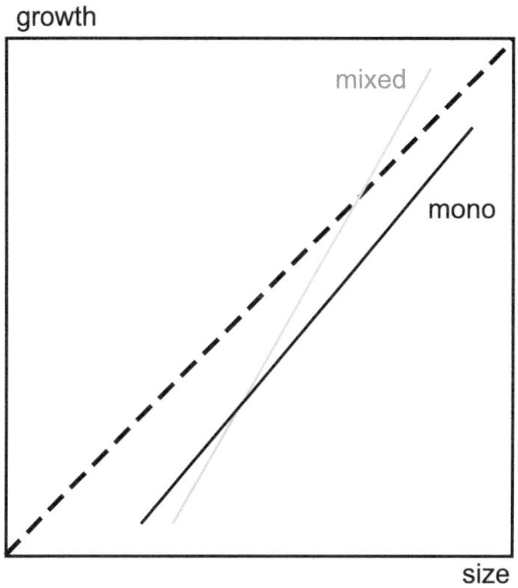


**Fig. 5.11** The slope of the iv–$v$ relationships in mixed and monospecific stands indicates how mixing may increase the asymmetry of competition and growth partitioning among the trees in a stand (Schwinning and Weiner 1998). In this schematic graph, mixing favours the growth of tall trees and reduces the growth of small trees

hierarchy within the population, and the stronger the concentration of growth on the tall trees at the expense of their small neighbours. Conversely, this means that steep slopes indicate relatively low growth or even drop out of small trees in favour of tall trees in the stand. The growth-size slope is suitable for indicating and further analysing interindividual growth allocation patterns and their dependency on species mixture. Figure 5.11 represents a reaction pattern where mixing favours the growth of tall trees and reduces the growth of small trees compared with monospecific stands.

### 5.4.2 Symmetry and Asymmetry of Competition

Comparison of the iv–$v$ relation in mixed-species stands with monocultures may reveal the modification of size asymmetry by species mixing. Of main interest is how the iv–$v$ relationship of the mixed stand as a whole behaves in relation to the monocultures. The overarching analysis of all 84 pairwise comparisons between the mixed-species stands and the monocultures introduced in Tables 5.1 and 5.2 yielded Rb $= 1.14 \pm 0.06$, $n = 84$, $p < 0.05$. This indicates a significant increase in the slope by 14% in mixed-species stands compared with monocultures as schematically shown in Fig. 5.11. An analogous analysis at the species level yielded Rb $- 1.11 \pm 0.05$, $n = 84$, $p < 0.05$. The overall mean slope is $b_m = 0.040$ for mixed stands and $b_p = 0.038$ for monocultures.

Behind these overall mean reaction patterns are species-specific behaviours (Table 5.3). While mixed stands of Norway spruce and European beech have significantly shallower slopes compared with the pure stands of these species, the

**Table 5.3** Slope (mean ± standard deviation) of the iv–$v$ relationship (iv $= a + b \times v$) in mixed-species stands versus monocultures of the 42 triplets of Norway spruce/European beech, Scots pine/European beech, Douglas-fir/European beech, and Norway spruce/Scots pine

| Species | | Slope $b$ of iv–$v$ relationship | | | $b$ mixed/$b$ mono 1 | | | | $b$ mixed/$b$ mono 2 | | | |
|---|---|---|---|---|---|---|---|---|---|---|---|---|
| 1/2 | $n$ | Mean mixed | Mean mono 1 | Mean mono 2 | Min | Max | Mean | SE | Min | Max | Mean | SE |
| N.sp./E.be | 11 | 0.036 ± 0.003 | 0.038 ± 0.004 | 0.035 ± 0.003 | 0.48 | 0.95 | 0.78*** | 0.04 | 0.58 | 1.14 | 0.85* | 0.06 |
| Sc.p./E.be | 7 | 0.041 ± 0.005 | 0.034 ± 0.007 | 0.048 ± 0.005 | 0.76 | 3.64 | 1.36 | 0.39 | 0.37 | 1.04 | 0.78* | 0.10 |
| D.-fir./E.be | 18 | 0.040 ± 0.004 | 0.039 ± 0.004 | 0.041 ± 0.008 | 0.69 | 2.07 | 1.28** | 0.09 | 0.50 | 2.50 | 1.37*** | 0.11 |
| N.sp./Sc.p. | 6 | 0.032 ± 0.002 | 0.034 ± 0.003 | 0.030 ± 0.002 | 0.67 | 1.42 | 0.95 | 0.13 | 0.81 | 1.28 | 1.02 | 0.08 |

For further explanation see caption of Table 5.1

mixture of Douglas-fir and European beech developed significantly steeper slopes than both the corresponding pure stands. Figure 5.12 illustrates for the mixture of Douglas-fir/European beech the mostly steeper slopes of the iv-v relationship in mixed-species stands (Fig. 5.12b, d, and e) compared with monocultures (Fig. 5.12a

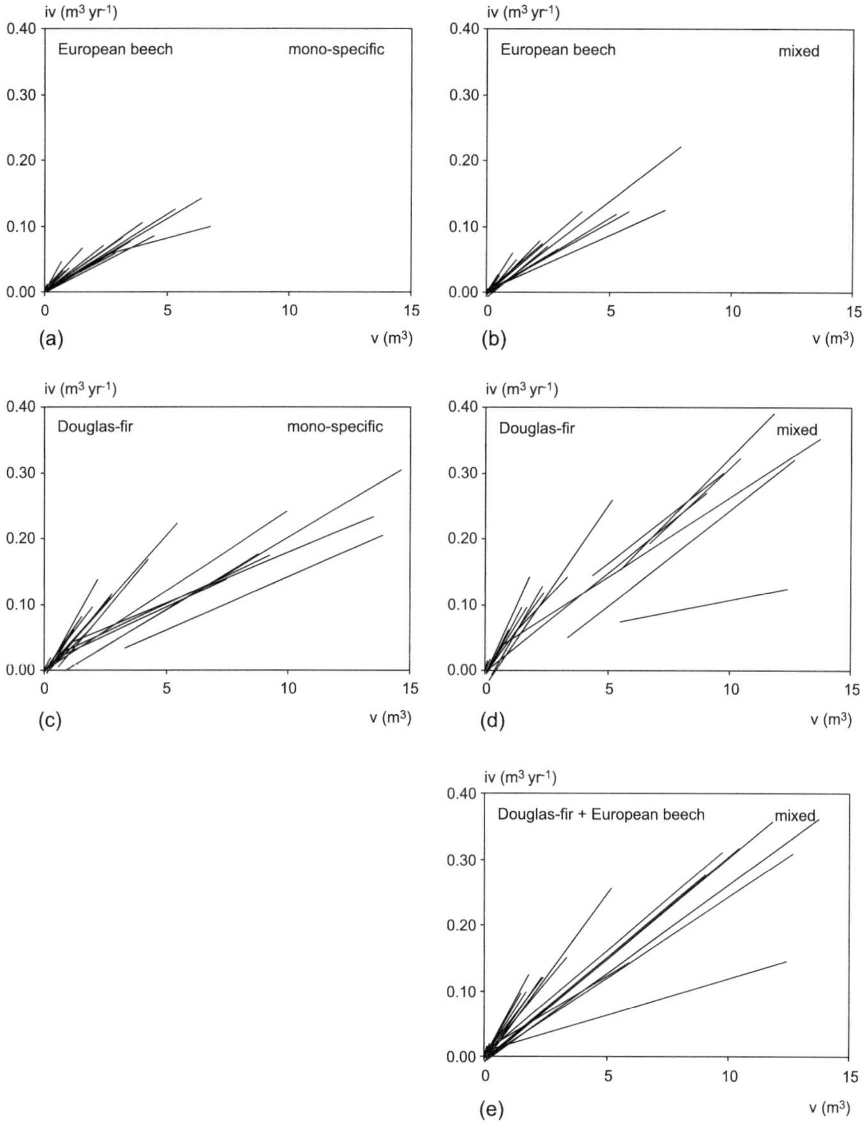


**Fig. 5.12** (**a**–**e**) Modification of the iv–*v* relationship by mixing Douglas-fir and European beech compared with the respective monocultures. The asymmetry of growth in the mixed stands (**b**, **d**, **e**) can differ considerably from the corresponding monocultures (**a** and **c**). The iv–*v* relationship is shown for each species in the mixed-species stands separately (**b** and **d**) and also for the mixed-species stands as a whole (**e**). Annual tree volume growth ($m^3$ $year^{-1}$), iv; tree volume ($m^3$), *v*

and b). Table 5.3 shows that the ratios between the slope in the mixed-species stands of Douglas-fir/European beech and the monocultures are significantly higher than 1.0.

The slope $b$ of the iv–$v$ relationship reveals the interindividual competition and growth partitioning between the trees in a stand (Schwinning and Weiner 1998; Wichmann 2001, 2002). The slope can be used to quantify and further analyse the size symmetry and asymmetry of competition, which is also called "mode of competition" by Hara (1993, 1992) and Yokozawa et al. (1998). The mostly steeper slope in mixed compared with monospecific stands indicates a stronger concentration of growth rates and resources on tall trees in the stand. Steep iv–$v$ relationships mean that larger individuals obtain a disproportionately higher share of resources and growth. Schwinning and Weiner (1998) expect this mode of growth-size relationship on high-quality sites where light is the limiting factor and can be pre-empted by the larger individuals. Shallow iv–$v$ relationships are assumed to prevail where below-ground resources (water and mineral nutrients) are a limiting factor, as these are mobile, diffuse quickly, and are difficult for larger individuals to pre-empt. Our finding of mostly steeper slopes in mixed compared with monospecific stands may corroborate the size-asymmetric competition and pre-emption of light by the tall trees at the expense of the growth of their small neighbours.

## 5.5 Modification of Tree Distribution by Removal Trees

### 5.5.1 Ratio Between the Diameter of the Removal and Total Stand for Characterising the Mode of Mortality

Based on the mean tree diameter of the removal stand, $d_{\text{removal}}$, and the remaining stand, $d_{\text{remain}}$, the ratio $d_{\text{rel}} = d_{\text{removal}}/d_{\text{remain}}$ characterises the size of the removal in relation to the remaining trees (Fig. 5.13). Note that the latter is a schematic figure with simplified assumptions about the mean diameter of the remaining and removal trees and about the shape of the distributions. The higher the $d_{\text{rel}}$ values, the taller the removed trees in relation to the remaining stand. Thinning from below or self-thinning befalls mainly small trees with $d_{\text{removal}} < d_{\text{remain}}$ and yields ratios of $d_{\text{rel}} < 1$. In the case of a schematic thinning, the mean sizes of the removal and remaining trees would be equal, and $d_{\text{rel}} \cong 1$ (range $d_{\text{rel}} = 0.9$–$1.1$). Thinning from above means tree elimination from the right side of the tree size distribution and can yield $d_{\text{rel}} > 1$. Selective thinning, which eliminates 1–2 of the strongest competitors of each future crop tree, may result in the range $d_{\text{rel}} = 0.8$–$1.2$. Comparison between a species $d_{\text{rel}}$ in the mixed stand and the neighbouring monoculture may reveal how mixing superimposes the self-thinning process in the monoculture which normally reduces the tree number from the left side of the tree size distribution.

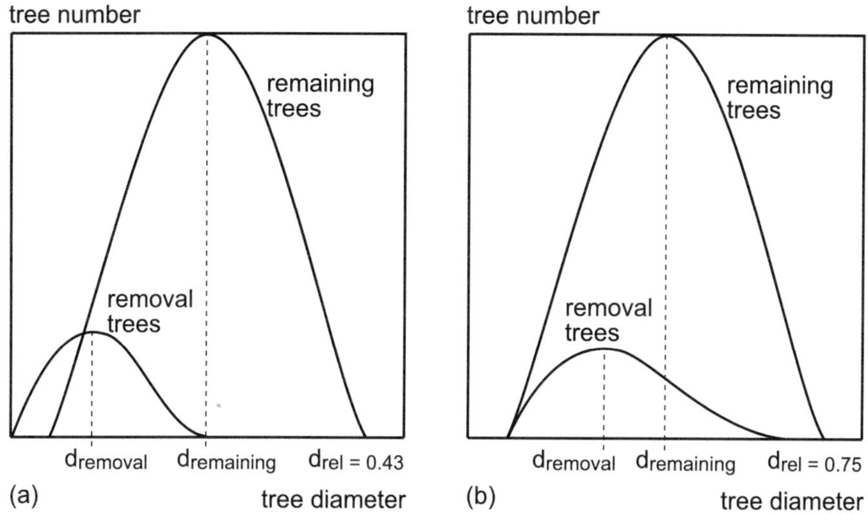


**Fig. 5.13** Schematic representation of changes in the removal ratio $d_{rel} = d_{removal}/d_{remaining}$ through mixing. (**a**) Restriction of the removals to the left side of the diameter distribution in the monospecific stand yields $d_{rel} = 0.43$, and (**b**) extension of the removals to the right side of the diameter distribution is reflected by $d_{rel} = 0.75$

## 5.5.2 Mode of Tree Mortality in Mixed Versus Monospecific Stands

For analysing the mode of mortality in mixed versus monospecific stands, unthinned or only moderately thinned experimental plots with long-term records of dead and removal trees are the most revealing database. For the following evaluation, a set of 11 fully stocked and at most moderately thinned stands was used, as this best reflects the species-specific mortality in mixed versus monospecific stands. The evaluation was based on more or less even-aged monospecific and mixed stands of European beech and sessile oak and European beech and Norway spruce located in Germany. Altogether we pooled data of 51 triplets of monospecific and mixed stands. The dataset represents the growing conditions of a rather broad time span (1905–2012), a range of stand ages (31–238 years) and mainly experiments from central and southeast Germany. Table 5.4 summarises basic characteristics of the plots; for more detailed information, see Matyssek et al. (2012, pp 243–271), Pretzsch (2009), Pretzsch et al. (2010), and Pretzsch et al. (2013a) who used the same experiments to analyse mixing effects at tree and stand level.

The significantly higher ratio $d_{rel}$ of beech in mixture compared with monocultures (Table 5.4, Fig. 5.14a) indicates that the associated tree species, Norway spruce and sessile oak, exert an alien-thinning effect from above. The mean $d_{rel}$ ratio of the removal beeches in the mixed stand is about 10% higher ($d_{rel} \cong 0.82$ versus $d_{rel} \cong 0.73$) than in monoculture, i.e. mortality reaches wider into the right branch of the tree size distribution in mixed stands compared with

**Table 5.4** Overview of the differences between mixed and monospecific stands in terms of size distribution, size-growth relationship, and mode of mortality

| | European beech | | | Norway spruce | | | sessile oak | | |
|---|---|---|---|---|---|---|---|---|---|
| Variable | mixed | mono | mixed/mono | mixed | mono | mixed/mono | mixed | mono | mixed/mono |
| $n$ | 68 | 68 | 68 | 41 | 41 | 41 | 32 | 32 | 32 |
| $d_{mean}$ (cm) | 26.5 | 29.1 | **0.92 ± 0.08** | 37.2 | 33.2 | **1.12 ± 0.03** | 25.0 | 24.6 | 1.03 ± 0.05 |
| $v_{mean}$ ($m^3$) | 0.89 | 1.05 | **0.82 ± 0.12** | 1.08 | 1.41 | **1.30 ± 0.08** | 0.62 | 0.59 | 1.13 ± 0.15 |
| $d_{min}$ (cm) | 14.6 | 15.2 | 1.08 ± 0.14 | 25.8 | 21.0 | **1.27 ± 0.11** | 17.7 | 17.0 | **1.07 ± 0.07** |
| $d_{max}$ (cm) | 49.0 | 43.6 | 0.96 ± 0.07 | 49.4 | 46.4 | **1.07 ± 0.04** | 35.9 | 33.7 | **1.08 ± 0.06** |
| $s_d$ (cm) | 74.7 | 79.3 | 1.05 ± 0.10 | 73.6 | 64.5 | **1.17 ± 0.09** | 46.5 | 41.5 | **1.13 ± 0.13** |
| $d_{rel}$ (cm $cm^{-1}$) | 0.82 | 0.73 | **1.20 ± 0.09** | 0.75 | 0.74 | 1.03 ± 0.10 | 0.78 | 0.79 | 1.0 ± 0.08 |

The columns 'mixed' and 'mono' report the species-specific arithmetic means of all n observations within the respective groups. The column 'mixed/mono', in contrast, shows the mean of the ratio resulting from the pairwise division of the characteristic of the mixed stand by the respective value of the neighbouring monoculture. Bold ratios indicate significant ($p < 0.05$) differences between the species behaviour in mixed versus monospecific stands

Number of stands analysed, $n$; arithmetic mean diameter, $d_{mean}$; arithmetic mean stem volume, $v_{mean}$; minimum and maximum tree diameter, $d_{min}$ and $d_{max}$; standard deviation of tree diameter, $s_d$; $d_{rel} = d_{removal}/d_{remaining}$

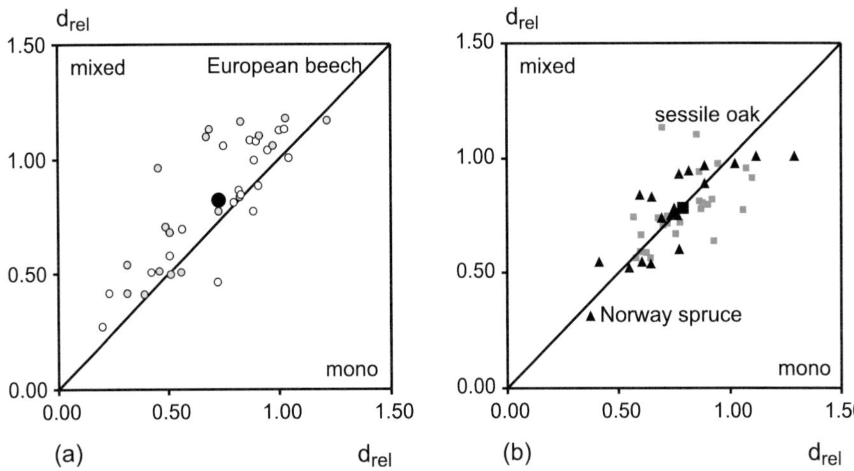


**Fig. 5.14** Ratio between the diameter of the removal and remaining stand, $d_{rel}$, for mixed versus monospecific stands for (**a**) beech and (**b**) oak and spruce. Observations close to the bisector line indicate similar behaviour for mixed and monospecific stands, while deviations indicate true mixing effects. (**a**) Response of beech to admixture of oak is represented by empty circles, its response to spruce by filled circles. The large black circle indicates the mean relationship for beech in mixed versus monospecific stands. (**b**) Response of oak to the admixture of beech is represented by rectangles and the response of spruce to beech by triangles. The large black rectangle and triangles in the centre, close to the 1:1 line, indicate the mean relationship for oak and spruce in mixed versus monospecific stands

monocultures. In the analysed even-aged mixed stands, Norway spruce and sessile oak are ahead in size growth and obviously able to slow down the growth and reduce the number of beeches during early stand development.

In contrast, the $d_{rel}$ of Norway spruce and sessile oak is not significantly modified by mixing (Fig. 5.14b). The mean tree size growth of Norway spruce and sessile oak can be fostered by mixing. However, their removal ratio is not significantly modified through the presence of beech. Their $d_{rel}$ values indicate a thinning-from-below effect which is rather equal in mixed and monospecific stands ($d_{rel} \cong 0.75$). Table 5.4 underlines that, in the case of European beech, $d_{rel}$ significantly ($p < 0.05$) exceeds 1.0 and indicates the mortality shifts from the smaller diameter classes to the taller trees in mixed stands.

The ratios between mixed and monospecific stand characteristics in Table 5.4 reflect that in the case of beech, size growth tends to be reduced in mixed compared with monospecific stands and in the case of spruce and oak, it is accelerated. The size variation can become wider (spruce and oak), the size inequality greater (beech), and the removal trees larger (beech) in mixed compared with monospecific stands.

## 5.6 Hierarchy of Tree Size, Tree Growth, and Growth Dominance

### 5.6.1 Gini Coefficients, $GC_v$ and $GC_{iv}$, and Growth Dominance Coefficient, GDC

#### 5.6.1.1 Coefficient by Gini and Curve by Lorenz for Characterising the Size and Growth Hierarchy

The coefficient by Gini and curve by Lorenz can be used to quantify the size or growth hierarchy between the trees in forest stands (see de Camino (1976); Kramer (1988, p 82)). We loan from economics the Gini coefficient, GC,

$$\mathrm{GC} = \frac{\sum_{i-1}^{n} \sum_{j=1}^{n} \left| x_i - x_j \right|}{2n(n-1) \times \bar{x}}$$

to quantify the relative distribution of stem volume ($GC_v$) and volume growth ($GC_{iv}$), respectively, between the trees in mixed versus monospecific stands. Variables $x_i$ and $x_j$ denote size or growth (or other tree characteristics) for the $i$'th and the $j$'th tree in the stand with $i = 1 \ldots n$ trees. GC = 0.0 applies for a very homogeneous distribution of the respective tree variable, e.g. maximum equality of size or growth distribution. The stronger the inequality of size or growth between the trees, the higher the GC (Fig. 5.15a and b). The curves of the cumulative distributions together with the sketched stands reflect the inequality of size which can also cause an inequality of growth.

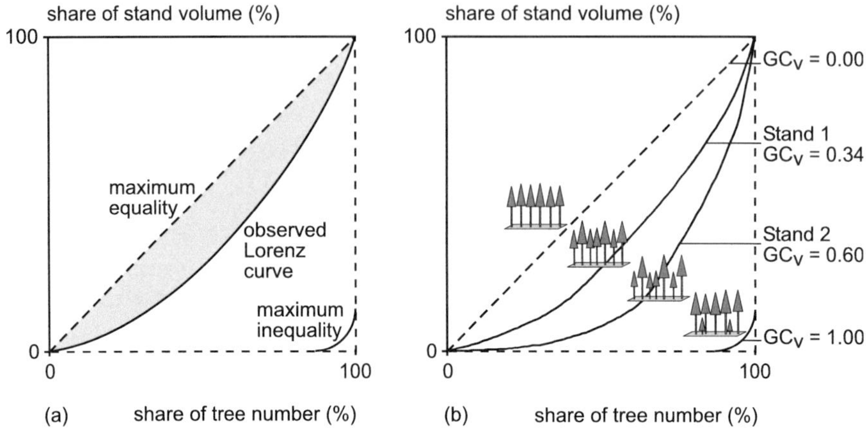


**Fig. 5.15** Schematic representation of the stem volume partitioning between the trees in a forest stand. (**a**) Lorenz curve applied to quantify the degree of equality in stem volume distribution. (**b**) Equal volume distribution in rather monolayered stands ($GC_v = 0.00$), increasing inequality in heterogeneous stands (Stand 1, $GC_v = 0.34$; Stand 2 $GC_v = 0.60$), and strong inequality ($GC_v$ approaching 1.0)

Application of these measures to mixed and monospecific stands can reveal how mixing modifies the hierarchy between the trees in a population, e.g. whether species mixing can favour the growth distribution towards small understorey trees compared with monocultures. The Lorenz curve (Fig. 5.15a), known for the analysis of income inequality in human populations, can be used to visualise the inequality of growth in forest stands. The larger the area between the bisector line (maximum equality) and the observed Lorenz curve, the greater the inequality, and the higher the GC. The GC is equivalent to the ratio of the grey-coloured area (area between the Lorenz curve and the bisector line is shown in Fig. 5.15a) to the total area of the square, multiplied by two.

#### 5.6.1.2 Curve of the Cumulative Tree Growth as Function of Cumulative Stem Volume: Growth Dominance Coefficient, GDC

The cumulative distribution of tree growth over stem volume combines information about size distribution and the respective growth distribution between the trees in a stand. For this purpose, the trees of the stand are ranked from smallest to largest volume, the cumulative volume of the trees is registered on the abscissa and the cumulative volume growth on the ordinate. The resulting curves illustrate how much smaller trees contribute to the total stand growth compared with taller trees (see Fig. 5.16). In this figure curve 1 indicates a growth dominance of tall trees, curve 3 a growth dominance of small trees, and curve 2, following the bisector line, indicates that all trees contribute to stand volume growth proportionally to their volume (Box 5.2).

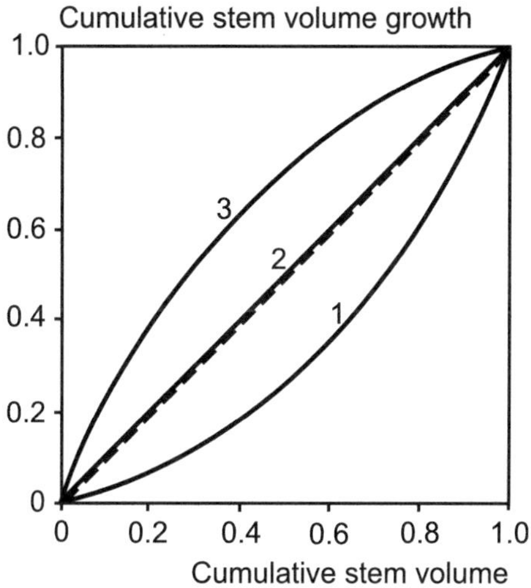


**Fig. 5.16** Schematic representation of the cumulative distribution of stem growth (ordinate) over the cumulative distribution of stem volume (abscissa) for forest stands with different competitive status of small compared with tall trees. The curves 1, 2, and 3 represent stands where small trees make a low, medium, and strong contribution to the total stand growth. Curves 1, 2, and 3 result in a growth dominance coefficient of GDC = 0.17, 0.0, and −0.12, respectively

**Box 5.2 Conceptual Model of the Change in Growth Dominance During Stand Development by Binkley et al. (2006)**

For a combined view on both the size and growth effect of individuals in a population, Binkley et al. (2006) proposed the analysis of the cumulative distribution of stem volume growth over stem volume (Box Fig. 5.2-1). According to the conceptual model by Binkley et al. (2006), growth dominance—quantified by the coefficient of growth dominance GDC—changes during stand development as follows: In the *open-grown young stage* (phase 1) suppression is low, and therefore individual tree growth is proportional to size (bisector line). Stand closure means the beginning of the *differentiation stage* (phase 2) when large trees grow overproportionally and continuously slow down or outcompete smaller neighbours. Due to this strong competitive pressure, the relative contribution of the stem growth of small trees may stay behind their portion to stem volume causing a convex curve (seen from below). In the *mature stage* (phase 3), growth may again become proportional to size as the growth of tall trees subsides and smaller trees benefit from interruptions of the canopy. In the *fragmentation and regeneration stage* (phase 4), smaller trees come up, benefit from the opened crown, and contribute overproportionally to the stand growth (based on Binkley et al. (2006, p 195, Fig. 2)).

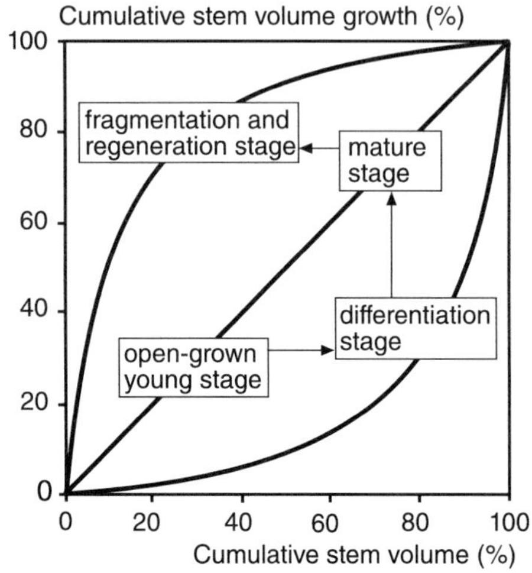


**Box Fig. 5.2-1** Different patterns of the cumulative distribution of stem volume growth ($y$-axis) over cumulative stem volume ($x$-axis) during mono-cohort forest stand development. In the *open-grown young stage*, suppression is low meaning that individual tree growth is proportional to size (*bisector line*). Stand closure represents the beginning of the *differentiation stage* when large trees grow overproportionally and continuously slow down or outcompete smaller neighbours. In the *mature stage*, growth may again become proportional to size as tall trees subside in growth and smaller ones benefit from interruptions of the canopy. In the *fragmentation and regeneration stage*, smaller trees come up, benefit from the opened crown, and contribute overproportionally to the stand growth (based on Binkley et al. (2006, p 195, Fig. 2))

We use this approach for the subsequent mixed-species stands analysis. When the curve in mixed stands is compared with monospecific stands, it indicates how the contribution of small and tall trees is modified by species mixing. Beyond this graphical representation, the relationship between cumulative tree growth and cumulative stem volume can be characterised by the difference between the Gini coefficients for cumulative growth ($GC_{iv}$) and the Gini coefficients for cumulative volume ($GC_v$). This difference ($GC_{iv}$-$GC_v$) is similar to the growth dominance coefficient GDC used by Binkley et al. (2006) and Pommerening et al. (2016). If the curve follows the bisectoral line, GDC = 0 (see line 2 in Fig. 5.16). If it runs above the bisectoral line, GDC < 0, and if the curve lies below the bisector line, GDC > 0 (see curves 3 and 1, respectively, in Fig. 5.16).

### 5.6.2 More Unequal Tree Sizes and Tree Growth Partitioning in Mixed Compared with Monospecific Stands but Similar Growth Dominance

The Gini coefficients $GC_v$ in Table 5.5 (upper section) indicate that the inequality is mostly significantly higher in mixed-species stands compared with monocultures ($GC_v$ mixed/$GC_v$ mono = 0.95–2.09). Figure 5.17a and b illustrates this general

**Table 5.5** Gini coefficients for volume, $GC_v$ and Gini coefficients for volume growth, $GC_{iv}$ (mean ± standard deviation) in mixed-species stands versus monocultures (from top to bottom)

| Species | | $GC_v$ for volume | | | $GC_v$ mixed/$GC_v$ mono 1 | | | | $GC_v$ mixed/$GC_v$ mono 2 | | | |
|---|---|---|---|---|---|---|---|---|---|---|---|---|
| 1/2 | *n* | Mean mixed | Mean mono 1 | Mean mono 2 | Min | Max | Mean | SE | Min | Max | Mean | SE |
| N.sp./E.be | 11 | 0.41 ± 0.02 | 0.30 ± 0.03 | 0.44 ± 0.03 | 0.91 | 2.24 | 1.47** | 0.14 | 0.63 | 1.33 | 0.95 | 0.06 |
| Sc.p./E.be | 7 | 0.37 ± 0.02 | 0.32 ± 0.04 | 0.39 ± 0.04 | 0.83 | 2.64 | 1.74** | 0.27 | 0.73 | 1.44 | 0.98 | 0.09 |
| D.-fir./E.be | 18 | 0.54 ± 0.02 | 0.38 ± 0.02 | 0.42 ± 0.03 | 0.64 | 3.37 | 1.79*** | 0.18 | 0.96 | 3.21 | 1.38** | 0.12 |
| N.sp./Sc.p. | 6 | 0.47 ± 0.02 | 0.31 ± 0.03 | 0.23 ± 0.02 | 0.71 | 1.84 | 1.29 | 0.17 | 1.50 | 2.81 | 2.09*** | 0.18 |
| Species | | $GC_{iv}$ for volume growth | | | $GC_{iv}$ mixed/$GC_{iv}$ mono1 | | | | $GC_{iv}$ mixed/$GC_{iv}$ mono2 | | | |
| 1/2 | *n* | Mean mixed | Mean mono 1 | Mean mono 2 | Min | Max | Mean | SE | Min | Max | Mean | SE |
| N.sp./E.be | 11 | 0.42 ± 0.02 | 0.35 ± 0.03 | 0.47 ± 0.03 | 0.78 | 1.69 | 1.24** | 0.08 | 0.72 | 1.28 | 0.92 | 0.05 |
| Sc.p./E.be | 7 | 0.40 ± 0.02 | 0.29 ± 0.05 | 0.45 ± 0.05 | 0.78 | 2.39 | 1.66* | 0.27 | 0.71 | 1.33 | 0.96 | 0.10 |
| D.-fir./E.be | 18 | 0.59 ± 0.02 | 0.39 ± 0.03 | 0.45 ± 0.03 | 0.72 | 2.54 | 1.64*** | 0.12 | 0.95 | 2.90 | 1.41*** | 0.10 |
| N.sp./Sc.p. | 6 | 0.49 ± 0.01 | 0.41 ± 0.04 | 0.28 ± 0.02 | 0.80 | 1.83 | 1.27 | 0.15 | 1.39 | 2.31 | 1.83*** | 0.18 |

Note that coefficients of $GC_v$ and $GC_{iv}$ close to 0.0 indicate equality of size and growth, respectively. Coefficients close to 1.0 indicate high inequality in size or size growth. For further explanation see caption of Table 5.1

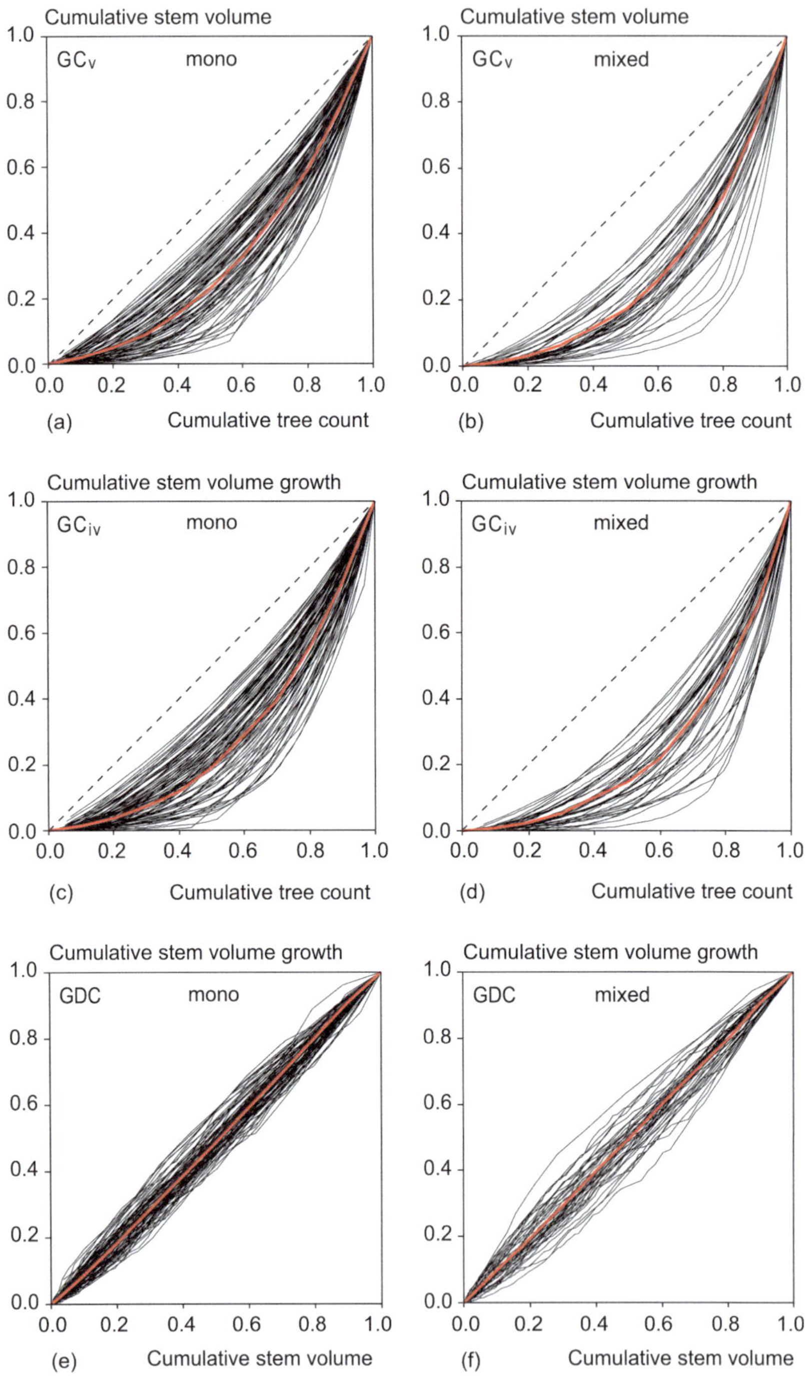


**Fig. 5.17** (**a**–**f**) Hierarchy of tree volume, stem volume growth, and growth dominance for 84 monocultures (**a**, **c**, **e**) and 42 mixed stands (**b**, **d**, **f**) of the triplet series of Norway spruce/European beech, Scots pine/European beech, Douglas-fir/European beech, and Norway

tendency. It shows the cumulative stem volume as a function of the cumulative tree count for all monocultures and mixed-species stands of the triplets. The mean curves (bold lines) indicate that the smallest 50% of the trees account for 25% of the total stand volume in monocultures and for just 15% in the mixed-species stands. This inequality in size indicates a high number of small trees and a restricted number of tall trees in mixed stands. In monocultures, in contrast, the trees are more equally sized.

The finding of higher inequality in mixed-species stands versus monocultures also applies for the growth partitioning among the trees within the stands, represented by $GC_{iv}$ (Table 5.5, lower section). This is illustrated by the cumulative stem volume growth as a function of the cumulative tree count in Fig. 5.17c and d. That both $GC_v$ and $GC_{iv}$ are mostly significantly higher in mixed-species stands than in monocultures indicates a more unequal size distribution and growth partitioning in mixed stands.

In contrast, the GDCs show only small and non-significant absolute differences between mixed and monospecific stands (Fig. 5.17e and f). The cumulative stem volume growth as a function of cumulative stem volume always lies close to the 1:1 line (GDC = 0). The ratios between GDCs in mixed versus monospecific stands vary strongly but do not differ significantly from GDC = 1.0.

### 5.6.3 Remarkable Growth-Structure Elasticity of Mixed-Species Stands

Both $GC_v$ and $GC_{iv}$ indicate a much more unequal partitioning of stem volume and stem volume growth among the trees in mixed compared with monospecific stands (Fig. 5.17a–d, Table 5.5). Mixed stands have many small trees on the one hand and a strong concentration of volume and growth on a restricted number of tall trees on the other hand. Monocultures have greater equality of size and growth distribution. In monocultures, the largest 20% of the trees account on average for 40% of the stem volume. In mixed stands this figure is 50% (see Fig. 5.17a versus b). In monocultures, the largest 20% of the trees account on average for 44% of the stand growth, whereas in mixed stands this figure is 52% (see Fig. 5.17c versus d).

**Fig. 5.17** (continued) spruce/Scots pine (for description of the triplet dataset, see Sect. 5.3.2 and Tables 5.1–5.3). (**a** and **b**) Cumulative stem volume, *v*, as a function of the cumulative tree count, *n*. On average, the Gini coefficients for monocultures ($GC_v = 0.36 \pm 0.01$) are significantly lower than for neighbouring mixed stands ($GC_v = 0.46 \pm 0.01$) of the same species. (**c** and **d**) Cumulative stem volume growth, iv, as a function of the cumulative tree count, *n*. On average, the Gini coefficients for monocultures ($GC_{iv} = 0.40 \pm 0.01$) are significantly lower than for neighbouring mixed stands ($GC_{iv} = 0.50 \pm 0.02$) of the same species. (**e** and **f**) Growth dominance coefficient, GDC, which is the cumulative stem volume growth, iv, as a function of cumulative stem volume, *v*, and results from $GC_{iv}$–$GC_v$. On average, GDCs of monocultures ($GDC = 0.02 \pm 0.01$) are not different to those of mixed-species stands ($GDC = 0.01 \pm 0.01$)

Table 5.5 showed that the difference between $GC_v$ and $GC_{iv}$ is rather similar for monospecific (on average GDC = $GC_{iv}$–$GC_v$ = 0.40–0.36 = 0.04) and mixed (on average GDC = $GC_{iv}$–$GC_v$ = 0.50–0.46 = 0.04) stands. So, the coefficients of growth dominance (GDC = $GC_{iv}$–$GC_v$) are also similar and near GDC = 0 (varying closely around the 1:1 line in Fig. 5.17e and f) for monospecific and mixed stands. As the stem volume growth partitioning is very similar to the stem volume partitioning, the relationship between cumulative growth and cumulative volume is rather proportional.

This reveals a remarkable growth elasticity of mixed-species stands. The reasons for this constant GDC, even under strongly varying stand structure, are probably the complementary ecological traits in mixed stands. A combination of light-demanding species in the upper canopy and shade-tolerant species in the lower canopy ensures that the relative contribution of the stem growth of small trees does not stay behind their portion of stem volume.

According to the conceptual model of growth dominance and the growth dominance coefficient, GDC, by Binkley et al. (2006), most of the 42 triplets are in phase 3, i.e. in the mature stage (Box 5.2). However, both $GC_v$ and $GC_{iv}$ indicate a much more unequal partitioning of stem volume and stem volume growth among the trees in mixed compared with monospecific stands and suggest strong differentiation and competition typical for phase 2. Mixing may release the competitive pressure in mixed stands through niche complementarity and thus result in a proportional relationship between cumulative growth and cumulative volume—although density, inequality, and size asymmetry are much higher than in monocultures. Species mixing seems to equalise the relationship between cumulative growth and cumulative volume and thereby guarantee a continuously low-growth dominance and size-proportional contribution of all trees to the stand growth.

## 5.7 Modification of the Crown Allometry and thereby the Canopy Space Filling in Mixed versus Monospecific Stands

The addressed changes in the stem size distribution reflect one important effect of mixing on the size-structure dynamics of forest stands. Beyond this, species mixing can also modify the tree height, the crown size, and thereby the canopy space filling (Fig. 5.1a–c).

### 5.7.1 From Stem Size Distribution to Crown Size Distribution

The mixing experiment Waldbrunn 105 of sessile oak and European beech (see also Fig. 5.6) can be used to illustrate an effect of mixing on the frequency distribution of crown variables that is even stronger than that on tree diameter. Behind this is the crown release effect and crown expansion in mixed or thinned stands which can result in a wide variability of tree allometry (see Chap. 6) and variation in wood quality parameters (see Chaps. 9 and 11).

Figure 5.18 is based on the tree diameter and crown diameter of the 87-year-old stands in 1989 (first surveyed 1935, 10 successive surveys). Till their present age, the trees on these plots have had sufficient time to adapt their stem and crown shapes to the intra- and interspecific neighbourhood in the monospecific and mixed stands, respectively. Note that the grey lines represent the frequency distributions of the trees in the mixed stands scaled up to the unit area of 1 ha by the mixing proportion of the respective species.

Mixing barely changes the diameter distribution of sessile oak (Fig. 5.18a) and slightly slows the diameter growth of smaller beeches (Fig. 5.18d). The crown diameter distribution of the trees growing in mixture indicates the length and diameters of branches and consequently the branch diameter. In comparison with the neighbouring monocultures, the crown diameter distribution in mixed stands is shifted to the left in the case of oak (Fig. 5.18b) and to the right in the case of beech (Fig. 5.18e).

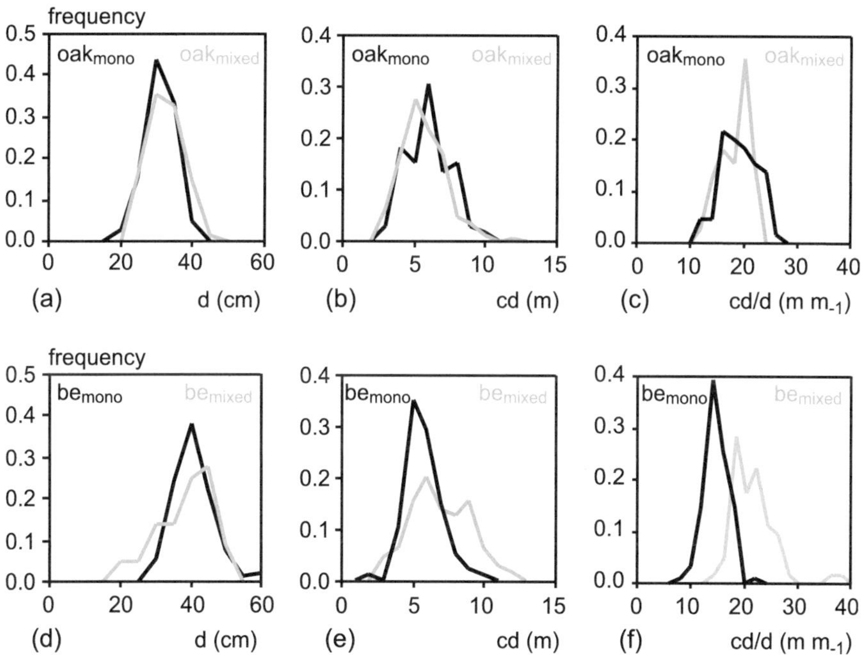


**Fig. 5.18** (**a–f**) Frequency distribution of tree attributes in mixed stands (*grey lines*) compared with monospecific stands (*black lines*) of sessile oak (**a–c**) and European beech (**d–f**) in the long-term experiment Waldbrunn 105 at age 87. In the case of oak, mixing only slightly modifies frequency distribution of tree diameter, *d*; crown diameter, cd; and the ratio cd/*d*, compared with monocultures. In the case of beech, mixing shifts and extends the frequency distributions of *d*, cd, and particularly of the ratio cd/*d*

The cd/*d* ratio (Fig. 5.18c and f) indicates how many times larger the crown diameter is in relation to the stem diameter. The cd/*d* ratio of a tree with cd = 5m and *d* = 50 cm would be 10 (cd/*d* = 500 cm/50 cm = 10), while a tree with the same diameter but a crown width of 10 m would yield cd/*d* = 20 (cd/*d* = 1000 cm/ 50 cm = 20). The higher the cd/*d* ratio, the broader the crown and the longer and thicker the branches. Note that crown size and branch diameter are closely associated with stem shape, wood strength, and stiffness (see Chap. 9). In the case of oak (Fig. 5.18c), mixing barely shifts the cd/*d* frequency distribution. However, the cd/d ratio distribution of beech is shifted considerably to the right and is much more extended in mixed compared with monospecific stands (Fig. 5.18f). The evaluation is based on 68 and 30 oaks and on 95 and 65 beeches in monospecific and mixed stands, respectively. The coefficient of variation ranges between 15 and 21% for *d*, 23 and 28% for cd, and 15 and 20% for cd/*d*. Especially in the case of beech, the coefficient of variation is high (between 15 and 33%). This example shows the considerable extent to which tree attributes such as crown diameter and cd/d ratio can change in mean and variation, although with regard to the stem diameter, there are only minor differences between mixed stands and monocultures.

In Chap. 6 we will further explore the effect of inter- versus intraspecific neighbourhood on the tree allometry and space occupation. For forest practice, the modification of the frequency distribution by mixing towards taller trees of the leading species (e.g. Norway spruce, sessile oak) and less small-sized and more uniform individuals in the case of beech may cause an improvement in the assortment yield, even if productivity at stand level remains unchanged. Further comparison of the frequency distribution of species in mixed and monospecific stands may be extended to proxies of wood quality such as ratios of *h*/*d* or cd/*d* which may decrease and cl/*h*, which may increase wood quality (tree height, *h*; tree diameter, *d*, crown diameter, cd; crown length, cl). Frequency distributions of the latter proxy variables of tree wood quality enable an integrated view of the effect of species mixing on both quality (quality aspects such as distortion, knottiness, wood density, stiffness, and strength) and quantity (number of trees with respective qualities) of the produced wood and potential wood products (Pretzsch and Rais 2016).

### 5.7.2 Canopy Space Filling in Mixed versus Monospecific Stands

In temperate forests the pre-emption of light by the tallest trees causes size-asymmetric competition (Schwinning and Weiner 1998; Wichmann 2002). Trees can acclimate to size-asymmetric competition through morphological shifts in order to remedy the light limitation (Grams and Andersen 2007). In monospecific stands, especially those that are rather homogeneous in terms of genetics and tree size, all individuals engage in similar behaviour to compete for the growing space and resources that are not sufficient for all. As the resource demands, physiological abilities, and structural variability of competing trees are more similar in monocultures, canopy structure remains mostly homogeneous, competition rather size asymmetric, and canopy depth low. In mixed stands, in contrast, the trees' elbowroom may be wider. If the crown shapes and the light ecology of the combined species complement each other, the trees might simply have more canopy space to occupy without mechanical abrasion or penetration of neighbouring crowns (Fish et al. 2006; Putz et al. 1984).

For closer analysis of the canopy space filling, Pretzsch (2014) used 253 crown maps in monocultures (n=87) and two-species ($n = 111$) and $\geq$ three-species ($n = 55$) mixed stands in Germany. All three groups comprise even-aged and uneven-aged stands of mainly Norway spruce, European beech, sessile and common oak, Scots pine, silver fir (*Abies alba* Mill.), and sycamore maple (*Acer pseudoplatanus* L.). The following analysis of the canopy space filling in mixed versus monospecific stands is based on combined measurement of tree positions and crown sizes (crown length and crown projection area) on long-term experimental plots. The crown measurements ($n = 35{,}728$ measured crowns) date back to the 1950s and cover a broad range of tree species growing in monospecific and mixed stands. For a detailed account of eight-radii crown measurements, see Pretzsch (2009, pp 115–118). Among the various measures for assessing forest canopies (Jennings et al. 1999), the ground coverage by crowns, CGC; the sum of the crown projection area, SCPA; and the crown engagement, CE, are of special interest (Box 5.3).

**Box 5.3 From Crown Projection Area of Individual Trees to the Canopy Ground Cover, CGC; the Sum of Crown Projection Areas, SCPA; and the Crown Engagement Ratio, CE**

The crown projection area of a tree, cpa, results from projection of four or eight crowns. In Box Fig. 5.3-1 the crown projection areas are represented by polygons. The crown projection area can be calculated via the mean crown radius $\bar{r}_q$, which corresponds to the quadratic mean of four or eight radial measurements $r_1, \ldots r_n$ ($\bar{r}_q = \sqrt{(r_1^2 + r_2^2 + \ldots + r_n^2)/n}$ ). The canopy ground cover, CGC, characterises the cover of the ground by crowns when looking vertically from above. This may be useful for estimating the throughfall of precipitation, deposition, etc. CGC is generally analysed using dot counts of crown maps. For this purpose, a grid is laid over the map. CGC is obtained from the number of dot points covered by the crown, $n$, divided by the total number of grid dot points, $p_{\text{total}}$ ($CGC = p_{\geq 1\text{-fold}}/p_{\text{total}}$). For the stand in Box Fig. 5.3-1 $p_{\geq 1\text{-fold}} = 93$, $p_{\text{total}} = 108$ applies, and CGC = 93/108 = 0.86; i.e. 86% of the stand area is covered by crowns.

However, the sun does not shine in a perpendicular direction into forest stands in our latitudes and the sum of crown projection area, SCPA (including the multiple layering), or the crown engagement ratio, CE, may be more relevant for production ecology than CGC. The ratio between the sum of crown projection areas, SCPA = $\sum$ csa, and the plot size, A, quantifies the

(continued)

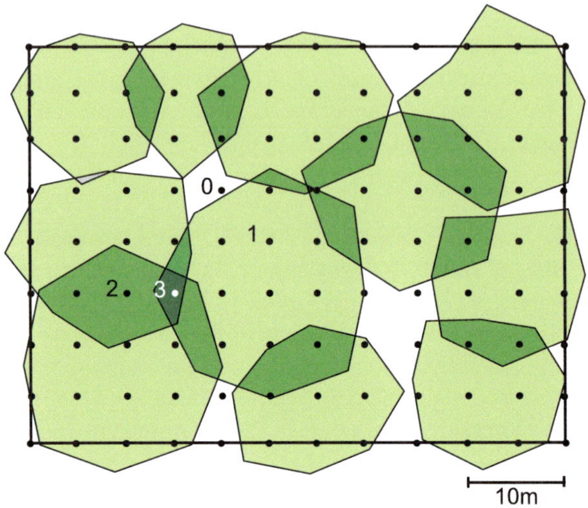


**Box Fig. 5.3-1** Crown map with uncovered area portions (white), one-, two-, and ≥threefold crown coverage (light grey, grey, and dark grey). In this example, 14% of the plot area is uncovered, 86% is covered ≥one - fold, and the sum of crown areas is SCPA = 2310 $m^2$ which is 105% of the plot area and results in a crown engagement of CE = 1.05. The numbers 0–3 in the map represent dots with zero-, one-, two-, and ≥threefold crown coverage

**Box 5.3** (continued)

crown engagement, CE, and can be derived in two ways: (1) by summing up the cross-section area of all trees on a plot and dividing by plot size A (then $CE = \sum csa/A$) and (2) based on the results of a dot count evaluation which distinguishes for each grid point between uncovered, onefold covered, twofold covered … *n*-fold covered ($CE = p_{1\text{-fold}} \times 1 + p_{2\text{-fold}} \times 2 + \ldots . + p_{n\text{-fold}} \times n/p_{total}$). In our example, the result is $CE = (75 \times 1 + 16 \times 2 + 2 \times 3)/108 = 1.05$, i. e. the sum of the crown cross-section area is 1.05-fold the stand area.

The stands have a mean plot size of 0.30 ha and stand ages of 16–283 years. Their crown measurements from 1951 to 2013 were used to calculate CGC, SCPA, and CE. All measures vary widely as the plots cover fully but also sparsely stocked stands. From plots with repeated crown surveys, we included only those with ≥10 years between the successive inventories in order to avoid autocorrelation between the measurements.

The 95% and 75% percentile regression lines in Fig. 5.19a show that the sum of crown area amounts to 150 and 100%, respectively, in monospecific stands and increases with species richness by up to 220% and 180% (95% and 75% percentile). That means that in the fully stocked mixed stands, tree species are so densely interlocked that the sum of the crown projection area can become more than twice

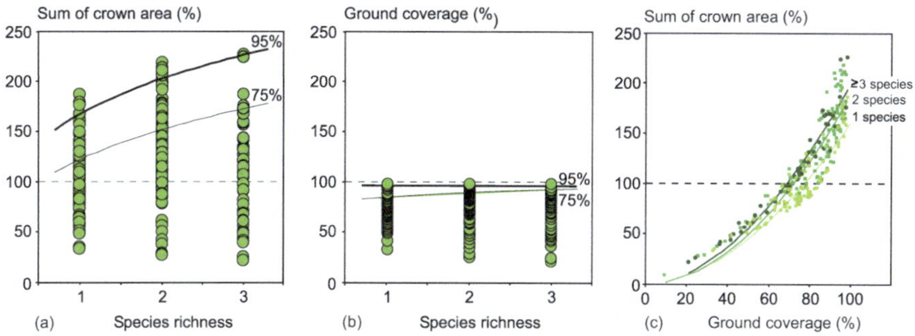


**Fig. 5.19** (**a**–**c**) Sum of crown projection area per unit area and ground coverage in mixed compared with monospecific stands. (**a**) Sum of crown projection area per unit area depending on tree species richness, (**b**) ground coverage depending on tree species richness, and (**c**) sum of crown projection area per unit area depending on ground coverage and tree species richness. Data are 253 crown maps of 187 plots belonging to 52 long-term experiments in even-aged and uneven-aged stands in Germany with mean plot size 0.29 ha, earliest and latest survey from 1951 and 2004, and minimum and maximum stand age of 16 and 283 years, respectively.The 95% and 75% quantile regression lines in Fig. 5.19a and b represent the sum of crown area and ground coverage for fully and loosely stocked stands. The OLS regression lines in Fig. 5.19c represent the mean sum of crown area per unit area depending on ground coverage for ≥three-, two-, and one-species stands (*upper*, *middle*, *lower line*, respectively)

the stand area. The relationship between ground coverage by crowns and tree species richness shown in Fig. 5.19b demonstrates that in most fully stocked mono- as well as polycultures, 5–10% of the stand area is uncovered by crowns. The frequently reported mechanical abrasion (Putz et al. 1984) resulting in crown shyness may cause the 5–10% uncovered area in both monospecific and mixed stands. The wide variation in ground coverage below the ceiling line (95% and 75% quantile regression lines) can be attributed to the broad range of thinning grades included in the dataset of 253 crown maps. The above finding that species richness may increase CE even when CGC is similar becomes even clearer in Fig. 5.19c. Especially in denser stands with CGC > 80%, the sum of crown projection area per unit area is about 25% higher in three-species stands compared with monocultures (upper versus lower line in Fig. 5.19c).

Studies in mixed-species stands by Kennel (1965), Pretzsch (2009, pp 267–269), and Preuhsler (1981) suggest that the sum of crown projection area per unit area is often much higher than the ground coverage due to an up to sevenfold crown overlap especially in polycultures of species with complementary ecology. When studies comparing crown coverage between stands only count once areas which have been covered twice or more often, they neglect that the sum of the crown areas and all related advantages such as structural heterogeneity, growth, and resilience may be underestimated (Assmann 1970, pp 102–107).

Based on long-term experimental plots in Germany in monospecific and mixed stands of spruce and beech ($n = 110$) as well as oak and beech ($n = 74$), the phenomenon of multiple crown coverage is analysed more closely (see Table 5.6).

**Table 5.6** Ground cover by crowns (mean ± standard error) in even-aged monocultures and mixed stands of Norway spruce, European beech, and sessile oak based on crown maps of unthinned and moderately thinned stands

| | | | | | onefold and multiple ground coverage | | | |
|---|---|---|---|---|---|---|---|---|
| Species | Stand type | n | Ground cover (%) | Ground | onefold (%) | twofold (%) | ≥ threefold (%) | Rel. sum of crown area (%) |
| Norway spruce | mono | 32 | 77 ± 2 | 23 ± 2 | 64 ± 1 | 12 ± 2 | 1 ± 1 | 91 ± 4 |
| European beech | mono | 25 | 83 ± 3 | 17 ± 3 | 50 ± 2 | 26 ± 3 | 7 ± 1 | 123 ± 7 |
| N. sp. and E. be | mixed | 53 | 88 ± 1 | 12 ± 1 | 48 ± 2 | 30 ± 1 | 10 ± 1 | 138 ± 4 |
| sessile oak | mono | 22 | 64 ± 4 | 36 ± 4 | 50 ± 2 | 12 ± 3 | 2 ± 1 | 81 ± 7 |
| European beech | mono | 25 | 83 ± 3 | 17 ± 3 | 50 ± 2 | 26 ± 3 | 7 ± 1 | 123 ± 7 |
| s. oak and E. be | mixed | 27 | 85 ± 3 | 15 ± 3 | 35 ± 2 | 30 ± 2 | 20 ± 3 | 156 ± 10 |

The uncovered and covered stand area adds up to 100%. In addition, columns 5–8 give the percentage of onefold, twofold, and ≥threefold crown cover and the relative sum of crown projection areas

The crown measurements cover the period from 1954 till the present and stand ages of between 26 and 207 years. In order to study species-specific behaviour in coping with crowding, only fully stocked and rather even-aged stands were included, in which all species were left unthinned or were only moderately thinned.

Table 5.6 shows that the ground coverage by crowns is on average only 64–83% in monocultures and 85–88% in mixture; i.e. in both kinds of stands, a considerable portion is uncovered by crown projection area. Of special interest is that about 40–50% of the mixed stand ground area is covered more than onefold with crown projection area. The respective range in monocultures is 13–33%. The more multiple coverage in mixed versus monospecific stands is also reflected by a relative sum of crown area of 138–156% in mixed stands compared with 81–123% in monocultures. Even in monocultures, beech achieves 83% crown cover with 33% more than onefold coverage, indicating its outstanding shade tolerance and crown plasticity.

Comparing monocultures and mixed stands reveals the 'overpacking' of the canopy space in mixed stands compared with neighbouring monocultures. The term overpacking is used as an analogy with overyielding (see Chap. 4, Box 4.1), which refers to the frequently observed superiority of mixed stands regarding productivity (Vilà et al. 2005; Pretzsch 2005). Particularly species combinations with complementary light ecology such as Norway spruce and European beech as well as sessile oak and European beech may lead to much denser and vertically layered canopies which in turn may cause higher light interception (Kelty 2006; Morin et al. 2011) and overyielding (Bauhus et al. 2004; Pretzsch et al. 2010, 2013b) compared with monocultures.

Many studies show relatively close relationships between the absorbed photosynthetically active radiation, APAR, and crown size, whether these are quantified in terms of leaf area, crown surface area, crown projection area, crown length, or crown width (Binkley et al. 2013; Forrester et al. 2012). These relationships will probably vary between species and growing conditions, but as long as they are known, crown size characteristics should be a reasonable proxy for light interception. This is further indicated by studies showing that when trees increase their leaf area, they often seem to simultaneously increase crown length or width rather than increasing leaf area density. For example, Forrester et al. (2013) found that thinning, pruning, and fertiliser application changed leaf area, crown length, and width, but not leaf area density. As the crown projection area is much easier to measure at individual tree level and is often available from past inventories of long-term plots, it can serve as a proxy for both leaf area and light interception. As crown morphology represents the holding fixture of leaves and the light interception and canopy structure results from tree-tree interaction, closer analysis of both may reveal species-specific behaviour in mixed compared with monospecific stands.

A combination of species with different crown shapes and albedos may decrease the reflection and light loss at the upper canopy layer by 5–10%. Particularly the rather notched canopies of conifers with light wells and low albedos reflect less light compared with deciduous forests (Otto 1994, p 213; Dirmhirn 1964, p 132).

Mixing light-demanding with shade-tolerant species may increase the light interception due to complementary, species-specific light compensation points (LCP) and light saturation points (LSP). For example, sessile and common oak represent light-demanding species with both high light saturation (LSP = 680 μmol $m^{-2}$ $s^{-1}$) and compensation points (LCP = 17 μmol $m^{-2}$ $s^{-1}$) which can make the most of the light supply in the upper canopy but, unlike European beech, have difficulty surviving in the deep shade (LCP and LSP for leaves and $P_{max}$ according to Ellenberg and Leuschner (2010, pp 103–105)). Beech combines a lower light saturation point (LSP = 460 μmol $m^{-2}$ $s^{-1}$) with a lower light compensation point (LCP = 13 μmol $m^{-2}$ $s^{-1}$) and is well equipped to forage for light beside or below oak or spruce (LSP and LCP = 750 and 20 μmol $m^{-2}$ $s^{-1}$). The filling of canopy space by species with complementary light ecology and the closer interlocking of their crowns enables foraging of full as well as weak light as well as a more complete light interception and often results in a lower light intensity at the forest floor of mixed compared with monospecific stands (Mitscherlich 1971, p 82).

#### 5.7.2.1 Temporal Change of Sum of Crown Projection Area in Mixed Versus Monospecific Stands

Figure 5.20a–c shows for monospecific and mixed stands of spruce and beech the change in the sum of crown projection area per unit area with progressing stand development (represented by the quadratic mean tree diameter, $d_q$, on the $x$-axis).

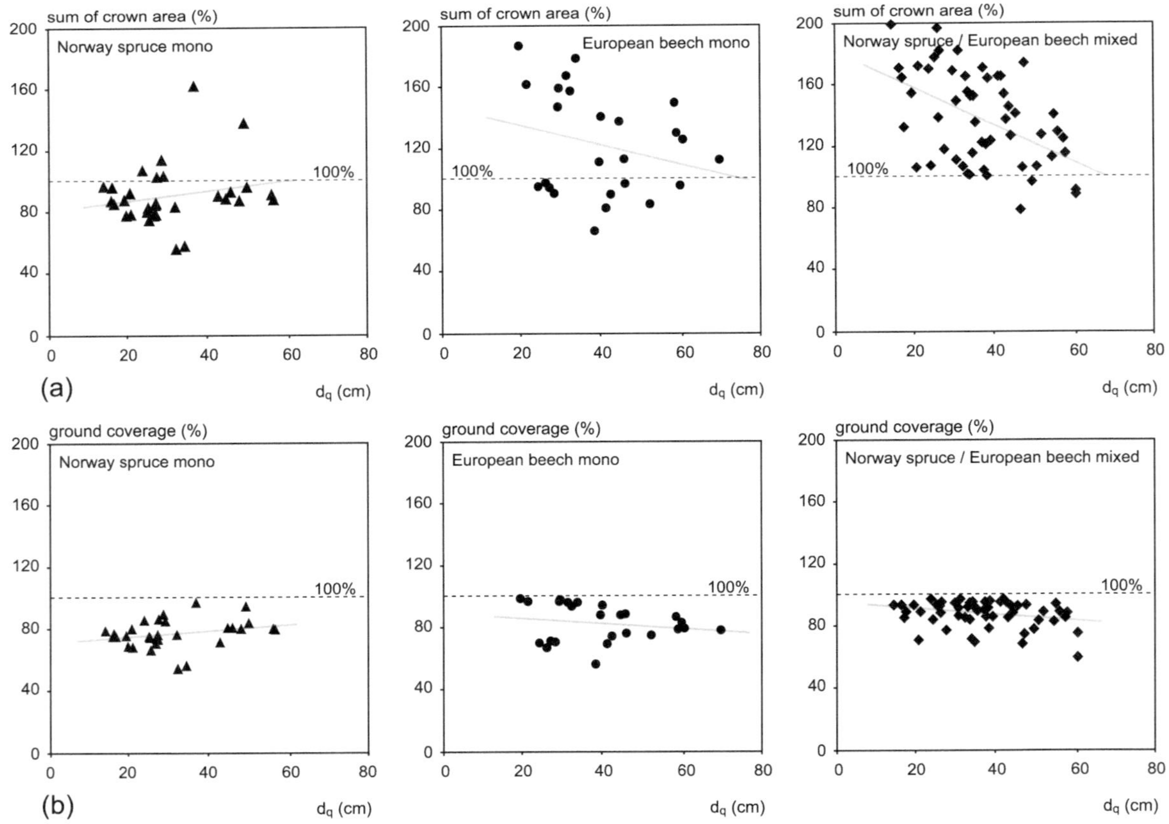


**Fig. 5.20** (**a–f**) Sum of crown projection area, SCPA, per unit area (%) and ground coverage, CGC, (%) in even-aged monospecific and mixed stands of Norway spruce and European beech in dependence on quadratic mean diameter, $d_q$. Sample size from left to right $n = 32$, 25, and 54. (**a–c**) Sum of crown area in monospecific and mixed stands of Norway spruce and European beech and (**d–f**) ground coverage by crowns for monospecific and mixed stands of Norway spruce and European beech. For further statistical characteristics, see Pretzsch (2014)

Compared with the respective monocultures (100% line), the mixture of spruce and beech results in values which start and remain much higher even in the advanced phase of stand development (Fig. 5.20c). Choi et al. (2001) found a similar trend in northern hardwood forests. The decrease with stand age may result from gaps between trees which increase with age and need longer to be covered by crowns again. The relative ground coverage starts lower and can decrease below 80% in the monocultures (Fig. 5.20d and e), whereas in mixed stands it remains at about 95% across the stand development (Fig. 5.20f).

### 5.7.3 Stand Density and Crown Density in Mixed and Monospecific Stands

In the following, we use the triplet data introduced in Sect. 5.3.2 to show the combined effect of both stocking density and canopy density. Furthermore, we analyse how those density and canopy characteristics affect the overyielding of mixed compared with monospecific stands. Stand density will be quantified by the standardised stand density index ($SDI_{1,2} = SDI_{1,(2)} + SDI_{(1),2} \times e_{2 \Rightarrow 1}$) introduced in Sect. 4.3.1. The ratio $RD_{1,2} = SDI_{1,2}/SDI_1$ reflects the relationship between the stand density in mixed versus monospecific stands. The quantification of the canopy density was based on the crown cross-section area of all trees on the monospecific and mixed-species plots of the triplets. The crown cross-section areas, cpa, of individual trees were added to get the sum of cpa per plot for the two monocultures of each triplet ($SCPA_1$, $SCPA_2$) and for the mixed stand ($SCPA_{1,2}$). As ratio between the mixed and monospecific stands regarding the sum of crown projection area, we used $RSCPA_{1,2} = SCPA_{1,2}/(SCPA_1 \times m_1 + SCPA_2 \times m_2)$ analogously to the RPA introduced in Chap. 4, Box 4.1.

#### 5.7.3.1 Stand Density and Canopy Density

The stand density index, SDI, in the mixed stands is mostly similar to or significantly higher than the SDI of the neighbouring monocultures (Fig. 5.21a). The pairwise comparison between the SDI in mixed-species stands and that in monocultures yielded a mean ratio of RD = 1.10 ± 0.06. The upper section of Table 5.7 shows that in mixed stands of Douglas-fir/European beech and Norway spruce/Scots pine there can be significantly higher stocking densities compared with monocultures ($SDI_{mixed}/SDI_{mono} = 1.53$ and 1.10, respectively).

The canopy density of mixed-species stands can be significantly higher because of the slightly higher stocking density and a considerably higher sum of crown projection area, caused by wider crown extension. Figure 5.21b illustrates that in most cases the observed sum of crown projection area per hectare in mixed stands,

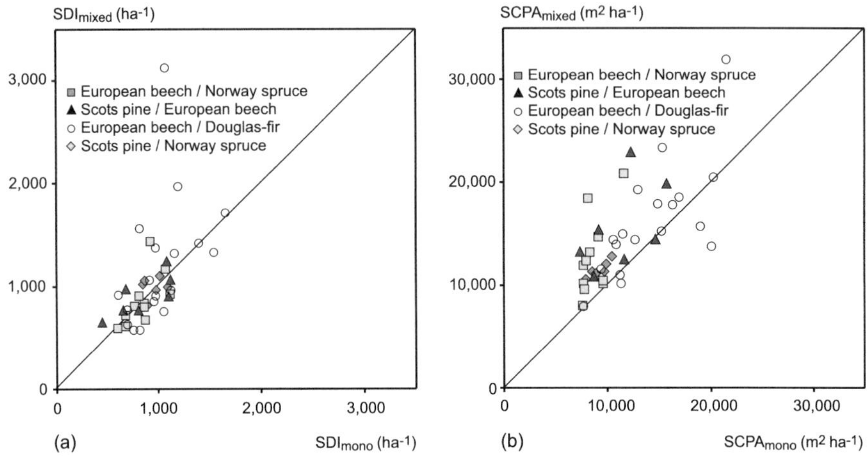


**Fig. 5.21** Observed density on 42 mixed stands compared with monospecific stands based on triplets of Norway spruce/European beech, Scots pine/European beech, Douglas-fir/European beech, and Norway spruce/Scots pine. (**a**) Observed stand density index in mixed stands compared with the weighted mean of both neighbouring monocultures. (**b**) Sum of crown projection area per hectare, $SCPA_{mixed}$, in mixed stands plotted over the values for monocultures, $SCPA_{pure}$. Observations above the 1:1 line indicate higher density of stocking or canopy cover in mixed stands compared with monocultures

$SCPA_{mixed}$, is much higher than the weighted mean of the two monocultures, $SCPA_{mono}$. SCPA values of 10,000–30,000 $m^2$ $ha^{-1}$ in mixed stands mean that a stand area of one hectare is covered between one and three times by tree crowns. The corresponding values for monocultures (7500–20,000 $m^2$ $ha^{-1}$) are distinctly smaller. On average, the sum of crown projection area amounts to 11,472–16,144 $m^2$ $ha^{-1}$ in the mixed stands of the triplets and 6933–17,095 $m^2$ $ha^{-1}$ in the monocultures (Table 5.7, lower section).

For the different species combinations, we found mean RSCPA values of 0.96–1.83, indicating canopies that are 4% lower and 83% denser in mixed-species stands than in monocultures. For all four species combinations separately and also for all triplets together, the RSCPA values mostly exceed the value of 1.0 ($p < 0.001$). The finding that the mean of RSCPA amounts to 1.16–1.83 for Norway spruce/European beech, 1.27–1.60 for Scots pine/European beech, 0.96–1.44 for Douglas-fir/European beech, and 1.06–1.51 for Norway spruce/Scots pine (Table 5.7, lower section) emphasises high canopy densities especially in mixtures with beech. Mixed-species canopies are on average 33% denser than monospecific canopies.

**Table 5.7** Stocking density, SDI, and sum of crown projection area, SCPA, in mixed-species stands versus monocultures of the 42 triplets of Norway spruce/European beech, Scots pine/European beech, Douglas-fir/European beech, and Norway spruce/Scots pine

| Species | | SDI (trees ha$^{-1}$) | | | RD = $SDI_{mixed}/SDI_{mono1}$ | | | | RD = $SDI_{mixed}/SDI_{mono2}$ | | | |
|---|---|---|---|---|---|---|---|---|---|---|---|---|
| 1/2 | *n* | Mean mixed | Mean mono 1 | Mean mono 2 | Min | Max | Mean | SE | Min | Max | Mean | SE |
| N.sp./E.be | 11 | 738 ± 59 | 846 ± 48 | 661 ± 50 | 0.50 | 1.46 | 0.89 | 0.08 | 0.52 | 2.09 | 1.17 | 0.12 |
| Sc.p./E.be | 7 | 863 ± 58 | 845 ± 100 | 761 ± 69 | 0.55 | 1.60 | 0.99 | 0.13 | 0.68 | 1.33 | 1.05 | 0.09 |
| D.-fir./E.be | 18 | 930 ± 108 | 1030 ± 69 | 675 ± 56 | 0.43 | 2.11 | 0.93 | 0.09 | 0.55 | 4.38 | 1.53* | 0.21 |
| N.sp./Sc.p. | 6 | 1046 ± 33 | 953 ± 39 | 1248 ± 235 | 0.99 | 1.21 | 1.10*** | 0.03 | 0.54 | 1.34 | 0.96 | 0.13 |
| Species | | SCPA | | | RSCPA = $SCPA_{mixed}/SCPA_{mono1}$ | | | | RSCPA = $SCPA_{mixed}/SCPA_{mono2}$ | | | |
| 1/2 | *n* | Mean mixed | Mean mono 1 | Mean mono 2 | Min | Max | Mean | SE | Min | Max | Mean | SE |
| N.sp./E.be | 11 | 12,625 ± 1183 | 6933 ± 361 | 11,097 ± 829 | 1.15 | 2.42 | 1.83*** | 0.14 | 0.78 | 2.13 | 1.16 | 0.11 |
| Sc.p./E.be | 7 | 15,566 ± 1636 | 10,224 ± 1219 | 12,588 ± 1162 | 1.03 | 2.23 | 1.60*** | 0.16 | 0.94 | 1.81 | 1.27* | 0.13 |
| D.-fir./E.be | 18 | 16,144 ± 1293 | 17,095 ± 1032 | 12,227 ± 1170 | 0.54 | 1.34 | 0.96 | 0.06 | 0.79 | 2.75 | 1.44*** | 0.11 |
| N.sp./Sc.p. | 6 | 11,472 ± 313 | 7659 ± 378 | 10,915 ± 486 | 1.40 | 1.72 | 1.51*** | 0.05 | 0.95 | 1.20 | 1.06 | 0.03 |

The ratio of stand density (RD = $SDI_{mixed}/SDI_{mono}$) and the ratio of sum of crown projection area (RSCPA = $SCPA_{mixed}/SCPA_{mono}$) quantify the stand density and crown area in mixed versus pure stands. For further explanation see caption of Table 5.1

## 5.8 Effect of the Size-Structure Dynamics on Overyielding at the Stand and the Species Level

The data of the 42 triplets with 126 plots in monospecific and mixed-species stands introduced in Sect. 5.3.2 can be applied to analyse relationships between overyielding at the stand level and characteristics at the size distribution level. The analysis of overyielding was based on the relative productivity, RPA, introduced in Chap. 4 (see Box 4.1). It quantifies the mixing effect on the stand productivity.

### 5.8.1 Overyielding at the Stand and Species Level

On average, the productivity of mixed-species stands on the 42 triplets was 104% of the weighted mean of the neighbouring monocultures. At the species level, the productivity in mixture is 107–118% of the monospecific stands, i.e. the species grow 7–18% more in mixed compared with monospecific stands. On average, European beech benefits the most from growing in mixture, Norway spruce benefits the least, and the other species lie in between. The minimum and maximum values show a broad variation of the mixing effects on both stand and species level.

In the mixture of Scots pine and European beech, both species interact in such a way that the total benefit amounts to 16%. The mixture of Douglas-fir and European beech can be particularly beneficial for beech; however, on average the benefit is 1% and marginal at stand level. The relative productivity at stand level can be different from the mean relative productivity at species level as the former considers the mixing proportions (Sterba et al. 2014) and the coupling between strong positive effects of one species with negative effects of the other.

### 5.8.2 Relationship Between Size Structure and Over- and Underyielding at the Stand Level

The analysis of how overyielding of the mixed stands of the 42 triplets emerges from the tree distribution level was based on the relative productivity, $RP_{1,2}$, of mixed stands versus monocultures (see Chap. 4, Box 4.1). To explain the mixing effect at the tree distribution level, the relative stand density, RD; the relative sum of crown projection area, RSCPA; the relative slope b of the iv–$v$ relationship, Rb; the relative Gini coefficient of $v$, iv, $RGC_{v}$ resp. $RGC_{iv}$; and the relative growth dominance coefficient, RGDC, were used. Similar to the relative productivity, RP, the latter six ratios represent the quotients between the distribution characteristics in mixed versus pure stands.

While all other ratios had only weak or non-significant effects on RP, the ratios RD, $RGC_{v}$, RSCPA, and Rb had a clearly positive effect. Two of the revealed

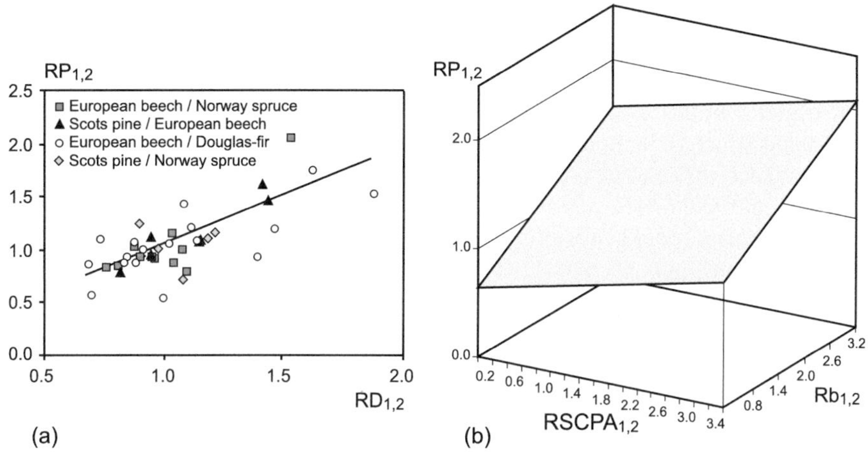


**Fig. 5.22** Dependency of overyielding at the stand level $RP_{1,2}$ from (**a**) the relative stand density, $RD_{1,2}$ ($RP = 0.20 + 0.81 \times RD$, $n = 42$, $R^2 = 0.52$, $p < 0.001$), and (**b**) the ratio between the sum of crown projection area in mixed and monospecific stands, $RSCPA_{1,2}$, and the ratio between the slope of the iv–$v$ relationship, $Rb_{1,2}$, in mixed and monospecific stands ($RP = -0.16 + 0.62 \times RSCPA + 0.36 \times Rb$, $n = 42$, $R^2 = 0.47$, $p < 0.001$)

relationships between the size structure and over- and underyielding at the stand are visualised in Fig. 5.22a and b.

### 5.8.3 Increase in Canopy Space Filling and Light Interception

The significantly positive effect of the relative sum of crown projection area, Gini coefficient $GC_v$, slope of the iv–$v$ relationship, and stocking density on the productivity of the mixed stands suggests that higher light interception and light-use efficiency can be essential causes of their overyielding. Kelty (1992) assumed that the intercepted light may also be used more efficiently when crowns or branches of shade-tolerant species replace more light-demanding species. He speculated that due to the replacement of less efficient organs, crown parts, or whole trees, particularly smaller trees growing in the shade of the lower crown may contribute to the overyielding of mixed versus monospecific stands. Following this assumption, the effect of beech on pine may be compared with thinning pine from below. By eliminating the slow and inefficiently growing understorey trees in monospecific stands, stand growth can be accelerated (Assmann (1970, pp 248–261); Pretzsch (2005)). However, in contrast to thinning, this does not only entail an elimination but also a replacement by even more efficiently growing individuals of a better adapted species.

## 5.9 Modification of Stand Structure in Monospecific and Mixed-Species Stands by Site Conditions

The triplet data for Scots pine and European beech introduced in Sect. 5.2 reveal how stand structure in monospecific and mixed-species stands can be modified by water availability. Based on the set of 32 triplets of mature and fully stocked monocultures and mixed stands of Scots pine and European beech located along a productivity and water availability gradient (annual mean temperature 6.0–10.5 °C und annual precipitation 520–1175 mm) through Europe, Pretzsch et al. (2016) examined how mixing modifies the stand structure in terms of stand density, horizontal tree distribution pattern, vertical stand structure, size distribution pattern, and variation in tree morphology. In particular, they analysed how site conditions modify these aspects of stand structure. For this typical mixture of a light-demanding and shade-tolerant species, they found that mixing significantly increases the above-mentioned aspects of structural heterogeneity compared with monocultures. Mixing effects such as an increase in stand density and diversification of vertical structure and tree morphology were caused by species identity (additive effects) but also by species interactions (multiplicative effects). The superior heterogeneity of mixed stands over monocultures increases from dry to moist sites (Fig. 5.23).

**Fig. 5.23** Stands of Scots pine and European beech (*centre*) have significantly higher structural heterogeneity than monocultures of Scots pine (*left*) and European beech (*right*). The superior heterogeneity of mixed stands over monocultures increases from dry to moist sites (from top to bottom).Canopies can be denser in mixed stands (*centre*) compared with monospecific stands (*left* and *right*) due to wider tree crown extension, multilayering, and higher stocking density. The complete canopy space filling may increase the light interception in mixed stands (after Pretzsch et al. 2016)

Mixed stands of Scots pine and European beech are more heterogeneous in structure and can carry more trees of a given size, and this effect increases with site productivity. The complementary light ecology of these species (pine being light demanding, beech being shade tolerant) increases the light interception or light-use efficiency to such an extent that not only stand productivity (Pretzsch et al. 2016) but also the stand density can be higher than in monocultures (Pretzsch and Biber 2016). For a more detailed analysis of the effect of species mixing on stand density, see Chap. 4, Sect. 4.3.2.

The finding by Pretzsch et al. (2016) that this tendency and vertical structuring increases with site productivity substantiates the assumption that greater light interception explains the increase in density and growth. On rich sites where water and nutrient supply are higher, the light complementarity might become more effective than on poor sites where other environmental conditions are limiting (Forrester and Albrecht 2014).

## 5.10 Scientific Evidence and Practical Relevance

### 5.10.1 Size Structure as Link Between Tree-Level and Stand-Level Research on Mixed Stands

The particular pattern of canopy space filling by crowns of different sizes can increase the productivity of mixed over monospecific stands. The combination of species with complementary ecological traits enables a coexistence of trees of variable sizes in different canopy layers. The vertical stratification can be coupled with a denser crown space filling, flooding of light also into deeper canopy layers, higher rates of survival of small trees, more complete light exploitation by crowns, and higher stand density. Even trees in deeper canopy layers contribute to stand growth proportionally to their stem volume. This reveals an amazing elasticity of tree growth over the different sizes and layers which is probably caused by adaptation via allometric plasticity, by optimised species-specific niche occupation (shade tolerant below light demanding), and a combination of small and tall trees (high-growth beside low-growth per tree stand area).

Even stands with similar mean tree dimensions (mean diameter, mean volume) or cumulative hectare-related characteristics (standing stock, biomass) can differ considerably in their size structure and the size-growth partitioning between the trees. These differences can strongly affect stand productivity via differences in resource acquisition, resource-use efficiency, and respiratory losses.

Analysis of mixing effects at the organ or tree level may provide insight into particular patterns and processes. However, their relevance for any overyielding at stand level requires that such analyses are available for trees of different sizes and social positions and that the size-specific patterns can be upscaled to the stand level taking into account the location, shape, and density of the frequency distribution.

As overyielding is strongly determined by a change in the location, shape, and density of the size and size-growth distribution, analyses based on a small number of dominant trees in mixed compared with monospecific stands are barely sufficient for explaining overyielding or underyielding effects at the stand level (Strigul et al. 2008; Webster and Lorimer 2003).

### 5.10.2 Consequences for Forest Management

The periodic annual growth and total yield of monocultures can be increased through removal of inefficient subdominant and suppressed trees, i.e. by thinning from below. Subdominant and suppressed trees in even-aged stands can have an unfavourable ratio between resource use and growth (Assmann 1970, pp 34–38). Their elimination through thinning leaves more water and nutrients to the remaining trees which use those resources more efficiently and can overcompensate for the growth loss of the removed trees and even increase the total stand growth by about 20–30% above the level of unthinned stands (Pretzsch 2005). Replacing inefficiently slow-growing and excessively respiring trees of one species with a more efficient and shade-tolerant second species in mixed-species stands may have a similar effect to thinning from below in monocultures. A precondition for this increase in efficiency is a combination of species with complementary light ecology, e.g. early-successional light-demanding above late-successional shade-tolerant species. Besides other reasons for leaving small trees (insurance against total stand loss by windthrow, resilience, regeneration), their contribution to stand growth may be used to maintain sustainable and continuous growth in monocultures (Sterba 1999) and particularly in mixed or uneven-aged stands where their share of stand productivity can be considerable.

The close relationship between stand density, canopy space filling, and overyielding (Fig. 5.22) indicates that part of the mixing effect may be a density effect. Higher stand densities might mean a greater light capture and additional increase in productivity. The latter effect would be rather density dependent and could be lost through thinning. The increase in stand density and light capture may be higher the more complementary the ecological niches and the resulting efficiency effects are; i.e. there might be a positive feedback between an increase in light-use efficiency and stand density. An increase in light-use efficiency brought about by combining complementary tree species in different layers would be rather density independent and could scarcely be eliminated by silviculture in the course of density reductions.

The SDI analyses showed that species mixing can increase the maximum stand density, probably because mixed stands capture more light and use the intercepted light more efficiently than neighbouring monocultures (see Sect. 4.3.2). Greater light interception by more extensive crowns, denser canopy space filling, and multilayering may only result in overyielding when the necessary below-ground resources are not growth limiting. However, increased light-use efficiency by a combination of species with complementary light ecology is likely to always result in higher growth as the available water and nutrients are simply used more efficiently compared with monocultures. European beeches in the shade of Scots pines, for instance, may use the intercepted light more efficiently for growth—i.e. consume less water and nutrients—due to a lower compensation point than would be needed by Scots pines growing under the same competitive constellation. So, the higher interception and use efficiency of light enable a higher stand density and more trees to survive and grow per unit area. Other silvicultural measures such as spacing, thinning, and fertilisation let stands grow faster but pass through the same allometric density trajectories (Long et al. 2004). Mixing, in contrast, can cause a long-lasting increase in the stock of resources in the system and may thereby increase the maximum stand density which means a growth and survival of more and/or taller trees compared with neighbouring monocultures in the same stand development phase.

If silvicultural guidelines for monocultures, e.g. regarding stand density or number of future crop trees, are applied to mixed-species stands without adaptation to their special structure, density, and productivity potential, this can cause a considerable loss of resources and productivity.

## 5.11 Synthesis

In temperate forests where the nutrient and water supply in the soil is sufficient for all trees so that they are mainly competing above ground for light, overyielding can emerge from tree and size-structure dynamics. On long-term experiments and temporally established triplets in Central Europe, the canopy space filling can be much denser in mixed compared with monospecific stands (Fig. 5.24a). The size distribution can be more right skewed and the kurtosis as well as the number of trees higher as in monocultures. In most of the mixed stands, there was a greater inequality in size and size growth than in monocultures (Fig. 5.24b) but similar growth dominance coefficients. The iv–*v* relationships indicate a stronger size asymmetry of competition in most of the mixed stands compared with neighbouring monocultures (Fig. 5.24c). Species mixing can cause a higher stocking density and significantly higher sum of crown projection area. Overyielding of mixed compared with monospecific stands increases with the density of stocking and canopy, inequality of size and size growth, and steepness of the size-growth relationship. These differences suggest that deeper entrance into the canopy, more complete interception, and higher use efficiency of light can be main causes of the overyielding of mixed stands (Fig. 5.24d).

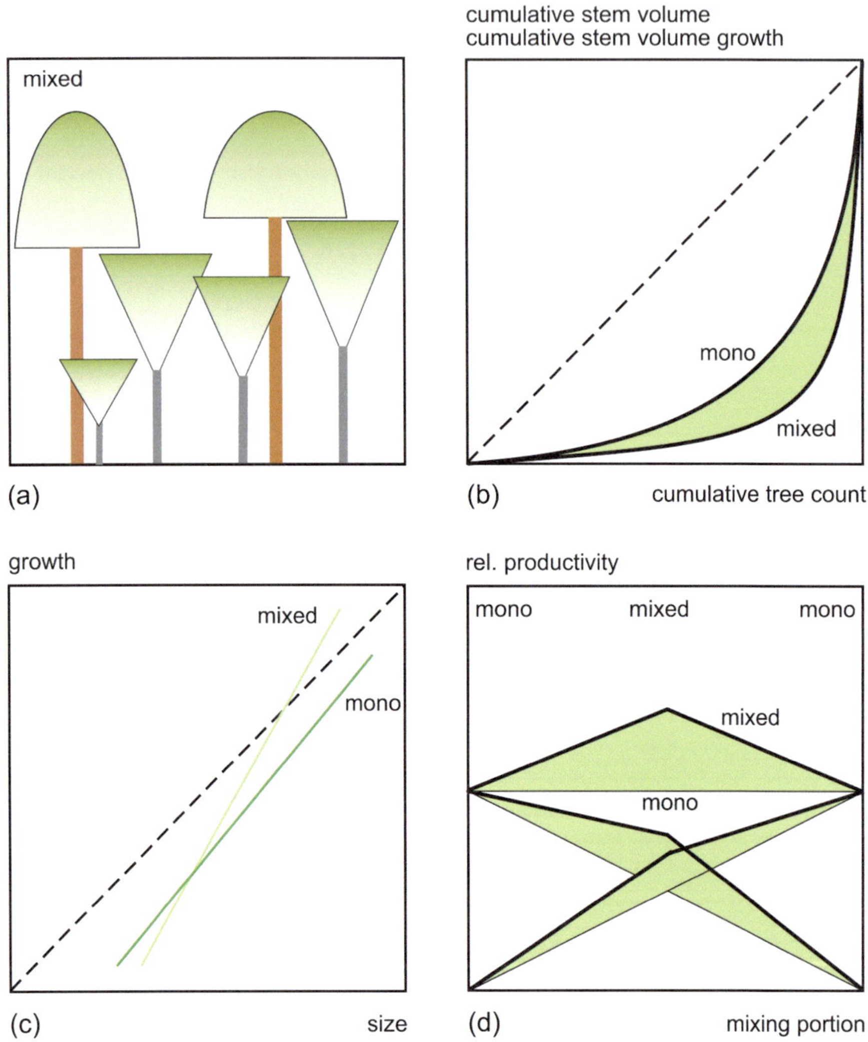


**Fig. 5.24** (**a**–**d**) Effect of species mixing on the stand structure dynamics and overyielding at the stand level. (**a**) Canopies can be denser in mixed than in monospecific stands due to wider tree crown extension, multilayering, and higher stocking density. (**b**) Mixed stands can have higher size inequality and also higher numbers of small compared with tall trees. This is indicated by a steeper increase in cumulative stem volume and cumulative stem growth over cumulative tree count and higher Gini coefficients. (**c**) In most cases, the iv–*v* relationships are higher in mixed compared with monospecific stands and indicate more size-asymmetric competition. (**d**) In most cases, mixed stands overyield neighbouring monocultures. The structural traits suggest that overyielding is partly based on deeper entrance of light into the canopy, more complete interception, and higher use efficiency of light

**Take-Home Messages**

1. The evolution of the size structure during stand development determines growth and yield as well as a broad spectrum of forest functions including biodiversity, protection, and recreation.
2. The slow but continuous feedback between the within-stand environment→tree growth→stand structure→within-stand environment results in the species' adaptation to the interspecific environment. Due to this continuous adaptation, the size distribution of mixed stands can increasingly deviate from those in neighbouring monocultures with proceeding stand development.
3. The total number of trees can be higher in mixed than in monospecific stands. In addition, the minimum tree size can be lower and the maximum size higher. The skewness and the kurtosis can also be higher.
4. In most cases, the iv–*v* relationships are higher in mixed than in monospecific stands and indicate more size-asymmetric competition.
5. Mixing of ecologically complementary species can increase the inequality of size and growth distribution between small and tall trees. This is indicated by a steeper increase in cumulative stem volume and cumulative stem growth over cumulative tree count and higher Gini coefficients.
6. The size heterogeneity brought about by mixing can increase the canopy space filling. The diversified canopy space ensures a growth elasticity of small and tall trees.
7. Canopies can be denser in mixed compared with monospecific stands due to wider tree crown extension, multilayering, and higher stocking density.
8. In many cases, mixed stands overyield neighbouring monocultures. The structural traits suggest that overyielding is based on deeper entrance of light into the canopy, more complete interception, and higher use efficiency of light.
9. The effect of the size-structure dynamics on the overyielding at the stand level suggests that common analyses at individual tree level should cover not just dominant trees but the whole size distribution. Upscaling the behaviour of dominant trees from tree to stand level remains questionable if it neglects size-specific reaction patterns and their frequency given that these strongly affect the stand-level productivity.
10. The part of the overyielding in complementary species mixture which is based on denser space filling and higher stocking density can be lost through silvicultural density reductions. In contrast, the increase in light-use efficiency should depend on the degree of complementarity in light ecology rather than on stand density.

## References

Ammer U, Schubert H (1999) Arten-, Prozeß- und Ressourcenschutz vor dem Hintergrund faunistischer Untersuchungen im Kronenraum des Waldes. Forstwissenschaftliches Centralblatt 118:70–87

Assmann E (1970) The principles of forest yield study. Pergamon Press, Oxford, New York

Avery TE, Burkhardt HE (1983) Forest measurements, 3rd edn. McGraw-Hill, New York

Bauhus J, van Winden AP, Nicotra AB (2004) Above-ground interactions and productivity in mixed-species plantations of *Acacia mearnsii* and *Eucalyptus globulus*. Can J For Res 34:686–694

Begon ME, Harper JL, Townsend CR (1998) Ökologie. Spektrum Akademischer Verlag, Heidelberg

Biging GS, Dobbertin M (1995) Evaluation of competition indices in individual tree growth models. For Sci 41:360–377

Binkley D, Kashian DM, Boyden S, Kaye MW, Bradford JB, Arthur MA, Fornwalt PJ, Ryna MG (2006) Patterns of growth dominance in forests of the Rocky Mountains, USA. For Ecol Manag 236(2):193–201

Binkley D, Campoe OC, Gspaltl M, Forrester DI (2013) Light absorption and use efficiency in forests: Why patterns differ for trees and stands. For Ecol Manag 288:5–13

Brus DJ, Hengeveld GM, Walvoort DJJ, Goedhart PW, Heidema AH, Nabuurs GJ, Gunia K (2011) Statistical mapping of tree species over Europe. Eur J For Res 131(1):145–157

Buongiorno J, Dahir S, Lu HC, Lin CR (1994) Tree size diversity and economic returns in uneven-aged forest stands. For Sci 40(1):83–103

Cannell MGR, Grace J (1993) Competition for light: detection, measurement, and quantification. Can J For Res 23:1969–1979

Choi J, Lorimer CG, Vanderwerker J, Cole WG, Martin GL (2001) A crown model for simulating long-term stand and gap dynamics in northern hardwood forests. For Ecol Manag 152 (1):235–258

Coggins SB, Coops NC, Wulder MA (2010) Estimates of bark beetle infestation expansion factors with adaptive cluster sampling. Int J Pest Manag 57(1):11–21

Condit R, Hubbell SP, Foster RB (1995) Mortality rates of 205 neotropical tree and shrub species and the impact of a severe drought. Ecol Monogr 65(4):419–439

Coomes DA, Allen RB (2007) Mortality and tree-size distributions in natural mixed-age forests. J Ecol 95(1):27–40

de Camino R (1976) Zur Bestimmung der Bestandeshomogenität. Allgemeine Forst- und Jagdzeitung 147:54–58

Dirmhirn I (1964) Das Strahlungsfeld im Lebensraum. Akademische Verlagsgesellschaft, Frankfurt am Main

Ellenberg H, Leuschner C (2010) Vegetation Mitteleuropas mit den Alpen in ökologischer, dynamischer und historischer Sicht. Ulmer, Stuttgart

Fish H, Lieffers VJ, Silins U, Hall RJ (2006) Crown shyness in lodgepole pine stands of varying stand height, density, and site index in the upper foothills of Alberta. Can J For Res 36 (9):2104–2111

Forrester DI, Albrecht AT (2014) Light absorption and light-use efficiency in mixtures of *Abies alba* and *Picea abies* along a productivity gradient. For Ecol Manag 328:94–102

Forrester DI, Lancaster K, Collopy JJ, Warren CR, Tausz M (2012) Photosynthetic capacity of *Eucalyptus globulus* is higher when grown in mixture with *Acacia mearnsii*. Trees Struct Funct 26:1203–1213

Forrester DI, Kohnle U, Albrecht AT, Bauhus J (2013) Complementarity in mixed-species stands of *Abies alba* and *Picea abies* varies with climate, site quality and stand density. For Ecol Manag 304:233–242

Freist-Dorr M (1992) Das Einzelbaumwachstum in langfristig beobachteten Mischbestandsversuchen, dargestellt am Beispiel der Eichen-Buchen-Versuchsfläche

Waldbrunn 105. Forstwissenschaftliches Centralblatt vereinigt mit Tharandter forstliches Jahrbuch 111(1):106–116
Grams TE, Andersen CP (2007) Competition for resources in trees: physiological versus morphological plasticity. Prog Bot 69:356–381
Hara T (1992) Effects of the mode of competition on stationary size distribution in plant populations. Ann Bot 69(6):509–513
Hara T (1993) Mode of competition and size-structure dynamics in plant communities. Plant Species Biol 8(2–3):75–84
Jennings SB, Brown ND, Sheil D (1999) Assessing forest canopies and understorey illumination: canopy closure, canopy cover and other measures. Forestry 72(1):59–74
Kelty MJ (1992) Comparative productivity of monocultures and mixed stands. In: Kelty MJ, Larson BC, Oliver CD (eds) The ecology and silviculture of mixed-species forests. Kluwer Academic Publishers, Dordrecht, pp 125–141
Kelty MJ (2006) The role of species mixtures in plantation forestry. For Ecol Manag. doi:10.1016/j.foreco.2006.05.011
Kennel R (1965) Untersuchungen über die Leistung von Fichte und Buche im Rein- und Mischbestand. Allgemeine Forst- und Jagdzeitung 136(149–161):173–189
Kennel R (1973) Die Bestimmung des Ertragsniveaus bei der Buche. Forstwissenschaftliches Centralblatt 92(1):226–234
Kramer H (1988) Waldwachstumslehre. Paul Parey, Hamburg, Berlin
Letcher BH, Priddy JA, Walters JR, Crowder LB (1998) An individual-based, spatially-explicit simulation model of the population dynamics of the endangered red-cockaded woodpecker, *Picoides borealis*. Biol Conserv 86:1–14
Long JN, Dean TJ, Roberts SD (2004) Linkages between silviculture and ecology: examination of several important conceptual models. For Ecol Manag 200:249–261
Matyssek R, Schnyder H, Oßwald W, Ernst D, Munch JC, Pretzsch H (2012) Growth and defence in plants. In: Ecological Studies, vol 220. Springer, Heidelberg, p 470
McKelvey K, Noon BR, Lamberson RH (1993) Conservation planning for species occupying fragmented landscapes: The case of the northern spotted owl. In: Kareiva PM, Kingsolver JG, Huey RB (eds) Biotic interactions and global change. Sinauer, Sunderland, MA, pp 424–450. (480 p)
MCPFE (1993) Resolution H1: General guidelines for the sustainable management of forests in Europe. Proc 2nd Ministerial Conference on the Protection of Forests in Europe. Helsinki, Finland
Mitscherlich G (1971) Wald, Wachstum und Umwelt, vol 2. Band. Waldklima und Wasserhaushalt. JD Sauerländer's Verlag, Frankfurt am Main
Morin X, Fahse L, Scherer-Lorenzen M, Bugmann H (2011) Tree species richness promotes productivity in temperate forests through strong complementarity between niches. Ecol Lett. doi:10.1111/j.1461-0248.2011.01691.x
Müller J, Hothorn T, Pretzsch H (2007) Long-term effects of logging intensity on structures, birds, saproxylic beetles and wood-inhabiting fungi in stands of European beech *Fagus sylvatica* L. For Ecol Manag 242(2):297–305
Müller J, Mehr M, Bässler C, Fenton MB, Hothorn T, Pretzsch H, Klemmt H-J, Brandl R (2012) Aggregative response in bats: prey abundance versus habitat. Oecologia 169(3):673–684
Noss RF (1990) Indicators for monitoring biodiversity: a hierarchical approach. Conserv Biol 4 (4):355–364
Otto HJ (1994) Waldökologie. Ulmer, Stuttgart
Peck JE, Zenner EK, Brang P, Zingg A (2014) Tree size distribution and abundance explain structural complexity differentially within stands of even-aged and uneven-aged structure types. Eur J For Res 133(2):335–346
Peltola H (1996) Model computations on wind flow and turning moment by wind for Scots pines along the margins of clear-cut areas. For Ecol Manag 83(3):203–215

Pommerening A, Brzeziecki B, Binkley D (2016) Are long-term changes in plant species composition related to asymmetric growth dominance in the pristine Białowieża forest? Basic Appl Ecol. doi:10.1016/j.baae.2016.02.002
Pretzsch H (2005) Stand density and growth of Norway spruce (*Picea abies* (L.) Karst.) and European beech (*Fagus sylvatica* L.). Evidence from long-term experimental plots. Eur J For Res 124(3):193–205
Pretzsch H (2006) Species-specific allometric scaling under self-thinning: evidence from long-term plots in forest stands. Oecologia 146(4):572–583
Pretzsch H (2009) Forest dynamics, growth and yield. Springer Verlag, Berlin
Pretzsch H (2014) Canopy space filling and tree crown morphology in mixed-species stands compared with monocultures. For Ecol Manag 327:251–264
Pretzsch H, Biber P (2010) Size-symmetric versus size-asymmetric competition and growth partitioning among trees in forest stands along an ecological gradient in central Europe. Can J For Res 40:370–384
Pretzsch H, Biber P (2016) Tree species mixing can increase maximum stand density. Can J For Res 46(10), 1179–1193. doi:10.1139/cjfr-2015-0413
Pretzsch H, Dieler J (2011) The dependency of the size-growth relationship of Norway spruce (*Picea abies* [L.] Karst.) and European beech (*Fagus sylvatica* [L.]) in forest stands on long-term site conditions, drought events, and ozone stress. Trees Struct Funct 25:355–369
Pretzsch H, Rais A (2016) Wood quality in complex forests versus even-aged monocultures. Review and perspectives. Wood Sci Technol 50(4):845–880
Pretzsch H, Schütze G (2014) Size-structure dynamics of mixed versus pure forest stands. For Syst 23(3):560–572
Pretzsch H, Schütze G (2015) Effect of tree species mixing on the size structure, density, and yield of forest stands. Eur J For Res 135(1):1–22
Pretzsch H, Block J, Dieler J, Dong PH, Kohnle U, Nagel J, Spellmann H, Zingg A (2010) Comparison between the productivity of pure and mixed stands of Norway spruce and European beech along an ecological gradient. Ann For Sci 67:712
Pretzsch H, Bielak K, Block J, Bruchwald A, Dieler J, Ehrhart H-P, Kohnle U, Nagel J, Spellmann H, Zasada M, Zingg A (2013a) Productivity of mixed versus pure stands of oak (*Quercus petraea* (Matt.) Liebl. and *Quercus robur* L.) and European beech (*Fagus sylvatica* L.) along an ecological gradient. Eur J For Res 132(2):263–280
Pretzsch H, Bielak K, Bruchwald A, Dieler J, Dudzińska M, Ehrhart H-P, Jensen AM, Johannsen VK, Kohnle U, Nagel J, Spellmann H, Zasada M, Zingg A (2013b) Mischung und Produktivität von Waldbeständen. Ergebnisse langfristiger ertragskundlicher Versuche [In German with English summary]. Allgemeine Forst- und Jagdzeitung 184:177–196
Pretzsch H, del Rio M, Schütze G, Ammer C, Annighöfer P, Avdagic A, Barbeito I, Bielak K, Brazaitis G, Coll L, Drössler L, Fabrika M, Forrester DI, Kurylyak V, Löf M, Lombardi F, Matovic B, Mohren F, Motta R, den Ouden J, Pach M, Ponette Q, Skzyszewski J, Sramek V, Sterba H, Svoboda M, Verheyen K, Zlatanov T, Bravo-Oviedo A (2016) Mixing of Scots pine (*Pinus sylvestris* L.) and European beech (*Fagus sylvatica* L.) enhances structural heterogeneity, and the effect increases with water availability. For Ecol Manag 373:149–166
Preuhsler T (1981) Ertragskundliche Merkmale oberbayerischer Bergmischwald-Verjüngungsbestände auf kalkalpinen Standorten im Forstamt Kreuth. Forstwissenschaftliches Centralblatt 100(1):313–345
Prodan M (1965) Holzmeßlehre. JD Sauerländer's Verlag, Frankfurt am Main
Putz FE, Parker GG, Archibald RM (1984) Mechanical abrasion and intercrown spacing. Am Midl Nat 112(1):24–28
Schütz JP (1989) Der Plenterbetrieb. Fachbereich Waldbau, ETH, 54 p
Schwinning S, Weiner S (1998) Mechanisms determining the degree of size asymmetry in competition among plants. Oecologia 113:447–455
Skov KR, Kolb TE, Wallin KF (2004) Tree size and drought affect ponderosa pine physiological response to thinning and burning treatments. For Sci 50(1):81–91

Sterba H (1999) 20 years target diameter thinning in the "Hirschlacke", forests of the monastery of Schlägl. Allgemeine Forst- und Jagdzeitung 170(9):170–175
Sterba H, del Río M, Brunner A, Condes S (2014) Effect of species proportion definition on the evaluation of growth in pure vs. mixed stands. For Syst 23(3):547–559
Strigul N, Pristinski D, Purves D, Dushoff J, Pacala S (2008) Scaling from trees to forests: tractable macroscopic equations for forest dynamics. Ecol Monogr 78(4):523–545
Valinger E, Lundqvist L, Bondesson L (1993) Assessing the risk of snow and wind damage from tree physical characteristics. Forestry 66(3):249–260
van Kuijk M, Anten NPR, Oomen RJ, van Bentum DW, Werger MJA (2008) The limited importance of size-asymmetric light competition and growth of pioneer species in early secondary forest succession in Vietnam. Oecologia 157:1–12
Vilà M, Inchausti P, Vayreda J, Barrantes O, Gracia C, Ibáñez JJ, Mata T (2005) Confounding factors in the observational productivity-diversity relationship in forests. In: Forest diversity and function. Springer, Berlin, Heidelberg, pp 65–86
Webster CR, Lorimer CG (2003) Comparative growing space efficiency of four tree species in mixed conifer–hardwood forests. For Ecol Manag 177(1):361–377
Weiner J (1990) Asymmetric competition in plant populations. Trends Ecol Evol 5(11):360–364
Weiner J, Thomas SC (1986) Size variability and competition in plant monocultures. Oikos 47:211–222
Wichmann L (2001) Annual variations in competition symmetry in even-aged Sitka spruce. Ann Bot 88:145–151
Wichmann L (2002) Competition symmetry. Chapter 7. In: Thesis P (ed) Modelling the effects of competition between individual trees in forest stands. Unit of Forestry, Copenhagen, pp 67–77
Yokozawa M, Kubota Y, Hara T (1998) Effects of competition mode on spatial pattern dynamics in plant communities. Ecol Model 106(1):1–16
Zeide B (1985) Tolerance and self-tolerance of trees. For Ecol Manag 13:149–166
Zeide B (1993) Analyses of growth equations. For Sci 39(3):594–616

# Individual Tree Structure and Growth in Mixed Compared with Monospecific Stands

6

Hans Pretzsch

**Abstract**

This chapter focuses on the shape, growth, and productivity of individual trees growing in inter- versus intraspecific environments. The individual tree senses and responds to the prevailing environmental conditions. The properties of the individuals determine the forest stand dynamics as individuals of different species interact with each other. Therefore, the level of the individual tree is most suitable for understanding competition, competition reduction through complementarity, and facilitation, which can result in the differences between structure dynamics and productivity of mixed compared with monospecific stands. The chapter shows how species mixing can modify the size development, persistence, and productivity of individual trees in mixed stands compared with members of the same species in neighbouring monospecific stands. Many of the beneficial tree mixing reactions result from complementary crown and root shape, spatially or temporally complementary resource exploitation, redistribution of resources, or modification of growth allocation and allometry introduced in this chapter.

## Contents

6.1 Principles of Individual Tree Ontogeny ..... 273
6.2 Interaction Between Trees in Inter- Versus Intraspecific Environments ..... 278
6.3 Basic Feedback Loop Between Growth, Structure, and Local Environment in Forest Stands ..... 282
6.4 Growth and Yield at the Tree Level ..... 284
6.4.1 Dominant Rather Than Open-Grown Trees as Reference ..... 284
6.4.2 Evidence of Facilitation and Competition Effects at the Tree Level ..... 286
6.4.3 Basic Growth Reaction Patterns in Intra- Versus Interspecific Neighbourhoods 288

H. Pretzsch (✉)
Chair for Forest Growth and Yield Science, Faculty of Forest Science and Resource Management, Technical University of Munich, Hans-Carl-von-Carlowitz-Platz 2, 85354 Freising, Germany
e-mail: Hans.Pretzsch@lrz.tum.de



H. Pretzsch et al. (eds.), *Mixed-Species Forests*, DOI 10.1007/978-3-662-54553-9_6

6.5 Crown Size and Morphology in Mixed Compared with Monospecific Stands ......... 290
6.5.1 Interspecific Variation in Crown Shape and Size ................................ 290
6.5.2 Intraspecific Variation in Crown Size ............................................ 292
6.5.3 Modification of Tree Allometry in Interspecific Local Environments .......... 296
6.5.4 From Pattern to Process of Crown Expansion in Interspecific Neighbourhoods 303
6.6 From Tree Morphology to Space Occupation ........................................... 311
6.6.1 Crown Projection Area, Tree Stand Area, and Degree of Crown Engagement 312
6.6.2 Dynamics of Canopy Gap Occupation .......................................... 313
6.7 Efficiency of Tree Crown Area and Tree Growing Area ................................ 316
6.7.1 Ratios for Comparing Tree Growth in Mixed Versus Monospecific Stands . . . 318
6.7.2 Elimination of the Size Effect from Efficiency Parameters ..................... 319
6.7.3 Transition from Tree Growth to Tree Productivity in Mixed Versus Monospecific Stands: An Example ............................................. 321
6.7.4 Crown Engagement cpa/sa as Link Between Tree and Stand Level Evaluation 322
6.8 Overview of Some Main Causes and Effects of Tree Species Mixing on Individual Tree Growth ........................................................................ 324
6.8.1 Complementary Exploitation of Crown and Root Space ........................ 324
6.8.2 Hydraulic Lift and Hydraulic Redistribution ................................... 326
6.8.3 Increased Availability of Mineral Nutrient Supply Through Deep-Rooting or Atmospheric $N_2$ Fixation ........................................................ 327
6.8.4 Temporal Complementarity of Niches, Resource Use, and Growth ............ 328
6.8.5 Modification of Growth Partitioning and Allometry of Trees in Inter- Versus Intraspecific Neighbourhoods ................................................... 329
Take-Home Messages ........................................................................ 331
References ..................................................................................... 332

Size growth and persistence are key requirements for a tree to compete and maximise fitness under the selective pressure in a forest stand. This chapter shows how species mixing can modify the size development, persistence, and productivity of individual trees in mixed stands compared with members of the same species in neighbouring monospecific stands. Living in association, both in intra- and interspecific neighbourhoods, entails pros and cons in comparison to solitary growth. An individual tree may be facilitated physically, e.g. by neighbours as they protect against storm breakage (von Lüpke and Spellmann 1997, 1999), sun scorch of bark (Assmann 1970), or snow sliding (Kuoch 1972; Mayer and Ott 1991, pp. 194–197; Strobel 1995). However, the meaning of neighbourhood is ambivalent, as it also entails competition when there are insufficient resources for all (Connell 1990). The interplay between facilitation and competition determines intraspecific and particularly interspecific coexistence (Larocque et al. 2013).

Mixing effects can be studied at various levels of organisation, e.g. at organ, individual, stand, or landscape level. Given that the individual tree forms the basic unit in forest stand ecology, this chapter focusses on the shape, growth, and productivity of individual trees growing in inter- versus intraspecific environments. The individual tree senses and responds to the prevailing environmental conditions. The properties of the individuals determine the forest stand dynamics as individuals of different species interact with each other (Smith and Smith 2009, p. 13). Therefore, the level of the individual tree is most suitable for understanding competition, competition reduction through complementarity, and facilitation, which can result in the differences between structure dynamics and productivity of mixed compared with monospecific stands discussed in Chaps. 4 and 5.

Understanding of the interspecific interactions between tree species can be improved through generalisation of the hologenome theory (Zilber-Rosenberg and Rosenberg 2008). According to this theory, a holobiont in a strict sense is a host organism in interaction with all associated microorganisms. More generally, all regular organisms which closely interact with each other (e.g. trees in mixed-species stands) can be perceived as holobiont-like systems and may also be studied as entities rather than focussing on the individual organisms (Matyssek and Lüttge 2012).

Tree species mixing may change the environmental conditions and thereby the physiology, morphology, and stand structure. As a result, the tree and stand growth also undergo changes. Because of the strong interaction between structural patterns and processes, the structure and morphology may reflect the results of mixing effects which are much more difficult to measure than the underlying physiological processes. To quantify mixing effects in this chapter, we compare the individuals' behaviour in terms of tree shape and growth in mixed versus monospecific stands. We use the trees' structure and growth in a monospecific stand as a reference for quantifying tree behaviour in neighbouring mixed stands.

Reaction patterns of shape, allometry, growth, and productivity at the individual tree level as presented in this chapter may contribute to understanding mixing effects. Further insight into the underlying mechanisms require analyses of the change in supply, capture, or use efficiency of resources in poly- versus monocultures (Binkley et al. 2004; Richards et al. 2010) as introduced in Chap. 3. We will set the growth of individual trees in relation to their required growing area (equivalent to tree stand area) and, in this way, link the tree level with the stand level and size distribution level (Chaps. 4 and 5).

## 6.1 Principles of Individual Tree Ontogeny

To explain inter- and intraspecific interactions between trees such as competition, competition reduction, and facilitation as well as their net effect on the growth of individual trees in mixed and monospecific stands, plant biology mostly starts with the ontogeny of a solitarily growing tree. The schematic Fig. 6.1a shows the unimodal course of the stem volume increment, ivs, and the s-shaped curve of

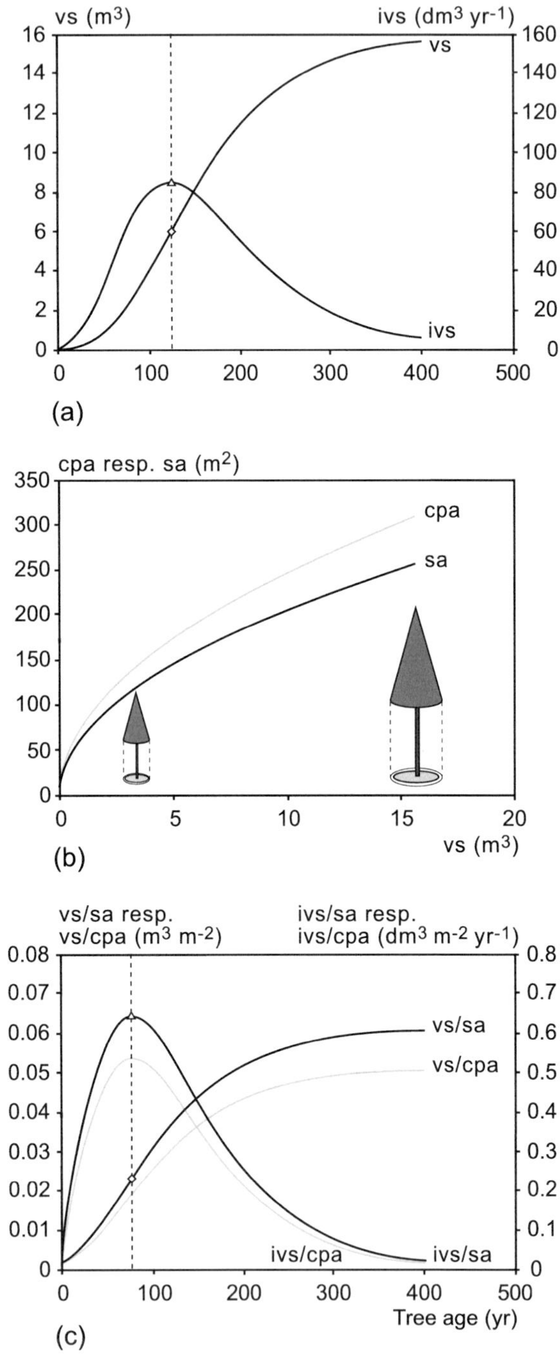


**Fig. 6.1** Schematic representation of the development of a solitarily growing individual tree. (**a**) Growth of stem volume, ivs, and development of volume yield, vs, over tree age. (**b**) Allometric dependency of crown projection area and tree stand area, sa, on stem volume, vs. In this example, we assume that tree stand area and crown projection area are equally scaled in relation to tree volume ($sa \propto cpa \propto v^{1/2}$). Just the allometric factors, $a$, in $cpa = a_1 \times vs^{1/2}$ and $sa = a_2 \times vs^{1/2}$, which indicate the tree shape, differ ($a_1 \neq a_2$). (**c**) Development of volume growth and volume yield per area. ivs/sa and vs/sa reflect the tree volume growth and yield per tree stand area (*bold lines*), while ivs/cpa and vs/cpa represent the tree growth and yield per crown projection area

stem volume, abbreviated to vs. These curves represent the characteristic individual tree ontogeny and were originally derived theoretically from basic relationships between tree assimilation and respiration (von Bertalanffy 1951). However, many works (e.g. Ryan and Yoder 1997; Yoder et al. 1994) show that it is rather nutrient limitation, genetic changes in meristem tissue, and hydraulic limitations that are behind the characteristic unimodal course of the stem volume increment, ivs, and the s-shaped curve of stem volume (see Boxes 6.1 and 6.2).

**Box 6.1 Characteristics of a Tree's Crown Morphology, Crown Extension, and Growing Area**

Description and analysis of tree structure, growth, and growth efficiency in this chapter are mainly based on the tree variables shown in Box Fig. 6.1-1. To compare tree morphology in mixed versus monospecific stands, the following ratios between tree organ sizes are frequently used:

$h/d$ = the $h$–$d$ ratio using $h$ (m) and stem diameter at breast height (cm) addresses tree stability.

cd/$d$ = the cd–$d$ ratio with crown diameter (m) and stem diameter (cm) addresses the crown extension.

$d_r/d_s$ = $d_r$–$d_s$ ratio with diameter of the main root, $d_r$, and stem diameter at breast height, $d_s$, quantifies the root-stem relationship. When based on increment cores from stem and main roots, this relationship provides insight into the allocation principle between stem and main root (Fig. 6.18).

cpa/sa = this ratio between crown projection area, cpa, and tree stand area, sa, is a measure for the interlocking of neighbouring crowns.

Most relationships between tree organ sizes change non-isometrically with tree size, i.e. when organ size $x$ increases, organ $y$ does not enlarge linearly ($y = a + \alpha \times x$) but according to the allometry equation ($y = a \times x^{\alpha}$) with $\alpha \neq 1$. This means that the above ratios between organs, e.g. $h/d$ or cd/$d$, show an ontogenetic drift with increasing size (see Box 6.3).

Raw density of wood (RDW) of species in boreal and temporal forests ranges mainly between 350 and 650 kg m$^{-3}$ (Pretzsch 2009, p 67). RDW can be used to transform stem volume, vs, to stem mass, ms (ms = vs × RDW).

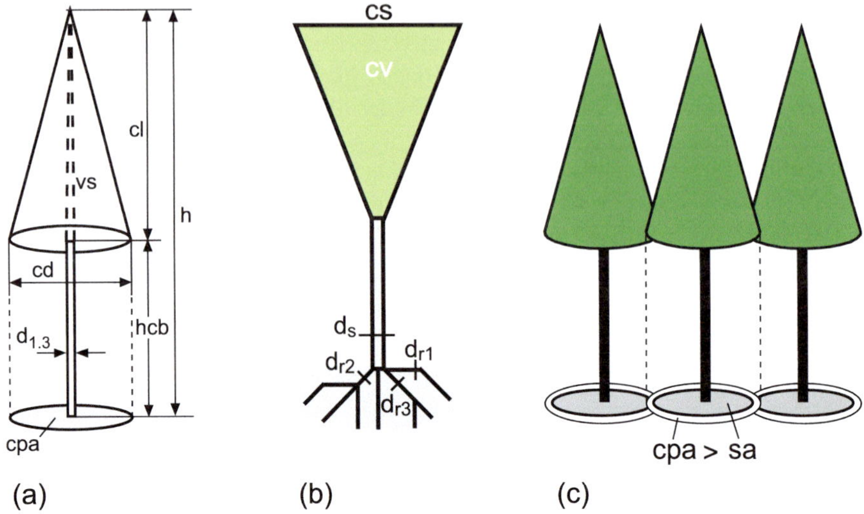


**Box Fig. 6.1-1** (**a**–**c**) Essential tree characteristics. (**a**) Tree height, h; crown length, cl; height to crown base, hcb; crown diameter, cd; tree diameter at breast height, $d_{1.3}$; and stem volume, vs. Variable cpa represents the crown projection area. (**b**) Crown volume, cv; crown surface area, cs; diameter of the stem, $d_s = d_{1.3}$; and diameter of the three tallest roots, $d_{r1} \ldots d_{r3}$. (**c**) Tree stand area equivalent to growing area, sa, and crown projection area, cpa. Particularly in densely packed stands, cpa is often greater than sa

**Box 6.2 The Rules by Kleiber (1947) and Rubner (1931) for Explaining the Ontogeny and the Growth and Yield Curves of Individual Trees**

According to these rules, the unimodal course of the net growth (Fig. 6.1a) results from the difference between the anabolistic term ($A = a \times v^{3/4}$) as minuend and the catabolistic term ($R = b \times v^1$) as subtrahend ($NG = a \times v^{3/4} - b \times v^1$). The variable $A$ represents the gross assimilation, $R$ the respiration, and NG the net growth of the tree. For the anabolistic term, the rule by Kleiber (1947) assumes that $A$ scales with 3/4 over plant mass or volume. According to Rubner's rule (1931), respiration increases linearly with mass or volume (plant mass $\propto$ plant volume). These two underlying allometric relationships are shown together in Box Fig. 6.2-1a and reflect that the net growth (grey area between the curves) is low for small trees, high for medium-sized trees, and low again for giants. This in turn reveals that plant size is a conflicting trait. It entails both supremacy over smaller neighbours and also low growth due to high expense for respiration. Intra- and interspecific interactions can affect both $A$ and $R$ and thus modify the trees' development. Box Figure 6.2-1b shows the characteristic unimodal curve of the net growth,

(continued)

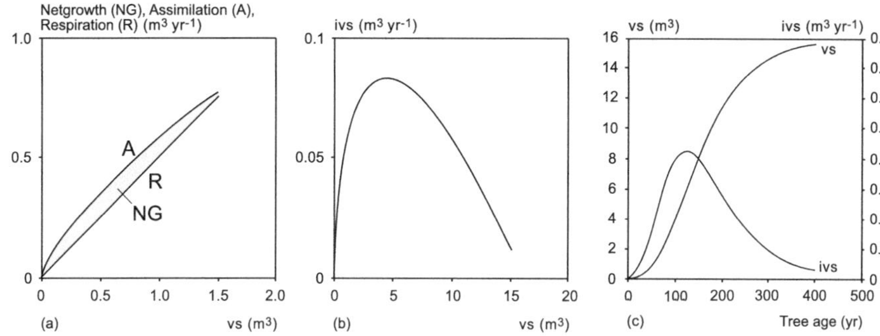


**Box Fig. 6.2-1** (**a–c**) Derivation of the unimodal growth and s-shaped yield development of a tree over age from the allometric rules by Kleiber (1947) and Rubner (1931). (**a**) Net growth (*grey* area between the curves, NG) is the difference between gross assimilation (*concave line*, $A = a \times v^{3/4}$) and respiration (*straight line*, $R = b \times v^1$). (**b**) Development of net plant growth, ivs = NG, over plant volume, vs. The latter curve results from the relationships between $A$ and vs and between $R$ and vs in Box Fig. 6.2-1a. (**c**) Course of plant volume, vs, and plant volume increment, ivs, over age, developed from the relationship between ivs and vs in Box Fig. 6.2-1b assuming an initial plant volume of 0.001 $m^3$

**Box 6.2** (continued)

ivs, over tree volume, vs, which results from that in Box Fig. 6.2-1a. Starting with the unimodal ivs–vs relationship in Box Fig. 6.2-1b and assuming an initial stem volume for plant age 1 (e.g. $v_1 = 0.001$ $m^3$), the course of volume increment, ivs, and the achieved volume, vs, can be derived (Box Fig. 6.2-1c). The development of ivs over age is the unimodal growth curve; the sigmoid development of vs over age results from integration of ivs.

While the shown unimodal course of the stem volume increment, ivs, and the s-shaped curve of stem volume, abbreviated to vs, are a good basis and reference for analysing any external effects such as intra- and interspecific competition and facilitation on tree growth, their causal explanation has been improved. Ryan and Waring (1992) showed that respiration of older and taller trees generally declines as growth declines (i.e. $R << b \times v^1$) because more than half of the respiration is growth respiration. Even maintenance respiration of big trees typically goes down (Yoder et al. 1994). Many works (e.g. Ryan and Yoder 1997) show that it is rather nutrient limitation, genetic changes in meristem tissue, and hydraulic limitations that are behind the characteristic unimodal course of the stem volume increment, ivs, and the s-shaped curve of stem volume. Although it is temptingly simple, the theoretical explanation of individual tree ontogeny by von Bertalanffy (1951) based on the rules by Kleiber (1947) and Rubner (1931) is increasingly called into question by empirical findings.

Figure 6.1a represents the growth through 'the plant's eye view', i.e. the development of growth and size. The production ecology represented by the 'forester's eye view' asks for the growth and volume per growing area, i.e. for the productivity. To transition from tree growth to tree growth per growing area, we apply the allometric relationship between stem volume and tree growing area ($sa \propto vs^{1/2}$) which is shown in Fig. 6.1b (West et al. 2009). By dividing the increment, ivs, and volume, vs, by the required growing area, we get the growth per unit area (productivity ivs/sa) and standing volume per unit area (standing stock vs/sa) of the tree (Fig. 6.1c).

Note that the transition from tree growth to tree productivity is based on growing area, sa, and not crown projection area, cpa. Unless cpa = sa, crown area productivity (ivs/cpa) differs from growing area productivity (ivs/sa) (Gspaltl et al. 2012). Particularly in mixed stands, cpa is often larger than sa as illustrated in Fig. 6.1b. In this case, ivs/cpa << ivs/sa, i.e. upscaling from tree to area level based on crown projection area underestimates tree productivity. The transition from tree growth to tree growth per area means that the time point of culmination and inflexion shifts towards younger tree ages as the required stand area of a tree increases with size (compare broken vertical lines in Fig. 6.1c and a).

The depicted unimodal and s-shaped development courses of a solitary tree in terms of growth per plant and growth per unit area (Fig. 6.1a and c) form the reference for the subsequent introduction to the effects of competition and facilitation.

## 6.2 Interaction Between Trees in Inter- Versus Intraspecific Environments

Suppose that of three equally sized trees established on similar sites, one tree grows solitarily, one in medium, and the third in maximum density with neighbours of the same species (Fig. 6.2a, trees 1–3). When we choose the solitary tree (tree 1) as a reference and set its growth rate to 1.0, what will the growth rate of the trees growing in medium and high stand density be and what does this depend on? On the one hand, the individual growing area becomes smaller from trees 1 to 3 so that the competition for resources (water, light, nutrients) increases. Helms (1998, p 34) defined competition as '... the extent to which each organism maximizes fitness by both appropriating contested resources from a pool not sufficient for all, and adapting to the environment altered by all participants...'. However, living in association with neighbours may also have advantages over solitary growth. For example, neighbours may protect against windthrow (Knoke and Hahn 2007; Griess and Knoke 2011), improve the soil conditions (Augusto et al. 2002; Heinsdorf 1999), or prevent nearest neighbours from sun scorch of bark or drought (Assmann 1970; Pretzsch et al. 2012b). The latter abiotic or biotic improvement of the neighbours' living conditions is called facilitation (Vandermeer 1992). Vandermeer (1992, p. 9) defined facilitation as '... the process in which two individual plants or two populations of plants interact in such a way that at least one exerts a positive effect on the other. Double facilitation is equivalent to

**Fig. 6.2** Schematic representation of the effect of competition and facilitation in (**a**) a monoculture and (**b**) a mixed stand on tree growth under solitary growing conditions (*left*) and in medium (*centre*) and high stand density (*right*). The outer hulls of the crown bodies (white envelopes) represent the current growth rate of the trees in question. From left to right (from trees 1 to 3 and 4 to 6, respectively) competition becomes stronger as does facilitation. As a result, trees 2 and 5 encounter medium density but can have higher growth rates (white envelopes) than trees 1 and 4 under solitary conditions

mutualism'. In Fig. 6.2a, the trees' growth rates (indicated by the white envelopes of trees 1–3) when growing in solitary conditions and in medium and high density reflect the net effect of competition and facilitation on tree growth. In relation to the growth of the solitary tree 1, the interaction between competition and facilitation results in a net facilitation under medium density (Fig. 6.2a, envelope tree 2 > envelope tree 1) and a net competition effect (Fig. 6.2a, envelope tree 3 < envelope tree 1) in the case of high density.

Analysing, understanding, modelling, and prognosticating monocultures often entail a thorough consideration of the effect of competition between neighbouring trees but mostly ignores facilitation. A plant may be facilitated by its neighbour through improvement of environmental factors (non-resource) or resource supply or a combination of both. Examples for improvement of environmental factors are slowing down the wind speed and avoidance of storm breakage, shading and avoidance of sunburn, providing a barrier against spread of insects, suppression of forest floor vegetation and competing weeds, and protection against snowslides or browsing. This kind of facilitation does not entail direct costs in terms of resources for the benefactor. In contrast, when facilitation is based on an improved resource supply, the benefactor may lose resources while the beneficiary gains.

Competition and facilitation act simultaneously. It is difficult to separate them. In field experiments in monospecific stands, the net effect of the interaction between facilitation and competition on tree structure and growth rather than their

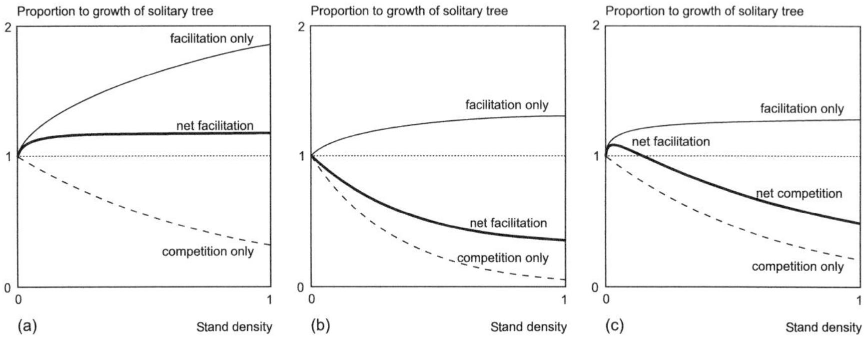


**Fig. 6.3** Schematic representation of the modification of solitary growth (1.0-line) through facilitation and competition in dependence on stand density in monocultures. The graphs show three essential patterns of how net facilitation and net competition result from the interplay between facilitation only and competition only. (**a**) Facilitation only has the upper hand in relation to competition only so that net facilitation is effective over the whole range of stand densities. (**b**) Competition only has the upper hand in relation to facilitation only so that net competition is effective over the whole range of stand densities. (**c**) The balance between competition only and facilitation only results in net facilitation under low stand density and net competition under high stand densities

particular effects is accessible through measurement. To simplify matters, the interplay between competition and facilitation is first illustrated for monospecific stand conditions. In Fig. 6.3 we set the growth in solitary conditions as the reference (1.0-line) and show three basic patterns of how facilitation and competition can modify this solitary growth. Remember that in terms of absolute growth, a solitary tree follows the trajectories shown in Fig. 6.1 and that we ask here how neighbourhood might modify this generic trajectory.

Figure 6.3a assumes that competition only increases from low to high stand densities and that facilitation also only increases with density. Given that in this theoretical case, the balance between the negative effect of competition and the positive effect of facilitation on growth is always positive, we consider net facilitation (facilitation only minus competition only) over the whole range of stand densities. Figure 6.3b represents the opposite case where competition only always has the upper hand and net competition prevails over the whole range of stand densities. These two reaction patterns with net facilitation or rather net competition over the whole range of densities tend to be exceptions. Figure 6.3c represents the most frequently observed reaction pattern, where net facilitation dominates under low densities and net competition is predominant in dense stands. In sparsely stocked stands, competition is still low enough to be compensated or even overcompensated by facilitation. Such low density conditions, where facilitation may have the upper hand, are frequently observed in grassland ecosystems which are too poor for the development of closed forests (Callaway and Walker 1997; Canham et al. 1994, 2004). But they are less relevant in forest ecosystems (Forrester 2013). As stand density increases, growth mostly declines, can hardly be balanced by facilitation, and causes mostly net competition. However, this consideration highlights that although not easy to detect and to separate from competition, facilitation may also play a role in forest stands with high stand density. The facilitative effect of medium density compared with high density or solitary growth can be revealed by Nelder trials (Nelder 1962) where the tree growth can only be at optimum between the extremes (Uhl et al. 2015).

The combined effect of competition and facilitation becomes even more complex in mixed stands where trees differ in terms of traits and niches. Suppose the above trees in solitary and medium- and high-density constellations grow in mixed as opposed to monospecific stands (Fig. 6.2b, trees 4–6). In this example, 50% of the neighbours belong to a second species, i.e. competition and facilitation become interspecific and modified compared with monospecific stand conditions.

Reduced competition and an even stronger facilitation caused by species mixing might result in an even more obvious growth maximum under medium stand density (tree 5, envelope tree 5 >> envelope tree 4) compared with growth rates under solitary conditions or growth rates under maximum density.

Figure 6.4a indicates this shift in the growth-density relationship due to a reduction in competition (e.g. by spatial or temporal niche complementarity) and an increase in facilitation (e.g. tapping nutrients in deep soil layers which are favourable for species 1) through admixture of trees of species 2. Such a reaction pattern can cause a widening of the density range with net facilitation (horizontal arrow). In other words, the range where net competition has the upper hand narrows. For mixed-species stands as in monospecific stands, a net competition which is normal for trees growing in medium or high stand density does not exclude the effect of facilitation. In both cases, competition merely has the upper hand

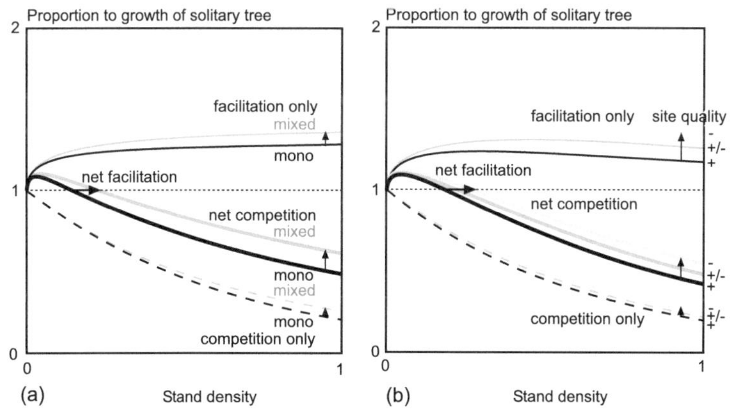


**Fig. 6.4** Schematic representation of the modification of solitary growth (1.0-line) through facilitation and competition in dependence on stand density in mixed-species stands. (**a**) Broadening of the range of net facilitation (*horizontal arrow* pointing right) because of additional facilitation and competition reduction through species mixing. (**b**) Increasing net facilitation (*horizontal arrow* pointing right) due to the increase in facilitation from mild (+) to harsh (−) environmental conditions

over facilitation. The relationship between the net tree growth in mixture and in monoculture (vertical arrows in Fig. 6.4a) reflects the mixing effect in terms of tree growth.

Facilitation effects are widely held to be strong under harsh environmental conditions and weaker under favourable conditions (Brooker et al. 2008; Callaway and Walker 1997). Accordingly, the density range with net facilitation may increase as site quality decreases (see Fig. 6.4b curves referred to as +, ±, −). Note that it is also possible for the net curve to remain the same although its components change in the case that the positive and negative changes in competition and facilitation cancel each other out.

While in plant ecology the net effect of competition and facilitation is mostly related to solitary plant growth (Callaway and Walker 1997; Belsky and Canham 1994; Forrester 2013), in closed forests we are interested in the mixing effects under mean or high stand density. In field experiments in monospecific stands, the net effect of the interaction between facilitation and competition rather than their specific effects are accessible through measurement. To complicate the situation, measurements of solitary plant or tree growth which are required as a reference are scarcely available, except from very rare experiments like the outer Nelder circles (Nelder 1962; Uhl et al. 2015) or spacing and thinning experiments, which include plots with open-growing trees. So, in most cases only growth of medium to high density plants is measured, and it is difficult to quantify whether net facilitation or competition has the upper hand because of the lack of reference measurements. The same restriction applies for quantification of net facilitation or net competition in mixed stands in relation to solitary plant growth (according to the definition of facilitation and competition). However, what can instead be easily detected in field experiments is the change in the net curve from monospecific to mixed stands (see upwards shift of the net curves in the centre of Fig. 6.4a and b marked with vertical arrows).

The relationship between the tree growth in mixed versus monospecific stands indicates the net mixing effect in closed stands. Of special interest is to what extent mixing increases or reduces net competition of trees in closed stands rather than whether the growth-density relationship shows net facilitation or net competition of a solitary tree. The former would mean an increase or decrease in productivity of the species in mixed versus monospecific stands. This practically relevant effect is represented by the relationship between mixed and monospecific stand performance in the right branch or even at the right end of the net growth-density relationship in Fig. 6.4.

## 6.3 Basic Feedback Loop Between Growth, Structure, and Local Environment in Forest Stands

The tree–tree interaction illustrated by an example in the previous section can be conceptualised by the basic feedback loop between functioning, structure, environment, and functioning in forest stands (Fig. 6.5). In order to explain the emergence

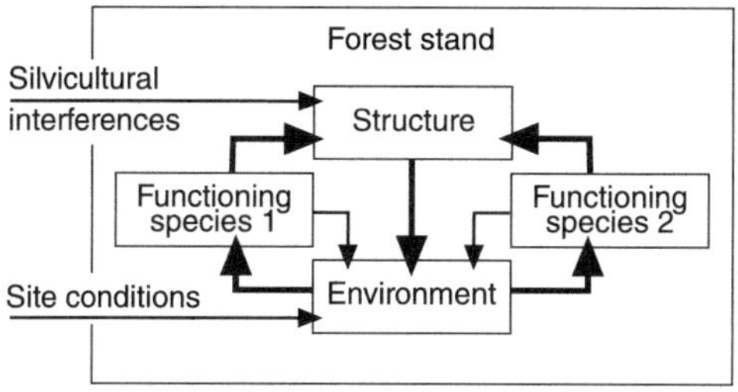


**Fig. 6.5** Feedback loop between stand structure, environmental conditions, and tree functioning in a two-species stand. The outer feedback loops (FSE loop) structure → environment → functioning → structure (*bold arrows*) are slow, and the inner loops (FS loop) environment → functioning → environment work faster. Further explanation is provided in the text

of intra- and interspecific interactions between neighbouring trees, a distinction between their functioning (e.g. germination, growth, defence, flowering, fruiting, senescence), their structure formation (e.g. body size, crown and root extension, crown space filling with foliage), and environment (resource supply, e.g. water, nutrient, or $CO_2$ supply) and environmental factors (e.g. temperature, salinity, browsing) is helpful.

The conceptual model in Fig. 6.5 clarifies essential plant–plant interactions in monospecific and mixed-species communities. Trees interact with their environment in two ways, via structure and functioning (Hari 1985). The feedback between functioning and environment (FE loop) can be very rapid and temporary, e.g. defence substances are exuded quickly after or nearly simultaneously to the injury or pathogen attack and are reduced after stress release (within-stand environment versus external environmental drivers). Reduction of atmospheric $CO_2$ concentration or soil water supply by roots and crowns of fast-reacting neighbours can immediately reduce photosynthesis or growth of slower-reacting trees.

The feedback between functioning, structure, and environment (FSE loop), in contrast, is slow and accumulative. The functioning changes the structure (e.g. tree size and structure) and, via tree and stand structures, the trees' environment. The crown structure, for instance, determines where the water and mineral nutrients from the crown periphery drip to and to what extent neighbours are shaded. The pattern of water drip and shading determines where the roots and crowns of neighbours grow and where they forage. This in turn determines their morphological structure. Trees develop and maintain their root, stem, and crown structure for decades and affect and adapt to their surroundings rather permanently.

The feedback between environment, growth, and structure may be relatively clear in monospecific stands where the species apply similar tricks and traits when appropriating contested resources, adapting to the environment, and modifying the stand structure. In mixed stands with two species as shown in Fig. 6.5, or with even more species, interactions and their effects on growth, structure, and environment may be more varied. Mixed-species modify their environment to their own benefit, e.g. by overtopping the neighbouring crowns or penetrating neighbouring roots in order to improve access to contested resources and acclimation to the environment altered by their neighbours. The principle feedback remains the same when neighbours belong to different species; however, their tricks and traits to modify their environment may be different, so that a broader range of reaction patterns, structures, and changes to the environment may occur.

The subsequent analyses of tree structure and tree growth in mixed versus monospecific stands address the two key elements of the feedback loop. Long-term experiments provide retrospective information on growth and structure but rarely time series on environmental variables such as light profile, nutrient supply, or water uptake of trees in mixed versus monospecific stands. Therefore, the effect of mixing on the local environment is less known and understood than the effect on structure and growth. However, future approaches will close this knowledge gap to enhance understanding of the feedback and underlying mechanism as a whole (Binkley et al. 2004; Pretzsch et al. 2015).

## 6.4 Growth and Yield at the Tree Level

Any changes to the local environmental conditions in mixed stands compared with monospecific stands primarily determine tree functioning in terms of growth (Fig. 6.5). This section will deal with these growth responses to mixing. Reduction or increase of the growth rate due to mixing can modify the tree and stand structure. In addition, the local environment may considerably modify the growth allocation in terms of tree morphology and thereby the tree and stand structure. Such structural changes which have a feedback effect on the local environment will be the subject of Sect. 6.5.

### 6.4.1 Dominant Rather Than Open-Grown Trees as Reference

Grassland plant ecology mostly uses solitarily growing plants as a reference for detecting facilitation or competition effects (Canham et al. 1994, 2004; Forrester 2014). When association with neighbours increases plant growth above the level of solitary plants, this indicates facilitation. When the opposite occurs, this serves as evidence of competition. Using open-grown trees as a reference for mixing effects such as facilitation or competition reduction is barely feasible given that these are rare in woodlands and, when situated in distant open land, are mostly incomparable in terms of site characteristics.

In Fig. 6.6 we therefore use dominant trees (d) in monocultures as a reference instead of open-grown trees (o). The former have lower growth rates than open-grown trees, but long-term experiments in monocultures provide a sound database for predicting the potential growth of dominant trees in forest stands in dependence on the site conditions. In the following, we use the unimodal diameter growth-tree diameter curve expected for dominant trees in monocultures as a reference for

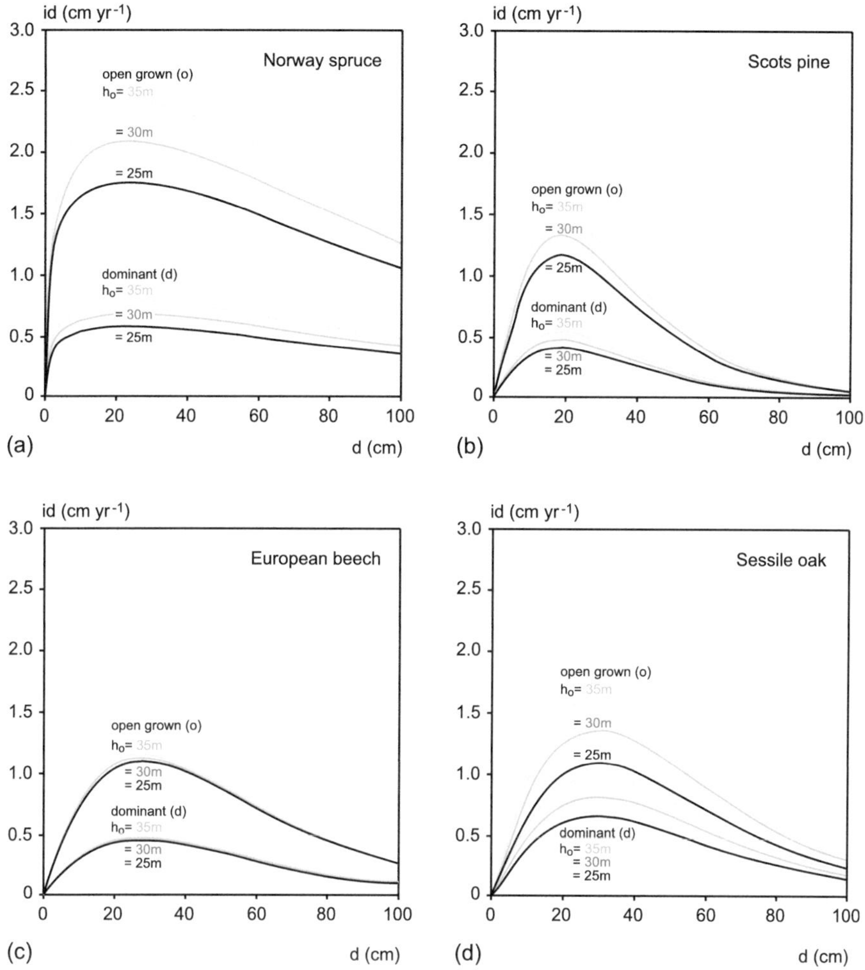


**Fig. 6.6** Tree diameter increment along the tree diameter and site spectrum covered by data from long-term experimental plots in southern Germany according to Pretzsch and Biber (2010). Model curves are given for trees growing under solitary conditions (*upper curve triplets*) and for dominant trees growing in stands (*lower curve triplets*). The *light grey*, *grey*, and *black lines* represent the diameter growth for fertile, mean, and poor sites (dominant height of 35 m, 30 m, and 25 m, respectively, at stand age 100 years)

revealing the competition and species admixture effect. For derivation of the species-specific unimodal potential diameter growth-diameter relationships, see Pretzsch and Biber (2010). Following the approach from Sect. 6.2, we use this potential curve (the curve for dominant trees in monospecific stands) as the 1.0-line to show any mixing effects. Of main interest is how stand density and mixing proportion lead growth away from this baseline.

## 6.4.2 Evidence of Facilitation and Competition Effects at the Tree Level

Figure 6.7 shows exemplarily for even-aged mixtures of Norway spruce/European beech, European beech/sessile oak, and Scots pine/Norway spruce how mixing can modify the individual tree diameter growth under ceteris paribus conditions, i.e. when all effects (tree size, stand density, or site conditions) except for species composition are statistically eliminated. The analysis is based on long-term mixing experimental plots in Bavaria (Pretzsch 2009). Established in the 1930s, they

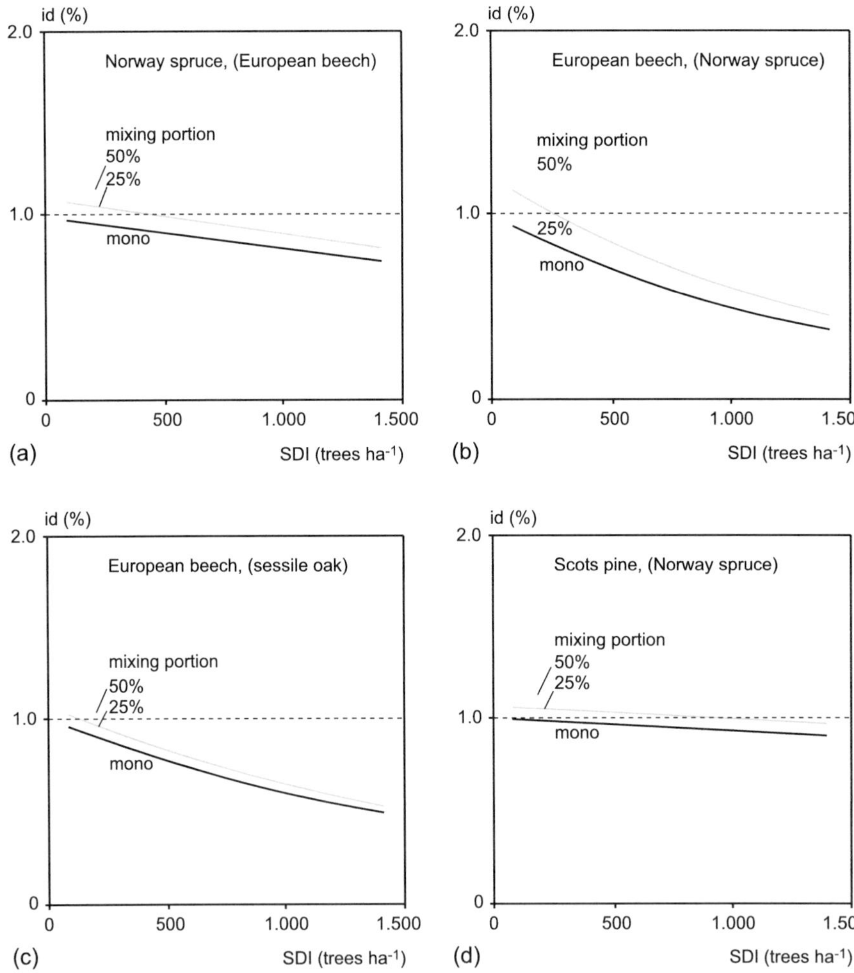


**Fig. 6.7** Relative tree diameter increment, id, of various tree species in dependence on the mixing portion (monospecific stand, 25%, 50% admixture of a second tree species) and the given local SDI on the $x$-axis. The 1.0-line serves as a reference and represents the growth of a dominant tree in monospecific stands of the respective species

mostly include reference plots in monospecific stands, provide individual tree data (tree position, tree size, growth rate) of several thousand trees in mixed and neighbouring monospecific stands, and thus provide information about mixing effects on tree growth.

The underlying model for the reaction patterns shown in Fig. 6.7 considers that growth of trees in mixed and monospecific stands may simply differ because of different size. Given space restrictions it is introduced here only briefly: the size effect is taken into consideration by the unimodal relationship between growth and size in monospecific stands (Fig. 6.6). This unimodal curve expected for dominant trees in monospecific stands is used as a reference to reveal the effects of species mixing and local stand density. Of main interest is how mixing proportion and stand density lead growth away from the baseline (1.0-line). Competition was quantified in the model using the local stand density index, i.e. the number of trees, $n_i$, within a defined radius around each tree scaled up to a hectare ($N_i = n_i/(r_i^2 \times \pi)$) and converted to the $SDI_i = N_i \times (25/d_q)^{-1.605}$ according to Reineke (1933). This total SDI was split into two components, the part containing the species equal to the species 1 (= species of the centre tree) and the rest. The mixing portion was therefore the ratio between the SDI portion of trees differing from the species of the centre tree and the total SDI. So a mixing portion of mix = 0 signifies a monoculture composed of species equal to the centre tree, whereas mix = 0.25 indicates that within the circle of radius r around the centre tree, 25% of trees are unequal to species 1. We considered two-species stands, so that mix = 0.25 or 0.50 means an admixture of 25 or 50%, respectively, of trees of species 2 in the vicinity of species 1. To show the tree reaction patterns at the tree level, we applied a model which considers the size effect on growth (id–d relationship) and the stand density effect (id-SDI relationship) so that the effect of inter- versus intraspecific environments on growth under ceteris paribus conditions can be eliminated. The model was fitted with individual tree data from *N.* spruce ($n$ = 10,160), E. beech ($n$ = 13,861), pine ($n$ = 2052), and oak ($n$ = 4960) growing in monospecific and mixed stands.

In all tree associations shown in Fig. 6.7, stand density reduces tree growth; the reduction is very strong in the case of beech and rather small in the cases of oak and pine. Even more variable is the mixing reaction between the species combinations investigated. In mixture of Norway spruce and European beech (Fig. 6.7a, b), both species show a reduction in competition and increase in growth compared with the monospecific stand (black solid line). Norway spruce benefits moderately from beeches in the neighbourhood (Fig. 6.7a), while European beech benefits strongly from Norway spruce (Fig. 6.7b). In the mixture with sessile oak, European beech again benefits considerably (Fig. 6.7c), while tree growth of oak is hardly affected (not shown). Scots pine is the beneficiary in the mixture with Norway spruce (Fig. 6.7d), and the benefit gained by Scots pine is partly at the expense of Norway spruce (not shown). Of the species considered, European beech experiences the strongest growth reduction through increased stand density index (SDI) but also the greatest benefit through admixture of other species like Norway spruce and sessile oak. The results suggest that mixing reactions vary depending on both the particular species investigated and the species which is added in its vicinity.

Important is the finding that the positive mixing effects in terms of growth are modified by stand density. The mixing effects are present to various extents all along the density range. However, they can be at maximum in sparsely stocked stands and decrease with increasing stand density (from left to right in Fig. 6.7). Under low density, facilitation and competition can raise the growth above the potential for dominant trees in monospecific stands (1.0-line), i.e. the net effect of mixing can be positive. Of course, at the latest when density gets very close to zero, there is no longer any facilitation or competition caused by trees. In dense stands, the increase from monoculture to 25% and 50% is still relevant, but the positive mixing effect cannot overcompensate the negative effects of density on tree growth, so the combined effect of mixing and density on tree growth is negative.

### 6.4.3 Basic Growth Reaction Patterns in Intra- Versus Interspecific Neighbourhoods

The last section showed that—depending on the mixing portion and the degree of stand density—living in association can increase or decrease the growth rate. This means that the unimodal course of individual tree growth is modulated depending on the prevailing local environmental neighbourhood conditions of the tree. Figure 6.8 is a schematic representation showing that facilitation and competition in inter- and intraspecific neighbourhoods can modify the volume growth in manifold ways. In contrast to an open-grown tree (o), dominant trees (d) in forest stands grow permanently under net competition. For the dominant tree (d) in Fig. 6.8a we assume a constant reduction in the solitary tree growth by 35% due to a sequestration of resources by neighbours. The trajectories of dominant trees (d) are more suitable as a reference for indicating mixing effects than are open-grown trees (o) (see Sect. 6.4.1). However, in Fig. 6.8b–e we include both open-grown and dominant trees as references.

Figure 6.8b shows how permanent facilitation, e.g. through wide spacing, can increase the growth of an individual tree (bold line) permanently above that of the solitarily open-grown tree (o). In this case, the balance between the respective positive and negative effects on the growth of competing and facilitating neighbours is positive. Trees growing in group structure in the alpine zone compete for light, but their neighbours also protect them against snow and wind so that they frequently benefit from being associated and can grow more than solitary trees. Forest management makes use of such a permanent facilitation by planting trees in groups or clusters known as 'Rotten' in the alpine zone (Strobel 1995) or 'Nester' in the lowlands (Saha et al. 2012).

The pattern in Fig. 6.8c (bold line) occurs frequently when a dominant tree benefits from neighbours of a different species through competition reduction (e.g. the effect of a rather translucent species) or facilitation (e.g. the effect of an atmospheric nitrogen fixing species or hydraulic water redistribution). In this case, the growth of a dominant tree might be accelerated permanently, and its superiority may extend over the whole tree age.

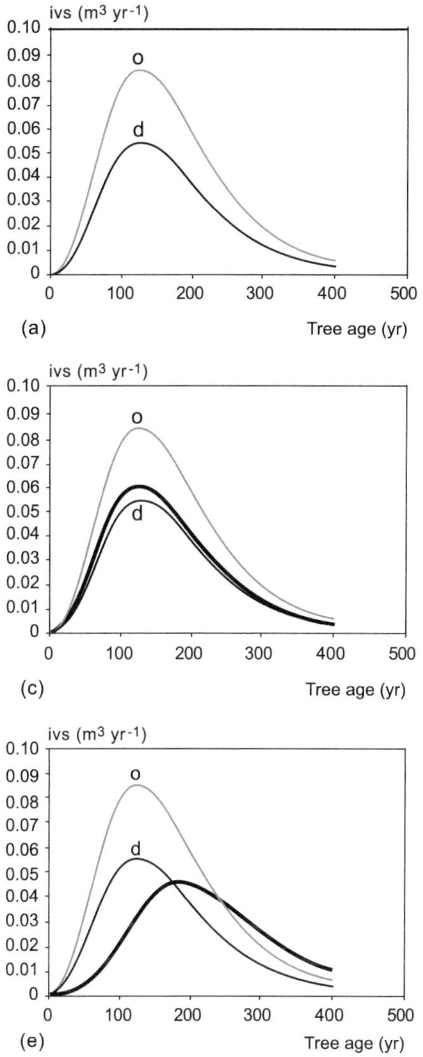


**Fig. 6.8** Generic patterns of tree volume growth in dependence on age, modified by competition and facilitation. The *bold lines* represent the response to competition and facilitation. The thin curves for open-grown trees (o) and dominant trees in a stand (d) serve as references. (**a**) Behaviour of an open-grown tree (o) and a dominant tree (d) in a forest stand. (**b**) Facilitation can increase tree growth beyond open-growth level. (**c**) Mixture can increase dominant tree growth through facilitation and competition reduction. (**d**) Mixture can increase and accelerate or (**e**) decrease and slow down growth in relation to dominant trees

The net effect of competition and facilitation may vary over time. Figure 6.8d (bold line) shows a frequently observed acceleration in growth through facilitation in the early state, followed by a reduction in the advanced state of plant development. So, tree growth exceeds the reference (d) in the early phase and falls behind in the late development phase. For example, in mixture with beech, larch growth can be accelerated in the juvenile state through competition reduction, whereas it can be slowed down by European beech in the mature phase due to the increasing height, crown extension, and overshading effects. This means an earlier culmination of volume growth due to the rather solitary conditions of an early-successional species when combined with a late-successional species.

Figure 6.8e (bold line) shows the opposite case. Neighbouring trees might reduce the growth level compared with dominant trees and delay the time of culmination. Suppression of Norway spruce by Scots pine is a common example of this growth response pattern in species mixing (Wiedemann 1951; Christmann 1939).

## 6.5 Crown Size and Morphology in Mixed Compared with Monospecific Stands

### 6.5.1 Interspecific Variation in Crown Shape and Size

Tree species can differ considerably in crown shape and size (Fig. 6.9). The allometric equation in logarithmic ($\ln(y) = a + \alpha \times \ln(x)$) or untransformed version ($y = e^a \times x^\alpha$) can be used to compare different species regarding their crown size and scaling of the crown to tree size (Niklas 2004). Factor a represents a multiplicative effect of species on the crown expansion. Species-specific differences of exponent $\alpha$ signify an exponential difference in the dynamic of the crown expansion with increasing size (Box 6.3).

To show the interspecific variation in tree growing area, we used the dataset compiled by Pretzsch and Dieler (2012) which includes 126 yield tables of 52 species, of which 30 are of angiosperm and 22 of gymnosperm taxonomy. Species included the genera *Abies*, *Acer*, *Alnus*, *Betula*, *Carpinus*, *Castanea*, *Cunninghamia*, *Eucalyptus*, *Fagus*, *Fraxinus*, *Juglans*, *Larix*, *Nothofagus*, *Picea*, *Pinus*, *Populus*, *Prunus*, *Pseudotsuga*, *Quercus*, *Robinia*, *Shorea*, *Thuja*, and *Tilia*. From these yield tables, we derived among others the mean tree diameter, $\bar{d}$, and

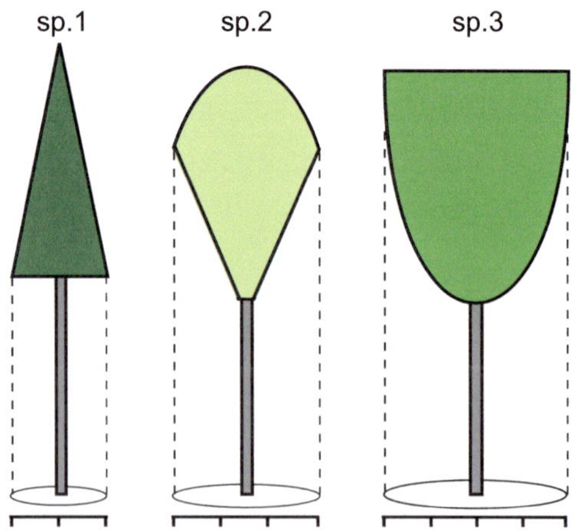


**Fig. 6.9** Tree crown shapes can resemble slim cones (sp. 1), opened umbrellas (sp. 2), or very wide bowls (sp. 3). Correspondingly, their crown radius, crown diameter, and projection area can vary considerably even for trees of equal stem diameter

tree stand area, $\overline{sa}$. The dataset includes only yield tables with self-thinning conditions or light and moderate thinning regimes where crown cross-section area, $\overline{cpa}$, is scaled proportionally to tree stand area, $\overline{sa}$, ($\overline{cpa} \cong \overline{sa}$). Note that light and moderate thinning permanently maintain a stand's canopy in a structure in which the tree crowns touch each other but barely overlap (Verein Deutscher Forstlicher Versuchsanstalten 1873, 1902).

Figure 6.10a shows the considerable variation in the allometry between stand area and tree diameter; the intercepts amount to $a = -1.96, -3.57, 2.47$ (mean, min, max) and the slopes to $\alpha = 1.47, 0.14, 2.33$ (mean, min, max). According to Fig. 6.10, angiosperms ($\ln(sa) = -1.33 + 1.36 \times \ln(d)$) need on average more growing area than gymnosperms ($\ln(sa) = -2.03 + 1.52 \times \ln(d)$). Angiosperms with a tree diameter of $d = 25$ cm, for example, occupy on average $sa = 19.51\,m^2$, while gymnosperms occupy just $sa = 16.90\ m^2$, i.e. significantly ($p < 0.05$) less tree stand area. The fact that species can differ in the intercept of their allometric crown relationship is represented, e.g. in yield tables, by species-specific tree numbers per ha at a given stand age (Assmann and Franz 1965) and by species-specific levels of the self-thinning lines (Pretzsch 2006).

The species-specific growing space requirement is relevant for calculation of the mixing proportions, stand density, and quantification of mixing effects on growth. If the species-specific growing space requirements are not considered and the calculation of the mixing proportions is simply based on mean tree stand area or tree numbers, this can result in flawed diagnoses of overyielding at the stand level and especially at the species level (see Chap. 4, Box 4.3).

Note that in this section the relationships between crown width and stem diameter are represented by the allometric relationship between crown projection area, cpa,

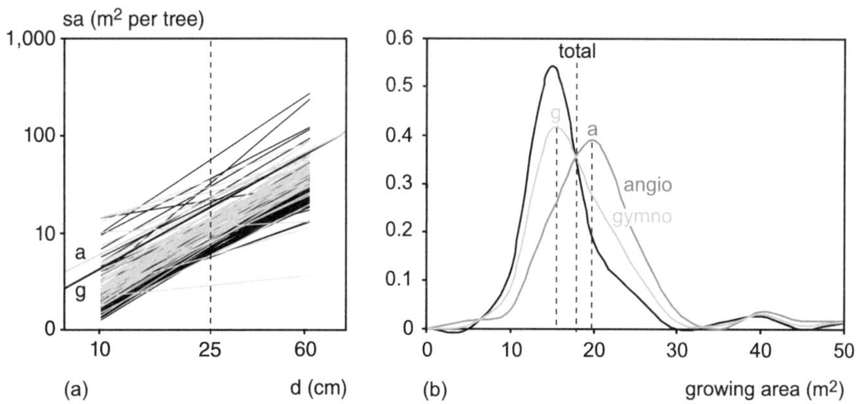


**Fig. 6.10** Overview of relationship between mean tree stand area, $\overline{sa}$, and quadratic mean tree diameter, $d_q$, in even-aged stands for 52 tree species of which 30 are of angiosperm (a and *grey*) and 22 of gymnosperm (g and *black*) taxonomy according to Pretzsch and Dieler (2012). (**a**) The mean $\overline{sa}$ -$d_q$-line of angiosperms differs significantly ($p < 0.05$) from the mean line of gymnosperms in its intercept, but not in its slope. (**b**) Mean tree stand area requirement for trees with $d = 25$ cm calculated on the basis of the species-specific $\overline{sa}$-$d_q$-allometries shown in (**a**). For further statistical characteristics, see Pretzsch (2014)

and stem diameter at breast height, d, in logarithmic ($\ln(\mathrm{cpa}) = a + \alpha \times \ln(\mathrm{d})$) or untransformed ($cpa = e^{a} \times d^{\alpha}$) version. When crown diameter, cd, or crown radius, cr, is required, they can be calculated as follows: $cd = 2 \times \sqrt{cpa/\pi}$ and $\mathrm{cr} = \sqrt{\mathrm{cpa}/\pi}$, respectively.

## 6.5.2 Intraspecific Variation in Crown Size

The crown size and shape of a tree are strongly determined by the local environment in the stand that has prevailed in the past and is now present. While maximum crown extension is achieved under solitary growing conditions, crown width decreases from open-grown (o) to dominant (d) and suppressed trees (s) (Fig. 6.11a). Stand density and competition can considerably modify the crown allometry, e.g. the cd–d relationship (Fig. 6.11b). The extent to which a tree may vary in crown extension in different environments is a crucial requirement for its success in coping with crowding.

In the following, we use 2346 crown projection area measurements of beech crowns in monospecific stands with spacing and thinning ranging from solitary growing conditions to moderate thinning and self-thinning, to illustrate the intraspecific variation (Fig. 6.12) (see Pretzsch 2014). The stand age ranges from 57 to 207 years; the surveys come from the experiments Arnstein 638, Gerolzhofen 627, Fabrikschleichach 15, Hain 27, Starnberg 91, Waldbrunn 105/106, and Zwiesel 111 in Bavaria in the years 1980–2004. Regressions for the upper 95% quantile ($\mathrm{cpa} = 3.03 \times d^{0.92}$), the group of moderately and heavily thinned stands ($\mathrm{cpa} = 1.05 \times d^{1.01}$) and the slightly and unthinned plots ($\mathrm{cpa} = 0.21 \times d^{1.36}$) describe the broad variation (Fig. 6.12, upper, middle, and lower regression line). According to these allometric equations, a beech with 25 cm stem diameter occupies 58 m$^2$ when

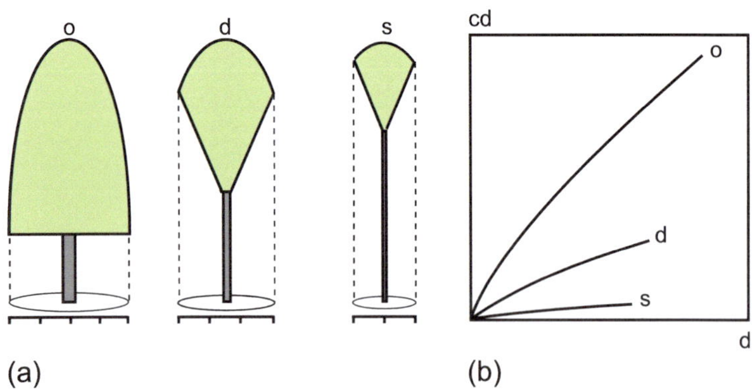


**Fig. 6.11** Schematic representation of the effect of stand density on the lateral crown expansion. (**a**) Crown width of an open-grown, solitary tree (o), a dominant tree growing under medium stand density (d), and in suppressed conditions under self-thinning conditions (s). (**b**) Allometric relationship between crown diameter, cd, and stem diameter, *d*, for open-grown, dominant, and suppressed trees

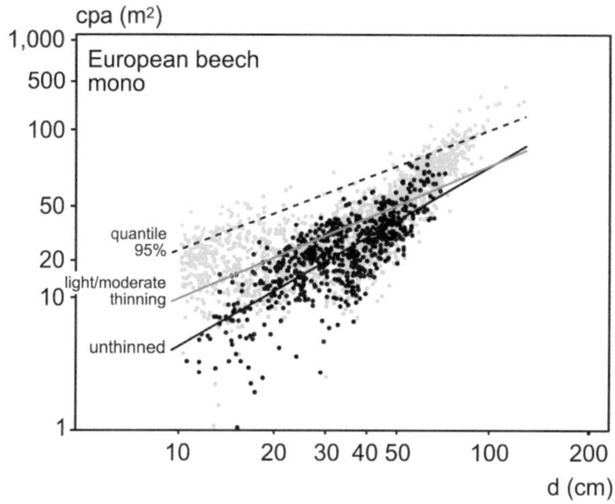


**Fig. 6.12** Allometric relationship between tree crown projection area, cpa, and stem diameter, *d*, of European beech derived from crown measurements on long-term experimental plots in Germany. As the database includes solitary trees as well as trees in thinned and unthinned stands of all ages, the ln(cpa) – ln(*d*) relationships can be derived for solitary trees (*upper line*: 95% quantile regression: $\ln(\text{cpa}) = 1.11 + 0.92 \times \ln(d)$), for trees in moderately heavily (B/C-grade, $n = 1685$) thinned stands (*middle line*: $\ln(\text{cpa}) = 0.05 + 1.01 \times \ln(d)$), and for trees in unthinned stands (A-grade, $n = 661$) (*lower line*: $\ln(\text{cpa}) = -1.59 + 1.36 \times \ln(d)$)

growing without lateral restriction, 27 $m^2$ under medium stand density, and 16 $m^2$ when growing in close to self-thinning conditions. This plasticity equips beech with high competitive strength in intra- and interspecific environments.

In terms of a species' acclimation to intra- and interspecific competition, the maximum and minimum of its crown extension are of special interest as these indicate the range of plasticity. Figure 6.13a shows exemplarily for maple the upper (95%) and lower (5%) quantile of the cpa–d allometry in addition to the mean relationship. The ranking of species in terms of their 95% quantile is different. Species such as lime tree and silver birch lose when ranked in terms of their 95% quantile, whereas rather plastic species such as beech and silver fir win in this ranking. This emphasises that deriving crown behaviour for solitary conditions from average behaviour in closed stands, especially from monocultures, is questionable. For example, beech, elm, maple, and silver fir profit considerably in crown expansion when released from competition in monospecific or mixed stands.

Figure 6.14 illustrates that crown plasticity can differ considerably between tree species. The data comes from crown measurements on long-term experimental plots in Germany and covers a broad range of tree ages and densities, in monospecific and mixed stands (Pretzsch et al. 2015; Pretzsch 2014). Based on the 95% and 5% quantile of the cpa–*d* allometry (Fig. 6.14, upper and lower lines), the following relative measure, CPL, for tree crown plasticity can be derived. Based on the lower quantile line ($\text{cpa}_{5\%} = a_{5\%} \times 25^{\alpha_{5\%}}$) and the upper one ($\text{cpa}_{95\%} = a_{95\%} \times 25^{\alpha_{95\%}}$), and using a reference stem diameter of 25 cm (where stand density in monolayered stands is rather high or even at maximum), CPL is formulated as follows:

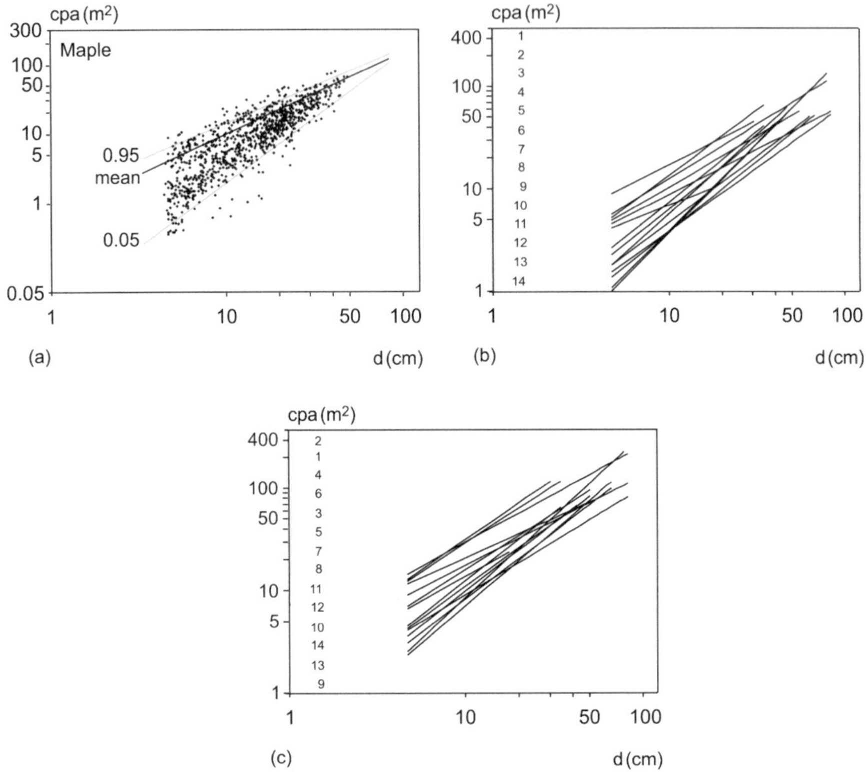


**Fig. 6.13** Crown projection area-stem diameter relationship of various tree species. (**a**) Example derivation of the upper (95% quantile), mean, and lower (5% quantile) ln(cpa)-ln($d$) relationship for sycamore maple (*Acer pseudoplatanus* L.). (**b**) Interspecific variation in the mean ln(csa) – ln ($d$) relationships. The model $\ln(\text{cpa}) = a + \alpha \times \ln(d)$ was fitted by OLS to the data and yielded the following parameters ($a$, $\alpha$) for mountain elm (1.30, 0.67), European beech (0.18, 1.02), lime tree (0.08, 0.98), sycamore maple (–1.96, 1.55), hornbeam (–0.25, 1.24), silver fir (0.37, 0.82), rowanberry tree (1.13, 0.40), black alder (–0.85, 1.22), silver birch (–1.69, 1.48), Scots pine (–1.41, 1.27), Norway spruce (–1,43, 1.20), European larch (–1.69, 1.32), sessile oak/ common oak (–2.48, 1.64), and common ash (–3.02, 1.79). (**c**) Interspecific variation in the 95% quantile of the ln (cpa) – ln($d$) relationships.[1]mountain elm (*Ulmus glabra* Huds.), [2]European beech (*Fagus sylvatica* L.), [3]lime tree (*Tilia cordata* Mill.), [4]sycamore maple (*Acer pseudoplatanus* L.), [5]hornbeam (*Carpinus betulus* L.), [6]silver fir (*Abies alba* Mill.), [7]rowanberry tree (*Sorbus aucuparia* L.), [8]black alder (*Alnus glutinosa* (L.) Gaertn.), [9]silver birch (*Betula pendula* Roth), [10]Scots pine (*Pinus sylvestris* L.), [11]Norway spruce (*Picea abies* (L.) Karst.), [12]European larch (*Larix decidua* Mill.), [13]sessile oak/common oak (*Quercus petraea* (Matt.) Liebl./*Quercus robur* L.), and [14]common ash (*Fraxinus excelsior* L.)

$\text{CPL} = \text{cpa}_{95\%,25}/\text{cpa}_{5\%,25} = (a_{95\%}/a_{5\%}) \times 25^{\alpha_{95\%}-\alpha_{5\%}}$. For a tree with a reference diameter of 25 cm, CPL indicates how wide a crown can range in solitary conditions in relation to maximum restriction. By setting the 95% in relation to the 5% width, any species-specific differences in shape and form (e.g. that beech crowns are a priori wider than spruces) are eliminated. CPL only indicates the relative potential for expansion which is of particular importance for competing in mixture.

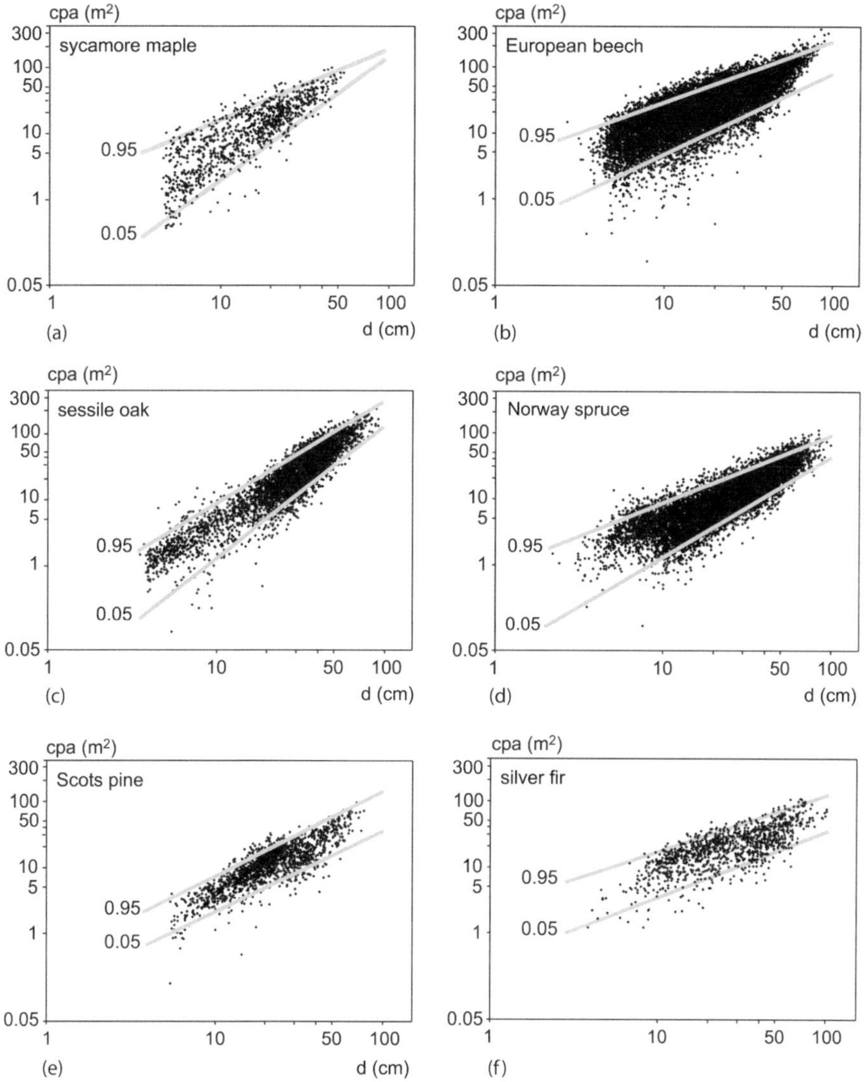


**Fig. 6.14** Allometric relationships between stem diameter, d, and crown projection area, cpa, for European beech ($n$ = 14,898), silver fir (*Abies alba* Mill.) ($n$ = 1079), sessile/common oak (*Quercus petraea* (Matt.) Liebl. and *Quercus robur* L.) ($n$ = 4485), Norway spruce ($n$ = 10,724), sycamore maple ($n$ = 942), and Scots pine ($n$ = 1609) in even-aged and uneven-aged stands. Range of crown dimensions measured on long-term experimental plots which cover dense as well as very sparsely spaced stands. The upper and lower lines represent the 95% and 5% quantile regression $\ln(\text{cpa}) = a + \alpha \times \ln(d)$. The width of the scattering and the distance between the 95% and 5% quantile regression represents the crown plasticity. For the statistical characteristics of the quantile regressions, see Pretzsch (2014)

Analyses of CPL values for various tree species in Europe revealed a maximum value for European beech of CPL = 5.1, i.e. upper crown area is more than fivefold compared to its lower crown area under strong competition. With respect to CPL, the species represented in Fig. 6.14 are ranked as follows: beech (CPL = 5.1) > silver fir (4.7) > oak (4.5) > spruce (4.2) > maple (4.0) > pine (3.7). Out of the set of 14 species represented in Fig. 6.13b and c, beech (5.1) has the highest value, whereas alder (2.8) and birch (2.6) have the lowest.

## 6.5.3 Modification of Tree Allometry in Interspecific Local Environments

### 6.5.3.1 Shift in Crown Allometry in Interspecific Compared with Intraspecific Environments

Species mixing can considerably reduce a tree's competition and increase its lateral crown expansion even when the stand density in mixture equals that in monoculture (Dieler and Pretzsch 2013). Of special relevance for stand dynamics, growing area efficiency and stand productivity are the behaviours of the lateral and vertical crown extension in mixed versus monospecific stands. This determines both a tree's growth and its space occupation and competition pressure on its neighbours.

We use data from Bavarian long-term experiments in monospecific and mixed stands of Norway spruce and European beech (Pretzsch and Schütze 2009) to scrutinise any shift in morphology caused by intra- versus interspecific competition. Figure 6.15 shows that the lateral and vertical crown extension of (a) Norway spruce and (b) European beech is higher in mixed than in monospecific stands.

Scaling between cpa and d is mostly considerably steeper as predicted by West et al. (2009) for the allometrically ideal plant ($\alpha_{\mathrm{cpa},\,d} = 4/3$ in $\mathrm{cpa} \propto d^{4/3}$). It is shallower in mixed than in monospecific stands but lies at a higher level in mixture (Fig. 6.15a). With increasing size, the collectives become more similar in this regard, i.e. in young and middle-aged stands, the species profit from interspecific competition but in mature stands, where crowns are less restricted, crown extensions become similar. According to allometric theory (Enquist et al. 2009; West et al. 2009), scaling between hcb and d should be $\mathrm{hcb} \propto d^{2/3}$. That applies to neither of the species investigated here; the exponents are mostly lower. The intercepts of trees in mixed stands differ considerably from those in monospecific stands; crowns are significantly longer in mixed compared with monospecific stands. The *h–d* allometry, in contrast, is only slightly modified by species mixing, and the observed scaling is close to $h \propto d^{2/3}$ as predicted by the allometric theory. Stem taper may be different in mixed versus monospecific stands; however, the *h–d* relationships are similar. Note that by comparing the species' behaviour based on their scaling size, differences are eliminated. This contrasts with less indicative and questionable comparisons based on ratios (see Box 6.3).

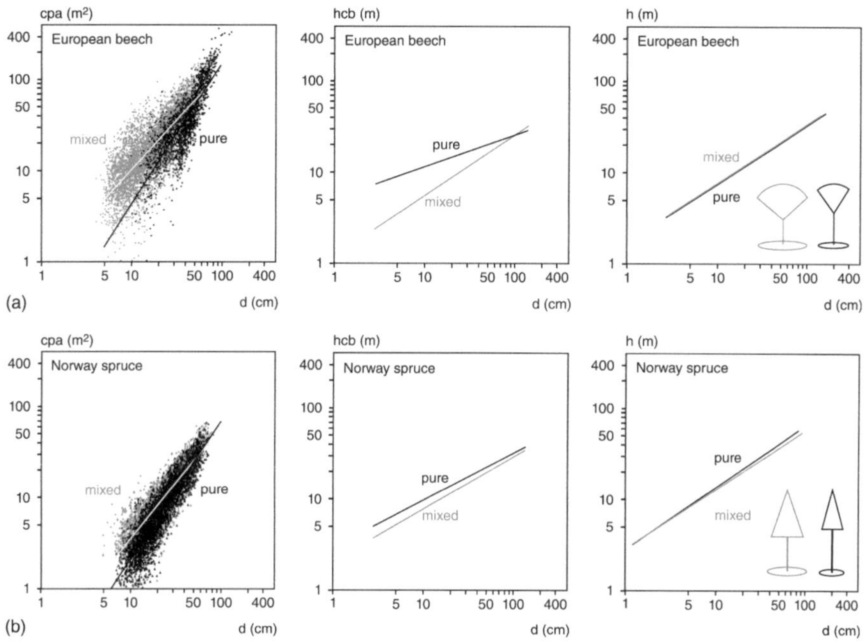


**Fig. 6.15** Crown allometry of (**a**) Norway spruce and (**b**) European beech in mixed stands (*grey*) compared with monospecific stands (*black*). Mixing significantly increases lateral and vertical crown extension in terms of the relationship between crown projection area, cpa, and tree diameter, d, and between height to crown base, hcb, and tree diameter, d. It hardly changes the relationship between tree height, *h*, and tree diameter, *d*. The respective equations are for spruce monoculture $\mathrm{cpa} = 0.05 \times d^{1.61}$, $\mathrm{hcb} = 3.04 \times d^{0.52}$, $h = 2.78 \times d^{0.68}$; for spruce in beech $\mathrm{cpa} = 0.09 \times d^{1.49}$, $\mathrm{hcb} = 2.19 \times d^{0.57}$, $h = 2.96 \times d^{0.64}$; for beech monoculture $\mathrm{cpa} = 0.24 \times d^{1.43}$, $\mathrm{hcb} = 5.14 \times d^{0.31}$, $h = 3.00 \times d^{0.64}$; and for beech in spruce $\mathrm{cpa} = 0.46 \times d^{1.29}$, $\mathrm{hcb} = 1.22 \times d^{0.67}$, $h = 3.09 \times d^{0.64}$

**Box 6.3 Allometric Relationship Between Tree Dimensions, Ratios, and Ontogenetic Drift**

As an example for size relationships between tree organs, Box Fig. 6.3-1 shows the allometric relationship between tree diameter, d in (cm), and crown diameter, cd in (m). It is linear at the ln–ln scale (Box Fig. 6.3-1a) and degressive with $\alpha_{\mathrm{cd},d} < 1.0$ at the logarithmic scale (Box Fig. 6.3-1b).

The allometric theory assumes $\mathrm{cd} \propto d^{2/3}$ (Enquist et al. 1998; West et al. 2009); however, our empirical evaluation based on crown measurement of $n = 3974$ Norway spruces yielded $\ln(\mathrm{cd}) = -1.24(\pm 0,018) + 0.75 (\pm 0.006) \times \ln(d)$, i.e. $\mathrm{cd} = 0.289 \times d^{3/4}$ with $\alpha_{\mathrm{cd},d} = 0.75$. The OLR regression was based on the ln-transformed values, according to Niklas (1994, p 331) RMA slope is $\alpha_{\mathrm{RMA}} = \alpha_{\mathrm{OLS}}/r_{y,x}$ with $r_{y,x} = r_{\mathrm{cd},d} = 0.95$ as correlation

(continued)

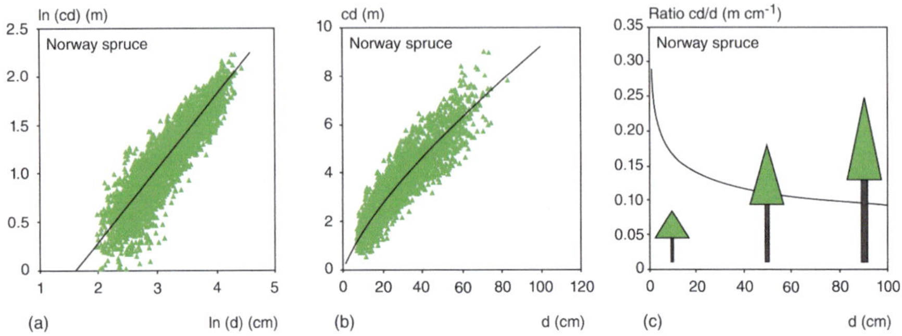


**Box Fig. 6.3-1** (**a**–**c**) Analysis of the allometric relationship between tree crown diameter, cd in (m), and stem diameter, d in (cm), and the resulting ratio cd/d for Norway spruce. (**a**) Based on $n = 3974$ crown records, OLR regression yields $\ln(\mathrm{cd}) = -1.24(\pm 0,018) + 0.75(\pm 0.006) \times \ln(d)$ ($R^2 = 0.82$). (**b**) The same relationship shown in linear grid and regression line $\mathrm{cd} = 0.289 \times d^{3/4}$. (**c**) Ontogenetic drift in the relationship between cd/d and tree size ($\mathrm{cd}/d = 0.289 \times d^{-1/4}$)

**Box 6.3** (continued)

coefficient between cd and d. In this case, $\alpha_{\mathrm{RMA}} = 0.75/0.95 = 0.79$ deviated even more from the slope for the allometrically ideal plant predicted by allometric theory ($\alpha_{\mathrm{RMA}} = 2/3 = 0.66$).

From $\mathrm{cd} = 0.289 \times d^{3/4}$, it follows that $\mathrm{cd}/d \propto d^{-1/4}$, i.e. the ratio decreases with tree size. Assmann (1970, p 112) used this ratio to compare crown morphology within and between different species. The non-isometric cd–*d* relationship shows that the ratio changes simply with size due to the ontogenetic drift of the tree size proportions. That means that comparison can be misleading when trees in mixed and monospecific stands differ in size. As the ratio decreases with size (Box 6.3), any size differences between the compared trees in mixed versus monospecific stands should be eliminated before comparison (see Sect. 6.7.2).

In this case, the observed cd/d ratio of the trees in mixture can be extrapolated to the size of neighbouring monocultures by $\mathrm{cd}/d' = \mathrm{cd}/d \times (d_p/d_m)^{-1/4}$. After this transformation, the adjusted observed ratio ($\mathrm{cd}/d'$) can be compared with the observation in neighbouring monocultures.

Analysis of cpa–*d* allometry of beech in monocultures compared with beech in mixture with Norway spruce, European larch, common ash, and sessile oak shows striking differences (Fig. 6.16). This analysis is based on densely stocked stands with no or only light thinning. Mixing matters even when the stand density is at maximum. Obviously a neighbouring European beech restricts the crown of a beech more than any other of the analysed species. For a beech with stem diameter 25 cm, the allometric equation shown in Fig. 6.16 (be) predicts a crown projection area of cpa = 17 $m^2$. Beeches with the same stem diameter achieve cpa = 25 $m^2$ when

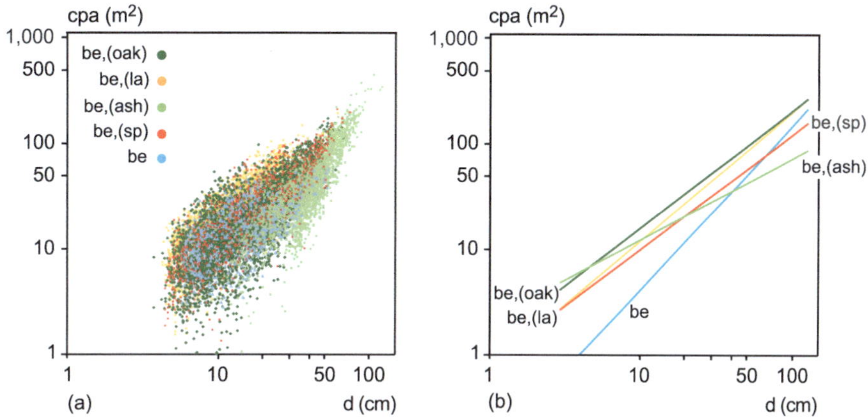


**Fig. 6.16** Allometric relationship between crown projection area, cpa, and the tree diameter, d, for European beech in monocultures (*blue*) and shift in the allometry when beech is mixed with Norway spruce, European larch, ash, and sessile oak. The respective equations are $cpa = 0.12 \times d^{1.54}$ for be; $csa = 2.17 \times d^{0.76}$ for be, (ash); $csa = 0.75 \times d^{1.21}$ for be, (la); $cpa = 0.84 \times d^{1.08}$ for be, (sp); and $cpa = 1.30 \times d^{1.10}$ for be, (oak)

mixed with ash, and 27, 37, or even 45 $m^2$ when mixed with spruce, larch, or oak, respectively. The ranking of neighbours regarding the effect on crown restriction is beech > ash > spruce > larch > oak. For European beech, mixing with each of the other species means 'competition reduction' in terms of crown extension in the sense of Kelty (1992) and Vandermeer (1992). In other words, a neighbouring Norway spruce, sessile oak, or Scots pine means a relief in crown restriction compared with a neighbouring beech. This is in accordance with findings by Pretzsch and Biber (2005) and Zeide (1985) that self-thinning is the highest in beech monocultures and much lower in monocultures of the other three species, with the ranking European beech > Norway spruce > Scots pine > sessile oak.

While allometric scaling theory predicts $cpa \propto d^{4/3}$ (i.e. $\alpha = 1.33$) for the allometrically ideal plant, the allometric exponent is at a maximum $\alpha = 1.54$ in beech monocultures and ranges between $\alpha = 0.76$ (be, (ash)) and $\alpha = 1.21$ depending on the species composition of the neighbours.

#### 6.5.3.2 Root Morphology and Root-Shoot Allometry

According to the optimal partitioning theory (McCarthy and Enquist 2007), the shape of the tree crown, root system, and the relationship between these depend highly on the resource supply of the plant. Part of the large variation in the root-shoot relationship of plants can be explained by this theory. It predicts that the limitation of a resource leads to the promotion of growth of the plant organ responsible for supplying that critical resource (Comeau and Kimmins 1989; Keyes and Grier 1981).

In Fig. 6.17, four examples indicate the behaviour of the root-shoot ratio. In the first example, where light, water, and nutrient conditions are favourable

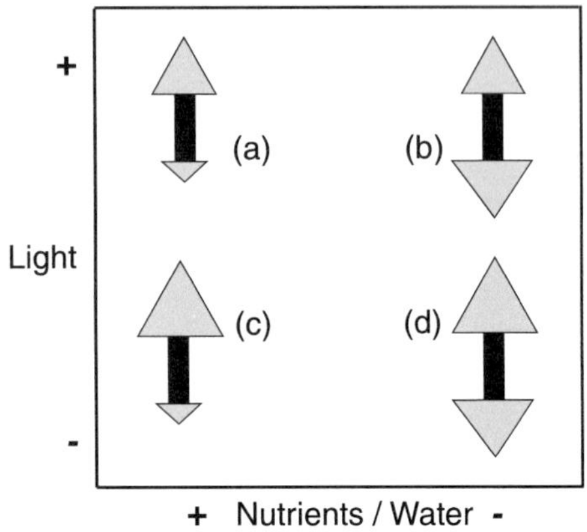


**Fig. 6.17** Partitioning of total plant biomass on shoot and root organs in relation to supply of nutrients and water ($x$-axis) and supply of light ($y$-axis). Limitation of nutrient and water supply causes a partitioning in favour of roots. Limitation of energy supply increases the investment of biomass in above-ground organs (By courtesy of Kimmins (1993, p 13))

(Fig. 6.17a), the tree develops a root-shoot ratio of 10:90. Under adequate light conditions, but with a water or nutrient deficiency (Fig. 6.17b), the tree invests more in root growth, especially fine roots. Thus, the root-shoot ratio increases in favour of roots to 45:55. With an adequate water and nutrient supply combined with critical light conditions, e.g. on nutrient-rich soils or for understorey trees (Fig. 6.17c), shoot growth is enhanced so that the tree root-shoot ratio may become 30:70. In the case of rich soils, trees in the upperstorey tend to allocate growth resources to extensive crown development, whereas trees in the understorey often invest in height growth to escape the shade (Oliver and Larson 1996). If light, water, and nutrient supply is limited (Fig. 6.17d), the root-shoot ratio may resemble the example in Fig. 6.17a of 10:90.

Changes in the partitioning between root and shoot growth can indicate a modification of environmental conditions through mixing as shown for the rather easily accessible crown growth. Intraspecific variation in root morphology due to species mixing, however, is more difficult to measure. Analysis of root-shoot allometry based on tree-ring analyses at increment cores from stem and coarse roots can reveal how the root-shoot relationships depend on site conditions (Pretzsch et al. 2012a, b) and silvicultural treatment (Pretzsch et al. 2014). For the methodology see Box 6.4. Figure 6.18 shows results of combined increment boring of coarse roots and stem at breast height applied in mixed spruce/beech stands versus monospecific spruce stands.

At both sample sites in the prealpine lowlands in Bavaria (SON 814 and ARN 815), the mixture with beech causes a significantly shallower root-shoot allometry of spruce compared with neighbouring monospecific stands. In the size range of 20–50 cm stem diameter, which is well backed with sample trees (right sections of the regression lines), spruces in mixture have a significantly lower root-stem growth

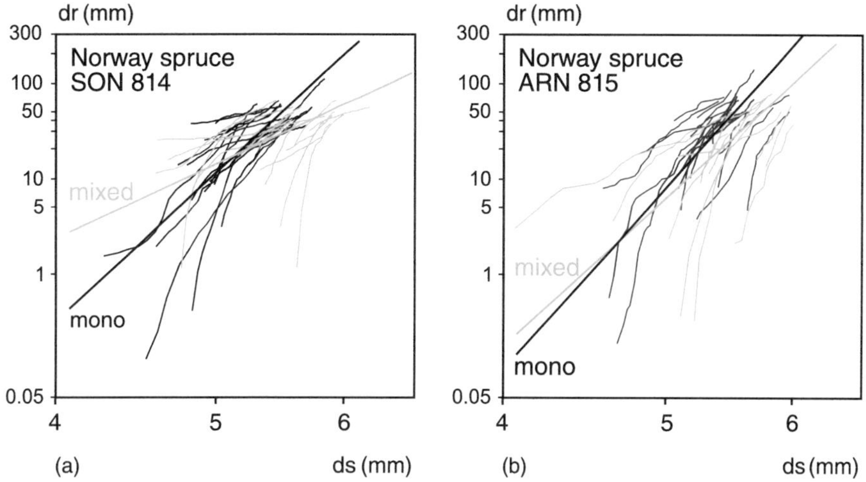


**Fig. 6.18** Development of coarse root diameter, dr, in relation to stem diameter, ds, of Norway spruce in mixture with European beech compared with neighbouring monospecific stands of Norway spruce on the long-term experimental plots (**a**) SON 814 and (**b**) ARN 815. The straight lines represent the allometric linear regression lines $\ln(dr) = a + \alpha \times \ln(ds)$ for spruce in mixed stands (*grey*) and monospecific stands (*black*)

than in monospecific stands. Coarse root growth is more sluggish in growth reaction than the resource-capturing fine roots. However, as in the case of the crown structure, coarse roots indicate the long-term dynamics as they represent the holding fixture and connecting pipes for fine roots. A cause of the repartitioning from root to stem growth in mixed stands may be a better supply of below-ground resources in mixed versus monospecific stands (Rothe 1997; Rothe and Binkley 2001) which triggers crown expansion to remedy above-ground resource limitation. Spatial niche separation between spruce roots in the upper soil horizon and beech roots in the deeper layer may be an additional reason (Rothe 1997; Wiedemann 1942, 1951). Decrease in root diameter growth entails shorter root length and a smaller influence zone of a tree. This may indicate a reduced investment in mechanical stabilisation and a decrease in the tree's growing area in mixed versus monospecific stands (see Sect. 6.6).

**Box 6.4 Analysing the Allometric Relationship Between Coarse Root and Stem Growth of Trees in Mixed Versus Monospecific Stands**

To analyse the allometric relationship between coarse root and stem growth increment, cores can be sampled from both stem at breast height and coarse roots in close distance from the stem axis (Pretzsch et al. 2014; Nikolova et al. 2011). For each trunk, two increment cores may be sampled from the cardinal direction north and east (Box Fig. 6.4-1, left). In addition, the trees' two to three biggest coarse roots can be sampled. Because of their rather elliptical

(continued)

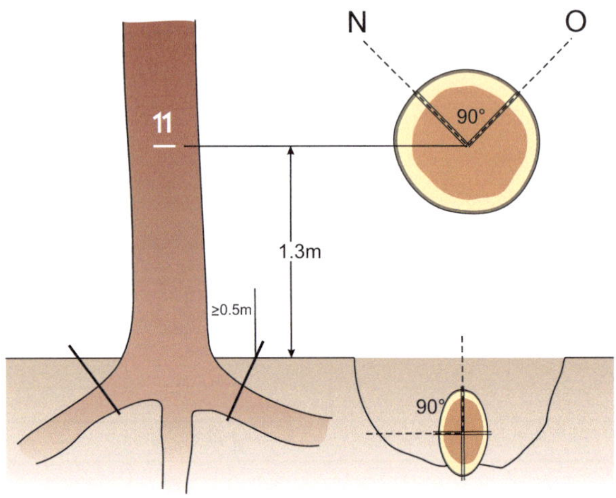


**Box Fig. 6.4-1** Principle of sampling increment cores from the stem and two main roots

**Box 6.4** (continued)

cross section, two increment cores may be taken from each root at a 45° angle from the vertical line. To enable the sampling, the selected roots need to be partly excavated (Box Fig. 6.4-1, right).

Based on tree-ring analysis, the diameter development of coarse roots may be plotted over the stem diameter development for each tree in a double-logarithmic scale. The grey trajectories in the background of Box Fig. 6.4-2 show such root-shoot allometries exemplarily for European beech and Douglas-fir grown in monospecific and mixed-species stands (Thurm et al. 2016).

The regression analysis of the individual coarse root diameter-stem diameter trajectories revealed that both European beech and Douglas-fir have reduced coarse root growth in mixed compared with monospecific stands (Box Fig. 6.4-2c versus a and f versus d). In both monocultures (Box Fig. 6.4-2a and d), the allometric relationship between coarse root growth and stem diameter growth is the steepest, i.e. the investment in roots compared with stem is maximal.

Certainly, the selected coarse roots represent just a small portion of the trees' whole root system, and their development is more sluggish and persistent than that of the ephemeral fine roots. But analogously to the stem which indicates the development of the crown and leaf area, the coarse root diameter reflects the activity of the whole root, since the coarse roots ultimately provide the basic structure and pipe system for the fine roots. So, the coarse root growth might be used as an integrative and non-specific indicator for the root system as a whole.

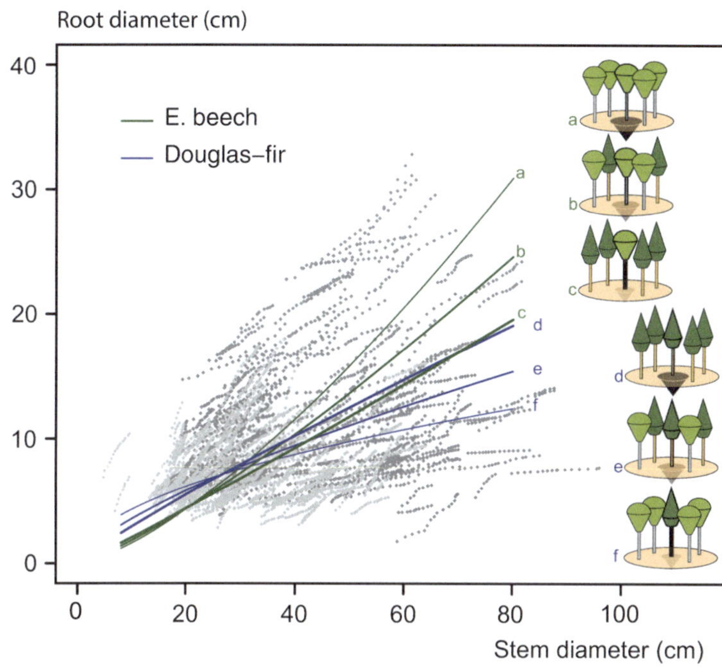


**Box Fig. 6.4-2** Compared with monospecific stands (**a** and **d**), in mixed-species stands the growth of coarse roots in European beech (**b** and **c**) and Douglas-fir (**e** and **f**) in relation to stem growth is reduced. In monocultures of European beech (**a**) and Douglas-fir (**d**), the allometric relationship between coarse root growth and stem diameter growth is steepest. The graphs show the results of increment boring at coarse roots and stem of European beech ($n = 85$) and Douglas-fir ($n = 90$) in 50–100-year-old monospecific and mixed-species stands on moist and nutrient rich sites in southern Germany. For the double-logarithmic allometric model, coarse root diameter $= f$ (stem diameter, mixing proportion, tree species), see Thurm et al. (2016)

### 6.5.4 From Pattern to Process of Crown Expansion in Interspecific Neighbourhoods

#### 6.5.4.1 Dynamics of Lateral Crown Expansion in Inter- Versus Intraspecific Environments

The change in intraspecific crown morphology in mixed versus monospecific stands becomes even more obvious by analysing successive crown maps of monospecific and mixed stands (Pretzsch 1992, 2009). Based on long-term plots with first crown measurements in the 1950s and successive measurements every 10–20 years, a 3D model delivers distances between neighbouring crowns (Fig. 6.19). The successive eight-radius crown measurements (Box 6.5) provide lateral crown growth for all eight directions. This enables derivation of functions for predicting lateral crown

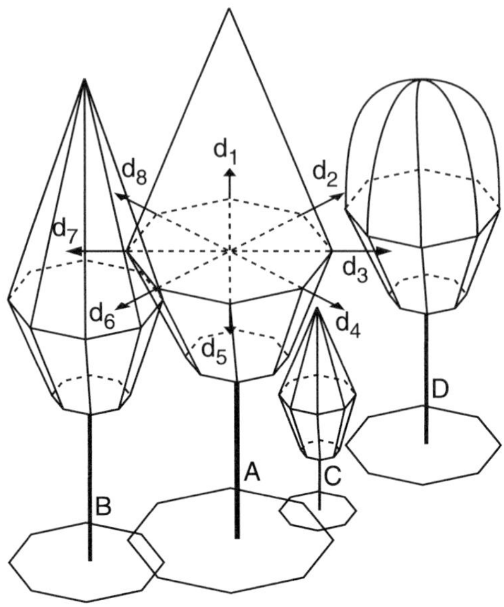


**Fig. 6.19** Determination of the lateral crown restriction of individual trees: distances from the crown perimeter of tree A to the neighbouring crowns of the trees B, C, and D are determined in eight directions $d_1 \ldots d_8$. When crowns intersect, the distance can become negative (see Fig. 6.20)

growth in dependence on the distance to neighbouring crowns in the respective directions (Fig. 6.20).

The functions describe the relative changes in crown radii of Norway spruce and European beech in relation to the crown distance to neighbours. A relative increment of crown radius of 1.0 means that a tree can expand its crown radius without lateral restriction and achieve the potential crown radial increment. A relative increment of 0.0 or less, by contrast, means stagnation and degeneration of the crown.

The curves in Fig. 6.20 show the decline in crown radial increment as the central tree crown becomes increasingly interlocked with neighbouring crowns as well as in the crown distance at which a crown dieback commences (below the dotted 0-line). Based on these distance-dependent crown growth functions, the crown cross-sectional development at a trial plot can be simulated for each individual tree in relation to its neighbourhood. Figure 6.21 shows the crown map for a 20 m × 20 m section of the experimental trial plot ZWI 111/3 in 1954, the year the simulation run starts based on the crown measurements in this year. The crown development of Norway spruce and European beech, at this time 60 and 80 years old, respectively, was simulated for a 50-year period at 5-year intervals. The interim results at times $t = 10$ years, $t = 30$ years, and $t = 50$ years (Fig. 6.21b, c, and d) show how the crowns expand and respond to increasing restriction by developing

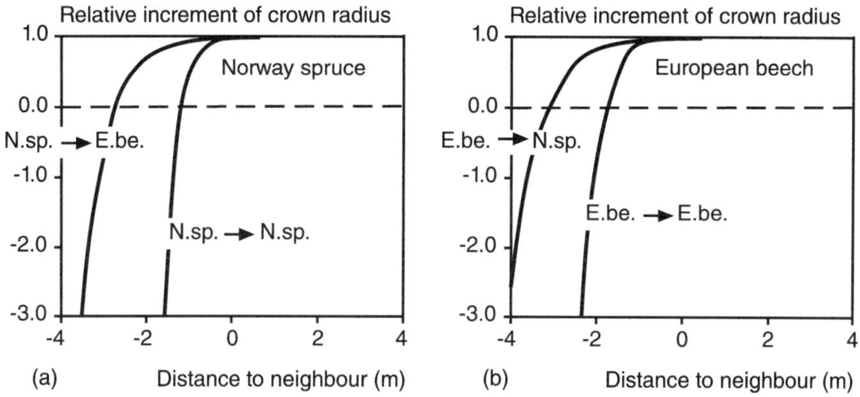


**Fig. 6.20** Influence of the distance to the neighbouring crown on the relative increment of the crown radius (Pretzsch 2009). (**a**) Radial crown increment of Norway spruce in a Norway spruce (N. sp → N. sp.) and a European beech (N. sp. → E. be.) neighbourhood. Functions: $icr_{sp\rightarrow sp} = 1.0-e^{-3.84\ (dist+1.2)}$ and $icr_{sp\rightarrow be} = 1.0-e^{-1.71\ (dist+2.7)}$. (**b**) Radial crown growth of European beech in a European beech (E. be. → E. be.) and a Norway spruce (E. be. → N. sp.) neighbourhood: $icr_{be\rightarrow be} = 1.0\text{-}e^{-2.56\ (dist+1.8)}$ and $icr_{be\rightarrow sp} = 1.0-e^{-1.49\ (dist+3.1)}$

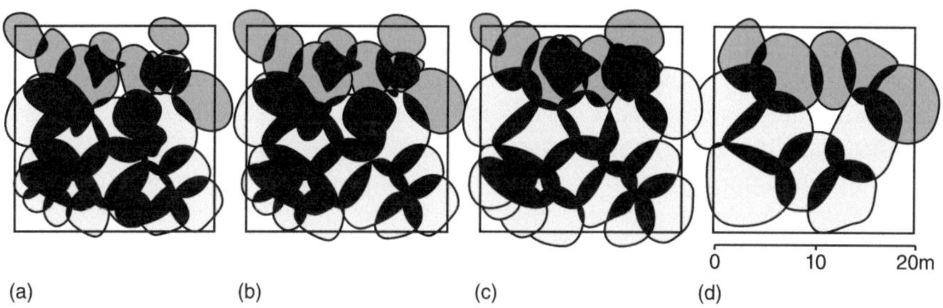


**Fig. 6.21** Crown cross-sectional area development on the experimental plot Zwiesel 111/3 based on simulation runs with the individual tree growth simulator SILVA 2.0 (Pretzsch 2009) starting with the crown measurement in the year 1954. The crown map for a 20 m × 20 m section of the experimental trial plot Zwiesel 111/3 is presented in a 50-year simulation run at the times (**a**) $t = 0$, (**b**) $t = 10$ years, (**c**) $t = 30$ years, and (**d**) $t = 50$ years

oval and eccentric crown shapes. To develop the crown perimeter, the eight crown radii are connected by a cubic spline (Pretzsch 2009).

The simulations and underlying functions (Fig. 6.20) reveal the following species-specific competition behaviour: when a Norway spruce grows towards another Norway spruce, the decrease in radial increment after an overlap of 1.0 m is almost linear. Yet, if it grows towards a European beech, a decrease in radial increment of the crown occurs much later. Effectively, this leads to a tighter interlocking of a Norway spruce crown with neighbouring European beech than with neighbouring Norway spruce (Pretzsch 2014). Similarly, the crown interlocking for a European beech growing towards a Norway spruce is also tighter than for European beech in monocultures. This finding confirms the species-specific behaviour of beech shown in Fig. 6.16. For Norway spruce and European beech, a reduction in lateral crown expansion of more than 5% only occurs when crowns interlock by 1.0–2.0 m, and the crowns recede only after the crowns overlap by 2.0–3.0 m. Thus, the crown growth dynamics of Norway spruce and European beech differ considerably from, e.g. the response of the light-demanding species European larch, which shows a clear reduction in the growth of side branches once the distances from branches to the neighbouring crown fall below 40 cm (Schütz 1989).

#### 6.5.4.2 Variability of Crown Projection Area in Inter- Versus Intraspecific Environments

Solitary trees achieve wide and, apart from a slight tendency towards ovality due to one-sided solar irradiation in northern or southern latitudes, rather circular crowns. The symmetry of their crowns indicates unimpeded lateral expansion or at least all-round homogeneous restriction by, for example, water, light, or nutrients (Møller and Swaddle 1997). When coping with crowding, crowns may adapt their lateral extension to their prevailing neighbourhood conditions and increasingly lose the symmetry typical for solitary growth. Species with higher crown plasticity can to some extent overcome their restriction by occupying emerging niches, penetrating neighbouring crowns, or even edging out neighbours. The eccentric crown expansion and crown asymmetry which enables occupation of additional space by directional lateral crown expansion should be distinguished from degeneration-caused asymmetry through die-off and mechanical abrasion of branches and crown parts due to overwhelming competition.

**Box 6.5 Measures for the Variability of Lateral Crown Expansion of Individual Trees in Intra- and Interspecific Environments**

The following measures of crown plasticity are based on common crown maps with eight-radius crown projection areas as shown in Box Fig. 6.5-1. The graph shows exemplarily a 400 $m^2$ sized section of the 80-year-old triplet Freising 813/2 of Norway spruce, European beech, and a mixture of both species close to Freising in southern Bavaria. The crown projection areas (in $m^2$ per crown) and their shape (divergence from circle) can be compared between mixed and neighbouring monocultures to reveal inter- versus intra-specific crown variability.

The following three measures address different aspects of the variability of lateral crown expansion (see Box Fig. 6.5-2):

The ratio cpa/sa between crown projection area, cpa, and stand area of an individual tree, sa, indicates the degree of crown engagement. A value of

(continued)

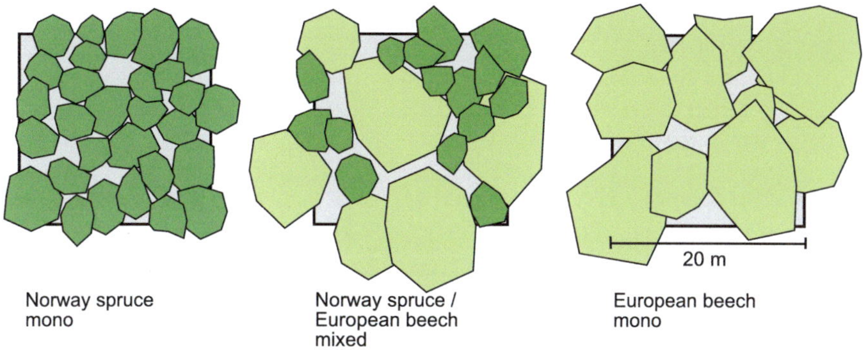


**Box Fig. 6.5-1** Crown maps with eight crown radii as basis for quantifying the variability of lateral crown expansion of individual trees. 400 $m^2$ sized sections of the 80-year-old triplet Freising 813/2 of Norway spruce (**a**), a mixed stand of Norway spruce and European beech (**b**), and European beech (**c**) close to Freising in southern Bavaria

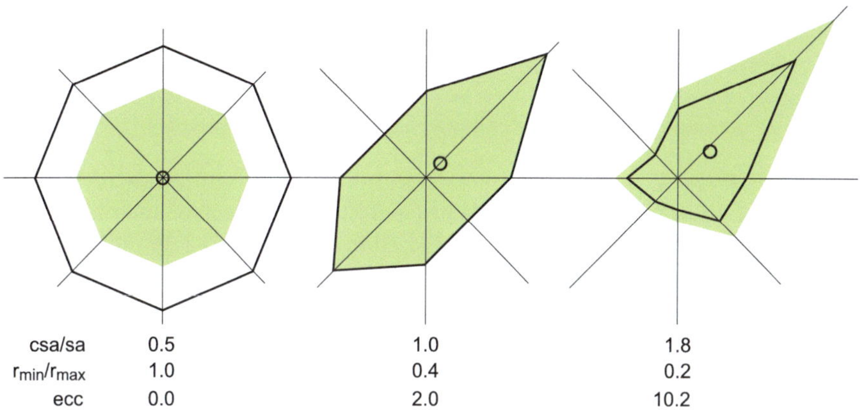


**Box Fig. 6.5-2** Crown projection area (*grey* area) and growing area (*black line*) of trees in (*left*) low density, (*centre*) medium density, and (*right*) high density with decreasing crown symmetry and measures for quantifying crown variability. cpa/sa, cpa-sa ratio represents crown extension in relation to stand area. $r_{min}/r_{max}$, crown rotundity in terms of the ratio between the largest to shortest radius. ecc, Crown eccentricity indicated by the standardised distance between the gravity centre of the crown projection area and the tree position

**Box 6.5** (continued)

cpa/sa = 1 would indicate identity of cpa and sa. High values (cpa/sa > 1) indicate wide crown expansion reaching beyond the trees' sa. Small cpa/sa values indicate a tree's suppression and crown recession (see Box Fig. 6.5-2 from left to right).

(continued)

**Box 6.5** (continued)

The ratio between the shortest and longest of the eight crown radii, $r_{min}/r_{max}$, indicates the crown rotundity. In Box Fig. 6.5-2 the rotundity decreases (from left to right) with stand density due to growing competitive pressure.

Crown eccentricity, ecc $= \sqrt{(x_s - x_g)^2 + (y_s - y_g)^2}/d_{1.3}$, is based on the Cartesian coordinates of the stem position, $x_s$ and $y_s$, and on the coordinates of the centre of gravity, $x_g$ and $y_g$, of the crown, calculated on the basis of the coordinates of the corner points of the crown projection area ($x_g = \sum_{i=1}^{8} x_i/8$, $y_g = \sum_{i=1}^{8} y_i/8$). Division of the distance $\sqrt{(x_s - x_g)^2 + (y_s - y_g)^2}$ by the stem diameter, $d_{1.3}$, eliminates size effects and makes different trees and species comparable. A measure of, for example, ecc = 0, 2.0, and 10.2 means that the gravity centre of the crown is perpendicular above the tree position (ecc = 0), twice (ecc = 2), or even more than ten times (ecc = 10) the stem diameter deviating from the tree position (see Box Fig. 6.5-2 left to right). Note that in Box Fig. 6.5-2 the trees' centres of gravity are represented by circles and the trees' stem foot positions by the origin of the coordinate system.

In the following, crown maps of monocultures and mixed stands of Norway spruce and European beech and sessile oak and European beech on long-term plots in Germany with eight-radius crown measurements are used to illustrate the differences between crown dynamics in intra- and interspecific environments. In order to quantify the variability of crown structures in those stands, the measures introduced in Box 6.5 will be applied.

Using the ratios, the crown engagement, rotundity, and eccentricity can be compared between mixed and neighbouring pure stands. The cpa/sa ratios in Table 6.1 reflect that, except for beech when mixed with oak, all three species can significantly expand their crowns laterally far beyond their stand area when growing in mixed compared with pure stands.

The ratio $r_{min}/r_{max}$ reveals that, compared with the other species, beech crown projection areas are mostly less circular, i.e. more jagged. Mixing significantly increases the rotundity in the case of spruce ($p < 0.001$), decreases the rotundity of beech ($p < 0.001$), and leaves the rotundity of the crowns of sessile oak unaffected.

The ecc values are the highest for beech, especially when growing in mixed stands. Values of ecc = 5.7 to 7.4 indicate that beeches have plastic crowns for resource capture even distant from their stem position. The measures show that crown morphology can significantly shift from intra- to interspecific competition and trigger the space occupation of the combined species in a species-specific way. Mixing increases the crown eccentricity of both beech ($p < 0.001$) and oak ($p < 0.001$) but fosters the centricity in the case of Norway spruce ($p < 0.001$).

**Table 6.1** Lateral crown expansion of individual trees in even-aged mixed versus pure forest stands of Norway spruce and European beech (above) and sessile oak and European beech (below)

| Species combination | N. sp. (E. be.) mixed | N. spruce pure | N. spruce mix/pure | E. be. (N. sp.) mixed | E. beech pure | E. beech mix/pure |
|---|---|---|---|---|---|---|
| *n* (Trees) | 4634 | 3623 | | 4845 | 3173 | |
| cpa/sa ($m^2\ m^{-2}$) | 1.22 (± 0.095)**a** | 0.98 (± 0.042)**a** | 1.24 | 1.84 (± 0.109)**c** | 1.32 (± 0.073)**c** | 1.39 |
| $r_{min}/r_{max}$ ($m\ m^{-1}$) | 0.51 (± 0.026)**c** | 0.43 (± 0.003)**c** | 0.78 | 0.36 (± 0.027)**c** | 0.38 (± 0.003)**c** | 1.02 |
| ecc ($cm\ cm^{-1}$) | 1.80 (± 0.020)**c** | 1.90 (± 0.030)**c** | 0.95 | 5.70 (± 0.060)**c** | 4.40 (± 0.060)**c** | 1.30 |

| Species combination | E. be. (s. oak) mixed | E. beech pure | E. beech mix/pure | s. oak (E. be.) mixed | s. oak pure | s. oak mix/pure |
|---|---|---|---|---|---|---|
| *n* (Trees) | 2326 | 3173 | 0.73 | 1959 | 2888 | |
| cpa/sa ($m^2\ m^{-2}$) | 1.48 (± 0.192) | 1.32 (± 0.073) | 1.12 | 1.39 (± 0.207)**a** | 0.90 (± 0.074)**a** | 1.54 |
| $r_{min}/r_{max}$ ($m\ m^{-1}$) | 0.35 (± 0.004)**c** | 0.38 (± 0.003)**c** | 0.88 | 0.38 (± 0.004) | 0.38 (± 0.003) | 0.88 |
| ecc ($cm\ cm^{-1}$) | 7.4 (± 0.110) **c** | 4.40 (± 0.060)**c** | 1.68 | 3.30 (± 0.050)**c** | 3.10 (± 0.040)**c** | 1.06 |

The data comes from crown projection measurements on fully stocked experimental plots in Bavaria/south Germany (for explanation of variables, see Box 6.5). Lowercase bold letters (**a** and **c**) behind the standard errors indicate that Welch two sample *t*-tests revealed significant differences (at level $p < 0.05$ and $p < 0.001$, respectively) between the species behaviour in mixed and pure stands

#### 6.5.4.3 Species Mixing and Inner Crown Properties

Species mixing can obviously modify the fractal dimension of the crown surface area and leaf area from a lower space filling (Euclidian scaling) towards higher space filling of the crown volume (fractal scaling) (Pretzsch and Dieler 2012). Tree crowns as described by Oldemann (1990), Purves et al. (2007) and Roloff (2001) lie somewhere along the continuum between the borderline cases of an umbrella-like crown with the majority of the leaf surface area allocated close to the convex hull (surface dimension $n = 2$) and a broom-like crown with leaf surface area distributed all over the crown space (surface dimension $n = 3$) (Zeide 1998).

Terrestrial laser scanning and subsequent skeletonisation of tree crowns based on the voxel patterns revealed significantly different inner crown structures in mixed versus monospecific stands of Norway spruce and European beech (Bayer et al. 2013). An example of the procedure from compilation of point clouds to derivation of inner crown properties is shown in Fig. 6.22a–d.

Although similar in stem diameter and tree height, beeches growing in mixture with spruce can have significantly more branches of first order and—because of stronger ramification—nearly twice the number of second and third order branches compared with those in monocultures (Bayer et al. 2013). The branch bending can

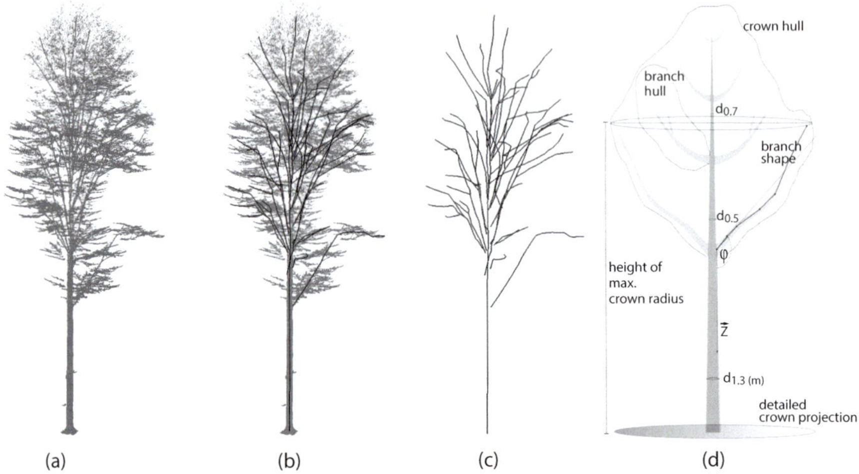


**Fig. 6.22** (**a–d**) TLidar scans in mixed-species stands of Norway spruce and European beech, skeletonisation, and measurement of crown periphery and inner structure. (**a**) Original voxel cloud compiled from various scans, (**b**) skeletonisation, (**c**) extracted branch pattern, (**d**) measurement of outer and inner crown characteristics (according to Bayer et al. 2013)

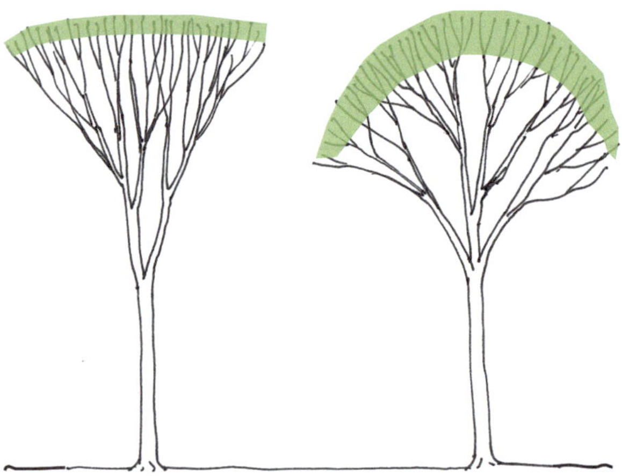

**Fig. 6.23** Schematic representation of (*left*) broom-like crown structure of adult European beech in even-aged monocultures and (*right*) bouquet-like crown structures in mixed stands with spruce (according to Arz (2013)). While individuals growing in monocultures develop steeper branches, those growing in mixture with Norway spruce have more branches, flatter branches, and nearly twice the number of second and third order branches (Bayer et al. 2013)

be higher and the branch angle flatter. Also, the sum of branch length and the crown volumes can be higher in mixed compared with monospecific stands. The schematic representation in Fig. 6.23 accentuates the broom-like crown shape of beeches when growing in intraspecific interaction (left) with most of the leaf area allocated close to the top of the convex hull. In interspecific environments, the crowns are rather bouquet-like (right) with the leaves reaching deeper into the crown volume (Arz 2013; Bayer et al. 2013).

## 6.6 From Tree Morphology to Space Occupation

All individuals in monocultures—particularly when they are relatively homogeneous in genetics and size—use similar strategies to compete for the growing space in a stand and for the same resources when these are not sufficient for all. When their demands and abilities are relatively similar (Fig. 6.24a), the function and structure of all competitors vary within a small range. Theoretically, this leads to a hexagonal pattern of tree distribution. Given that all are equipped with rather similar tricks and traits, neighbouring trees are in a stalemate rather than outwitting each other. That is why many species, when growing in monocultures, have tree crown projection areas that are rather similar to growing area. Their crowns barely interlock and 5–10% of the stand area is not covered by crowns at all (see Chap. 5, Fig. 5.19 and Table 5.6).

Compared with monocultures, the trees' elbowroom in mixed stands can be wider. If the crown shapes of the combined species complement each other (Fig. 6.24b), the trees might simply have more space to occupy without mechanical abrasion or penetration of the neighbours' crowns, while mixtures with similar

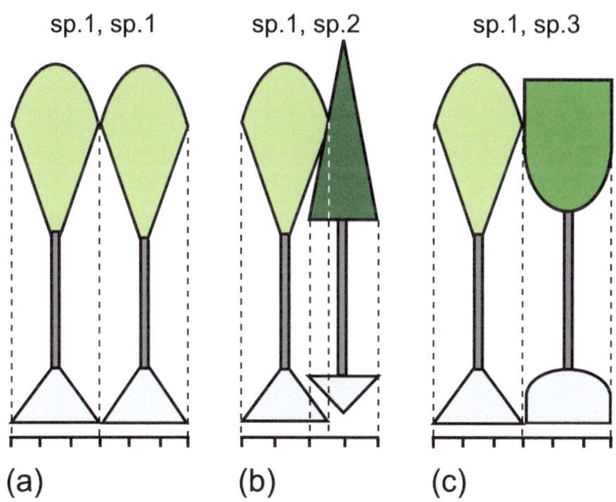


**Fig. 6.24** Crown and root shapes of tree species can complement or prevent each other from filling the canopy and soil space. Schematic representation of beech in monoculture (**a**), mixing of spruce and beech (**b**), and mixing of beech and oak (**c**). Packing density can be increased compared with monoculture of beech through the complementary structure of spruce and beech (**b**) but not through the rather obstructive structure of beech and oak (**c**)

crown shapes (Fig. 6.24c) cause less competition reduction. Beyond these differences in species-specific structure, the species-specific strategies for space occupation through, e.g. crown penetration due to different light compensation points and lateral expansion due to change in branch angle may result in competition release and wider crown expansion than in monoculture of one species and crown restriction and regression compared with the monospecific stands for the other species.

### 6.6.1 Crown Projection Area, Tree Stand Area, and Degree of Crown Engagement

Regarding the crown as a tool for resource sequestration, any crown and root plasticity may be relevant for tree and stand productivity. Suppose that if species combined in mixture retained the same morphology as they have in monoculture, the mixing effect may simply be additive. Any crown structures emerging in mixture, e.g. a plastic acclimation or complementarity of spatial niches, may cause higher resource capture and multiplicative mixing effects such as overyielding or underyielding in mixed versus monospecific stands.

The crown projection area, cpa, of a tree of a given size indicates its reach for resource acquisition, light interception, and shading of neighbours. In contrast, the tree stand area, sa, represents its available growing area. So the ratio between cpa and sa, introduced as the degree of crown engagement in Box 6.5 indicates how wide crowns reach beyond their growing area, fill the canopy space, and forage for light in their local environment.

Table 6.1 shows exemplarily for monospecific and mixed stands of spruce, beech, and oak that the cpa/sa ratio is mostly close to cpa/sa = 1.0 in monocultures, i.e. the tree crown area is similar to its growing area. In mixture, trees can overcome the lateral crown restriction by expanding and interlocking more closely. In total, the cpa/sa value increases by 24 and 39%, respectively, when spruce and beech are mixed, and by 12 and 54%, respectively, when beech and oak grow in mixture (see Table 6.1, cpa/sa values). This increase in the crown expansion beyond growing area can be coupled with a reduction in the mean growing area per tree in mixed versus monospecific stands (Sect. 6.4).

Figure 6.25a is a schematic representation showing that in monocultures crown projection area is rather similar to stand area, whereas in mixed stands the crowns often exceed the tree stand area. Due to the higher stand density in mixed compared with monospecific stands, the mean tree stand area decreases (see Chap. 4, Sect. 4.3.2). So, in mixed stands the cpa-d allometry can shift upwards and the sa-d allometry downwards, and the cpa/sa relationship, i.e. the crown engagement, often rises in mixed compared with monospecific stands (see $cpa_m/sa_m > cpa_p/sa_p$ in Fig. 6.25b).

Tree size is an ambiguous trait. As trees become larger, they tend to gain more privileged access to resources, occupy more space, and have a greater repression effect on neighbours. However, size can also mean higher maintenance costs and susceptibility to windthrow, drought, or bark beetle attacks. When trees expand far

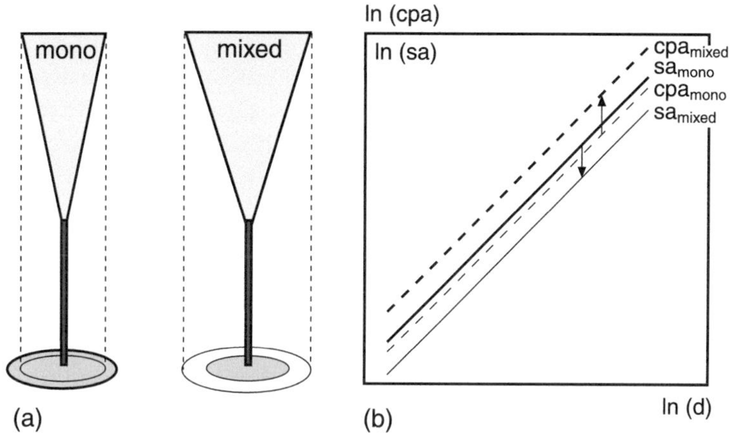


**Fig. 6.25** Schematic representation of the crown shape of beech monocultures (p) compared with mixed stands (m). While crown projection area is frequently rather similar to or just slightly lower than growing area in monocultures, mixing can considerably increase crown projection area, cpa, and slightly reduce tree stand area, sa. As a result, the cpa/sa ratio which indicates the degree of crown engagement can rise significantly when trees grow in interspecific environments

beyond their stand area by spreading their crowns without much additional investment in tree diameter or tree mass, this can mean a higher resource acquisition without the drawbacks of higher resource investment. Section 6.6 will show that this can increase the growth per stand area and thereby the tree and stand productivity.

### 6.6.2 Dynamics of Canopy Gap Occupation

Crown plasticity is species specific. We quantified it by CPL in Sect. 6.5.2 and saw that, for instance, CPL of beech >> CPL of Scots pine. A species' crown plasticity is relevant for its persistence and continuous extension in closed stands; it enables crown competition through penetration or even edging out of neighbouring crowns in the course of stand development. But plasticity also means a strong capacity for first access to unoccupied space in the early stand phase or spontaneous occupation of gaps caused by mortality or thinning. During stand development, various kinds of disturbances (e.g. tree mortality due to competition, storm damage, bark beetle attacks, and thinnings) may cause openings in the canopy and gaps in the root space. The quicker and wider a species' crown extension, the greater the benefit it gains from erratic gaps. At the stand level, plastic crowns which can quickly occupy gaps can mean continuous cover of the ground and high resilience of productivity in the case of disturbances.

The following rough estimation underlines the relevance of gap occupation for interspecific competition and stand productivity. The roughly 3% annual rate of tree loss due to mortality or thinning means a gap area of 300 $m^2$ per hectare and year.

A continuation of the drop-out process of trees over a 100-year rotation period can cause a transient gap area of about 30,000 $m^2$, i.e. 3.0 ha. In other words, a stand's given growing area is repeatedly contested by neighbours, occupied by trees, released after mortality or thinning, and finally captured by the most competitive neighbours. Starting with some thousand (when planted) or million trees (when naturally regenerated), capture and recapture of growing area due to competition ultimately leave just a few hundred trees per hectare in the mature stand. Again and again trees drop out and release space for the remaining neighbours, which can occupy it faster and more completely the more plastic their crowns are.

Particularly in more or less monolayered stands, this ongoing mortality and subsequent closure through crown extension causes a continuous opening up of the canopy. Crown size and accordingly gap size increases with tree age. As larger gaps need more time to be recaptured and closed by neighbours and because older trees' lateral shoot growth is rather slow, the crown re-coverage becomes more sluggish with increasing stand age. The ongoing gap dynamics are probably one additional reason why even in unthinned monolayered monocultures, 5–15% of the stand area remains uncovered by crowns (Chap. 5, Table 5.6). In temperate forests, where individual tree growth is mainly light limited, gaps are closed more rapidly through crown extension above ground than through tree root extension below ground (Bauhus 2009). Species with higher crown plasticity can achieve more additional canopy space through faster and more extensive gap occupation equivalent with additional resource capture and competitive strength. Particularly gaps caused abruptly by thinning favour those species capable of wide and fast crown extension. Additional growing space caused by mortality emerges more continuously and is more gradually occupied by neighbours already in advance of a tree's death.

Time series of TLidar measurements before and after gap cuttings (2006, 2008, 2010, and 2012) in monospecific and mixed stands of Norway spruce and European beech reveal the species-specific strategies for occupying crown gaps caused by thinning (Bayer et al. unpublished). Figure 6.26 illustrates the initial gap size (non-black areas in the centre), the gap filling till 2008 (dark grey) and 2012 (light grey), and the still uncovered area in 2012 (white). The progressing occupation of gap area (in $m^2$) is much slower in the monospecific spruce stand (left) compared with beech (middle). This finding corresponds with the much lower growth resilience to stand density reduction through thinning of spruce compared with beech (Pretzsch 2005b). The gap in the mixed stand of spruce and beech is closed relatively quickly; again beech contributes much more than spruce. On average, the area occupation in the 6-year period of spring 2006 to spring 2012 amounts to 0.8 $m^2$ $year^{-1}$ for spruce and 2.8 $m^2$ $year^{-1}$ for beech.

Figure 6.27a quantifies the dynamic of gap closure for the canopy cover of the ground. While spruce in the monospecific stand (sp) closes about 5 $m^2$ during the 6-year period (2006–2012), beech (be) and the mixed stand (sp, be) close more than 15 $m^2$, i.e. about three times this area. Figure 6.27b reveals the contribution of both species to closing the gap in the mixed stand in terms of their sum of crown area which has grown into the gap since 2006. Again, beech (17.5 $m^2$) is superior to

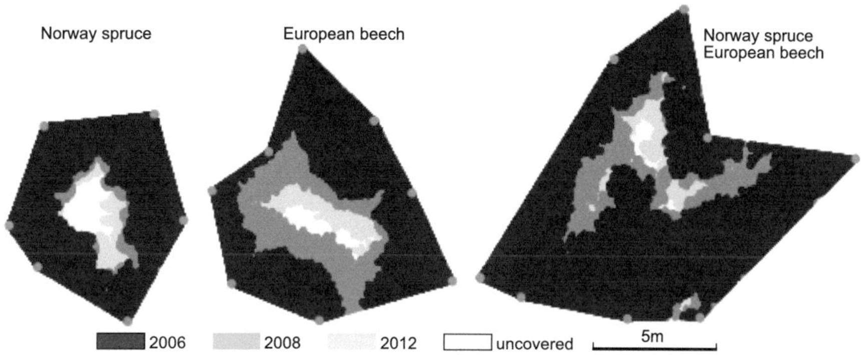


**Fig. 6.26** Progressing canopy closure in monocultures of Norway spruce (*left*), European beech (*centre*), and a mixed stand of Norway spruce and European beech (*right*) after cutting gaps in 2006 (experimental area FRE 813/1). TLidar scans show the initial state of the canopy in 2006 (*black*), the progressing area occupation until 2008 (*dark grey*) and 2012 (*light grey*), and the still open part in 2013 (*white*)

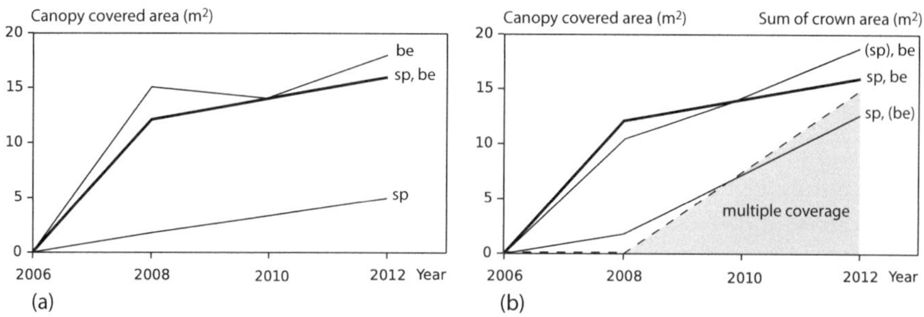


**Fig. 6.27** Progressing gap area occupation on the experimental plot FRE 813/1 from 2006 (closed gap area = 0 m$^2$) to 2012 for monospecific and mixed stands of Norway spruce and European beech. (**a**) Progressing gap closure in monospecific spruce (sp), monospecific beech (be), and mixed stand of both species (sp, be) in terms of canopy cover of the ground. (**b**) Canopy cover for the mixed plot of spruce and beech (*bold line*), sum of crown projection area for N. spruce (sp, (be)) and E. beech ((sp), be), and multilayering on the mixed plot (hatched area)

spruce (10 m$^2$), and large portions of the canopy covered area are multilayered (hatched area). The high multiple crown coverage in mixed stands indicates the complementary use of light by spruce and beech.

TLidar reveals that beech is superior in terms of both the absolute area occupation and the velocity of gap occupation. This continuous contest for space occupation is rather unspectacular, but in view of the above-stated total area of about 3 ha during a rotation, it is highly relevant for a species' survival and spread. Efficient space occupation does not necessarily mean superior productivity but signifies an increase in dominance and persistence through restraining, penetrating, and suppressing the other species in a kind of permanent guerrilla warfare. Even in stands without active silvicultural interference, this contest for space and resources can gradually change the mixing proportion in favour of those species which are best equipped for continuous gap dynamics.

A superior crown plasticity and gap occupation of species such as European beech, maritime pine (*Pinus pinaster* Ait.), or lenga (*Nothofagus pumilio* [Poepp. U. Endl.] Krasser), respectively, is probably one essential reason for their dominance in Central European, Mediterranean, and South American forest ecosystems.

## 6.7 Efficiency of Tree Crown Area and Tree Growing Area

Productivity means growth per time and unit area and indicates how efficiently a tree uses a given growing area for volume or biomass growth. So far we have compared tree morphology and growth in mixed stands versus monocultures without any relation to the required growing area. The following transmission to growth per tree stand area yields tree and stand productivity.

Dividing the annual volume increment of a tree ($m^3$ $year^{-1}$) by its crown projection area or stand area ($m^2$) (see Box 6.6) yields a measure of the tree's area use efficiency in terms of volume growth. In this case, area is used as a surrogate variable for the tree's resource supply.

Note that indices which set the growth in relation to the consumption of light, water, or nitrogen and water follow the same concept and are much more informative because they are directly related to the resources. Although they reflect the efficiency of light, water, and nutrient use, they are still rarely available for trees in mixed and monospecific stands.

Efficiency parameters based on growing area or growing space were frequently used to compare the performance of trees of different sizes or social classes in the stand, at different ages, or in different environments (Gspaltl et al. 2012; Pretzsch and Schütze 2009; Sterba et al. 1993; Sterba and Amateis 1998). When calculated for trees in mixed stands and neighbouring monospecific stands, efficiency parameters can contribute to quantifying and understanding the emergence of mixing effects.

**Box 6.6 Crown Projection Area, cpa, and Stand Area, sa, of Trees**

The crown projection area of a tree, cpa, results from projection of four or eight crowns. In Box Fig. 6.6-1a, the crown projection area is represented by a circle with the mean crown radius circle with a radius $\bar{r}_q$, which corresponds to the quadratic mean of four or eight radial measurements $r_1$, $\ldots r_n$. The growing area—also called stand area—of a tree, sa, is a substitute for its resource supply (Box Fig. 6.6-1a). It results from dividing the

(continued)

**Box 6.6** (continued)

whole-stand area (e.g. the plot area of 1 ha) by the number of living stocking trees. The sum of the tree stand areas of all trees is the stand area ( $AA = \sum_{i=1}^{n} sa_i$ ). The simplest method for obtaining the mean stand area, $\overline{sa} = AA/n$, is to divide the stand area by the number of trees, $n$. Other methods such as the circle segment method by Alemdag (1978), the rastering methods by Faber (1981, 1983) and Nagel (1985), and the growing area polygons by Brown (1965), Jack (1968), Fraser (1977), and Pelz (1978) make additional assumptions about the trees such as their diameter, crown size, or growth rate in order to partition the total stand area between them (see Pretzsch (2009)).

The cpa/sa relationship can become greater in mixed stands and the degree of crown closure and therefore the crown closure and the sum of crown projection areas higher than in monocultures (Box Fig. 6.6-2). Wider crown extension may result in denser canopy space filling and higher tree growth per unit area.

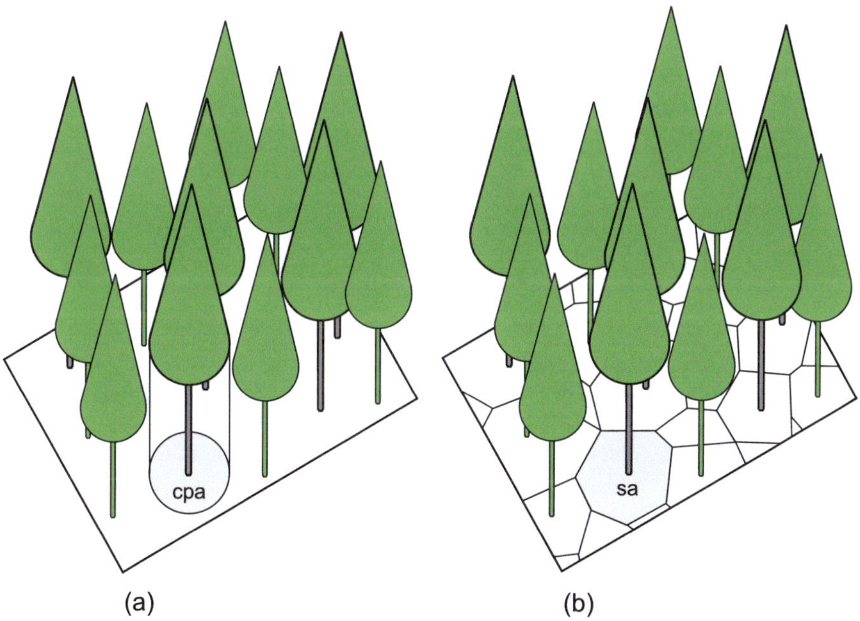


**Box Fig. 6.6-1** Measures of crown extension and growing area. (**a**) Crown projection area, cpa, and (**b**) stand area, sa, of an individual tree

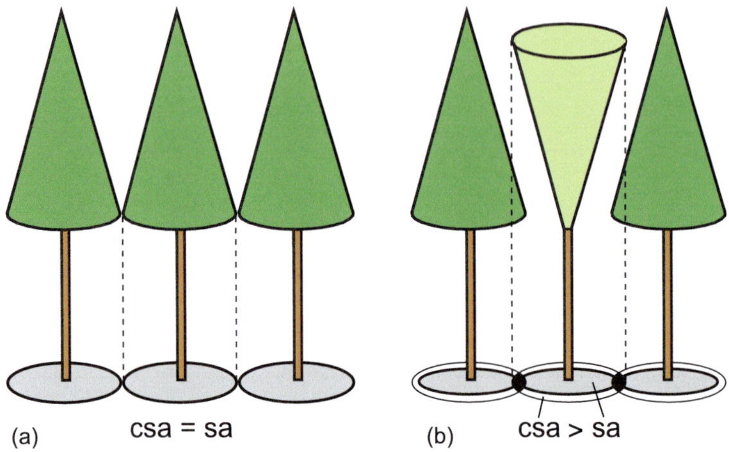


**Box Fig. 6.6-2** Relationship between tree growing area, sa, and crown projection area, cpa. (**a**) Crown projection area, cpa, is often equal to or slightly smaller than tree stand area, sa, in monocultures. (**b**) Because of structural or functional complementarity, crown projection area is often larger than stand area in mixed stands. This results in multiple crown coverage (black intersections of crowns)

### 6.7.1 Ratios for Comparing Tree Growth in Mixed Versus Monospecific Stands

The ratios presented in Table 6.2 are useful for analysing efficiency of growth at the tree level and for the transition from growth per tree to growth per stand unit area (Box 6.6). The ratios can be quantified for trees in monospecific and mixed stands. Comparing these can reveal, for example, whether overyielding results from higher stand density, higher growth rate per tree, or from a combination of both.

The ratios $E_{\mathrm{cpa}/v}$ and $E_{\mathrm{iv/cpa}}$ indicate the efficiency of the crown in terms of crown space occupation and exploitation without considering the required tree stand area, sa, i.e. they reflect aspects of plant ecology. $E_{\mathrm{cpa}/v}$ ($\mathrm{m}^2\ \mathrm{m}^{-3}$) indicates how much crown area a tree occupies per unit of tree volume. $E_{\mathrm{iv/cpa}}$ ($\mathrm{cdm\ year}^{-1}\ \mathrm{m}^{-2}$) reflects the growth efficiency of the tree's crown projection area. The ratios $E_{\mathrm{cpa/sa}}$, $E_{\mathrm{iv/sa}}$, and $E_{v/\mathrm{sa}}$ relate the structure and growth of the tree to its required growing space. They thereby enable the transition from tree growth to tree productivity. $E_{\mathrm{cpa/sa}}$ ($\mathrm{m}^2\ \mathrm{m}^{-2}$) reflects the degree of crown extension beyond the tree stand area and is equivalent to the degree of crown engagement. $E_{\mathrm{iv/sa}}$ ($\mathrm{cdm\ year}^{-1}\ \mathrm{m}^{-2}$) represents the efficiency of stand area in terms of tree volume growth ($\mathrm{m}^3\ \mathrm{year}^{-1}$) per tree stand area ($\mathrm{m}^2$) and is equivalent to the tree productivity. $E_{v/\mathrm{sa}}$ ($\mathrm{m}^3\ \mathrm{m}^{-2}$) indicates the stand density in terms of tree volume per tree growing area (Pretzsch and Schütze 2005). We use the above ratios to compare tree development in mixed versus monospecific stands. The ratio $rE_{\mathrm{iv/csa}} = E_{\mathrm{iv/cpa},m}/E_{\mathrm{iv/cpa},p}$ represents how trees perform with regard to crown growth efficiency in mixed versus monospecific stands. Most of the ratios change

**Table 6.2** Efficiency ratios for characterising individual tree growth in relation to tree size, crown projection area, and tree stand area

| Variable | Ratio | Unit | Characteristic |
|---|---|---|---|
| $E_{\text{cpa/v}}$ | cpa/v | $m^2\ m^{-3}$ | Efficiency of crown space occupation |
| $E_{\text{iv/cpa}}$ | iv/cpa | $m^3\ year^{-1}\ m^{-2}$ | Crown area growth efficiency |
| $E_{\text{cpa/sa}}$ | cpa/sa | $m^2\ m^{-2}$ | Degree of crown engagement |
| $E_{\text{iv/sa}}$ | iv/sa | $m^3\ year^{-1}\ m^{-2}$ | Stand area growth efficiency |
| $E_{\text{v/sa}}$ | v/sa | $m^3\ m^{-2}$ | Stand density in terms of volume |

To compare the efficiencies between different groups of trees, the $E$-values may be allometrically adjusted in order to eliminate size effects (see also Sect. 4.2.1 and Box 4.2)

with tree size and require allometric adjustment before being applied to compare trees in mixed versus monospecific stands when there are differences in tree size, e.g. in tree diameter or tree volume (Dirnberger and Sterba 2014).

### 6.7.2 Elimination of the Size Effect from Efficiency Parameters

Efficiency ratios such as $E_{\text{iv/cpa}}$ or $E_{\text{iv/sa}}$ are useful for comparing tree growth in mixed versus monospecific stands. However, these ratios change with tree size. If the size of trees in mixed and monospecific stands differs, such size-dependent changes in the efficiency parameters should be eliminated. Allometric theory (Enquist et al. 1998; West et al. 2009) and empirically derived allometric exponents (Pretzsch and Dieler 2012) are available for such corrections. If these are not undertaken, then mixing effects might be confounded with size effects as shown in the following example.

Supposing that Norway spruce volume increment, iv, scales with an exponent of $\alpha_{\text{iv},\,v}=0.42$ and cpa with $\alpha_{\text{csa},\,v}=0.62$ to total plant volume, then the efficiency of crown area, $E_{\text{iv, csa}}=\text{iv/csa}$ scales with $\alpha_{\text{iv/cpa},\,v}=-0.20$ to plant volume (as $\text{iv}\propto v^{0.42}$ and $\text{cpa}\propto v^{0.62}$ $E_{\text{iv, cpa}}=\text{iv/cpa}\propto v^{0.42-0.62}\propto v^{-0.20}$) as shown in Fig. 6.28 at (a) linear and (b) double-logarithmic scale. Suppose a tree of volume $v_m=1.0\ \text{m}^3$ growing in the mixed stand has a crown area efficiency of $E_{iv/cpa,\,m}=1.25\ \text{cdm year}^{-1}\ \text{m}^{-2}$ and should be compared with a tree of volume $v_p=3.0\ \text{m}^3$ and crown area efficiency of $E_{\text{iv/cpa},\,p}=1.20\ \text{cdm year}^{-1}\ \text{m}^{-2}$ in the neighbouring monocultures. The relative efficiency of the tree in the mixed stand is then $rE_{\text{iv/cpa}}=E_{\text{iv/cpa},\,m}/E_{\text{iv/cpa},\,p}=1.25/1.20=1.04$, which is equivalent to a 4% increase through mixing. In this case it is obvious that the size growth of the tree (or a mean tree) in the mixed stand is behind the monoculture ($v_m<<v_p$), and part of its superiority is related to its younger allometric state. If we apply $\alpha_{\text{iv/cpa},\,v}=-0.20$ and estimate which $E'_{\text{iv, cpa},\,m}$ the tree in the mixed stand would have if it was the same size as the reference tree in the monoculture ($v_p=3.0\ \text{m}^3$), we get $E'_{\text{iv/cpa},\,m}=(\text{iv}_m/\text{cpa}_m)\times(v_p/v_m)^{-0.20}$ in the example $E'_{\text{iv/cpa},\,m}=(1.25)\times(3/1)^{-0.20}=1.00$ and the relative efficiency calculated on the basis of

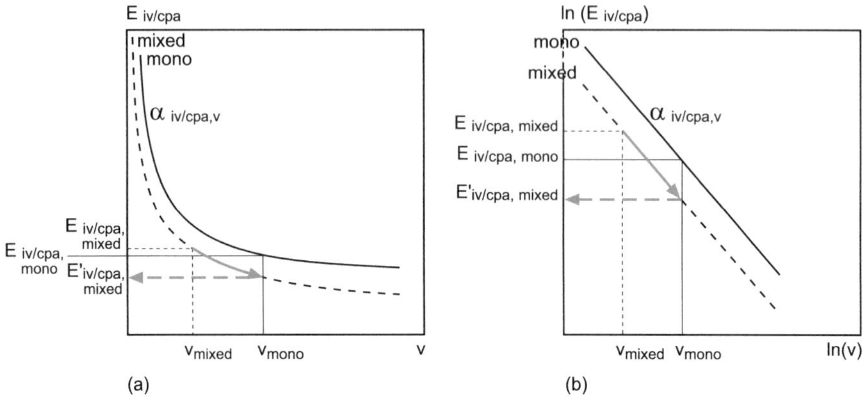


**Fig. 6.28** Elimination of size effects from efficiency parameters such as $E_{iv/cpa}$ prior to their application for comparing tree growth in mixed versus monospecific stands, illustrated at (**a**) linear and (**b**) logarithmic scale. Allometric correction is recommendable when the compared trees in mixed and monospecific stands differ in size, as represented by their stem volume, $v_m$ and $v_p$, respectively. The adjustment starts with the observed $E_{iv/cpa}$ value for the tree with volume $v_m$ in mixture ($E_{iv/cpa,\ m}$) and uses the allometric slope $\alpha_{iv/csa,\ v}$ to extrapolate up to the volume $v_p$ (solid arrow from $v_m$ to $v_p$). The read off value is the adjusted efficiency $E^{'}_{iv/cpa,\,m}$ which can be compared with the reference value $E_{iv/cpa,\,r}$ for the monoculture in order to calculate the real mixing effect $rE = E^{'}_{iv/cpa,\,m}/E_{iv/cpa,\,p}$

$rE'_{iv/cpa} = E'_{iv/cpa,m}/E_{iv/cpa,p} = 1.00/1.20 = 0.83$. In other words, the real mixing effect is −17% as opposed to +4%; the different sizes create the illusion of a positive mixing effect.

In such cases, one should either refrain from comparison at the tree level or eliminate the size effects as shown. Based on the uncorrected mixing effect $rE_{iv,\,cpa} = 1.04$ and the 'real' mixing effect $rE^{'}_{iv,\,cpa} = 0.83$, we can calculate the size effect $S = rE_{iv,\,cpa}/rE^{'}_{iv,\,cpa} = 1.04/0.83 = 1.25$. The observed uncorrected mixing effect $rE_{iv,\,csa} = 0.83 \times 1.25 = 1.04$ can be split into a real negative mixing effect ($rE^{'}_{iv,\,cpa} = 0.83$) and a positive size effect ($S = 1.25$). Part of the size effect might have been caused in the past by mixing but should not be confounded with present mixing effects.

For the allometric adjustment in this chapter, the following exponents for Norway spruce and European beech, $\alpha_{iv/v,\,v} = -0.58, -0.41$, $\alpha_{cpa/v,\,v} = -0.38, -0.31$, $\alpha_{iv/cpa,\,v} = -0.20, -0.10$, and $\alpha_{sa/v,\,v} = -0.38, -0.31$, were specially derived from trees on long-term experiments. See Pretzsch and Dieler (2012) for empirically derived allometric scaling relationships of other tree species or Enquist et al. (1998, 2009), and West et al. (2009) for general scaling exponents of allometrically ideal plants.

### 6.7.3 Transition from Tree Growth to Tree Productivity in Mixed Versus Monospecific Stands: An Example

The transition from growth of the individual tree to the productivity of the mean tree and the stand productivity of mixed versus monospecific stands is illustrated for the age series NOR 811, FRE 813, and SON 814 of monospecific and mixed forest stands of Norway spruce and European beech in southern Germany. The age series provide triplets of neighbouring plots which are different in terms of species composition (spruce monoculture, mixed spruce/beech, beech monoculture) and relatively similar in all other characteristics such as site conditions, stand age, silvicultural treatment, etc. The 12 triplets of plots cover 36–161-year-old stands with dominant height of 25–44 m at age 100, standing volume stock of 55–1.041 $m^3$ $ha^{-1}$, and current annual volume growth of 2–32 $m^3$ $ha^{-1}$ $year^{-1}$. The relatively even-aged and monolayered monospecific and mixed stands are fully stocked, unthinned, or only slightly thinned and represent good to optimal site conditions for both spruce and beech. For more details about the plots, see Pretzsch and Schütze (2009) and Pretzsch et al. (2010).

As the considered stands are more or less even aged and the size characteristics are normally distributed, this is the easiest way to explain the transition from the tree level to the stand level. In these stands, the mean tree is a good representative of the respective cohort. In order to demonstrate how mixing can modify the growth and growth efficiency at the individual tree level, we calculated the mean ratios cpa/v, iv/cpa, cpa/sa, iv/sa, and v/sa introduced in Table 6.2 for all plots of the 12 triplets. The mean ratios between the mean tree volume in mixed versus monospecific stands of $v_m/v_r = 0.85$ for beech and $v_m/v_r = 1.04$ for spruce indicate that beech is 15% smaller and spruce 4% larger in mixed stands versus the neighbouring monoculture of the same triplet. The effects of these size differences were eliminated through allometric adjustment; the adjusted ratios are marked by a stroke (e.g. $E'$ mixed).

$E_{cpa/v}$ in Table 6.3 reflects that with a given tree volume, $v$, trees in mixed stands develop a wider crown than those in monospecific stands. The table shows an

**Table 6.3** Refined analysis and allometric adjustment of tree growing space efficiency and productivity of (a) Norway spruce and (b) European beech growing in mixed and monospecific stands

| | Norway spruce | | | European beech | | |
|---|---|---|---|---|---|---|
| Efficiency ratios E | E′ mixed | E mono | E′ mix/E mono | E′ mixed | E mono | E′ mix/E mono |
| $E_{cpa/v}$ ($m^2$ $m^{-3}$) | 13.96 | 11.83 | 1.18 | 38.33 | 26.44 | 1.45 |
| $E_{iv/cpa}$ (cdm $year^{-1}$ $m^{-2}$) | 2.92 | 3.04 | 0.96 | 1.32 | 1.66 | 0.80 |
| $E_{cpa/sa}$ ($m^2$ $m^{-2}$) | 1.01 | 0.95 | 1.06 | 2.21 | 1.31 | 1.69 |
| $E_{iv/sa}$ (cdm $year^{-1}$ $m^{-2}$) | 2.96 | 2.93 | 1.01 | 2.17 | 1.89 | 1.15 |
| $E_{v/sa}$ ($m^3$ $m^{-2}$) | 0.07 | 0.08 | 0.96 | 0.06 | 0.05 | 1.13 |

For each of the ratios, the table presents allometrically adjusted ratios for mixed stands, observed values for monocultures, and the ratios between mixed and monospecific stands increase of 18% for spruce and 45% for beech. Beech is already superior to spruce in terms of crown extension in monospecific stands (26.44/11.83 = 2.2) but even more so in mixed stands (38.33/13.96 = 2.8). On average, both species benefit from mixture in $E_{cpa/v}$, which indicates that the benefit of one species is not at the expense of the other but rather that their strategies are complementary. The $E_{iv/cpa}$ values indicate that mixing reduces the crown area efficiency. This reduction amounts to 4% for the rather vertically expanding spruce and 20% for laterally expanding beech. In contrast to $E_{cpa/v}$, the crown efficiency $E_{iv/cpa}$ is about twofold for spruce compared with beech. Mixing increases the crown engagement $E_{cpa/sa}$ in spruce by just 6% ($E'_{cpa/sa,m}/E_{cpa/sa,p} = 1.06$) but in beech by 69% ($E'_{cpa/sa,m}/E_{cpa/sa,p} = 1.69$). In both mixed and monospecific stands of beech, the crown engagement $E_{csa/sa}$ is superior to spruce. Mixing hardly changes the stand area efficiency $E_{iv/sa}$ (= productivity) of spruce ($E'_{iv/sa,m}/E_{iv/sa,p} = 1.01$) but increases that of beech by 15% ($E'_{iv/sa,m}/E_{iv/sa,p} = 1.15$). So, productivity of spruce is generally higher in both monospecific and mixed stands, but beech benefits more from mixing than spruce. Note that especially for beech, efficiency measures based on crown area ($E_{iv/cpa}$) may indicate a negative mixing effect. However, due to the multiple crown coverage of beech as indicated by its high $E_{cpa/sa}$ values, it can achieve an overyielding in inter- as opposed to intraspecific environments. The wider extension and denser interlocking of beech crowns enables a 13% higher stand density of beech in mixed stands as indicated by $E_{v/sa}$ ratios of $E'_{v/sa,m} = 0.06$ compared with $E'_{v/sa,p} = 0.05$ in monoculture.

### 6.7.4 Crown Engagement cpa/sa as Link Between Tree and Stand Level Evaluation

For linking tree to stand level evaluation, the overlap and interlocking of crowns is crucial. From $E_{iv/sa} = \text{iv/sa}$, $E_{iv/csa} = \text{iv/csa}$, and $E_{sa/csa} = \text{sa/csa}$, it follows that $E_{iv/sa} = E_{iv/csa} \times E_{csa/sa}$. The latter equation conceptualises individual tree productivity as a product of individual tree crown efficiency and crown engagement. Consider further that the mixing effects in terms of crown efficiency and crown engagement can be expressed as $rE_{iv/csa} = E_{iv/csa,m}/E_{iv/csa,p}$ and $rE_{cpa/sa} = E_{cpa/sa,m}/E_{cpa/sa,p}$, respectively. The relative productivity of trees in mixed versus monospecific stands is then $rE_{iv/sa} = rE_{iv/cpa} \times rE_{cpa/sa}$, i.e. the relative effect of mixing on productivity is the product of the effect of mixing on crown efficiency and crown engagement.

Figure 6.29 illustrates the efficiency ratios for the 12 triplets in monospecific and mixed stands of Norway spruce and European beech introduced in the previous section (see also Table 6.3). In Fig. 6.29a the mean crown efficiency of beech in mixture is plotted over the efficiency in neighbouring monocultures. Observations close to the 1.0-line would indicate equality of efficiency. Points above or below this bisectoral line indicate a plus or minus, respectively, as compared with monospecific stands. In the case of beech, crown efficiency is reduced through mixing, but crown expansion increased to such an extent that productivity is 10–20% higher

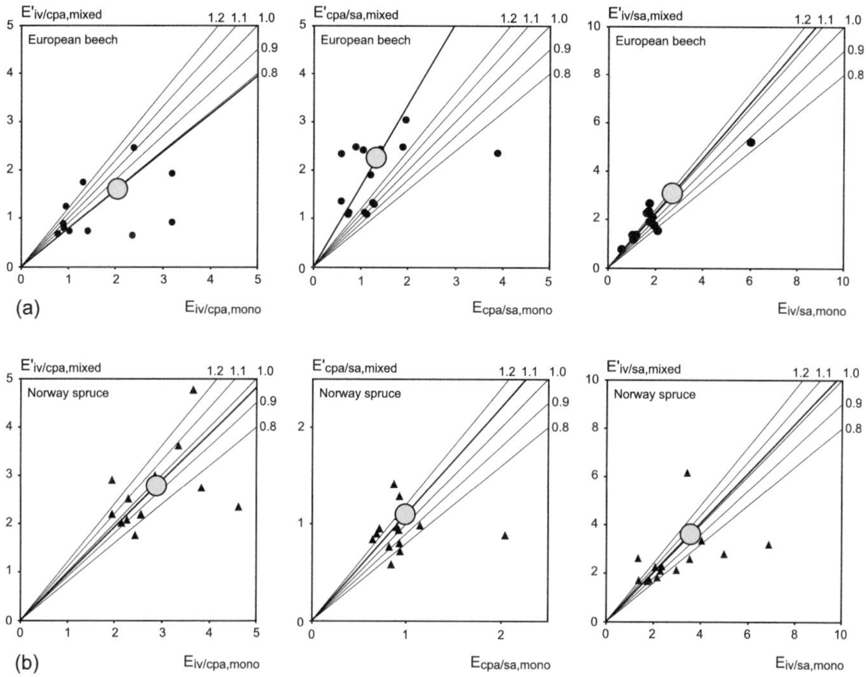


**Fig. 6.29** Ratios for refined analysis of tree growing space efficiency and productivity of (**a**) European beech and (**b**) Norway spruce growing in mixed versus monospecific stands. The following efficiency parameters are shown for each species $E'_{\mathrm{iv/csa},m}$ vs. $E_{\mathrm{iv/cpa},p}$ (left), $E'_{\mathrm{cpa/sa},m}$ vs. $E_{\mathrm{cpa/sa},p}$ (centre), $E'_{\mathrm{iv/sa},m}$ vs. $E_{\mathrm{iv/sa},p}$(right). Small symbols represent means for the included mixed versus monospecific stand plots; large symbols represent mean values over all 12 triplets. In the case that the behaviours of mixed and monospecific stands were equal, the observations would lie on the bisectoral lines

in mixed compared with neighbouring monospecific stands (Fig. 6.29a, from left to right). For Norway spruce, the reactions to mixing are similar but less pronounced (Fig. 6.29b, from left to right). When coping with interspecific competition, both species benefit to a different extent from a release of lateral crown restriction. Their wider crown spread and overlap can overcompensate the reduction of crown efficiency and result in a higher tree and stand productivity. Fully stocked mixed stands can differ slightly from monocultures in terms of the ground cover ratio, but the main difference is their higher sum of crown area due to two- to fourfold crown overlaps (see also Sect. 5.7.2). As crown maps are rare and crown areas difficult to measure, the multiple coverage was often neglected and counted only once in analyses (e.g. Assmann 1970, p. 105), but this seems to be crucial for over-/underyielding and should be considered more thoroughly in future.

Figure 6.30 illustrates the complementarity of Norway spruce and European beech in terms of space occupation with a section of the mixed stand SON 814/8 (age 80, survey 2004) which is one of the 12 triplets (spruce monoculture, spruce/beech mixed, beech monoculture) used for this example analysis. While spruce

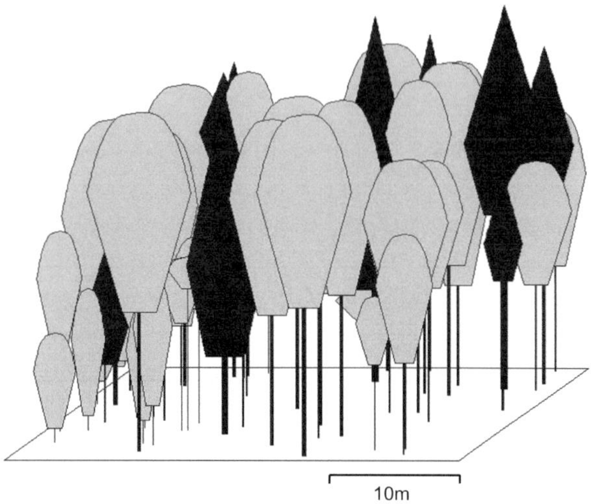


**Fig. 6.30** Section of the mixed stand SON 814/8 (age 80, survey 2004) belonging to one of the 12 triplets (spruce monoculture, spruce/beech mixed, beech monoculture) of this study. The section illustrates the complementarity of Norway spruce and European beech in terms of space occupation

tends to expand vertically, beech fills the canopy space through lateral expansion and can increase the stand density.

## 6.8 Overview of Some Main Causes and Effects of Tree Species Mixing on Individual Tree Growth

Many of the beneficial tree mixing reactions introduced in this chapter result from complementary crown and root shape, spatially or temporally complementary resource exploitation, redistribution of resources, or modification of growth allocation and allometry. This section summarises the main causes and effects of tree species mixing on individual tree growth. There are many more causes of beneficial mixing effects, such as mechanical stabilisation, reduction of crown shyness, mechanical abrasion of parts of the crown by wind and storm, barrier building against biotic and abiotic disturbances, holobiontic interactions via mycorrhizal networks, or common use of root channels for exploitation of deeper soil layers. As the effects of these are so difficult to disentangle, the following overview of cause and effect is still rather phenomenological, qualitative, and mostly restricted to two-species mixtures. Many ongoing research projects strive to generalise and better quantify the effects and to extend the research to mixtures with more than two species.

### 6.8.1 Complementary Exploitation of Crown and Root Space

Any changes in tree crown morphology and canopy structure in interspecific versus intraspecific environments suggest a modification of the trees' space occupation, resource capture, and productivity. Of special interest is a wider crown extension as this indicates a competition reduction and an increase in light interception and may also contribute to overyielding of mixed compared with monospecific stands (Chap. 4). Changes in crown structure and allometry underline that the mixing effect goes beyond a merely additive effect where the species' behaviour in terms of structure and growth remains similar to that in the monoculture (Fig. 6.31a–c).

In temperate forests where tree growth is primarily limited by light, canopy space filling may be denser and light interception higher in mixed than in monospecific stands. Especially when their crown shapes are complementary, the crown expansion may be wider (Fig. 6.31d), the tree packing density higher (Fig. 6.31e), or both reactions may be combined (Fig. 6.31f). Similar complementarity with

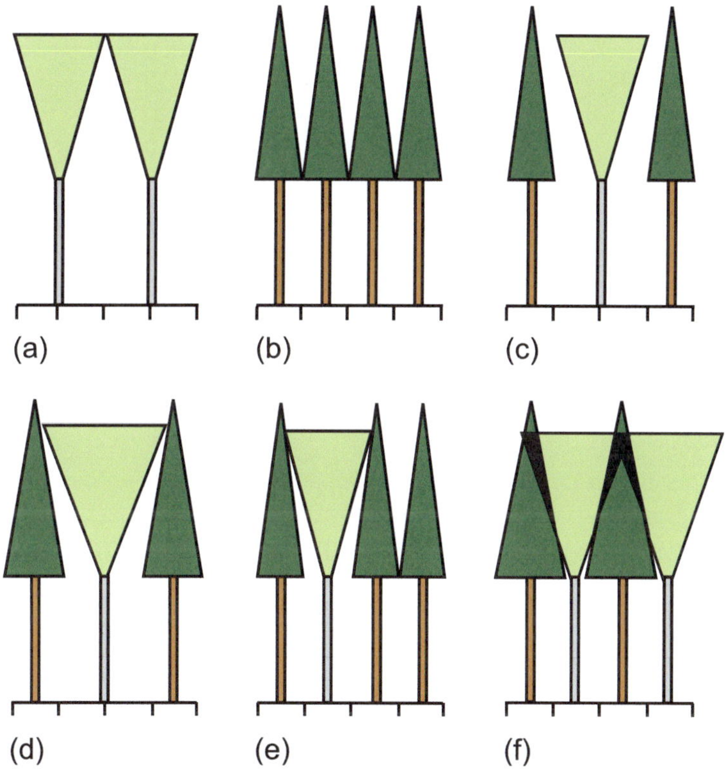


**Fig. 6.31** Potential effects of species mixing on the canopy and root space filling when tree crowns have complementary shapes. When two species with complementary crown shapes (**a** and **b**) retain the same shape and space occupation in mixed-species stands—as is common for monospecific stands—the space filling in mixture may be simply additive (**c**). However, in many cases, species mixing enables a wider crown expansion (**d**), higher stand density (**e**), or a combination of both (**f**). The cases (**d**)–(**f**) indicate a multiplicative mixing effect through denser canopy space filling (Pretzsch 2014)

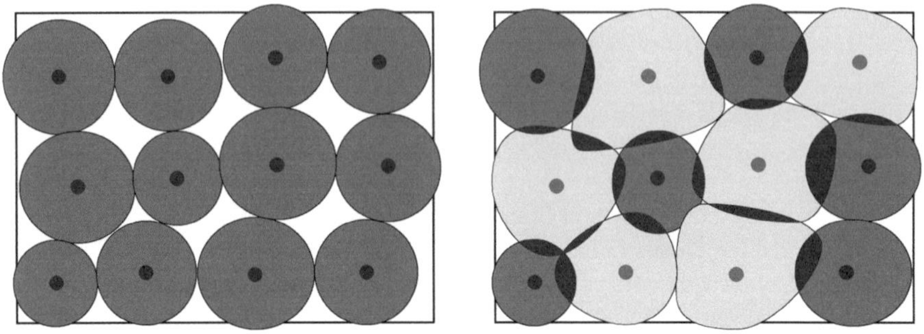


**Fig. 6.32** Schematic representation of canopy pattern in monospecific versus mixed stands. (**a**) Rather circular crowns in monospecific stand with about 5–10% uncovered space in between, a crown cover of 90–95%, and sum of crown areas below 100%. (**b**) More irregular crown shapes in mixed stand with scarce uncovered space in between, crown cover close to 100%, and multiple crown overlap resulting in a sum of crown areas of greater than 100%

regard to their root morphology may increase the benefit of inter- compared with intraspecific neighbourhoods.

Because of the complementarity in crown shape (e.g. a combination of ▼-shaped with ▲-shaped crowns) and resource demand (e.g. combination of light-demanding and shade-tolerant species), tree crowns may expand more widely in inter- compared with intraspecific environments. On top of that, mixing can change inner crown properties such as branch length, branch angle, number of branches, and their straightness. The cpa/sa relationship can become greater in mixed stands (Fig. 6.32b) and the degree of crown closure—and therefore the crown closure and the sum of crown projection areas—may be higher than in monocultures (Fig. 6.32a). Larger crowns and denser canopy space filling can increase the tree and stand growth.

Analogously, a complementary root morphology may benefit inter- compared with intraspecific neighbourhoods in the root space.

### 6.8.2 Hydraulic Lift and Hydraulic Redistribution

Especially in dry years and on dry sites where water is the limiting factor, processes in the root space become decisive for tree growth and survival. However, outcomes under drought conditions may turn into benefits for some of the associates. One such phenomenon is the well-described 'hydraulic lift' or 'hydraulic redistribution' of water in woody plants, particularly observable in arid conditions, through deeply extending root systems (Fig. 6.33a), which release water at night into surface-near soil horizons of low water potential (Fig. 6.33b) as described in detail by Caldwell and Richards (1989) and Prieto et al. (2012). Once released, such water represents

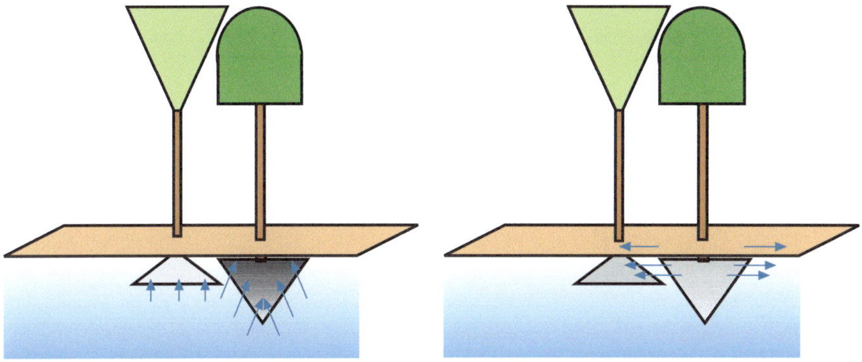


**Fig. 6.33** If shallow-rooting species are mixed with deep-rooting species, they can benefit from hydraulic redistribution. (**a**) During the day when stomata are opened, both species take up water for transpiration. (**b**) Dry soil conditions may extract water from the deep-reaching roots, especially during the night when water flow is reduced. Shallower rooting trees may benefit from this redistribution in terms of water supply and growth (Caldwell and Richards 1989)

an external storage capacitance for the deep-rooting plant, but concomitantly enables survival of neighbouring shallow-rooting plants as the profiteers. The resulting benefit is not an active support provided by the deep-rooting plant, but occurs indirectly as an emergent feature of the specific species association under a specific site scenario. In such a case, competition turns into facilitation, which denotes an indirectly mediated, functional emergence between neighbouring plants. One or several of the associates may benefit with regards to water availability and growth.

### 6.8.3 Increased Availability of Mineral Nutrient Supply Through Deep-Rooting or Atmospheric $N_2$ Fixation

In mixed stands one tree species may benefit from another species which can exploit mineral nutrients in deeper root layers (Fig. 6.34a) or can fix atmospheric $N_2$ (Fig. 6.34a). In both cases, mineral nutrients are procured by a benefactor species and partly transferred to the other through litter fall. Thus, the additional nutrients are socialised among the neighbouring trees of the benefactor independently of their species. An additional benefit may come from root channels abandoned by one species and used by the other for penetrating compacted soil layers and reaching resources in deeper and otherwise inaccessible soil layers (Bauhus and Messier 1999; Gaiser 1952; Puhe 2003; Stone and Kalisz 1991).

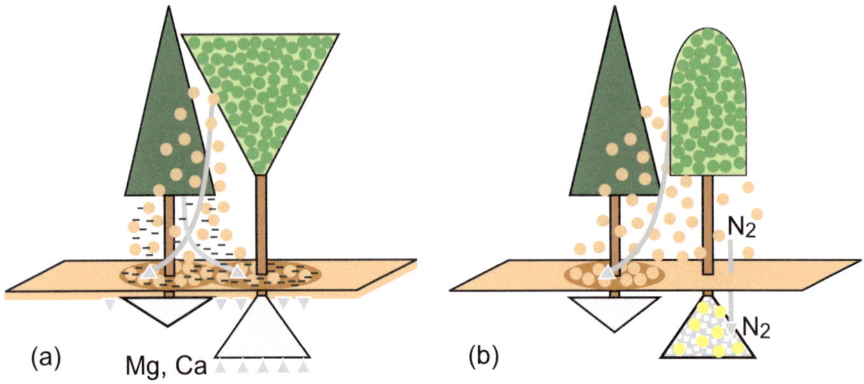


**Fig. 6.34** In assemblages of tree species with different behaviours of mineral nutrient exploitation, one species may facilitate the other through socialisation of nutrient supply via leaf and needle turnover. (**a**) Shallower rooting species may benefit from deep-rooting neighbours because they act as a 'mineral nutrient pump' (Rothe 1997; Rothe and Binkley 2001). (**b**) Species with the potential for atmospheric $N_2$ fixation may facilitate neighbouring species that lack this ability, especially when growing on N-limited sites (Forrester et al. 2006, 2007)

### 6.8.4 Temporal Complementarity of Niches, Resource Use, and Growth

Many studies show clear differences in the course of growth between the species across their lifetimes (Assmann 1970, p 45) but also within a single year (Pretzsch 2005a; Rötzer et al. under review; Schober 1950/51). This asynchrony in growth entails a temporal diversification in resource uptake which can release stress.

For instance, Norway spruce may profit from a released competition for water in early spring when European beech is still leafless and lives from the reserves of the previous year (Fig. 6.35). In this period, photosynthesis and growth of spruce may profit from higher temperatures and light supply till beech sprouts in April/May. Afterwards, European beech may profit from the mixture in relation to the mono-specific stand as the admixed Norway spruces have shallower roots. During the main growing season, scattered European beeches in the mixture with Norway spruce may have greater water availability than in monocultures of beech as the interspecific competition with neighbouring spruces is lower than the intraspecific competition with beech (Rötzer et al. under review). This may explain why beeches in interspecific environments grow much better than those beside beeches, especially in drought years (Goisser et al. 2016; Rötzer et al. under review).

Analogously, temporal diversification of space occupation during stand development may relax resource competition. It occurs, for example, in species assemblages of shade-tolerant and shade-intolerant, shallow- and deep-rooting, and early- and late-successional tree species. For given site conditions, species 1 and 2 in Fig. 6.36 may access resources in different spaces or at different times.

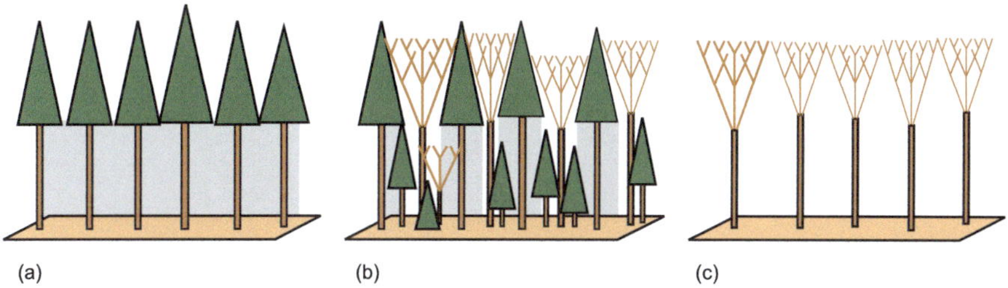


**Fig. 6.35** Temporal complementarity of resource use and growth when mixing evergreen species (**a**) with deciduous tree species (**c**). In mixtures of both species (**b**), conifers can start growing before the sprouting of their deciduous neighbours. This asynchrony may promote smaller conifers in particular and enrich stand structure and stand productivity (Schober 1950/51)

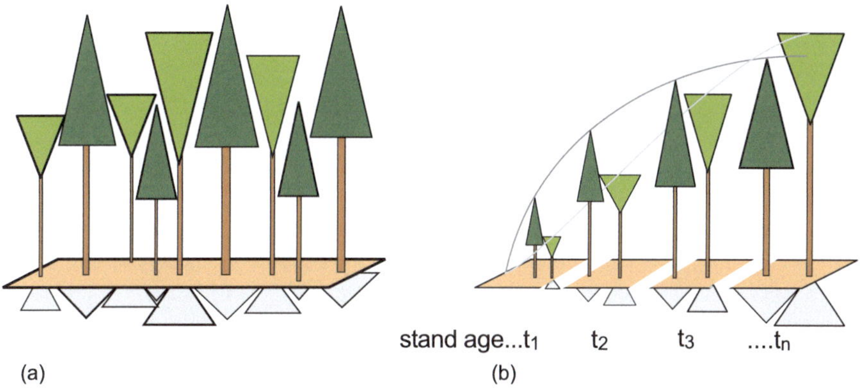


**Fig. 6.36** Complementary occupation and packing density in the canopy and root space found in one stand development phase (**a**) may change over time (**b**). In the mixture shown on the right (**b**), conifer is initially ahead but later lags behind in height growth. However, due to the species-specific crown development, complementary space occupation may result in all stand development phases

## 6.8.5 Modification of Growth Partitioning and Allometry of Trees in Inter- Versus Intraspecific Neighbourhoods

In order to acclimatise to an interspecific neighbourhood, trees species may change their growth partitioning in mixed-species stands compared with monospecific stands. Most of the comparisons between the productivity of mixed and pure stands are based on stem volume production (Liang et al. 2016; Pretzsch et al. 2015). Comparison based on total biomass production may produce different results, as tree species mixing can change stem-crown allometry (Bayer et al. 2013), root-shoot relationship (Thurm et al. 2016), and also tree-ring width and wood density (Zeller 2016) as illustrated in Fig. 6.37.

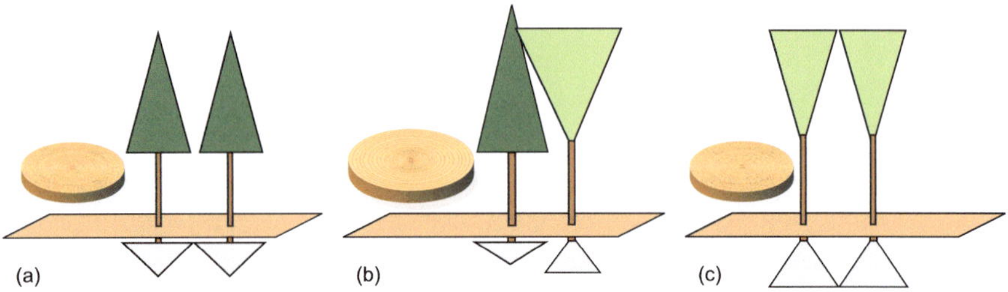


**Fig. 6.37** Tree species mixing may modify the growth allocation and allometry in mixed (**b**) compared with monospecific stands (**a** and **c**). Mixing may modify the crown size, the root-shoot relationship, the tree-ring width, and wood density. So, comparison between mixed and monospecific stands based on the stem volume growth may be misleading (Zeller 2016)

An increase in crown in relation to stem size in mixed-species stands as reported by Dieler and Pretzsch (2013) and Pretzsch (2014) would mean that the overyielding would be even higher when calculated for the total above-ground volume of mixed versus pure stands. However, the decrease of root in relation to shoot growth as reported by Thurm et al. (2016) and reported in Sect. 6.5.3 as well as the decrease of wood density revealed by Zeller (2016) would lower the overyielding of mixed versus pure stands when calculated for the total biomass growth. Ongoing works on within-tree growth partitioning in mixed versus pure stands will clarify how mixed stand productivity compares with that in pure stands when calculated for the stem volume productivity, which is of primary interest for forestry. They will also clarify how mixed stand productivity compares with pure stands based on their total biomass production, which is relevant for ecosystem understanding and carbon balance.

The presented tree properties emerging when growing in interspecific in contrast to intraspecific neighbourhoods go beyond an acceleration or reduction of growth velocity, i.e. a quicker tree and stand development. Rather, they can change the trees' structure and space requirement. The presented tree properties emerging in interspecific in contrast to intraspecific competition indicate complementarity in use of above- and below-ground resources.

Beyond the increased productivity, the wider extension of tree crowns and multiple crown overlap in mixture can increase the growth resilience after disturbances, as gaps are smaller or can be closed more quickly. Properties emerging in mixed but not in monospecific stands such as increase in crown size, change in the crowns' space filling pattern, reduction in growing area requirement, and increase in tree productivity are crucial for appropriate estimation of maximum density, silvicultural treatment, and dynamic modelling of mixed-species stands. Because of their size, firm position, and longevity, trees are the founder species in ecosystems and determine the living conditions of many ecosystem characteristics, functions, and services. So, the effects of mixing on tree and furthermore on stand structure have far-reaching effects on most ecosystem properties.

## Take-Home Messages

1. Interspecific difference in tree size causes well-known additive mixing effects such as higher or lower tree numbers per unit area in mixed versus monospecific stands due to different growing space requirements. Interspecific functional and structural complementarity (light-demanding vs. -shade-tolerant, early- vs. late-successional, shallow vs. deep-rooting, deciduous vs. evergreen species) may cause multiplicative mixing effects such as multiple crown layering, denser canopy space filling, and eventually overyielding.
2. The feedback between environment, growth, and structure enables a structural acclimation of trees to their local environment. Their size and longevity improve the capacity to increasingly acclimate morphologically with progressing stand development. Mixing can foster the variability of both the local environment and the tree structure and growth.
3. Species mixing can trigger emergent properties of individual tree structure and growth which may result in an increase in resource capture and overyielding of mixed compared with monospecific stands.
4. Depending on the net effect of competition and facilitation by neighbours, the unimodal increment curve of individual trees can be modified in different ways; e.g. its level can be raised or lowered and its rhythm can be accelerated or slowed down.
5. Due to the complementarity of species and a reduction of competition, crowns may expand more widely in length and width in interspecific compared with intraspecific environments.
6. Mixing can generate allometric relationships, crown projection areas, and inner crown properties which differ significantly from respective properties in monospecific stands.
7. Wider crown extension and multiple crown overlap in mixed stands may indicate a more complete canopy space filling and light capture. Wider reach of the crowns combined with lower growing space requirements can increase the tree productivity in mixed versus monospecific stands by 10–20%.
8. By relating the achieved tree growth to growing area instead of plant or crown size, we proceed from plant biology to production ecology.
9. Tree characteristics change non-linearly with tree size. Therefore, comparison of structure or growth of trees in mixed versus monospecific stands should consider and eliminate any size differences between trees in mixed and monospecific stands.
10. The beneficial tree mixing reactions introduced in this chapter result mainly from complementary crown and root shape, spatially or temporally complementary resource exploitation, redistribution of resources, or modification of growth allocation and allometry.

## References


Alemdag IS (1978) Evaluation of some competition indexes for the prediction of diameter increment in planted white spruce. Can For Serv, Ottawa, Canada

Arz MAO (2013) Strukturelle Kronenanalyse von Fichte (*Picea abies* [L.] Karst.) und Buche (*Fagus sylvativa* L.) im Rein- und Mischbestand. Kombination von terrestrischen Laserscan- und Zuwachsdaten

Assmann E (1970) The principles of forest yield study. Pergamon Press, Oxford, New York

Assmann E, Franz F (1965) Vorläufige Fichten-Ertragstafel für Bayern. Forstwissenschaftliches Centralblatt 84, 1/2, Sonderdruck, Verlag Paul Parey, Hamburg/Berlin

Augusto L, Ranger J, Binkley D, Rothe A (2002) Impact of several common tree species of European temperate forests on soil fertility. Ann For Sci 59:233–253

Bauhus J (2009) Rooting patterns of old-growth forests: is aboveground structural and functional diversity mirrored belowground? In: Cea W (ed) Old-growth forests, ecological studies, vol 207. Springer, Berlin, pp 211–229. doi:10.1007/978-3-540-92706-8_10

Bauhus J, Messier C (1999) Soil exploitation strategies of fine roots in different tree species of the southern boreal forest of eastern Canada. Can J For Res 29(2):260–273

Bayer D, Seifert S, Pretzsch H (2013) Structural crown properties of Norway spruce and European beech in mixed versus pure stands revealed by terrestrial laser scanning. Trees 27(4): 1035–1047

Belsky AJ, Canham CD (1994) Forest gaps and isolated savanna trees. Bioscience 44(2):77–84

Binkley D, Stape JL, Ryan MG (2004) Thinking about efficiency of resource use in forests. For Ecol Manag 193:5–16

Brooker RW, Maestre FT, Callaway RM, Lortie CL, Cavieres LA, Kunstler G, Michalet R (2008) Facilitation in plant communities: the past, the present, and the future. J Ecol 96(1):18–34

Brown GS (1965) Point density in stems per acre, vol 38. New Zealand Forest Research Note, Wellington, New Zealand

Caldwell MM, Richards JH (1989) Hydraulic lift: water efflux from upper roots improves effectiveness of water uptake by deep roots. Oecologia 79(1):1–5

Callaway RM, Walker LR (1997) Competition and facilitation: a synthetic approach to interactions in plant communities. Ecology 78(7):1958–1965

Canham CD, Finzi AC, Pacala SW, Burbank DH (1994) Causes and consequences of resource heterogeneity in forests: interspecific variation in light transmission by canopy trees. Can J For Res 24(2):337–349

Canham CD, LePage PT, Coates KD (2004) A neighborhood analysis of canopy tree competition: effects of shading versus crowding. Can J For Res 34(4):778–787

Christmann (1939) Ertragstafel für den Kiefern-Fichten-Mischbestand. In: Wiedemann E (ed) Ertragstafeln wichtiger Holzarten bei verschiedener Durchforstung sowie einiger Mischbestandsformen. Schaper Verlag, Hannover, p 100

Comeau PG, Kimmins JP (1989) Above- and below-ground biomass and production of Lodgepole pine on sites with differing soil moisture regimes. Can J For Res 19:447–454

Connell JH (1990) Apparent versus "real" competition in plants. In: Grace JB, Tilman D (eds) Perspectives on plant competition. Academic Press, San Diego, pp 9–26

Dieler J, Pretzsch H (2013) Morphological plasticity of European beech (*Fagus sylvatica* L.) in pure and mixed-species stands. For Ecol Manag 295:97–108

Dirnberger GF, Sterba H (2014) A comparison of different methods to estimate species proportions by area in mixed stands. For Syst 23(3):534–546

Enquist BJ, Brown JH, West GB (1998) Allometric scaling of plant energetics and population density. Nature 395:163–165

Enquist BJ, West GB, Brown JH (2009) Extension and evaluations of a general quantitative theory of forest structure and dynamics. Proc Natl Acad Sci U S A 106(17):7046–7051

Faber PJ (1981) Die Standflächenschätzung über den Distanzfaktor. In: Proc Dt Verb Forstl Forschungsanst. Sek Ertragskd, Soest, pp 87–95

Faber PJ (1983) Concurrentie en groei van de bomen binnen een opstand (Konkurrenz und Wachstum der Bäume in einem Waldbestand). Pijksinstituut voor onderzoek in de bos- en landschapsbouw "De Dorschkamp", vol 18(1). Uitvoerig verslag, Wageningen

Forrester DI (2013) The spatial and temporal dynamics of species interactions in mixed-species forests: from pattern to process. For Ecol Manag. doi:10.1016/j.foreco.2013.10.003

Forrester DI (2014) A stand-level light interception model for horizontally and vertically heterogeneous canopies. Ecol Model 276:14–22

Forrester DI, Bauhus J, Cowie AL, Vanclay JK (2006) Mixed-species plantations of Eucalyptus with nitrogen-fixing trees: a review. For Ecol Manag 233:211–230

Forrester DI, Bauhus J, Cowie AL, Mitchell PA, Brockwell J (2007) Productivity of three young mixed-species plantations containing N2-fixing *Acacia* and non-N2-fixing *Eucalyptus* and *Pinus* trees in Southeastern Australia. For Sci 53(3):426–434

Fraser AR (1977) Triangle based probability polygons for forest sampling. For Sci 23(1):111–121

Gaiser RN (1952) Root channels and roots in forest soils. Soil Sci Soc Am J 16(1):62

Goisser M, Geppert U, Rötzer T, Paya A, Huber A, Kerner R, Bauerle T, Pretzsch H, Pritsch K, Häberle K-H, Matyssek R, Grams TEE (2016) Does belowground interaction with *Fagus sylvatica* increase drought susceptibility of photosynthesis and stem growth in *Picea abies*? For Ecol Manag 375:268–278

Griess VC, Knoke T (2011) Growth performance, windthrow, and insects: meta-analyses of parameters influencing performance of mixed-species stands in boreal and northern temperate biomes. Can J For Res 41:1141–1158

Gspaltl M, Sterba H, O'Hara KL (2012) The relationship between available area efficiency and area exploitation index in an even-aged coast redwood (*Sequoia sempervirens*) stand. Forestry 85(5):567–577

Hari P (1985) Theoretical aspects of eco-physiological research. In: Tigerstedt PMA, Puttonen P, Koski V (eds) Crop physiology of forest trees. Helsinki University Press, Helsinki, pp 21–30. (336 p)

Heinsdorf D (1999) Das Revier Sauen – Ein Beispiel für erfolgreichen Waldumbau, Schriftenreihe des Ministeriums für Ernährung, Landwirtschaft und Forsten. vol Band VI. Ministerium für Ernährung, Landwirtschaft und Forsten, Brandenburg, Landesforstanstalt Eberswalde

Helms JA (1998) The dictionary of forestry. The Society of American Foresters, Bethesda, MD

Jack WH (1968) Single trees sampling in evenaged plantations for survey and experimentation. In: 14th IUFRO Congress, München, pp 379–403

Kelty MJ (1992) Comparative productivity of monocultures and mixed stands. In: Kelty MJ, Larson BC, Oliver CD (eds) The ecology and silviculture of mixed-species forests. Kluwer Academic, Dordrecht, pp 125–141

Keyes MR, Grier CC (1981) Above-and below-ground net production in 40-years-old Douglas-fir stands on low and high productivity sites. Can J For Res 11:599–605

Kimmins JP (1993) Scientific foundations for the simulation of ecosystem function and management in FORCYTE-11. Forestry Canada, Northern Forestry Centre, Edmonton, Alberta

Kleiber M (1947) Body size and metabolic rate. Physiol Rev 27(4):511–541

Knoke T, Hahn A (2007) Baumartenvielfalt und Produktionsrisiken: Ein Forschungseinblick und-ausblick – Diversity of tree species and risk: A research insight and outlook. Schweizerische Zeitschrift fur Forstwesen 158(10):312–322

Kuoch R (1972) Zur Struktur und Behandlung von subalpinen Fichtenwäldern. Schweiz Z Forstwes 123:77–89

Larocque GR, Luckai N, Adhikary SN, Groot A, Bell FW, Sharma M (2013) Competition theory-science and application in mixed forest stands: review of experimental and medelling methods and suggestions for future research. Environ Rev 21:71–84

Liang J, Crowther TW, Picard N, Wiser S, Zhou M, Alberti G, Schulze E-D, McGuire AD, Bozzato F, Pretzsch H, de-Miguel S, Paquette A, Hérault B, Scherer-Lorenzen M, Barrett CB, Glick HB, Hengeveld GM, Nabuurs G-J, Pfautsch S, Viana H, Vibrans AC, Ammer C, Schall P, Verbyla D, Tchebakova N, Fischer M, Watson JV, HYH C, Lei X, Schelhaas M-J,

Lu H, Gianelle D, Parfenova EI, Salas C, Lee E, Lee B, Kim HS, Bruelheide H, Coomes DA, Piotto D, Sunderland T, Schmid B, Gourlet-Fleury S, Sonké B, Tavani R, Zhu J, Brandl S, Vayreda J, Kitahara F, Searle EB, Neldner VJ, Ngugi MR, Baraloto C, Frizzera L, Bałazy R, Oleksyn J, Zawiła-Niedźwiecki T, Bouriaud O, Bussotti F, Finér L, Jaroszewicz B, Jucker T, Valladares F, Jagodzinski AM, Peri PL, Gonmadje C, Marthy W, O'Brien T, Martin EH, Marshall AR, Rovero F, Bitariho R, Niklaus PA, Alvarez-Loayza P, Chamuya N, Valencia R, Mortier F, Wortel V, Engone-Obiang NL, Ferreira LV, Odeke DE, Vasquez RM, Lewis SL, Reich PB (2016) Positive biodiversity-productivity relationship predominant in global forests. Science 354(6309). doi:10.1126/science.aaf8957

Matyssek R, Lüttge U (2012) Gaia. The planet holobiont. Nova Acta Leopoldina NF 114(391): 325–344

Mayer H, Ott E (1991) Gebirgswaldbau Schutzwaldpflege. Gustav Fischer Verlag, Stuttgart, New York

McCarthy MC, Enquist BJ (2007) Consistency between an allometric approach and optimal partitioning theory in global patterns of plant biomass allocation. Funct Ecol 21:713–720

Møller AP, Swaddle JP (1997) Asymmetry, developmental stability, and evolution. Oxford Serie in ecology and evolution. Oxford University Press, Oxford

Nagel J (1985) Wachstumsmodell für Bergahorn in Schleswig-Holstein. Universität Göttingen, 124p

Nelder JA (1962) New kinds of systematic designs for spacing experiments. Biometrics 18(3): 283–307

Niklas KJ (1994) Plant allometry. Univ Chicago Press, Chicago, IL

Niklas KJ (2004) Plant allometry: is there a grand unifying theory? Biol Rev 79:871–889

Nikolova PS, Zang C, Pretzsch H (2011) Combining tree-ring analyses on stems and coarse roots to study the growth dynamics of forest trees: a case study on Norway spruce (*Picea abies* [L.] H. Karst). Trees 5:859–872

Oldemann RAA (1990) Forests: elements of silvology. Springer, Berlin

Oliver CD, Larson BC (1996) Forest stand dynamics: Update edition. Wiley, New York

Pelz DR (1978) Estimating individual tree growth with tree polygons. Blacksburg, VA

Pretzsch H (2005a) Diversity and productivity in forests. In: Scherer-Lorenzen M, Körner C, Schulze E-D (eds) Forest diversity and function, Ecological studies 176. Springer, Berlin, pp 41–64

Pretzsch H (2005b) Stand density and growth of Norway spruce (*Picea abies* (L.) Karst.) and European beech (*Fagus sylvatica* L.). Evidence from long-term experimental plots. Eur J For Res 124(3):193–205

Pretzsch H (2006) Von der Standfächeneffizienz der Bäume zur Dichte-Zuwachs-Beziehung des Bestandes. Beitrag zur Integration von Baum- und Bestandesebene. Allgemeine Forst- und Jagdzeitung 177:188–199

Pretzsch H (2009) Forest dynamics, growth and yield. Springer, Berlin

Pretzsch H (2014) Canopy space filling and tree crown morphology in mixed-species stands compared with monocultures. For Ecol Manag 327:251–264

Pretzsch H, Biber P (2005) A re-evaluation of Reinekes rule and stand density index. For Sci 51 (4): 304–320

Pretzsch H, Biber P (2010) Size-symmetric versus size-asymmetric competition and growth partitioning among trees in forest stands along an ecological gradient in central Europe. Can J For Res 40:370–384

Pretzsch H, Dieler J (2012) Evidence of variant intra- and interspecific scaling of tree crown structure and relevance for allometric theory. Oecologia 169(3):637–649. doi:10.1007/s00442-011-2240-5

Pretzsch H, Schütze G (2005) Crown allometry and growing space efficiency of Norway Spruce (*Picea abies* [L.] Karst) and European Beech (*Fagus sylvatica* [L.]) in pure and mixed stands. Plant Biol 7(6):628–640

Pretzsch H, Schütze G (2009) Transgressive overyielding in mixed compared with pure stands of Norway spruce and European beech in Central Europe: evidence on stand level and explanation on individual tree level. Eur J For Res 128:183–204

Pretzsch H, Block J, Dieler J, Dong PH, Kohnle U, Nagel J, Spellmann H, Zingg A (2010) Comparison between the productivity of pure and mixed stands of Norway spruce and European beech along an ecological gradient. Ann For Sci 67:712

Pretzsch H, Biber P, Uhl E (2012a) Coarse root-shoot allometry of *Pinus radiata* modified by site conditions in the Western Cape province of South Africa. Southern Forests 74(4):237–246

Pretzsch H, Schütze G, Uhl E (2012b) Resistance of European tree species to drought stress in mixed versus pure forests: evidence of stress release by inter-specific facilitation. Plant Biol 15:483–495

Pretzsch H, Heym M, Pinna S, Schneider R (2014) Effect of variable retention cutting on the relationship between growth of coarse roots and stem of black spruce (*Picea mariana* (Mill.) Britton). Scand J For Res 29(3):222–233

Pretzsch H, Forrester DI, Rötzer T (2015) Representation of species mixing in forest growth models. A review and perspective. Ecol Model 313:276–292

Prieto I, Armas C, Pugnaire FI (2012) Water release through plant roots: new insights into its consequences at the plant and ecosystem level. New Phytol 193(4):830–841

Puhe J (2003) Growth and development of the root system of Norway spruce (*Picea abies*) in forest stands—a review. For Ecol Manag 175(1):253–273

Purves DW, Lichstein JW, Pacala SW (2007) Crown plasticity and competition for canopy space: a new spatially implicit model parameterized for 250 North American tree species. PLoS One 2(e870). doi:10.1371/journal.pone.0000870

Reineke LH (1933) Perfecting a stand-density index for even-aged forest. J Agric Res 46:627–638

Richards AE, Forrester DI, Bauhus J, Scherer-Lorenzen M (2010) The influence of mixed tree plantations on the nutrition of individual species: a review. Tree Physiol 30(9):1192–1208

Roloff A (2001) Baumkronen. Verständnis und praktische Bedeutung eines komplexen Naturphänomens. Ulmer, Stuttgart

Rothe A (1997) Einfluß des Baumartenanteils auf Durchwurzelung, Wasserhaushalt, Stoffhaushalt und Zuwachsleistung eines Fichten-Buchen-Mischbestandes am Standort Höglwald. Forstl Forschungsber München 163:174

Rothe A, Binkley D (2001) Nutritional interactions in mixed species forests: a synthesis. Can J For Res 31:1855–1870

Rubner M (1931) Die Gesetze des Energieverbrauchs bei der Ernährung, vol 16/18. Proc preuß Akad Wiss Physik-Math Kl, Berlin, Wien

Ryan MG, Waring RH (1992) Maintenance respiration and stand development in a subalpine lodgepole pine forest. Ecology 73(6):2100–2108

Ryan MG, Yoder BJ (1997) Hydraulic limits to tree height and tree growth. Bioscience 47(4): 235–242

Saha S, Kuehne C, Kohnle U, Brang P, Ehring A, Geisel J, Leder B, Muth M, Petersen R, Peter J, Ruhm W, Bauhus J (2012) Growth and quality of young oaks (*Quercus robur* and *Quercus petraea*) grown in cluster plantings in central Europe: a weighted meta-analysis. For Ecol Manag 283:106–118

Schober R (1950/51) Zum jahreszeitlichen Ablauf des sekundären Dickenwachstums. Allgemeine Forst- und Jagdzeitung 122:81–96

Schütz JP (1989) Zum Problem der Konkurrenz in Mischbeständen. Schweiz Z Forstwes 140(12): 1069–1083

Smith TM, Smith RL (2009) Elements of ecology. Pearson International Edition, 7th edn. Benjamin Cummings, San Francisco

Sterba H, Amateis RL (1998) Crown efficiency in a loblolly pine (*Pinus taeda*) spacing experiment. Can J For Res 28(9):1344–1351

Sterba H, Andrae F, Pambudhi F (1993) Crown efficiency of oak standards as affected by mistletoe and coppice removal. For Ecol Manag 62(1):39–49

Stone EL, Kalisz PJ (1991) On the maximum extent of tree roots. For Ecol Manag 46(1–2):59–102

Strobel GW (1995) Rottenstruktur und Konkurrenz im subalpinen Fichtenwald

Thurm EA, Biber P, Pretzsch H (2016) Tree growth is partitioned in favor of the stem on the expense of the roots for Douglas-fir (*Pseudotsuga menziesii* (Mirb.) Franco) and European beech (*Fagus sylvatica* L.) if growing in mixture and under humid conditions. Trees. doi:10.1007/s00468-016-1512-4

Uhl E, Biber P, Ulbricht M, Heym M, Horváth T, Lakatos F, Gál J, Steinacker L, Tonon G, Ventura M, Pretzsch H (2015) Analysing the effect of stand density and site conditions on structure and growth of oak species using Nelder trials along an environmental gradient: experimental design, evaluation methods, and results. For Ecosyst 2(1):17

Vandermeer J (1992) The ecology of intercropping. Cambridge University Press, Cambridge

Verein Deutscher Forstlicher Versuchsanstalten (1873) Anleitung für Durchforstungsversuche. In: von Ganghofer A (ed) Das Forstliche Versuchswesen, vol 2. Schmid'sche Buchhandlung, Augsburg, pp 247–253

Verein Deutscher Forstlicher Versuchsanstalten (1902) Beratungen der vom Vereine Deutscher Forstlicher Versuchsanstalten eingesetzten Kommission zur Feststellung des neuen Arbeitsplanes für Durchforstungs- und Lichtungsversuche. Allgemeine Forst- und Jagdzeitung 78:180–184

von Bertalanffy L (1951) Theoretische Biologie: II. Band, Stoffwechsel, Wachstum, 2nd edn. A Francke AG, Bern

von Lüpke B, Spellmann H (1997) Aspekte der Stabilität und des Wachstums von Mischbeständen aus Fichte und Buche als Grundlage für waldbauliche Entscheidungen. Forstarchiv 68: 167–179

von Lüpke B, Spellmann H (1999) Aspects of stability, growth and natural regeneration in mixed Norway spruce-beech stands as a basis of silvicultural decisions. In: Olsthoorn AFM, Bartelink HH, Gardiner JJ, Pretzsch H, Hekhuis HJ, Franc A (eds) Management of mixed-species forest: silviculture and economics, vol 15. IBN Scientific Contributions, pp 245–267

West GB, Enquist BJ, Brown JH (2009) A general quantitative theory of forest structure and dynamics. Proc Natl Acad Sci U S A 106(17):7040–7045

Wiedemann E (1942) Der gleichaltrige Fichten-Buchen-Mischbestand. Mitt Forstwirtsch u Forstwiss 13:1–88

Wiedemann E (1951) Ertragskundliche und waldbauliche Grundlagen der Forstwirtschaft. Frankfurt am Main

Yoder BJ, Ryan MG, Waring RH, Schoettle AW, Kaufmann MR (1994) Evidence of reduced photosynthetic rates in old trees. For Sci 40(3):513–527

Zeide B (1985) Tolerance and self-tolerance of trees. For Ecol Manag 13:149–166

Zeide B (1998) Fractal analysis of foliage distribution in loblolly pine crowns. Can J For Res 28: 106–114

Zeller L (2016) Tree ring width and wood density in mixed versus pure stands of Scots pine and European beech. TUM, 39 p

Zilber-Rosenberg I, Rosenberg E (2008) Role of microorganisms in the evolution of animals and plants: the hologenome theory of evolution. FEMS Microbiol Rev 32(5):723–735

# Ecological Stability of Mixed-Species Forests

# 7

Jürgen Bauhus, David I. Forrester, Barry Gardiner, Hervé Jactel, Ramon Vallejo, and Hans Pretzsch

**Abstract**

In many parts of the world, forests are likely to face novel disturbance regimes as a result of global change processes, and there is concern that the capacity of forest ecosystems to withstand, recover from, or adapt to these novel disturbance regimes may decline. Creation and maintenance of species-diverse forests is seen as an important option to adapt forests to uncertain future disturbances.

J. Bauhus (✉)
Chair of Silviculture, Faculty of Environment and Natural Resources, University of Freiburg, Tennenbacherstr. 4, 79106 Freiburg, Germany
e-mail: juergen.bauhus@waldbau.uni-freiburg.de

D.I. Forrester
Swiss Federal Institute for Forest, Snow and Landscape Research WSL, Zürcherstrasse 111, 8903 Birmensdorf, Switzerland
e-mail: david.forrester@wsl.ch

B. Gardiner
UMR 1391 ISPA, Bâtiment C2, INRA Centre de Bordeaux Aquitaine, 71 Avenue Edouard Bourleaux, CS 20032, 33882 Villenave d'Ornon cedex, France
e-mail: barry.gardiner@bordeaux.inra.fr

H. Jactel
Entomologie Forestière & Biodiversité, Forest Entomology & Biodiversity, UMR BIOGECO – INRA, 69, route d'Arcachon, 33612 CESTAS cedex, France
e-mail: herve.jactel@inra.fr

R. Vallejo
Department of Evolutionary Biology, Ecology and Environmental Sciences, Faculty of Biology, University of Barcelona, Av. Diagonal, 643, 08028 Barcelona, Spain
e-mail: vvallejo@ub.edu

H. Pretzsch
Chair for Forest Growth and Yield Science, Faculty of Forest Science and Resource Management, Technical University of Munich, Hans-Carl-von-Carlowitz-Platz 2, 85354 Freising, Germany
e-mail: Hans.Pretzsch@lrz.tum.de



H. Pretzsch et al. (eds.), *Mixed-Species Forests*, DOI 10.1007/978-3-662-54553-9_7

However, it is not known whether benefits of mixed-species forests consist mainly of risk spreading among tree species that have different susceptibility to various stressors and disturbance agents or whether they also have emergent properties resulting from interactions among species, which increase the resistance and resilience of participating species or the entire ecosystem. Here we review the evidence for the effects of tree diversity on the resistance and resilience of forests in relation to a number of abiotic (drought, wind, fire) and biotic (insect herbivores, pathogens) stress and disturbance factors. For the abiotic disturbances, damage or reduction in ecosystem function can be reduced, compared with monocultures of susceptible or less resilient species, when more resistant or resilient species are mixed with less susceptible and less resilient species. However, storm, fire, or drought damage to individual species may not be reduced in mixtures when compared to monocultures. The stress or disturbance impacts may even be aggravated for one or more species in some mixtures, as is shown for drought. There is more evidence for beneficial diversity effects in relation to biotic disturbance agents. Mixing tree species reduces the impact of insect herbivores on individual susceptible tree species in the majority of cases, where the community is dominated by specialist herbivores. However, the opposite effect may occur with generalist herbivores, which can be promoted by tree diversity. Similarly, tree diversity can reduce the impact of specialist pathogens on host tree species, whereas there is little evidence for positive influences in the case of generalist pathogens. In most cases, tree species diversity dilutes the impact of disturbance agents and, owing to different susceptibility of species to specific disturbances, insures against a complete damage or loss. In addition, mixing tree species can reduce temporal variation in growth and stabilise productivity. However, there is little evidence for true, positive diversity effects, where diversity leads to an increase in the resistance and resilience of component species in mixed-species communities. From an economic point of view, mixing might help to reduce risk for a more vulnerable and valuable species, even if there are no benefits for the admixed species. However, forest managers should be aware that mixtures do not provide universally higher resistance or resilience in relation to disturbances than monocultures. In most cases, it depends to a large extent on the attributes of the species in mixture in relation to the specific disturbances.

## Contents

7.1 Introduction ........ 339
7.2 Drought ........ 343
7.2.1 Mixing Effects on Water Use and Water-Use Efficiency ........ 345
7.2.2 Responses to Drought Stress ........ 346
7.3 Storm Damage ........ 351
7.3.1 Susceptibility of Trees and Stands to Storm Damage ........ 352
7.3.2 Do Tree Species Mixtures Reduce the Vulnerability to Storms? ........ 355
7.4 Fire Risk ........ 356
7.5 Other Abiotic Disturbances ........ 361
7.6 Pest Insects ........ 361

7.7 Pathogens .......... 364
7.8 Interactions Between Stress and Disturbance Factors .......... 368
7.9 Can Tree Species Diversity Stabilise Productivity Over Time? .......... 369
Take-Home Messages .......... 373
References .......... 374

## 7.1 Introduction

We have seen in previous chapters that tree species diversity may enhance ecosystem functioning and that there is an expectation that it may also increase ecological stability. So far in this book, the focus has been on the effects of tree species diversity on the productivity of forest stands. The productivity of forests and their general functioning can be substantially impaired by stress and disturbance. Recent exceptionally strong disturbances or novel combinations of different types of disturbances are pushing forest ecosystems beyond their resilience threshold (Millar and Stephenson 2015). It is feared that owing to these unprecedented disturbances, forest health and hence the provision of ecosystems services may be severely impaired (Trumbore et al. 2015).

In this chapter we will explore the question of whether or not, and to what extent, mixed-species forests may be less affected by environmental stress and ecological disturbances. One advantage of tree species diversity is to spread the risk. Because different species are not equally susceptible to particular biotic or abiotic stress and disturbance factors, the likelihood that some trees will escape infection, infestation, windthrow, or fire damage increases with the number of tree species in a forest stand. These surviving trees can then maintain, at least to some extent, the structure and the functioning of the forest and facilitate the recovery from stress and disturbance (Jactel et al. 2009).

Environmental stress and ecological disturbances are natural phenomena in forest ecosystems. Whereas the concept of stress relates to individual organisms, disturbance refers to populations, communities, and ecosystems. Both stress and disturbance can cause a deviation from the 'normal' functioning of individuals, populations, communities, and ecosystems. One of the difficulties in defining and quantifying stress and disturbance is therefore in describing what we mean by normal, because there are no constant conditions in natural systems. Any unfavourable condition or substance that affects or blocks a plant's metabolism, growth, or development can be regarded as stress (Lichtenthaler 1998). Here we will focus on strong or chronic stress events in plants that eventually reduce cell activity and plant growth to a minimum. Intense or long-lasting stress can lead to serious damage in plants ultimately causing the death of cells, organs, or the entire organism, if the stress-coping or repair mechanisms of plants are overtaxed (Lichtenthaler 1998). We exclude in our further discussion those stress events such as shading that can be largely compensated for by acclimation, the physiological adjustment to changing environmental conditions. Natural stress may be caused by many abiotic and biotic factors such as very low or very high levels of irradiance, temperature, and water availability, mineral nutrient deficits, browsing, and pests and pathogens. In addition, there are many direct anthropogenic stress factors caused, for example, by air pollution or herbicides; these will not be dealt with here.

When stress events are so strong that the stress-coping mechanisms of plants are overburdened and mortality of organs or entire plants occurs, we speak of disturbances. This distinction may not always be easy. For example, droughts typically cause stress in plants, but severe droughts may lead to widespread mortality and subsequent changes in the structure and composition of vegetation communities. A classical definition of disturbance by Pickett and White (1985) is: 'A disturbance is any relatively discrete event in time that disrupts ecosystem, community, or population structure and changes resources, substrate availability, or the physical environment'.

This and similar definitions have been critiqued by many ecologists (e.g. Van der Maarel 1993), because there is no timescale involved and because the disruption is not defined. Instead the disruption is related to the concept of fluctuation of environmental conditions, measured as statistical variation. Yet, the above definition can be regarded as an absolute definition of disturbance because it requires measurement of the disruption in structure and the changes in resources, substrates, and environments (White and Jentsch 2001). The ecosystem property or process that is disrupted or changed through disturbance may simply be the biomass that is partially or completely destroyed, as proposed by Grime (1979). The destruction of biomass is a variable that can be easily measured across a wide range of ecosystems. However, the magnitude of disturbance (see Fig. 7.1) could be measured in many other ways as well, for example, the loss of species, the changes in nutrient cycling, and so on.

Ecological disturbances can range from small-scale, caused by the death of a single tree, to a large-scale, catastrophic, and stand replacing event such as those caused by windstorms, intense fires, and outbreaks of pest insects. Disturbance regimes to which particular forests are exposed are typically characterised by the type of disturbance (e.g. fire, windthrow, flooding, insect outbreak, etc.), its intensity, the spatial extent, frequency of recurrence, and itsrandomness (Pickett and White 1985; Attiwill 1994).

Disturbances are omnipresent, they are an inherent and unavoidable element of ecosystems and their dynamics and are important for ecosystem function and integrity (White and Jentsch 2001). Native forest species have evolved with the prevailing disturbance regimes and have developed mechanisms to resist disturbances and to recover from disturbances (Attiwill 1994). Hence, forest management approaches have been developed to consider and emulate the natural disturbance regimes to minimise the impact on biodiversity (e.g. Seymour and White 2002).

At the same time, in managed forests, particularly in intensively managed semi-natural forests, plantations, or urban forests, ecological disturbances are commonly viewed as accidents because they can destroy lives, capital and infrastructure, and can impair the provision of ecosystem services. For that reason, risk management strategies have been developed in forestry, for example, to reduce the incidence or

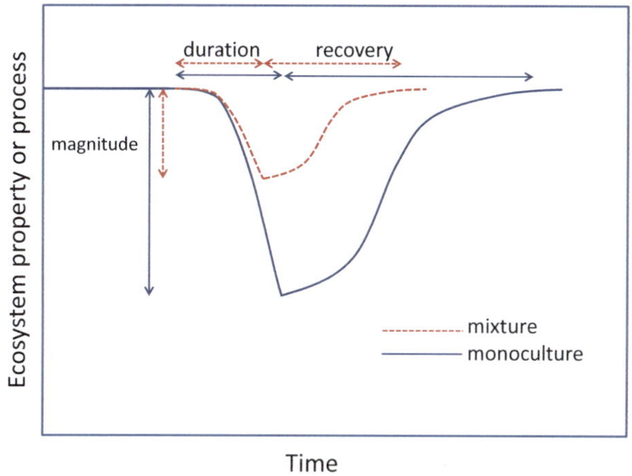


**Fig. 7.1** Conceptual illustration of hypotheses related to the response of monospecific and mixed-species forests to disturbance or environmental stress. In mixtures, the magnitude of the disturbance may be smaller, its duration shorter, and the recovery from disturbance faster than in monocultures. The **magnitude**, as understood here, is not the magnitude of the disturbance factor, which may be measured as the wind force or the fire intensity, but the deviation in the ecosystem property or process from a pre-disturbance steady-state level. If we subtract the magnitude of deflection from the pre-disturbance state, we obtain a measure of **resistance**; the smaller the magnitude of deflection in relation to a certain disturbance force, the higher the resistance. Their **duration** of disturbance is likely to be the same in monocultures and mixtures in the case of most abiotic disturbance agents, e.g. the duration of a storm, flood or fire. However, it is conceivable that fires might last for a shorter period of time, if mixed forests provide less flammable fuel than some monospecific forests. In the case of biotic disturbances agents such as pest insects, the outbreak might last for a shorter period in mixtures than in monocultures. In this figure, a complete **recovery** of the ecosystem property or process following disturbance is depicted. This may not necessarily be the case. The difference between the level of the ecosystem property or process attained following the recovery period and the pre-disturbance period can be quantified as **resilience** (see also Box 7.1)

severity of the effects of fires or storms (Tàbara et al. 2003; Gardiner and Quine 2000). There is also the fear that some disturbance regimes might change as a result of global change, and hence forest ecosystems might experience either new types of disturbances or more intense or more frequent disturbances than they have experienced historically (Dale et al. 2001). These new disturbance regimes might have unprecedented consequences for forest ecosystem functioning and the provision of goods and services. Therefore strategies have been developed to maintain or increase the ecological stability of forests in the face of global change (e.g. Millar et al. 2007; Seppälä et al. 2009). These strategies typically comprise options to increase the resistance and resilience of forest ecosystems in the face of disturbances and their adaptability with regard to long-term changes in environmental stresses and disturbance regimes (Millar et al. 2007; Bolte et al. 2009). Maintaining and developing mixed-species forests is a common element of forest management adopting these strategies (Bauhus et al. 2013). One important assumption underpinning these strategies is that in mixtures, the magnitude of the disturbance may be smaller, its duration shorter, and the recovery from disturbance faster or more complete than in monocultures (Fig. 7.1). On the one side, an increasing number of functionally different species increases the probability that some of these are less strongly affected by external stressors or disturbances. In addition, the probability increases that one or some species can take over the function of another that may have been strongly impaired or lost from the system owing to the disturbance or new conditions (Walker et al. 1999; Yachi and Loreau 1999). On the other side, through interactions between species, the detrimental effect of stress or disturbance may be buffered for some species through the changes to ecosystem properties and processes caused by the presence of other species (Fig. 7.2). For example, slow-growing and frost-sensitive tree species can be

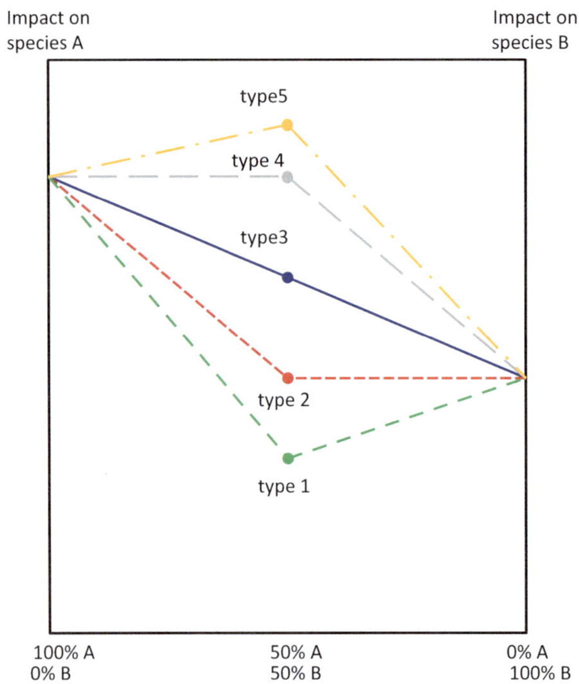


**Fig. 7.2** Possible impacts of stress and disturbances on mixed-species forests illustrated for two monocultures of tree species A (*left*) and B (*right*) and their 1:1 mixture (*centre*). The impact is depicted here as the stress or disturbance incidence for individual trees of each species. In monoculture, species B is less susceptible to the disturbance agent than species A. For the mixed stand, five different outcomes are possible: (1) Trees of both species benefit from mixing in a synergistic way. The disturbance impact in mixture is lower than in monoculture for both species. The reduction in impact is more pronounced for the more susceptible species A. (2) The impact is reduced for the more susceptible species A, but not for species B. (3) The disturbance impact is reduced in the susceptible species A, but at the expense of increased damage of species B. (4) Mixing does not reduce the disturbance damage in trees of the more susceptible species, but increases the damage in the less susceptible species. (5) The effect of mixing is antagonistic; it increases the damage level of disturbance in both companion species

protected by frost-tolerant, fast-growing pioneer species (Örlander and Karlsson 2000) (response type 2 in Fig. 7.2). In addition, the effects of species-specific disturbance agents such as specialist herbivores or pathogens may be simply diluted in mixtures (Castagneyrol et al. 2014a) (also response type 2 in Fig. 7.2). For some specific mixtures, the reduced susceptibility to stress or disturbance for one or the other participating tree species has been demonstrated, e.g. *Picea abies* in Europe (Neuner et al. 2015). However, it has not been established whether these theoretical benefits of mixed-species forests apply to all types of different mixtures and stress or disturbance factors. Using a modelling approach for a temperate forest landscape in Central Europe, Pedro et al. (2015) could demonstrate that increasing tree species diversity could buffer the impacts of disturbances on the forest carbon cycle, which was used as a measure of ecosystem functioning. In their model system, which comprised up to 13 tree species of different successional stages as well as different types and intensities and frequencies of disturbances, the beneficial effects of tree diversity were more pronounced in young successional stages than in old-growth stages. Whereas the mitigating influence of diversity on the disturbance impact varied with forest age, it generally had a positive influence on the resilience. In addition, the positive effects of diversity were mostly driven by the functional diversity of participating tree species, and they became more pronounced with increasing frequency of disturbance. The advantage of this modelling approach is that it could cover spatial (2500 ha) and temporal scales (500 years) that are impossible to capture in experiments or retrospective analyses. At the same time, the modelling approach offers only a few hints as to the underlying mechanisms responsible for the diversity effects on disturbance impacts and resilience.

In the following we will therefore review current knowledge on the interactions between the ecological disturbances or environmental stress and tree species diversity. We will focus in particular on the influence of tree species diversity on resistance, resilience, and ecological stability of forests in relation to stress and disturbance factors that are likely to change in their frequency, intensity, extent, and interactions with each other with global change (Dale et al. 2001). These comprise abiotic factors such as storms, drought, and fire as well as biotic factors such as insect pests, fungal pathogens, and herbivory. The focus will be on processes that change ecosystem resistance and resilience in relation to different stress and disturbance factors in mixtures when compared to monocultures.

## 7.2 Drought

The frequency and severity of drought and heat stress are predicted to increase during the twenty-first century (IPCC 2013) with drastic consequences for forest ecosystem functioning and mortality (e.g. Allen et al. 2010). Despite the fact that tree mixtures are commonly advocated to adapt forests to climate change, including in areas where more frequent and severe droughts are predicted (Reif et al. 2010), the effect of tree species mixtures on drought stress in trees and on forest hydrological cycles has received only little attention. This is surprising, since we have seen above that mixed-species forests are often more productive than their monospecific counterparts and forest evapotranspiration usually increases with productivity (Law et al. 2002).

Therefore we might ask: how can mixtures of tree species influence the drought tolerance of individual species in mixture or the whole tree community through either increasing their resistance or resilience to drought (Fig. 7.1). In situations when mixtures are more productive than monocultures of the companion species, it is likely that they will also use more water (see Chap. 3, Forrester et al. 2010; Kunert et al. 2012). The few studies that have compared the water use and water-use efficiency of mixtures compared with monocultures have all shown that when the growth of a given species increases in mixture, the transpiration and/or water-use efficiency of that species will also increase (Forrester 2015). On dry sites or during drought periods, an increase in transpiration may in consequence increase water stress for some or all species (response type 4 or 5 in Fig. 7.2). However, transpiration is only one of several water-related processes than can influence the evapotranspiration and water stress of forests (see Chap. 3). There are a range of interspecific differences that influence water availability or use and potentially influence resistance and resilience to drought as well (see Chap. 3). For a given level of productivity, these could have a positive effect in mixtures compared with monocultures.

Considering the disturbance responses depicted in Fig. 7.1, mixed-species stands might be considered more tolerant to drought, if they lead to less intense drought stress, shorter periods of stress, and a faster recovery from stress for the whole community when compared to monocultures. In this case, the drought impact should follow the response types 1 or 2 in Fig. 7.2. The magnitude and the duration of drought effects (slower growth or higher mortality) may be reduced if resource partitioning or facilitation between species occurs, thereby increasing the water availability or improving the microclimatic conditions for some or all species. The mechanisms leading to resource partitioning or facilitation such as stratification of root systems among species or hydraulic lift have been explained in Chap. 3. However, they may be effective only in moderate droughts. During severe droughts, when there is no more plant available soil water in the entire rooting depth of the soil profile, these mechanisms will not offer any benefits as has been also observed for the effects of thinning on drought response of individual trees (e.g. Sohn et al. 2013). Different response patterns of trees in mixtures compared with monocultures, with regard to drought events, may therefore be partially explained by the severity of droughts.

There are few studies that have compared water use and drought stress of trees in mixtures to those in monocultures or along gradients in tree diversity (Anders et al. 2006; del Río et al. 2017; Forrester et al. 2010, 2016; Grossiord et al. 2014a, b, c; Kunert et al. 2012; Metz et al. 2015; Lebourgeois et al. 2013; Pretzsch et al. 2014; Thurm et al. 2016; Schume et al. 2004). The majority of these studies have been carried out at a single site or have followed a design that does not permit extrapolation of results to a larger population: confidence levels 3–4 according to Fig. 2.1 (in Chap. 2). Some of these studies have been carried out in planted experiments at one or two sites (Forrester et al. 2010; Grossiord et al. 2014d; Kunert et al. 2012; Moore et al. 2011), and others have been carried out in semi-natural forests at a large number of sites, to capture a larger proportion of the variation within the population of interest representing confidence level 2 (Forrester et al. 2016; Grossiord et al. 2014c; Lebourgeois et al. 2013; Pretzsch et al. 2013).

### 7.2.1 Mixing Effects on Water Use and Water-Use Efficiency

A recent review found that while the effects of mixing on transpiration were variable (increases, decreases, and no differences were all reported), there was still a general pattern such that when the growth of a given species increased in mixture, the transpiration and/or water-use efficiency of that species also increased (Forrester 2015). In mixtures of *Eucalyptus globulus* and *Acacia mearnsii*, water use was higher than in monocultures, although the water-use efficiency of the two companion species had also increased, partially compensating the higher transpiration (Forrester et al. 2010). Similarly, water use in mixed stands of *Picea abies* and *Fagus sylvatica* was substantially higher than in monospecific *P. abies* stands (Schume et al. 2004). In mixed boreal forests comprising *Betula pendula*, *P. abies*, and *Pinus sylvestris*, soil water availability decreased and drought stress increased with tree species diversity in dry years but not in wet years (Grossiord et al. 2014b). In Mediterranean mixed oak forests of *Quercus petraea* and *Q. cerris*, the former species showed no differences in transpiration, whereas transpiration of the latter species during summer drought was more strongly reduced in mixture than in monoculture (Grossiord et al. 2014a) (response type 4 in Fig. 7.2). In a pot experiment with hardwood seedlings, transpiration was significantly higher, on average 11%, in the mixtures than in the monocultures when water supply was ample (Lübbe et al. 2015). In a second treatment with restricted water supply, the diversity effect on transpiration was slightly smaller. The overyielding in transpiration was largely attributable to a species identity effect, here the presence of two species with large leaf areas and sapwood areas.

In the above cases, we therefore have evidence for negative diversity effects and selection effects, where only particular species with certain morphology and physiology are disadvantaged or advantaged in mixture.

However, in mixtures and monocultures of *Pinus sylvestris* and *Fagus sylvatica*, groundwater recharge was higher under mixtures (Anders et al. 2006) than under *P. sylvestris*, despite a higher water use of trees in mixture. In this case, the improved soil moisture and drainage in mixture was the result of suppression of an understorey comprising dwarf shrubs and grasses with high interception and transpiration by the shade-casting beech. This study shows that interactions in mixed stands can be complex and go beyond the direct influence of one tree species on the other.

### 7.2.2 Responses to Drought Stress

In a study of the sensitivity of radial growth of *Abies alba* along an altitudinal and moisture gradient in the Vosges mountains, this species showed reduced sensitivity (increased resistance) to summer drought at the driest sites when mixed with *F. sylvatica* but not when mixed with *P. abies* (Lebourgeois et al. 2013). The authors assumed that this response may be attributable to changes in interception and below-ground competition. However, they did not report how radial growth of the tree species community responded to drought years at the stand level. When comparing the drought response in terms of radial growth of *F. sylvatica*, *P. abies*, and *Q. petraea* in mixed and monospecific stands in Bavaria, positive effects of mixtures were only found for *F. sylvatica* (Pretzsch et al. 2013) (see also Box 7.1). *F. sylvatica* was more resistant and resilient when mixed with *Q. petraea* than in monospecific stands. Possible explanations for this phenomenon were that the asynchronous stress reaction pattern of *F. sylvatica* and *Q. petraea* and a facilitation of *F. sylvatica* by hydraulic lift of water by *Q. petraea* might reduce the drought stress for *F. sylvatica* (see Box 7.1 for a more detailed description of this case study). Similar results were found for *F. sylvatica* when radial growth during and after extreme drought events was compared between monospecific and heterospecific neighbourhoods, where companion species were either *P. sylvestris*, *Fraxinus excelsior*, *Acer pseudoplatanus*, *Quercus* sp., *Tilia cordata*, or *P. abies* (Metz et al. 2015).

**Box 7.1 Indices for Resistance, Resilience, and Recovery of Tree Growth in Relation to Drought Stress**

Experiments on the effect of species mixing and species diversity on the stress response of trees are mostly based on young stands and hardly transferable to mature trees. For example, experiments regarding drought stress on mature trees in mixed versus monospecific stands are rare. Instead, dendroecological studies have been employed to compare the drought response in mixed versus monospecific stands based on increment cores or stem analysis. In Central Europe, the extremely dry years 1976 and 2003 have been used to assess the drought response of trees to extreme stress (e.g. Kohler et al. 2010). These years may be regarded as forerunners of future climatic conditions (Schär et al. 2004), and they have also been employed to quantify the drought response of trees in mixed and monospecific stands, as will be outlined below.

The basis of the model examples for stress reactions in mixed versus monospecific stands are long-term time series of annual radial or basal area growth of trees in mixed and neighbouring monospecific stands with comparable conditions regarding site quality, tree age, stand density, and other attributes. To quantify indices of resistance, resilience, and recovery in relation to stress, the model of Lloret et al. (2011) was adopted (Box Fig. 7.1-1).

(continued)

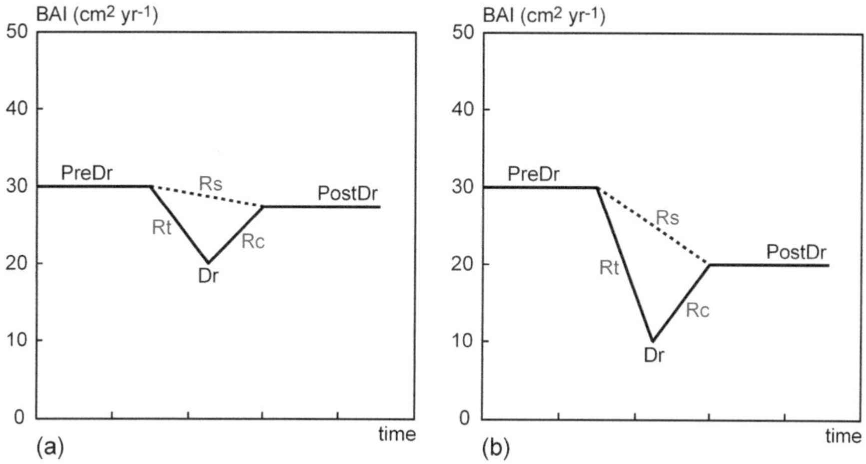


**Box Fig. 7.1-1** Response of radial growth (basal area increment, BAI, in $cm^2$ per year) in relation to two different stress events. The response pattern is characterised by three different phases (see also Fig. 7.1): (**a**) growth in the period before drought, PreDr; (**b**) growth during the drought period, Dr; and (**c**) growth after the drought period, PostDr (modified after Lloret et al. 2011). Indices, for resistance, Rt = Dr/PreDr; recovery, Rc = PostDr/Dr; and resilience, Rs = PostDr/PreDr were used for characterisation of the stress response patterns. In the graphs, Rt, Rc, and Rs are represented by the gradient of decline from PreDr to Dr, the increase from Dr to PostDr, and the difference in level of PreDr and PostDr, respectively. **Resistance** quantifies the decrease from the pre-drought period to the drought period. Rt = 1 means a complete resistance; the further the value falls below Rt = 1, the lower the resistance. **Recovery** describes the growth reaction after the drought period. Rc = 1 indicates a persistence at the low-growth level even after the drought period, values of Rc < 1 indicate a further decline, and values Rc > 1 represent a recovery from the level reached during the drought period. **Resilience** represents the ratio between post-drought and pre-drought increment. $Rs \geq 1$ indicates a full recovery or even an increase after the episodic stress, while values of Rs < 1 indicate a growth decline and low resilience. (**a**) Tree with low-growth decline during drought (PreDr = 30, Dr = 20, PostDr = 27.5) indicated by high resistance, Rt = 0.67, and resilience, Rs = 0.92, and medium recovery, Rc = 1.38. (**b**) Tree with strong growth decline during drought (PreDr = 30, Dr = 10, PostDr = 20) indicated by low resistance, Rt = 0.33, and considerable recovery Rc = 2.00, which results in a medium resilience, Rs = 0.67

**Box 7.1** (continued)

*Drought Tolerance of Trees in Mixed Versus Monospecific Stands*

Drought stress can be modified distinctly by species mixing. Pretzsch et al. (2013) showed that based on tree-ring measurements on increment cores from 559 trees of Norway spruce (*Picea abies* [L.] Karst.), European beech (*Fagus sylvatica* [L.]), and Sessile oak (*Quercus petraea* (Matt.) Liebl.) in southern Germany, with half of them sampled in monospecific and half in mixed stands. Indices for resistance, recovery, and resilience were applied to

(continued)

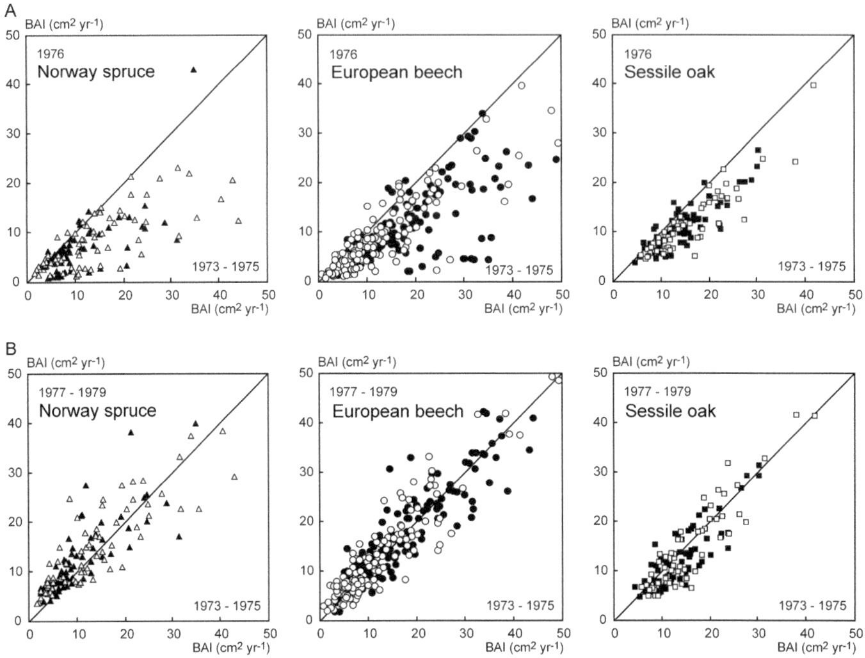


**Box Fig. 7.1-2** Resistance (**a**) and resilience (**b**) of basal area increment (BAI in $cm^2$ per year) in *P. abies* (Norway spruce), *F. sylvatica* (European beech), and *Q. petraea* (Sessile oak) in 1976. (**a**) Relationship between annual growth in 1976 and growth in the 3-year-reference-period before drought. (**b**) Relationship between annual growth in the 3-year-period after the drought (BAI in $cm^2$ per year) and the growth in the 3-year-reference-period before drought. The filled symbols represent individuals in pure stands, the empty symbols individuals in mixed stands

**Box 7.1** (continued)

quantify the tree's growth reaction to drought stress in the years 1976 and 2003. The temperatures from May to August lay 5–6 °C above the long-term mean and the precipitation in February to April and in summer from July to September were less than 50% of the normal level. In 1976 the drought started in spring and continued to late summer.

Growth reductions during the 1976 drought occurred in all three species, in *Q. petraea* to a lesser extent than in *P. abies* and *F. sylvatica*, as is indicated by the resistance values below 1 (Box Fig. 7.1-2a).

Trees growing in monospecific stands (empty symbols) showed a lower drought resistance than trees in mixed stands. *P. abies* and *F. sylvatica* were more resilient in relation to drought (Rs > 1) when compared to *Q. petraea* (RS ≅ 1.0) (Box Fig. 7.1-2b).

(continued)

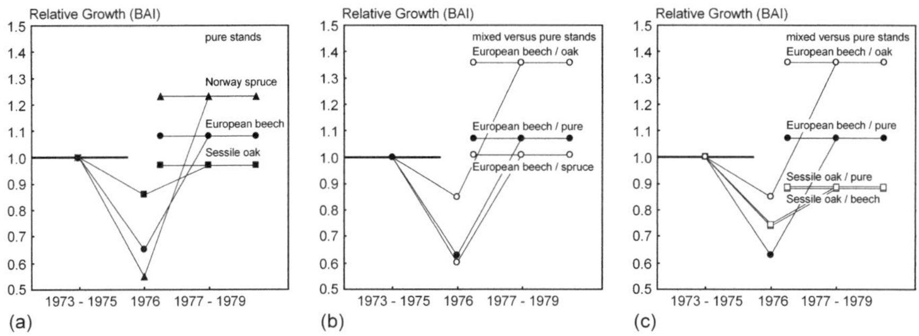


**Box Fig. 7.1-3** Species-specific stress reactions caused by the drought in 1976 in relation to the mean growth level in the 3-year period before drought (1973–1975) (reference line = 1.0). (**a**) *P. abies* (Norway spruce), *F. sylvatica* (European beech), and *Q. petraea* (Sessile oak) in monospecific stands. (**b**) European beech in monospecific and mixed stands. (**c**) European beech and Sessile oak in monospecific and mixed stands. The courses represent the growth in the dry year 1976 and in the recovery period (periodical mean of 1977–1979) in relationship to the growth in the reference period (periodical mean of 1973–1975)

**Box 7.1** (continued)

In comparison to *F. sylvatica* and *Q. petraea*, the growth of *P. abies* was less drought resistant and fell most distinctly from 1973–1975 to 1976 (Box Fig. 7.1-3). However, the stronger the growth decline of a species, the stronger its recovery from 1976 to the period 1977–1979. Hence the ranking of recovery from drought (*Q. petraea* < *F. sylvatica* < *P. abies*) was opposite to the ranking of drought resistance (*P. abies* < *F. sylvatica* < *Q. petraea*). Characteristic differences between the performances in mixed and monospecific stands occurred only in *F. sylvatica* (Box Fig. 7.1-3b). In addition, drought stress in *F. sylvatica* was reduced in mixture with *Q. petraea*, whereas growth performance of *Q. petraea* did not differ between mixed and monospecific stands (Box Fig. 7.1-3b).

The reduced drought stress of *F. sylvatica* when mixed with *Q. petraea* may be partially attributable to hydraulic lift of water by the deep-rooting *Q. petraea* which facilitates the more shallow-rooting *F. sylvatica*. Caldwell et al. (1998) list *Quercus* among other genera which exhibit hydraulic lift in the field, and they stress that all deep-rooting plants with low resistance to water loss from roots might provide this facilitative feature. According to Dawson (1993) hydraulic lift is the passive movement of water from deep, moist soil layers through the root system upwards. Under extremely dry conditions, the water potential in upper soil layers is lower than in the root system and the roots loose water to the soil. Shallower rooting species like *F. sylvatica* or *P. abies* can benefit from this increase of soil moisture both directly as their water supply increases and indirectly as the nutrient

(continued)

**Box 7.1** (continued)
availability increases with soil moisture. Körner (2002) stresses that the indirect effect of water surplus via nutrient availability on plant growth may be much more relevant as the direct effect. The presence of *Q. petraea* not only increased the resistance of *F. sylvatica* during the drought years but also causes a stronger recovery and resilience. Hence the growth of *F. sylvatica* in mixture with *Q. petraea* was higher in the years after the stress event when compared to before the event. The higher resilience in mixture may be attributable to better maintenance of the root system during the drought period compared to the monospecific stands, where *F. sylvatica* tends to suffer organ losses (Rennenberg et al. 2004). However, none of the processes that are referred to here to explain the phenomenon were actually measured, and hence we can only speculate about their contribution.

A large-scale study comprising 160 forest stands representing gradients of tree species diversity across six major European forest types provided ambiguous results regarding the effect of tree species diversity on drought exposure of these forest ecosystems (Grossiord et al. 2014b, c; Forrester et al. 2016). In this study, soil drought exposure was assessed in each forest stand by calculating the species-level and community-level increase in carbon isotope composition of latewood from a wet to a dry year ($\Delta\delta^{13}$CS). This signal indicated that community-level drought exposure decreased with increasing tree species diversity only in two forest types (temperate beech, thermophilous deciduous), increased with diversity in one forest type (boreal), and was not correlated with diversity in the other three forest types (hemi-boreal, mountainous beech, Mediterranean) (Grossiord et al. 2014b, c). In the three forest types where diversity effects were significant, the patterns were driven by single species. In the temperate beech forests, stress declined in mixtures only because *F. sylvatica* was less stressed. In the thermophilous deciduous forests, stress declined in mixtures only because *Castanea sativa* was less stressed, and in the boreal forests, stress increased in mixtures because *P. abies* was more stressed (Forrester et al. 2016). Therefore information about the drought response of a single species can provide very little information about the response of other species or the whole community. By focussing on the two-species mixtures, it was found that the diversity effects on a given species were usually not significant because of divergent results between different types of mixtures. While there were many two-species mixtures where both species were less water stressed compared with their monocultures, there were also many where both species were more stressed compared with their monocultures (Forrester et al. 2016). Furthermore, the response of a given mixture was different in the different regions.

The above results indicate that there is no general pattern between tree species diversity and drought response of forests. The observed drought impacts correspond mostly with the patterns 2, 3, and 4 in Fig. 7.2. The different patterns observed appear to be influenced by selection effects that are based on the properties of particular species in mixture, for example, the presence of deep-rooting species, and by local environmental conditions. All the above studies about drought stress speculated that the negative or positive diversity effects resulted from interactions that improved or diminished water availability for the interacting species, although many of these studies did not actually measure the processes behind these patterns. In addition, we also do not know whether the unmeasured processes that are proposed to be important in some studies are actually relevant. For example, *Q. petraea* redistributed water from deeper to more shallow regions of the soil in mixtures with *F. sylvatica*, but there was no evidence that *F. sylvatica* actually used that water (Zapater et al. 2011).

Based on the assembled evidence, we can conclude that managing forest ecosystems with high tree species diversity may not necessarily lead to improved drought tolerance, and there maybe even trade-offs between the use of tree species diversity for other functions and the susceptibility to drought stress. When it comes to predicting water stress, species identity appears to be more important than species diversity.

## 7.3 Storm Damage

Storm damage in forests ranges from the loss of leaves or needles and branches to the fall and breakage of individual trees and entire stands. The extent of damage is related on the one side to the strength and turbulence of the wind force and on the other side to the predisposition of trees and stands. Wind damage is one of the most important and widespread disturbances to forests. For example, in Europe, storms are the most important abiotic damaging agent and contribute to more than half of all recorded damage (Schelhaas et al. 2003). In the following we will focus on severe wind damage, although mechanical stress through abrasion between crowns of different species may also play an important role in shaping mixed stands and influence their productivity (e.g. Hajek et al. 2015).

Windthrow is, in most forested regions, a much more common phenomenon than breakage, because the resistance of stems to break is typically higher than the resistance to uprooting under unfrozen soil conditions (e.g. Peltola et al. 2000).

At any particular site, severe storm damage is a highly infrequent event. This low frequency is largely attributable to the temporal dynamics of storm susceptibility of trees and forests in relation to age and height development. Some authors have proposed critical tree heights (e.g. 20 m) below which the risk of windthrow is relatively low and above which it increases rapidly (Ruel 1995). However, this critical height varies with soil conditions, tree species, and wind speed (Cremer et al. 1982). Typically trees acclimate to their wind environment as best they can. It is when trees are unable to acclimate fully (e.g. shallow rooting depth stopping root penetration) and when there are sudden changes in wind environment (e.g. following thinning) that trees become especially vulnerable to storm events.

The susceptibility of trees to windthrow and stem breakage depends on a variety of factors, some of which are related to site conditions and tree species identity or

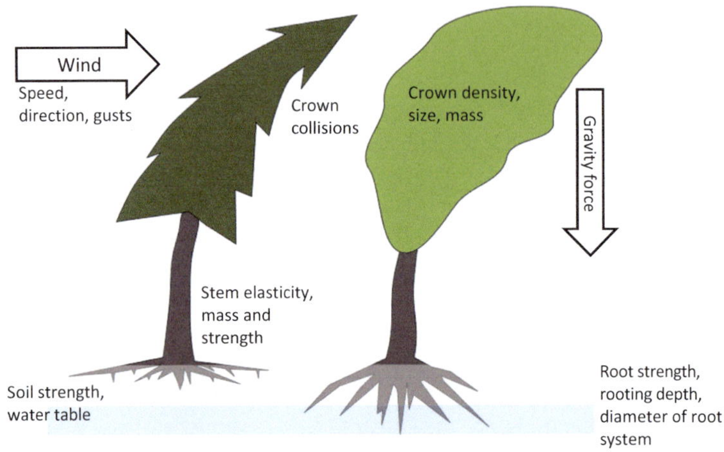


**Fig. 7.3** Factors influencing the vulnerability of trees to wind damage (adapted from Peltola 2006 and Kanimura in Peltola et al. 2013). In mixed-species forests, the magnitude of many of these factors may be changed (see text for explanations)

may be influenced by stand composition and management (Peltola 2006; Hanewinkel et al. 2013) (Fig. 7.3). In the context of this chapter, we will first discuss the factors which may be altered in mixed-species forests when compared to monocultures. Then we will review the evidence for effects of tree species diversity on susceptibility of forests to storms. This will be followed by a discussion about whether or not mixed or diverse forests might recover more quickly from storm disturbances.

### 7.3.1 Susceptibility of Trees and Stands to Storm Damage

Resistance to windthrow and stem breakage can be addressed at the individual tree and at the stand level. Under a constant wind load, the mechanical stability of a tree against windthrow is determined by the wind and gravity forces (see Fig. 7.3 and Peltola 2006). For a given wind force, the bending moment therefore increases with the size, density, and mass of tree crowns and with the mass and elasticity of the stem as well as with tree height. The latter two factors increase the displacement of the tree tip. Thus tall trees with a dense and heavy crown (e.g. conifers when compared to deciduous trees during the leafless period) would be more susceptible. Importantly, all the evidence indicates that trees with large crowns are under ceteris paribus conditions less susceptible to wind damage (e.g. Cremer et al. 1982; Dunham and Cameron 2000). The resisting forces comprise the anchorage of the tree, which is determined by the weight of the root-soil plate, the root strength and the soil shear strength, as well as the wood strength and the elasticity and thickness of stems (Peltola 2006; Ruel 1995). Therefore, trees with extensive and deep root systems that are firmly anchored in the soil have a greater resistance to uprooting than trees with shallow or small, in terms of lateral extent, root systems (e.g. Coutts 1983). Thus site conditions that influence the vertical extent of root systems, for example, through certain soil characteristics such as water-logged horizons, have a large influence on the susceptibility to windthrow. Likewise, slender trees (high height-to-diameter ratio) with low wood strength would be more prone to uprooting or breakage than sturdy trees with high wood strength growing in the same stand. Under dynamic wind loading, trees sway irregularly and their oscillatory motions result in large bending moments imposed on the stems and root systems, thus increasing the probability of tree failure (Peltola et al. 2013).

At the level of forest stands, factors that increase wind speed or turbulence, such as canopy roughness or the presence of gaps, can increase the risk of windthrow, if trees are not acclimated to this condition. The behaviour of wind turbulence over forests is of particular importance since trees fall or break under the action of wind gusts rather than an average wind load (Brunet 2013). Stand density can affect wind stability in two directions. On the one side, individual trees will become more stable in widely spaced stands, where they have a lower height-to-diameter ratio and a larger root system (e.g. Cremer et al. 1982). On the other side, stand density influences the wind forces applied to trees and the dissipation of wind energy (Ruel 1995). In widely spaced, open stands that permit a deeper wind penetration, the forces acting on individual trees can be much higher than in dense, closely spaced stands (Ruel 1995).

Within forest stands, interactions between trees may reduce their susceptibility to damage. This is very obvious at forest edges where sturdy trees that have developed large root systems protect trees in the interior of the stand (Peltola et al. 2013). Large, well-anchored trees within stands may have similar effects on other trees located downwind. In addition, the swaying of a tree may be reduced through the collision with crowns of neighbouring trees that offer mutual support in this way (Rudnicki et al. 2003). This mechanism is likely stronger in more heterogeneous stands in which trees, owing to differences in dimensions and hence natural sway frequencies, swing in a more asynchronous fashion than homogenous stands (Moore and Maguire 2004).

In addition, trees have very effective damping systems involving modal coupling between stem and branches that help dissipate energy (Spatz and Theckes 2013). Also, anchorage of trees may be increased through interlocking and grafting among root systems of neighbouring trees, which increase the effective root plate (Cremer et al. 1982; Ruel 1995). However, there appears to be no study that has quantified this effect (Loehle and Jones 1990). Since root grafting occurs mostly between conspecific trees and interspecific connections are extremely rare (e.g. Eis 1972), a possible advantage of combined roots systems should be lower in mixtures than in monocultures of those species that develop root grafts.

It is obvious from this list of factors influencing tree and stand-level stability that tree species identity and the mixing of species can have a strong influence on the susceptibility to windthrow. A direct comparison of species is difficult, because of a lack of experiments that permit the assessment of damage among a wide range of different tree species under ceteris paribus conditions. However, through statistical modelling of extensive storm damage following major storm events, some general patterns regarding the major factors explaining the extent of damage including tree species properties have emerged (e.g. Albrecht et al. 2012; Canham et al. 2001; Rich et al. 2007). Tree species with a high susceptibility to windthrow or stem breakage typically have one or more of the following characteristics:

- **Shallow root systems**, either genetically determined or through the influence of soil conditions.
- **High crown density** leading to higher drag in the case of storms. This is one reason why conifers are typically less resistant to winter storms than deciduous species (Hanewinkel et al. 2013).
- Greater average **tree height** (e.g. Albrecht et al. 2012).
- **Low wood strength** (e.g. Putz et al. 1983).

Comparatively wind-firm species would have the opposite characteristics. The interesting question is, now, whether it is possible to change the susceptibility and resistance to storms through the mixing of species at the species or the stand level. Theoretically, the following situations may occur when we mix species:

- The effect of interspecific interactions (competition and facilitation) on tree allometry can differ from that of intraspecific interactions (e.g. Pretzsch and Schütze 2009). Hence, trees of a given species may become more slender or sturdy and have different crown lengths in mixtures, which may increase or decrease their stability. A stabilising effect of mixtures at the community level might occur, if interspecific competition in mixtures was lower than intraspecific competition in monocultures for the majority of species.
- Owing to different height growth dynamics or final heights, tree species that are taller than the others will have their crowns more exposed to the wind force. Individual trees which are taller than their neighbours are subjected to higher wind loads because wind speeds increase with height above ground and within and above the canopy space of forests (Cremer et al. 1982).
- Mixed stands may have a higher or lower stand density than monocultures (see Chaps. 4 and 8) or may have a higher or lower wind penetration through mixing of species with different crown densities (e.g. evergreen and deciduous species in winter). This may change the wind loading that acts on individual trees.
- In mixtures of species that differ in stem taper, elasticity, crown density, or root anchorage and hence exhibit different swaying characteristics (Moore and Maguire 2004), there may be more mutual support through stand structure than in monospecific stands of the species.
- Below-ground competition might lead to shallower or deeper root systems in one or more species. For example, root systems of *P. abies* were shallower when the species was mixed with *F. sylvatica* than in monospecific stands (Schmid and Kazda 2002), yet in another study the reverse trend, deeper roots of *P. abies* in mixture with *F. sylvatica*, was observed (Kuhr 2000).

The above list of possible outcomes of mixing tree species on a number of factors that influence tree and stand stability in relation to wind shows that it is difficult to disentangle mixing effects from other effects that influence the variables representing stability, e.g. stand density, site conditions, etc. In addition, there may also be indirect influences through silvicultural practices that differ in mixtures, where thinning may be more or less frequent or intense or where the production cycle length may be shorter or longer, thus leading to different final tree heights. It is therefore not surprising that the evidence for possible effects of tree species mixtures on resistance to storm damage is extremely variable and often contradictory. The main problem is that most evidence comes from analyses of damage to forests following storms, so that there is always the confounding influence of other factors such as site and soil characteristics and previous forest management. In the following we will review the evidence for effects of mixed species stands on the magnitude of storm damage.

### 7.3.2 Do Tree Species Mixtures Reduce the Vulnerability to Storms?

Only few studies conclude that mixed-species stands have a higher overall stability than monospecific stands (e.g. Slodicák 1995; Schütz et al. 2006; Valinger and Fridman 2011). However, this conclusion is often based on the comparison with monospecific stands of the more vulnerable tree species. In the study of Schütz et al. (2006), the admixture of broadleaved tree species reduced the damage in *P. abies* stands caused by the storm Lothar in the Swiss Midlands. Many more studies, reviewed in Knoke et al. (2008), have provided similar evidence of the benefit of admixing broadleaved tree species in *P. abies* stands (response type 2 in Fig. 7.2). Similarly, in conifer plantations on water-logged soils, the rooting depth and hence the stability of *P. sitchensis* can be improved through mixing with the deep-rooting *Pinus contorta*, which lowers the water table (Pyatt 1993). In contrast, admixture of *P. abies* to broadleaved species did not increase stability of those stands. In addition, while the stability of the more vulnerable tree species may not actually be improved, the probability of total stand damage may still be reduced through the presence of less susceptible species. In some cases, the windthrow risk to a more vulnerable species may not be reduced when mixed with a more stable one, as was found for *P. abies* when mixed with *F. sylvatica* in situations where, owing to differences in growth rates and final heights, the crowns of *P. abies* were more exposed to wind than in monocultures (Lüpke and Spellmann 1997).

Other studies based on evaluations of forest inventory data (Colin et al. 2008, 2009) do not provide evidence of positive effects of mixtures on stand stability. Thus, while it may be advantageous to mix a less resistant tree species with a more resistant one to avoid total stand damage in cases of severe storms, it is most likely that the damage to those mixed stands will be reduced only in proportion to the share of the stable species (Lüpke and Spellmann 1997) (response type 2 in Fig. 7.2). There is no evidence of positive interactions that lead to greater stability in relation to wind disturbance for all species in mixed stands, which would correspond with response type 1 in Fig. 7.2. It thus appears that tree species identity is more important in predicting wind damage to mixed stands than species richness (Dhôte 2005). Based on rankings of the vulnerability of different tree species to wind damage (e.g. Colin et al. 2009; Mason and Valinger 2013) and knowledge of the compatibility of species under different site conditions, mixed stands may be created that offer a high degree of storm resistance.

One important question is whether there is any benefit in terms of windthrow risk of having mixed-species stands as opposed to having a mixture of patches of monocultures of different species at the level of landscapes. The evidence to date does not suggest there is any benefit.

As discussed above, most evidence for differences in vulnerability to storm damage between mixtures and monocultures is based on damage assessments following storms, where it is difficult to control for all the confounding factors. These assessments have therefore yielded contradictory results regarding the relative wind stability of different tree species and the effects of species mixing. Since there has been no kind of proper replicated experiment addressing tree species identity and diversity effects, the level of confidence based on the ranking of Binkley and Menyailo (2005) (Fig. 2.1 in Chap. 2) does not exceed 4.

Regarding the resilience of mixed-species forests following storm events when compared to monocultures, there is even less evidence. It is conceivable that mixed semi-natural forests with a greater variation in stand conditions provide more opportunities to develop advance regeneration and respond with a greater variety of propagules to disturbance (e.g. Brang 2001; Bauhus et al. 2013), which would be the most important prerequisite for the resilience of forests. There are many descriptions of the high tree species diversity in the natural regeneration after windthrow (e.g. Drouineau et al. 2000). However, there has been no study that has quantified the resilience, for example, in terms of biomass accumulation, following wind disturbance in relation to the pre-disturbance diversity or structure of forests; the same is true for other types of disturbances. See also Mason (2002) for a more extensive discussion of the question, whether structurally diverse stands are less susceptible to storm damage and recover better from it than even-aged stands. The persistence of at least a partial cover of trees following catastrophic storms may however be an important advantage for the recovery of stands (Dhôte 2005).

## 7.4 Fire Risk

The other main widespread abiotic risk to forests is posed by fires. Fires are a prevalent type of natural and man-made disturbance in most forest types and climates, and fire disturbance is expected to increase with climate change in most of these regions (Bond et al. 2005; Pausas and Vallejo 1999; Van der Werf et al. 2008; Flannigan et al. 2006; Peterson et al. 2005). In contrast to other disturbances (e.g. wind) where the intensity and frequency is independent of plant and ecosystem

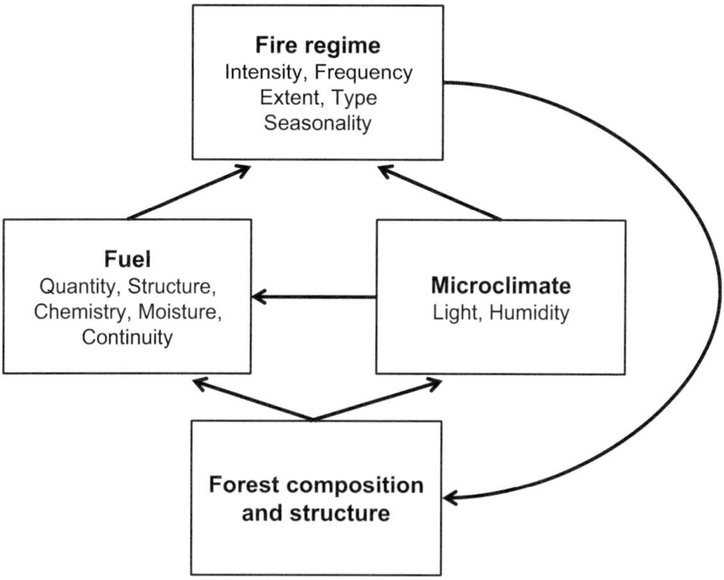


**Fig. 7.4** Conceptual presentation of possible effects of tree species mixtures on fire regimes at the stand level. The tree species composition might influence fuel conditions and microclimate; see text for examples. In addition to the factors listed here, the fire regime is influenced by factors such as climate and topography

properties, in the case of fire, the disturbance itself is shaped by ecosystem and forest species features (fuels) (Fig. 7.4). Hence, fire intensity or severity depends on fuel composition (dead and alive) and structure, in addition to heat sources (ignitions), weather conditions (wind, air humidity), and topography. Additionally, species respond differently in their adaptations to resist and to regenerate after fire.

The main direct effect of fires on vegetation is via the killing of plant tissues through heat. During (low-intensity) ground fires, surface fine roots and the cambium of thin-barked trees as well as seedlings and saplings may be affected, whereas in (high-intensity) crown fires, all foliage as well as the cambium of stems and branches may be affected. Species that have co-evolved with fire have developed a range of structures and mechanisms to cope with this disturbance (e.g. Gill 1975). The different adaptations and responses to fire have formed the basis of classifications of plant functional types. The classifications are based on plant functional traits describing the capacity to protect above-ground or below-ground tissues from fire and the capacity to regenerate through sprouts, suckers, or seeds following fires (Allen 2008). The underlying mechanisms comprise the protection of the cambium and adventitious buds through thick bark and by positioning them below ground, the protection of seeds in cones and woody capsules as can be found in many pines and eucalypts, and the effective shedding of dead branches to prevent fuel ladders (e.g. Noble and Slatyer 1980; Gill 1997; Fernandes et al. 2008). Tree species that have adapted to fires are often classified as either resprouters or seeders. The former invest in protective tissues and hence have the capacity to resprout after fires from buds that have been protected from the heat under thick bark or below ground (e.g. Gill 1997). Many eucalypt species and some oaks such as *Quercus suber* are characterised by thick bark. Seeders, by contrast, may be completely killed by the fires but regenerate from seeds that have been protected either in the canopy or in the soil. Some tree species, for example of the genera *Pinus* and *Eucalyptus*, produce litter that decomposes slowly and is highly flammable, hence promoting fires (Ormeno et al. 2009). Many coniferous species are highly flammable due to the high content of resins and essential oils (e.g. Dimitrakopoulos and Papaioannou 2001). This plant trait does not only increase the probability of fire occurrence but also the severity of fire (González et al. 2007). Other plant properties that influence the severity of fire are the litter production rate which influences the accumulation of fuels, in particular fine fuel, and the retention of dead branches on the stem, which functions as a fuel ladder and can increase the probability of crown fires as well as fire temperatures and heat release (Schwilk 2003; Taylor and Fonda 1990).

There are of course also many tree species from forests where fires are extremely rare, which show no such adaptations to fire. In addition to the resistance of individual species or trees to fires, resistance at the level of forest communities or ecosystems can be related to the flammability of the vegetation and fuels. Thus forests comprising species that are not adapted to fire may still be highly resistant to it because they are highly inflammable [e.g. rainforests with little understorey and litter that has a high ash content and decomposes fast (Bowman 2000)]. For example, tree species with low flammability of litter have been commonly used as green fire breaks within plantation landscapes of species with higher flammability (Johnson 1975).

Owing to these differences among species, forest composition affects wildfire hazard, the resistance to fires, and post-fire regeneration (resilience). Mixing tree species with different properties regarding fuel conditions, effects on the microclimate, fire resistance, and resilience may on the one side lead to intermediate conditions, when compared to monospecific stands (Fig. 7.4). For example, the probability of fire occurrence in northwestern Portugal was intermediate in mixed stands comprising conifers and deciduous hardwoods when compared to conifer forests, where it was higher, and compared to deciduous hardwoods, where it was lowest (Moreira et al. 2001). On the other side, tree species may interact and therefore create stand conditions that differ from the expected value, based on a mixing proportion of species, with regard to the probability of fire occurrence and the severity of fire. The following nonadditive effects may occur, when we mix different tree species (see also Fig. 7.4):

- **Changes in fuel conditions.** If mixtures are more productive, they will produce more litter and hence higher fuel loads, unless litter decomposition is accelerated at the same time. Mixing different litters may lead to nonadditive effects regarding their fuel behaviour (when fuel behaviour of mixtures cannot be predicted from that of respective monocultures). Admixing species with highly decomposable litter of low flammability to species with litter of high flammability might reduce the flammability of litter fuel and change the fuel bed properties. The horizontal variation in species with different litter properties might also influence fire spread. For example, in pine-oak forests of the southeastern USA, where regular prescribed burning to maintain low fuel loads is being practised, it may be more desirable to favour those understorey oaks with litter burning characteristics (e.g. intensity, flaming duration, mass loss) similar to that of the dominant pines to facilitate fire spread (Kane et al. 2008). Differences in litter flammability between species might of course also be used from a management perspective to reduce fire spread. However, fuel properties might also interact in the opposite way such that highly flammable fuels increase the flammability of other fuel types in mixture (de Magalhaes and Schwilk 2012).

Mixtures with species that have a high concentration of fuels around the tree base and also a high density of ladder fuels might increase the probability of crown fires in species that would otherwise not or rarely experience these. There may be interactions with other disturbance types leading to inputs of high amounts of fuel, e.g. after insect outbreaks (Fleming et al. 2002). If the packing of crowns (of flammable species) is higher in mixtures than in monocultures, this may also sustain crown fires to a higher degree (Agee 1996).

- **Changes in microclimate**. In mixtures that sustain a higher leaf area index than monocultures of the participating species, the abundance of shrubs and grasses, which present fine fuels, can be reduced. A higher leaf area in mixtures may also lead to a higher relative humidity of the air below the canopy and therefore higher fuel moisture contents and thus lower flammability when compared to more open stands.

For many of the processes depicted in Fig. 7.4, we have no experimental evidence regarding the influence of species mixtures. The ability to predict fire behaviour on the basis of plant functional traits is still rudimentary. One example of such an analysis is the study by Schwilk and Caprio (2011) in mixed forests of the Sierra Nevada in California, where they found that fire severity was greater in stands inhabited by species with longer leaves than in stands containing more short-leaved species. This effect was attributed to a lower litter density when long leaves were present and was also driven by the presence of pines.

Fire behaviour of mixed-species vegetation includes a complex set of processes and interactive influences of factors that are not well understood (Blauw et al. 2015). We have some evidence that the flammability and other fire properties of fuel in species mixtures can deviate substantially from those that we would expect based on the fuel properties of component species, although there appear to be only two studies that have examined this experimentally (Van Altena et al. 2012). For example, in Californian mixed-conifer forests, the flammability of litter mixtures from eight dominant tree species was nonadditive in the sense that it was higher than can be expected from single-species litters owing to the disproportionally large influence of the most flammable litter component in mixtures (de Magalhaes and

Schwilk 2012). In this case, the admixing of certain species with highly flammable litter may actually incur a disproportionally large increase in fire risk for the tree community (response type 4 in Fig. 7.2).

However, these indications on mixing effects of fuels are based on laboratory combustion experiments. For a number of reasons, the assessment of fuel flammability in the laboratory cannot be extrapolated to field conditions and wildfires with natural fuel-complex structures and different heat transfer processes (Fernandes and Cruz 2012). As is the case for wind disturbance, there are no proper replicated field experiments with different tree species and mixtures to assess the significance of these possible interactions. Owing to limited knowledge of interactions among tree species and between different tree species combinations and the understorey, fire behaviour in mixed-species stands remains extremely difficult to predict. Based on the scant evidence, it appears that the effect of mixing species on resistance towards fire is mostly based on species identity effects. The few nonadditive effects that have been identified indicate that the fire risk for the community might increase proportionally more with the admixture of species that produce highly flammable litter (response types 3–5 in Fig. 7.3). Since significant fire events are typically not confined to individual forest stands, it would be important to assess whether the mixing of tree species at the level of stands has advantages over mixing monospecific stands at the landscape level for reducing the risk of ignition, fire spread, fire severity, and other important aspects of fire behaviour that determine options for fire control and the damage through fire disturbance.

The resilience of forest communities following fire disturbance depends largely on the demographic mechanisms that allow plant populations to recover after disturbances, which are expressed in the functional traits of the participating species (Allen 2008; Pausas et al. 2004b). Trees of the different species can either survive fires through protection of living tissues above ground and/or below ground or they are killed by fire. Those that are killed might regenerate from seed stored in the canopy or in the soil, or they are—at least temporarily—lost from the site.

From the perspective of maintenance of forest cover and ecosystem functioning, it would be advantageous to manage mixed-species forests with tree species that can either survive or regenerate from seed on site to facilitate a quick recovery at little cost (e.g. Puerta-Piñero et al. 2012). It appears that forest recovery through species that resprout is less affected by changes in fire frequencies, whereas species that recruit from seed might not have sufficient time between fires to reproduce (e.g. Pausas and Lloret 2007). This can be especially critical for young monospecific pine plantations that are vulnerable to fire until they reach maturity in fire-prone regions, i.e. 15–20 years depending on site quality (Pausas et al. 2004a).

In Mediterranean forest, maintaining species such as oaks that recover after fire by basal resprouts (e.g. *Quercus* ilex) or epicormic resprouts (*Quercus suber*) might thus reduce the immaturity risk of recurrent, short-interval fires, in pines such as *P. halepensis*, *P. nigra*, or *P. sylvestris*. Mixed forests of both genera would therefore increase the resilience when compared to the more vulnerable pines, but not when compared to monospecific oak stands (response type 2 in Fig. 7.2). Whether the species-specific resilience is affected by the mixing of tree species has not been investigated. This might be the case if mixing tree species leads to a higher incidence of crown fires that kill the canopy-stored seeds or if in mixtures the fuel conditions, e.g. fuel load, change in a way that leads to significantly hotter fires that exceed the protective mechanisms of resprouting species. At this stage, it appears that forest community resilience in relation to fire is mainly driven by species identity effects and not by nonadditive effects of mixtures.

## 7.5 Other Abiotic Disturbances

There are a number of other abiotic disturbances and stress factors that can cause substantial damage in forests ecosystems. These include uprooting or breaking of trees and branches under heavily loads of snow and ice (e.g. Nykänen et al. 1997; Rhoads et al. 2002) and flooding. Unlike storm and fire, these disturbances are typically limited to certain parts of the landscape or to altitudes where these events can occur. Similar to the other abiotic disturbance agents discussed above, there are large differences among tree species regarding their resistance and resilience in relation to these stresses and disturbances (e.g. Nykänen et al. 1997; Warrillow and Mou 1999). Likewise there has been no examination of nonadditional effects of tree mixtures. As with the other abiotic disturbances, we assume that resistance and resilience of forest communities depends on the particular combination of tree species and that the main effect of mixing is the spreading of risk.

## 7.6 Pest Insects

Insect herbivory can have profound impacts on forest ecosystem productivity and composition (Attiwill 1994). Worldwide, insect herbivores are responsible of ca. 8% of annual loss of foliar biomass (Kozlov et al. 2015). Even minor but chronic damage to long-lived plants like trees can result in significant growth reductions (Zvereva et al. 2012). In contrast to the abiotic disturbances discussed above, there is clear evidence for the effects of tree mixtures on the resistance of forests to insect pests from experiments. The hypothesis that mixed forests are more resistant to pest insects than tree monocultures is supported by two meta-analyses which have shown that in the majority of cases, a particular tree species is less prone to damage by a given pest insect when grown in mixture when compared to monoculture (Jactel and Brockerhoff 2007; Castagneyrol et al. 2014a). Thus we can have a high level of confidence in these results (see Fig. 2.1 in chapter 2). Recently, insect defoliation was monitored on a network of 208 forest plots sampled along two orthogonal gradients of increasing tree species richness and latitudes in Europe (Baeten et al. 2013). This study revealed a positive effect of tree species richness on resistance to insect herbivores, irrespective of the climate (Guyot et al. 2016).

Three main mechanisms have been proposed to explain this pattern of associational resistance, where the mixing of herbivore-susceptible plant species with herbivore-resistant plant species can reduce the herbivory in the susceptible species (Tahvanainen and Root 1972; Barbosa et al. 2009) (response type 2 in Fig. 7.2). The first two mechanisms are bottom-up, i.e. a result of tree-tree interactions. The resource concentration hypothesis (Tahvanainen and Root 1972) predicts that specialist herbivores are more likely to find, remain, and reproduce on their host plants and hence achieve higher loads when their food plants grow in dense patches in monospecific stands rather than scattered in mixtures (Otway et al. 2005). Hence, in tree species mixtures, the occurrence of host tree species is more diluted among other associated tree species providing fewer feeding resources (or breeding substrate, or carrying capacity) to specialised insect herbivores, thus leading to reduced opportunity to build up epidemic populations and generate damage (Vehviläinen et al. 2006; Castagneyrol et al. 2013).

Secondly, trees from the focal species (host trees) are also less apparent, i.e. more difficult to find and colonise by specialist insect herbivores. Reduced tree apparency may be due to disruption of host-finding cues such as visual cues, when trees are hidden by taller neighbours (e.g. Dulaurent et al. 2011, Castagneyrol et al. 2013, 2014b), or olfactory cues, when volatile chemical stimuli of non-host plants interfere with the host-finding ability of specialised insect herbivores searching for host species (e.g. Stanton 1983; Zhang and Schlyter 2004; Jactel et al. 2011). These two mechanisms, which represent response type 2 in Fig. 7.2, apply to specialised herbivores and their interactions with host plants. It can be expected that the effect of tree species diversity increases with the host specificity of insect herbivores. A recent meta-analysis found that damage by monophagous (feeding on tree species within a given genus) or oligophagous insects (feeding on different genera within a given family) was reduced by 42% and 15%, respectively, whereas there was no significant reduction in damage by polyphagous herbivore insects that feed on several tree families (Castagneyrol et al. 2014a). In some cases, insect damage may be even higher in mixed forests, which is a case of associational susceptibility (response type 4 or 5 in Fig. 7.2). This may occur when generalist herbivores spill over onto neighbouring species after depletion of the favoured host plant (White and Whitham 2000) and when polyphagous herbivores benefit from dietary mixing, e.g. through complementary uptake of nutrients or dilution of toxins by feeding on different plant species (Bernays et al. 1994; Unsicker et al. 2008). For example, in a highly diverse Chinese subtropical forest, where herbivory appeared to be dominated by polyphagous insects, folivory in seedlings increased with tree species richness (Schuldt et al. 2010).

The third main mechanism is the increasing top-down regulation of insect herbivores (primary consumers) by parasitoids and predators (secondary consumers) with increasing tree species diversity, also known as the natural enemies hypothesis (Root 1973) (response type 1 or 2 in Fig. 7.2). Here, tree species richness increases the abundance and diversity of insect herbivores which in turn provide more feeding resources to generalist predators (Castagneyrol and Jactel 2012) that can develop more stable populations and then are ready to prey on pest populations when they reach an epidemic level (Jactel et al. 2006). Mixed forests are commonly also structurally more complex and thus can provide predators and parasitoids with a broader range of nesting sites or refuges against adverse conditions (e.g. Kaitaniemi et al. 2007; Muiruri et al. 2015). For example, individual numbers of most families of parasitoid wasps were statistically significantly higher in mixed *Pinus sylvestris* stands with deciduous trees resulting from forest conversion activities when compared to monospecific pine stands (Jäkel and Roth 2004). Lastly, mixtures can provide specialist parasitoids with alternative feeding resources such as pollen, nectar or honeydew (Dulaurent et al. 2011) which increase their fitness.

Owing to these different mechanisms, different effects of tree species diversity on different insect herbivores may be seen in the same forest stand (e.g. Plath et al. 2012). For example, in sapling mixtures of *Pinus sylvestris* and *Betula pendula*, a lower proportion of European pine sawfly (*Neodiprion sertifer*) larvae and eggs survived on the *P. sylvestris* in mixtures than in monocultures. This was explained with the higher abundance of ants in mixtures. However, higher ant abundance also led to a higher number of ant-tended aphid colonies on *P. sylvestris* in mixtures when compared to monocultures (Kaitaniemi et al. 2007). Thus, it may be impractical to regard associational resistance or associational susceptibility in relation to individual insect herbivore species, and instead it should be seen as a net effect of tree mixtures or diversity on herbivore damage of the whole system (community), equivalent to over- or underyielding in relation to productivity in mixtures. Most studies, however, assessed the associational resistance for focal tree species or focal pest species or both, but not the community-level damage through herbivory from the community of insect herbivores. In addition, most studies compared monocultures with mixtures but did not examine 'true' tree species diversity effects, for which there is only little evidence, i.e. a significant correlation between tree species richness/diversity and insect herbivory. However, only recently a significant negative correlation between tree species richness and overall insect defoliation was observed in European mature forests (Guyot et al. 2016). For example, defoliation of chestnut trees (*Castanea sativa*) owing to the introduced Asian chestnut gall wasp (*Dryocosmus kuriphilus*) declined with increasing tree species richness from chestnut monocultures to four species mixtures in Italian hardwood forests (Guyot et al. 2015).

In most cases, the identity of species in mixed tree communities was more important than tree species richness per se. In particular the association of more contrasting functional types and of phylogenetically distinct tree species (i.e. mixtures of broadleaved and conifer trees), results in greater associational resistance to insect herbivory (Jactel and Brockerhoff 2007; Castagneyrol et al. 2014a).

From studies in agriculture, it is known that reduction in herbivore abundance through crop species diversity depends strongly on the spatial scale of mixing species and can decrease with increasing plot size (Bommarco and Banks 2003). However, the effects of tree diversity on herbivory have been mostly studied at a single spatial scale of forest plots or stands, and thus transferability to other scales may be limited. Also, to our knowledge, there are no studies testing the effect of tree diversity on the temporal dynamics of forest insect populations and on the resilience of mixed forests following pest outbreaks. Therefore it is not possible to predict the long-term effect of tree diversity on herbivory and thus its influence on productivity over longer periods.

## 7.7 Pathogens

Like herbivores, pathogens (biotic agents, mostly fungi or bacteria, causing a disease) are an integral part of forest ecosystems, but they can sometimes produce undesirable effects through substantial loss of ecosystem health, which may incur significant economic impacts (Ayres and Lombardero 2000).

In contrast to many abiotic disturbances, pathogens, like many insect herbivores, may selectively eliminate the less vigorous or genetically unfit individuals of a population (Castello et al. 1995). The impact of pathogens on tree species and forest ecosystems can be enormous and has increased in some regions owing to the introduction of exotic pathogens (Lovett et al. 2006). Well-known examples comprise the substantial decline or almost complete loss of tree species such as the American chestnut (*Castanea dentata*) through the introduced fungus *Cryphonectria parasitica* (Anagnostakis 1987); jarrah (*Eucalyptus marginata*) on wet sites, where it is affected by the introduced root pathogen *Phytophthora cinnamomi* (Davison and Tay 1987); and white pines by the white pine blister rust (*Cronartium ribicola*) (MacDonald and Hoff 2001). But also native pathogens such as the root rot *Heterobasidion annosum* may cause substantial damage in managed conifer stands (Woodward et al. 1998). Pathogens may affect the leaves or needles, shoots, stems and roots of trees.

As for the other disturbance agents, the question we address here is whether mixed-species forests offer greater resistance to pathogens. Unlike for the abiotic disturbances, the impact of pathogens is typically prolonged and may last until all host species have been killed, which can take many decades depending on the aggressiveness and spread of the pathogen in the community. Some of the mechanisms that might confer a higher resistance of mixed stands are the same as for disturbance by pest insects. For specialised pathogens, the host species occur more diluted in mixtures and offer fewer contact points between infected and healthy individuals (Fig. 7.5). The evidence of the majority of studies on the relationships between diversity and disease risk provided evidence for dilution effects, whereby pathogen transmission and disease incidence decrease with increasing species diversity (Keesing et al. 2010). In addition, in mixed stands more antagonists of the pathogen species may be found. For example, in grassland diversity experiments, plant diversity contributed to plant community resistance against pathogens through an indirect soil feedback by fostering beneficial bacterial communities (Latz et al. 2012). Amplification effects, where increased diversity leads to an intensification of disease risk in a community, may occur when pathogen transmission is not a function of host density, for example in generalist pathogens. In addition, mixing tree species may change stand microclimate and soil moisture conditions and thus alter conditions for pathogens.

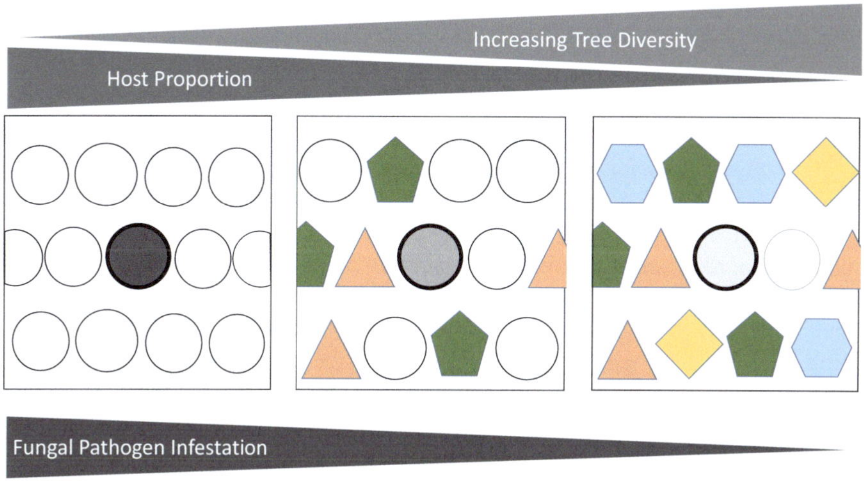


**Fig. 7.5** Conceptual illustration of the effect of neighbourhood tree species diversity on fungal pathogen infestation (adapted from Hantsch et al. 2014a, CC BY-NC-ND 3.0). Fungal pathogens of an individual host tree (in the centre of plates marked with a *fat circular black line*) are expected to decrease with increasing tree species diversity in the immediate neighbourhood; as indicated by the declining intensity of *grey* of the host tree. With increasing tree species diversity, the susceptible species becomes increasingly diluted in the neighbourhood (community), and pathogen transmission is thought to decrease between individuals of the host species. Different symbols represent different tree species

In the following we will review the evidence for beneficial effects of mixtures for pathogens in different plant compartments and provide examples for the above mechanisms.

**Root Pathogens** The root rot caused by *Heterobasidion annosum* has been called the 'most limiting single disease in coniferous plantations throughout the northern hemisphere' (Shea 1971) and is in many regions in Europe and North America the economically most important pathogen (Woodward et al. 1998). For its spread, the fungus relies on tree wounds for spore infections or root contacts (grafts). Its movement through free soil is very limited owing to antagonists. The disease is more prevalent in managed than in natural forests owing to the many possibilities for infection created through cut stumps and logging damage. Since the biology of the fungus suggests that the incidence of root rot can be reduced through wider spacing between individual trees of susceptible species and creation of soil patches with antagonistic properties, silvicultural control of root rot through mixed stands has been studied extensively and the experience has been reviewed by Korhonen et al. (1998). The susceptibility of tree species to *H. annosum* differs greatly and, in addition, individual tree species have different susceptibilities to the different known intersterility groups of *H. annosum* (Korhonen et al. 1998). Many studies have reported reduced incidence of diseased trees in mixtures when compared to monocultures, where the positive effect has been attributed to reduced numbers of root contacts and grafts of the susceptible tree species or to antagonistic soil properties beneath resistant tree species. However, there have also been several studies that have indicated no beneficial effects of mixtures or even infection also of the admixed, presumably more resistant species. Where crown and root competition increases for a particular species in mixture, this may also increase its susceptibility to *H. annosum*. This may in turn increase the spread of the fungus in mixed stands (Schönhar 1990). Korhonen et al. (1998) concluded that the effect of resistant admixed tree species on the incidence of *H. annosum* infestation in susceptible conifers is in most cases beneficial but small, following pattern 2 in Fig. 7.2. Apparently, at least 20–30% of evenly distributed resistant trees are required in mixtures to significantly reduce the decay frequency in susceptible conifers (e.g. Piri et al. 1990). There have been also reported cases conforming with patterns 3, 4, and 5 in Fig. 7.2 (Korhonen et al. 1998). Obviously the effect of mixing on root rot infestation depends not only on the species combination but also on site properties, management practices and stand history. Most results have been obtained from unreplicated case studies established in existing forests, where confounding factors have not been controlled. Hence the level of confidence based on the ranking of Binkley and Menyailo (2005) (Chap. 2) that we can place on this finding is between 4 and 5.

Other studies provide interesting results in relation to other root pathogens. The experimental removal of paper birch (*Betula papyrifera*) from 10- to 15-year-old Douglas-fir (*Pseudotsuga menziesii*) plantations to increase conifer productivity, increased mortality in the conifer species after 2 years due to root disease with increasing paper birch thinning intensity where the pathogen *Armillaria ostoyae* was already present in untreated stands (Baleshta et al. 2005). However, from this study it was not clear whether the increased infection in Douglas-fir resulted from a decreased dilution of root contacts between conifers or from an increased growth substrate for the fungus through the provision of cut birch stumps, which also acted as a host. In addition, it had been proposed that population sizes of fluorescent pseudomonad bacteria, which were antagonistic towards *A. ostoyae* and four times higher in birch soil than in Douglas-fir soil, may be responsible for this effect (DeLong et al. 2002). Three years later there was no difference in Douglas-fir mortality between the different levels of birch removal (Baleshta et al. 2015), lending support to the assumption that the temporarily increased mortality was attributable to the proliferation of the fungus on cut birch stumps.

More evidence for a dilution effect of tree species mixtures on *Armillaria* was obtained in an experiment with susceptible conifers and largely resistant hardwood species, which were planted at four densities and in several mixtures on recently logged sites (Gerlach et al. 1997). Here, mortality in the three most affected species (*Abies balsamea*, *Larix laricina*, *Picea mariana*) was reduced by 75% when grown in aspen-rich plots when compared to conifer-rich plots. Since seedling mortality was related to species identity, planting density, and proportion of conifers, the mixing response followed that of option 2 in Fig. 7.2. Beneficial effects of mixing susceptible and non-susceptible tree species at the time of planting may however disappear in the long term, if the spatial scale of mixtures does not prevent root contact of larger trees (Morrison et al. 2014).

**Leaf Pathogens** Also for foliar fungal pathogens, the degree of infestation in susceptible plants may be influenced by the neighbourhood, i.e. the level of species diversity among adjacent host and non-host plant species (Fig. 7.5) (Hantsch et al. 2014a). In relation to fungal foliar pathogens, the associational resistance (Tahvanainen and Root 1972; Jactel et al. 2006), which was introduced above in relation to herbivory, may also be attributable to higher structural heterogeneity and hence higher microclimatic variability in mixed-species neighbourhoods. Since microclimatic conditions such as moisture and sunlight are of critical importance for foliar pathogen reproduction, growth and survival (e.g. Blakeman 1993), variations in crown size, structure, and density that come with species-diverse neighbourhoods may be an important determinant of fungal infestation of a particular focal tree.

The effects of tree species diversity and species identity in the immediate neighbourhood of host trees has been analysed in controlled tree diversity experiments. In one of these young experiments (tree heights between 1.5 and 3 m) with *Tilia cordata* and *Quercus petraea* as the focal host species that were mixed with up to four additional tree species, high neighbourhood tree diversity reduced both the total fungal species richness and fungal infestation rate on leaves of *T. cordata*, but for *Q. petraea* only fungal infestation was reduced (Hantsch et al. 2014a). In addition, the infestation by three host-specialised fungal pathogen species was reduced. These beneficial effects were attributed to tree species diversity in the neighbourhood; they were independent of the proportion of conspecific host trees. However, fungal species richness and infestation were also affected by the proportion of particular non-host tree species in the neighbourhood (Hantsch et al. 2014a). In a similar study conducted in another tree diversity experiment (BIOTREE, Scherer-Lorenzen et al. 2007), where trees were between 1 and 7 m tall at the time of sampling, the effect of functional diversity on foliar fungal pathogen richness and load was tested in plots that consisted of four different tree species and differed in functional diversity of the community (Hantsch et al. 2014b). The different combinations of 4 species were drawn from a pool of 16 different tree species of the temperate zone. In this case, no effect of functional diversity on pathogen richness or load was observed, although the community-level pathogen richness and load were positively influenced by the proportion of susceptible tree species at the plot level (Hantsch et al. 2014b). Since some of the functional traits used in the design of this experiment also relate to crown attributes, canopy heterogeneity may not be an important determinant of foliar fungal infestation at the community level. The beneficial effects of tree species mixtures described above may also occur at the level of within-species genetic diversity, where individual clones differ in susceptibility to a particular pathogen. For example, yield of *Salix burjatiea* short-rotation coppice plantations in Ireland was consistently higher and rust infections by *Melampsora* sp. were lower in clonal mixtures than in any of the component clones grown in monoculture (McCracken and Dawson 1998).

Other studies on pathogen infestation have shown that, even in the case of a generalist pathogen such as *Phytophthora ramorum*, plant species diversity can reduce the risk of infection (Haas et al. 2011). Although this pathogen, which has killed millions of oak (*Quercus* spp.) and tanoak (*Notholithocarpus densiflorus*) trees in coastal areas of California and Oregon, has over 100 plant species classified as hosts, the risk of infection is diluted with increasing species diversity because most alternative hosts have a lower competency than the main hosts bay laurel (*Umbellularia californica*) and *N. densiflorus*.

In summary, tree species diversity may reduce susceptibility of individual species to pathogens and hence increase resistance of the whole tree community. This effect may simply result from negative density effects (Fig. 7.5), but more complex mechanisms such as through development and maintenance of more diverse and abundant antagonists may also operate in species-diverse forests. In case of generalist pathogens, the effect of tree species diversity is likely much lower. A more detailed discussion about the relationship between tree species diversity and impact of fungal pathogens can be found in Pautasso et al. (2005). Regarding the evidence, most of the older studies are not based on experiments but on observational studies and comparisons of existing forest stands that differ in tree species richness. Hence, there is often no control over important variables such as stand density or site conditions (c.f. Koricheva et al. 2006). Most tree diversity experiments are still young and few have employed different spatial scales of mixtures. Hence the results obtained from these experiments can only be indicative of the effects of tree diversity on the influence of pathogens across forest ages and larger spatial scales; the same is true for insect herbivores.

## 7.8 Interactions Between Stress and Disturbance Factors

Above we have discussed the effect of tree species diversity on resistance and resilience in relation to different individual disturbance factors. Often different stress and disturbance factors interact resulting in compound effects that are much stronger than those created by a single disturbance factor (Millar and Stephenson 2015; Dale et al. 2001). For example, a meta-analysis on damage by forest insects and pathogens on drought stress showed that primary pests and pathogens living on foliage caused more damage to drought-stressed trees than to control trees, irrespective of stress severity (Jactel et al. 2012). The same analysis showed that damage by secondary agents increased with stress severity. However, the impact of primary damaging agents living in wood was significantly lower in water-stressed trees. Likewise fires are typically promoted by droughts, which create more highly flammable fuel (Dale et al. 2001). Trees felled by windstorms provide breeding material for damaging agents such as bark beetles leading to rapid increase in population size and subsequent damage in remaining standing trees (e.g. Schroeder and Lindelöw 2002).

We have seen above that the evidence for the effect of tree diversity on the influence of individual stress and disturbance factors on trees and forests is in many situations scant. One obvious problem is that disturbances such as storms or fires are so variable and operate at spatial and temporal scales that are extremely difficult to capture within an experimental design, let alone designs that also control for tree species diversity. This applies even more so for the effects of combined disturbance factors. In future, we might gain some insights from recently established tree diversity experiments, where combined disturbances such as water stress and insect herbivory may be observed or can actually be experimentally manipulated within the scale of the experiment.

## 7.9 Can Tree Species Diversity Stabilise Productivity Over Time?

Stress and disturbances lead to reductions in productivity. The effects on productivity of disturbances that remove a significant proportion of the biomass through mortality of trees are not well quantified, firstly because experiments or long-term plots providing pre-disturbance productivity data were not designed to capture the variation in disturbance, secondly because this would have to be quantified in the long term to capture the whole recovery period, and thirdly because many forest growth experiments were established for a different purpose and hence were often simply abandoned following disturbance, when the original research question could no longer be addressed. Disturbance impacts on productivity are also difficult to quantify from repeated large-scale forest inventories because inventory periods are so long (typically around 10 years), and the spatial coverage is so extensive that multiple disturbances, which may also be interacting, may not be captured. Hence, changes in productivity cannot be linked to particular disturbance events. In contrast, stress events that have only caused variation in productivity over time but not widespread mortality, e.g. through drought stress or herbivory, may have been recorded on long-term monitoring plots through repeated measurements or can be traced back through analysis of growth rings (e.g. Jucker et al. 2014). Thus we can address the question, whether species mixing/tree species diversity can stabilise productivity over time.

Tree diversity may stabilise productivity through a combination of processes including species asynchrony and species interactions, as has been found for grasslands, where community-level studies have shown that inter-annual fluctuations in primary productivity were lower in diverse communities compared to species-poor ones (Hector et al. 2010; Hautier et al. 2014). Species asynchrony can be thought of as a form of temporal complementarity among species (Hautier et al. 2014). Their asynchronous growth patterns in responses to temporal variation in environmental conditions reflect niche differences among species. For example, evergreen and deciduous temperate tree species with different times of leaf flush and onset of spring growth might be differently affected by spring droughts. Species interactions such as facilitation or reduced competition between heterospecific and conspecific neighbours can enable a higher proportion of the site's resources to be used and an increase in the efficiency with which the acquired resources are used (see Chap. 2, Forrester 2014). Recent evidence suggests that the same processes also contribute to reduction in stress, expressed as a temporal reduction in growth rates, and stabilisation of productivity over time in forests (e.g. Jucker et al. 2014; Río et al. 2013). For example, Jucker et al. (2014) analysed the relationship between tree species richness and above-ground wood production across five different European forest ecosystem types in Mediterranean, temperate, and boreal regions. Stability of productivity was defined as the ratio between the temporal mean of productivity ($\mu$) and the temporal standard deviation of productivity ($\sigma$). Thus, processes which lead to an increase in mean productivity and/or to a reduction in variance through increased resistance and resilience in relation to stress and disturbance increase stability. Productivity was derived from tree-ring analysis in stands with species richness ranging from one to four or five. Growth asynchrony was calculated following the approach proposed by Loreau and de Mazancourt (2008). This community-level variable of species asynchrony increases as the average pairwise temporal correlation in radial growth among species becomes stronger. In this study, tree species richness stabilised productivity in four out of five forest types. This was partially attributable to less variable productivity of trees in mixtures when compared to monocultures, which was likely the result of reduced competition and possibly also facilitation in heterospecific neighbourhoods. The asynchronous growth responses of the different tree species to climate decreased the inter-annual fluctuations in productivity in mixed-species stands. The asynchrony effect, however, levelled off quickly with increasing tree species richness. In addition, the overyielding of mixtures when compared to monocultures further contributed to increased stability of productivity (Jucker et al. 2014).

Using a different approach, Río et al. (2013) provide evidence that nonadditive mixing effects, overyielding and underyielding, depend on the growth conditions in respective years (Box 7.2 and Fig. 7.6). In years with favourable growth conditions, there was an increase in interspecific competition in mixed stands, whereas in years with adverse growth conditions, interspecific interaction was dominated by the processes' facilitation and/or reduced competition, resulting in lower growth reduction than in respective monospecific stands. These shifts agree with the general pattern predicted by the stress-gradient hypothesis (see Chap. 2) but using a temporal stress gradient instead of a spatial one. Obviously, these different effects of mixing also lead to a temporal stabilisation of productivity in mixtures. However, it is important to consider the temporal scale. In the long-term successional development of forest, competitive and facilitative interactions among tree species may change with time, and hence beneficial effects of mixtures may be restricted to certain periods (e.g. Cavard et al. 2011; Lasky et al. 2014).

The observed patterns highlight the importance of considering temporal variation when analysing interspecific interactions. Given the stabilising effect of tree mixtures on the temporal variability of growth, the promotion of mixed stands in forest management can improve the forest resilience in terms of growth in the face of climate change with increasingly frequent stress events (Fig. 7.7).

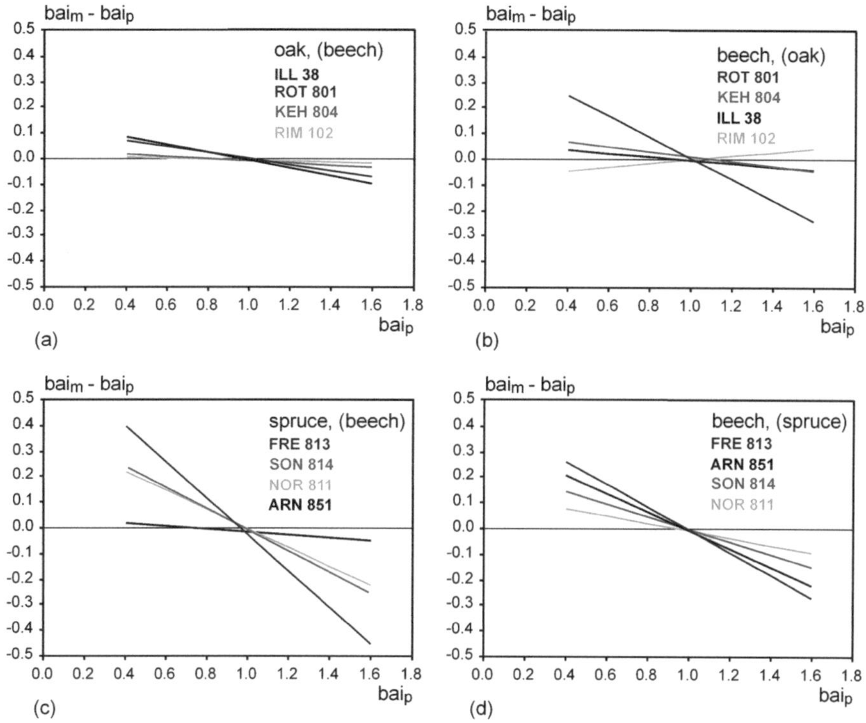


**Fig. 7.6** Tree growth benefits from mixtures in years with adverse growth conditions and is reduced, relative to monocultures, in years with favourable growth conditions. The mixing effects are shown in terms of relative tree productivity (mixed versus pure stand) on the *y*-axis in dependence on the annual productivity in the pure stand on the *x*-axis. Here, mean basal area growth indices (BAI) were calculated on the basis of the same tree-ring measurements in mixed (m) and pure (p) stands of *Picea abies* (Norway spruce), *Fagus sylvatica* (European beech), and *Quercus petraea* (Sessile oak) in southern Germany that were presented in Box 7.1. The regression lines represent the relationship between mixing effect and annual growth conditions for sites of different productivity (Río et al. 2013)

**Box 7.2 Detecting Temporal Variation of Mixing Effects on Tree Growth by Comparing the Annual Growth of Mean Tree Growth Rates in Mixed Versus Neighbouring Monospecific Stands**

Analysis of the annual variation of mixing effects can be based on time series of annual basal area increments of trees of a given species in mixed and neighbouring monospecific stands. When based on triplets of neighbouring plots comprising a monospecific stand of species 1, of species 2, and of a mixed stand of both, a sample of dominant trees is cored in order to get the retrospective growth for, for example, the last $n = 20$–$40$ years. By using only dominant trees, mixing effects can be separated from effects related to tree dimension and position in relation to the main canopy.

(continued)

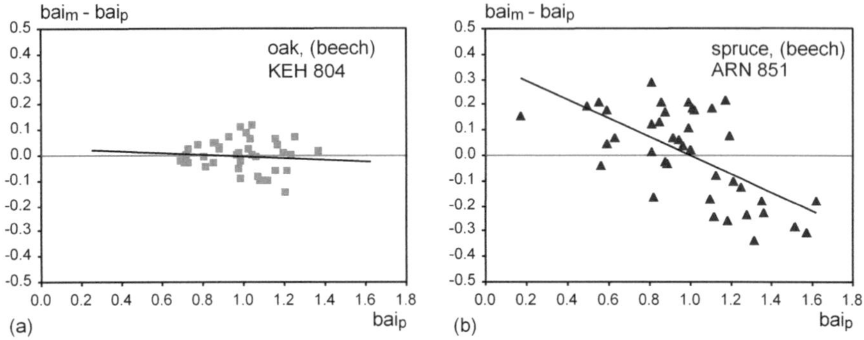


**Box Fig. 7.2-1** Analysis of the temporal variation of mixing effects using records of annual basal increment from increment cores. The difference between the annual mean tree growth in mixed and monospecific stands ($iba_m$–$iba_p$) is plotted against the mean tree growth in pure stands for a 40-year observation period. The slope of the regression line indicates how the mixing effect varies between years of high and low growth. (**a**) The observed growth rates for the oaks at plot Kelheim 804 and the fitted straight line show that inter-annual variations in growth conditions have no effect on the mixing effect. (**b**) The negative slope of the fitted straight line indicates that *P. abies* trees at the plot Arnstein 851 show positive mixing effects in low-growth years ($iba_m$–$iba_p > 0$) and negative mixing effects in high-growth years ($iba_m$–$iba_p < 0$)

**Box 7.2** (continued)

For species 1 the average basal area growth can be derived for monospecific stands ($iba_1$) and for mixed stands ($iba_{1,(2)}$) for each of the last n years. The difference $iba_{1,(2)} - iba_1$for every year indicates whether the mean tree in the mixed stand is superior ($iba_{1,(2)} - iba_1 > 0$), inferior ($iba_{1,(2)} - iba_1 < 0$), or equal ($iba_{1,(2)} - iba_1 = 0$) to the mean tree of the monospecific stand. The difference $iba_{1,(2)} - iba_1$ is plotted against the $iba_1$ values of the respective year and a linear model ($iba_{1,(2)} - iba_1 = a + b \times iba_1$) is fitted to the data (Box Fig. 7.2-1). A slope of $b = 0$ indicates no effect of the variation in growth conditions of respective years; $b < 0$ indicates a positive mixing effect in low-growth years and, if $b > 0$, a positive effect of mixing in high-growth years. The same evaluation can be performed for $iba_2$ and $iba_{(1),2}$ of species 2.

Río et al. (2013) eliminated any differences in the level and long-term trend between the time series of the trees from mixed and monospecific stands. Appropriate steps for such elimination are (1) the smoothing by moving average or spline, (2) indexing of the annual growth rates in relation to the smooth component, and (3) using the indexed values for further evaluation. The evaluation in this chapter (see Box Fig. 7.2-1) uses indexed growth rates based on moving averages over 10 years (Río et al. 2013). Notice that any systematic over- or underyielding effects between mixed and monospecific stands (differences between the mean level of $iba_{1,(2)}$ and $iba_1$) are eliminated by this indexing.

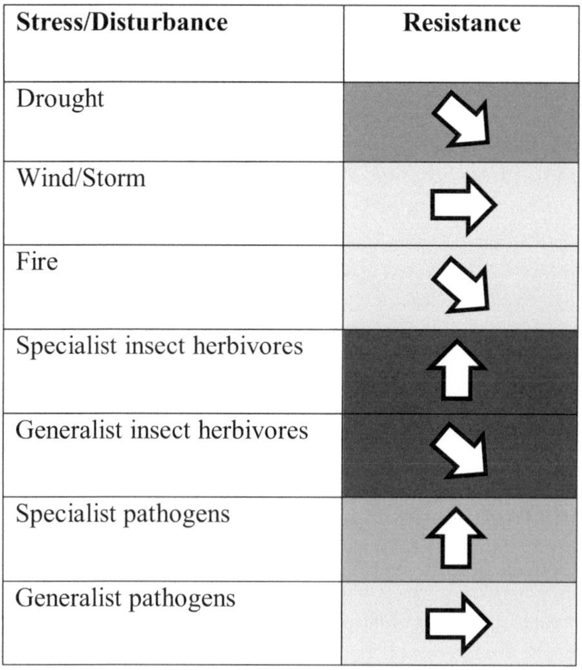


**Fig. 7.7** Overview of effects of mixing tree species and tree species diversity on the resistance of the entire tree community to different stress and disturbance factors. Effects of mixing at the tree community level result from comparisons with monocultures of all participating species. *Arrows* pointing downward indicate that antagonistic effects prevail, *upward arrows* indicate synergistic effects, and *arrows* pointing sideways indicate no interactions between species, where the community-level response is likely to be intermediate in mixtures when compared to monocultures. *Diagonal arrows* indicate intermediate effects. The colour intensity of the box indicates the level of evidence for that finding, from light grey = low evidence to dark grey = high level of evidence

## Take-Home Messages

1. Creation and maintenance of species-diverse forests is seen as an important option to adapt forests to uncertain future disturbances. However, the currently available evidence suggests that mixed-species forests are not always more resistant to ecological disturbances when compared to monospecific forests and that there is not a clear relationship between resistance to disturbance and tree species diversity; this depends on the type of disturbance. There is little evidence for true, positive diversity effects, where diversity leads to an increase in the resistance and resilience of component species in mixed-species communities.

2. For abiotic disturbances (drought, wind, fire), damage or reduction in ecosystem function can be reduced, compared with monocultures of susceptible or less resilient species, when more resistant or resilient species are mixed with less susceptible and resilient species. However, storm, fire, or drought damage to the individual tree species may not be reduced in mixtures when compared to monocultures. The stress or disturbance impacts may even be aggravated for one or more tree species in some mixtures, as is shown for drought.
3. There is more evidence to show that tree diversity reduces the effects of biotic disturbance agents (insect herbivores, pathogens). The impact of insect herbivores on individual susceptible tree species is reduced in mixtures in the majority of cases, where the insect herbivore community is dominated by specialists, but the opposite effect may occur with generalist herbivores, which can be promoted by tree diversity and thus cause greater damage in mixtures. Similarly, tree diversity can reduce the impact of specialist pathogens on host tree species, whereas there is little evidence for positive influences in the case of generalist pathogens.
4. In general terms, tree species diversity dilutes the impact of disturbance agents and, owing to different susceptibility of species to specific disturbances, insures against a complete damage or loss. This has the advantage that the forest structure and ecosystem functioning are maintained to some extent, and thus the post-disturbance recovery is facilitated. Mixing tree species can also reduce temporal variation in growth and stabilise productivity.
5. Forest managers should be aware that mixtures do not provide universally higher resistance or resilience in relation to disturbances than monocultures. In most cases, it depends to a large extent on the attributes of the species in mixture in relation to the specific disturbances.

## References

Agee JK (1996) The influence of forest structure on fire behavior. In: Proceedings of the 17th annual forest vegetation management conference, pp 16–18

Albrecht A, Hanewinkel M, Bauhus J, Kohnle U (2012) How does silviculture affect storm damage in forests of south-western Germany? Results from empirical modeling based on long-term observations. Eur J For Res 131:229–247

Allen HD (2008) Fire: plant functional types and patch mosaic burning in fire-prone ecosystems. Prog Phys Geogr 32:421–437

Allen CD, Macalady AK, Chenchouni H et al (2010) A global overview of drought and heat-induced tree mortality reveals emerging climate change risks for forests. For Ecol Manage 259:660–684

Anagnostakis SA (1987) Chestnut blight: the classical problem of an introduced pathogen. Mycologia 79:23–37

Anders S, Müller J, Augustin S, Rust S (2006) Die Ressource Wasser im zweischichtigen Nadel-Laub-Mischbestand. In: Fritz P (ed) Ökologischer Waldumbau in Deutschland: Fragen, Antworten, Perspektiven. Oekom Verlag, pp 152–183

Attiwill PM (1994) The disturbance of forest ecosystems: the ecological basis for conservative management. For Ecol Manage 63:247–300
Ayres MP, Lombardero MJ (2000) Assessing the consequences of global change for forest disturbance from herbivores and pathogens. Sci Total Environ 262:263–286
Baeten L, Verheyen K, Wirth C et al (2013) A novel comparative research platform designed to determine the functional significance of tree species diversity in European forests. Perspect Plant Ecol Evol Syst 15:281–291
Baleshta KE, Simard SW, Guy RD, Chanway CP (2005) Reducing paper birch density increases Douglas-fir growth rate and Armillaria root disease incidence in southern interior British Columbia. For Ecol Manage 208:1–13
Baleshta KE, Simard SW, Roach WJ (2015) Effects of thinning paper birch on conifer productivity and understory plant diversity. Scand J For Res 30:699–709
Barbosa P, Hines J, Kaplan I et al (2009) Associational resistance and associational susceptibility: having right or wrong neighbors. Annu Rev Ecol Evol Syst 40:1–20
Bauhus J, Puettmann KJ, Kuehne C (2013) Close-to-nature forest management in Europe: does it support complexity and adaptability of forest ecosystems? In: Messier C, Puettmann KJ, Coates KD (eds) Managing forests as complex adaptive systems: building resilience to the challenge of global change. Routledge, The Earthscan Forest Library, pp 187–213
Bernays EA, Bright KL, Gonzalez N, Angel J (1994) Dietary mixing in a generalist herbivore: tests of two hypotheses. Ecology 75:1997–2006
Binkley D, Menyailo O (2005) Gaining insights on the effects of tree species on soils. In: Binkley D, Menyailo O (eds) Tree species effects on soils: implications for global change. Kluwer Academic Publishers, Dordrecht, pp 1–16
Blakeman JP (1993) Pathogens in the foliar environment. Plant Pathol 42:479–493
Blauw LG, Wensink N, Bakker L et al (2015) Fuel moisture content enhances non-additive effects of plant mixtures on flammability and fire behavior. Ecol Evol 5:3830–3841
Bolte A, Ammer C, Löf M et al (2009) Adaptive forest management in central Europe: climate change impacts, strategies and integrative concept. Scand J For Res 24:473–482
Bommarco R, Banks JE (2003) Scale as modifier in vegetation diversity experiments: effects on herbivores and predators. Oikos 102:440–448
Bond WJ, Woodward FI, Midgley GF (2005) The global distribution of ecosystems in a world without fire. New Phytol 165:525–538
Bowman DM (2000) Australian rainforests: islands of green in a land of fire. Cambridge University Press, Cambridge, 344 p
Brang P (2001) Resistance and elasticity: promising concepts for the management of protection forests in the European Alps. For Ecol Manage 145:107–119
Brunet Y (2013) Susceptibility to wind damage. In: Gardiner B, Schuck A, Schelhaas M-J, Orazio C, Blennow K, Nicoll B (eds) Living with storm damage to forests: what science can tell us, vol 3. European Forest Institute, Joensuu, pp 25–30
Caldwell MM, Dawson TE, Richards JH (1998) Hydraulic lift: consequences of water efflux from the roots to plants. Oecologia 113:151–161
Canham CD, Papaik MJ, Latty EF (2001) Interspecific variation in susceptibility to windthrow as a function of tree size and storm severity for northern temperate tree species. Can J For Res 31:1–10
Castagneyrol B, Jactel H (2012) Unravelling plant-animal diversity relationships: a meta-regression analysis. Ecology 93:2115–2124
Castagneyrol B, Giffard B, Péré C, Jactel H (2013) Plant apparency, an overlooked driver of associational resistance to insect herbivory. J Ecol 10:418–429
Castagneyrol B, Jactel H, Vacher C et al (2014a) Effects of plant phylogenetic diversity on herbivory depend on herbivore specialization. J Appl Ecol 51:134–141
Castagneyrol B, Régolini M, Jactel H (2014b) Tree species composition rather than diversity triggers associational resistance to the pine processionary moth. Basic Appl Ecol 15:516–523

Castello JD, Leopold DJ, Smallidge PJ (1995) Pathogens, patterns, and processes in forest ecosystems. BioScience 45:16–24
Cavard X, Bergeron Y, Chen HYH et al (2011) Competition and facilitation between tree species change with stand development. Oikos 120:1683–1695
Colin F, Brunet Y, Vinckler I, Dhôte J-F (2008) Résistance aux vents forts des peuplements forestiers, et notamment des mélanges d'espèces. Rev For Fr LX:191–205
Colin F, Vinkler I, Rou-Nivert P, Renaud J-P, Hervé J-C, Bock J Piton B (2009) Facteurs de risques de chablis dans les peuplements forestiers: les leçons tirées des tempêtes de 1999. In: Birot Y, Landmann G, Bonhême I (eds) La forêt face aux tempêtes. Editions Quae, p 177–228
Coutts MP (1983) Root architecture and tree stability. Plant Soil 71:171–188
Cremer KW, Borough CJ, McKinnell FH, Carter PR (1982) Effects of stocking and thinning on wind damage in plantations. NZ J For Sci 12:244–268
Dale VH, Joyce LA, McNulty S et al (2001) Climate change and forest disturbances: climate change can affect forests by altering the frequency, intensity, duration, and timing of fire, drought, introduced species, insect and pathogen outbreaks, hurricanes, windstorms, ice storms, or landslides. BioScience 51:723–734
Davison EM, Tay FCS (1987) The effect of waterlogging on infection of *Eucalyptus marginata* seedlings by *Phytophthora cinnamomi*. New Phytol 105:585–594
Dawson TE (1993) Hydraulic lift and water use by plants: implications for water balance, performance and plant-plant interactions. Oecologia 95:565–574
de Magalhaes RMQ, Schwilk DW (2012) Leaf traits and litter flammability: evidence for non-additive mixture effects in a temperate forest. J Ecol 100:1153–1163
del Río M, Schütze G, Pretzsch H (2013) Temporal variation of competition and facilitation in mixed species forests in Central Europe. Plant Biol 16:166–176
del Río M, Pretzsch H, Ruíz-Peinado R, Ampoorter E, Annighöfer P, Barbeito I, Fabrika M (2017) Species interactions increase the temporal stability of community productivity in *Pinus sylvestris–Fagus sylvatica* mixtures across Europe. J Ecol. doi:10.1111/1365-2745.12727
DeLong RL, Lewis KJ, Simard SW, Gibson S (2002) Fluorescent pseudomonad population sizes baited from soils under pure birch, pure Douglas-fir, and mixed forest stands and their antagonism toward *Armillaria ostoyae* in vitro. Can J For Res 32:2146–2159
Dhôte JF (2005) Implication of forest diversity in resistance to strong winds. In: Scherer-Lorenzen M, Körner C, Schulze ED (eds) Forest diversity and function – temperate and boreal systems, Ecological studies, vol 176. Springer, Berlin, Heidelberg, Germany, pp 291–308
Dimitrakopoulos AP, Papaioannou KK (2001) Flammability assessment of Mediterranean forest fuels. Fire Technol 37:143–152
Drouineau S, Laroussinie O, Birot Y et al (2000) Joint evaluation of storms, forest vunerability and their restoration. European Forest Institute Discussion Paper 9, Joensuu, Finland, 38 p
Dulaurent AM, Rossi JP, Deborde C et al (2011) Honeydew feeding increased the longevity of two egg parasitoids of the pine processionary moth. J Appl Entomol 135:184–194
Dunham RA, Cameron AD (2000) Crown, stem and wood properties of wind-damaged and undamaged Sitka spruce. For Ecol Manage 135:73–81
Eis S (1972) Root grafts and their silvicultural implications. Can J For Res 2:111–120
Fernandes PM, Cruz MG (2012) Plant flammability experiments offer limited insight into vegetation–fire dynamics interactions. New Phytol 194:606–609
Fernandes PM, Vega JA, Jiménez E, Rigolot E (2008) Fire resistance of European pines. For Ecol Manage 256:246–255
Flannigan MD, Amiro BD, Logan KA et al (2006) Forest fires and climate change in the 21st century. Mitig Adapt Strat Glob Chang 11:847–859
Fleming RA, Candau JN, McAlpine RS (2002) Landscape-scale analysis of interactions between insect defoliation and forest fire in central Canada. Clim Change 55:251–272
Forrester DI (2014) The spatial and temporal dynamics of species interactions in mixed-species forests: from pattern to process. For Ecol Manage 312:282–292

Forrester DI (2015) Transpiration and water-use efficiency in mixed-species forests versus monocultures: Effects of tree size, stand density and season. Tree Physiol 35:289–304
Forrester DI, Theiveyanathan S, Collopy JJ, Marcar NE (2010) Enhanced water use efficiency in a mixed *Eucalyptus globulus* and *Acacia mearnsii* plantation. For Ecol Manage 259:1761–1770
Forrester DI, Bonal D, Dawud S et al (2016) Drought responses by individual tree species are not often correlated with tree species diversity in European forests. J Appl Ecol 53:1725-1734 doi:10.1111/1365-2664.12745
Gardiner BA, Quine CP (2000) Management of forests to reduce the risk of abiotic damage – a review with particular reference to the effects of strong winds. For Ecol Manage 135:261–277
Gerlach JP, Reich PB, Puettmann K, Baker T (1997) Species, diversity, and density affect tree seedling mortality from Armillaria root rot. Can J For Res 27:1509–1512
Gill AM (1975) Fire and the Australian flora: a review. Aust For 38:4–25
Gill MJ (1997) Eucalypts and fire: interdependent or independent? In: Williams J, Woinarski J (eds) Eucalypt ecology: individuals to ecosystems. Cambridge University Press, Cambridge, pp 151–167
González JR, Trasobares A, Palahí M, Pukkala T (2007) Predicting tree survival in burned forests in Catalonia (North-East Spain) for strategic forest planning. Ann For Sci 64:733–742
Grime JP (1979) Plant strategies and vegetation processes. Wiley, Chichester
Grossiord C, Granier A, Ratcliffe S et al (2014a) Tree diversity does not always improve resistance of forest ecosystems to drought. Proc Natl Acad Sci U S A 111(41):14812–14815
Grossiord C, Gessler A, Granier A et al (2014b) Interspecific competition influences the response of oak transpiration to increasing drought stress in a mixed Mediterranean forest. For Ecol Manage 318:54–61
Grossiord C, Granier A, Gessler A et al (2014c) Does drought influence the relationship between biodiversity and ecosystem functioning in boreal forests? Ecosystems 17:394–404
Grossiord C, Gessler A, Granier A et al (2014d) Impact of interspecific interactions on the soil water uptake depth in a young temperate mixed species plantation. J Hydrol 519:3511–3519
Guyot V, Castagneyrol B, Vialatte A et al (2015) Tree diversity limits the impact of an invasive forest pest. PLoS One 10:e0136469
Guyot V, Castagneyrol B, Vialatte A et al (2016) Tree diversity reduces pest damage in mature forests across Europe. Biol Lett 12:20151037
Haas SE, Hooten MB, Rizzo DM, Meentemeyer RK (2011) Forest species diversity reduces disease risk in a generalist plant pathogen invasion. Ecol Lett 14:1108–1116
Hajek P, Seidel D, Leuschner C (2015) Mechanical abrasion, and not competition for light, is the dominant canopy interaction in a temperate mixed forest. For Ecol Manage 348:108–116
Hanewinkel M, Albrecht A, Schmidt M (2013) Influence of landscape and within stand characteristics on wind damage. In: Gardiner B, Schuck A, Schelhaas M-J, Orazio C, Blennow K, Nicoll B (eds) Living with storm damage to forests: What Science Can Tell Us, vol 3. European Forest Institute, Joensuu, pp 39–46
Hantsch L, Bien S, Radatz S et al (2014a) Tree diversity and the role of non-host neighbour tree species in reducing fungal pathogen infestation. J Ecol 102:1673–1687
Hantsch L, Braun U, Haase J et al (2014b) No plant functional diversity effects on foliar fungal pathogens in experimental tree communities. Fungal Div 66:139–151
Hautier Y, Seabloom EW, Borer ET et al (2014) Eutrophication weakens stabilizing effects of diversity in natural grasslands. Nature 508:521–525
Hector A, Hautier Y, Saner P et al (2010) General stabilizing effects of plant diversity on grassland productivity through population asynchrony and overyielding. Ecology 91:2213–2220
IPCC (2013) Climate change 2013: The physical science basis. IPCC, Cambridge, UK
Jactel H, Brockerhoff EG (2007) Tree diversity reduces herbivory by forest insects. Ecol Lett 10:835–848
Jactel H, Menassieu P, Vetillard F et al (2006) Tree species diversity reduces the invasibility of maritime pine stands by the bast scale, *Matsucoccus feytaudi* (Homoptera: Margarodidae). Can J For Res 36:314–323

Jactel H, Nicoll BC, Branco M et al (2009) The influences of forest stand management on biotic and abiotic risks of damage. Ann For Sci 66:1–18
Jactel H, Birgersson G, Andersson S, Schlyter F (2011) Non-host volatiles mediate associational resistance to the pine processionary moth. Oecologia 166:703–711
Jactel H, Petit J, Desprez-Loustau ML et al (2012) Drought effects on damage by forest insects and pathogens: a meta-analysis. Glob Chang Biol 18:267–276
Jäkel A, Roth M (2004) Conversion of single-layered Scots pine monocultures into close-to-nature mixed hardwood forests: effects on parasitoid wasps as pest antagonists. Eur J For Res 123:203–212
Johnson VJ (1975) Hardwood fuel-breaks for north eastern United States. J For 73:588–589
Jucker T, Bouriaud O, Avacaritei D, Coomes DA (2014) Stabilizing effects of diversity on aboveground wood production in forest ecosystems: linking patterns and processes. Ecol Lett 17:1560–1569
Kaitaniemi P, Riihimäki J, Koricheva J, Vehviläinen H (2007) Experimental evidence for associational resistance against the European pine sawfly in mixed tree stands. Silva Fennica 41:259–268
Kane JM, Varner JM, Hiers JK (2008) The burning characteristics of southeastern oaks: discriminating fire facilitators from fire impeders. For Ecol Manage 256:2039–2045
Keesing F, Belden LK, Daszak P et al (2010) Impacts of biodiversity on the emergence and transmission of infectious diseases. Nature 468:647–652
Knoke T, Ammer C, Stimm B, Mosandl R (2008) Admixing broadleaved to coniferous tree species: a review on yield, ecological stability and economics. Eur J For Res 127:89–101
Kohler M, Nägele G, Sohn S, Bauhus J (2010) Can drought tolerance of Norway spruce (*Picea abies* (L.) Karst.) be increased through thinning? Eur J For Res 129:1109–1118
Korhonen K, Delatour C, Greig BJW, Schönhar S (1998) Silvicultural Control. In: Woodward S, Stenlid J, Karjalainen R, Hüttermann A (eds) *Heterobasidion annosum*: biology, ecology, impact and control. CAB International, pp 283–313
Koricheva J, Vehviläinen H, Riihimäki J et al (2006) Diversification of tree stands as a means to manage pests and diseases in boreal forests: myth or reality? Can J For Res 36:324–336
Körner C (2002) Ökologie. In: Sitte P et al (eds) Strasburger Lehrbuch für Botanik. Spektrum Akademischer Verlag, Heidelberg, pp 930–932
Kozlov MV, Lanta V, Zverev V, Zvereva EL (2015) Global patterns in background losses of woody plant foliage to insects. Glob Ecol Biogeogr 24:1126–1135
Kuhr M (2000) Grobwurzelarchitektur in Abhängigkeit von Baumart, Alter, Standort und sozialer Stellung. PhD thesis, University of Göttingen
Kunert N, Schwendenmann L, Potvin C, Hölscher D (2012) Tree diversity enhances tree transpiration in a Panamanian forest plantation. J Appl Ecol 49:135–144
Lasky JR, Uriarte M, Boukili VK et al (2014) The relationship between tree biodiversity and biomass dynamics changes with tropical forest succession. Ecol Lett 17:1158–1167
Latz E, Eisenhauer N, Rall BC et al (2012) Plant diversity improves protection against soil-borne pathogens by fostering antagonistic bacterial communities. J Ecol 100:597–604
Law BE, Falge E, Guc L et al (2002) Environmental controls over carbon dioxide and water vapor exchange of terrestrial vegetation. Agric For Meteorol 113:97–120
Lebourgeois F, Gomez N, Pinto P, Mérian P (2013) Mixed stands reduce Abies alba tree-ring sensitivity to summer drought in the Vosges mountains, western Europe. For Ecol Manage 303:61–71
Lichtenthaler HK (1998) The stress concept in plants: an introduction. Ann NY Acad Sci 851 (1):187–198
Lloret F, Keeling EG, Sala A (2011) Components of tree resilience: effects of successive low-growth episodes in old ponderosa pine forests. Oikos 120:1909–1920
Loehle C, Jones RH (1990) Adaptive significance of root grafting in trees. Funct Ecol 4:268–271
Loreau M, de Mazancourt C (2008) Species synchrony and its drivers: neutral and non-neutral community dynamics in fluctuating environments. Am Nat 172:48–66

Lovett GM, Canham CD, Arthur MA et al (2006) Forest ecosystem responses to exotic pests and pathogens in eastern North America. BioScience 56:395–405
Lübbe T, Schuldt B, Coners H, Leuschner C (2015) Species diversity and identity effects on the water consumption of tree sapling assemblages under ample and limited water supply. Oikos 125:86–97
MacDonald GI, Hoff RJ (2001) Blister rust: an introduced plague. In: Tomback DF, Arno SF, Keane RE (eds) Whitebark pine communities, ecology and restoration. Island Press, Washington, DC, pp 193–220
Mason WL (2002) Are irregular stands more windfirm? Forestry 75:347–355
Mason WL, Valinger E (2013) Managing forests to reduce storm damage. In: Gardiner BA, Schuck A, Schelhaas M-J, Orazio C, Blennow K, Nicoll B (eds) Living with storm damage to forests. What Science Can Tell Us, vol 3. EFI, Joensuu, Finland, pp 87–96
McCracken AR, Dawson WM (1998) Short rotation coppice willow in Northern Ireland since 1973: development of the use of mixtures in the control of foliar rust (*Melampsora* spp.) Eur J For Pathol 28:241–250
Metz J, Annighöfer P, Schall P et al (2015) Site adapted admixed tree species reduce drought susceptibility of mature European beech. Glob Chang Biol 22:903–920
Millar CI, Stephenson NL (2015) Temperate forest health in an era of emerging mega-disturbance. Science 349:823–826
Millar CI, Stephenson NL, Stephens SL (2007) Climate change and forests of the future: managing in the face of uncertainty. Ecol Appl 17:2145–2151
Moore JR, Maguire DA (2004) Natural sway frequencies and damping ratios of trees: concepts, review and synthesis of previous studies. Trees 18:195–203
Moore GW, Bond BJ, Jones JA (2011) A comparison of annual transpiration and productivity in monoculture and mixed-species Douglas-fir and red alder stands. For Ecol Manage 262:2263–2270
Moreira F, Rego FC, Ferreira PG (2001) Temporal (1958–1995) pattern of change in a cultural landscape of northwestern Portugal: implications for fire occurrence. Landsc Ecol 16:557–567
Morrison DJ, Cruickshank MG, Lalumière A (2014) Control of laminated and Armillaria root diseases by stump removal and tree species mixtures: amount and cause of mortality and impact on yield after 40 years. For Ecol Manage 319:75–98
Muiruri EW, Rainio K, Koricheva J (2015) Do birds see the forest for the trees? Scale-dependent effects of tree diversity on avian predation of artificial larvae. Oecologia 180:619–630
Neuner S, Albrecht A, Cullmann D et al (2015) Survival of Norway spruce remains higher in mixed stands under a dryer and warmer climate. Glob Chang Biol 21:935–946
Noble IR, Slatyer RO (1980) The use of vital attributes to predict successional changes in plant communities subject to recurrent disturbances. Vegetatio 43:5–21
Nykänen ML, Peltola H, Quine C et al (1997) Factors affecting snow damage of trees with particular reference to European conditions. Silva Fenn 31:193–213
Örlander G, Karlsson C (2000) Influence of shelterwood density on survival and height increment of *Picea abies* advance growth. Scand J For Res 15:20–29
Ormeno E, Cespedes B, Sanchez IA et al (2009) The relationship between terpenes and flammability of leaf litter. For Ecol Manage 257:471–482
Otway SJ, Hector A, Lawton JH (2005) Resource dilution effects on specialist insect herbivores in a grassland biodiversity experiment. J Anim Ecol 74:234–240
Pausas JG, Lloret F (2007) Spatial and temporal patterns of plant functional types under simulated fire regimes. Int J Wildland Fire 16:484–492
Pausas JG, Vallejo VR (1999) The role of fire in European Mediterranean ecosystems. In: Chuvieco E (ed) Remote sensing of large wildfires. Springer, Berlin, pp 3–16
Pausas JG, Bladé C, Valdecantos A (2004a) Pines and oaks in the restoration of Mediterranean landscapes of Spain: new perspectives for an old practice – a review. Plant Ecol 171:209–220
Pausas JG, Bradstock RA, Keith DA, Keeley JE (2004b) Plant functional traits in relation to fire in crown-fire ecosystems. Ecology 85:1085–1100

Pautasso M, Holdenrieder O, Stenlid J (2005) Susceptibility to fungal pathogens of forests differing in tree diversity. In: Scherer-Lorenzen M, Körner C, Schulze E-D (eds) Forest diversity and function: temperate and boreal systems, Ecological Studies 176. Springer, Berlin, pp 263–289

Pedro MS, Rammer W, Seidl R (2015) Tree species diversity mitigates disturbance impacts on the forest carbon cycle. Oecologia 177:619–630

Peltola HM (2006) Mechanical stability of trees under static loads. Am J Bot 93:1501–1511

Peltola H, Kellomäki S, Hassinen A, Granander M (2000) Mechanical stability of Scots pine, Norway spruce and birch: an analysis of tree-pulling experiments in Finland. For Ecol Manage 135:143–153

Peltola H, Gardiner B, Nicholl B (2013) Mechanics of wind damage. In: Gardiner B, Schuck A, Schelhaas M-J, Orazio C, Blennow K, Nicoll B (eds) Living with storm damage to forests: What Science Can Tell Us, vol 3. European Forest Institute, Joensuu, pp 31–38

Peterson DL, Johnson MC, Agee JK, et al (2005) Forest structure and fire hazard in dry forests of the Western United States. PNW-GTR-628, U.S. Department of Agriculture, Forest Service, Pacific Northwest Research Station, Portland, Oregon, 30 p

Pickett STA, White P (1985) The ecology of natural disturbance and patch dynamics. Academic Press, Orlando, FL

Piri T, Korhonen K, Sairanen A (1990) Occurrence of Heterobasidion annosum in pure and mixed spruce stands in southern Finland. Scand J For Res 5:113–125

Plath M, Dorn S, Riedel J et al (2012) Associational resistance and associational susceptibility: specialist herbivores show contrasting responses to tree stand diversification. Oecologia 169:477–487

Pretzsch H, Schütze G (2009) Transgressive overyielding in mixed compared with pure stands of Norway spruce and European beech in Central Europe: evidence on stand level and explanation on individual tree level. Eur J For Res 128:183–204

Pretzsch H, Schütze G, Uhl E (2013) Resistance of European tree species to drought stress in mixed versus pure forests: evidence of stress release by inter-specific facilitation. Plant Biol 15:483–495

Pretzsch H, Rötzer T, Matyssek R et al (2014) Mixed Norway spruce (Picea abies [L.] Karst) and European beech (Fagus sylvatica [L.]) stands under drought: from reaction pattern to mechanism. Trees 28:1305–1321

Puerta-Piñero C, Brotons L, Coll L, González-Olabarría JR (2012) Valuing acorn dispersal and resprouting capacity ecological functions to ensure Mediterranean forest resilience after fire. Eur J For Res 131:835–844

Putz FE, Coley PD, Lu K et al (1983) Uprooting and snapping of trees: structural determinants and ecological consequences. Can J For Res 13:1011–1020

Pyatt DG (1993) Multi-purpose forests on peatland. Biodivers Conserv 2:548–555

Reif A, Brucker U, Kratzer R et al (2010) Waldbewirtschaftung in Zeiten des Klimawandels – Synergien und Konfliktpotenziale zwischen Forstwirtschaft und Naturschutz. Naturschutz und Landschaftsplanung 42:261–266

Rennenberg H, Seiler W, Matyssek R et al (2004) Die Buche (Fagus sylvatica L.) – ein Waldbaum ohne Zukunft im südlichen Mitteuropa? Allg Forst- und Jagdztg 175:210–224

Rhoads AG, Hamburg SP, Fahey TJ et al (2002) Effects of an intense ice storm on the structure of a northern hardwood forest. Can J For Res 32:1763–1775

Rich RL, Frelich LE, Reich PB (2007) Wind-throw mortality in the southern boreal forest: effects of species, diameter and stand age. J Ecol 95:1261–1273

Root RB (1973) The organisation of a plant-arthropod association in simple and diverse habitats: the fauna of collards, *Brassica olacea*. Ecol Monogr 43:95–124

Rudnicki M, Lieffers VJ, Silins U (2003) Stand structure governs the crown collisions of lodgepole pine. Can J For Res 33:1238–1244

Ruel JC (1995) Understanding windthrow: silvicultural implications. For Chronicle 71:434–445

Schär C, Vidale PL, Lüthi D et al (2004) The role of increasing temperature variability in European summer heatwaves. Nature 427:332–335
Schelhaas MJ, Nabuurs GJ, Schuck A (2003) Natural disturbances in the European forests in the 19th and 20th centuries. Glob Chang Biol 9:1620–1633
Scherer-Lorenzen M, Schulze E-D, Don A et al (2007) Exploring the functional significance of forest diversity: a new long-term experiment with temperate tree species (BIOTREE). Perspect Plant Ecol Evol Syst 9:53–70
Schmid I, Kazda M (2002) Root distribution of Norway spruce in monospecific and mixed stands on different soils. For Ecol Manage 159:37–47
Schönhar S (1990) Ausbreitung und Bekämpfung von Heterobasidion annosum in Fichtenbeständen auf basenreichen Lehmböden. Allg Forstzeitschrift 36:911–913
Schroeder LM, Lindelöw Å (2002) Attacks on living spruce trees by the bark beetle *Ips typographus* (Col. Scolytidae) following a storm-felling: a comparison between stands with and without removal of wind-felled trees. Agric For Entomol 4:47–56
Schuldt A, Baruffol M, Böhnke M et al (2010) Tree diversity promotes insect herbivory in subtropical forests of south-east China. J Ecol 98:917–926
Schume H, Jost G, Hager H (2004) Soil water depletion and recharge patterns in mixed and pure forest stands of European beech and Norway spruce. J Hydrol 289:258–274
Schütz JP, Götz M, Schmid W, Mandallaz D (2006) Vulnerability of spruce (Picea abies) and beech (Fagus sylvatica) forest stands to storms and consequences for silviculture. Eur J For Res 125:291–302
Schwilk DW (2003) Flammability is a niche construction trait: canopy architecture affects fire intensity. Am Nat 162:725–733
Schwilk DW, Caprio AC (2011) Scaling from leaf traits to fire behaviour: community composition predicts fire severity in a temperate forest. J Ecol 99:970–980
Seppälä R, Buck A, Katila P (eds) (2009) Adaptation of forests and people to climate change. A Global Assessment Report. IUFRO World Series vol 22, Helsinki, 224 p
Seymour RS, White AS (2002) Natural disturbance regimes in northeastern North America—evaluating silvicultural systems using natural scales and frequencies. For Ecol Manage 155:357–367
Shea KR (1971) Disease and insect activity in relation to intensive culture of forests. In: XV IUFRO congress, Gainesville, Proceedings. Florida, USA, pp 109–117
Slodicák M (1995) Thinning regime in stands of Norway spruce subjected to snow and wind damage. In: Couts MP, Grace J (eds) Wind and trees. Cambridge University Press, Cambridge, pp 436–447
Sohn J, Gebhardt T, Ammer C et al (2013) Mitigation of drought by thinning: short-term and long-term effects on growth and physiological performance of Norway spruce (*Picea abies*). For Ecol Manage 308:188–197
Spatz HC, Theckes B (2013) Oscillation damping in trees. Plant Sci 207:66–71
Stanton ML (1983) Spatial patterns in the plant community and their effects upon insect search. In: Ahmad S (ed) Herbivorous insects: host seeking behavior and mechanism. Academic Press, New York, pp 125–156
Tàbara D, Saurí D, Cerdan R (2003) Forest fire risk management and public participation in changing socioenvironmental conditions: a case study in a Mediterranean region. Risk Anal 23:249–260
Tahvanainen JO, Root RB (1972) The influence of vegetational diversity on the population ecology of a specialized herbivore, *Phyllotreta cruciferae* (Coleoptera: Chrysomelidae). Oecologia 10:321–346
Taylor KL, Fonda RW (1990) Woody fuel structure and fire in subalpine fir forests, Olympic National Park, Washington. Can J For Res 20:193–199
Thurm EA, Uhl E, Pretzsch H (2016) Mixture reduces climate sensitivity of Douglas-fir stem growth. For Ecol Manage 376:205–220
Trumbore S, Brando P, Hartmann H (2015) Forest health and global change. Science 349:814–818

Unsicker SB, Oswald A, Köhler G, Weisser WW (2008) Complementarity effects through dietary mixing enhance the performance of a generalist insect herbivore. Oecologia 156:313–324
v Lüpke B, Spellmann H (1997) Aspekte der Stabilität und des Wachstums von Mischbeständen aus Fichte und Buche als Grundlage für waldbauliche Entscheidungen. Forstarchiv 68:167–179
Valinger E, Fridman J (2011) Factors affecting the probability of windthrow at stand level as a result of Gudrun winter storm in southern Sweden. For Ecol Manage 262:398–403
Van Altena C, van Logtestijn RS, Cornwell WK, Cornelissen JH (2012) Species composition and fire: non-additive mixture effects on ground fuel flammability. Front Plant Sci 3:63
Van der Maarel E (1993) Some remarks on disturbance and its relations to diversity and stability. J Veg Sci 4:733–736
Van Der Werf GR, Randerson JT, Giglio L et al (2008) Climate controls on the variability of fires in the tropics and subtropics. Global Biogeochem Cycles 22:3. doi:10.1029/2007GB003122
Vehviläinen H, Koricheva J, Ruohomäki K et al (2006) Effects of tree stand species composition on insect herbivory of silver birch in boreal forests. Basic Appl Ecol 7:1–11
Walker B, Kinzig A, Langridge J (1999) Plant attribute diversity, resilience, and ecosystem function: the nature and significance of dominant and minor species. Ecosystems 2:95–113
Warrillow M, Mou P (1999) Ice storm damage to forest tree species in the ridge and valley region of southwestern Virginia. J Torrey Bot Soc 126:147–158
White PS, Jentsch A (2001) The search for generality in studies of disturbance and ecosystem dynamics. In: Progress in botany. Springer, Berlin, Heidelberg, pp 399–450
White JA, Whitham TG (2000) Associational susceptibility of cottonwood to a box elder herbivore. Ecology 81:1795–1803
Woodward S, Stenlid J, Karjalainen R, Hüttermann A (eds) (1998) *Heterobasidion annosum*: biology, ecology, impact and control. CAB International, Wallingford, UK, 589 p
Yachi S, Loreau M (1999) Biodiversity and ecosystem productivity in a fluctuating environment: the insurance hypothesis. Proc Natl Acad Sci U S A 96:57–64
Zapater M, Hossann C, Bréda N et al (2011) Evidence of hydraulic lift in a young beech and oak mixed forest using 18O soil water labelling. Trees 25:885
Zhang Q-H, Schlyter F (2004) Olfactory recognition and behavioural avoidance of angiosperm non-host volatiles by conifer bark beetles. Agric For Entomol 6:1–19
Zvereva EL, Zverev V, Kozlov MV (2012) Little strokes fell great oaks: minor but chronic herbivory substantially reduces birch growth. Oikos 121:2036–2043

# Modelling Mixed-Species Forest Stands

# 8

Hans Pretzsch, Thomas Rötzer, and David I. Forrester

**Abstract**

The chapter first describes common models for monospecific stands and then the environmental conditions, processes, and structures that need to be included in forest growth models that are to be applied to mixed-species forests, how these different processes are incorporated into models, and the strengths and weaknesses of tree-level and stand-level approaches. The chapter gives an introduction to empirical models, process-based models, and hybrid models, which are a combination of the former two groups. Empirical models describe the system behaviour statistically, not the structure and mechanistic functioning of the system. Process-based models describe the trees and stand development on the basis of the underlying structure, within-stand environment, and functioning. Hybrid models represent a compromise between empirical and process-based models; they may bridge knowledge gaps of processes using statistical relationships. The chapter focuses on models that start at the individual tree level and scale up to the stand level or models that start and finish at the stand level.

## Contents

8.1 Models for Forest Science and Forest Management ..... 385
8.2 From Early Experience Tables to Process-Based Computer Models ..... 387
8.2.1 Overview of Stand Growth Models ..... 387
8.2.2 From Empirical to Process-Based and Hybrid Approaches ..... 389
8.3 Integration of Mixing Effects in the Algorithms of Forest Growth Models ..... 396

H. Pretzsch (✉) • T. Rötzer
Chair for Forest Growth and Yield Science, Faculty of Forest Science and Resource Management, Technical University of Munich, Hans-Carl-von-Carlowitz-Platz 2, 85354 Freising, Germany
e-mail: Hans.Pretzsch@lrz.tum.de; Thomas.Roetzer@lrz.tum.de

D.I. Forrester
Swiss Federal Institute for Forest, Snow and Landscape Research WSL, Zürcherstrasse 111, 8903 Birmensdorf, Switzerland
e-mail: david.forrester@wsl.ch



H. Pretzsch et al. (eds.), *Mixed-Species Forests*, DOI 10.1007/978-3-662-54553-9_8

8.3.1 Integration of Mixing Effects in Empirical Models ..... 396
8.3.2 Integration of Mixing Effects in Process-Based Models ..... 398
8.4 Functioning, Structure, and Within-Stand Environment as the Main Interacting Units in Mixed-Species Stands and Models ..... 399
8.4.1 Temporal and Spatial Resolutions for Modelling the Effects of Species Mixing on the Functioning, Environment, and Structure ..... 400
8.4.2 Functioning ..... 406
8.4.3 Structure ..... 410
8.4.4 Within-Stand Environment ..... 411
8.5 Model Selection: Which Spatial and Temporal Resolutions? ..... 412
8.5.1 Using the Individual Tree Hybrid Model SILVA to Simulate Mixed-Species Stands ..... 413
8.5.2 Application for Scenario Analysis at the Stand Level: A Monospecific *P. abies* Stand Versus a *P. abies-F. sylvatica* Mixed Stand ..... 414
8.5.3 Confidence in the Performance of a Process-Based Model as a Prerequisite for Application in Forest Practice ..... 420
8.5.4 Using a Stand-Level Process-Based Model to Develop Silvicultural Regimes for Mixtures ..... 422
Take-Home Messages ..... 423
References ..... 425

Models are useful tools to summarise existing knowledge about forest dynamics into a form that can be used by researchers and foresters. Importantly, they can be used to predict the spatial and temporal variability in mixing effects in terms of growth and other ecosystem functions and the interactions between different environmental factors. Empirical data for many of these situations is often non-existent. Despite this potential and the growing body of empirical evidence showing that mixing effects are not simply an additive function of each monoculture, many forest growth models, as well as related concepts and theory, are still based on monospecific stands. This chapter first describes common models for monospecific stands and then the environmental conditions, processes, and structures that need to be included in forest growth models that are to be applied to mixed-species forests, how these different processes are incorporated into models, and the strengths and weaknesses of tree-level and stand-level approaches.

Forest growth models are often divided into three groups, empirical models, process-based models, and hybrid models, which are a combination of the former two groups. All are considered in this chapter; however, the focus is on the latter two groups because while empirical models can provide accurate predictions of mixing effects for the environmental conditions that existed in the forests where they were developed, they are unlikely to provide reliable predictions of mixing effects for other environmental conditions resulting from changes in climate, management or disturbance regimes, or novel species combinations, and this is often precisely what researchers or managers are interested in. It is important to note that in reality, all models are empirical to some extent. However, we use the terms empirical, process-based, and hybrid to broadly divide models into groups based on the proportion of process-based components that are in the model.

Three main forest characteristics that need to be considered in forest models are the environmental conditions within the stand (e.g. vertical light profile, rooting space, humus layer), the functioning (e.g. photosynthetic rate, light use, growth), and the tree and stand structure (e.g. crown and stem shape, shoot and root morphology). Each of these is influenced by and influences the others and can be considered at different spatial and temporal resolutions.

Stand-level models often seek to simplify the functions, environmental attributes, and structural attributes, thereby avoiding potential problems when scaling from the individual tree or leaf levels up to the stand level, while also simplifying model parameterisation and making it easier to interpret and use the models. Most of the functions, environmental attributes, and structural attributes that are important in mixed-species forests can be modelled at the level of the stand. One exception is horizontal heterogeneity, such as when the species are mixed row-by-row or are clumped due to the gap dynamics of the stand, as opposed to more homogeneous structures of tree-by-tree mixing.

Tree-level models are often able to incorporate functions, environmental attributes, and structural attributes at the spatial and temporal resolutions at which they actually occur, including the horizontal heterogeneity. These models can be useful for examining how the variability in the properties of individual trees, as well as the horizontal spatial structural and environmental heterogeneity within a stand, can influence the stand-level dynamics. However, their actual predictions of stand-level variables can be less reliable than those of stand-level models due to the propagation of errors when upscaling. Tree-level and stand-level approaches are complementary and can be substantiated by further empirical research into mixed-species forests. This validation will be a critical next step because despite the interest in modelling mixed-species forests, there are only a very few examples where the predicted mixing effects have been compared with empirical measurements of mixing effects (Forrester and Tang 2016).

This chapter describes some of the main approaches used to model forest stand dynamics. It focuses on models that start at the individual tree level and scale up to the stand level or models that start and finish at the stand level.

## 8.1 Models for Forest Science and Forest Management

A model results when a real system, for example, a forest stand, has been abstracted and reproduced biometrically. The model, or system model, collects, organises, and aggregates reliable knowledge bases in a system algorithm. When a system model is converted into a useful computer program, a forest growth simulator is created that can reproduce the behaviour of the system and perform scenario and prediction computations. Through the organisation, synthesis, and utilisation of forest growth knowledge, forest models and growth simulators can expand both the basic and applied knowledge required in forest management and forest science equally. Thus, model development and system simulation are of paramount importance for forest growth research.

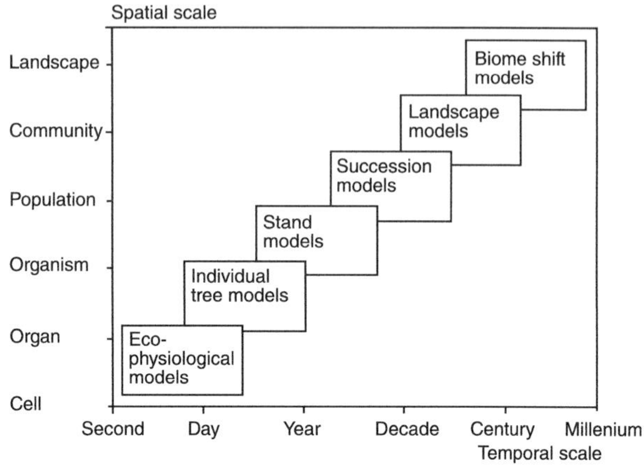


**Fig. 8.1** The spatial and temporal aggregation of modelling processes and structures increases from eco-physiological process models to management models and to succession, landscape, and biome shift models (modified from Battaglia and Sands (1998))

The aim and purpose of a model, as well as the state of knowledge about the system considered, determines the degree of complexity necessary, or possible, and the temporal and spatial resolution of the model (Battaglia and Sands 1998). The temporal scale may range from seconds to thousands of years, the spatial scale from the cell and mineral surfaces up to the vegetation zone, resulting in eco-physiological process models, individual tree models, yield tables, gap models, landscape, and biome shift models (Fig. 8.1).

As in many science branches, in forest science there is a trend from holism towards reductionism. Forest research moves further and further from stand to tree, from tree form and structure to allometry, from allometry to primary and secondary metabolism, to proteome, transcript, and gene level. As the level of detail in forest science increases, however, the link to forest practice requires that the practically relevant level is still considered in experiments as well as in models. A more and more detailed analysis of stands and ecosystems is not necessarily a forward movement in the knowledge base relevant to ecosystem management. If forest management makes use of any scientific knowledge at all, then it is the integrated knowledge at the stand or landscape level, rather than fragmented details on a small temporal and spatial scale, as interesting as the results of such reductionism are. Contrary to the reductionism in science, system management often requires an integrated and a more holistic view of the system in question.

Forest growth models are simplifications and generalisations of actual forest dynamics. The degree of simplification depends on the purpose of the model and the depth of knowledge about the actual system. Thoroughly validated forest growth models are very valuable tools for forest research and practice because they provide instant predictions of forest growth dynamics. Given the long lifespan of trees, it often takes decades or even centuries to obtain empirical data about the effect of a given treatment on growth, and experiments that were established decades or centuries ago are rare and may not contain the same treatments that are of interest today.

## 8.2 From Early Experience Tables to Process-Based Computer Models

### 8.2.1 Overview of Stand Growth Models

Before forest scientists understood even some of the basic processes beyond tree growth, considerable empirical knowledge had accumulated from pure observation. The growth relationships derived from observation and not primary causes made a large contribution to today's understanding of how trees grow and the way they respond under certain growing conditions in the forest stand. The progression from stand-oriented growth models like the monospecific stand yield tables from Schwappach (1893) and Wiedemann (1932) to stand simulators for management purposes and process-based models as research tools and the development of succession, landscape, and biome shift models (Fig. 8.1) reflect the advance in forest ecosystem knowledge, the change in the aims of forest modelling, and the development of theories of forest dynamics.

With a history of over 250 years, yield tables for monospecific stands may be considered the oldest models in forestry science and forest management. They represent the development of important stand-level parameters (stem number, mean height, mean diameter, basal area, form factor, cumulative annual increment, total production, and mean annual increment) in tabular form in 5-year intervals, separately for defined treatments and yield classes (Pretzsch 2001). From the earlier experience tables (von Cotta 1821; Paulsen 1795) at the end of the eighteenth century, which were based on estimations or limited measurements, to the first standardised yield tables (Gehrhardt 1909; Schwappach 1893), based on long-term observations, and subsequently to the computer-supported yield table models (Assmann and Franz 1963; Schmidt 1971) and yield tables produced by stand simulators (Franz 1968; Hradetzky 1972), models of this generation have become a decisive information base for sustainable volume production.

Early experience tables estimated the site fertility class based on the standing stem volume and stand age. The site fertility class was then applied to estimate stand growth and yield. More developed yield tables used stand height and age to estimate the site fertility class, and then the site fertility class and stand age were used to read stand growth and yield variables off the yield tables. Later, tables used site index or vegetation type as surrogate variables to assess the site fertility. Only very few yield tables were compiled for mixed-species stands (Bonnemann 1939; Wiedemann 1942), and such stands are of very limited practical relevance due to the small number of mixing proportions, site classes, and thinning grades they include.

In the 1960s, a second generation of models was initiated, which, in addition to stand-level data, also produced stem number frequencies and size classes to enable improved predictions of assortment yield and financial net return. Differential equation models, where the state variables of the model are updated in discrete time intervals (Moser 1972), distribution extrapolation models (Clutter and Bennett 1965), and stochastic evolution models (Suzuki 1971) served this purpose by abstracting the development dynamics of even-aged homogeneous monospecific stands by shifting a stem number-diameter distribution along the time axis. Although initially developed for monospecific stands, such models can also be statistically fitted to mixed stand datasets and applied for prediction of mixed stand dynamics (Liang et al. 2005; Nguyen et al. 2012).

Individual tree models employ a much higher resolution in the system abstraction and modelling (Newnham 1964; Ek and Monserud 1974; Nagel 1996; Pretzsch et al. 2002; Wykoff et al. 1982). They divide the stand into a mosaic of individual trees and model their interactions as a spatial-temporal system. The level of description is identical to the level of biological observation, and the information unit in the model (individual tree) is equally the basic unit of the stand. As individual tree models contain feedback loops between stand structure and growth, they have greater complexity and flexibility than their precursors. We define position-dependent and position-independent individual tree models as approaches in which stand competition is modelled with and without consideration of the spatial distribution pattern (stem coordinates, distances between tree pairs, crown parameters), respectively. Pretzsch (2001) reviews the most relevant competition indices, which form the core of such models, as they control individual tree increment. Stand-level data required in forestry can be produced by summarising and aggregating the data of all tree individuals in a stand (Pukkala 1987; Sterba et al. 1995). By considering the spatial constellation of the individual trees and their growth response to intra- and interspecific competition, they pave the way to modelling and simulating the dynamics of even- and uneven-aged mono- and mixed-species stands (Pretzsch et al. 2015).

Small area or gap models reproduce the growth of single trees in forest patches (e.g. of 100 $m^2$ area) in relation to the prevailing mean growing conditions at the site (Shugart 1984; Leemans and Prentice 1989). In these models, the relationships between environmental conditions and growth are described partially statistically and partially eco-physiologically; as a result they adopt an intermediate course between statistically based single-tree models and eco-physiologically oriented process models. They are used to investigate the competition and succession in close-to-nature forests. Individual tree and gap models estimate increment at the tree or stand level, using independent variables such as tree and stand characteristics as well as environmental factors. They were primarily built for modelling succession processes and natural dynamics in monospecific and mixed stands (Bugmann 2001; Shugart 1984) but were improved concerning the representation of the stand structure (Lindner et al. 1997) and further developed for application for climate change (Keane et al. 2001) and forest management (Huth and Tietjen 2007; Kammesheidt et al. 2001) scenario analyses.

Eco-physiological process-based models predict tree and stand development from primary processes such as photosynthesis, respiration, or carbon allocation (Bossel 1994; Mäkelä and Hari 1986; Mohren 1987). They are based on basic physical, chemical, and eco-physiological relationships as much as possible and seek statistical support only to bridge the gap in knowledge where necessary. These models predict the primary production at the individual tree level (Grote and

**Table 8.1** Examples of within-stand environmental (E), functioning (F), and structural (S) conditions that can change at different temporal and spatial resolutions as a result of species interactions (modified from Pretzsch et al. (2015))

| | Temporal resolution | |
|---|---|---|
| Spatial resolution | Short term (minute, hour, day) | Long term (year, decade, lifetime) |
| Point, parts of stand, tree | E: light flecks, wind speed<br>F: photosynthesis, hydraulic redistribution<br>S: cell structure, tree ring development | E: root channels, root layers<br>F: tree mortality, leaf turnover<br>S: tree crown plasticity |
| Whole stand, ecosystem | E: canopy climate ($CO_2$ and $H_2O$ exchange)<br>F: water translocation<br>S: current height growth, bud break | E: soil fertility, nutrient supply<br>F: gap dynamic, regeneration<br>S: standing stock |

Pretzsch 2002) or stand level (Landsberg 1986, 2003) and also provide information about carbon, nitrogen, and water cycles, thereby supporting a comprehensive understanding and management of ecosystems. Estimations of primary productivity form the basis of such model approaches. With the exception of some process-based or hybrid stand-level models (e.g. 3-PG; Landsberg and Sands (2010)), eco-physiologically based models have served as research tools more than management tools due to the large demand for initialisation data and the collation and processing of time series of growth determinants by a powerful computer. However, in the future they will become increasingly involved in practical uses; in particular, the integration of structure and function will be of major practical relevance (Kurth 1999; Pretzsch et al. 2008). The increasing demand for information about forest ecosystems and the desire to understand and predict the responses of forest ecosystems to disturbances requires a degree of complexity provided only in eco-physiological process or hybrid models. Eco-physiological models which consider the environmental conditions (Pretzsch et al. 2015), source and/or sink-driven tree functioning (Hayat et al. 2017), and tree and stand structure (Rötzer et al. 2009) have the highest potential to implicitly cover the effect of intra- and interspecific interactions on tree and stand growth (Table 8.1).

## 8.2.2 From Empirical to Process-Based and Hybrid Approaches

### 8.2.2.1 Progression from Describing to Understanding the Site-Growth Relationship

The attempt to model and predict stand productivity, growth, or yield for a given forest stand and the corresponding experiments for generating a suitable database date back to the eighteenth century (Assmann 1970; Pretzsch and Rötzer 2015). However, the approaches have become more and more process based, spatially explicit, and more focused on the primary resource and environmental variables.

Initially, in the eighteenth century, the standing stem volume of forest stands was used to classify a given stand into a site quality system (Pressler 1877):

Site fertility class $= f$ (standing volume),

and then the site fertility class could be applied to estimate growth and yield, e.g.

Volume growth $= f$ (site fertility class).

This classification was circular; the standing volume must be estimated before determining the site fertility class, after which projections of the future volume productivity could be calculated. This approach only made sense as long as light and moderate thinnings were common. With the change to more intensive management concepts in the nineteenth century, the thinning component of total production increased so that standing volume became an increasingly poorer indicator of site fertility class.

As the relationship between stand age and stand height correlates closely with total stand production (Eichhorn 1902) and is less dependent on thinning intensity, it provides an alternative to the former approach (Sect. 4.4). Thus, the use of age and mean height, introduced by Baur (1877) and de Perthuis de Laillevault (1803),

Site fertility class $= f$ (mean height, stand age)

for the classification of stand growth became established despite some initial reluctance (Heyer 1845). Estimates of stand growth and yield result from

Stand growth $= f$ (site fertility class, stand age).

With the intensification of thinning from below, which significantly influences calculations of mean height, another change was made in the mid-twentieth century towards the use of dominant height as an indicator of site fertility (Assmann 1970). The idea of using stand volume or height growth as a 'phytometer' for the productivity of a site has continued to the present day. However, this approach is again being questioned as forest practice increasingly implements thinning from above in management regimes to encourage structurally diverse mixed stands.

This calls into question how indicative dominant height is of the site conditions because site conditions may be reduced below the site-specific dominant height by thinning from above or a suppression phase. The more a stand deviates from an even-aged, mono-layered structure, the greater the influence of density and competition on the relationships between age and height, and the less suitable the age-height relationship becomes as an indicator of site fertility. Especially in highly structured mixed stands, mean and dominant heights are unsuitable as site fertility indicators.

Heyer (1845) demanded that yield studies not be directed exclusively towards the volume yield, but towards the investigation and measurement of 'primary site factors', such as temperature, nutrient supply, radiation, etc. One step in this direction was made by Cajander (1926). He developed a classification system for boreal forests that facilitated the growth and yield estimation from the forest floor vegetation as the indicator (species lichens, mosses, grasses, herbs, shrubs). While this method became standard in the rather uniform and undisturbed boreal forests, the heterogeneity of forest types and human influences makes this approach inadequate for Central European forests.

However, the core of Cajander's idea was to combine locally available indicator variables for classifying a particular stand with growth and yield information deduced from site-specific yield tables. The increasing availability of site information and growth and yield data from inventories led to an increased meshing of locally acquired information about site, growth, and yield with general growth and yield relationships deduced from models. Moosmayer and Schöpfer (1972), Wykoff et al. (1982), and Wykoff and Monserud (1988) developed relationships between site conditions and growth at the tree or stand level by regression analysis:

Volume growth $= f$ (stand attributes, site characteristics).

As independent variables they used metric information (e.g. annual precipitation, mean temperature, slope, exposition), ordinal (e.g. levels of nutrition supply, levels of water supply), and nominal (e.g. growth region, degree of disturbance of topsoil by machines) variables. To estimate the potential height growth, volume growth, and yield in the forest growth simulator SILVA, Kahn (1994) used a set of nine metric site variables.

Experimental field plots, monitoring plots, and chamber experiments enable the study of metabolic, physiological, and growth processes as well as the factors affecting growth (environmental conditions, resource supply) at increasingly more refined spatial and temporal resolution. For example, temperature measured per day, hour, or minute and radiation recorded separately for different wavelengths can be used to estimate net primary production, according to the following approach:

Net primary productivity $= f$ (leaf area, radiation, temperature, nutrient, water).

Simultaneously recorded assimilation rate, respiration, height, and diameter also enable a refined estimation of gross production (gC $\text{min}^{-1}$), (cm $\text{day}^{-1}$), (mm $\text{day}^{-1}$). The development of inventory monitoring and innovative regionalisation tools, which deliver all the relevant variables for such primary variable-based approaches that Heyer (1845) already had in mind, is in progress. Knowledge about the site-growth relationship forms the backbone of forest growth models, and hence the availability of site variables is decisive for the applicability of these models. To overcome missing knowledge about the relationships between the primary causes affecting tree and stand development (environmental conditions, resource supply), forest research has developed a rich experience in the application of 'surrogate variables' or 'proxy variables' (Oliver and Larson 1996; Zeide 2003). Examples for the application of a 'surrogate variable' or 'proxy variable' include the use of age-height relations to estimate stand growth, the area available for growth or growing space of a tree to estimate its resource supply, or the application of

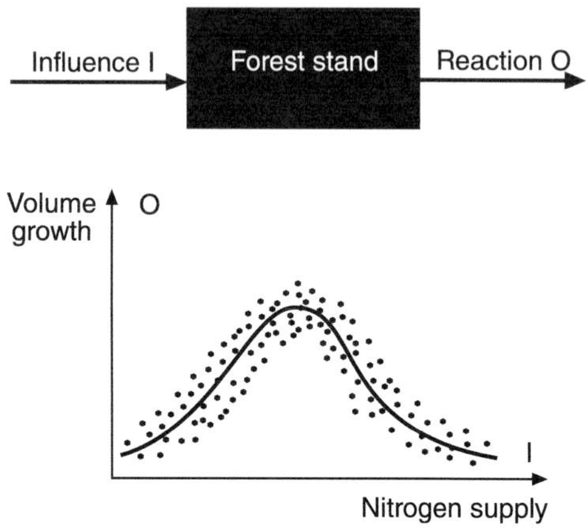


**Fig. 8.2** Empirical model or black-box approaches are based on statistical correlations, e.g. between age and size or nitrogen supply and growth response

competition indices to estimate the height and diameter increment of an individual tree in relation to resource supply (Chap. 6). In all cases, the primary factors remained unsolved and were replaced by easy-to-measure auxiliary variables that reflected the hidden relationships. However, today many of the primary factors are much easier to measure, and the use of these proxy variables may therefore decline and be replaced with more direct and accurate variables.

#### 8.2.2.2 Advancement of Model Approaches

Based on records of the resource supply, environmental conditions (causes) on experimental plots, and the growth responses of individual trees and whole stands as determined from measurements of diameter, height, and crown growth (effects), forest growth research derives stochastic relationships between cause and effect and draws conclusions about the behaviour of the system.

Depending on their approach to modelling the relationship between causes and effects and between environmental conditions and growth, models are often classified as empirical, process based, or hybrid.

*Empirical Models* Early modelling approaches in particular did not really consider the underlying processes (e.g. photosynthesis, respiration, transpiration, nutrient uptake, allocation, senescence, and mortality). Such approaches are described as empirical or black-box approaches since we remain in the dark, to a large extent, about the inner structure of the system and the underlying causal relationships. Only the system behaviour is explained statistically, not the structure and function of the system. The description of the relationship determined by regression analysis between the nitrogen inputs in a forest stand and the wood production provides an example (Fig. 8.2). When the modelling approach includes the system elements and their interactions, we speak of a white-box approach. Here, the effect of the nitrogen on photosynthesis, allocation, and decomposition processes would be described so that the stem wood production would be obtained from the underlying eco-physiological processes (Fig. 8.4).

Early forest growth research pursued the traditional black-box approach with the aim of predicting the behaviour of the system. Its main interest was to describe, analyse, and predict accurately the overall behaviour of biological systems, be it tree growth, stand development, or age-class rotation at the forest estate level.

Yield tables and other empirical models are based on regression analytically adapted growth curves between age and tree or stand size, e.g. age and mean stand height ($h$) for different site conditions ($h = 20, 25, \ldots 40$ m at age 100). If age, $a_1$, and height, $h_1$, at a given age are known, the height increment in the following period (triangular field) can be read off. Analogously, diameter, basal area, and volume increment can also be deduced from the initial variables age and height (Fig. 8.3).

*Process-Based Models* The relationships statistically derived for stable systems cannot be transferred directly to other ecological conditions. If one assumes that the physiological response to environmental changes is rapid and consistent, the establishment of short-term experiments may possibly be sufficient to forecast long-term growth responses. In the view of an increasing instability in forest growth due to atmospheric pollution, increased $NO_x$ deposition, atmospheric $CO_2$ concentrations, and climate change, forest growth research cannot restrict itself to statistical links between system variables at high levels of aggregation, but must understand and predict the responses of forest ecosystems. Hence, the development of eco-physiological process models has gained momentum as they model forest growth on the basis of physiological cause-effect chains and interactions between the system processes even though little or nothing is known about their impacts in the long term.

In such models, processes need to be included at a higher temporal and spatial resolution to gain an understanding of system behaviour and to make realistic predictions. Often, black-box approaches are modified to grey-box or white-box approaches when they are supported by a deeper knowledge of structure and

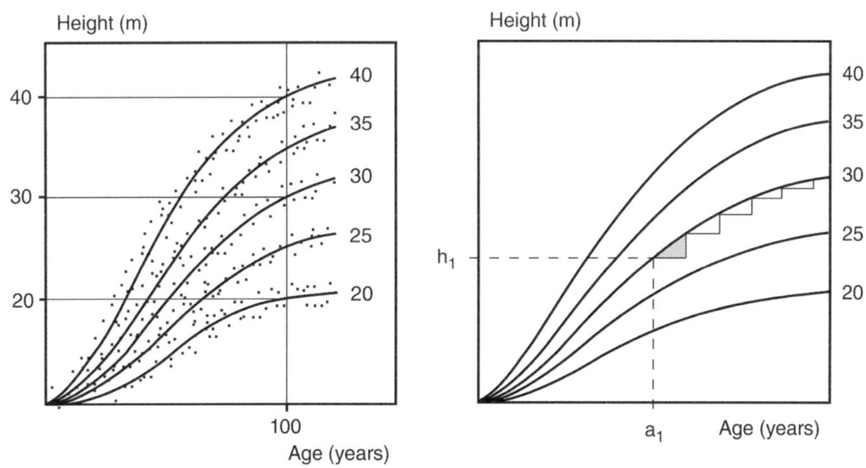


**Fig. 8.3** Principle of statistical growth models represented schematically. Explanation in the text

process. The transition in terminology from black-grey-white reflects how processes and structures not yet understood or considered are gradually resolved and included in the system model. The statistical relationships are then replaced by process knowledge from subordinate levels in the hierarchy and explained in biological terms.

Process-based models describe, among other things, the carbon, nitrogen, and water cycle and stand growth on the basis of the biological-physical-chemical processes such as incoming radiation, transpiration, photosynthesis, respiration, allocation, ageing, and mortality (Fig. 8.4). This requires the input of the initial structure, differentiated information on environmental conditions, and resource supply. However, such models deliver a broad set of output variables that managers need in order to decide the future direction of forest development, long-term planning, and environmental policy. While the empirically based models deliver only a small number of the criteria and indicators needed in forest ecosystem management (forest resources, vitality and stability, production and regeneration, biological diversity, protection, socio-economic aspects), mechanistic models have at least the potential to provide the remaining information.

*Hybrid Models* These represent a compromise between empirical and process-based models and bridge any knowledge gaps of processes by statistical relationships. They attempt to combine the strengths of empirical and process-based models. The hybrid models introduced, for example, by Kimmins (1993), Landsberg (2003), and Pretzsch et al. (2002, 2008) may present a compromise between empirical models, which provide only limited output variables, and mechanistic models, which are parameterised from intense measurements at only a few sites. These models contain functions that estimate the productivity (biomass, wood volume) in relation to primary factors (e.g. precipitation, leaf nitrogen content, temperature, radiation). This can be done on the basis of resource-use efficiencies

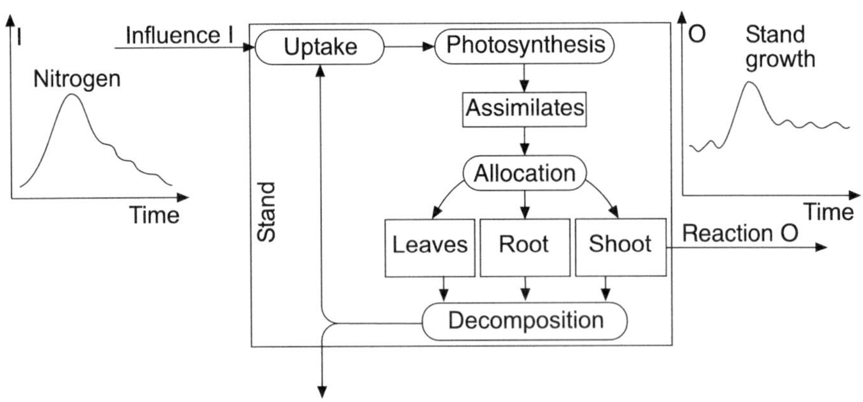


**Fig. 8.4** If, for example, the response of a forest stand to nitrogen input in separate individual processes and structures is determined to understand a system, one speaks of a grey-box or white-box system depending on the degree of resolution

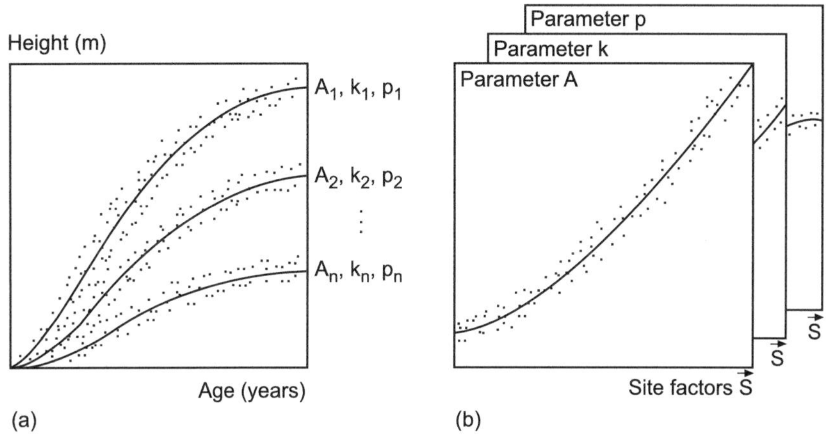


**Fig. 8.5** The principle of the site productivity module of the hybrid model SILVA 2.2 represented schematically. Mainstays are (**a**) the parameterisation of height-age curves which yield the curve parameters *A*, *k*, and *p* and (**b**) the derivation of statistical relationships between the curve parameters *A*, *k*, and *p* and the vector $\vec{s}$ representing site factors. Using this system of equations, the height-age curves and other characteristics of any sites within the range of the covered site conditions can be estimated

(e.g. nitrogen-use efficiency or energy-use efficiency) or by assessing the growth curve parameters in relation to the site condition variables (Fig. 8.5). For a good performance, the initial values and calibration of the internal estimator functions of such models must be backed by a regionally extensive network of inventory data on biomass, soil conditions, climate, and growth.

Figure 8.5 illustrates the principle of hybrid models with the example of the growth simulator SILVA (Kahn 1994; Pretzsch et al. 2002). The model was calibrated from long-term time series of site-specific experimental plots (height and yield development), which have been adjusted by regression analysis, and statistically related to the respective site variables $\vec{s}$. The resulting site-specific age-height curve parameters (*A*, *k*, *p*) (Fig. 8.5a) build an estimation system for growth predictions based on regional site conditions (Fig. 8.5b). The start-up and control variables of the model are generated from regional-specific inventory data (stand structure, height, site conditions, etc.). Application of hybrid models requires the definition and supply of standardised initial variables (e.g. standing volume, spatial stand structure) and relevant driving variables (e.g. proxy variables like code for the respective eco-region, altitude, and more primary information like temperature, precipitation, radiation, nutrient supply) indicating the environmental conditions and resource supply. To assure accuracy and acceptance of such models in the future, a set of evaluation variables (e.g. biomass growth and yield) needs to be defined (Mette et al. 2009; Pretzsch 2002), and a standard set of output variables (e.g. indicators for forest resources, vitality and stability, production and regeneration, biological diversity, protection, socio-economic aspects) specified to meet the information requirements for sustainable forest ecosystem management.

## 8.3 Integration of Mixing Effects in the Algorithms of Forest Growth Models

The structures of algorithms used to predict the growth of mixed-species forests can be divided into four main groups (Fig. 8.6a–d).

### 8.3.1 Integration of Mixing Effects in Empirical Models

Empirical models for mixed-species stands are mostly based on one of the following three concepts shown in Fig. 8.6a–c.

When there is no information about mixtures and species interactions, then the mixtures are modelled as the weighted mean of the monocultures (Fig. 8.6a). Monospecific stand models, such as yield tables or diameter distribution models, are selected for each species and then submodels for site indices and thinning are selected. Growth and yield predictions are provided for each species at specified ages, and these are used to calculate a weighted mean for each species based on the

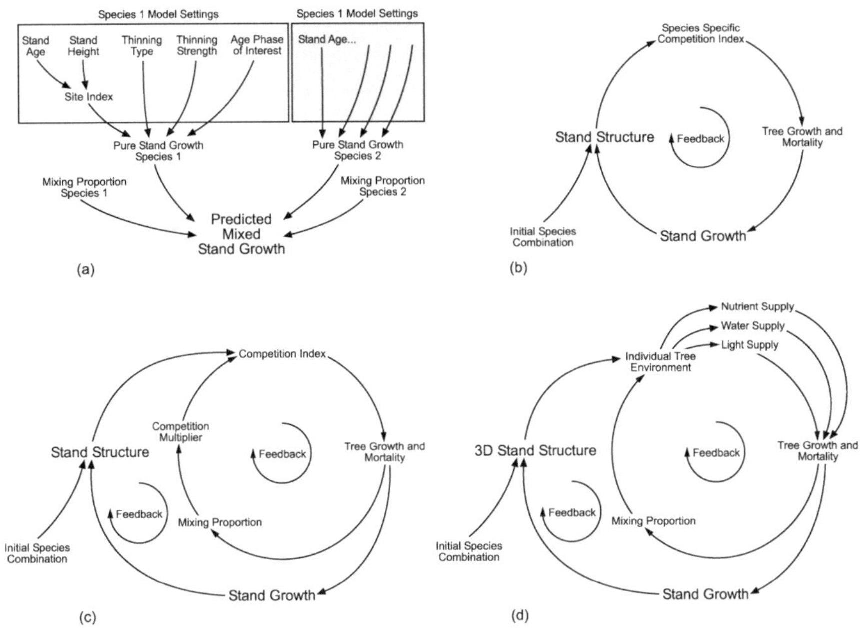


**Fig. 8.6** Main principles for deriving and predicting mixed-species forest growth by models. (**a**) Deriving the growth of mixed-species stands as a weighted mean of monocultures using models of the respective species. (**b**) Indirect incorporation of mixing effects in individual tree growth models by integration of species-specific competition indices. (**c**) Direct incorporation of mixing effects using multipliers affecting growth rates and stand density. (**d**) Process-based representation of mixing effects by incorporating within-stand environmental conditions, species-specific structures, and resource uptake and availability

mixing proportion (e.g. of basal area) that the given species contributes to the stand. That is, mixing proportions ($m_1$, $m_2$, ..., $m_n$) are used to calculate the expected productivity ($p_{mix}$) of the mixture as a weighted mean of the monoculture productivity (e.g. $p_{mix} = p_1 \times m_1 + p_2 \times m_2 + \ldots + p_n \times m_n$). The same approach is used to calculate stand characteristics such as tree number, basal area, standing volume, basal area growth, etc. An even simpler approach uses parameter sets that apply to whole mixtures, as opposed to parameters for each individual species. Several studies have used these approaches for mixed-species forests (Nightingale et al. 2008; Peters et al. 2013; Kramer and Akça 1995). However, while these simplifications avoid the need for models that consider species interactions, they assume that the species interactions have no significant effect on forest growth and functioning or that these effects do not change from site to site or as stands develop. Chapters 3–6 showed that the productivity of mixtures can often vary from that expected based on monocultures. Mixtures may be 50% more productive than the weighted mean monoculture productivity when they contain nitrogen-fixing species (Forrester et al. 2006) and 20–30% more productive in other mixtures (Pretzsch et al. 2013). Chapters 3 and 4 also showed that these mixing effects vary along spatial and temporal gradients in resource availability or climatic conditions. Relying on the weighted mean of monocultures ignores these mixed-species dynamics.

Many empirical models account for the vertical layering structure of forests using variables such as tree height, height to crown base, and crown width. Stand-level models use a 2D structure, while tree-level models can apply a 3D structure by including tree coordinates to define the horizontal dimension. The 2D or 3D structures are then used to calculate competition indices for each tree or species as a proxy for the availability of resources (Köhler and Huth 1998; Pukkala et al. 2009; Pretzsch et al. 2002; Hasenauer 1994). The individual trees' or species growth and probability of survival for the next period are then determined using these competition indices, and the size of all trees or species and the stand structure can be updated (Fig. 8.6). The feedback between stand structure and tree growth via the competition indices means that species interactions can have a strong influence on stand development. That is, the competition indices will be influenced by the development of canopy stratification and stand density and will consequently influence the growth of each species, some of which may even be outcompeted. The total stand growth is calculated as the sum of all individual trees or species. The temporal resolution of these models is often 1 or 5 years. In these models, competition is modelled in relation to available space rather than in terms of resource availability or environmental conditions. For example, competition for light might be predicted using competition indices calculated from tree heights and crown sizes, whereas competition for soil resources might be predicted using competition indices calculated from tree diameters, root mass, or rooting depths. If data from monocultures are used to parameterise these models, the mixing effects will relate to species-specific growing space, whereas when data from mixtures is also used, additional complementarity effects may be indirectly represented.

A third algorithm structure used in empirical models for mixed-species stands (Fig. 8.6c) predicts mixing effects directly by including multipliers. This can only be done when the growth differences between mixtures and monocultures, at different stand densities, are known. This extends the classic tree-level modelling approaches that model the relationship between competition indices and growth and consider maximum stand density using self-thinning lines. This modifier approach (Fig. 8.6c) has been applied analogously to model the effects of site conditions, insect damage, and fertiliser application, even when the processes are not yet understood (Komarov et al. 2003; Monserud and Sterba 1996; Wykoff et al. 1982).

The direct calculation of mixing effects using empirical multipliers (Fig. 8.6c) will enable the simulation of mixing effects in any growth model, and the predictions are likely to be relatively accurate within the parameter space used to calculate the multipliers. However, this approach is only possible when these effects are known and these effects are often only known for a restricted range of conditions. In reality, there are countless combinations of species, silvicultural treatments, soil conditions, and climatic conditions in any given region. Furthermore, foresters may be interested in combinations of these factors that do not currently exist, especially when the goal is to establish new mixed-species plantations.

Process-based models (Fig. 8.6d) can potentially provide robust extrapolations to untested environmental conditions, silvicultural regimes (Weiskittel et al. 2010), and species combinations and proportions. The process-based modelling approach (Fig. 8.6d) is therefore the focus of the subsequent sections.

### 8.3.2 Integration of Mixing Effects in Process-Based Models

Process-based models predict the effects of competition and facilitation between trees by considering the actual resources and environmental factors instead of using proxies such as competition indices (Fig. 8.6d). Competition for resources is simulated for each individual tree or cohort by calculating the availability and uptake of resources (Grote and Pretzsch 2002; Kellomäki and Väisänen 1997; Kimmins et al. 1990a, b, 1999; Rötzer et al. 2009; Forrester and Tang 2016). The effect of species interactions on growth is predicted by feedbacks between species-specific spatial structures, environmental conditions including resource supply, and species-specific physiology in terms of growth and mortality (Fig. 8.6d). All of these can be influenced by the species mixing.

The most important functions, environmental conditions, and structural attributes to include in a model will be those that have the largest effects on growth or any other output variables of interest (Pretzsch et al. 2015). The following sections discuss how various functions, environmental conditions, and structural attributes have been incorporated into forest growth models and the temporal and spatial scales at which they are usually included. Nevertheless, regardless of the components that are included in a model, or the spatial and temporal resolutions, it is critical that all components of the model have been validated against empirical observations, as opposed to only comparing observed and predicted growth. Without thorough evaluations there cannot be much confidence in the predictions from any given model.

## 8.4 Functioning, Structure, and Within-Stand Environment as the Main Interacting Units in Mixed-Species Stands and Models

The potential of mixed-species forests to provide higher levels of ecosystem services than monocultures has been mentioned in all chapters. However, to make the most of this potential and to increase their attractiveness to foresters, it is important to develop reliable models that can predict when and where mixtures are most likely to provide higher levels of ecosystem services and when they may not. Based on the above introduction to forest stand growth models in general, we now focus on the environmental conditions, processes, and structures that need to be included in forest growth models that are to be applied to mixed-species forests.

In order to integrate mixing effects into forest growth models, we first consider that the three main interacting units of forest stand dynamics include the *functioning* of trees and stands, the stand and tree *structure*, and the *environmental conditions* within the stand (Fig. 8.7). The environmental conditions within the stand might include vertical gradients in light, temperature, and soil moisture. The functioning could include transpiration, photosynthesis, and growth. The structures could include the stem, root, and crown structures. There are also strong feedbacks between all three units. For example, the canopy structure determines the light conditions within the stand, the light supply drives photosynthesis, and growth, allocation, and the standing biomass determine the tree and stand structures (see Figs. 6.15 and 6.16).

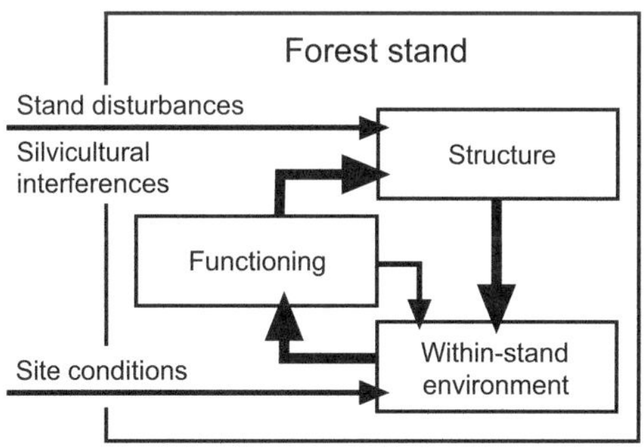


**Fig. 8.7** Schematic representation of the connections between the within-stand environment, functioning, and structure. The species within the stand can slowly modify their environment via structure (feedback circle represented by *bold arrows*) or quickly modify their environment via functioning (*thin arrow*). External factors, such as disturbances, silvicultural interferences, and site conditions, influence the structure and within-stand environment of the stand and thereby its functioning (Pretzsch et al. 2015)

### 8.4.1 Temporal and Spatial Resolutions for Modelling the Effects of Species Mixing on the Functioning, Environment, and Structure

The effects of mixing tree species on the stand environment, functioning, and structure can be modelled at different *spatial* and *temporal resolutions* (Table 8.1). At high resolutions, changes in the environmental, functioning, or structural conditions may be modelled at the tree, organ, or cell levels and in time steps of minutes and hours. This approach is sometimes referred to as 'bottom-up'. A high-resolution approach integrates the effects of species interactions with high spatial and temporal resolution. However, it can produce questionable output when scaling up from the tree to the stand level. Any errors in organ- or tree-level predictions may be propagated when upscaling and calculating stand-level patterns. Emergent system properties may be neglected.

Lower-resolution, stand-level approaches summarise stand dynamics in monthly, annual, or sometimes 5-year steps. This approach is sometimes referred to as 'top down'. Lower-resolution models reflect the effects of species interactions without as much spatial or temporal detail for causal drivers. That is, organ- and tree-level approaches often consider processes at the resolutions on which they actually occur. In contrast, stand-level approaches can show whether those processes have significant effects on the long-term system dynamics. For the purposes of this chapter, species interactions are classified in terms of their effects on forest growth using spatial scales of points, part of the stand, tree, whole stand, and ecosystem (Table 8.1). Temporal scales include minutes, hours, days, months, years, decades, and lifetime of a tree. All of the processes described in this chapter and throughout this book can be defined in terms of the spatial and temporal resolutions at which they occur. This indicates how they might be incorporated into models.

An overview of frequently reported processes or stand attributes that can be influenced by mixing tree species is shown in Table 8.2. These are defined in terms of their temporal and spatial resolution as well as their effects on the stand environment, functioning, and structure. In addition, the requirements and approaches that can be used to reproduce these mechanisms in growth models are discussed. For example, to simulate the water translocation in the soil by a deep-rooted species (or tree) to a shallow-rooted species (or tree), the soil layers have to be linked and, in the case of tree-to-tree translocation, horizontal differences in soil water availability as effected by different individual trees may also need to be quantified. This requires a sophisticated water balance module with a low spatial scale (tree) and a short temporal scale (day) to be able to simulate the processes at the scales on which they occur.

The mechanistic reproduction of mixing effects in models requires a description of the relevant processes, such as the distribution of light or the water cycle. To reproduce some mixing effects, such as light flecks, growth models may need to be based on individual trees. However, the effects of mixing tree species on the environment, function, or structure cannot be grouped in terms of a specific cycle

**Table 8.2** Potential changes in stand structure, functioning, and within-stand environment that can result from mixing tree species in forests

| Temporal scale | Spatial scale | Effect on | Factor influenced by the mixing of species | Prerequisites and possible mechanistic reproduction | | Required description |
|---|---|---|---|---|---|---|
| | | | | Individual tree models | Cohort (stand) models | |
| (a) Short-term effects (hour, day, month) of tree species mixing on stand structure, functioning, and within-stand environment | | | | | | |
| Short (hour, day, month) | High (organ, tree) | Environment | Light flecks | Highly resolved spatial structure of the model | Not possible | l |
| | | | Wind speed reduction | Change of wind speed for calculation of transpiration, leaf temperature | Change of wind speed for calculation of transpiration, leaf temperature | o |
| | | | Water balance | Individual tree transpiration, soil water availability, canopy rainfall interception by each tree | Transpiration by each cohort, soil water availability, canopy rainfall interception by each cohort | w |
| Short (hour, day, month) | High (organ, tree) | Functioning | Light absorption and photosynthesis | Light gradient for individual tree influenced by neighbouring trees, and changes in inter-tree shading due to changes in stand density | Light availability reduction based on competing cohorts, and changes in inter-tree shading due to changes in stand density | l |
| | | | Hydraulic lift | Linked soil layers for individual trees | Linked water availability for each cohort | w |
| | | | Water uptake and drought stress | Exchange of water in the rooting zone, linked soil layers for individual trees, within canopy climatic gradients | Drought index, linked water availability for different cohorts, within canopy climatic gradients | w |

(continued)

**Table 8.2** (continued)

| Temporal scale | Spatial scale | Effect on | Factor influenced by the mixing of species | Prerequisites and possible mechanistic reproduction | | Required description |
|---|---|---|---|---|---|---|
| | | | | Individual tree models | Cohort (stand) models | |
| (a) Short-term effects (hour, day, month) of tree species mixing on stand structure, functioning, and within-stand environment | | | | | | |
| Short (hour, day, month) | High (organ, tree) | Structure | Cell structure | Highly spatially resolved models | Cohort-based predictions | s |
| | | | Ratio of early to late wood | Highly spatially resolved models | Cohort-based predictions | s |
| | | | Phenology (temporal partitioning of growing space) | Species-specific simulation of bud burst, leaf fall, and length of growing season | Cohort-specific simulation of phenology | p |
| Short (hour, day, month) | Low (tree, cohort) | Environment | Matter balance and partitioning | Different matter balances (C, N, $H_2O$) have to be predicted for each individual tree | Different matter balances (C, N, $H_2O$) have to be predicted for each cohort | n |
| Short (hour, day, month) | Low (tree, cohort) | Functioning | Allelopathy | Inhibition of germination and growth through chemicals, simulation of allelochemicals or carbohydrates | Simulation of allelochemicals or carbohydrates | o |
| | | | Water translocation from species to species | Linked soil layers for individual trees | Linked water availability for cohorts | w |
| Short (hour, day, month) | Low (tree, cohort) | Structure | Current height growth | Simulated on tree basis | Influence of neighbouring cohorts on height growth | s |

| Temporal scale | Spatial scale | Effect on | Factor influenced by the mixing of species | Prerequisites and possible mechanistic reproduction | | Required description |
|---|---|---|---|---|---|---|
| | | | | Individual tree models | Cohort (stand) models | |
| (b) Long-term effects (hour, day, month) of tree species mixing on stand structure, functioning, and within-stand environment | | | | | | |
| Long (>= year) | High (organ, tree) | Environment | Root channels | Linked root sections for individual trees, change of soil characteristics | Change of soil characteristics depending on species | s |
| | | | Root layers and zones | Partitioning of root space, linked root sections for individual trees | Root space for different species, link between cohorts | s |
| Long (>= year) | High (organ, tree) | Functioning | Tree mortality | Species-specific simulation of mortality depending on neighbouring trees | Species-specific simulation of mortality depending on neighbouring cohorts | s |
| | | | Leaf turnover (dependent on nutrient cycle, water balance, light) | Simulation of nutrient and water cycles with linkages between trees | Simulation of nutrient and water cycles with linkages between cohorts | n |
| Long (>= year) | High (organ, tree) | Structure | Tree crown and root plasticity | Crown and root architecture of each tree must be known | Mean crown and root architecture of each species must be known | s |
| Long (>= year) | Low (tree, cohort) | Environment | Soil fertility | Nutrient cycle has to be considered | Nutrient supply of the cohorts must be considered | n |
| | | | Soil water capacity | Change of soil characteristics must be possible | Change of soil characteristics must be possible | w |
| | | | N-fixation | Nitrogen cycle has to be regarded, N-fixing tree species parameterised | Nitrogen supply of the cohorts has to be simulated, N-fixing tree species parameterised | n |
| | | | Barrier effects against pests | Pests have to be considered | Pests have to be considered | o |

(continued)

**Table 8.2** (continued)

| Temporal scale | Spatial scale | Effect on | Factor influenced by the mixing of species | Prerequisites and possible mechanistic reproduction | | Required description |
|---|---|---|---|---|---|---|
| | | | | Individual tree models | Cohort (stand) models | |
| (b) Long-term effects (hour, day, month) of tree species mixing on stand structure, functioning, and within-stand environment | | | | | | |
| Long (>= year) | Low (tree, cohort) | Functioning | Gap dynamic (tree establishment and disturbances) | Simulated on individual tree basis | Not possible | s |
| Long (>= year) | Low (tree, cohort) | Structure | Standing stock, carbon partitioning | Initial spatial structure and biomass distribution of the trees must be known | Initial biomass distribution of cohorts must be known | s |
| | | | Size growth and ontogeny (species with different life cycles, feedback tree size-resource acquisition) | Competition for space or resources between individual trees | Competition between cohorts for space or resources, allometric adjustment | s |
| | | | Wood quality | Quality parameters such as stem straightness, branchiness, or knot distribution have to be simulated | Quality parameters such as stem straightness, branchiness, or knot distribution have to be simulated | o |
| | | | Third party trade-offs (1. species inhibits soil vegetation-> reduced resource consumption-> increase of growth for 2. species) | Simulation of soil vegetation, resource uptake by individual trees | Cohorts for soil vegetation, resource uptake by each cohort | o |

| | | | | | | |
|---|---|---|---|---|---|---|
| | | | Stability and protection (against extremes such as wind throw, snow load, frost winter drought) | Stability parameters have to be simulated | Stability parameters have to be simulated | s |
| | | | Mechanistic branch abrasion and crown shyness | Module for these processes is needed, dependent on wind speed | Module for these processes is needed, dependent on wind speed | s |

The factors are listed starting from high resolution to low resolution as well as their classification and prerequisites for their mechanistic reproduction in individual tree and cohort models. Many of these are described in previous chapters, particularly Chaps. 3 and 6. The cycles and processes that have to be simulated are also indicated (l, light; n, nutrient cycle; p, phenology; s, spatial structure; w, water cycle; o, others). Please note that regeneration dynamics are excluded or process chain. That is, within the water cycle, mixing effects at fine spatial resolutions (organ or tree level) affect the stand environment at coarser spatial resolutions (e.g. stand precipitation, interception). They also affect the functioning, such as hydraulic redistribution or water uptake during drought (Sect. 6.8). At even coarser temporal resolutions, such as several years, species mixing might also change the soil water holding capacity.

## 8.4.2 Functioning

### 8.4.2.1 Light and Phenology

The absorption of photosynthetically active radiation is fundamental for tree growth, and the predictions of light absorption are therefore critical in process-based forest growth models. Light absorption can be predicted relatively easily for homogeneous, closed canopies, such as even-aged monocultures. However, light absorption predictions are much more difficult for mixtures because they often have vertically and horizontally heterogeneous canopies.

Many process-based tree-level models predict the light absorption by individual tree crowns using information about tree positions, tree heights, crown sizes, and leaf optical properties. Comparisons with empirical data have shown that these calculations are often relatively accurate (le Maire et al. 2013; Charbonnier et al. 2013; Ligot et al. 2014). Other tree-level models use simplified crown descriptions (SORTIE; Canham et al. 1999) or competition indices (SILVA; Pretzsch et al. (2002)) to quantify light competition.

Stand-level models predict light absorption of mixed-species forests using similar inputs such as tree heights, crown sizes, and leaf optical properties, except that these are species-specific means, rather than actual values for each individual tree. Stand-level models also ignore information about horizontal tree positions, so the distance between trees is assumed to be uniform. Despite these simplifications, stand-level light absorption model predictions can compare well with those of tree-level models for a wide range of species and stand structures (Forrester 2014b; Forrester et al. 2014).

The seasonality of light absorption (e.g. Baldocchi et al. 1984; Hertel et al. 2012) is also often considered in process-based forest growth models. This seasonality can be accounted for by modelling phenological phases such as bud break and leaf fall. Interestingly, of the 53 models reviewed by Pretzsch et al. (2015), only 33% simulated seasonal changes within the canopy as a function of structure on environmental conditions. Only about 19% of the models could potentially simulate the effects of species interactions on phenology, and only 13% already simulated mixing effects on phenology in mixtures. An example of such a process-based model is BALANCE, which calculates bud break for each species and leaf senescence for each individual tree (Rötzer et al. 2004, 2010). Most models use much simpler approaches to describe seasonality, such as by simply differentiating species as evergreen or deciduous and/or using long-term means for the start and end of the growing season.

#### 8.4.2.2 Water Balance

The water balance and the water pools and fluxes that influence it are clearly going to impact forest functioning. Even in moist stands, there can be periodic droughts that significantly influence tree growth or survival, and therefore the water balance is often considered in process-based models. Water-related species interactions can be incorporated by modelling the soil water balance and species-specific fluxes such as the proportion of rainfall intercepted by the canopy, soil evaporation, and tree transpiration. Different species will often occupy different vertical positions within the canopy and possibly also different positions within the soil. As a result they may be exposed to contrasting environmental conditions. This can be accounted for when models consider the vertical profiles of vapour pressure deficit, radiation, and soil moisture. The contrasting responses by different species to a given environmental condition can be quantified by defining the physiology of each species, such as the relationship between vapour pressure deficit and stomatal conductance. Models that consider phenology of leaf production can also prevent transpiration during the leafless periods.

Water-related processes such as hydraulic redistribution and mycorrhizal symbioses might also influence species interactions. However, as mentioned in Chap. 3, there is very little empirical information about the impact of differences in mycorrhizal communities or mycorrhizal connections between differing tree species on transpiration (Simard et al. 2012), and this process is rarely incorporated into forest growth models. While many studies have examined hydraulic redistribution, which can amount from 0.1 to 1 mm per day in forest ecosystems (Neumann and Cardon 2012), there is very little information for mixed-species forests, so it is difficult to determine whether this process is worth including in models (Goisser et al. 2016; Pretzsch et al. 2014).

#### 8.4.2.3 Mineral Nutrients

Light and water dynamics can be modelled in the same way in different forests, simply by assigning different physiological and allometric parameters to each species. In contrast, mineral nutrient dynamics are much more complex, and only a small number of nutrient models are usually not general enough to apply to a wide range of soil and forest types. Furthermore, different nutrients are important in different regions or even on different sites within a given region. Therefore several different nutrients may need to be modelled. While many nutrient models consider nitrogen, fewer account for other nutrients such as P or K, let alone interactions between different nutrients (Pretzsch et al. 2015), which limits their applicability given the importance of nutrients such as P or K in many tropical ecosystems (Laclau et al. 2013).

As a result of this difficulty to model nutrients, several models do not include dynamic calculations of nutrient availability or use. Instead, nutrient dynamics are ignored or simplified or empirical approaches are used (e.g. MAESPA, 3-PG; Duursma and Medlyn (2012); Forrester and Tang (2016)). Alternatively, a forest growth model might be coupled with another model that considers nutrients (Xenakis et al. 2008).

Some of the largest mixing effects reported have occurred in stands where nitrogen-fixing species facilitated the growth of non-nitrogen-fixing species. However, this facilitative effect of greater nitrogen availability is generally not included dynamically in forest growth models due to a lack of knowledge about the dynamics of nitrogen fixation in forests, or how nitrogen fixation is influenced by species interactions. While nitrogen fixation has been measured in mixtures at a few points in time (Binkley et al. 1992; Forrester et al. 2007; Bouillet et al. 2008), there is very little information about how rates of nitrogen fixation or the proportion of nitrogen derived from the atmosphere are influenced by species interactions (Van Kessel et al. 1994).

Accelerated rates of litter decomposition have been found in mixtures compared with monocultures in about 50% of cases, with no significant effect or antagonistic effects in other cases (Rothe and Binkley 2001; Gartner and Cardon 2004). However, the effects of these differences in decomposition on growth are rarely quantified so it is difficult to validate and justify their incorporation into models. Nevertheless nutrient cycling and its effect on nutrient availability have been incorporated into models by accounting for litter chemical and physical properties and forest floor microclimate. This is generally done using a stand-level approach (even in tree-level models) where soil properties are assumed to be uniform throughout the stand.

#### 8.4.2.4 Maximum Density and Mortality

The mortality functions in most forest growth models are empirical, even in process-based models. This is because the processes that drive mortality are not well understood. Furthermore, trees do not necessarily die immediately after the event that triggers their death, and it is difficult to develop mortality functions that link different causes of mortality including drought, density-dependent mortality, pests, and diseases (McDowell et al. 2011). Another difficulty is that the mortality functions used in models are often developed using monocultures, or from mixtures where there is no consideration about whether interspecific competition influenced the function. This reflects a paucity of empirical data required to develop these functions and also a need to develop functions that incorporate the effects of interspecific interactions. While several models include algorithms for the intra- and interspecific effects on tree mortality (Pukkala et al. 1998; Monserud and Sterba 1999; Yao et al. 2001; Zhao et al. 2004), the effects of stand density (Chap. 4) and canopy packing density (Chap. 5) may also need to be considered in mortality models.

The self-thinning line, commonly represented by the relationship between the logarithm of the tree number per hectare and the logarithm of the mean tree diameter (Fig. 8.8, left), quantifies the maximum stand density of fully stocked stands and the progressing tree number reduction due to competition-driven mortality. Tree species mixing may alter the slope, intercept, or both of the self-thinning line as niche complementarity can modify canopy packing and stand density. Any shifts in the self-thinning line mean a change of the mortality process in mixed-species stands compared to monocultures.

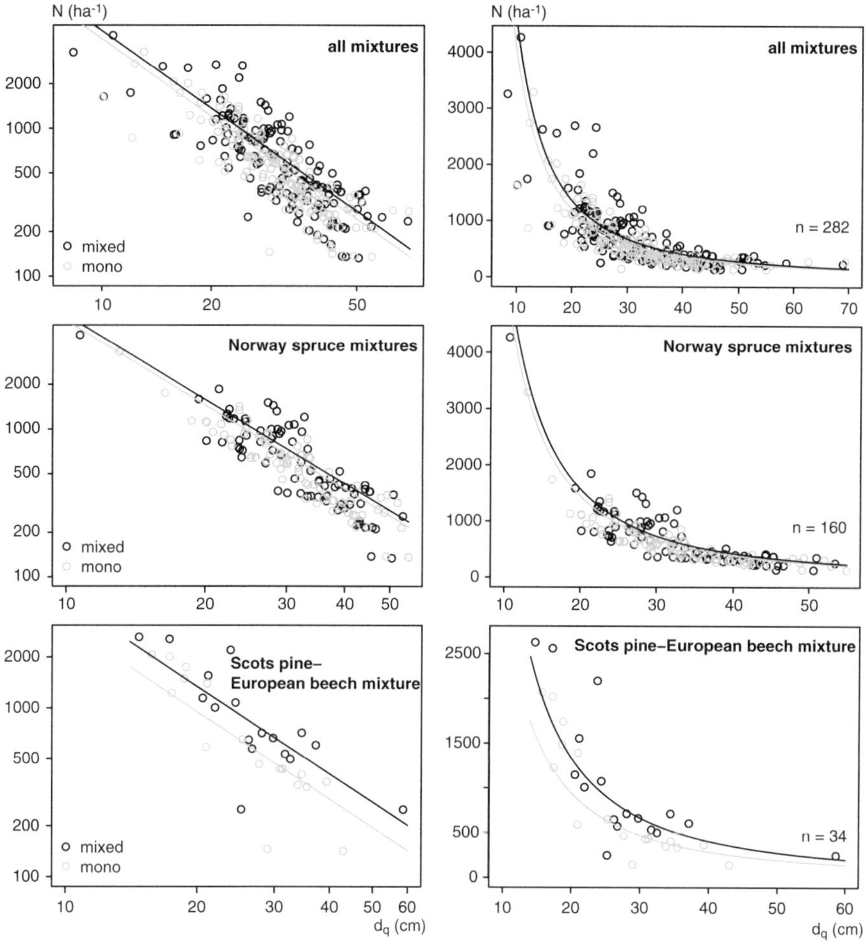


**Fig. 8.8** Increase of the maximum stand density by tree species mixing indicated by an upward shift of the relationship between tree number per hectare and mean tree diameter in double-logarithmic (*left*) and linear (*right*) representation. For all species assemblages, the self-thinning line of the mixed stands (*black lines*) lies above the weighted mean self-thinning line of the two monocultures (*grey lines*). Data from Pretzsch and Biber (2016)

Based on triplets of fully stocked and unthinned long-term plots of mixed-species stands and neighbouring monocultures, Pretzsch and Biber (2016) revealed over-density of mixed-species stands on the basis of the mean diameter-tree number relationship in double-logarithmic (left) and linear (right) representation (Fig. 8.8). Most mixtures included Norway spruce (*Picea abies*). The evaluation showed that in mixed stands, the maximum density was on average 16.5% higher than in neighbouring pure stands. Species mixtures with Norway spruce exceeded densities of monospecific stands by 8.8% on average. Individual species mixtures showed a significant density effect of +29.1% for Norway spruce mixed with European larch

(*Larix decidua*) and +35.9% for Scots pine (*Pinus sylvestris*) in association with European beech (*Fagus sylvatica*). The self-thinning line of all the mixed stands (black) lay above the weighted mean self-thinning line of the two monocultures (grey). The level of the self-thinning was raised by species mixing, whereas the slope remained the same. This shows that mixing can reduce tree mortality of the total stand so that their self-thinning line can be significantly higher.

#### 8.4.2.5 Carbon Partitioning

The partitioning of carbon between roots, stems, and leaves and the resulting tree allometry influence the ability of trees to acquire and compete for resources. Partitioning depends on environmental conditions including soil and climatic characteristics, and stand density, and how these are influenced by neighbouring plants (Litton et al. 2007). Since all of these can be influenced by species interactions, it is not surprising that carbon partitioning can differ between mixed and monospecific stands (see Box 3.1).

Carbon partitioning is part of a feedback loop that will affect all other processes considered by forest growth models and is therefore an important model component. Many models modify partitioning and allometry in response to soil water and nutrient availability and a few models also include effects of stand density (Pretzsch et al. 2015). Species interactions can therefore influence partitioning when they influence soil water and nutrient availability or stand density.

As an example of the importance of partitioning and allometry, tree allometry influences the stand structure and therefore the environmental conditions within the stand and is therefore critical for predictions of light absorption. As explained above, light models generally require information about crown sizes, e.g. leaf area, crown diameter, and crown length, and these are often calculated using allometric equations. These equations can vary between mixtures and monocultures (Pretzsch 2014; Forrester and Albrecht 2014; Forrester et al. 2017). For example, the relationship between crown projection area and stem diameter for European beech can vary depending on whether European beech is growing in monocultures or in mixtures with Norway spruce, European larch, *F. excelsior* (European ash), and *Q. petraea* (sessile oak) (Fig. 6.16). These differences can have significant effects on the amount of light absorbed by individual crowns (Forrester and Albrecht 2014). Despite this, growth models generally do not yet allow for intra-specific changes in allometry in response to interspecific competition.

### 8.4.3 Structure

Three broad aspects of stand structure are considered: horizontal, vertical, and their variability such as in terms of diameter distributions (Table 8.2a, b). Tree-level models automatically account for each of these structural aspects whether they are empirical or process based. In empirical tree-level models, the structure is used when calculating competition indices (Fig. 8.6b). In process-based tree-level models, the local structure and the resulting environmental conditions are considered when modelling different leaf- or tree-level processes and interactions. The vertical and horizontal variability in structure is recorded in tree-level models by keeping track of individual tree positions and diameter or height distributions.

In stand-level models, the vertical structure is quantified using the mean height (and height to crown base) of each species. This is generally done in all mixed-species stand-level models, regardless of whether they are process based or empirical. This vertical heterogeneity is important with regards to competition for light (Forrester 2014a; Forrester et al. 2014). Occasionally, the canopy structures are simplified even further such that all of the leaf area is positioned at the top of the crown, rather than distributed between the top and bottom of the crown (Bugmann 2001). This has been shown to produce biased estimates of light absorption such that taller species gain an unrealistic competitive advantage (Forrester et al. 2014), which will be exacerbated as the stand develops. Models that position all leaf area at the top of a crown are unlikely to reproduce reliable mixing effects even if all other processes are predicted accurately. Another simplification used in stand-level models is to divide the crowns and canopy into vertical layers, each of which is allocated a given portion of the leaf area, often assuming a uniform vertical distribution of leaf area within the crown.

The horizontal structure is required to model any factors that influence stand density, such as thinning or natural disturbances. Tree-level models consider the horizontal structure of a stand in terms of the positions of all individual trees, whereas stand-level models are restricted in their ability to account for horizontal structure, and assume that there is a uniform horizontal distance between all trees and species. Therefore, while stand-level models can account for horizontal heterogeneity by changing stand density (trees per ha, basal area, leaf area), they cannot account for heterogeneous horizontal stand structures such as row-by-row or group-by-group mixtures or gaps where a given cohort or species develops from seedlings under the influence of an older and taller cohort that surrounds the gap.

In contrast to above-ground stand structure, the below-ground stand structure is much harder to quantify. Therefore, while much is known about how above-ground tree allometry or above-ground stand structure varies between mixtures and monocultures (Forrester and Albrecht 2014; Bauhus et al. 2004; Pretzsch 2014), far less is known about how species interactions influence below-ground allometry and stand structure (Schume et al. 2004; Laclau et al. 2013; Thurm et al. 2016). This has inhibited the development of modelling approaches to account for below-ground functions and environmental conditions.

### 8.4.4 Within-Stand Environment

In many mixtures, each species will occupy different positions within the stand due to differences in height growth and rooting depth. As a result, they may experience different local environmental conditions within the stand and this may need to be accounted for in models (Table 8.2a, b). Generally, if the processes and structures are modelled realistically, then the environmental conditions within the stand are probably adequately accounted for. For example, the water balance is determined by the structural and functional characteristics of the stand, and light availability is influenced by the vertical and horizontal distributions of leaves.

In contrast to process-based models, empirical tree-level models usually summarise all environmental conditions using competition indices. This can result in accurate predictions of mixing effects (e.g. BWIN PRO; Nagel (1999)) for the environmental conditions and stand structures of the forests where the empirical data was collected to build the models. However, empirical models are unlikely to provide reliable predictions of mixing effects for different environmental conditions that result from changes in climate, management, or disturbance regimes. Empirical models may also be insensitive to inter-annual variability in climatic conditions that can strongly effect forest productivity, especially in short rotations (Stape et al. 2004).

## 8.5 Model Selection: Which Spatial and Temporal Resolutions?

Different forest growth models operate at a range of spatial and temporal resolutions, and this influences the complexity and speed of the calculations, the complexity of input requirements and model usability, and whether any upscaling is required. Therefore, when selecting a model, one of the most important considerations is the spatial and temporal resolution that is required. Similarly, when developing a model, it is important to consider the resolution at which each process, structure, and environmental condition should be integrated.

An appropriate model resolution depends on the purpose and the desired scale of the outputs that are required (Battaglia and Sands 1998) more than on the resolution at which the function or environmental condition occurs. That is, even though many functions in mixed-species forests occur at high spatial (leaf or tree) or temporal (seconds, hours) resolutions, they can usually be summarised using calculations at the stand level. In recognition of this, several recent studies have used stand-level models for mixed-species forests (Härkönen et al. 2010; Forrester and Tang 2016) in contrast to the many tree-level models that had previously been developed (Pretzsch et al. 2002, 2015).

If the aim is to understand how the characteristics of each species influence their interactions and stand-level variables (e.g. growth, transpiration) or the relative dominance of each species, then stand-level models are useful because they avoid complex higher-resolution calculations, detailed and expensive physiological data or tree-level input information, and associated scaling problems (Härkönen et al. 2010; Medlyn 1998; Battaglia and Sands 1998). Stand-level models may therefore suit forest managers if they require relatively simple models to parameterise and use.

Stand-level models are capable of modelling many forest management practices, such as thinning (Landsberg et al. 2005) or mixing effects (Forrester and Tang 2016), which are influenced by the vertical and horizontal structural attributes and environmental conditions described above. Process-based stand-level models such as TRIPLEX, 4C, and 3-PG have also been used to provide additional detail about stand structure by incorporating empirical equations that describe the development of the diameter distributions (Wang et al. 2011; Lasch et al. 2005; Landsberg et al. 2005; Forrester and Tang 2016) or by using hierarchical modelling approaches (e.g. CABALA; Battaglia et al. (2015)). The diameter distributions can also indicate how much biomass, merchantable volume, etc. is removed from the stand following disturbances.

On the other hand, tree-level models can show how the properties of individual trees influence forest growth. They can also show how horizontal structural heterogeneity, such as row-by-row or group-by-group mixing, can influence stand dynamics, as opposed to the tree-by-tree mixing that stand-level models are restricted to. That is, tree-level models are important when the aim of modelling is to understand how this variability influences the development and a characteristic of whole stands (Grimm 1999). Tree-level models can be used to determine which processes and interactions occurring at the organ and tree levels are the most important determinants of the stand-level patterns.

### 8.5.1 Using the Individual Tree Hybrid Model SILVA to Simulate Mixed-Species Stands

As a first example, we use the individual tree hybrid model SILVA. It requires as start-up and control parameters initial stand parameters, site variables, and silvicultural prescriptions (Pretzsch et al. 2002, 2008). The simulation results comprise estimates of growth and yield as well as ecological and socio-economic indicators (Pretzsch 2009, pp. 516–551). To begin, data for the dimensions and positions of the individual trees and site conditions are needed, and the site-growth model is adapted to the given site conditions (Fig. 8.9). Any missing dimensions or positions such as tree heights, crown dimensions, or stem coordinates are complemented by means of a stand structure generator (Pretzsch 1997). Once the start-up and control parameters have been compiled, the simulation run can proceed. The actual growth prognosis is carried out in 5-year cycles. The user may determine the number of cycles, i.e. the length of the simulation period used. Each cycle comprises four steps. The first step is to quantify the three-dimensional growth arrangement for each individual tree via a competition index. In a second step, the trees to be removed are identified according to the thinning regime defined by the user. Then, the competition index determined previously is used to control the changes in the size of all trees in the stand. The fourth step involves the use of a mortality model to determine which trees did not survive due to competition effects. Steps 1–4 are repeated until the entire prediction period has been covered.

Each tree may be classified into timber grades according to user-defined specifications, since the stem dimensions are known. The economic development at the tree and stand level can be followed because wood prices and harvesting costs are taken into account. For the ecological evaluation, spatial stand structure at

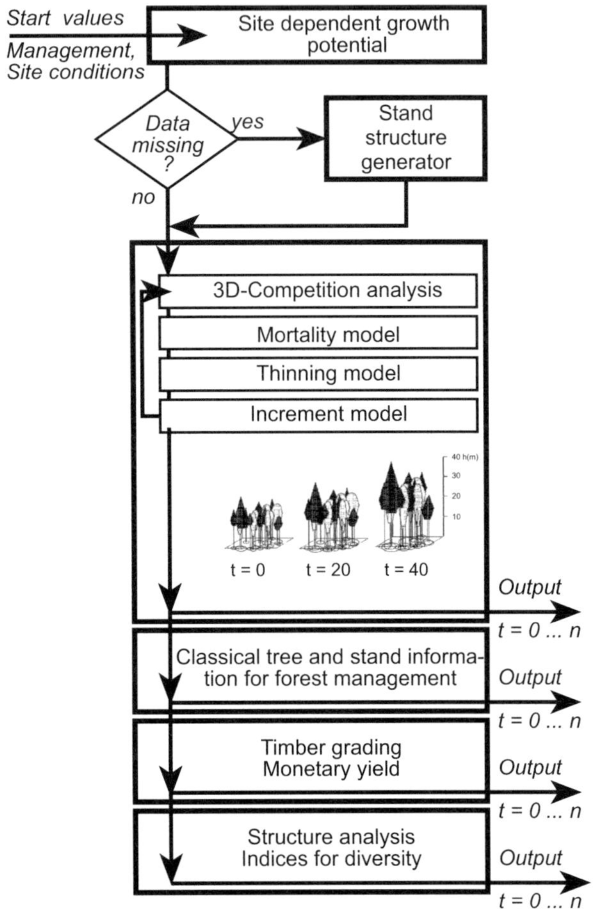


**Fig. 8.9** Overview of the prediction algorithm in the growth simulator SILVA 2.2. Variable $t$ denotes the simulation time in years, running from 0 to $n$

different scales is characterised by ecological standard indices. Moreover, results of the total biomass production of trees are included. Apart from the numerical data, the visualisation module provides vivid impressions of the three-dimensional forest structure at the stand and landscape levels (see Chap. 10).

### 8.5.2 Application for Scenario Analysis at the Stand Level: A Monospecific *P. abies* Stand Versus a *P. abies-F. sylvatica* Mixed Stand

A simulation run requires the stem coordinates of all trees at the initial point at time $t_0$. If these are unavailable, then all the coordinates are produced with the structure generator STRUGEN (Pretzsch 1993, 1997). This supplementation of missing values, and the generation of initial structures, can deliver realistic initialisation

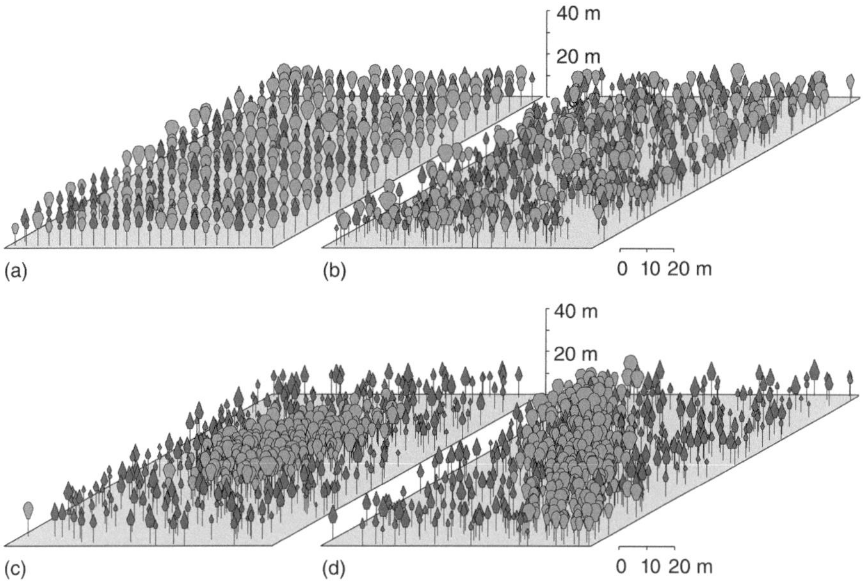


**Fig. 8.10** Generation of initial stand structure of Norway spruce-European beech mixed stands by the structure generator STRUGEN with (**a**) single mixtures at plant spacing 4 × 4 m, (**b**) single mixtures with random distribution, (**c**) mixture in patches with a given diameter, and (**d**) mixture in stripes (see Pretzsch 1993, 1997)

data for the simulation for a wide spectrum of stand structural configurations (cf. Fig. 8.10). Agreement between the actual and generated initial structures in this case does not mean that a particular tree in the actual and generated stand must be situated at the same position, but rather that the qualities that characterise the structure correspond in the real and generated stand in terms of species mixture, the vertical structure, and the horizontal distribution patterns.

For the estimation of the further stand development, the spatial growth arrangement is quantified for each individual tree. The following comparison of a pure *P. abies* stand and a *P. abies*-*F. sylvatica* mixed stand illustrates the use of the growth model SILVA at the stand level. The model was initialised to develop *P. abies* and *F. sylvatica* on a moderately moist site with poor nutrient status in the growth region 'Oberbayerisches Tertiärhügelland'. The *P. abies* site index is about 34 m at age 100 years after Assmann and Franz (1965), and *F. sylvatica* falls into site class II with moderate thinning after Schober (1967). Initially the monospecific stand comprised 1,400 planted *P. abies* trees per hectare (Norway spruce aged 20 years), whereas the mixed stand consisted of 1000 planted *P. abies* trees and 600 naturally regenerated *F. sylvatica* trees per hectare (*P. abies* aged 20 years; *F. sylvatica* aged 25 years; cf. Fig. 8.11). A moderate selection thinning was carried out in both stands controlled by the relevant species-specific tree number-tree height curve.

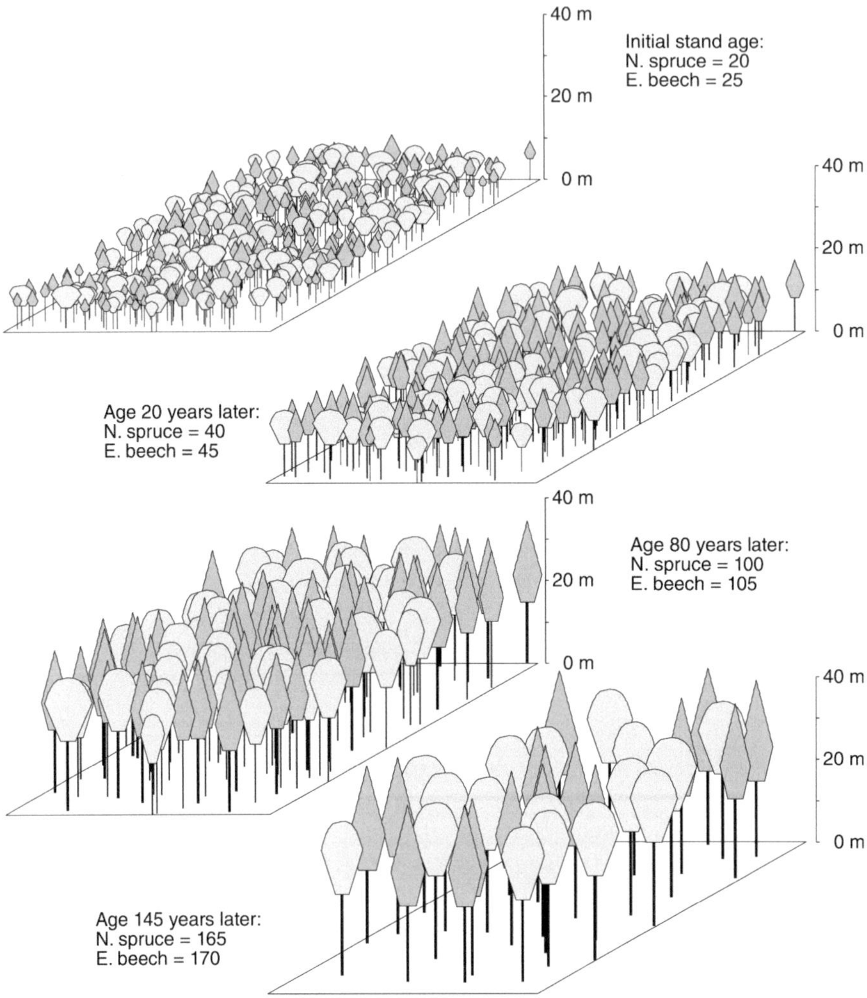


**Fig. 8.11** The growth simulator SILVA 2.2 models the development of monospecific and mixed stands in 5-year steps: a Norway spruce-European beech mixed stand in the growth region 'Oberbayerisches Tertiärhügelland' in south Bavaria is simulated from 20 to 165 years of age; the dominant height site index of Norway spruce is about 34 m after Assmann and Franz (1965), and European beech falls into site class II.0 according to Schober (1967), with moderate thinning; a heavy selection thinning has been conducted in the stands

In the following we highlight some key information from the extensive range of available output variables, which includes performance of growth and yield at the tree and stand level, crown development, stem quality, assortment structures, carbon and nutrient contents, nutrient export due to harvest, harvesting costs, value development of the stand and individual trees, net return, stability, structural stand diversity, and scenic value.

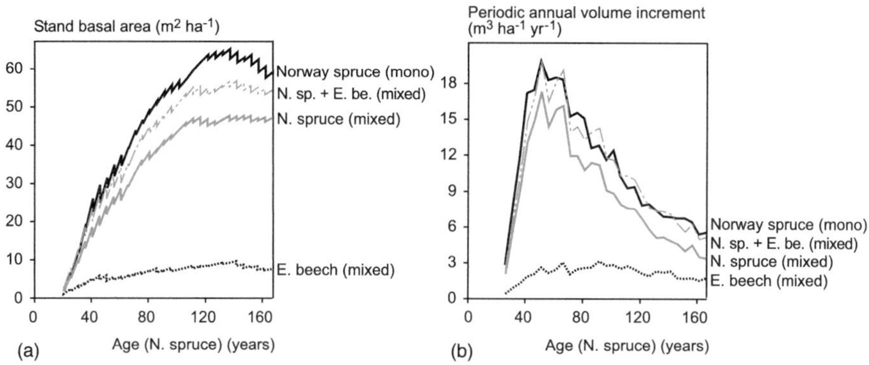


**Fig. 8.12** Results of the variant study on monospecific Norway spruce stand versus a mixed Norway spruce-European beech stand. (**a**) This shows the basal areas of the stands and (**b**) the mean periodic annual volume increment for the 145-year prediction period

In this example, simulations cover a period of 145 years and Fig. 8.12 provides an example of the development of basal area and mean periodic annual volume increment. Thinning is repeated several times in the first half of the prediction period because the volume growth of Norway spruce at this stage of stand development is high. Figure 8.12b shows that the annual volume increment at young stand ages rises more rapidly in the pure (full black line) than in the mixed stand (grey dotted line). The admixture of European beech causes a slight decrease in culmination height and an increase in the time required for the culmination of the mean periodic annual volume increment due to the competitive strength of European beech. With advancing age, European beech may even cause the volume increment of the mixed stand to repeatedly exceed that of the pure stand, despite the later culmination age of European beech (see Sect. 4.1.2).

For the assessment of financial yield, the mean annual value increment $MAI_{value}$ (Fig. 8.13a) is considered as the most appropriate characteristic (i.e. financial yield at respective age plus accumulated intermediate financial yield, divided by stand age). In contrast, the mean periodic annual value increment $PAI_{value}$ (Fig. 8.13b) corresponds to the difference between the total value achieved in two consecutive periods divided by the duration of these periods.

SILVA simulates the stand as a mosaic of individual trees and calculates, for example, the $MAI_{value}$ at the tree level analogous to the stand level (Fig. 8.14). In the mixed stand, the $MAI_{value}$ of the remaining Norway spruce with diameters of 50–60 cm culminates with $MAI_{value}$ of 0.75–2.0 € per year per tree, while European beech, at 0.5–0.75 € per year per tree, is still on an upward trajectory. That is, in view of its suitability for analysing and optimising modern harvesting scenarios based on the promotion of individual trees, the model presents a planning tool, which facilitates the calculation of growth and yield characteristics at both the tree and stand level, and provides information for analysing the trade-off between both views.

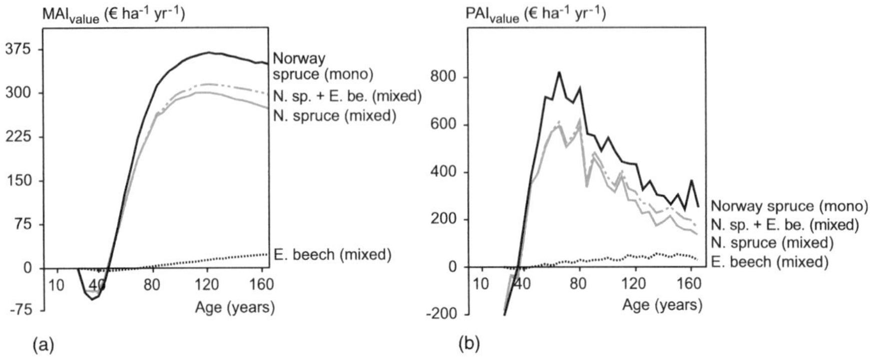


**Fig. 8.13** Results of the variant study on a pure Norway spruce stand versus a mixed Norway spruce-European beech stand. (**a**) This shows the development of the mean annual value increment $MAI_{value}$ and (**b**) of the mean periodic annual value increment $PAI_{value}$ for the 145-year prediction period

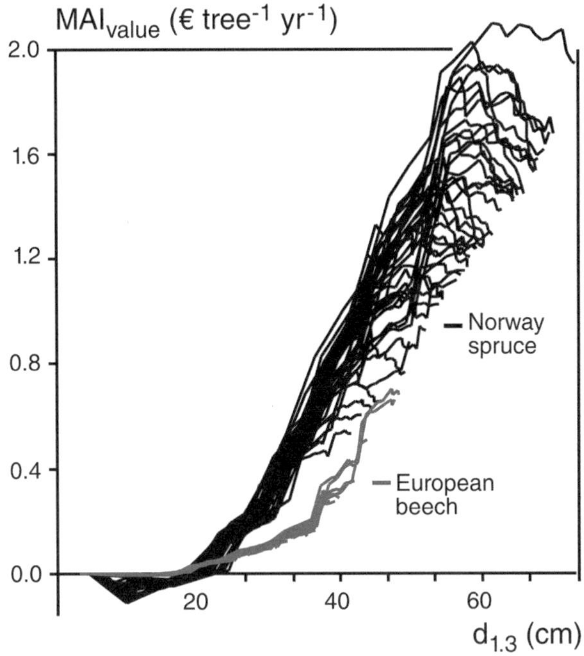


**Fig. 8.14** Development of the mean annual value increment $MAI_{value}$ for individual Norway spruce (*black solid line*) and European beech (*grey solid line*) trees in the mixed stand

SILVA models the stand in 5-year time steps as a spatial-temporal system of individual trees and structural indices, which provides indicator values for habitat and species diversity that can be calculated at any stage of the simulation run. The *R* Index from Clark and Evans (1954) for the horizontal tree distribution pattern, the *A* Index from Pretzsch (1998) for the vertical species profile, and *S* Index from

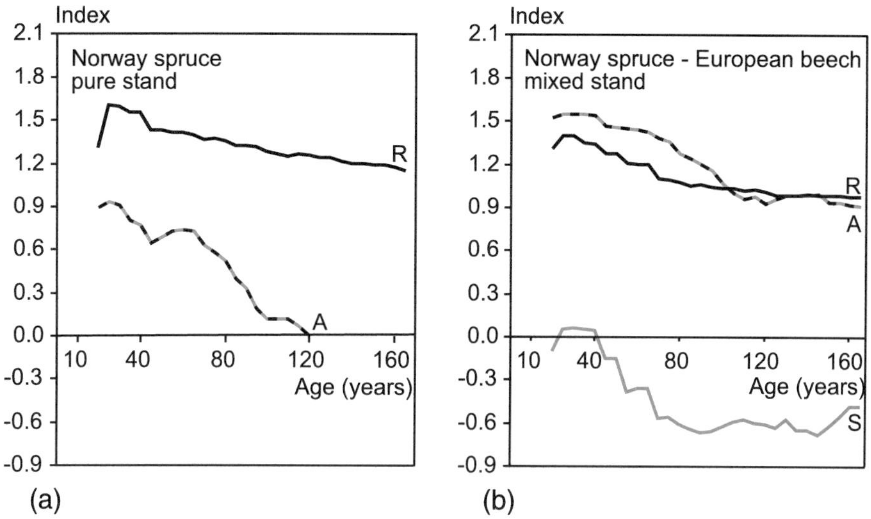


**Fig. 8.15** Changes in structure in the (**a**) monospecific Norway spruce stand and (**b**) the Norway spruce-European beech mixed stand over the 145-year simulation period, expressed by the structure parameter *R* by Clark and Evans (1954), *A* by Pretzsch (1998), and *S* by Pielou (1977)

Pielou (1977) for species intermingling all show how stand structures in a pure *P. abies* stand and a *P. abies-F. sylvatica* mixed stand (Fig. 8.15) are modified by age and thinning.

In monospecific and mixed stands, the selection thinning converts the initially more regular distribution ($R = 1.1$–$1.4$) into a random to clumped one ($R = 0.9$–$1.0$) by maintaining the lower and middle components of the stand. As expected, the selection thinning of the monospecific *P. abies* stand, which is still highly structured during the pole phase, reduces the vertical structure with increasing age. The *A* Index declines almost linearly. In mixed stands *A* is about 1.5 in the initial phase and decreases to only 0.9 during the life of the stand. This is due to the species mixture present at the beginning as well as to the selection thinning, which enables *F. sylvatica* to survive in the lower and intermediate positions within the stand. The S Index values indicate that an intensive individual tree mixture develops in the mixed stand over the prediction period as a result of the tending measures in the pole and sapling stages and ensures a continuous heterogeneous mosaic of interspecific and intraspecific competition between *P. abies* and *F. sylvatica* until the stand reaches maturity.

Figure 8.16 shows how various results may be summarised graphically in order to compare the simulation results for monospecific and mixed stands. The structural indices and the $h_q/d_q$ value indicate the structure of a stand with a dominant height of 30 m; similar profiles can be produced at any time during the simulation run. The MAI of volume and $\text{MAI}_{\text{value}}$ of the monospecific Norway spruce stand on the site selected appear to be superior to that of the Norway spruce-European beech mixed stand for the thinning regime modelled and for the assumptions made about

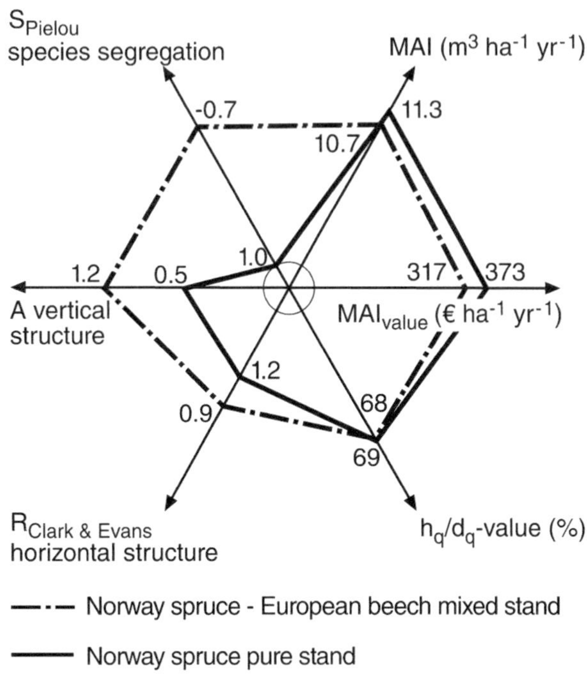


**Fig. 8.16** Profile of the most important output parameters for monospecific and mixed stands (structural indices *S*, *A*, *R*, $h_q/d_q$-values, maximal $MAI_{value}$, and maximal MAI) facilitates weighing up and optimisation of ecological and economic aspects of stand management

harvesting costs, wood prices, and risks. Yet the structure and stability characteristics indicate that the monospecific stand is inferior.

### 8.5.3 Confidence in the Performance of a Process-Based Model as a Prerequisite for Application in Forest Practice

Thorough evaluations of any process-based models should include an assessment of all processes that are included in the model (Grimm 1999; Weiskittel et al. 2010). This is because it is easy to obtain very good fits to observed growth data for the wrong physiological reasons simply by 'tuning' some of the parameters (Sands 2004). Although this is stating what is probably obvious, many process-based or hybrid models have surprisingly only been tested for their ability to predict growth and not for the other processes that they calculate to actually predict growth. If the processes are not calculated correctly, then the growth predictions may become erroneous as soon as the conditions differ from those where the model was 'tuned'. Before using a process-based model, it is important to check that all components have been tested against empirical data or at least against the predictions of other models that are known to be accurate. Ideally, the models should have been tested in the whole range of forest types, stand structures, climates, etc., where they are actually going to be used. An example of a process-based (or hybrid) forest growth model that has been thoroughly tested for all of its components, all in a wide range of forest types, is 3-PG (Landsberg and Sands 2010), which is freely available (Fig. 8.17). Thorough testing has given confidence in the performance of 3-PG, which is now one of the most widely used forest growth models in the world.

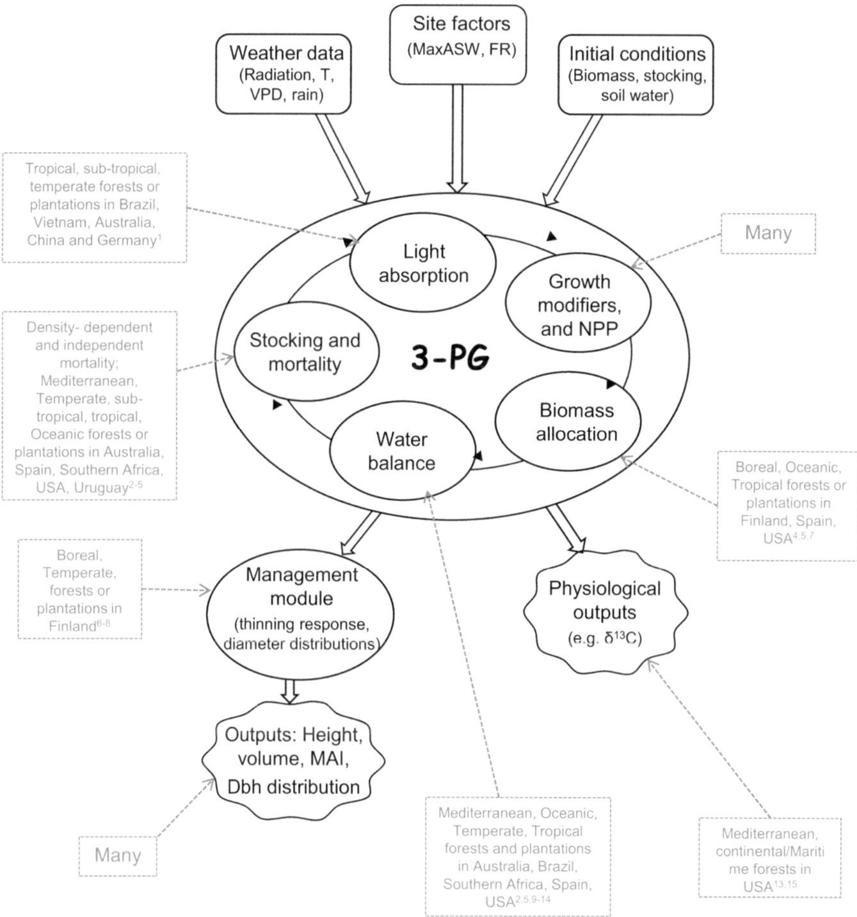


**Fig. 8.17** The main components of the 3-PG model (modified from Landsberg and Sands (2010)). The biological core of 3-PG comprises five submodels for predicting light absorption, growth modifiers and NPP, biomass allocation, stem number and stem mortality, and the soil water balance. A management submodel converts state variables into outputs often required by foresters. The input requirements (weather, site factors, initial conditions) depend on the data availability and the desired outputs. All submodels have been tested against empirical data from a wide range of forests; a few examples are cited in the *grey boxes*: [1]Forrester et al. (2014), [2]Dye (2001), [3]Sands and Landsberg (2002), [4]Gonzalez-Benecke et al. (2014), [5]González-García et al. (2016), [6]Rodríguez et al. (2002), [7]Landsberg et al. (2005), [8]Miehle et al. (2009), [9]Almeida et al. (2004), [10]Almeida et al. (2007), [11]Almeida et al. (2010), [12]Feikema et al. (2010), [13]Wei et al. (2014a), [14]Almeida and Sands (2015), [15]Wei et al. (2014b)

### 8.5.4 Using a Stand-Level Process-Based Model to Develop Silvicultural Regimes for Mixtures

A simple framework is presented in Fig. 3.4 to describe the spatial and temporal dynamics of species interactions. In reality, there will be several different types of interactions occurring simultaneously in a given mixture, and the framework is too simple to describe the effects of all these simultaneously occurring interactions along gradients of different resources and climatic variables. This is where process-based models can be useful. Process-based models can be used by foresters to give a first impression of how a given type of mixture will develop under a specific combination of site conditions and silvicultural treatments.

Figures 8.18 and 8.19 are examples of how the 3-PG$_{mix}$ model was used to design silvicultural systems for subtropical plantations in China. The simulated complementarity effect changed with age and soil fertility (Fig. 8.18). This indicated sites where mixtures could provide the most benefits, in terms of productivity, compared with monocultures. It also indicated when different silvicultural treatments might need to be applied to manage competition. Complementarity also varied with rainfall and stand density (Forrester and Tang 2016), and this information could be used to refine the site conditions and silviculture most appropriate for this species composition. After selecting a preferred species combination and site condition, the 3-PG$_{mix}$ model was used to test different silvicultural regimes (Fig. 8.19). It is important to note that while models can give an impression of how a stand might develop, they are no replacement for actual empirical data and experiments. The models simply represent best estimates based on the current knowledge about the functioning of forests.

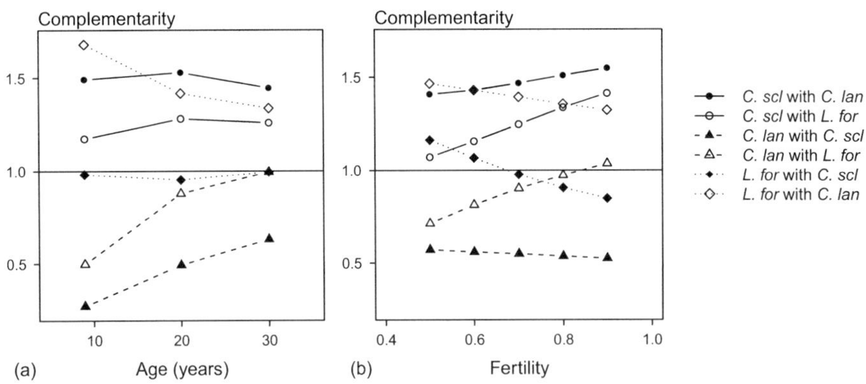


**Fig. 8.18** The complementarity effect (calculated for net primary productivity using Eq. 3.3) simulated using 3-PG$_{mix}$ for two-species mixtures containing *Castanopsis sclerophylla* (C. scl), *Cunninghamia lanceolata* (C. lan), and *Liquidambar formosana* (L. for) along the age and fertility gradients in the subtropics of China. When complementarity is >1, the trees of the given species grow faster in mixture than in monoculture; if complementarity is <1, then trees of the given species grow slower in the mixture than in monoculture. Modified from Forrester and Tang (2016)

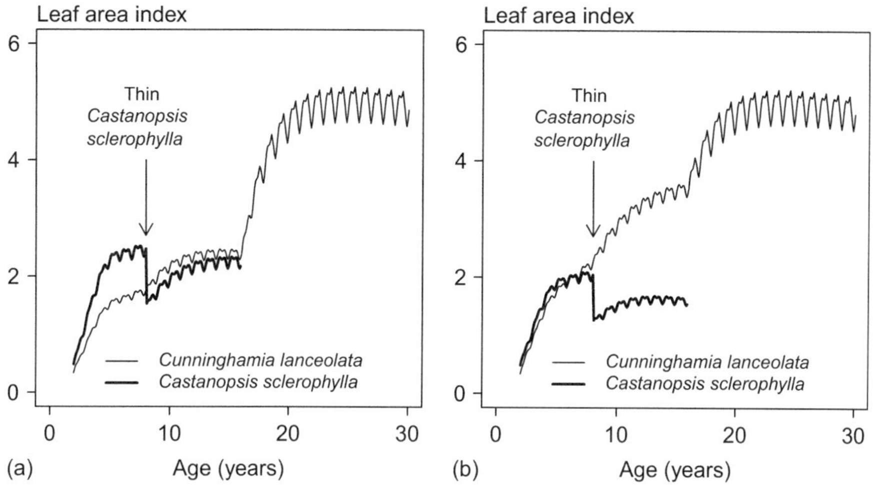


**Fig. 8.19** Using the 3-$PG_{mix}$ model to develop a silvicultural regime for a 1:1 mixture of *Cunninghamia lanceolata* and *Castanopsis sclerophylla*. Both species are planted at 500 trees per ha, and at age 8 years, *C. sclerophylla* is thinned to 300 trees per ha and then thinned out completely at age 15 years. In (**a**) the site is slightly more favourable to *C. sclerophylla*, while at (**b**) both species have very similar early leaf area index development. The wavy pattern of each line reflects the seasonality of leaf production

## Take-Home Messages

1. Models are useful tools to summarise existing knowledge about forest dynamics into a form that can be applied by researchers and foresters. Although mixed-species stands are on the advance in forestry, many forest growth models, as well as related concepts and theory, are still based on monospecific stands.
2. Models provide a framework for integration of existing knowledge about important functional, structural, and within-stand environmental attributes that influence the spatial and temporal dynamics of mixing effects.
3. Forest growth models are often divided into three groups, empirical models, process-based models, and hybrid models, which are a combination of the former two groups. Empirical models describe the system behaviour statistically, not the structure and mechanistic functioning of the system. Process-based models describe the trees and stand development on the basis of the underlying structure, within-stand environment, and functioning. Hybrid models represent a compromise between empirical and process-based models; they may bridge knowledge gaps of processes by statistical relationships.
4. Empirical models for mixed-species stands are mostly based on one of the following three concepts: the modelling of mixed stands as a weighted mean of the monocultures, the modelling of mixed stands on the basis of the tree-tree competition for growing space, or the consideration of the mixture effects by

multipliers which quantify emergent properties in mixed compared with pure stands. Process-based approaches model the competition and facilitation between trees considering the actual resources and environmental factors instead of using proxies such as competition indices.

5. Empirical models can provide accurate predictions of forest growth under the conditions where they were developed, but they are unlikely to reliably predict mixing effects for other environmental conditions such as when there are changes in climate, management or disturbance regimes, or for novel species combinations.
6. Process-based models incorporate the environmental conditions within the stand (e.g. vertical gradients in light, temperature, soil moisture), the functioning (e.g. photosynthetic rate, light-use efficiency, growth), and the tree and stand structure (e.g. crown and stem shape, shoot and root morphology, stand density, diameter distributions) when predicting the behaviour of mixed-species stands.
7. Stand-level models simplify the expressions of functions and environmental and structural attributes, thereby avoiding potential scaling issues, simplifying parameterisation, and facilitating the interpretation of outputs and the use of the models. However, they cannot account for horizontal heterogeneity such as when species are clumped.
8. Tree-level models often operate at higher spatial and temporal resolutions that are closer to the actual resolutions of functions and environmental and structural attributes. They are useful for examining how variability in the properties of individual trees, as well as the horizontal spatial heterogeneity of a stand, can influence the stand-level dynamics. However, there can be scaling problems when going from the tree to the stand level.
9. An appropriate model resolution depends first of all on the aim of the model and the scale of the outputs that are required. If the aim is to understand how the characteristics of each species influence their interactions and stand-level variables (e.g. growth, transpiration) or the relative dominance of each species, then process-based stand-level models are useful because they avoid complex higher-resolution calculations and tree-level input information. On the other hand, empirical or hybrid tree-level models can show how horizontal structural heterogeneity, such as row-by-row or group-by-group mixing, can influence stand dynamics, as opposed to the tree-by-tree mixing that stand-level models are restricted to.
10. Further understanding and model development regarding mixed-species stands requires long-term research into structure (e.g. organ and tree morphology, root-shoot allometry, size structure of the stand), within-stand environment (e.g. resource supply, environmental factors), and functioning (e.g. tree growth, partitioning between trees, mortality) at the organ, tree, and stand level. The obtained findings will support model parameterization and validation.

## References


Almeida AC, Sands PJ (2015) Improving the ability of 3-PG to model the water balance of forest plantations in contrasting environments. Ecohydrology. doi:10.1002/eco.1661

Almeida AC, Landsberg JJ, Sands PJ (2004) Parameterisation of 3-PG model for fast-growing *Eucalyptus grandis* plantations. For Ecol Manag 193:179–195

Almeida AC, Soares JV, Landsberg JJ, Rezende GD (2007) Growth and water balance of *Eucalyptus grandis* hybrid plantations in Brazil during a rotation for pulp production. For Ecol Manag 251:10–21

Almeida AC, Siggins A, Batista TR, Beadle C, Fonseca S, Loos R (2010) Mapping the effect of spatial and temporal variation in climate and soils on *Eucalyptus* plantation production with 3-PG, a process-based growth model. For Ecol Manag 259:1730–1740

Assmann E (1970) The principles of forest yield study. Pergamon Press, Oxford, NY

Assmann E, Franz F (1963) Vorläufige Fichten-Ertragstafel für Bayern. Institut für Ertragskunde der Forstl Forschungsanstalt München:103

Assmann E, Franz F (1965) Vorläufige Fichten-Ertragstafel für Bayern. Forstwissenschaftliches Centralblatt 84, 1/2, Sonderdruck edn. Verlag Paul Parey, Hamburg

Baldocchi DD, Hutchinson BA, Matt DR, McMillen RT (1984) Seasonal variations in the radiation regime within an oak-hickory forest. Agric For Meteorol 33:177–191

Battaglia M, Sands PJ (1998) Process-based forest productivity models and their application in forest management. For Ecol Manag 102:13–32

Battaglia M, Bruce J, Latham R, O'Grady A, Greenwood A (2015) Process-based size-class distribution model of trees within forest plantations: a hierarchical modeling approach. For Ecol Manag 344:63–72

Bauhus J, van Winden AP, Nicotra AB (2004) Above-ground interactions and productivity in mixed-species plantations of *Acacia mearnsii* and *Eucalyptus globulus*. Can J For Res 34:686–694

Baur F (1877) Die Fichte in Bezug auf Ertrag, Zuwachs und Form. Springer, Berlin

Binkley D, Sollins P, Bell R, Sachs D, Myrold D (1992) Biogeochemistry of adjacent conifer and alder-conifer stands. Ecology 73:2022–2033

Bonnemann A (1939) Der gleichaltrige Misehbestand von Kiefer und Buche. Mitt aus Forstwirtschaft und Forstwissenschaft 10:439–483

Bossel H (1994) TREEDYN3 forest simulation model. Reihe B, vol 35. Ber Forschungszentrum Waldökosysteme, Univ Göttingen

Bouillet JP, Laclau JP, Goncalves JLM, Moreira MZ, Trivelin PCO, Jourdan C, Silva EV, Piccolo MC, Tsai SM, Galiana A (2008) Mixed-species plantations of *Acacia mangium* and *Eucalyptus grandis* in Brazil 2: nitrogen accumulation in the stands and biological $N_2$ fixation. For Ecol Manag 255:3918–3930

Bugmann H (2001) A review of forest gap models. Clim Change 51(3–4):259–305

Cajander AK (1926) The theory of forest types. Acta For Fenn 29:108

Canham C, Coates KD, Bartemucci P, Quaglia S (1999) Measurement and modeling of spatially explicit variation in light transmission through interior cedar-hemlock forests of British Columbia. Can J For Res 29:1775–1783

Charbonnier F, le Maire G, Dreyer E, Casanoves F, Christina M, Dauzat J, Eitel JUH, Vaast P, Vierling LA, Roupsard O (2013) Competition for light in heterogeneous canopies: application of MAESTRA to a coffee (*Coffea arabica* L.) agroforestry system. Agric For Meteorol 181:152–169

Clark PJ, Evans FC (1954) Distance to nearest neighbour as a measure of spatial relationships in populations. Ecology 35(4):445–453

Clutter JL, Bennett FA (1965) Diameter distributions in old-field slash pine plantations, Georgia, vol 13. USDA Southeastern Forest Exp Station, Asheville, NC

von Cotta H (1821) Hülfstafeln für Forstwirte und Forsttaxatoren. Arnoldische Buchhandlung, Dresden

Duursma RA, Medlyn BE (2012) MAESPA: a model to study interactions between water limitation, environmental drivers and vegetation function at tree and stand levels, with an example application to [$CO_2$] × drought interactions. Geosci Model Dev 5:919–940

Dye PJ (2001) Modelling growth and water use in four *Pinus patula* stands with the 3-PG model. S Afr For J 191:53–63

Eichhorn F (1902) Ertragstafeln für die Weißtanne. Verlag Julius Springer, Berlin

Ek AR, Monserud RA (1974) Trials with program FOREST: growth and reproduction simulation for mixed species even- or uneven-aged forest stands. In: Fries J (ed) Growth models for tree and stand simulation. Res notes, 30th edn. Royal College of Forestry, Stockholm, pp 56–73 (379 p)

Feikema PM, Morris JDR, Beverly C, Collopy JJ, Baker TG, Lane PNJ (2010) Validation of plantation transpiration in south-eastern Australia estimated using the 3PG+ forest growth model. For Ecol Manag 260:14–22

Forrester DI (2014a) The spatial and temporal dynamics of species interactions in mixed-species forests: from pattern to process. For Ecol Manag 312:282–292

Forrester DI (2014b) A stand-level light interception model for horizontally and vertically heterogeneous canopies. Ecol Model 276:14–22

Forrester DI, Albrecht AT (2014) Light absorption and light-use efficiency in mixtures of *Abies alba* and *Picea abies* along a productivity gradient. For Ecol Manag 328:94–102

Forrester DI, Tang X (2016) Analysing the spatial and temporal dynamics of species interactions in mixed-species forests and the effects of stand density using the 3-PG model. Ecol Model 319:233–254

Forrester DI, Bauhus J, Cowie AL, Vanclay JK (2006) Mixed-species plantations of Eucalyptus with nitrogen-fixing trees: a review. For Ecol Manag 233:211–230

Forrester DI, Schortemeyer M, Stock WD, Bauhus J, Khanna PK, Cowie AL (2007) Assessing nitrogen fixation in mixed- and single-species plantations of *Eucalyptus globulus* and *Acacia mearnsii*. Tree Physiol 27:1319–1328

Forrester DI, Guisasola R, Tang X, Albrecht AT, Dong TL, le Maire G (2014) Using a stand-level model to predict light absorption in stands with vertically and horizontally heterogeneous canopies. For Ecosyst 1:17

Forrester DI, Benneter A, Bouriaud O, Bauhus J (2017) Diversity and competition influence tree allometry – developing allometric functions for mixed-species forests. J Ecol. doi:10.1111/1365-2745.12704

Franz F (1968) Das EDV-Programm STAOET – zur Herleitung mehrgliedriger Standort-Leistungstafeln. München (Unpublished manuscript)

Gartner TB, Cardon ZG (2004) Decomposition dynamics in mixed-species leaf litter. Oikos 104:230–246

Gehrhardt E (1909) Ueber Bestandes-Wachstumsgesetze und ihre Anwendung zur Aufstellung von Ertragstafeln. Allgemeine Forst- und Jagdzeitung 85:117–128

Goisser M, Geppert U, Rötzer T, Paya A, Huber A, Kerner R, Bauerle T, Pretzsch H, Pritsch K, Häberle K-H, Matyssek R, Grams TEE (2016) Does belowground interaction with *Fagus sylvatica* increase drought susceptibility of photosynthesis and stem growth in *Picea abies*? For Ecol Manag 375:268–278

Gonzalez-Benecke CA, Jokela EJ, Cropper WP Jr, Bracho R, Leduc DJ (2014) Parameterization of the 3-PG model for *Pinus elliottii* stands using alternative methods to estimate fertility rating, biomass partitioning and canopy closure. For Ecol Manag 327:55–75

González-García M, Almeida AC, Hevia A, Majada J, Beadle CL (2016) Application of a process-based model for predicting the productivity of *Eucalyptus nitens* bioenergy plantations in Spain. GCB Bioenergy 8(1):194–210

Grimm V (1999) Ten years of individual-based modelling in ecology: what have we learned and what could we learn in the future? Ecol Model 115:129–148

Grote R, Pretzsch H (2002) A model for individual tree development based on physiological processes. Plant Biol 4:167–180

Härkönen S, Pulkkinen M, Duursma R, Mäkelä A (2010) Estimating annual GPP, NPP and stem growth in Finland using summary models. For Ecol Manag 259:524–533
Hasenauer H (1994) Ein Einzelbaumwachstumssimulator für ungleichaltrige Kiefern- und Buchen- Fichtenmischbestände. Forstl Schr Univ Bodenkultur, Wien
Hayat H, Hacket-Pain A J, Pretzsch H, Rademacher TT, Friend AD (2017) Modelling tree growth taking into account carbon source and sink limitations. Front Plant Sci. doi: 10.3389/fpls.2017.00182
Hertel C, Leuchner M, Rötzer T, Menzel A (2012) Assessing stand structure of beech and spruce from measured spectral radiation properties and modeled leaf biomass parameters. Agric For Meteorol 165:82–91
Heyer G (1845) Wedekinds Neue Jahrb. 30:1–127
Hradetzky J (1972) Modell eines integrierten Ertragstafel-Systems in modularer Form. Universität Freiburg
Huth A, Tietjen B (2007) Management strategies for tropical rain forests: results of ecological models and requirements for ecological–economic modelling. Ecol Econ 62(2):207–215
Kahn M (1994) Modellierung der Höhenentwicklung ausgewählter Baumarten in Abhängigkeit vom Standort. Forstl Forschungsber München 141:221
Kammesheidt L, Köhler P, Huth A (2001) Sustainable timber harvesting in Venezuela: a modelling approach. J Appl Ecol 38(4):756–770
Keane RE, Austin M, Field C, Huth A, Lexer MJ, Peters D, Solomon A, Wyckoff P (2001) Tree mortality in gap models: application to climate change. Clim Change 51(3–4):509–540
Kellomäki S, Väisänen H (1997) Modelling the dynamics of the forest ecosystem for climate change studies in the boreal conditions. Ecol Model 97:121–140
Kimmins JP (1993) Scientific foundations for the simulation of ecosystem function and management in FORCYTE-11. Forestry Canada, Northern Forestry Centre, Edmonton
Kimmins JP, Comeau PG, Kurz W (1990a) Modelling the interactions between moisture and nutrients in the control of forest growth. For Ecol Manag 30:361–379
Kimmins JP, Scoullar KA, Apps MJ, Kurz WA (1990b) The FORCYTE experience: a decade of model development. In: Proceedings of Symposium on Forestry Canada
Kimmins JP, Mailly D, Seely B (1999) Modelling forest ecosystem net primary production: the hybrid simulation approach used in FORECAST. Ecol Model 122:195–224
Köhler P, Huth A (1998) The effects of tree species grouping in tropical rainforest modelling: simulations with the individual-based model FORMIND. Ecol Model 109:301–321
Komarov AS, Chertov OG, Zudin SL, Nadporozhskaya MA, Mikhailov AV, Bykhovets SS, Zudina E, Zoubkova E (2003) EFIMOD 2 – the system of simulation models of forest growth and elements cycles in forest ecosystems. Ecol Model 170:373–392
Kramer H, Akça A (1995) Leitfaden zur Waldmeßlehre. JD Sauerländer's Verlag, Frankfurt am Main
Kurth W (1999) Die Simulation der Baumarchitektur mit Wachstumsgrammatiken. Wissenschaftlicher Verlag, Berlin
Laclau J-P, Nouvellon Y, Reine C, Gonçalves JLM, Krushe AV, Jourdan C, le Maire G, Bouillet J-P (2013) Mixing *Eucalyptus* and *Acacia* trees leads to fine root over-yielding and vertical segregation between species. Oecologia 172:903–913
Landsberg JJ (1986) Physiological ecology of forest production. Academic Press, London
Landsberg JJ (2003) Modelling forest ecosystems: state of the art, challenges, and future directions. Can J For Res 33:385–397
Landsberg JJ, Sands PJ (2010) Physiological ecology of forest production: principles, processes and models. Elsevier, Amsterdam
Landsberg J, Mäkelä A, Sievänen R, Kukkola M (2005) Analysis of biomass accumulation and stem size distributions over long periods in managed stands of *Pinus sylvestris* in Finland using the 3-PG model. Tree Physiol 25:781–792

Lasch P, Badeck F-W, Suckow F, Lindner M, Mohr P (2005) Model-based analysis of management alternatives at stand and regional level in Brandenburg (Germany). For Ecol Manag 207:59–74

Leemans R, Prentice IC (1989) FORSKA, a general forest succession model. Meddelanden fran Växtbiologiska Institutionen 2. Inst Ecol Bot, Univ Uppsala, Uppsala

Liang J, Buongiorno J, Monserud RA (2005) Growth and yield of all-aged Douglas-fir western hemlock forest stands: a matrix model with stand diversity effects. Can J For Res 35 (10):2368–2381

Ligot G, Balandier P, Courbaud B, Claessens H (2014) Forest radiative transfer models: which approach for which application? Can J For Res 44:385–397

Lindner M, Sievänen R, Pretzsch H (1997) Improving the simulation of stand structure in a forest gap model. For Ecol Manag 95(2):183–195

Litton CM, Raich JW, Ryan MG (2007) Carbon allocation in forest ecosystems. Glob Chang Biol 13:2089–2109

le Maire G, Nouvellon Y, Christina M, Ponzoni FJ, Gonçalves JLM, Bouillet J-P, Laclau J-P (2013) Tree and stand light use efficiencies over a full rotation of single- and mixed-species *Eucalyptus grandis* and *Acacia mangium* plantations. For Ecol Manag 288:31–42

Mäkelä A, Hari P (1986) Stand growth model based on carbon uptake and allocation in individual trees. Ecol Model 33:205–229

McDowell NG, Beerling DJ, Breshears DD, Fisher RA, Raffa KF, Stitt M (2011) The interdependence of mechanisms underlying climate-driven vegetation mortality. Trends Ecol Evol 26:523–532

Medlyn BE (1998) Physiological basis of the light use efficiency model. Tree Physiol 18:167–176

Mette T, Albrecht A, Ammer C, Biber P, Kohnle U, Pretzsch H (2009) Evaluation of the forest growth simulator SILVA on dominant trees in mature mixed Silver fir–Norway spruce stands in South-West Germany. Ecol Model 220(13):1670–1680

Miehle P, Battaglia M, Sands PJ, Forrester DI, Feikema PM, Livesley SJ, Morris JD, Arndt SK (2009) A comparison of four process-based models and a statistical regression model to predict growth of *Eucalyptus globulus* plantations. Ecol Model 220:734–746

Mohren GMJ (1987) Simulation of forest growth, applied to Douglas fir stands in the Netherlands. PhD Thesis, Agricultural University, Wageningen

Monserud RA, Sterba H (1996) A basal area increment model for individual trees growing in even- and uneven-aged forest stands in Austria. For Ecol Manag 80:57–80

Monserud RA, Sterba H (1999) Modeling individual tree mortality for Austrian forest species. For Ecol Manag 113:109–123

Moosmayer HU, Schöpfer W (1972) Beziehungen zwischen Standortsfaktoren und Wuchsleistung der Fichte. Allgemeine Forst- und Jagdzeitung 143(10):203–215

Moser JW (1972) Dynamics of an uneven-aged forest stand. For Sci 18:184–191

Nagel J (1996) Anwendungsprogramm zur Bestandesbewertung und zur Prognose der Bestandesentwicklung. Forst und Holz 51(3):76–78

Nagel J (1999) Konzeptionelle Überlegungen zum schrittweisen Aufbau eineswaldwachstumskundlichen Simulations systems für Nordwestdeutschland, vol 128. Forstl Fak Univ Göttingen u Niedersächs Forstl Versuchsanst, Frankfurt am Main

Neumann RB, Cardon ZG (2012) The magnitude of hydraulic redistribution by plant roots: a review and synthesis of empirical and modeling studies. New Phytol 194:337–352

Newnham RM (1964) The development of a stand model for Douglas-fir. Forestry Facuty, University of British Columbia, Vancouver

Nguyen TT, Biber P, Pretzsch H (2012) Analysis and management of stand dynamics of Vietnamese dipterocarp forests by applying a dynamic growth model. Ann For Sci 69(5):581–601

Nightingale JM, Hill MJ, Phinn SR, Davies ID, Held AA, Erskine PD (2008) Use of 3-PG and 3-PGS to simulate forest growth dynamics of Australian tropical rainforests I. Parameterisation and calibration for old-growth, regenerating and plantation forests. For Ecol Manag 254:107–121

Oliver CD, Larson BC (1996) Forest stand dynamics. Wiley, New York
Paulsen JC (1795) Kurze praktische Anleitung zum Forstwesen. Verfaßt von einem Forstmanne, Detmold
de Perthuis de Laillevault R (1803) Traité de l'aménagement et de la restauration des bois et forêts de la France. Madame Huzard, Paris
Peters EB, Wythers KR, Bradford JB, Reich PB (2013) Influence of disturbance on temperate forest productivity. Ecosystems 16:95–110
Pielou EC (1977) Mathematical ecology. Wiley, New York
Pressler M (1877) Forstliche Zuwachs-, Ertrags- und Bonitierungs-Tafeln mit Regeln und Beispielen, 2nd edn. Selfpublishing, Tharandt
Pretzsch H (1993) Analyse und Reproduktion räumlicher Bestandesstrukturen. Versuche mit dem Strukturgenerator STRUGEN, vol 114. Schr Forstl Fak Univ Göttingen u Niedersächs Forstl Versuchsanst. JD Sauerländer's Verlag, Frankfurt am Main
Pretzsch H (1997) Analysis and modeling of spatial stand structures. Methodological considerations based on mixed beech-larch stands in Lower Saxony. For Ecol Manag 97:237–253
Pretzsch H (1998) Structural diversity as a result of silvicultural operations. Lesnictví-Forestry 44 (10):429–439
Pretzsch H (2001) Modellierung des Waldwachstums. Blackwell Wissenschafts-Verlag, Berlin
Pretzsch H (2002) Application and evaluation of the growth simulator SILVA 2.2 for forest stands, forest estates and large regions. Forstwissenschaftliches Centralblatt 121:28–51
Pretzsch H (2009) Forest dynamics, growth and yield. Springer, Berlin
Pretzsch H (2014) Canopy space filling and tree crown morphology in mixed-species stands compared with monocultures. For Ecol Manag 327:251–264
Pretzsch H, Biber P (2016) Tree species mixing can increase maximum stand density. Can J For Res 46:1179–1193. doi:10.1139/cjfr-2015-0413
Pretzsch H, Rötzer T (2015) Indicating forest ecosystem and stand productivity: from deductive to inductive concepts. In: Laroque GR (ed) Ecological forest management handbook, Series of applied ecology and environmental management. CRC Press, Taylor & Francis Group, p 604
Pretzsch H, Biber P, Dursky J (2002) The single tree-based stand simulator SILVA: construction, application and evaluation. For Ecol Manag 162:3–21
Pretzsch H, Grote R, Reineking B, Rötzer T, Seifert S (2008) Models for forest ecosystem management: a European perspective. Ann Bot 101:1065–1087
Pretzsch H, Bielak K, Bruchwald A, Dieler J, Dudzińska M, Ehrhart HP, Jensen AM, Johannsen VK, Kohnle U, Nagel J, Spellmann H, Zasada M, Zingg A (2013) Species results from long-term experiments. Allgemeine Forst- und Jagdzeitung 184:177–196
Pretzsch H, Rötzer T, Matyssek R, Grams TEE, Häberle KH, Pritsch K, Kerner R, Munch JC (2014) Mixed Norway spruce (*Picea abies* [L.] Karst) and European beech (Fagus sylvatica [L.]) stands under drought: from reaction pattern to mechanism. Trees 28:1305–1321
Pretzsch H, Forrester DI, Rötzer T (2015) Representation of species mixing in forest growth models. A review and perspective. Ecol Model 313:276–292
Pukkala T (1987) Simulation model for natural regeneration of pinus sylvestris, picea abies, betula pendula and betula pubescens. Silva Fenn 21:207–216
Pukkala T, Miina J, Kurttila M, Kolström T (1998) A spatial yield model for optimizing the thinning regime of mixed stands of *Pinus sylvestris* and *Picea abies*. Scand J For Res 13:31–42
Pukkala T, Lähde E, Laiho O (2009) Growth and yield models for uneven-sized forest stands in Finland. For Ecol Manag 258:207–216
Rodríguez R, Espinosa M, Real P, Inzunza J (2002) Analysis of productivity of radiata pine plantations under different silvicultural regimes using the 3PG process based model. Aust For 65:165–172
Rothe A, Binkley D (2001) Nutritional interactions in mixed species forests: a synthesis. Can J For Res 31:1855–1870

Rötzer T, Grote R, Pretzsch H (2004) The timing of bud burst and its effect on tree growth. Int J Biometeorol 48:109–118

Rötzer T, Seifert T, Pretzsch H (2009) Modelling above and below ground carbon dynamics in a mixed beech and spruce stand influenced by climate. Eur J For Res 128:171–182

Rötzer T, Leuchner M, Nunn AJ (2010) Simulating stand climate, phenology, and photosynthesis of a forest stand with a process-based growth model. Int J Biometeorol 54:449–464

Sands PJ (2004) Adaptation of 3-PG to novel species: guidelines for data collection and parameter assignment. CRC for Sustainable Production Forestry

Sands PJ, Landsberg JJ (2002) Parameterisation of 3-PG for plantation grown *Eucalyptus globulus*. For Ecol Manag 163:273–292

Schmidt A (1971) Wachstum und Ertrag der Kiefer auf wirtschaftlich wichtigen Standorteinheiten der Oberpfalz. Forstl Forschungsber München 1:187

Schober R (1967) Buchen-Ertragstafel für mäßige und starke Durchforstung. In: Schober R (ed) Die Rotbuche 1971, vol 43/44. JD Sauerländer's Verlag, Frankfurt am Main, p 333

Schume H, Jost G, Hager H (2004) Soil water depletion and recharge patterns in mixed and pure forest stands of European beech and Norway spruce. J Hydrol 289:258–274

Schwappach A (1893) Wachstum und Ertrag normaler Rotbuchenbestände. Verlag Julius Springer, Berlin

Shugart HH (1984) A theory of forest dynamics. The ecological implications of forest succession models. Springer, New York

Simard SW, Beiler KJ, Bingham MA, Deslippe JR, Philip LJ, Teste FP (2012) Mycorrhizal networks: mechanisms, ecology and modelling. Fungal Biol Rev 26:39–60

Stape JL, Ryan MG, Binkley D (2004) Testing the utility of the 3-PG model for growth of *Eucalyptus grandis*×*urophylla* with natural and manipulated supplies of water and nutrients. For Ecol Manag 193:219–234

Sterba H, Moser M, Monserud RA (1995) Prognaus – Ein Waldwachstumssimulator für Rein- und Mischbestände. Österreich Forstzeitg 5:19–20

Suzuki T (1971) Forest transition as a stochastic process. Mitt Forstl Bundesversuchsanst Wien 91:69–86

Thurm EA, Biber P, Pretzsch H (2016) Tree growth is partitioned in favor of the stem on the expense of the roots for Douglas-fir (*Pseudotsuga menziesii* (Mirb.) Franco) and European beech (*Fagus sylvatica* L.) if growing in mixture and under humid conditions. Trees. doi:10.1007/s00468-016-1512-4

Van Kessel C, Farrell RE, Roskoski JP, Keane KM (1994) Recycling of the naturally-occurring $^{15}$N in an established stand of *Leucaena leucocephala*. Soil Biol Biochem 26:757–762

Wang W, Peng C, Zhang SY, Zhou X, Larocque GR, Kneeshaw DD, Lei X (2011) Development of TRIPLEX-management model for simulating the response of forest growth to pre-commercial thinning. Ecol Model 222:2249–2261

Wei L, Marshall JD, Link TE, Kavanagh KL, Du E, Pangle RE, Gag PJ, Ubierna N (2014a) Constraining 3-PG with a new $\delta^{13}$C submodel: a test using the $\delta^{13}$C of tree rings. Plant Cell Environ 37:82–100

Wei L, Marshall JD, Zhang J, Zhou H, Powers RF (2014b) 3-PG simulations of young ponderosa pine plantations under varied management intensity: why do they grow so differently? For Ecol Manag 313:69–82

Weiskittel AR, Maguire DA, Monserud RA, Johnson GP (2010) A hybrid model for intensively managed Douglas-fir plantations in the Pacific Northwest, USA. Eur J For Res 129:325–338

Wiedemann E (1932) Die Rotbuche 1931. Mitt Forstwirtsch u Forstwiss 3(1):189

Wiedemann E (1942) Der gleichaltrige Fichten-Buchen-Mischbestand. Mitt Forstwirtsch u Forstwiss 13:1–88

Wykoff WR, Monserud RA (1988) Representing site quality in increment models: a comparison of methods. In: Ek AR, Shifley SR, Burk TE (eds) Proceedings of IUFRO conference, Minneapolis, MN, Aug 1987. Gen Tech Rep NC-120, pp 184–191

Wykoff WR, Crookston NL, Stage AR (1982) User's guide to the stand prognosis model. US Forest Service, Ogden, UT

Xenakis G, Ray D, Mencuccini M (2008) Sensitivity and uncertainty analysis from a coupled 3-PG and soil organic matter decomposition model. Ecol Model 219:1–16

Yao X, Titus SJ, MacDonald SE (2001) A generalized logistic model of individual tree mortality for aspen, white spruce, and lodgepole pine in Alberta mixed wood forests. Can J For Res 31:283–291

Zeide B (2003) The U-approach to forest modeling. Can J For Res 33:480–489

Zhao D, Borders B, Wilson M (2004) Individual-tree diameter growth and mortality models for bottom land mixed-species hardwood stands in the lower Mississippi alluvial valley. For Ecol Manag 199:307–322

# Silvicultural Options for Mixed-Species Stands

9

Jürgen Bauhus, David I. Forrester, Hans Pretzsch, Adam Felton, Patrick Pyttel, and Adam Benneter

**Abstract**

Silvicultural approaches for mixed-species stands typically aim at creating and maintaining stand compositional and structural diversity to provide the desired ecosystem function. An important aim of regeneration and tending is to develop stable mixtures that facilitate the anticipated interactions among the participating tree species and at the same time require little silvicultural input to maintain the composition of compatible tree species. To ensure survival and vigorous growth of trees in mixture in the long term, it is often necessary to separate species in space and time during the regeneration phase. Weaker competitors can be given a head start through advance regeneration or may be regenerated in patches that protect the majority of individuals from interspecific competition. In managed forests, tree species in mixture may also have a service function. For example, nurse crops may be used to facilitate establishment of

J. Bauhus (✉) • P. Pyttel • A. Benneter
Chair of Silviculture, Faculty of Environment and Natural Resources, University of Freiburg, Tennenbacherstr. 4, 79106 Freiburg, Germany
e-mail: juergen.bauhus@waldbau.uni-freiburg.de; Patrick.Pyttel@waldbau.uni-freiburg.de; adam.benneter@waldbau.uni-freiburg.de

D.I. Forrester
Swiss Federal Institute for Forest, Snow and Landscape Research WSL, Zürcherstrasse 111, 8903 Birmensdorf, Switzerland
e-mail: david.forrester@wsl.ch

H. Pretzsch
Chair for Forest Growth and Yield Science, Faculty of Forest Science and Resource Management, Technical University of Munich, Hans-Carl-von-Carlowitz-Platz 2, 85354 Freising, Germany
e-mail: Hans.Pretzsch@lrz.tum.de

A. Felton
Southern Swedish Forest Research Centre, Institutionen för Sydsvensk Skogsvetenskap, Box 49, 23053 Alnarp, Sweden
e-mail: adam.felton@slu.se



H. Pretzsch et al. (eds.), *Mixed-Species Forests*, DOI 10.1007/978-3-662-54553-9_9

sensitive tree species and to accelerate ecosystem recovery following disturbances. Trainer species are typically used to improve the quality development of stems of economically important species. Whereas the regeneration of mixtures has received considerable attention in research, there is little specific information regarding the thinning of mixed-species stands, which are often treated like an agglomerate of monospecific stands of the companion species. In the absence of quantitative information on optimal tree spacing, stand density, or the growth response to thinning in mixtures, crop tree thinning approaches are recommended. These facilitate the maintenance of desired species proportions as well as stand stability and development of quality attributes in crop trees through simple indicators. Although it is often assumed that the development of stem quality in mixtures is inferior to that in monospecific stands, there is so far little solid evidence to support this assumption. Rather, it appears that stem quality depends on specific interactions between species and the structural heterogeneity of stands.

## Contents


9.1 Management for Heterogeneity or Homogeneity ........ 434
9.2 Types of Silvicultural Mixtures ........ 438
9.2.1 Introduction to Different Types of Silvicultural Mixtures ........ 438
9.2.2 Mixtures to Increase Productivity or Economic Performance of the Tree Crop 438
9.2.3 Admixing Species with Service Functions: Trainer Trees and Nurse Crops . . . 440
9.3 Creating and Maintaining Long-Term Mixtures ........ 445
9.3.1 Why Are Stable Mixtures Desirable? ........ 445
9.3.2 How Can We Predict the Compatibility of Species and Develop Silvicultural Prescriptions? ........ 451
9.3.3 Regeneration Methods for Mixed-Species Forests ........ 459
9.4 Tending and Thinning in Mixtures ........ 470
9.4.1 How Does Thinning in Mixtures Differ from that in Monocultures? ........ 470
9.4.2 The Purpose of Thinning ........ 471
9.4.3 Conventional Thinning Prescriptions ........ 473
9.4.4 Considerations for Thinning Prescriptions for Mixed-Species Stands ........ 474
9.4.5 Crop Tree Approaches to Thinning ........ 478
9.4.6 Thinning in Uneven-Aged Stands ........ 483
9.5 Tree Quality in Mixtures ........ 487
Take-Home Messages ........ 489
References ........ 490


## 9.1 Management for Heterogeneity or Homogeneity

Silviculture is the deliberate manipulation of forest structure, composition, and dynamics to provide ecosystem goods and services on a sustainable basis (Bauhus and Pyttel 2015). The extent to which forests are managed for the various goods and services depends on the objectives of the landowner (Nyland 2002). The creation and maintenance of forest stands comprising a higher degree of species-mixing promises the delivery of a range of ecosystem goods and services at a higher level in comparison to monospecific forests (Gamfeldt et al. 2013). Especially at a time when society demands of forests the provision of recreation and 'living space' services at an ever-increasing rate, management decisions must meet ever higher standards regarding multifunctionality. For a wide range of European forests, it has been shown that biodiversity can indeed promote multiple ecosystem functions, when the management goal is to provide moderate levels of many ecosystem functions simultaneously. This has been coined the 'jack-of-all-trades' effect, which is caused by averaging of the effects of individual species on ecosystem functions in diverse forest stands (van der Plas et al. 2016). If the goal is to provide particular ecosystem functions at a very high level, this may be better achieved with monospecific or low-diversity forests. However, the majority of the global forest area worldwide is not dedicated exclusively to either production or conservation and is thus likely to be managed for multiple functions (FAO 2016). Silvicultural systems to develop and maintain forests providing multiple ecosystem goods and services would therefore benefit from considering the use of tree species mixtures. Unsurprisingly, the use of mixed-species forests has emerged as a central element of alternative silvicultural approaches that contrast with more traditional silvicultural methods aimed at even-aged and monospecific forest stands (Puettmann et al. 2015).

Forest management takes place at different spatial scales, ranging from the enterprise or property to management units and individual stands. In this chapter, we will focus mostly on the stand scale, which is the scale at which silvicultural activities are considered and applied. The sum of all management activities that take place at the stand level from regeneration to harvesting of trees is called a silvicultural system. All silvicultural decisions that may be taken reflect the goals and aspirations of forest owners, embedded in the context of societal aspirations that are expressed in the form of legislation, certification, and other elements of forest governance. Commonly, silvicultural systems are named after the regeneration method such as clear-felling, shelterwood, seed tree system, single tree selection, etc., because the type and intensity of harvesting and the residual stand structure that remains are important determinants for all the other management activities that form a silvicultural system. Silvicultural systems aim at improving the forest stand conditions to optimise the yield of goods and services. Historically, the focus has been on the delivery of desired commodity products at predictable qualities and quantities at regular intervals and over long periods (Nyland 2002). To ensure this, desired species have to be regenerated in the preferred composition. Subsequently, the regrowing stands are being tended to maintain or change that species composition, to ensure the desired development of tree quality and dimensions, to promote stand stability and reduce risks such as windthrow, to maintain productivity, to facilitate income through thinning of trees, and to provide flexible options for the future stand development. Finally, trees are being harvested either individually (single tree selection), in groups (group selection, irregular group shelterwood), gaps, strips, shelterwoods, or clear cuts, to name some important silvicultural systems (Matthews 1991). The removal of mature trees opens up growing space for the establishment or release of a new cohort of trees. The establishment and subsequent growth of trees may be further promoted through specific site preparation practices such as removal of competing vegetation or the scarification of soils. In the following, we will look at these different phases of silvicultural systems in more detail to address the question of which practices may be used to establish and maintain mixed-species forests and how these practices may differ from those that are commonly used in monospecific forests.

To help understand the goals and challenges of managing mixed-species forests, it helps to contrast it initially against silvicultural systems used for monospecific, even-aged plantation forests (Fig. 9.1, left panel). The management goal commonly pursued with industrial plantations is typically the efficient production of a homogenous commodity product. The high growth rates of tree species used in fast-growing plantations warrant high inputs of labour and energy. These are used for tree breeding programmes, soil preparation, fertilisation, planting, pruning, forest protection measures, and monitoring programmes. Important aims of these practices are high yields and also homogenous growth of the crop so that silvicultural practices as well as the processing of logs can be standardised as much as possible. For example, if all trees are very similar in size, they can be pruned at the same time, and they respond similarly to thinning and can be effectively handled within a narrow processing chain when harvested in similar dimensions.

**Fig. 9.1** Contrasting silvicultural systems and forest types. *Left panel*: harvesting and re-establishment of a clonal crop of *Eucalyptus* hybrid (*urophylla* x *grandis*) for production of pulp wood in short rotations (ca. 7 years) in southern Brazil (Photo: D.I. Forrester). *Right panel*: mixed-species, uneven-aged seminatural mountain forest in southern Germany, where trees are harvested according to target diameters (achieved after 80 to 140 years) and the stand is regenerated naturally (Photo: J. Bauhus)

In contrast, most mixed-species stands can be found in seminatural forests (Fig. 9.1, right panel). In most situations, the tree species have experienced little or no genetic improvement. They are often regenerated naturally, which means that tree individuals establish at different times and at variable spacing and species combinations. The soil has not been ploughed or ripped and thus retains its high spatial heterogeneity in physical and chemical properties. The species differ in physiological requirements, competitive strength, growth rates, potential products, etc. Further, the tree species may be harvested at different dimensions and different tree ages, and the products might go to different consumers and industries. In mixed-species forests, particularly in those comprising species with strongly different growth rates and a high diversity in tree dimensions and stand structures, the silvicultural management becomes inevitably more challenging. Management interventions such as pruning of trees, thinning, or harvesting, which are typically related to tree dimension, may have to be spread over longer periods in order to adapt to the staggered development of trees to ensure that individual trees or neighbourhoods are treated or harvested at the appropriate developmental stage. For example, target diameter harvesting in monospecific and even-aged European beech (*Fagus sylvatica*) forests may easily proceed over a period of 40 years or more (e.g. Tarp et al. 2005). In mixed-species forests with an even greater variation in tree age, tree size, and growth rates, this process might take considerably longer. This greater variation requires either close monitoring of tree development or fairly accurate models of stand and tree development to enable the timing of management interventions, which will be more frequent but typically less intensive than in monospecific forests. In addition, where stands are undergoing regular repeated thinning and trees are harvested using selection systems in structurally diverse stands, avoiding damage to retained trees and advance regeneration becomes an enormous challenge. For example, in the German state of Baden-Württemberg, one in four trees is damaged through harvesting operations (Nill 2011). There are many more ways in which management complexity increases in mixed forests. As we have seen in previous chapters, owing to interactions among species, they may develop differently in mixtures when compared to monospecific stands, and knowledge about these interactions is required as well.

One could go on to list additional differences, but it should be clear already that management complexity is inherently much higher in compositionally and structurally more diverse forests, and management outcomes are therefore less predictable (Puettmann et al. 2009). However, increasing complexity of mixed-species forests does not only pose challenges to management, but they also represent opportunities to manage systems that are more resilient and adaptable to an uncertain future, including the changing aspirations of society (Puettmann et al. 2009). Hence, management for heterogeneity in stand structure and composition has advanced to become an important approach in forestry (Puettmann et al. 2015).

## 9.2 Types of Silvicultural Mixtures

### 9.2.1 Introduction to Different Types of Silvicultural Mixtures

Mixed-species forests can develop in many different ways. They can be naturally mixed, comprising the tree species of a native forest community, and they can be the result of a deliberate combination of species through artificial regeneration, or might have originated from the combination of these, for example, the enrichment of native forests with additional tree species or the manipulation of native tree communities to promote the more desirable species. Since silviculture comprises deliberate manipulations of forest composition and structure, we will deal here only with those types of mixtures that have resulted from silvicultural treatments.

The forest management goals that may be pursued with silvicultural mixtures cover a wide range of objectives which will be discussed below.

### 9.2.2 Mixtures to Increase Productivity or Economic Performance of the Tree Crop

Although it is now well established (see preceding chapters; Liang et al. 2016) that mixing tree species leads in the majority of cases to increased productivity when compared to monospecific stands of the companion species, there are not many cases, where increased biomass production is a declared objective of the management of mixtures. This has likely to do with the fact that transgressive overyielding, where the mixture is also more productive than the most productive monoculture, is not so often observed. Hence, where the focus of management is on biomass production, this may still be better achieved with monospecific stands, for example, for the production of pulpwood with *Eucalyptus* species.

Yet, there are also examples where the admixture of other species to eucalypts, in particular nitrogen-fixing tree species, can be used to increase biomass production. In this case, the admixture of N-fixing species may have the additional objective of improving the N nutrition or to replace mineral fertiliser. For this type of mixture, both species are usually planted at about the same time at total stand densities of about 800 to 1600 trees per ha. Species are mixed tree-by-tree or row-by-row, often in 1:1 mixtures, which tend to show the highest complementarity effects (Fig. 9.2). Rotation lengths vary from as short as 7 years to about 20 years, depending on the region and the desired end products. Alternative thinning regimes have been proposed, depending on these objectives, including thinning the N-fixing species for fuelwood or pulpwood, while retaining the slash on site to provide another pulse of N as the foliage mass decomposes. A case study of growth dynamics of this system is described in Sect. 4.2.2. The *Eucalyptus* trees are very shade intolerant and therefore need to obtain the upper canopy position within this mixture. The N-fixing species can occupy the upper canopy position during the first 4–5 years (Forrester et al. 2004), while rates of N fixation are high, but the *Eucalyptus* then needs to obtain a dominant position to avoid becoming suppressed.

**Fig. 9.2** Five-year-old mixed plantations of a tropical *Eucalyptus* hybrid (*urophylla* x *grandis*) with the tropical *Acacia mangium* in southern Brazil. Both species are used for pulpwood production. The shade-intolerant eucalypt occupies the overstorey, whereas the acacia occupies the midstorey level (Photo: D.I. Forrester)

Examples of this system were reviewed in Forrester et al. (2006). See also Binkley et al. (2003) and Bouillet et al. (2013). While the facilitative effect of N fixation plays a major role in the complementarity effect of this mixture, the mixtures have also been shown to have greater light absorption, transpiration, light-use efficiency, and water-use efficiency (Binkley et al. 1992; Forrester et al. 2010, 2012). However, on very productive sites, where N is not a growth limiting factor, mixing of N-fixing tree species to fast-growing eucalypts may not lead to a biomass productivity that is greater than that of the eucalypt monoculture (see Fig. 9.1, left panel).

More commonly, the objective of species mixing is the improved economic performance of mixtures when compared to one or more monospecific stands of companion species. Often, species that are economically more attractive owing to superior volume or value production are added to a forest matrix of less valuable native species. For example, in Central Europe, stands of the dominant native species, European Beech (*Fagus sylvatica*), typically provide lower income to forest owners than comparable stands (on the same site) of conifer species such as Norway spruce (*Picea abies*) or Douglas-fir (*Pseudotsuga menziesii*) (e.g. Möhring et al. 2008). In the past, this has led to both the conversion of native forests to monospecific conifer stands and also to the admixture of conifer species to beech. More recently, it has been demonstrated that mixed *F. sylvatica*-*P. abies* stands are also economically more favourable than pure *P. abies* stands when productions risks are being considered (Griess and Knoke 2013 and Chap. 11). However, these results are specific to this particular species combination and may not be extrapolated to other hardwood-conifer mixtures. Whereas older mixed stands of *F. sylvatica* with conifer species (*P. abies*, *P. menziesii*, *Abies* sp., *Larix* sp.) have often been developed through enrichment planting of conifers in young natural regeneration of *F. sylvatica*, young mixtures are often the result of conversion of former pure conifer stands to increase naturalness and ecological stability of these forests (e.g. Löf et al. 2010).

### 9.2.3 Admixing Species with Service Functions: Trainer Trees and Nurse Crops

#### 9.2.3.1 Trainer Species

The use of admixed species that provide a servicing function within silvicultural systems has a long tradition. A classic example is the use of trainer species to improve the quality development of a valuable main crop species, as can be seen in the mixtures of *Quercus petraea and Q. robur* with shade-tolerant species such as *F. sylvatica* or *Carpinus betulus*. In a typical setting, the oaks in the overstorey, which are grown for veneer logs or saw logs, will be embedded in a matrix of beech in the mid- and understorey (Fig. 9.3). High-value logs must be of large dimensions (>40 cm mid-diameter, preferably larger) and free of knots and other defects. Typical target diameters (DBH of 70–80 cm) can only be achieved in long production cycles of 160–240 years (von Lüpke 1998). The beech trees are used as trainers to shade the oak stems and thus help maintain thin primary branches as well as prevent the emergence and development of epicormic branches. The beech subcanopy is not required for natural self-pruning, which can also be achieved in dense monospecific oak stands. Once the beech stratum has reached a height at which it effectively shades the valuable lower stems of oaks, it facilitates the crown thinning in oaks. Without the trainer trees, they may develop abundant epicormic branches on the lower stem following crown thinning. Thinned oaks can then develop large crowns that facilitate a continuously high diameter increment. Given the long production cycles required for oak, it is necessary to control the

**Fig. 9.3** Mixed *Quercus petraea-Fagus sylvatica* stand near Müllheim, southern Germany. Trees marked with 'O' are the dominant oaks, those marked with 'B' are beech trees in the midstorey and understorey. The oaks are grown for high value veneer or saw logs. The beech trees are used as trainers to shade the oak stems and thus prevent the emergence and development of epicormic branches. This function of beech trees is particularly important when the canopy of oaks is interrupted in the process of thinning or harvesting. In addition, the beech stratum shades the forest floor and suppresses the development of ground vegetation which could become an impediment to the natural regeneration when the main crop of trees is harvested (Photo: J. Bauhus)

competition from beech, which otherwise grows into the crowns of the oaks, which leads to the dieback of the lower branches. For this purpose, beech trees need to be regularly removed.

The volume yield of mixed oak-beech stands in Central Europe is on average 30% higher than for respective monospecific stands of the two species. Both species benefit in terms of their productivity from these mixtures, but the complementarity effects are stronger on poor sites and decline or even become negative on very productive sites (Pretzsch et al. 2013, 2016). Unfortunately, there is no quantitative information of the mixing effect on the financial outcome. The establishment of these mixtures is often difficult, since on most sites, *F. sylvatica* is far more competitive than *Q. petraea* and *Q. robur* (von Lüpke 1998). When mixtures are regenerated naturally, the more shade-tolerant beech grows faster than the light-demanding oak seedlings and saplings at all light levels below 60% of photosynthetically active radiation in the open (Ligot et al. 2013). This situation requires constant intervention in the seedling and sapling stage to ensure a desired proportion of oaks in the developing stand. As this process can be very costly, the regeneration of the two species is sometimes separated in time and beech is planted under oaks, once they have reached pole dimensions.

The function of trainer species may also be assumed by tall understorey species. For example, in wet eucalypt forests of Tasmania, *Pomaderris apetala* had a strong facilitative effect on tree form and branch dimensions of the valuable hardwood *Acacia melanoxylon* and at the same time a strong competitive effect on growth. When *P. apetala* was removed in circles of 3.6 and 7.2 m around *A. melanoxylon*, diameter growth of the latter increased by 0.2 and 0.7 cm per year, respectively. However, removal of the trainer species also reduced the branch-free bole length at age 11 by 1.6 m and led to more and stronger branches (Jennings et al. 2003).

#### 9.2.3.2 Nurse Crops

The other important servicing function of species that are admixed to a main crop is that of a nurse. Nurse crops are typically mixtures of at least two tree species, commonly combining shade-intolerant, but fast-growing and stress-tolerant species in the overstorey with slow-growing shade-tolerant and stress-sensitive species in the understorey, where the former ameliorate the site conditions to improve survival or growth of the latter. There is therefore growing interest in nurse crop systems to rapidly restore or accelerate the recovery of large disturbed areas in terms of productivity and other ecosystem functions. A typical example are nurse crops with pioneer species such as *Populus* or *Betula* that can reduce the risk of frost damage by sheltering slower-growing target species such as *Picea* or *Abies* which may suffer from late frost events in large forest openings (Lundmark and Hallgren 1987; Örlander 1993). For example, the canopy of a birch nurse crop may increase temperatures near the ground during spring frost events by 4–6 °C (Geiger 1950). In addition, pioneer trees can also play a major role in the restoration of ecosystem functioning in terms of water transpiration and nutrient cycling thus benefitting the slower-growing tree species (e.g. Man and Lieffers 1999; Stark et al. 2015). This may be useful following forest disturbances, when, owing to high mineralisation rates of organic matter and reduced uptake by vegetation of water and nutrients, high rates of nitrate leaching accompanied by leaching of cations such as calcium (Ca), potassium (K), and magnesium (Mg) may occur, leading to soil acidification and a reduction in soil fertility (e.g. Vitousek and Melillo 1979; Bauhus und Bartsch 1995). The effect of these disturbances on nutrient cycling depends largely on the development of the vegetation and its uptake demand (Jerabkova et al. 2011). The rapid growth and rate of canopy closure of pioneer tree species tends to result in a higher nutrient and water uptake, thus reducing leaching, soil temperatures, and mineralisation rates, when compared to slower-growing tree species (e.g. Stark et al. 2015). Hence, these temporary mixtures are typically employed during the initial years of stand development after large-scale disturbances, clear-felling, or the afforestation of agricultural land, when the microclimatic conditions are extreme (Drouineau et al. 2000; Keenan et al. 1995; Gardiner et al. 2004).

Since the main purpose of these nurse crop systems is to facilitate the establishment of sensitive tree species, the fast-growing pioneers are typically removed once that goal has been achieved and before the target tree species start to suffer from interspecific competition from nurse trees (e.g. Shepperd and Jones 1985). In many situations, the nurse crop provides additional biomass to that of the main tree species, in particular when the latter is shade tolerant (e.g. Lundqvist et al. 2014; Nord-Larsen and Meilby 2016). However, in many cases, the nurse trees have simply been killed and not commercially harvested, because it was feared that their felling and extraction might damage the established regeneration of target species. This might be regarded as a typical example of a case where technical capabilities for tree harvesting and/or existing extraction systems do not (yet) match the increasing stand complexity found in those systems. To ensure that the biomass of nurse trees is utilised and not wasted, new approaches involve the establishment of widely spaced nurse crops for the purpose of woody biomass production (Unseld and Bauhus 2012). Here, nurse crops are managed in relatively short rotation periods employing wide tree spacings that still provide the desired nurse effect and at the same time enable harvesting of nurse trees with minimal damage to the advanced regeneration (Fig. 9.4).

For example, temporary nurse crop mixtures of *Populus* sp. or *Betula* sp. with *Quercus robur* have been found to have similar total productivity to *Quercus* monocultures in the long term; however, in the nurse systems, about 60–100 Mg $ha^{-1}$ of biomass could be harvested after 20 years, whereas the first harvest from the *Quercus* monocultures could not occur until at least age 50 (Stark et al. 2015).

Other applications of nurse crop systems aim at improved nutrition of the main species. One well-known example is the use of other conifer species such as *Pinus* sp. or *Larix* sp. to avoid nitrogen deficiency and growth stagnation in *Picea sitchensis* (Carlyle and Malcolm 1986). Apparently, the other conifers facilitate establishment of mycorrhizal associations in *P. sitchensis* when established on heathlands. However, in these systems, the selection of the appropriate nurse species can also have beneficial effects on stem quality (branchiness) of Sitka spruce (Cameron and Watson 1999).

Nurse crops may also be used to reduce the risk of fire damage through their suppressive effect on grasses and hence reducing the fuel load. For instance, on sites

**Fig. 9.4** (**a**) *Left panel*: Systematic design of planted nurse crops for afforestation or restoration of disturbed sites. Fast-growing pioneer species might be planted in rows, where skidding tracks will be located later (here depicted as *double rows*). This may not only make use of the beneficial effect of nurse trees but also provide an early income from the first thinning operation and protect the soil on skid tracks against compaction. (**b**) *Right panel*: Fully mechanised harvesting of an aspen nurse crop with a feller-buncher system. This significantly reduces damage to the advance regeneration of main tree-crop species when compared to manual felling and forwarding by horse (Weich 2012) (Photo: T. Weich)

where dry tropical forest was restored, the damage to native tree seedlings through two successive ground fires was significantly reduced when these grew under a nurse crop of *Leucaena leucocephala* when compared to open conditions, where the fuel load, consisting mostly of Guinea grass (*Urochloa maxima*), was more than ten times higher than beneath the nurse crop (Santiago-García et al. 2008). In this case, the interaction between nurse crop and the main tree species is rather indirect.

Tropical pines have been used as nurse crops to restore the species composition of native tropical forests with late-successional tree species (Ashton et al. 1997) or to establish plantations of valuable tropical hardwoods (Dordel et al. 2010). In addition to the beneficial effects on the regeneration of the main tree species, the use of conventional plantation species as nurse crops in restoration plantings has the added benefit that the revenue from harvesting the nurse trees reduces the costs of restoration efforts.

To ensure that trainer or nurse trees meet their functional role in these types of mixtures, the spatial arrangement and density has to be such that the interactions with trees of the main crop are optimised, typically through a fine-grained mixture. However, the optimal density of trainer and nurse trees appears to be relatively low in most situations, because competitive effects outweigh facilitative effects at high densities.

## 9.3 Creating and Maintaining Long-Term Mixtures

### 9.3.1 Why Are Stable Mixtures Desirable?

We have seen above that some mixtures are only temporary to achieve specific purposes (e.g. nurse crops). In situations, however, where mixtures are cultivated to achieve an overall higher and more stable provision of ecosystem goods and services, it is desirable to maintain mixtures over the whole production cycle of even-aged forests or to maintain them permanently in uneven-aged forests. For this purpose, it is advantageous to combine species that are highly compatible and complementary (see glossary) in terms of growth dynamics, regeneration requirements, and other interactions such as those in relation to pests and diseases or stress and disturbance. Mixtures of tree species that are highly compatible will help reduce the frequency and intensity of silvicultural interventions to regulate the desired stand development. For example, in tree species mixtures with similar growth rates and shade tolerance, there will be less need to favour or promote one or several species through repeated thinning than in situations where light-demanding species grow slower or have a lower end height than competing shade-tolerant species (e.g. Kerr 2004, and see Boxes 9.1 and 9.2). In situations, where certain species are affected by diseases such as soilborne pathogens, admixing disease-resistant species to reduce the root contact between susceptible species can reduce damage rates and help maintain mixing proportions (e.g. Lindén and Vollbrecht 2002). Combining tree species that have a positive influence on the quality development of one or more crop species (e.g. through reduced branch dimensions or more rapid crown lift) will benefit the value of the final product (e.g. Cameron and Watson 1999). Choosing compatible tree species for mixtures will thus not only assist in achieving the desired mixing effects but will also help reduce management costs and/or improve economic benefit. As will be discussed further below, compatibility between species may be further enhanced through temporal and spatial separation of species in mixture.

**Box 9.1 Risks and Opportunities from Mixing Native and Non-native Tree Species**

The use of introduced tree species in mixed-species stands may be particularly appealing in regions and forest types where there is a limited diversity of native tree species and where introduced species provide a preferred portfolio of ecosystem goods and services for forest owners or society in general. Introduced tree species are those which occur outside their natural range (Richardson and Rejmanek 2011), a notably encompassing concept which includes regional to intercontinental introductions. The use of introduced tree species in mixed-species production forests raises similar potential opportunities and ecological concerns as when introduced tree species are

(continued)

**Box 9.1** (continued)
used in monocultures. These include the potential for (1) increased productivity, (2) a better fit of physiological amplitude to future climatic conditions, and (3) greater resistance to the prevailing disturbance regime, including pests and pathogens. However, their use can also result in (4) reduced stand-level biodiversity, (5) adverse impacts on sensitive ecosystems from the escape and establishment of invasive tree species, (6) facilitation of pest and pathogen outbreaks, and (7) hybridisation with and genetic dilution of native tree species populations (Vor et al. 2015; Felton et al. 2013, 2016; Keane and Crawley 2002; Liu and Stiling 2006). Here we will not discuss these opportunities and risks of introduced species in general terms. Instead, we focus on the question to what extent some of these potentials and risks may be reduced or enhanced in mixed-species stands when combined with specific companion species.

**Production potentials:** Few introduced species have been purposefully admixed into stands composed of otherwise native species. In most situations, they were planted in monospecific stands. Hence, there are also only few studies that have specifically addressed the effect of admixing introduced and native tree species.

Douglas-fir (*Pseudotsuga menziesii*) has a long history of introduction in Europe and other parts of the world (Spellmann et al. 2015). Early mixed-species stands originated in beech and mixed-mountain forests managed using shelterwood cutting, in which Douglas-fir was used to fill in gaps lacking sufficient natural regeneration. As a result, mixed stands of Douglas-fir/European beech, or Douglas-fir/silver fir/European beech, and other species combinations have developed, in which Douglas-fir, owing to its superior growth and terminal height, often becomes a canopy emergent. In these mixtures, Douglas-fir grows faster than it does in conspecific monocultures, and this is largely attributed to its dominant position within the mixture (Thurm and Pretzsch 2016). It may also recover more rapidly from droughts in the mixtures. In contrast, growth of *F. sylvatica* was not significantly affected in the mixtures; except that following droughts, its recovery was slowed down (Thurm et al. 2016).

A classic example for the facilitation of an introduced species through a native species is provided by mixtures of *Picea sitchensis* and the native *Pinus sylvestris* in Scotland. Here, the afforestation of heathlands with monospecific *P. sitchensis* plantations often resulted in poor, stunted growth combined with nitrogen deficiency. When *P. sitchensis* was planted together with *Pinus sylvestris*, however, it was able to establish successfully without N fertilisation (Taylor and Tabbush 1990). The reason may be found in the forest floor of mixed stands, where increased N mineralisation led to greater accumulation of N in above-ground biomass in 15-year-old mixtures of

(continued)

**Box 9.1** (continued)

*P. sitchensis* with *Pinus sylvestris* (205 kg $ha^{-1}$) compared to pure *P. sitchensis* stands (68 kg $ha^{-1}$) (Miller et al. 1986, cited in Williams 1992). The enhanced availability of N beneath mixtures, which greatly increased productivity, was attributed to higher proteolytic activity of mycorrhizal fungi associated with pine roots (Ryan and Alexander 1992). In other words, the presence of the native Scots pine facilitated greater activity of mycorrhizal fungi and the possible establishment of mycorrhizal associations in *P. sitchensis*.

When considering the production potential of tree mixtures between native and introduced tree species, it would be most interesting to compare these mixtures with mixed stands comprising only native species in the same experiment, in particular where the performance of introduced species can be compared to those of native species with similar traits, e.g. congeneric species (e.g. Wein et al. 2016). This would permit a separation of the effects of functional traits of introduced species from that of species origin (native versus exotic). Unfortunately, such experiments are still very young and do not yet provide reliable results on the medium to long-term growth patterns (e.g. Tobner et al. 2014).

**Biodiversity impacts**: Introduced tree species can act as obstacles to biodiversity within a production stand when the environments and resources they provide diverge from that which native forest-dependent species were evolved to exploit (Felton et al. 2016). They may also enhance native biodiversity, if they provide habitat and resources that may have been lost from forests through the loss of native species (Mitchell et al. 2016; Felton et al. 2013). Outcomes for biodiversity can be expected to vary depending on the specific context assessed, including the choice of introduced tree species and its phylogenetic proximity to native species; the complementarity of the tree species combined in terms of their functional traits, structures, and resources; the proportion and number of the different trees used in mixtures; the extent of understorey vegetation retained; how the mixtures are planted (e.g. intimate versus patch); the taxonomic response group assessed; the metric considered (e.g. richness, abundance, community composition); and due to regional variation in the availability and ecological requirements of source populations.

Furthermore, evaluating the implications for biodiversity from the use of introduced tree species in mixed-species stands depends to a large extent on the baseline reference condition considered. As we have seen for the assessment of other stand-level properties and processes in mixtures, e.g. productivity, monocultures of all component species are required for a comprehensive assessment of mixing effects. For example, the majority of studies conducted seem to indicate that the use of introduced tree species in

(continued)

**Box 9.1** (continued)

mixtures with native tree species will often support a distinct community composition and possibly increased species richness, from that of introduced monocultures (Simmons and Buckley 1992; Gjerde and Saetersdal 1997; Donald et al. 1998; Hausner et al. 2002), but there are exceptions (Martinez et al. 2009; Oxbrough et al. 2012). In contrast, when mixtures involving one or more introduced tree species are compared to mixtures solely comprised of native tree species, they may be relatively depauperate (Schuldt and Scherer-Lorenzen 2014), depending on the proportion of introduced tree species (Simmons and Buckley 1992; Gjerde and Saetersdal 1997). Unfortunately much of our present knowledge is limited by the paucity of relevant studies, the comparison of sites that differ not only in species composition but also in management intensities, limited replication, or confounding tree species compositions and proportions.

**Invasive tree species**: Of substantial concern is the potential for introduced tree species to become invasive and adversely affect or even transform sensitive ecosystems (Essl et al. 2011; Krumm and Vítková 2016). Approximately 60% of tree species identified as invasive are used in forestry (Haysom and Murphy 2003), and forestry is the second most common source of introduced woody plant species worldwide (Richardson and Rejmanek 2011). It is also important to emphasise that an introduced tree species which escapes and spreads into the landscape can in turn upgrade the scale of other risks, such as pest and pathogen transfer to endemic tree species (Engelmark et al. 2001; Richardson and Rejmanek 2004). Multiple studies have demonstrated a link between propagule pressure, and the risk of an introduced tree species becoming invasive (e.g. Krivanek et al. 2006; Simberloff 2009; Essl et al. 2011), and have emphasised the importance of vegetation cover and disturbance regimes in determining the susceptibility of a given landscape to invasion (e.g. Ledgard 2002; Richardson and Rejmanek 2004; Orellana and Raffaele 2010; Nunez et al. 2011). The use of introduced tree species in mixtures thereby raises the potential of using the accompanying native tree species to ameliorate such risks. At the stand scale, the intermixing of an introduced tree species with native tree species may simply dilute such pressure of invasiveness by limiting the number of the introduced species' propagules dispersing from a given stand area. More targeted reductions in dispersal may perhaps be achieved by planting the native tree species component along stand boundaries or other at-risk edges (Engelmark et al. 2001). In addition, mixtures may be designed to prevent the establishment of introduced tree species. For example, light-demanding species, or those that require mineral soil contact to germinate and establish as seedlings, may be effectively controlled in a matrix of shade-tolerant species that cast dense shade or provide a thick forest floor layer. (Box Fig. 9.1-1). Natural regeneration of such species could be further controlled through the

(continued)

**Box Fig 9.1-1** Mixture of *Pseudotsuga menziesii* with *Fagus sylvatica*. The different traits of the tree species and associated structures (e.g. smooth bark *versus* rough bark, leaves *versus* needles) provide niches for faunal species that shape distinct communities (e.g. Goßner and Ammer 2006). In a matrix of native *F. sylvatica* with widely spaced *P. menziesii*, the specific broadleaved tree-related fauna might be maintained. In the shade of *F. sylvatica*, regeneration of *P. menziesii* is suppressed (Photo: Cecilia Malmqvist)

**Box 9.1** (continued)

use of silvicultural systems that create or use only small disturbances, e.g. single tree or group selection. For example, in mixed mountain forests in the southern Black Forest with a large proportion of mature Douglas-fir (ca. 20% of forest area), the contribution of the species to the regeneration layer (combined natural and artificial regeneration) was less than 4%, and this proportion was declining, most probably as a result of insufficient light availability (Steinmetz and Bauhus 2016). These forests are managed in selection systems involving small gap disturbances, in which Douglas-fir is accompanied by shade-tolerant species such as *Fagus sylvatica*, *Abies alba*, and *Picea abies*. In these circumstances, most Douglas-fir regeneration occurred beneath its own canopy, with far less regeneration beneath these other species. However, if it were mixed with species such as *Quercus petraea* or *Pinus sylvestris*, which have a more open canopy and grow on drier sites, its regeneration might be more abundant (Knoerzer 1999).

While the extent of tree species invasion observed in Europe is less than that observed in the southern hemisphere (e.g. Carrillo-Gavilán and Vilà 2010; Krumm and Vítková 2016), the increased and widespread use of introduced tree species in Europe, coupled with the abiotic and biotic disturbance events associated with anthropogenic climate change (Higgins and Richardson 1998; Richardson and Rejmanek 2004), may change this in the future.

(continued)

**Box 9.1** (continued)

**Hybridisation**: The use of introduced tree species can also result in the genetic dilution of native congenerics via hybridisation (McKay et al. 2005; Goto et al. 2011), with associated risks for wild tree species populations via processes including genetic assimilation or outbreeding depression (e.g. Levin et al. 1996; Vanden Broeck et al. 2004). Whereas the use of hybrid tree species in mixtures is likely to be limited, in those cases where such mixtures do occur [e.g. the use of hybrid larch (*Larix marschlinsii*) in combination with oak (*Quercus robur*) or European beech (*Fagus sylvatica*)], border trees may be of only limited effectiveness at reducing introgression, due to the far greater dispersal capacity of their pollen (Koivuranta et al. 2012). Nevertheless, the use of mixtures could still be argued to dilute pollen sources. Specific concerns can arise with respect to hybridisation risks involving rare or locally endangered tree species, such as introgression and gene swapping between European black poplar (*Populus nigra*) and the introduced *Populus* x *Canadensis* (*P. nigra* and *P. deltoides*). Several studies have already identified exotic genes from *P. deltoides* occurring within wild European black poplar seedlings in Belgium, Germany, and the Netherlands (e.g. Arens et al. 1998; Vanden Broeck et al. 2004; Ziegenhagen et al. 2008). In some cases, the risks of such events are projected to increase due to climatic change (Koivuranta et al. 2012).

**Pests and pathogens**: The pests and pathogens hosted by introduced tree species can threaten a region's biodiversity and ecosystems. The use of introduced tree species in Europe has been associated with pest and pathogen introductions, as in the co-introduction with Douglas-fir of Swiss needle cast fungi (*Phaeocryptopus gaeumannii*), Douglas-fir needle blight (*Rhabdocline pseudotsugae*), the Douglas-fir woolly aphid *Gilletteella cooleyi* and the seed cone wasp *Megastigmus spermotrophus* (Hansen et al. 2000; Goßner et al. 2005). Tree species that are introduced into forests without congeneric species may experience less severe outbreaks of native pests than tree species introduced into forests with congeneric species. However, if introduced tree species lacking congeneric species are accompanied by co-introduced pests, they may experience more damaging outbreaks than in their native environments (Roques et al. 2006). For example, the introduced fir seed chalid (*Megastigmus spermotrophus*) tends to cause more damage to *P. menziesii* in regions naturally lacking this tree genus than in its native range (Roques et al. 2006). In either regard, the juxtaposition of introduced tree species in close proximity to native tree species within the stand could plausibly reduce the risk of pest or pathogen outbreak by the mechanism described in Chap. 7. Still, the wrong tree species mixture can override such benefits and actually create better conditions for pest and pathogen outbreaks (Jactel and Brockerhoff 2007).

(continued)

**Box 9.1** (continued)

The complicating factor of having so many possible variables at play in mixed-species stands involving introduced tree species limits the widespread applicability of even well-designed studies. Furthermore, as several of the issues of concern will vary with climate, the insights we do have can possess a 'use-by-date'. One of the most critical questions regarding the admixing of introduced species to native forest communities is whether this process is reversible (see Felton et al. 2013). And this depends to a very large degree on the ecology and functional traits of the introduced species, for example, its seed dispersal, its ability to regenerate vegetatively through root suckers and coppice, its competitiveness in comparison to associated species, etc.

It is important to point out, however, that the question of biodiversity impacts is not a well-developed field of study relative to the size of the question and the results of any study thus far conducted will by necessity be context specific. Furthermore, whereas biodiversity impacts at the stand level are in many ways reversible, landscape-level impacts from pests, pathogens, hybridisation, and the escape of invasive trees may not be.

### 9.3.2 How Can We Predict the Compatibility of Species and Develop Silvicultural Prescriptions?

The most reliable way to predict species compatibility and to design silvicultural regimes for mixed-species forests is to use empirical data from the same site and tree species combination. However, there are many situations where this information is not available, for example, for afforestation or where new species combinations are established. So it is often necessary to rely on forest growth models or comparisons of species traits. Nevertheless, there are empirical data sources that may be available, and these differ in ways that can greatly influence the conclusions that can be derived from them (Forrester and Pretzsch 2015). Curves for the height development with age indicate at what stage of growth and how intensively tree crowns of different species are likely to interact when grown together (see also Fig. 9.5). This information can be obtained from yield tables for monospecific stands or site-specific inventory data. With additional information about the relative shade tolerance of the species participating in mixture, a basic assessment of likely crown competition and necessary thinning interventions can be made. Another simple source of information are forests where the species of interest occur together or with species that are very similar in their traits to those that are intended to be admixed. These may provide insights regarding the stand structures potentially required for maintaining these mixtures. However, they are less likely to provide information about how they can be established and about the silvicultural interventions that they will require for their maintenance. Planted experiments at single sites can show how the species interact in a more controlled environment in terms of climatic and edaphic conditions, stand density, and age.

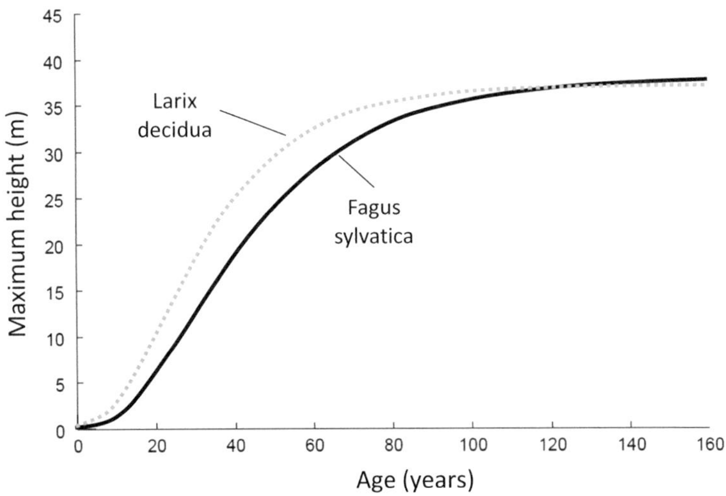


**Fig. 9.5** Development of maximum heights of European larch (*Larix decidua*) and European beech (*Fagus sylvatica*) in mixed stands in Lower Saxony (after Guericke 2001). The curves show that the very shade-intolerant larch initially has a superior height growth and that it is taller than beech up to the age of 120 years; after that point in time beech is slightly taller than larch. At no particular age is larch more than 7 m taller than beech. Considering that beech will grow into the crowns of larch, which leads to the shading and recession of the green crown from below, this height differential is not enough to produce vigorous larch trees with long crowns in a fine-grained single tree mixture of larch in a beech matrix, unless the beech component is frequently thinned to promote larch. Therefore, it is recommended (a) to focus this type of stands on sites, where larch has a comparative growth advantage, (b) to regenerate larch (in gaps) in advance of beech, (c) to admix larch to a beech matrix in the form of groups (20–40 m diameter) and patches (40–80 m diameter), and (d) to concentrate the thinning and pruning of larch on the early development phase in pole-sized trees to ensure the development of vital trees with large crowns and large diameters (Guericke 2001) (The maximum heights show a slightly higher curve but very similar pattern when compared to the top heights. Data were generated from 30 plots ranging in age from 34 to 158 years)

However, they again provide no information about how this species combination will perform on other sites, at different stand densities, and in future developmental phases. This underscores the value of replicated experiments of forest plots across site gradients, which can indicate the conditions where the mixture performs well and where it is not suitable. Forest inventories may also provide this information (e.g. Toïgo et al. 2015). As has been shown in Chap. 4 (Fig. 4.15f), the ranking of mixtures and monocultures can change as stands develop. Therefore, long-term plots are extremely valuable when considering species compatibility and the appropriate silvicultural regimes to maintain and manage mixtures. The value of such experiments depends of course on the breadth and detail of information that has been recorded and stored in the process. Unfortunately, the silvicultural input (e.g. in terms of man hours) that was required to achieve or maintain a certain structure or composition is often not recorded. This greatly restricts the development of recommendations for forestry practice.

Unfortunately, it is extremely rare to have long-term datasets for experiments that are also replicated across the range of site conditions of interest. Therefore, one of the most effective ways to predict species compatibility and to design suitable silvicultural regimes is to use forest growth models that are capable of predicting mixing effects and responses to silviculture. This can be aided by using the framework described in Chap. 3 (see Sect. 3.6.1 and Fig. 3.4).

Forest growth models require more information than simple alternative approaches, such as comparisons of species traits (e.g. shade tolerance), but there are many advantages. One is that they simultaneously consider the effects of many different traits and processes, which is important because the effect of one trait or process can counter the effect of another. If the model is process based (see Chap. 8), they can also be used to examine how the species compatibility and complementarity may change along spatial and temporal gradients. They can even be used to quantify how large any given trait difference might need to be to have a significant effect on growth or other simulated forest functions.

Forest growth models can also be used to examine how different silvicultural treatments will influence species compatibility and complementarity. Stochastic elements can be incorporated into the model in terms of the climatic conditions or disturbances and species-specific physiological parameters. The model simulations can then be used to identify critical development stages and intervention points during the rotation.

Often, forest growth models and empirical data from experiments or forest plots on tree height development, canopy layering, stand density, etc. are not available for the species of interest. In that case, species compatibility can be assessed by considering the species traits and whether or not the interactions expected to result from the differences in traits or growth dynamics will suit the site conditions. While comparing traits and growth dynamics is in principle a simple task to perform, it will only provide an indication of what might happen and how management complexity might increase with increasing compositional and structural diversity. However, thanks to the large research effort of the past decades, the accuracy of predictions is likely to improve now that our understanding of mixtures has expanded.

The approach described here is based on the framework described in Chap. 3 (Sect. 3.6.1 and Fig. 3.4). This framework illustrates how we can consider the climatic, edaphic, or biotic conditions of a site and then consider which types of species interactions might improve those conditions. Simultaneously, we need to consider whether any of the species in mixture is at risk of being suppressed by others. Most often, the goal is to avoid suppression of species in mixture. However, there are situations in which it is actually desirable that one species eventually overtops another one in the process of stand development. On sites highly susceptible to windthrow, where thinning of Sitka spruce was regarded as too risky because it destabilised stands, mixtures with 'sacrificial species' such as pine or larch were established, so that the spruce could gradually expand its crowns without thinning and gain dominance to form the final crop (Cameron and Watson 1999). Likewise, where trainer species are employed to ensure stem quality of more valuable species, these should commonly be overtopped by the main crop species (see Sect. 9.2.3 on trainer species).

A good indicator of the chance of suppression can be obtained by comparing the development of height over age in relation to shade tolerance (Fig. 9.5). In closed forests, species that are shade intolerant need to occupy the upper canopy to avoid suppression. If shade-intolerant species are overtopped or experience strong crown competition by more tolerant species for too long, then the mixture is unlikely to be successful, as has been shown in many mixtures containing N-fixing species (Forrester et al. 2006), or is often the case for mixtures such as those depicted in Fig. 9.5. When the height-over-age curves indicate a potential risk of suppression, it may still be possible to use that species composition by carefully considering the establishment or design of the mixture. For example, the establishment of the shade-tolerant or eventually taller species could be delayed until the shade-intolerant or slower-growing species has established itself, the species could be spatially separated by regenerating them in groups or larger spatial units (see Sect. 9.3.3), or spacing and thinning could be used to control competition (see Haines and DeBell 1980; FAO 1992; Kelty and Cameron 1995). This can be aided and done more effectively when considering that different types of species interactions may extend over different distances. For example, competition for light may come mainly from the nearest neighbours, whereas litterfall travels further and could influence nutrient cycling under trees beyond the nearest neighbours (Kelty and Cameron 1995; Ewers et al. 1996; Ferrari and Sugita 1996). Therefore, competition for light for the less shade-tolerant species could be reduced while maintaining facilitative effects of nutrient cycling by planting groups or multiple rows of a single species instead of mixing trees within a row or between single rows (Fig. 9.6). It is important to consider that these effects of competing neighbours on crown development and eventually tree vigour and quality are not only the result of competition for light but can also be driven by physical interaction through abrasion of buds and twigs (Hajek et al. 2015), particularly in taller trees that sway more strongly in wind. Hence, also, physical crown properties may be important traits to evaluate compatibility between tree species, but we have so far very little information about the magnitude of this effect for different species.

It should also be noted that height-over-age curves of a given species often change with site and species composition. Therefore, the curves used should apply to similar site and stand conditions, and if they are from monocultures, it should be expected that the actual height trajectories may differ slightly in even-aged mixtures. When height curves of each species are available from different sites, it will be possible to determine which sites are more and which ones are less suitable for the particular mixture (based on their height growth dynamics).

In uneven-aged forests, where there is no close relationship between tree age and height, this approach cannot be used.

The comparison of height-over-age curves and shade tolerance among tree species provides useful indicators of potential competition for light (Fig. 9.5). Comparison of nutrient dynamics can be a useful indicator of potential success of mixtures on nutrient-limited sites. Many examples of nutrient-related interactions were provided in Sect. 3.3. Two of the most obvious nutrient-related interactions can result when one species is capable of N fixation or when there are strong differences in the rates of nutrient cycling, which may be related to litter chemistry

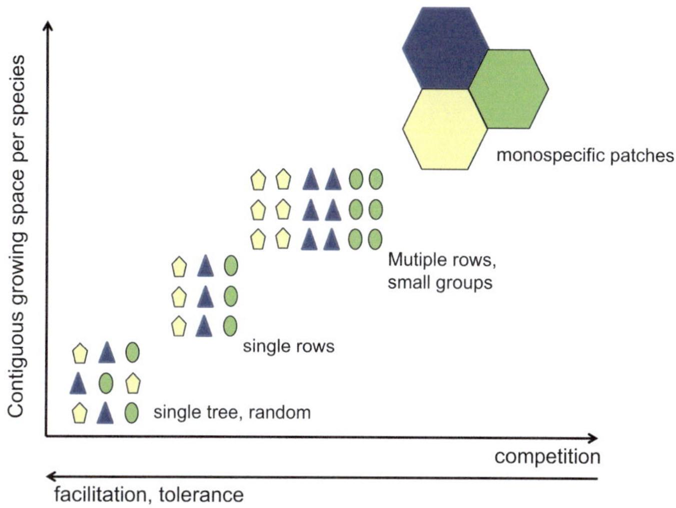


**Fig. 9.6** The higher the intensity of competition between different tree species or the lower the expected facilitation or tolerance among companion species, the greater is the need for initial spatial separation among species to avoid interspecific competition. Thus, individual trees of species that are (initially) weaker competitors have a greater probability to establish and survive, when they experience only or mostly intraspecific competition. At a later stage of stand development, initially monospecific patches of 4–7 m radius may be occupied by just one individual tree

and decomposability (Rothe and Binkley 2001). To estimate the effect of species interactions, it is also useful to know the actual rates of nutrient cycling processes. For example, N fixation rates of 5 kg per ha per year may not be sufficient to change growth in a companion species, whereas rates of 50 or 100 kg per ha per year may lead to very large facilitative effects (Forrester et al. 2006). Rates of nutrient cycling in one species that are, for example, double that of the other species may not be very useful if both species are not limited in their growth by that particular nutrient or if both are limited. Similarly, based on the framework described in Fig. 3.4, mixtures at sites where other resources such as water are limiting factors are unlikely to benefit from mixing species that improve nutrient availability.

Comparisons of water-related traits among tree species, such as rooting depth, can also be useful, but have so far not received much attention in mixtures. Traits that may be useful for comparison are listed in Sect. 3.5.2. As indicated above for nutrients, it is important to consider whether differences in any given trait are likely to result in large enough differences in water availability or within-stand climate in order to have a significant effect on interactions of those species.

When mixtures are being designed to withstand stress and disturbances, then traits relating to the specific disturbances will need to be considered. These were discussed in Chap. 7, which also indicated that in many cases it is not clear how different the traits actually need to be before they have a significant effect on the resistance or resilience of mixtures in relation to the particular stress or disturbance. That is, the actual differences in traits cannot easily be linked to the strength of the interaction, for example, if the rooting depth differs by 1 m, will this reduce the probability of windthrow of the shallow-rooted species by 1%, 10%, or 50%? This contrasts with differences in height or N fixation where there are many studies that indicate how much mixing effects could change depending on the differences in height or rates of N fixation. With regard to disturbances, the information base for selecting compatible species is probably best in relation to pests and diseases, where we have good knowledge about susceptibilities and resistances of different species. For example, phylogenetic distance between host and non-host species appears to be valid selection criterion for the development of mixtures with higher resistance to specialist herbivores (Castagneyrol et al. 2014).

The traits mentioned above are those that are often the most obvious to consider because of their direct, and often measurable, influence on the relevant process. It is important to realise that simply matching species with different functional traits is not necessarily the best strategy. On the one hand, designing biodiversity experiments has logically been done by combining species with contrasting traits, assuming that this will result in diversity effects (Scherer-Lorenzen et al. 2007; Bruelheide et al. 2014). However, this does not mean that doing this in practice will lead to complementarity; many of the treatments in biodiversity experiments do not show any complementarity despite containing species with contrasting traits. This illustrates that we do not know the most important functions and functional traits in every case. On the other hand, there are mixtures that show significant complementarity effects, although they consist of species that are very similar with only minor differences in traits that could have led to those mixing effects [e.g. mixtures of *Eucalyptus grandis* and *Eucalyptus pilularis*, which are shade-intolerant, deep-rooted, and fast-growing species with similar leaf and crown morphology (Forrester and Smith 2012)].

**Box 9.2 Maintaining Rare European Tree Species Through Historical Forest Management Regimes**

The large majority of tree species is typically uncommon or rare. These species are often called scattered, sporadic, secondary, or minor tree species, which make up only a very small proportion of the forest area or biomass, typically less than 1% (Hemery et al. 2010). In Europe, many rare tree species belong to the families of *Rosaceae*, *Aceraceae*, *Ulmaceae*, *Tiliaceae*, and some others. Through their diversity in traits, they contribute significantly to ecosystem functioning (Carrari et al. 2015). In particular the family of *Rosaceae* comprises many secondary trees species that are regarded as rare or sporadic (Carrari et al. 2015). Across Europe, the genus *Sorbus* is represented with several rare species ranging from thermophilous deciduous forests of southern, western, and Central Europe (*S. torminalis*, *S. domestica*, *S. aria*) to temperate and boreal forests (*S. aucuparia*) (San-Miguel-Ayanz et al. 2016). These species, many of which are considered drought tolerant, are likely to become increasingly important to maintain forests at the rear end of a climate change induced northward shift of forests and to increase the resilience of mixed forests on warmer and drier sites (Hemery et al. 2010; Kunz et al. 2016).

(continued)

**Box 9.2** (continued)

Minor species almost always occur in mixed-species stands, either as scattered individuals or in small patches. They have specific site requirements and habitats, and, in addition, their limited distribution may be caused by history of forest use that has selected against these species (Hemery et al. 2010), low reproductive capacity (which may also be caused by reduced genetic variation) or dispersal ability, and low competitive strength. For example, in Central Europe, the relatively higher occurrence of *Sorbus* species on dry and warm slopes is determined by the lower competitiveness of other species, mainly beech (Paganová 2008; Ewald et al. 1994). In addition, past silvicultural management has further influenced their abundance.

Silvicultural options to promote rare species must be based on solid knowledge of their physiology, ecology, and response to management interventions. Unfortunately, the knowledge base for most of these rare species is poor, as can be seen from the small number of scientific papers dedicated to them. Most of the available information is based on observations, but there are hardly any experiments. Here, we present results from some studies that have addressed the question on how two of these rare species may be maintained in mixed forests.

*Sorbus domestica* L. is one of the rarest tree species in Central Europe (Enescu et al. 2016), being classified as endangered in Austria, Germany, and Switzerland. Since a large proportion of the few recorded individuals occur in aged oak coppice forests, it has been widely assumed that its rarity results from transformation of coppice forests to high forests during the past 80 years. In order to test this hypothesis and to reconstruct stand development and growth-related competition processes between *S. domestica* and other tree species (mainly *Quercus petraea* (Matt.) Liebl.), a regional inventory was carried out in >40-year-old, former oak coppice forests in the federal state of Rhineland-Palatinate (southwest Germany, Pyttel et al. 2013a). As a result of this inventory, only 147 trees were found. Out of these, 46 trees were selected for detailed analysis of their competitive status. For that purpose, their age (determined from annual growth rings) and height were compared to that of *Quercus petraea* trees in the immediate neighbourhood.

Growth patterns of *S. domestica* and neighbouring oaks were found to be similar. Across all sites, early height growth in relation to tree age followed comparable trajectories for both tree species observed (Box Fig. 9.2-1). However, with increasing tree age, the interspecific competition by neighbouring oaks on *S. domestica* increased, as can be seen in their height growth patterns. Comparable patterns were found regarding other growth parameters like DBH, crown projection area, and height of the green crown base. The crown class of most observed service trees ranged between dominant and intermediate, whereas most oaks were found to be codominant.

(continued)

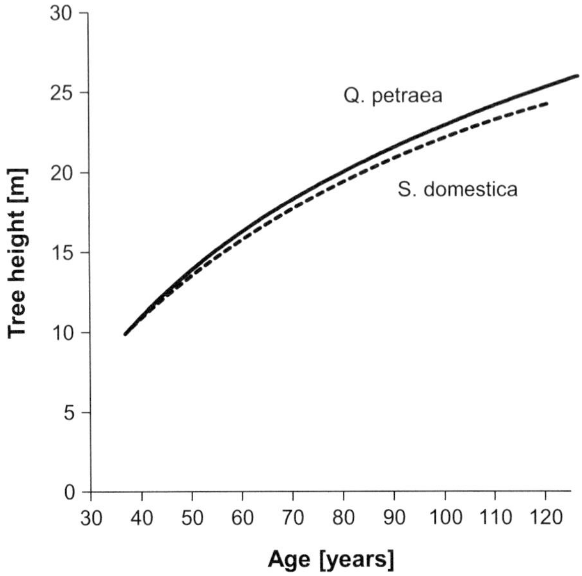


**Box Fig. 9.2-1** Height growth in relation to age of *S. domestica* (*dashed line*, $N = 72$) and *Q. petraea* (*solid line*, $N = 121$) growing in aged coppice forests in southwest Germany

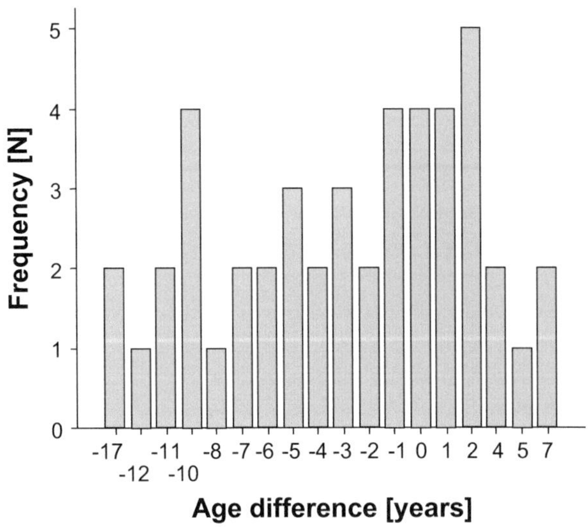


**Box Fig. 9.2-2** Age differences between *S. domestica* and neighbouring *Q. petraea* growing in aged oak coppice forests in southwest Germany ($N = 46$)

**Box 9.2** (continued)

Dendrochronological data support the hypothesis that coppicing promotes the establishment of new *S. domestica* cohorts. Almost all observed trees were established ± 5 years before or after the neighbouring oaks, which regenerated from coppice shoots (Box Fig. 9.2-2). Obviously, after the

(continued)

**Box 9.2** (continued)
available growing space was fully reoccupied by oak stump sprouts and seedlings of various tree species including *S. domestica*, no further service tree individuals were established.

These growth patterns suggest that *S. domestica* is a light-demanding species that regenerates in these forests only after disturbances and is not able to establish in closed oak forests. In oak-dominated forests of Central Europe, direct sunlight is necessary for the development of the service tree in both juvenile and adult phases (Brütsch and Rotach 1993). Since coppicing is no longer practiced, populations of *S. domestica* are commonly declining in forests that are managed according to a continuous cover approach. To preserve this species, the resumption of coppicing or comparable silvicultural measures should be considered.

Different considerations apply to the wild service tree (*Sorbus torminalis*). Like the former, *S. domestica* is found more often in (aged) oak coppice forests. However, in contrast to *S. domestica*, the species can tolerate the shade of the subcanopy formed by oak in all stages of development (Pyttel et al. 2013b). In the second story of aged oak coppice forests, the species typically shows little or no height and radial growth and broad, flat crowns. Under such light conditions, the species is still able to persist and to regenerate vegetatively through root suckers (Pyttel et al. 2013b). To which extent suckers contribute to the survival of the mother trees is not known. Hence, the species may not be directly dependent on frequent coppice cuts. However, flowering and fruiting and subsequent generative regeneration are dependent on access of crowns to direct light. Establishment of seedlings is facilitated in open forest spaces (Drapier 1993). Hence, coppicing and heavy conversion cuttings leaving *S. torminalis* as standards or seed trees are suitable options to maintain a genetically diverse population of *S. torminalis*.

These two examples illustrate that the ecological requirements of rare tree species may differ greatly, even within the same genus, and that silvicultural practices need to consider their different physiological and ecological requirements. Clearly, more research is needed to provide the necessary information base for the conservation of these rare species in managed and unmanaged forests.

### 9.3.3 Regeneration Methods for Mixed-Species Forests

Regeneration of mixtures differs in many ways to that of monospecific stands and therefore needs to consider different aspects and silvicultural approaches. The silvicultural aim of regenerating mixtures is to create and maintain compositional and structural heterogeneity of forest stands. This is ideally done with compatible tree species suitable to be managed in mixtures requiring low silvicultural tending efforts for maintaining the desired structure and composition. However, since there are very few combinations of tree species that are perfectly compatible, a range of silvicultural approaches to regeneration of mixtures have been developed to ease competition between particular species and thus reduce the risk of costly subsequent tending operations or of losing one or more desired species altogether. It should be kept in mind that the success of silvicultural efforts to regenerate mixed forests depends on many factors including the mode and timing of reproduction of the participating tree species; soil and site conditions and the seedbed types they provide, with or without disturbance; the species-specific response to canopy openings, which can range from small gaps to large clear cuts; the early interactions among tree species and ground vegetation; and last but not least the effects of pests, pathogens, and browsing. Hence, the outcome of silvicultural activities, which act in concert with long-term site conditions and short-term environmental fluctuations, is not entirely predictable. Since we cannot cover all possible species combinations here, the main principles will be illustrated further below with the help of some examples. Before, we will briefly cover the main aspects of forest regeneration.

Trees and forest stands may be regenerated artificially or naturally. Natural regeneration may be through seedlings that establish from natural seed fall or vegetatively through suckers or coppice shoots that emerge from superficial roots (e.g. in *Populus tremula* or *P. tremuloides*) or cut stumps, respectively. Natural regeneration from seeds is the most prominent of these forms, although in some regions, regeneration of forests through coppicing is still widely practiced. Natural regeneration is likely the preferred process when the species composition and the genetic make-up of mature trees conform to the production and other stand management goals. It is often the preferred choice, when natural forest communities are to be maintained (Larsen and Nielsen 2007) or when the expected yield and financial return do not justify high investments at the time of regeneration. The expenses incurred at the time of regeneration naturally take a long time to be recovered in commercial thinning or harvesting operations. Natural regeneration is typically also employed, when dense regeneration layers with more than 5000 or 10,000 seedlings $ha^{-1}$ are to be established and the costs of planting that many trees would be prohibitive. Such high seedling numbers are typically preferred for slow-growing hardwood species such as oaks or European beech, where dense stand conditions at the sapling and pole stage are required for the self-pruning process and stem quality development in terms of straight boles (e.g. Kuehne et al. 2013).

In contrast, artificial regeneration of forest stands through planting or sowing is applied in situations where (a) the current species or species combination does not match targets for the future stand composition; (b) the genetic make-up of the current stand is unsatisfactory, for example, through wrong selection of provenances in the past; (c) a tight control over the spatial arrangement of trees and species is desired (e.g. to avoid overstocking or to avoid early onset of interspecific competition; (d) seed production to facilitate natural regeneration is unreliable; and/or (e) establishment conditions are so adverse that they can only be overcome by tall seedlings (e.g. competition from understorey vegetation, ground frost, or browsing).

The spacing of planted seedlings depends typically on their growth rates, which determine the time during which they are exposed to harsh growing conditions near the ground and the time until canopy closure, their planting height, and on the quality of the planting material.

Sowing is a less common form of artificial regeneration and typically restricted to situations where (a) seed availability is not a problem; (b) the species to be regenerated develops a tap root, which is to be maintained (the development of tap roots is restricted or impossible in seedlings that have been lifted from nurseries); (c) the initial competition from understorey vegetation is not so strong as to impede the establishment of seedlings from seed; and/or (d) sowing is a much cheaper option than planting.

All of these approaches and combinations between them are available for the regeneration of mixed-species forests. As has been mentioned above, to create stable mixtures, it is often necessary to reduce competition between species. This may be achieved through some form of temporal or spatial (see Fig. 9.6) separation of the participating tree species or a combination of both.

#### 9.3.3.1 Provision of Seeds and Microsites to Facilitate Diverse Regeneration

The importance of the regeneration niche for the maintenance of species-rich plant communities has been discussed in the ecological literature for a long time (e.g. Grubb 1977). According to Grubb, all stages in the regeneration cycle are potentially important for the coexistence of species. These include (a) the production of viable seed, (b) their dispersal in space and time, (c) germination, (d) establishment, and (e) further development. In the following, we will address these different issues in relation to the establishment and maintenance of mixed-species forests.

Regeneration of species-rich stands requires the timely availability of seeds of all species to be regenerated. Where there are no mother trees of the desired species within the expected dispersal distance from the site to be regenerated, or where the genetic make-up of mature trees is not satisfactory, regeneration of these species must be through artificial means, e.g. sowing or planting. In other cases, it is typically desirable to regenerate the next generation from the locally adapted population of species and to include all desired species of the existing community in the future stand. This may be relatively easy, if all participating species can respond at the same time to harvesting and associated canopy opening through regeneration from seed (held in the canopy or in the seed bank) or from a seedling bank or through vegetative means. However, in many natural and man-made tree communities, this simultaneous availability of seeds, seedlings, or sprouts is not given. This may not be a problem, if the window of opportunity to establish regeneration is long enough for all species to regenerate. However, in many situations, this window of opportunity following harvesting may be very short before the seedbed is no longer receptive or before competing understorey vegetation has occupied the site to impede further establishment of seedlings (e.g. Neyland et al. 2009).

For example, seed availability might be highly irregular, in particular in masting tree species (Kelly 1994). In some cases, synchronised masting even across tree genera has been reported because different species respond to the same environmental triggers, e.g. for *Quercus robur*, *Carpinus betulus*, and *Acer platanoides* in eastern Poland (Pucek et al. 1993). However, in more species-rich temperate deciduous forest community with 16 principal tree species in Japan, reproduction was not or only partially synchronised across different species (Shibata et al. 2002). With only one masting species, the canopy openings may be tuned to the reproduction events of this species. In stands with several masting species, canopy openings may be tuned to the reproduction events of the most infrequently reproducing or the most valuable species. For example, in oak (*Quercus robur*)-dominated mixed stands with ash (*Fraxinus excelsior*) and hornbeam (*Carpinus betulus*), harvesting and regeneration are typically aligned with mast years of oak, which is also the least competitive of the three species. In order to regenerate several tree species with asynchronous reproduction events from natural seed fall, silvicultural approaches would have to aim at allocating patches of the stand to different species at different points in time for regeneration or employ also artificial regeneration to complement the species composition.

In contrast to masting tree species, pioneer species typically have always some seed available to colonise new sites, also at greater distance. Therefore, if canopy openings are large enough, have some receptive seedbed, and reproducing mother trees occur within the forest matrix, wind-dispersed pioneer tree species can be integrated into the regeneration layer of future stands (e.g. Huth and Wagner 2006).

Many tree species have specific requirements for germination and establishment. Germination of seeds typically requires adequate moisture, temperature, and for some species also light or other triggers such as heat (Baskin and Baskin 2014). In forests, typical substrate types available for germination of seeds include litter layer, moss cover, dead wood, exposed mineral soil, etc. For many species, exposed, loose mineral soil appears to be a favourable substrate for germination because (a) the contact with moist soil facilitates imbibition of seeds, (b) seeds may be slightly covered and hence protected from desiccation and predators, (c) the radicle can immediately get into contact with water and nutrients, and (d) they are less exposed to pathogens than within the forest floor. In moist environments, organic material such as decomposed forest floor can also be a favourable substrate (e.g. Oleskog and Sahlén 2000). Some species, in particular conifers such as *Picea* or *Abies*, regenerate well on decomposed logs or stumps (Szewczyk and Szwagrzyk 1996), which provide seedlings with a moist substrate and with an exposed position that reduces shading by ground vegetation and the length of snow cover. Mounds created by upturned root plates are another favourable microsite for regeneration in forests (Putz 1983). They provide exposed mineral soil as a germination substrate and at the same time also lift up seeds from the soil seed bank. In tree species that have evolved with fire as an ecosystem disturbance, burned mineral soil often provides the most favourable seedbed (e.g. Neyland et al. 2009; Charron and Greene 2002). It is impossible to describe here all the combinations of substrate or seedbed type and species requirements for successful germination. However, it appears that the regeneration of diverse forest also requires a diversity of substrate types to provide establishment opportunities and safe microsites that facilitate the regeneration of all species. Silvicultural approaches for regeneration of mixed-species forests might therefore (a) maintain existing microsites (downed dead wood, pit, and mound topography); (b) create microsites where necessary, e.g. higher stumps; and (c) provide variable seedbed conditions through partial soil scarification or patchwise burning of forest floor material. This heterogeneity in substrate may also provide for some spatial and temporal separation of tree species in the regeneration layer.

#### 9.3.3.2 Temporal Separation of Species in the Regeneration Process

The temporal separation typically aims at providing some species that are weaker competitors with a growth advantage over other species. For example, if juvenile growth is slower in a particular species and hence it runs a risk of being overgrown, the advance regeneration of this species represents a head start, reducing the intensity and frequency of subsequent silvicultural interventions for maintaining this species. If this slower-growing species is more shade tolerant than the companion species, it may be possible to provide it with this relative growth advantage by disadvantaging the other species through regulation of light availability. This is a principle that has been applied in a number of classical silvicultural systems (Matthews 1991).

This so-called advance regeneration consists most commonly of natural regeneration, but may also be artificial for the reasons outlined above. Advance regeneration is a common approach for the establishment of shade-tolerant species, in particular those that need the protection of the canopy, for example, against frost. This advance regeneration is also an insurance against disturbances in the late stages of stand development. For example, the very shade-tolerant *Abies alba* can be regenerated naturally in the deep shade of the initial stages of shelterwood or strip-shelterwood regeneration methods of stands, where it is mixed with *Picea abies* and other species (Fig. 9.7). At low light intensities, the other species either do not establish or grow so slowly that it is possible to develop a height advantage in *A. alba* before the stand is opened up more, and other species are regenerated or assume higher growth rates. Owing to this initial height advantage, it is subsequently easier to maintain *A. alba* in mixture with *P. abies*, which has higher height growth rates under higher levels of light availability that occur in late stages of the shelterwood with sparse canopy or after the canopy has been removed. Similar approaches may be used for advance regeneration of other shade-tolerant species such as *Fagus sylvatica*. For example, in mixed mountain forests of Romania,

**Fig. 9.7** Natural advance regeneration of Silver fir (*Abies alba*) in a Norway spruce (*Picea abies*)-dominated stand. The oldest saplings are ca. 2–3 m tall and ca. 20 years old. Natural regeneration of the other admixed species (*P. abies* and *Pinus sylvestris*) is insignificant at this stage. This height growth advantage of Silver fir is considered sufficient to initiate the natural regeneration of the companion species through gradual removal of the canopy. This temporal separation of species establishment through advance regeneration of the most shade-tolerant species is only possible where the mature forest stand is sufficiently resistant against disturbances such as windthrow to maintain the required canopy density (Photo: J. Bauhus)

Stancioiu and O'Hara (2006) compared the height growth of advance regeneration (0.5–5.0 m tall) of *A. alba*, *F. sylvatica*, and *P. abies* in relation to light availability above the regeneration, which was measured as a percent of the light above the canopy (PACL). In these forests, height growth of *F. sylvatica* and *A. alba* did not increase further beyond a relative light availability of ca. 25 and 65%, respectively, whereas height growth of *P. abies* continued to increase over the whole gradient of measured light availability up to full light. The regeneration of shade-tolerant *A. alba* and *F. sylvatica* clearly outcompeted that of *P. abies* in low light environments that were characterised by low light availability of less than 20–35% (PACL) and a basal area (BA) >30 $m^2\,ha^{-1}$. At intermediate stand densities (PACL = 35–70%; BA = 15–35 $m^2\,ha^{-1}$), the three tree species grew at comparable rates. At high light intensities (PACL > 80–90%; BA < 15–20 $m^2\,ha^{-1}$), *P. abies* tended to outgrow the more shade-tolerant species (Stancioiu and O'Hara 2006).

Similar silvicultural strategies can be employed to maintain mixed-species boreal forests composed of shade-tolerant conifers such as *Abies balsamea* and *Picea abies* and shade-intolerant species such as *Populus tremuloides*, *Betula papyrifera*, and *Pinus banksiana* (Messier et al. 1999; Harvey et al. 2002). Here, the shade-intolerant species typically dominate early successional forest following stand replacing disturbances such as fire or clear-felling. The shade-tolerant conifers establishing and growing in the understorey of these forests may be released through partial harvesting or overstorey removal to create mixed stands with a larger component of these species (Harvey et al. 2002).

In even-aged mixed regeneration of *F. sylvatica*, *Fraxinus excelsior*, and *Acer pseudoplatanus*, which may originate following shelterwood harvesting, the latter two species are initially similarly shade tolerant as beech in deep shade, but grow fast at higher light availability (Petriţan et al. 2009). However, beyond the pole stage of stand development, the growth rate of these species becomes slower than that of *F. sylvatica*, and it is therefore recommended to maintain a light availability of at least 20% of full light to provide these noble hardwood species with a sufficient height growth advantage (Petriţan et al. 2009).

To apply this silvicultural approach in practice, managers need to know how the growth relationships between competing species change with light availability or stand density and adjust the canopy light transmittance accordingly (Fig. 9.8). Where this information is not available, the close monitoring of annual shoot length and height increments of the different species can be used as an indicator. In addition, the ability of understorey trees to grow and survive at different levels of shade and competition from the overstorey is the result of complex interactions between the availability of light, nutrients, and water and the acclimation potential and developmental plasticity of plants, which changes with ontogenetic development (Givnish 1988). Hence, the competitive relationship between seedlings and saplings in the understorey of forests can change as they grow (e.g. Kneeshaw et al. 2006; Messier et al. 1999). To maintain a certain competitive balance between species, i.e. to maintain a shade-intolerant species in mixture, the canopy light

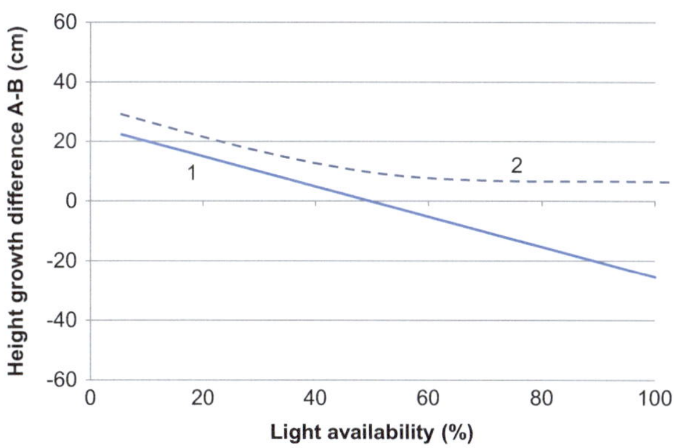


**Fig. 9.8** Annual height growth differences (cm) between two combinations of seedlings or saplings of a shade-tolerant (A) and a shade-intolerant (B) species in relation to light availability. For species combination 1, advance regeneration of the shade-tolerant species will be favoured at low light availability (<50%), whereas at high light availability, advance regeneration of the less shade-tolerant species will be advantaged. For species combination 2, the more shade-tolerant species exhibits higher rates of height growth than the less shade-tolerant species at all light levels. In this case, species B cannot be cultivated together with species A without continuous silvicultural intervention or through spatial separation

transmittance needs to be adjusted accordingly, for example, through continued target-diameter harvesting or progression of shelterwood removal.

The above examples show that careful management of light availability in silvicultural systems can be used to establish advance regeneration of shade-tolerant and comparatively slow-growing tree species to develop mixtures with faster-growing and less shade-tolerant species. Vice versa, canopy conditions may be adjusted to provide competitive advantages to shade-intolerant species, which might be eventually overgrown by shade-tolerant species with superior height growth rates later in stand development. However, there are also silviculturally very difficult species combinations, where a faster-growing species is more competitive over the whole range of light availability (pattern 2 in Fig. 9.8). In this situation, it is necessary to separate the species spatially (Fig. 9.6).

Another type of temporal separation of species may also be achieved through controlled disturbances. Where regeneration of fire-tolerant species is difficult to achieve owing to competition from fire-intolerant species, prescribed fires that kill the latter but not the former provide a competitive advantage to the fire-tolerant species. For example, regeneration of oaks in the eastern USA has been very difficult for a number of reasons, among them the competition of shade-tolerant understorey species such as *Acer rubrum* (Arthur et al. 2012). In contrast to oak seedlings and saplings, which have the ability to resprout from meristematic tissues of their large root system following fire-induced death of the shoot, many companion species have a much lower probability of survival (Brose and Van Lear 1998; Arthur et al. 2012). Fire thus decreases abundance of competitors and increases light availability to resprouting oaks, in particular if the fire follows reductions in canopy density through harvesting. However, oak seedlings must be of sufficient vigour and size to respond with rapid growth and take advantage of this window of opportunity created by disturbance (Brose 2010).

#### 9.3.3.3 Spatial Separation of Species in the Regeneration Process

Combining tree species with strongly different growth rates into long-term stable mixtures may be also achieved through initial spatial separation of monospecific groups or patches of species. This ensures that the majority of individuals in these monospecific groups or patches experience only intraspecific competition and not competition from interspecific neighbours (Fig. 9.6). This initial spatial separation of species is only of temporary nature, if the grain size of monospecific patches is small. For example, where monospecific patches occupy the space of the crown of mature trees of that species, an intrinsic mixture of species will be achieved in the second half of production cycles, when one crop tree has been liberated from its competitors in that patch. For the effective protection of weak competitors, single-species patches should occupy at least the space of the crown of a mature tree. In case of large growth and height differences, the patches should be accordingly larger.

The control over the spatial patterning of mixtures and the temporal dynamics of species interactions is greatest with artificial regeneration through planting. Combinations of different forms of regeneration for the establishment of mixed-species forests are also possible. For example, low-density planting of widely spaced clusters of seedlings in a matrix of natural regeneration has been employed successfully as a cheap alternative to establish mixed hardwood forests in large areas following windthrow or other disturbances (Saha et al. 2013, 2016). This ensures both the establishment of a minimum number of trees from the desired species and the maintenance of natural successional processes. As has been shown for young (10–20-year-olds) stands of *Quercus* species (*Q. petraea* and *Q. robur*) cultivated in the form of groups (20–25 trees planted at 1 m spacing), the quality of oaks (stem and crown form) was higher in the interior of groups, where they experienced mostly intraspecific competition than at the edge of groups, where they also experienced competition from fast-growing species that had established in the matrix between groups (Saha et al. 2014). The probability of oak trees to develop into future crop trees was highest in the interior of groups. Hence, some spatial separation of species may also aid quality development.

In some cases, a combination of spatial and temporal separation of species has been successfully employed to create mixtures, for example, when advance regeneration has been established in gaps (Fig. 9.9). This has been used to introduce tree species that were not present in the existing forest as is often the case when a more natural tree species composition is to be restored from secondary forests of species not native to the site. For example, in north-eastern Germany, oaks (*Q. petraea* and *Q. robur*) were established through planting in patch cuts of pine (*P. sylvestris*) ca. 10 years prior to regeneration of the pine matrix (Bilke 2004). Through the prior establishment, oaks in patches did not suffer from competition of the fast-growing

**Fig. 9.9** Mixed stands of pine (*Pinus sylvestris*) and oak (*Quercus petraea* and *Q. robur*). Groups of oaks were established in expansive monospecific pine stands in north-eastern Germany through advance regeneration in gaps of ca. 30 m diameter. In the areal image, oaks are visible through their yellow to red autumn foliage. Gaps were cut and planted with oak ca. 10 years before natural regeneration of the pine matrix. In older stands, oak has spread from the original groups into the adjacent pine stand (Photo: Gernod Bilke)

pine, not even at the edge of patches. However, quality of oak trees was lower at the edges, where they could temporarily develop larger crowns, when compared to oaks growing in the interior of patches. When these mixed pine-oak stands were fenced to protect natural oak regeneration from browsing, oaks could spread into the pine matrix (Bilke 2004) (Fig. 9.9). Hence, these patches of admixed species (Fig. 9.9) may be regarded as inoculum of conifer monocultures. Similar approaches have been used to reintroduce native *Abies alba* and *Fagus sylvatica* into monocultures of *Picea abies*.

In natural regeneration of mixtures, spatial separation of species may be achieved through attention to microsites and heterogeneity of light conditions that can be used to promote or disadvantage species in a desired pattern. Gaps create heterogeneous growing conditions in forests (Fig. 9.10). The environmental factors in gaps differ strongly from closed forests, and they vary considerably within larger gaps (e.g. Bazzaz 1996; Bauhus and Bartsch 1995). In gaps, more light and

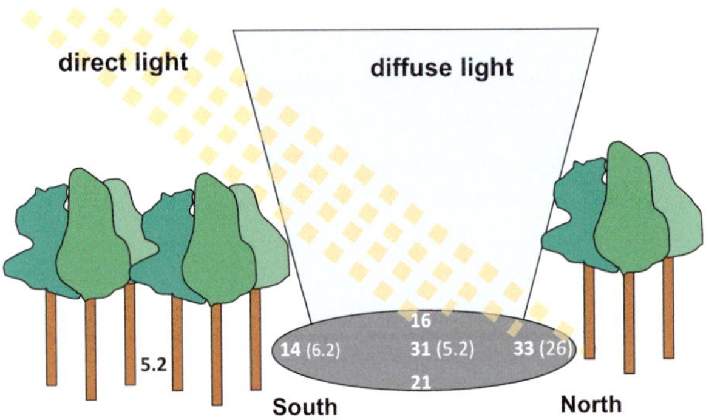


**Fig. 9.10** Heterogeneity of environmental conditions within canopy gaps. Diffuse light is highest in the centre of gaps. In larger gaps, e.g. where the gap diameter is greater than the height of surrounding trees, a considerable proportion of the gap (on the far side from the equator, here 'north' for the northern hemisphere) receives direct light, and total radiation over the course of the year is highest here. In the southern part of the gap, radiation levels are less than half of that in the northern section, but still higher than in the surrounding forest. Soil moisture is also highest in the centre of gaps. Within gaps, growing conditions for shade-intolerant tree species are often best in the central and northern gap section. However, growing conditions are also influenced by roots of surrounding trees and competing vegetation, which may benefit more than the seedlings from increased light availability. The values for relative light availability (of photosynthetically active radiation) in this figure are from gaps of ca. 25 m diameter that were cut into a 33 m tall stand dominated by *Fagus sylvatica* with a small proportion (11%) of six other hardwood species (Wagner 1999). The bold figures represent the light levels above the mixed-species regeneration. These were 33% in the north, 31% in the gap centre, 14% in the south, 21% in the east, and 16% in the west compared to 5.2% in the mature stand. The light received by beech seedlings is presented in brackets for selected parts of the gap. Owing to competing vegetation including the much taller seedlings of other, faster-growing tree species (mostly *Fraxinus excelsior*) in the gap centre, beech seedlings received not more light than in the closed stand. Beech seedlings were taller than those of ash at the edge of gaps, which was attributed to their higher tolerance to shade and also to root competition from surrounding trees (Wagner 1999)

commonly also more precipitation than in the surrounding stand are received by the regeneration layer. In addition, removal of mature trees increases soil water and nutrient availability with increasing distance from the edge of the forest into the gap (e.g. Bauhus and Bartsch 1995, 1996). Hence, gaps are characterised by gradients in resource availability. Owing to species differences in the ability of seedlings to exploit these resources and to tolerate competition and shade, distinct spatial patterns of species distributions can emerge. For example, growth and mortality of three *Acer* and four *Betula* species differed substantially among different locations within gaps leading to partitioning of this growing space (Bazzaz 1996). Competition among species was governed by a trade-off between their abilities to grow rapidly in high light environments and persistence in low light environment. It is thus logical that the ranking of height growth rates within seedling communities changes with gap size (Hibbs 1982). Hence, fast-growing, shade-intolerant species tend to dominate in large gaps and in the parts of gaps receiving most diffuse and direct radiation, whereas shade-tolerant species are more likely to occupy small gaps and the darker sections of large gaps. However, growth and mortality may be differently affected. For example, although the mortality of birches was highest in the northern part of gaps, this is also where survivors grew taller than at the southern edge (Bazzaz 1996). However, in other tree communities, the differentiation in height growth between species in relation to available light may be much less pronounced, and hence there may be no gap partitioning among species (e.g. Coates and Burton 1999; Coates 2002). This may be related to the physiological and morphological plasticity of species.

In addition, the competition between tree species will be influenced by the mode of regeneration. Species with the ability to sprout from roots and stumps have an advantage over those that establish from seeds (e.g. Dietze and Clark 2008). The same applies to species with the ability to maintain seedling banks in the understorey. These already established seedlings can respond quicker to canopy openings than species that need more light for germination and establishment.

Further, the concomitant development of ground or shrub vegetation may reduce light availability to desired tree seedlings within gaps to the extent that gap centres are not the sites most suitable for tree regeneration (e.g. Diaci et al. 2012; Kneeshaw and Prevost 2007). However, species patterning depends strongly on the silvicultural process of gap creation. Gaps may be cut directly through the simultaneous removal of a group of trees or they may be created gradually through reducing the canopy density of the gap area and also through extending the perimeter into the forest matrix. The latter is also known as 'Femelschlag' and may actually favour also regeneration of shade-tolerant species in the centre of gaps where these develop as advance growth before the canopy is fully removed.

Hence, gap-based silvicultural approaches provide a powerful tool to influence the spatial patterning of tree species through manipulating the size and speed of canopy cover removal (e.g. Kneeshaw and Prevost 2007). Regenerating mixed-species forests through creation of gaps of different size and at different rates should offer regeneration opportunities to all species, if light requirements of the most shade-intolerant species are met in the largest gaps and if harvesting coincides with the availability of seeds or advance regeneration. In addition, gap-based approaches to regeneration may follow the size distribution of the gaps in natural forests to emulate natural disturbance regimes and thus also maintain ecosystem diversity and function (Coates and Burton 1997). Spatial partitioning of tree species in the regeneration phase through environmental gradients is also achievable through other silvicultural systems that create heterogeneous conditions such as strip felling (Matthews 1991), where shade-tolerant species can regenerate as advance growth in the interior of stands, species of medium shade tolerance near the edge, and shade-intolerant species at the exterior edge, where the canopy has already been removed.

#### 9.3.3.4 The Effect of Browsing on Regeneration of Tree Species Mixtures

Browsing by ungulates influences seedling height and form, and it can affect tree survival rates, in particular of small seedlings. Browsing pressure on trees and shrubs may also increase penetration of light to the ground and thus promote more vegetation at ground level, which may in turn affect regeneration of tree species (Gill and Beardall 2001). It has been claimed that high population densities of ungulates represent the most important impediment to more tree species-diverse forests and hence also adaptation to climate change (Ammer et al. 2010). The problem of excessive browsing appears to be most prevalent in parts of Europe and North America. For these regions, there are several examples that show that deer browsing is selective and hence leads to changes in trees species composition and less diverse tree communities (e.g. Kittredge and Ashton 1995; Kuiters and Slim 2002). For example, in Europe and North America, oaks are more sensitive to browsing than their companion species such that continuous browsing leads to their decline in mixed forests (Gill and Beardall 2001; Kuiters and Slim 2002; Götmark et al. 2005; Rossell et al. 2007). With regard to moose browsing, tree mixtures in boreal forests do not seem to offer associational resistance. In a tree diversity experiment in Finland, moose browsing increased with the number of tree species and with the presence of the preferred tree species, *Betula pendula* (Vehviläinen and Koricheva 2006). Since strong browsing pressure reduces tree species richness and diversity (Gill and Beardall 2001), the development and maintenance of mixed-species forests is not possible or severly impeded without some forms of control of ungulate populations or protection of regeneration.

## 9.4 Tending and Thinning in Mixtures

### 9.4.1 How Does Thinning in Mixtures Differ from that in Monocultures?

The tending of forest stands comprises all silvicultural activities between the successful establishment of regeneration and the harvesting of trees. The most important tending practices include thinning and pruning and to some extent fertilisation. The latter is typically applied in intensively managed forests, such as fast-growing plantations, where the replenishment of nutrients being removed with biomass during tending and thinning is essential to maintain site productivity. In most other forests, it is of little practical relevance and therefore shall not be discussed further here.

Depending on management goals and the tree species present, thinning and pruning may also be achieved through natural processes. Self-thinning is the result of competition-induced mortality that occurs when the stand density approaches the site's carrying capacity for biomass (Westoby 1984). In this process, larger and dominant trees expand their growing space at the expense of smaller, suppressed, or intermediate-sized trees. In even-aged stands, natural mortality occurs at very high rates when stands and trees are in their most dynamic phase of growth since the concomitant crown expansion is driving this process. Self-thinning is highest following canopy closure in the subsequent pole size and early-mature growth stages (Oliver and Larson 1990). In mixtures, density-dependent mortality results from self-thinning through increasing density of conspecific individuals and from alien thinning through an increase in density of heterospecific individuals (see Chap. 5). Owing to the variable interactions among species depending on site conditions, the spatial patterning (grain) of mixture, and other factors, density-dependent mortality in mixtures is less predictable than in monocultures.

## 9.4.2 The Purpose of Thinning

For a number of reasons, forest managers rarely rely on self-thinning to regulate stand density, because stand management goals are accomplished more easily or to a higher degree through active thinning. Thinning is carried out for a number of reasons; here we focus on those that are most relevant in the context of mixed-species stands:

- To regulate tree species composition. During the early dynamic phase of tree growth, differences in the competitiveness of species, owing to their height growth rates, shade tolerance or other attributes, can lead quickly to the loss of inferior species from the stand. Thinning can increase the growing space and thus the probability to survive of individuals of less competitive species and thus help maintain tree diversity. Given that adaptation of forests to climate change through regeneration processes is such a slow process, forest management might want to ensure that no species are unnecessarily lost from existing mixed-species forest stands through inadequate control of competition through thinning. Hence, one goal of thinning in mixed-species stands should be the promotion of rare and possibly competitively inferior species that provide a potential for adaptation to future conditions.
- To concentrate site productivity on desired stems and species. Without thinning, trees that may eventually die, trees of poor shape, or trees of commercially undesirable species take up growing space and consume site resources such as

water and nutrients that can be directed towards vigorous, high-quality trees of desired species. Thus, thinning can dramatically improve the economic result of forest management (e.g. Cao et al. 2008).

- To reduce the time needed for reaching desired target dimensions of trees. Increasing the growing space of trees accelerates their crown expansion and consequently their stem diameter growth (e.g. Ginn et al. 1991).
- To reduce stress from competition and to increase tree health and vigour and thus reduce susceptibility to pests and pathogens. When trees have better access to site resources and can fix more carbon, they are less stressed and have more resources such as stored starch to respond to stress or disturbance. Consequently, well-thinned stands are less susceptible to secondary pests and pathogens that attack weakened or stressed individuals (e.g. Neumann and Minko 1981). Also, heavily thinned stands recover better from drought than unthinned or only moderately thinned stands (Sohn et al. 2016).
- To increase individual tree stability. The physical stability of trees depends among other things on the extent and depth of their root system, the length of their crown, and the thickness of the stem. These tree attributes are promoted by thinning (Cameron 2002).
- To increase or decrease the complementarity effects. The interaction between stand density and complementarity was described in Sect. 3.5.3. The balance between different interactions could be manipulated. For example, lower density and greater space between the trees could reduce competition for light and water while still allowing facilitative interactions that result from litterfall and nutrient cycling.
- To create or maintain spatial patterns among species. The vertical structure of the stand can be manipulated by thinning to ensure that shade-tolerant species are prevented from growing up into the canopy of a less tolerant species.
- To create variable stand density and canopy cover conditions to provide important habitat features across multiple scales and thus enhance habitat quality for certain species that require a variety of conditions within forest stands (Wilson and Puettmann 2007; Ares et al. 2010). Traditionally, thinning prescriptions were aimed at creating homogenous conditions for tree growth across the entire stands, either for all trees or at least for future crop trees. Where biodiversity is of concern, the aim may be to create heterogeneous conditions that comprise unthinned patches as well as gaps. This heterogeneity may be more easily achieved with mixed-species forests, especially where the mixing pattern is in the form of groups or patches. Owing to differences in growth dynamics and canopy stand densities, heterogeneous stand conditions are almost inevitable.

Typically, thinning regimes for forest stands have been described by the type of thinning, its frequency, intensity, and commencement. Thinning from above, or crown thinning, concentrates on the removal of the strongest competitors to future crop trees. In contrast, thinning from below starts at the other end of the diameter range with the removal of suppressed and intermediate trees (West 2006). In highly uneven-aged forests, where we find large mature trees intermingled with seedlings, saplings, pole-sized and early-mature trees, it becomes difficult to distinguish thinning from harvesting and initiation or promoting of regeneration.

The list above contains already several aspects on how thinning in mixed-species stands may differ from thinning in monocultures. Additional aspects will be covered further below. Before we discuss thinning prescriptions for mixed-species stands, we will briefly address conventional approaches for thinning prescriptions for monospecific stands.

### 9.4.3 Conventional Thinning Prescriptions

An important question is how these aspects that differ between monospecific and mixed stands might translate into thinning prescriptions. Typical thinning prescriptions for monocultures follow the temporal development of simple stand measures such as tree numbers or basal area to guide thinning interventions (see Fig. 9.11). For example, in even-aged conifer plantations, thinning interventions may be guided by empirically derived curves depicting an optimal number of trees at a given age or, preferably, stand height. If the latter is used, differences in site quality are already accounted for (Fig. 9.11). These curves describe a desired development of crop trees, in terms of their crown size, height-to-diameter ratio, tree quality (e.g. branch dimension) and diameter increment, which are all influenced by the growing space of individual trees (e.g. Mäkinen and Hein 2006). This desired development of crop trees aims at reaching a certain target diameter as soon as possible with vigorous trees that have a high physical stability, while maintaining a desired stand volume or biomass productivity. These prescriptions are typically based on evidence from many thinning trials with different thinning intensities carried out under a range of site conditions (e.g. Mäkinen and Isomäki 2004). In monospecific stands with more complex structures, for example, in European beech forests with a large and highly variable number of trees in the mid- und understorey, which serve the quality development of crop trees but have little influence on their growth rate, it would be impractical to guide thinning by stem numbers. In this case, thinning interventions can be guided by tree quality criteria for the onset of thinning (e.g. the minimum branch-free bole length as a result of a self-pruning process) and a combination of numbers of crop trees and basal area for a given age or height of crop trees (e.g. Altherr 1971; Klädtke 2002) (Fig. 9.11b).

For mixed-species forests, this type of empirical evidence from spacing and thinning trials does largely not exist. In addition, as we have seen in previous chapters (Chaps. 4 and 5), crown packing, allometries, tree interactions, and growth rates and patterns in mixtures can differ from monocultures. Hence, any attempt to derive prescriptions for mixtures based on the prescriptions for monospecific stands of the component species carries the risk to be inaccurate, if not flawed.

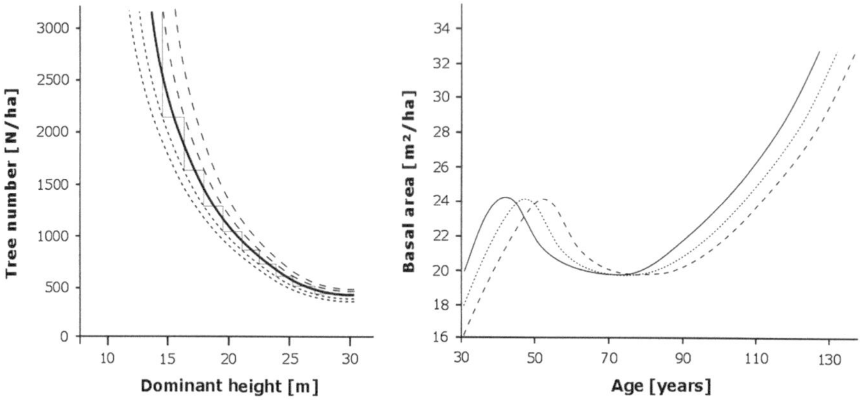


**Fig. 9.11** Classic stand tending concepts for crop tree thinning approaches in monospecific forests: (**a**) even-aged, single-layered conifer stand consisting of *Picea abies* (Abetz 1975) and (**b**) even-aged hardwood stand of *Fagus sylvatica* with midstorey and understorey trainer trees of the same species (Altherr 1971). In the case of even-aged *P. abies* stands (**a**) thinning, which removes a prescribed number of competitors of crop trees (crown thinning) per treatment, follows the height development of stands. Here the *dotted lines* represent 2 and 3 m height intervals at which thinning is carried out. Tree numbers are then reduced to those prescribed by the corresponding line below the *solid line*, which represents the average tree numbers. A typical pattern for the reduction of tree numbers without additional natural mortality is depicted by the *grey staircase-shaped line*. This type of thinning guideline is based on empirical information on the relationships between tree height and stand density on the one side, and physical tree stability and tree quality (stem taper, branch dimensions) on the other side. In the case of *F. sylvatica* (**b**), thinning follows the development of basal area over age; the *different curves* depict different yield classes, decreasing from left to right. Initially, the stands are kept close to facilitate self-pruning of trees. Once they have reached a desired branch-free bole length of 8–10 m, which corresponds to a stand basal area of ca. 24 $m^2$, crown thinning to liberate the crowns of future crop trees through removal of competitors commences and continues for ca. 40 years, after which the remaining number of final crop trees continues to grow until they have reached target dimensions. This type of thinning prescription is based on empirical knowledge of the relationships between stand density, tree height (age is here a proxy), and crop tree diameter growth and quality. In mixed stands, these curves for optimal tree numbers at a given height, or optimal basal area at a given age, will likely be different, in most cases higher than in monospecific forests

### 9.4.4 Considerations for Thinning Prescriptions for Mixed-Species Stands

It has been shown that interactions between tree species in mixture may depend on stand density (Forrester 2014). At very low stand densities, any complementary effects between species may be weak because the trees are spatially too far apart to interact effectively. At very high densities, competitive effects may outweigh complementary effects (e.g. Hunt et al. 1999). In between these extremes, changing stand density through thinning may increase or reduce complementary effects. The direction of change will likely depend on the type of resources that change in reaction to changes in stand density (Forrester 2014). For example, where water is the most limiting resource, thinning may increase complementary effects between species. In contrast, where complementarity effects depend strongly on differences

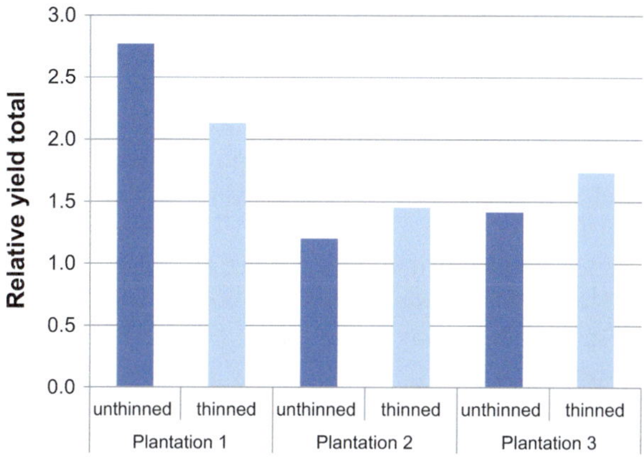


**Fig. 9.12** Relative yield totals of three different types of mixtures in native hardwood plantations in Costa Rica (data from Piotto et al. 2003). In plantation one and two, one of the four originally planted species had not succeeded. The relative yield totals were calculated on the volume indices provided for unthinned and thinned pure and mixed stands. The expected yield was calculated on the assumption that each species contributed equally to the plantation area (1/3 in stands with three species and 1/4 in stands with four species)

in shade tolerance and light use efficiency between species, decreasing stand density may reduce these effects. In addition, changes in stand density affect individual species in mixtures differently. For example, with increasing stand density in mixtures of *Pinus sylvestris* and *Fagus sylvatica*, complementary effects increased for the latter but decreased for the former (Condés et al. 2013). In mixtures of *Picea abies* and *Abies alba*, the effect of stand density on complementarity was also influenced by measures of site quality. Thus, even within a given type of mixture, different thinning approaches (intensity and/or method) may have to be developed to benefit from mixing effects or at least to have no adverse effect. Yet other studies have shown that different thinning histories had no effect on complementarity effects or the lack thereof for certain mixtures (Lu et al. 2016). Thinning in different types of mixed-species plantations of tropical hardwoods in Costa Rica had no consistent effect on complementarity (Piotto et al. 2003) (Fig. 9.12). Whereas the relative yield total was reduced by thinning in one species combination, it increased slightly in two other types of mixtures. Thus, changing stand density through thinning in these plantation systems may either intensify, reduce, or have no effect on the interactions between species.

Owing to differences in physiology, vitality, dimension, and relative canopy position in pre-thinning stands, different species respond differently to thinning, and considering these differences improves prediction of stand-level thinning responses for variables such as mortality or basal area increment (e.g. Kariuki 2008; Kuehne et al. 2016; Zhao et al. 2014). In many situations, it is difficult to disentangle the effect of species identity from other important variables. For example, thinning in a 12-year-old mixed-species hardwood stand of *Quercus rubra*, *Q. alba*, and *Juglans nigra* led to an increase in diameter growth in the two oak species, where *J. nigra* did not show a significant response (Gauthier and Jacobs 2010). The authors attributed this lack of a response in black walnut to the superior crown position that this species occupied in mixture. Here, thinning did not obviously improve access to light by crowns of this species. In general terms, there are surprisingly few studies that have documented the species-specific response to thinning of mixed-species stands when compared to the vast number of studies on thinning effects. There are very few planted experiments that have examined mixtures at different stand densities (e.g. Garber and Maguire 2004; Piotto et al. 2003) and few studies that analysed changes in resource use or leaf traits of individual tree species following thinning in mixed stands (e.g. Chiang and Brown 2010; Miesel 2012).

In addition, the thinning response of individual trees and species in mixture may differ from that in monoculture. It has been shown in a number of studies that the competition experienced by a subject tree depends on the species identity of its neighbours. For example, growth of individual *Picea abies* trees was reduced more by a conspecific neighbour than by a competing *Pinus sylvestris* tree in mixed stands of these two species (Pukkala et al. 1994). Likewise, the diameter growth of individual *P. sylvestris* was less affected by heterospecific neighbours of similar dimensions. Similar interactions have been reported for other species combinations for which the competitive influence of neighbours was attributed to individual species (e.g. Forrester et al. 2013; del Río et al. 2014). Therefore, the outcome of thinning at the level of individual trees as well as tree communities in mixed-species stands can depend strongly on the selection of trees to be removed or retained. While this information has been already considered in simulation models, the required inputs to drive such models (species, tree diameters, heights, and ages, and coordinates of all trees) make it difficult to translate into practical forest management (Pukkala et al. 1998).

Silvicultural guidelines such as stand density management diagrams (SDMD) often use maximum stand density as an upper boundary and reference line for regulating different density levels. Any deviation from the weighted mean line based on pure stands means that mixed stands can carry more or less trees of a given size per unit area. This is of course highly relevant for the assessment and regulation of stand density and yield and represents fundamental information for forest growth models. For example, for common two-species combinations of tree species used in Central European forests, the maximum stand density was on average 16.5% higher than the weighted mean density of neighbouring pure stands (Pretzsch and Biber 2016). This analysis was based on triplets of pure and mixed stands, which were unthinned or just slightly thinned and fully stocked and thus approximately at maximum stand density. Significant increases in maximum stand density when compared to monospecific plots were observed for *Picea abies* mixed with *Larix decidua* (29.1%) and for *Pinus sylvestris* in association with *Fagus sylvatica* (35.9%), which represent mixtures of shade-tolerant with shade-intolerant species. Hence, if density in these stands would be managed according to the proportion of the species and their stand density in monospecific stands, the overyielding observed in dense mixtures may not be achieved. Therefore, knowledge of any

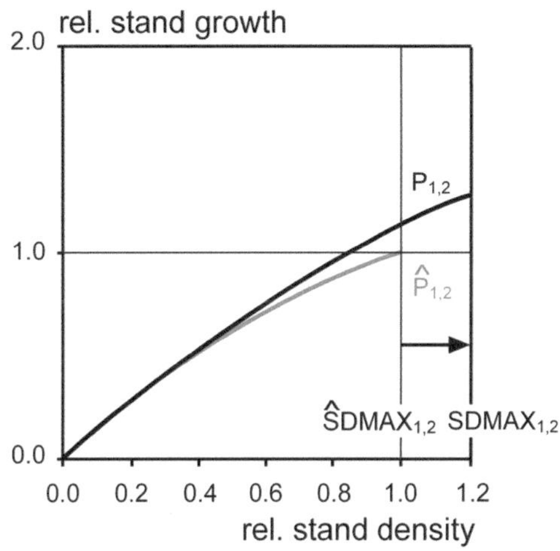


**Fig. 9.13** Effect of increased maximum stand density (SDMAX) in mixed-species stands on stand-level productivity. Regulation of mixed stand density from the potential-level $SDMAX_{1,2}$ to the level $\widehat{S}DMAX_{1,2}$, which represents the expected stand density (weighted mean) based on stand densities achieved in monospecific stands of the participating species, may reduce productivity losses from $p_{SDMAX_{1,2}}$ to $p_{\widehat{S}DMAX_{1,2}}$ due to suboptimal stocking

increase of the maximum stand density by species mixing is relevant for development of appropriate silvicultural guidelines. If thinning guidelines for mixed stands simply adopt the target curves for stand basal area or tree number developed for pure stands, this may result in suboptimal stand densities and thereby losses of stand productivity.

If the increase of the carrying capacity of mixed compared to monospecific stands is neglected, and mixed stand density is reduced to the level expected for monospecific stands, this can cause significant stand-level productivity losses as shown in Fig. 9.13. This can be avoided when silvicultural guidelines take into consideration the species' interaction in terms of density and productivity. This conceptual illustration does not imply that we recommend managing mixed or monospecific stands at the maximum stand density level. Their density might be regulated down to a certain percentage of the maximum density at which an optimum is achieved for stand and tree-level growth, stability, and quality of individual trees. This optimum has to be defined through experimentation and modelling. However, in mixed-species stands, the 100% level may be higher in terms of tree numbers or basal area, and thus also the absolute number of crop trees for moderate or heavy thinning may be higher than in monospecific stands.

Other differences between mixed and monospecific stands relate to the timing of thinning operations in relation to the development of trees and their quality. For example, in even-aged hardwood stands, the timing of the first commercial thinning would typically consider the quality development of trees, which may be quantified as the branch-free bole length following a process of self-pruning (e.g. Spiecker et al. 2009). In mixed stands comprising species of different growth rates and shade tolerance, it is unlikely that all species will reach target qualities and dimensions at the same time. For example, the diameter growth response to intensity of crop tree thinning in an 8-year-old sapling stands of *Acer saccharum*, *Betula alleghaniensis*, and *B. papyrifera* differed between tree species. Likewise, the stem height to the crown base decreased from the latter to the former species (Voorhis 1990). While faster-growing species have already reached the stage to be promoted by crown thinning, slow-growing ones may require more time. Hence, a single thinning operation in such a situation would represent a compromise regarding the dimension and quality development of different species in mixed stands. Alternatively, repeated thinning interventions tailored to the development of individual species would have to be carried out; these may also follow different thinning methods and intensities. Whether a more sophisticated, species-specific approach provides superior outcomes will depend on the additional gain in tree growth and quality on the one side and the additional costs of repeated thinning entries on the other side. One often experienced problem with crop tree thinning in mixed-species stands is that it does not follow the dynamic growth of individual species and trees sufficiently well, because thinning is delayed for too long. This often results in a large proportion of crop trees that have not been liberated from competition early enough (Ammann 2013). These trees may have good stem form and long branch-free bole lengths, but they lack the crown size and hence growth potential to reach the target diameter at a desired age or tree height (Ammann 2013). This leads to increased production risks associated with age-related problems, for example, formation of discoloured heartwood in some tree species (e.g. *Acer pseudoplatanus*; Kadunc 2007) or height-related problems such as windthrow (Albrecht et al. 2012) and eventually reduced economic performance of such stands. One option to avoid this problem is to combine thinning with pruning to facilitate early concentration of site resources on the most vigorous trees and to increase their diameter growth while maintaining a desirable stem quality. This way, crop tree thinning can commence before the desired branch-free bole length has been achieved through the process of self-pruning. The extensive experience with pruning of conifers and hardwood species (O'Hara 2007) provides a good knowledge base to facilitate this early liberation of crop trees to thus provide more flexibility with thinning of mixed-species stands.

### 9.4.5 Crop Tree Approaches to Thinning

The problems with the transfer of empirically derived guidelines from monospecific to mixed-species stands will increase with the degree of mixing and interactions between species, for example, from patchwise to a fine-grained tree-wise mixture. The difficulty will further increase with increasing tree species diversity and variability in tree dimensions. One possible alternative for tending of mixed-species forests may be seen in crop tree approaches, which are strongly guided by the development of individual trees.

A crop tree approach to manage mixed-species stands focusses on the development of selected, individual trees, and their immediate neighbourhoods, which influence this development through inter- and intraspecific interactions (Fig. 9.14). For the purpose of tending in fully stocked stands, these neighbourhoods can be defined as the growing space (crown projection area) of one crop tree when it has reached target dimensions. Different tree species have different target tree dimensions and different ratios of DBH to crown diameter. Within species, target tree dimensions may also differ between trees of different quality, because trees of lower quality may be harvested earlier as their current value increment is comparatively low. In mixtures, the species-specific ratio of DBH to crown diameter may be different from monospecific stands, and it may vary also with the species identity of neighbours. Hence, the spacing of crop trees in mixed stands needs to be variable to account for this. The development of stands is then the result of the development of these neighbourhoods and also patches that may remain untreated, because they do not contain crop trees. Crop tree thinning approaches in mixtures must then be guided by the following information:

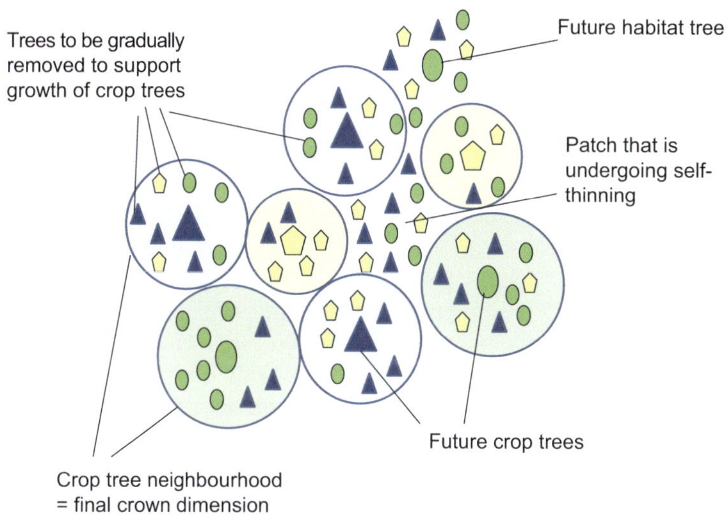


**Fig. 9.14** Spatial arrangement of crop trees and their neighbourhoods. *Large symbols* in the *centre of circled areas* represent crop trees. *Circled areas* represent the final crown dimension of crop trees of that particular species. The size of these neighbourhoods depends on the target size of crop trees and their crown-to-stem diameter ratio. Through thinning, all other competing trees within these crop tree neighbourhoods will be gradually removed. Small trees within these neighbourhoods that, owing to their height and dimension, do not compete with crop trees may be left; they may also serve a particular purpose such as shading the stems of crop trees. Patches without a crop tree may remain unthinned. The conceptual model depicted here assumes that the minimum distance between crop trees is based on their final crown size. This is typically the case where light is the most limiting resource. If water is most limiting, the circle defining the growing space of crop trees may be considerably larger than the crown projection area to provide crop trees with sufficient access to exploitable soil volume

- Desired species proportions (based on planting design or outcome of natural regeneration). The numbers of selected crop trees per species should then reflect these proportions.
- Preferred spatial and temporal mixing pattern based on knowledge of interactions among the species in mixture (single tree to patches, which may change from coarse to fine grain during stand development) (see Fig. 9.6).
- The relative growth rates and value increment of neighbouring trees of different species to facilitate selection of the most appropriate crop tree. For example, Zhao et al. (2014) developed an interesting approach to guide thinning in mixtures of *Tsuga heterophylla* and *Pseudotsuga menziesii*, where the latter is affected to different degrees by the Swiss needle cast fungus (*Phaeocryptopus gaeumannii*). Based on models of diameter growth taking into account a variety of stand and tree-level variables including an expression of disease severity (retention of foliage), they developed a simple model to predict the ratio of basal area increment between two adjacent individuals of the two species in relation to the initial DBH difference between these individuals and the disease severity. The resulting charts present a tool for practitioners in the field for the selection of the more promising crop tree. This approach is based on an excellent database, however, and relates to two-species mixtures. With more species, it would become quickly impractical unless species can be aggregated to functional types.
- Species-specific target diameters of future crop trees determine the minimum distance between crop trees, through the relationship between stem and crown dimensions (Fig. 9.14). For example, the distance between a pair of crop trees of *Picea abies* (with a typical DBH to crown diameter ratio of 1:15 and a target diameter of 45 cm) and *Fagus sylvatica* (with a typical DBH to crown diameter ratio of 1:20 and a target diameter of 65 cm) would be 9.88 or ca. 10 m (45 cm × 15 = 6.75 m, 65 cm × 20 = 13 m, (6.75 + 13)/2 = 9.88 m). Our information on tree allometry and hence the DBH-to-crown diameter ratio, which is a very useful practical measure, is largely derived from monospecific stands. However, this ratio is influenced by the species identity of immediate tree neighbours and by site quality (e.g. Dieler and Pretzsch 2013, Chap. 6). At this stage, we still lack information on how allometry or more specifically the DBH-to-crown diameter ratio is influenced for individual trees species in the most important mixtures and their developmental stages and associated stand structures. It is likely impractical to develop specific allometries for all these possible situations. A promising solution to this problem may be in general allometric functions that consider, in addition to DBH, the species composition and competition in the tree neighbourhood and the relative height of the target tree in relation to its neighbours (Forrester et al. 2016). While such an approach permits the development of general equations for a given species to be applied to a wide range of species compositions and stand structures, it does still require a translation into practical thinning guidelines. For example, foresters need to know how a species-specific DBH-to-crown-diameter ratio of crop trees of a

given tree species is modified depending on the relative position of the target tree and the type of tree neighbours. Since neighbourhood density and composition are actually controlled through thinning, species interactions are likely to have less effect on tree allometry of crop trees than on trees from unthinned mixed patches. Hence, the development of crop trees in regularly thinned stands may differ less between mixed and monospecific stands than the development of trees of similar dimension in unthinned mixed stands. However, this must be regarded as a hypothesis that needs to be tested. In the absence of better information, indicators such as height-to-diameter ratios or relative crown length, which are related to the growth rate, physical stability and quality of trees derived from monospecific stands, may also be used to guide thinning in mixed stands (Fig. 9.15).

- The onset of thinning in neighbourhoods of target trees may differ considerably depending on tree species and production goals. In dense patches of conifers, one

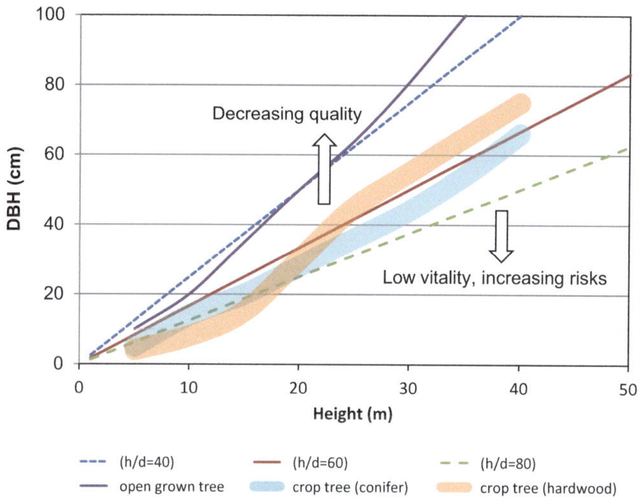


**Fig. 9.15** Range of height-to-diameter (*h*/*d*) ratios and possible growth trajectories for crop trees (broad bands) [adapted from Abetz and Klädtke (2002)]. The trajectories of the height-to-diameter ratios over tree height represent a growing space for crop trees that indicate the period of time to reach target diameters and the associated stem qualities. At higher *h*/*d* ratios (>80), trees would take too long to reach the target diameter. In this case, they may be already very tall and thus susceptible to windthrow and other height-related risks when reaching the target diameter (e.g. Albrecht et al. 2012). In addition they may be relatively old and thus prone to discoloration of heartwood and susceptible to fungal diseases, in particular in short-lived hardwood species. At very low *h*/*d* ratios, in particular in the juvenile phase, trees will develop very large crowns and thus thick branches, strong stem taper, and therefore stems of low timber quality (e.g. Mäkinen and Hein 2006). For hardwoods that undergo a phase of self-pruning facilitated through high stand density in the juvenile phase, growing space is only substantially increased once the lower part of the stem has reached quality targets (e.g. a branch-free bole length of 6–10 m) (Röhrig et al. 2006). Subsequent enlargement of crowns does not decrease but increase stem quality through the higher log dimensions that can be achieved

aim is often to thin very early to facilitate development of individual tree stability through large, symmetric root systems (Coutts et al. 1999). In dense natural regeneration, pre-commercial thinning may be carried out to space trees at a distance that facilitates development of individual tree stability until the first commercial thinning. The time of first commercial thinning is then often determined by average tree dimension, at which the thinning can be conducted without incurring a deficit through the silvicultural operation. In many hardwood species, growing space of individual trees is initially restricted to facilitate self-pruning and development of straight stems. Thinning to promote crop trees is then commenced, when future crop trees have reached a certain quality (e.g. branch free bole length of 6–10 m) (e.g. Spiecker et al. 2009). If pruning can be employed to grow trees of similar quality, then the time of first thinning can be brought forward, which has the added advantage that the phase of rapid crown expansion and stem growth can be used to develop target diameters in crop trees in shorter time.

- The number of trees used for retention purposes and timing of their marking/exclusion from harvesting. The conservation of biodiversity through some form of retention of structural elements such as dead wood and habitat trees has become an important aspect of management approaches for multifunctional forests (Gustafsson et al. 2012). Commonly, habitat trees are only marked for retention when stands enter the harvesting phase, when there may be only or mostly crop trees left. That means that habitat trees may have been treated and received investments into their development like a regular crop tree. It also means that trees that could have developed into very valuable habitat trees, e.g. very rich in microhabitat structures or with rare types of microhabitats, might have already been removed through thinning (Vuidot et al. 2011). Expectedly, the continuous removal of such trees during the stand tending phase has a negative influence on forest biodiversity (e.g. Müller et al. 2007). From an economic perspective as well as a biodiversity conservation perspective, it may therefore make sense to leave a certain number of trees with a good potential to develop rare or many microhabitat features in unthinned patches already early in the tending phase.

Crop tree approaches to thinning of even-aged mixed-species stands may be regarded as an interim solution until we have developed more sophisticated approaches that consider the specific interactions between different tree species and how these are influenced by tree size, age, stand density, and site quality. However, since there are so far very few experiments to provide a quantitative basis for new thinning approaches for mixtures, it may be necessary to employ this interim solution for a considerable time into the future. Crop tree approaches have the advantage of great silvicultural flexibility. They can be easily adjusted to accommodate different targets for mixing proportions and species-specific tree dimensions, growth rates, quality, and stability. The approach is further suited to concentrate management efforts on the most valuable trees, in terms of economic importance as well as other functions (e.g. as habitat trees). An important drawback is that trade-offs between achievements of goals for the development of individual trees (e.g. optimising diameter growth) and stand-level goals (e. g. optimising biomass productivity) have hardly been quantified for mixed-species stands. In addition, this silvicultural flexibility of crop tree approaches to adjust to the uncertain development of mixtures can best be realised in intensively managed forests with frequently recurring tending operations. In the most dynamic phase of tree growth, when crowns expand rapidly in height and width (in some cases several meters in height), it may be necessary to return every 2–5 years to enlarge the growing space of crop trees and to maintain desired mixing proportions. In most situations, such a high input can be afforded only if the system produces high value products.

### 9.4.6 Thinning in Uneven-Aged Stands

The above points apply to thinning approaches for even-aged stands. In uneven-aged forests, the situation is somewhat more difficult. Here, thinning may not only be used to control the overall stocking and species composition or to foster growth of individual trees but also to reduce tree numbers in certain age classes or strata to ensure desired regeneration processes and long-term sustained yield (O'Hara 2014). Thinning in uneven-aged forests is an important part of stocking control and the (re-)allocation of growing space to cohorts and individual trees. In these systems, harvesting of trees, typically according to target diameters, and possible regeneration treatments complete the measures of stocking control. In most situations, the management approach aims at maintaining or perpetuating a relatively constant stand structure in terms of tree species and size distributions over time because it is this structural complexity that provides a large variety of ecosystem functions and services (O'Hara 2014; Schütz 2001). Recent studies have demonstrated that the diversity of tree sizes has a similar function to the diversity of tree species. For example, stand structure, through the diversity of tree dimensions, had a positive influence on stand-level productivity in temperate conifer-dominated mountain forests as well as in boreal forests (Dănescu et al. 2016; Zhang and Chen 2015); see also Chap. 12. To maintain this stand structural diversity, it is important to remove the appropriate number of trees from different diameter classes. If too few trees are removed in recurrent harvesting and thinning operations, there is the risk that regeneration and growth of small trees, in particular of less shade-tolerant species, will be impeded in overstocked stands and that the stand might develop towards a structure with less vertical heterogeneity. If too many trees are removed, the growth potential of the site may not be fully utilised, and hence yield may decline. The maintenance of a relatively constant stand structure and harvested wood volumes across cutting cycles as well as the presence of sufficient regeneration are therefore important indicators of the sustainable management of these forests (O'Hara 2014).

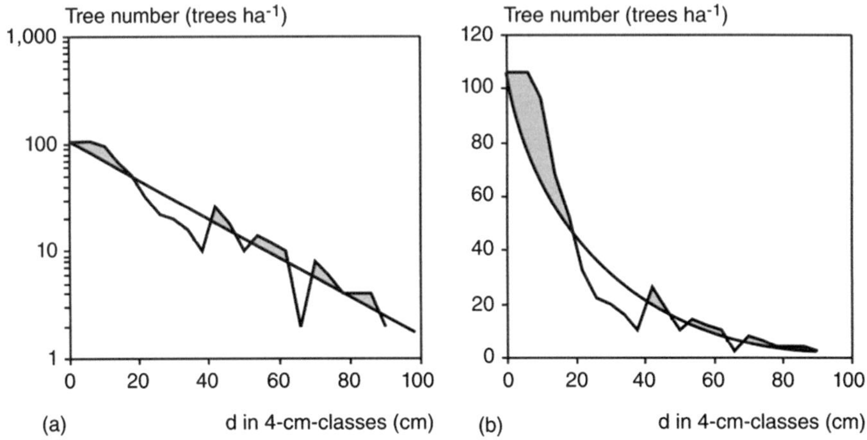


**Fig. 9.16** Silvicultural prescriptions using target diameter distributions for a Plenterforest with *Abies alba*, *Picea abies*, and *Fagus sylvatica*. The observed diameter distribution (*black solid jagged line*) for a selection forest (Freyung 129, plot 1) and corresponding target equilibrium curves are displayed (**a**) on a semilogarithmic scale and (**b**) *linearly grey areas* between the observed tree numbers and the target curve indicate in which diameter classes thinning should be carried out. The *white areas* between the observed tree numbers and the target curve indicate understocked diameter classes

There is a variety of different approaches to stocking control in uneven-aged forests, which are described and discussed in detail by O'Hara and Gersonde (2004). To illustrate the difficulties with these approaches in relation to the maintenance and optimisation of tree species composition, we will focus in the following on the use of diameter distributions to guide thinning in these forests. Although many different diameter distributions can be found in unmanaged and managed uneven-aged forests (e.g. Westphal et al. 2006), a 'reverse J' shaped or a negative exponential curve between tree diameter and frequency have become model distributions for many uneven-aged forests (Fig. 9.16) (Pretzsch 1985).

By sorting the trees in a stand into diameter classes with a width of 5 or 10 cm, for example, a stem number-diameter distribution is obtained in which the mean diameter, variation, skewness, and excess of the frequency distribution reflect the underlying stand structure. Assmann (1961) described the diameter classes as numerical tree classes because the trees within a class, like the tree classes defined by Kraft, are comparable in size and social status.

Therefore, thinning interventions can be tailored to maintain the diameter distribution. By comparing the actual diameter distribution with the desired, programmed distribution, the treatment is regulated. When the number of trees in particular diameter classes exceeds the desired tree number, which is predicted by the modelled diameter distribution, it will be reduced in the next thinning to match the modelled distribution. Figure 9.16 provides an example of this method for the silvicultural treatment of the selection forest experiment, Freyung 129. An equilibrium curve is obtained from the function

$$n = k \cdot e^{-a \cdot d},$$

which describes a declining exponential distribution that can be represented as a line in the semilogarithmic coordinate system (Fig. 9.16a)

$$\ln(n) = \ln(k) - a \cdot d,$$

where $n$ represents tree number per diameter class, $d$ diameter class, $k$ intercept, $a$ the slope of the line, and ln the natural logarithm. The presentation of the diameter distributions in a semilogarithmic coordinate system (Fig. 9.16a) shows clearly whether the stem numbers are above or below the reference curve. The curve parameters $k$ and $a$ can be determined from the target diameter $d_t$ and the corresponding stem number $n_t$ in this diameter class, and the mean diameter of the first diameter class $d_i$ and the corresponding stem number $n_i$ respectively. The curve parameter '$a$' reflects the process of recruitment:

$$a = \frac{\ln(n_t) - \ln(n_i)}{d_t - d_i},$$

where $k$ is determined by

$$k = n_i \cdot e^{a \cdot d_i}.$$

The aim of maintaining this diameter distribution is to achieve a steady state where the ingrowth is equal to the tree volume removed through thinning, harvesting, and natural mortality (O'Hara 2014).

Schütz (2001) suggests a simple algorithm for determining equilibrium curves for selection forests to obtain the desired distributions. These approximate exponential distributions as a rule, although the empirically derived curves often deviate somewhat from an exponential function because periodic diameter increment varies with tree size: it is extremely low, often only 1–2 mm per year in small diameter classes in typical Plenterforests and increases in larger diameter classes, in particular once trees are no longer overtopped. This results in so-called 'rotated sigmoid' curves (Schütz 2001). No matter how the reference curve is developed, its application is rather simple. In our example (Fig. 9.16), intensified thinning in the medium-sized diameter classes (40–60 cm DBH) and harvesting of large trees (>80 cm DBH) will accelerate growth of small trees (<20 cm DBH) to fill the gap in the 20–40 cm diameter classes. This system obviously depends on comprehensive inventories to control the diameter distribution and growing stock (Schütz 2001).

While the application of these equilibrium curves to guide thinning is relatively simple and has a long tradition in some parts of the world, for example, the Plenterforests in Central Europe, the tool may not be easily transferred to other

**Fig. 9.17** A Plenterforest dominated by *Picea abies* and *Abies alba*. Harvesting of a large Silver fir tree in the foreground has opened up growing space for advance regeneration and trees from smaller diameter classes. Within these diameter classes, trees will be eventually thinned to ensure adequate growth rates and progression through diameter classes (Photo: J. Bauhus)

forest types and more species-diverse stand conditions. As is indicated by Fig. 9.16, thinning follows strictly the stocking levels in diameter classes regardless of tree species distribution. Neither different regeneration requirements nor different growth rates of the tree species or interactions between the species are taken explicitly into account. In typical Plenterforests, the most important tree species *Abies alba*, *Picea abies*, and *Fagus sylvatica* are functionally similar regarding their shade tolerance and can establish and grow under similar light conditions (Fig. 9.17), although tree quality of *F. sylvatica* suffers more under heavy shade than that of the conifers. If forests are composed of tree species with a range in shade tolerances including shade-intolerant ones, a strict target-diameter harvesting of single trees might cause a loss of light-demanding species in the regeneration phase. Hence, for the purpose of also maintaining shade-intolerant species in the next generation, one might have to deviate from a strict diameter harvesting approach to also harvest patches of mature trees, opening up large enough gaps for regeneration. In addition, substantial differences among species in growth and mortality rates across tree dimensions may require species-specific stocking levels for diameter classes. For example, in light-demanding species such as eucalypts, old and slow-growing trees with very limited potential to respond to release can accumulate in smaller diameter classes (Bauhus et al. 2002). This can only be picked up through repeated inventories. In addition, in mixed stands with inherently different growth rates, assortment dimensions, and assortment prices of different tree species, relative value increment or another tree selection rule is likely more promising than using only diameter to select the trees for removal (Pukkala et al. 2015). Also in uneven-aged forests, crop tree approaches to thinning may be applied. However, the variables that guide this approach to maintain tree vitality (e.g. relative crown length, height-to-diameter ratio) must not only be species specific but also specific to each cohort or tree stratum in these forests. Other variables such as the distance between crop trees remain the same within cohorts or strata. However, between strata, these differences can be very irregular, i.e. a small future crop tree may be promoted at close distance to a mature crop tree that is close to reaching its target diameter (see Fig. 9.17)

## 9.5 Tree Quality in Mixtures

The commercial value of forest production, which is in most cases in the main interest of forest owners, has so far received little attention in previous biodiversity research in forests, where the focus has been on biomass production (Knoke 2008). The commercial value of harvested logs for solid wood products and veneer is largely determined by the quality of timber. Consequently, most forest owners with economic management objectives will manage forests with higher tree species diversity only if this improves or at least does not reduce the production of high-quality wood.

The quality of stem wood is determined by a wide range of internal and external properties. In the case of timber, the quality is largely determined by its intended end use (e.g. van Leeuwen et al. 2011). In order of their market value, solid logs may be used for (1) veneer production, (2) furniture and other visual uses and (3) construction use. Timber for other end uses such as pulp and paper, fibreboards, and fuel make up a large portion of the total wood consumption but yield much lower revenues per unit of volume. While knowledge of the intrinsic wood properties (e.g. density, colour, cell perimeter and coarseness, microfibril angle, fibre length) allows a more sophisticated assessment of the value of the final product, it is costly to obtain and onerous to study, because sampling is necessarily destructive. Therefore, stem quality is commonly assessed on the basis of external quality assessment proxies such as the presence of branches, knots, branch wounds and stem deformations, curvature, taper, epicormic branching, defects caused by mechanical damage, irregular growth, or pests and pathogens (e.g. Thies et al. 2004; Drobyshev et al. 2007; Wimmer and Johansson 2014; Dănescu et al. 2015). Many of these properties may be influenced in mixed when compared to monospecific stands.

Whereas the development of timber quality in even-aged monocultures has been intensively studied for many commercially important species (e.g Erickson and Harrison 1974; Zobel 1984; von Lüpke 1998), there is very little information for mixed-species stands. The latter tend to have more heterogeneous growing conditions owing to differing growth dynamics, morphology, and habitus of the participating species (Pretzsch and Rais 2016). The higher structural heterogeneity in mixtures may increase the variability in stem and crown properties such as taper, straightness, or form regularity of all or some species in the mixture (Pretzsch 2006; Jucker et al. 2015). For example, fast-growing species in a matrix of slower-growing species may quickly develop large crowns, which they keep above the main canopy layer. Trees of this species may then develop larger branches on the lower stem than they would in monospecific stands with the same original stand density. Similarly, species with high crown plasticity can more quickly grow into canopy gaps created through thinning than those with a lower plasticity or more rigid crown architecture (e.g. Dordel et al. 2010). Hence, crown form may become more easily irregular for these morphologically plastic species in mixtures than in monocultures. And crown dimensions, through their relationship with branch length, branch diameter, and consequently the size of knots, have a strong influence on stem quality (e.g. Maguire et al. 1999).

In mixtures of species differing in crown architecture and apical dominance, mixed stands may also exhibit more curved and leaning trees, if one species is forced to grow into unoccupied or unshaded growing space to compete for light, especially for light-demanding species dominated by species with low light transmittance (Grotta et al. 2004) or those with higher resistance to mechanical branch abrasion (e.g. Putz et al. 1984; Hajek et al. 2015). This may also lead to irregular crown shapes (e.g. Seidel et al. 2011; Brisson 2011) and associated variable branch dimensions, where branches exposed to more light are longer and thicker (Kint et al. 2010). Asymmetric crowns may lead to asymmetric stem cross sections and reaction wood. For example, crown imbalance and compression wood in Sitka spruce was significantly higher in mixtures with other conifers than in monospecific stands (Watson and Cameron 1995).

Since light exposure has been found to be associated with epicormic branch occurrence and persistence (e.g. Blum 1963; Takiya et al. 2010), differences in light transmittance between species may also promote epicormic branching in spots with higher light exposure compared to more homogeneous pure stands.

Yet tree mixture may also be purposefully employed to improve tree quality. There is some knowledge on the effects of specific tree species interactions on timber quality, such as facilitation of self-pruning of lower stem sections of valuable hardwoods through subdominant shade-tolerant species in the neighbourhood (e.g. von Lüpke 1998; Jennings et al. 2003; Saha et al. 2014). Here the shade-tolerant species are being used as trainer trees. In addition to controlling lower branches on crop trees through mixtures with trainer trees, admixed nurse trees can be used to control overall branch development and crown shape. The main difference between trainer and nurse species is, in addition to their different silvicultural purpose, their height relative to that of the species they are mixed with. Nurse crop species can be used to shade the stems and crowns of shorter and more shade-tolerant species. Examples include the nursing of *Quercus* species by *Populus* sp. and *Betula* sp. (Stark et al. 2015) or by *Alnus glutinosa*. (Nord-Larsen and Meilby 2016); of *Picea abies* by *Betula pendula* (Tham.); of *Fagus sylvatica* and *Picea sitchensis* by *Pseudotsuga menziesii* and *Larix kaempferi* (Nord-Larsen and Meilby 2016); of *Toona ciliata* by tropical *Pinus* species (Dordel et al. 2010); or of *Acacia melanoxylon* by *Pinus* or *Eucalyptus*

(Medhurst et al. 2003). When *Picea sitchensis* was mixed with hybrid larch (*Larix x eurolepis* Henry) as a nurse crop, it developed larger branches than in monocultures, whereas it developed smaller branches, when it was mixed with *Pinus contorta* as a nurse crop, indicating that the competitive strength of nurse crop species is important to control branch development in the main crop species (Cameron and Watson 1999).

Pretzsch and Rais (2016) reviewed the limited number of comparative studies on timber quality in mixed versus pure stands. For that purpose, they used morphological and structural properties of trees as proxies for their timber quality in mixed stands. These proxies such as stem shape and crown morphology were based on well-known relationships between morphology and timber quality based on many studies in monospecific stands. However, in their review, they did not strictly separate between mixed-species and uneven-aged stands with complex structures. They reported that knottiness was mostly stronger in complex stands, that wood density was not significantly influenced and that eccentricity of crowns, bending of stems, and irregular width of tree rings were more frequent in forests with complex than in those with simple structures. Thus, structural complexity was more important for tree quality than species identity or mixing.

**Take-Home Messages**

1. Silvicultural management approaches for mixed-species stands are inevitably more challenging than those developed for monospecific stands, and the complexity typically increases with the compositional and structural diversity of the system to be created and maintained. Hence, the outcome of silvicultural activities, which act in concert with long-term site conditions and short-term fluctuations of the abiotic and biotic environment, is not entirely predictable. Most research on regeneration of mixtures has focussed on the response of the different participating species to be regenerated, not on their interaction in the process.
2. An important aim of regeneration and tending is to develop stable mixtures that facilitate the anticipated interactions among the participating tree species and at the same time require little silvicultural input to maintain the composition of compatible tree species. Information about compatibility of species may be gained from experience in natural and cultivated systems or from species traits such as height growth pattern and shade tolerance.
3. To ensure survival and vigorous growth of trees in mixture in the long term, it is often necessary to separate species in space and time during the regeneration phase. Weaker competitors can be given a head start through advance regeneration or may be regenerated in patches that protect the majority of individuals from interspecific competition.
4. In managed forests, tree species in mixture may also have a service function. Nurse crops may be used to facilitate establishment of sensitive tree species and to accelerate ecosystem recovery following disturbances. Trainer species

are typically used to improve the quality development of stems in the economically important species.

5. Whereas the regeneration of mixtures has received considerable attention in research, there is little specific information regarding the thinning of mixed-species stands. These are often treated like an agglomerate of monospecific stands of the companion species. In the absence of quantitative information on optimal stand density, tree spacing, or the growth response to thinning in mixtures, crop tree thinning approaches are recommended. These facilitate the maintenance of desired species proportions as well as stand stability and development of quality attributes in crop trees through simple indicators.
6. Whether stem quality may be influenced by species mixture is extremely important for the economic result of forest management and thus also for a wider acceptance of mixed-species forests by forest owners. There is little evidence to support the notion that the development of stem quality in mixtures is reduced when compared to monospecific stands. Stem quality appears to depend on specific interactions between species and the structural heterogeneity of stands. Hence, the outcome for stem quality depends on the particular species combination and forest structure.

## References

Abetz P (1975) Eine Entscheidungshilfe für die Durchforstung von Fichtenbeständen. Allg Forstz 30:666–667

Abetz P, Klädtke J (2002) The target tree management system. Forstwiss Centralbl 121:73–82

Albrecht A, Hanewinkel M, Bauhus J, Kohnle U (2012) How does silviculture affect storm damage in forests of south-western Germany? Results from empirical modeling based on long-term observations. Eur J For Res 131:229–247

Altherr E (1971) Wege zur Buchen-Starkholzproduktion. Bericht zur 15. Hauptversammlung des Baden-Württembergischen Forstvereins, pp 123–127

Ammann P (2013) Erfolg der Jungwaldpflege im Schweizer Mittelland? Analyse und Folgerungen (Essay). Schweiz Zeitschr Forstw 164:262–270

Ammer C, Vor T, Knoke T, Wagner S (2010) Der Wald-Wild-Konflikt-Analyse und Lösungsansätze vor dem Hintergrund rechtlicher, ökologischer und ökonomischer Zusammenhänge. Universitätsverlag, Göttingen

Arens P, Coops H, Jansen J, Vosman B (1998) Molecular genetic analysis of black poplar (*Populus nigra* L.) along Dutch rivers. Mol Ecol 7:11–18

Ares A, Neill AR, Puettmann KJ (2010) Understory abundance, species diversity and functional attribute response to thinning in coniferous stands. For Ecol Manag 260:1104–1113

Arthur MA, Alexander HD, Dey DC et al (2012) Refining the oak-fire hypothesis for management of oak-dominated forests of the eastern United States. J For 110:257–266

Ashton PMS, Gamage S, Gunatilleke IAUN, Gunatilleke CVS (1997) Restoration of a Sri Lankan rainforest: using Caribbean pine Pinus caribaea as a nurse for establishing late-successional tree species. J Appl Ecol 34:915–925

Assmann E (1961) Waldertragskunde. Organische Produktion, Struktur, Zuwachs und Ertrag von Waldbeständen. BLV Verlagsgesellschaft, München, Bonn, Wien, 490 p

Baskin CC, Baskin JM (2014) Seeds: ecology, biogeography, and, evolution of dormancy and germination, 2nd edn. Elsevier, San Diego, CA, 1600 p

Bauhus J, Bartsch N (1995) Mechanisms of carbon and nutrient release and retention within beech forest gaps. I. Microclimate, water balance and seepage water chemistry. Plant Soil 168-169:579–584

Bauhus J, Bartsch N (1996) Fine root growth in beech (*Fagus sylvatica* L.) forest gaps. Can J For Res 26:2153–2160

Bauhus J, McElhinny C, Alcorn P (2002) Stand structure and tree growth in uneven-aged spotted gum (*Corymbia maculata*) forests: some implications for management. Forestry 75:451–456

Bauhus J, Pyttel P (2015) Managed forests. In: Peh KSH, Corlett RT, Bergeron Y (eds) Routledge handbook of forest ecology. Routledge, Oxon, pp 75–90

Bazzaz FA (1996) Plants in changing environments: linking physiological, population and community ecology. Cambridge University Press, Cambridge, 320 p

Bilke G (2004) Waldumbau in Nordostdeutschland durch Eichennachzucht in Mortzfeldtschen Löchern. Dissertation, University of Freiburg, 208 p

Binkley D, Dunkin KA, DeBell D, Ryan MG (1992) Production and nutrient cycling in mixed plantations of *Eucalyptus* and *Albizia* in Hawaii. For Sci 38:393–408

Binkley D, Senock R, Bird S, Cole TG (2003) Twenty years of stand development in pure and mixed stands of *Eucalyptus saligna* and N-fixing *Facaltaria moluccana*. For Ecol Manag 182:93–102

Blum BM (1963) Excessive exposure stimulates epicormic branching in young northern hardwoods. USDA Forest Service, Research Note: 1–6

Bouillet JP, Laclau JP, De Moraes Gonçalves JL et al (2013) Eucalyptus and Acacia tree growth over entire rotation in single- and mixed-species plantations across five sites in Brazil and Congo. For Ecol Manag 301:89–101

Brisson J (2011) Neighborhood competition and crown asymmetry in *Acer saccharum*. Can J For Res 31:2151–2159

Brose PH (2010) Long-term effects of single prescribed fires on hardwood regeneration in oak shelterwood stands. For Ecol Manag 260:1516–1524

Brose PH, Van Lear DH (1998) Responses of hardwood advance regeneration to seasonal prescribed fires in oak-dominated shelterwood stands. Can J For Res 28:331–339

Bruelheide H, Nadrowski K, Assmann T et al (2014) Designing forest biodiversity experiments: general considerations illustrated by a new large experiment in subtropical China. Methods Ecol Evol 5:74–89

Brütsch U, Rotach P (1993) Der Speierling (*Sorbus domestica*) in der Schweiz: Verbreitung, Ökologie, Standortsansprüche, Konkurrenzkraft und waldbauliche Eignung. Schweiz Z Forstwes 144:967–991

Cameron AD (2002) Importance of early selective thinning in the development of long-term stand stability and improved log quality: a review. Forestry 75:25–35

Cameron AD, Watson BA (1999) Effect of nursing mixtures on stem form, crown size, branching habit and wood properties of Sitka spruce (*Picea sitchensis* (Bong.) Carr.) For Ecol Manag 122:113–124

Cao T, Valsta L, Härkönen S et al (2008) Effects of thinning and fertilization on wood properties and economic returns for Norway spruce. For Ecol Manag 256:1280–1289

Carlyle JC, Malcolm DC (1986) Nitrogen availability beneath pure spruce and mixed larch and spruce stands growing on a deep peat. Plant Soil 93:105–112

Carrari E, Ampoorter E, Coppi A, Selvi F (2015) Diversity of secondary woody species in relation to species richness and cover of dominant trees in thermophilous deciduous forests of central Italy. Scand J For Res 31:484–494

Carrillo-Gavilán MA, Vilà M (2010) Little evidence of invasion by alien conifers in Europe. Divers Distrib 16:203–213

Castagneyrol B, Jactel H, Vacher C et al (2014) Effects of plant phylogenetic diversity on herbivory depend on herbivore specialization. J Appl Ecol 51:134–141

Charron I, Greene DF (2002) Post-wildfire seedbeds and tree establishment in the southern mixedwood boreal forest. Can J For Res 32:1607–1615

Chiang JM, Brown KJ (2010) The effects of thinning and burning treatments on within-canopy variation of leaf traits in hardwood forests of southern Ohio. For Ecol Manag 260:1065–1075
Coates KD (2002) Tree recruitment in gaps of various size, clearcuts and undisturbed mixed forest of interior British Columbia, Canada. For Ecol Manag 155:387–398
Coates KD, Burton PJ (1997) A gap-based approach for development of silvicultural systems to address ecosystem management objectives. For Ecol Manag 99:337–354
Coates KD, Burton PJ (1999) Growth of planted tree seedlings in response to ambient light levels in northwestern interior cedar-hemlock forests of British Columbia. Can J For Res 29:1374–1382
Condés S, Rio MD, Sterba H (2013) Mixing effect on volume growth of Fagus sylvatica and Pinus sylvestris is modulated by stand density. For Ecol Manag 292:86–95
Coutts MP, Nielsen CCN, Nicoll BC (1999) The development of symmetry, rigidity and anchorage in the structural root system of conifers. Plant Soil 217:1–15
Dănescu A, Ehring A, Bauhus J et al (2015) Modelling discoloration and duration of branch occlusion following green pruning in Acer pseudoplatanus and Fraxinus excelsior. For Ecol Manag 335:87–98
Dănescu A, Albrecht A, Bauhus J (2016) Structural diversity promotes productivity of mixed, uneven-aged forests in southwestern Germany. Oecologia 182:319–333
del Río M, Condés S, Pretzsch H (2014) Analyzing size-symmetric vs. size-asymmetric and intra- vs. inter-specific competition in beech (Fagus sylvatica L.) mixed stands. For Ecol Manag 325:90–98
Diaci J, Adamic T, Rozman A (2012) Gap recruitment and partitioning in an old-growth beech forest of the Dinaric Mountains: Influences of light regime, herb competition and browsing. For Ecol Manag 285:20–28
Dieler J, Pretzsch H (2013) Morphological plasticity of European beech (*Fagus sylvatica* L.) in pure and mixed-species stands. For Ecol Manag 295:97–108
Dietze MC, Clark JS (2008) Changing the gap dynamics paradigm: vegetative regeneration control on forest response to disturbance. Ecol Monogr 78:331–347
Donald PF, Fuller RJ, Evans AD, Gough SJ (1998) Effects of forest management and grazing on breeding bird communities in plantations of broadleaved and coniferous trees in western England. Biol Conserv 85:183–197
Dordel J, Simard SW, Bauhus J, Seely B, Pozas LJ, Prescott C, Hampel H (2010) Trade-offs among establishment success, stem morphology and productivity of underplanted *Toona ciliata*: effects of nurse-species and thinning density. For Ecol Manag 259:1846–1855
Drapier N (1993) Ecologie de l'Alisier torminal *Sorbus torminalis* (L.) Crantz. Rev For Fr 65:229–242
Drobyshev I, Linderson H, Sonesson K (2007) Relationship between crown condition and tree diameter growth in southern Swedish oaks. Environ Monit Assess 128:61–73
Drouineau S, Laroussinie O, Birot Y et al (2000) Joint evaluation of storms, forest vulnerability and their restoration. Technical Report. European Forest Institute, Joensuu. http://northerntosia.org/files/attachments/dp_09.pdf
Enescu CM, de Rigo D, Houston Durrant T, Caudullo G (2016) *Sorbus domestica* in Europe: distribution, habitat, usage and threats. In: San-Miguel-Ayanz J, de Rigo D, Caudullo G, Houston Durrant T, Mauri A (eds) European atlas of forest tree species. Office of European Union, Luxembourg, p 178;e01e816+
Engelmark O, Sjoberg K, Andersson B et al (2001) Ecological effects and management aspects of an exotic tree species: the case of lodgepole pine in Sweden. For Ecol Manag 141:3–13
Erickson HD, Harrison AT (1974) Douglas-fir wood quality studies part I: effects of age and stimulated growth on wood density and anatomy. Wood Sci Technol 8:207–226
Essl F, Mang T, Dullinger S et al (2011) Macroecological drivers of alien conifer naturalizations worldwide. Ecography 34:1076–1084
Ewald C, Zander M, Jander A (1994) Die Elsbeere (*Sorbus torminalis* [L.] Crantz) in Brandenburg. Der Wald 44:232–235

Ewers B, Binkley D, Bashkin M (1996) Influence of adjacent stand on spatial patterns of soil carbon and nitrogen in *Eucalyptus* and *Albizia* plantations. Can J For Res 26:1501–1503
FAO (1992) Mixed and pure forest plantations in the tropics and subtropics. FAO Forestry paper 103 (based on the work of TJ Wormald). In: Food and Agriculture Organization of the United Nations, FAO of the UN, Rome, Italy, p 151
FAO (2016) Global forest resources assessment 2015 - How are the world's forests changing? 2nd edn. FAO, Rome, 44 p
Felton A, Boberg J, Björkman C, Widenfalk O (2013) Identifying and managing the ecological risks of using introduced tree species in Sweden's production forestry. For Ecol Manag 307:165–177
Felton A, Gustafsson L, Roberge JM et al (2016) How climate change adaptation and mitigation strategies can threaten or enhance the biodiversity of production forests: insights from Sweden. Biol Conserv 194:1–20
Ferrari JB, Sugita S (1996) A spatially explicit model of leaf fall in hemlock-hardwood forests. Can J For Res 26:1905–1913
Forrester DI (2014) The spatial and temporal dynamics of species interactions in mixed-species forests: from pattern to process. For Ecol Manag 312:282–292
Forrester DI, Pretzsch H (2015) Tamm review: on the strength of evidence when comparing ecosystem functions of mixtures with monocultures. For Ecol Manag 356:41–53
Forrester DI, Smith RGB (2012) Faster growth of Eucalyptus *grandis* and *Eucalyptus pilularis* in mixed-species stands than monocultures. For Ecol Manag 286:81–86
Forrester DI, Bauhus J, Khanna PK (2004) Growth dynamics in a mixed-species plantation of *Eucalyptus globulus* and *Acacia mearnsii*. For Ecol Manag 193:81–95
Forrester DI, Bauhus J, Cowie AL, Vanclay JK (2006) Mixed-species plantations of *Eucalyptus* with nitrogen fixing trees: a review. For Ecol Manag 233:211–230
Forrester DI, Theiveyanathan S, Collopy JJ, Marcar NE (2010) Enhanced water use efficiency in a mixed *Eucalyptus globulus* and *Acacia mearnsii* plantation. For Ecol Manag 259:1761–1770
Forrester DI, Lancaster K, Collopy JJ et al (2012) Photosynthetic capacity of *Eucalyptus globulus* is higher when grown in mixture with *Acacia mearnsii*. Trees Struct Funct 26:1203–1213
Forrester DI, Kohnle U, Albrecht AT, Bauhus J (2013) Complementarity in mixed-species stands of Abies alba and Picea abies varies with climate, site quality and stand density. For Ecol Manag 304:233–242
Forrester DI, Benneter A, Bouriaud O, Bauhus J (2016) Diversity and competition influence tree allometric relationships – developing functions for mixed-species forests. J Ecol. doi:10.1111/1365-2745.12704
Gamfeldt L, Snäll T, Bagchi R et al (2013) Higher levels of multiple ecosystem services are found in forests with more tree species. Nat Commun 4:340
Garber SM, Maguire DA (2004) Stand productivity and development in two mixed-species spacing trials in the central Oregon cascades. For Sci 50:92–105
Gardiner ES, Stanturf JA, Schweitzer CJ (2004) An afforestation system for restoring bottomland hardwood forests: biomass accumulation of Nuttall oak seedlings interplanted beneath eastern cottonwood. Restor Ecol 12:525–532
Gauthier MM, Jacobs DF (2010) Ecophysiological responses of black walnut (*Juglans nigra*) to plantation thinning along a vertical canopy gradient. For Ecol Manage 259:867–874
Geiger R (1950) Das Klima der bodennahen Luftschicht: ein Lehrbuch der Mikroklimatologie. Vieweg Verlag, Braunschweig
Gill RMA, Beardall V (2001) The impact of deer on woodlands: the effects of browsing and seed dispersal on vegetation structure and composition. Forestry 74:209–218
Ginn SE, Seiler JR, Cazell BH, Kreh RE (1991) Physiological and growth responses of eight-year-old loblolly pine stands to thinning. For Sci 37:1030–1040
Givnish TJ (1988) Adaptation to sun and shade: a whole-plant perspective. Aust J Plant Physiol 15:63–92

Gjerde I, Saetersdal M (1997) Effects on avian diversity of introducing spruce *Picea spp.* plantations in the native pine *Pinus sylvestris* forests of western Norway. Biol Conserv 79:241–250

Goßner M, Ammer U (2006) The effects of Douglas-fir on tree-specific arthropod communities in mixed species stands with European beech and Norway spruce. Eur J For Res 125:221–235

Goßner M, Gruppe A, Simon U (2005) Aphidophagous insect communities in tree crowns of the neophyte Douglas fir [*Pseudotsuga menziesii* (Mirb.) Franco] and Norway spruce (*Picea abies* L.) J Appl Entomol 129:81–88

Götmark F, Berglund Å, Wiklander K (2005) Browsing damage on broadleaved trees in semi-natural temperate forest in Sweden, with a focus on oak regeneration. Scand J For Res 20:223–234

Goto S, Iijima H, Ogawa H, Ohya K (2011) Outbreeding depression caused by intraspecific hybridization between local and nonlocal genotypes in *Abies sachalinensis*. Restor Ecol 19:243–250

Griess VC, Knoke T (2013) Bioeconomic modelling of mixed Norway spruce – European beech stands: economic consequences of considering ecological effects. Eur J For Res 132:511–522

Grotta AT, Gartner BL, Radosevich SR (2004) Influence of species proportion and timing of establishment on stem quality in mixed red alder – Douglas-fir plantations. Can J For Res 34:863–873

Grubb PJ (1977) The maintenance of species-richness in plant communities: the importance of the regeneration niche. Biol Rev 52:107–145

Guericke M (2001) Untersuchungen zur Wuchsdynamik von Mischbeständen aus Buche und Europäischer Lärche (*Larix decidua* Mill.) als Grundlage für ein abstandsabhängiges Einzelbaumwachstumsmodell. PhD thesis, University of Göttingen

Gustafsson L, Baker SC, Bauhus J et al (2012) Retention forestry to maintain multifunctional forests: a world perspective. Bioscience 62:633–645

Haines SG, DeBell DS (1980) Use of nitrogen-fixing plants to improve and maintain productivity of forest soils. Proceedings: impact of intensive harvesting on young nutrition cycling, pp 279–303

Hajek P, Seidel D, Leuschner C (2015) Mechanical abrasion, and not competition for light, is the dominant canopy interaction in a temperate mixed forest. For Ecol Manag 348:108–116

Hansen EM, Stone JK, Capitano BR et al (2000) Incidence and impact of Swiss needle cast in forest plantations of Douglas-fir in coastal Oregon. Plant Dis 84:773–778

Harvey BD, Leduc A, Gauthier S, Bergeron Y (2002) Stand-landscape integration in natural disturbance-based management of the southern boreal forest. For Ecol Manag 155:369–385

Hausner VH, Yoccoz NG, Strann KB, Ims RA (2002) Changes in bird communities by planting non-native spruce in coastal birch forests of northern Norway. Ecoscience 9:470–481

Haysom K, Murphy S (2003) The status of invasiveness of forest tree species outside their natural habitat: a global review and discussion paper. In: Forest Health and Biosecurity Working paper FBS/3E. FAO, Rome

Hemery GE, Clark JR, Aldinger E et al (2010) Growing scattered broadleaved tree species in Europe in a changing climate: a review of risks and opportunities. Forestry 83:65–81

Hibbs DE (1982) Gap dynamics in a hemlock–hardwood forest. Can J For Res 12:522–527

Higgins S, Richardson D (1998) Pine invasions in the southern hemisphere: modelling interactions between organism, environment and disturbance. Plant Ecol 135:79–93

Hunt MA, Unwin GL, Beadle CL (1999) Effects of naturally regenerated *Acacia dealbata* on the productivity of a *Eucalyptus nitens* plantation in Tasmania, Australia. For Ecol Manag 117:75–85

Huth F, Wagner S (2006) Gap structure and establishment of Silver birch regeneration (*Betula pendula* Roth.) in Norway spruce stands (*Picea abies* L. Karst.) For Ecol Manag 229:314–324

Jactel H, Brockerhoff EG (2007) Tree diversity reduces herbivory by forest insects. Ecol Lett 10:835–848

Jennings SM, Wilkinson GR, Unwin GL (2003) Response of blackwood (*Acacia melanoxylon*) regeneration to silvicultural removal of competition in regrowth eucalypt forests of north-west Tasmania, Australia. For Ecol Manag 177:75–83

Jerabkova L, Prescott CE, Titus BD et al (2011) A meta-analysis of the effects of clearcut and variable-retention harvesting on soil nitrogen fluxes in boreal and temperate forests. Can J For Res 41:1852–1870

Jucker T, Bouriaud O, Coomes DA (2015) Crown plasticity enables trees to optimize canopy packing in mixed-species forests. Funct Ecol 29:1078–1086

Kadunc A (2007) Factors influencing the formation of heartwood discolouration in sycamore (*Acer pseudoplatanus* L.) Eur J For Res 126:349–358

Kariuki M (2008) Modelling the impacts of various thinning intensities on tree growth and survival in a mixed species eucalypt forest in central Gippsland, Victoria, Australia. For Ecol Manag 256:2007–2017

Keane RM, Crawley MJ (2002) Exotic plant invasions and the enemy release hypothesis. Trends Ecol Evol 17:164–170

Keenan R, Lamb D, Sexton G (1995) Experience with mixed species rainforest plantations in North Queensland. Commonwealth For Rev 74:315–321

Kelly D (1994) The evolutionary ecology of mast seeding. Trends Ecol Evol 9:465–470

Kelty MJ, Cameron IR (1995) Plot designs for the analysis of species interactions in mixed stands. Commonwealth For Rev 74:322–332

Kerr G (2004) The growth and form of ash (*Fraxinus excelsior*) in mixture with cherry (*Prunus avium*), oak (*Quercus petraea* and *Quercus robur*), and beech (*Fagus sylvatica*). Can J For Res 34:2340–2350

Kint V, Hein S, Campioli M, Muys B (2010) Modelling self-pruning and branch attributes for young *Quercus robur* L. and *Fagus sylvatica* L. trees. For Ecol Manag 260:2023–2034

Kittredge DB, Ashton PMS (1995) Impact of deer browsing on regeneration in mixed stands in southern New England. North J Appl For 12:115–120

Klädtke J (2002) Wachstum großkroniger Buchen und waldbauliche Konsequenzen. Forstarchiv 73:211–217

Kneeshaw DD, Prevost M (2007) Natural canopy gap disturbances and their role in maintaining mixed-species forests of central Quebec, Canada. Can J For Res 37:1534–1544

Kneeshaw DD, Kobe RK, Coates KD, Messier C (2006) Sapling size influences shade tolerance ranking among southern boreal tree species. J Ecol 94:471–480

Knoerzer D (1999) Zur Naturverjüngung der Douglasie im Schwarzwald. Dissertationes Botanicae 306

Knoke T (2008) Mixed forests and finance - Methodological approaches. Ecol Econ 65:590–601

Koivuranta L, Latva-Karjanmaa T, Pulkkinen P (2012) The effect of temperature on seed quality and quantity in crosses between European (*Populus tremula*) and hybrid aspens (*P. tremula* x *P. tremuloides*). Silva Fennica 46:17–26

Krivanek M, Pysek P, Jarosik V (2006) Planting history and propagule pressure as predictors of invasion by woody species in a temperate region. Conserv Biol 20:1487–1498

Krumm F, Vítková L (eds) (2016) Introduced tree species in European forests: opportunities and challenges. European Forest Institute, 423 p

Kuehne C, Kublin E, Pyttel P, Bauhus J (2013) Growth and form of *Quercus robur* and *Fraxinus excelsior* respond distinctly different to initial growing space: results from 24-year-old Nelder experiments. J For Res 24:1–14

Kuehne C, Weiskittel AR, Wagner RG, Roth BE (2016) Development and evaluation of individual tree-and stand-level approaches for predicting spruce-fir response to commercial thinning in Maine, USA. For Ecol Manag 376:84–95

Kuiters AT, Slim PA (2002) Regeneration of mixed deciduous forest in a Dutch forest-heathland, following a reduction of ungulate densities. Biol Conserv 105:65–74

Kunz J, Räder A, Bauhus J (2016) Effects of drought and rewetting on growth and gas exchange of minor European broadleaved tree species. Forests 7:239

Larsen JB, Nielsen AB (2007) Nature-based forest management – Where are we going? Elaborating forest development types in and with practice. For Ecol Manag 238:107–117
Ledgard N (2002) The spread of Douglas-fir into native forests. NZ J For 47:36–38
Levin DA, Francisco Ortega J, Jansen RK (1996) Hybridization and the extinction of rare plant species. Conserv Biol 10:10–16
Liang J, Crowther TW, Picard N et al (2016) Positive biodiversity–productivity relationship predominant in global forests. Science 354(6309):aaf8957
Ligot G, Balandier P, Fayolle A et al (2013) Height competition between *Quercus petraea* and *Fagus sylvatica* natural regeneration in mixed and uneven-aged stands. For Ecol Manag 304:391–398
Lindén M, Vollbrecht G (2002) Sensitivity of *Picea abies* to butt rot in pure stands and in mixed stands with *Pinus sylvestris* in southern Sweden. Silva Fennica 36:767–778
Liu H, Stiling P (2006) Testing the enemy release hypothesis: a review and meta-analysis. Biol Invasions 8:1535–1545
Löf M, Bergquist J, Brunet J et al (2010) Conversion of Norway spruce stands to broadleaved woodland–regeneration systems, fencing and performance of planted seedlings. Ecol Bull 53:165–173
Lu H, Mohren GM, den Ouden J et al (2016) Overyielding of temperate mixed forests occurs in evergreen–deciduous but not in deciduous–deciduous species mixtures over time in the Netherlands. For Ecol Manag 376:321–332
Lundmark T, Hallgren J-E (1987) Effects of frost on shaded and exposed spruce and pine seedlings planted in the field. Can J For Res 17:1197–1201
Lundqvist L, Mörling T, Valinger E (2014) Spruce and birch growth in pure and mixed stands in Sweden. For Chron 90:29–34
Maguire DA, Johnston SR, Cahill J (1999) Predicting branch diameters on second-growth Douglas-fir from tree-level descriptors. Can J For Res 29:1829–1840
Mäkinen H, Hein S (2006) Effect of wide spacing on increment and branch properties of young Norway spruce. Eur J For Res 125:239–248
Mäkinen H, Isomäki A (2004) Thinning intensity and growth of Scots pine stands in Finland. For Ecol Manag 201:311–325
Man R, Lieffers VJ (1999) Are mixtures of aspen and white spruce more productive than single species stands? For Chron 75:505–513
Martinez A, Carlos Iturrondobeitia J, Goldarazena A (2009) Effects of some ecological variables on carabid communities in native and non native forests in the Ibaizabal basin (Basque Country: Spain). Ann For Sci 66:1–14
Matthews JD (1991) Silvicultural systems. Oxford University Press, Oxford, 284 p
McKay JK, Christian CE, Harrison S, Rice KJ (2005) "How local is local?" – a review of practical and conceptual issues in the genetics of restoration. Restor Ecol 13:432–440
Medhurst JL, Pinkard EA, Beadle CL, Worledge D (2003) Growth and stem form responses of plantation-grown *Acacia melanoxylon* (R. Br.) to form pruning and nurse-crop thinning. For Ecol Manag 179:183–193
Messier C, Doucet R, Ruel JC et al (1999) Functional ecology of advance regeneration in relation to light in boreal forests. Can J For Res 29:812–823
Miesel JR (2012) Differential responses of *Pinus ponderosa* and *Abies concolor* foliar characteristics and diameter growth to thinning and prescribed fire treatments. For Ecol Manag 284:163–173
Miller HG, Alexander CE, Cooper J et al (1986) Maintenance and enhancement of forest productivity through manipulation of the nitrogen cycle. Unpublished Report to the EC. Contract No. BOS-093 UK
Mitchell RJ, Pakeman RJ, Broome A et al (2016) How to replicate the functions and biodiversity of a threatened tree species? The case of *Fraxinus excelsior* in Britain. Ecosystems 19:573–586
Möhring B, Leefken G, Gutsche C (2008) Economic valuation of beech forests. Ergebnisse angewandter Forschung zur Buche, Beiträge aus der NW-FVA 3:327–343

Müller J, Hothorn T, Pretzsch H (2007) Long-term effects of logging intensity on structures, birds, saproxylic beetles and wood-inhabiting fungi in stands of European beech *Fagus sylvatica* L. For Ecol Manag 242:297–305

Neumann FG, Minko G (1981) The sirex wood wasp in Australian radiata pine plantations. Aust For 44:46–63

Neyland M, Hickey J, Beadle C et al (2009) An examination of stocking and early growth in the Warra silvicultural systems trial confirms the importance of a burnt seedbed for vigorous regeneration in *Eucalyptus obliqua* forest. For Ecol Manag 258:481–494

Nill M (2011) Rindenschäden durch Holzernte in Baden-Württemberg – Ursachen und Prognose. Forstliche Versuchs- und Forschungsanstalt Baden-Württemberg. Schriftenreihe Freiburger forstliche Forschung, Bd. 50

Nord-Larsen T, Meilby H (2016) Effects of nurse trees, spacing, and tree species on biomass production in mixed forest plantations. Scand J For Res 31:592–601

Nunez MA, Moretti A, Simberloff D (2011) Propagule pressure hypothesis not supported by an 80-year experiment on woody species invasion. Oikos 120:1311–1316

Nyland RD (2002) Silviculture: concepts and applications, 2nd edn. McGraw-Hill

O'Hara KL (2014) Multiaged silviculture: managing for complex forest stand structures. Oxford University Press, New York, NY, 213 p

O'Hara KL (2007) Pruning wounds and occlusion: a long-standing conundrum in forestry. J For 105:131–138

O'Hara KL, Gersonde RF (2004) Stocking control concepts in uneven-aged silviculture. Forestry 77:131–143

Oleskog G, Sahlén K (2000) Effects of seedbed substrate on moisture conditions and germination of Scots pine (*Pinus sylvestris*) seeds in a mixed conifer stand. New For 20:119–133

Oliver CD, Larson BC (1990) Forest stand dynamics. McGraw-Hill, New York, 467 p

Orellana IA, Raffaele E (2010) The spread of the exotic conifer *Pseudotsuga menziesii* in *Austrocedrus chilensis* forests and shrublands in northwestern Patagonia, Argentina. NZ J For Sci 40:199–209

Örlander G (1993) Shading reduces both visible and invisible frost damage to Norway spruce seedlings in the field. Forestry 66(1):27–36

Oxbrough A, French V, Irwin S et al (2012) Can mixed species stands enhance arthropod diversity in plantation forests? For Ecol Manag 270:11–18

Paganová V (2008) Ecological requirements of wild service tree (*Sorbus torminalis* (L.) Crantz) and service tree (*Sorbus domestica* L.) in relation with their utilization in forestry and landscape. J For Sci 54:216–226

Petriţan AM, von Lüpke B, Petriţan IC (2009) Influence of light availability on growth, leaf morphology and plant architecture of beech (*Fagus sylvatica* L.), maple (*Acer pseudoplatanus* L.) and ash (*Fraxinus excelsior* L.) saplings. Eur J For Res 128:61–74

Piotto D, Montagnini F, Ugalde L, Kanninen M (2003) Growth and effects of thinning of mixed and pure plantations with native trees in humid tropical Costa Rica. For Ecol Manag 177:427–439

Pretzsch H (1985) Die Fichten-Tannen-Buchen-Plenterwaldversuche in den ostbayerischen Forstämtern Freyung und Bodenmais. Forstarchiv 56:3–9

Pretzsch H (2006) Species-specific allometric scaling under self-thinning: evidence from long-term plots in forest stands. Oecologia 146:572–583

Pretzsch H, Biber P (2016) Tree species mixing can increase maximum stand density. Can J For Res 46:1179–1193

Pretzsch H, Rais A (2016) Wood quality in complex forests versus even-aged monocultures. Review and perspectives. Wood Sci Technol 50:845–880

Pretzsch H, del Rio M, Schütze G, Ammer C, Annighöfer P, Avdagic A, Barbeito I, Bielak K, Brazaitis G, Coll L, Drössler L, Fabrika M, Forrester DI, Kurylyak V, Löf M, Lombardi F, Matovic B, Mohren F, Motta R, den Ouden J, Pach M, Ponette Q, Skzyszewski J, Sramek V, Sterba H, Svoboda M, Verheyen K, Zlatanov T, Bravo-Oviedo A (2016) Mixing of Scots pine

(*Pinus sylvestris* L.) and European beech (*Fagus sylvatica* L.) enhances structural heterogeneity, and the effect increases with water availability. For Ecol Manage 373:149–166
Pretzsch H, Bielak K, Block J et al (2013) Productivity of mixed versus pure stands of oak (*Quercus petraea* (Matt.) Liebl. and *Quercus robur* L.) and European beech (*Fagus sylvatica* L.) along an ecological gradient. Eur J For Res 132:263–280
Pucek Z, Jędrzejewski W, Jędrzejewska B, Pucek M (1993) Rodent population dynamics in a primeval deciduous forest (Białowieża National Park) in relation to weather, seed crop, and predation. Acta Theriol 38:199–232
Puettmann KJ, Coates D, Messier C (2009) A critique of silviculture: managing for complexity. Island Press, Washington, 188 p
Puettmann KJ, McG Wilson S, Baker S et al (2015) Silvicultural alternatives to conventional even-aged forest management – what limits global adoption? For Ecosyst 2:8. doi:10.1186/s40663-015-0031-x
Pukkala T, Vettenranta J, Kolström T, Miina J (1994) Productivity of mixed stands of *Pinus sylvestris* and *Picea abies*. Scand J For Res 9:143–153
Pukkala T, Miina J, Kurttila M, Kolström T (1998) A spatial yield model for optimizing the thinning regime of mixed stands of *Pinus sylvestris* and *Picea abies*. Scand J For Res 13:31–42
Pukkala T, Lähde E, Laiho O (2015) Which trees should be removed in thinning treatments? For Ecosyst 2:1–12
Putz FE (1983) Treefall pits and mounds, buried seeds, and the importance of soil disturbance to pioneer trees on Barro Colorado Island, Panama. Ecology 64:1069–1074
Putz FE, Parker GG, Archibald RM (1984) Mechanical abrasion and intercrown spacing. Am Midl Nat 112:24–28
Pyttel P, Weber T, Bauhus J (2013a) Regeneration und Wachstum des Speierlings (*Sorbus domestica*) in durchgewachsenen Niederwäldern. In: Becker G, Bauhus J, Konold W (eds) Schutz durch Nutzung: Ein Raum-Zeit-Konzept für die multifunktionale Entwicklung der Stockausschlagwälder in Rheinland-Pfalz. Culterra 62, Freiburg, pp 100–105
Pyttel P, Kunz J, Bauhus J (2013b) Growth, regeneration and shade tolerance of *Sorbus torminalis* in aged oak coppice forests. Trees Struct Funct 27:1609–1619
Richardson DM, Rejmanek M (2004) Conifers as invasive aliens: a global survey and predictive framework. Divers Distrib 10:321–331
Richardson DM, Rejmanek M (2011) Trees and shrubs as invasive alien species – a global review. Divers Distrib 17:788–809
Röhrig E, Bartsch N, v Lüpke B (2006) Waldbau auf ökologischer Grundlage. Eugen Ulmer, Stuttgart, Germany, 479 p
Roques A, Auger-Rozenberg MA, Boivin S (2006) A lack of native congeners may limit colonization of introduced conifers by indigenous insects in Europe. Can J For Res 36:299–313
Rossell CR Jr, Patch S, Salmons S (2007) Effects of deer browsing on native and non-native vegetation in a mixed oak-beech forest on the Atlantic coastal plain. Northeast Nat 14:61–72
Rothe A, Binkley D (2001) Nutritional interactions in mixed species forests: a synthesis. Can J For Res 31:1855–1870
Ryan EA, Alexander IJ (1992) Mycorrhizal aspects of improved growth of spruce when grown in mixed stands on heathlands. In: Read DJ, Lewis DH, Fitter AH, Alexander IJ (eds) Mycorrhizas in ecosystems. CAB International, Wallingford, pp 237–245
Saha S, Kuehne C, Bauhus J (2013) Tree species richness and stand productivity in low-density cluster plantings with oaks (*Quercus robur* L. and *Q. petraea* (Mattuschka) Liebl.) Forests 4:650–665
Saha S, Kuehne C, Bauhus J (2014a) The impact of intra- and interspecific competition on growth and stem quality of oaks (*Quercus robur* L. and *Q. petraea* (Mattuschka) Liebl.) established in cluster plantings. Ann For Sci 71:381–393
Saha S, Kuehne C, Bauhus J (2014b) The impact of intra- and interspecific competition on growth and stem quality of oaks (*Quercus robur* L. and *Q. petraea* (Mattuschka) Liebl.) established in cluster plantings. Ann For Sci 71:381–393

Saha S, Kuehne C, Bauhus J (2016) Lessons learned from oak cluster planting trials in central Europe. Can J For Res. doi:10.1139/cjfr-2016-0265
San-Miguel-Ayanz J, de Rigo D, Caudullo G, Houston Durrant T, Mauri A (2016) European atlas of forest tree species. Office of the European Union, Luxembourg, p e01e816+
Santiago-García RJ, Colon SM, Sollins P, Van Bloem SJ (2008) The role of nurse trees in mitigating fire effects on tropical dry forest restoration: a case study. Ambio 37:604–608
Scherer-Lorenzen M, Schulze ED, Don A et al (2007) Exploring the functional significance of forest diversity: a new long-term experiment with temperate tree species (BIOTREE). Persp Plant Ecol Evol Syst 9:53–70
Schuldt A, Scherer-Lorenzen M (2014) Non-native tree species (*Pseudotsuga menziesii*) strongly decreases predator biomass and abundance in mixed-species plantations of a tree diversity experiment. For Ecol Manag 327:10–17
Schütz JP (2001) Plenterwald und weitere Formen strukturierter und gemischter Wälder. Parey, Berlin, 207 p
Seidel D, Leuschner C, Müller A, Krause B (2011) Crown plasticity in mixed forests—quantifying asymmetry as a measure of competition using terrestrial laser scanning. For Ecol Manag 261:2123–2132
Shepperd WD, Jones JR (1985) Nurse crop. In: DeByle N, Winokur R (eds) Aspen: ecology and management in the Western United States. General technical report RM-119. USDA Forest Service, Rocky Mountain Forest and Range Experiment Station, Fort Collins, CO, pp 181–188
Shibata M, Tanaka H, Iida S et al (2002) Synchronized annual seed production by 16 principal tree species in a temperate deciduous forest, Japan. Ecology 83:1727–1742
Simberloff D (2009) The role of propagule pressure in biological invasions. Annu Rev Ecol Evol Syst 40:81–102
Simmons EA, Buckley GP (1992) Ground vegetation under planted mixtures of trees. In: Cannell MGR, Malcolm DC, Robertson PA (eds) The ecology of mixed-species stands of trees. Blackwell Scientific Publications, Oxford, pp 211–231
Sohn JA, Saha S, Bauhus J (2016) Potential of forest thinning to mitigate drought stress: a meta-analysis. For Ecol Manag 380:261–273
Spellmann H, Weller A, Brang P et al. (2015) Douglasie (Pseudotsuga menziesii (Mirb.) Franco). In: Vor T, Spellmann H, Bolte A, Ammer C (eds) (2015) Potenziale und Risiken eingeführter Baumarten: Baumartenportraits mit naturschutzfachlicher Bewertung. Universitätsverlag Göttingen. p 187-217
Spiecker H, Hein S, Makkonen-Spiecker K, Thies M (eds) (2009) Valuable broadleaved forests in Europe. EFI Research Report 22, Brill, Leiden, Boston, Köln, 256 p
Stancioiu PT, O'Hara KL (2006) Regeneration growth in different light environments of mixed species, multiaged, mountainous forests of Romania. Eur J For Res 125:151–162
Stark H, Nothdurft A, Bauhus J (2015) Forest restoration with *Betula ssp.* and *Populus ssp.* nurse crops increases productivity and soil fertility. For Ecol Manag 339:57–70
Steinmetz A, Bauhus J (2016) Naturverjüngung der Douglasie im Stadtwald Freiburg – invasiv? AFZ/Der Wald 71:25–28
Szewczyk J, Szwagrzyk J (1996) Tree regeneration on rotten wood and on soil in old-growth stand. Plant Ecol 122:37–46
Takiya M, Koyama H, Umeki K et al (2010) The effects of early and intense pruning on light penetration, tree growth, and epicormic shoot dynamics in a young hybrid larch stand. J For Res 15:149–160
Tarp P, Buongiorno J, Helles F et al (2005) Economics of converting an even-aged Fagus sylvatica stand to an uneven-aged stand using target diameter harvesting. Scand J For Res 20:63–74
Taylor CMA, Tabbush PM (1990) Nitrogen deficiency in Sitka spruce plantations. Forestry Commission Bulletin 89, HMSO, London, 20 p
Thies M, Pfeifer N, Winterhalder D, Gorte BGH (2004) Three-dimensional reconstruction of stems for assessment of taper, sweep and lean based on laser scanning of standing trees. Scand J For Res 19:571–581

Thurm EA, Pretzsch H (2016) Improved productivity and modified tree morphology of mixed versus pure stands of European beech (Fagus sylvatica) and Douglas-fir (*Pseudotsuga menziesii*) with increasing precipitation and age. Ann For Sci 73:1047–1061

Thurm EA, Uhl E, Pretzsch H (2016) Mixture reduces climate sensitivity of Douglas-fir stem growth. For Ecol Manag 376:205–220

Tobner CM, Paquette A, Reich PB et al (2014) Advancing biodiversity–ecosystem functioning science using high-density tree-based experiments over functional diversity gradients. Oecologia 174:609–621

Toïgo M, Vallet P, Perot T et al (2015) Overyielding in mixed forests decreases with site productivity. J Ecol 103:502–512

Unseld R, Bauhus J (eds) (2012) Energie-Vorwälder – Alternative Bewirtschaftungsformen zur Steigerung der energetisch nutzbaren Biomasse im Wald durch Integration von schnellwachsenden Baumarten. Freiburger Forstliche Forschung, Berichte, Heft 91

van der Plas F, Manning P, Allan E et al (2016) 'Jack-of-all-trades' effects drive biodiversity-ecosystem multifunctionality relationships in European forests. Nat Commun 7:11109

Van Leeuwen M, Hilker T, Coops NC et al (2011) Assessment of standing wood and fiber quality using ground and airborne laser scanning: a review. For Ecol Manag 261:1467–1478

Vanden Broeck A, Storme V, Cottrell J et al (2004) Gene flow between cultivated poplars and native black poplar (*Populus nigra* L.): a case study along the river Meuse on the Dutch–Belgian border. For Ecol Manag 197:307–310

Vehviläinen H, Koricheva J (2006) Moose and vole browsing patterns in experimentally assembled pure and mixed forest stands. Ecography 29:497–506

Vitousek PM, Melillo JM (1979) Nitrate losses from disturbed forests: patterns and mechanism. For Sci 25:605–619

von Lüpke B (1998) Silvicultural methods of oak regeneration with special respect to shade tolerant mixed species. For Ecol Manag 106:19–26

Voorhis NG (1990) Precommercial crop-tree thinning in a mixed northern hardwood stand. USDA Forest Service, Northeastern Forest Experiment Station, 4 p

Vor T, Spellmann H, Bolte A, Ammer C (eds) (2015) Potenziale und Risiken eingeführter Baumarten: Baumartenportraits mit naturschutzfachlicher Bewertung. Universitätsverlag, Göttingen

Vuidot A, Paillet Y, Archaux F, Gosselin F (2011) Influence of tree characteristics and forest management on tree microhabitats. Biol Conserv 144:441–450

Wagner S (1999) Ökologische Untersuchungen zur Initialphase der Naturverjüngung in Eschen-Buchen-Mischbeständen. Schriften aus der Forstl Fak der Univ Göttingen u der Niedersächs Forstl Versuchsanst 129:262 p

Watson B, Cameron A (1995) Some effects of nursing species on stem form, branching habit and compression wood content of Sitka spruce. Scott For 49:146–154

Weich T (2012) Optimierte Verfahren zur Beerntung gepflanzter Vorwälder. In: Unseld R, Bauhus J (eds) Energie-Vorwälder – Alternative Bewirtschaftungsformen zur Steigerung der energetisch nutzbaren Biomasse im Wald durch Integration von schnellwachsenden Baumarten. Freiburger Forstliche Forschung, Berichte, Heft 91, pp 130–146

Wein A, Bauhus J, Bilodeau-Gauthier S et al (2016) Tree species richness promotes invertebrate herbivory on congeneric native and exotic tree saplings in a young diversity experiment. PLoS One 11(12):e0168751

West PW (2006) Growing plantation forests. Springer, Berlin

Westoby M (1984) The self-thinning rule. Adv Ecol Res 14:167–226

Westphal C, Tremer N, von Oheimb G et al (2006) Is the reverse J-shaped diameter distribution universally applicable in European virgin beech forests? For Ecol Manag 223:75–83

Williams BL (1992) Nitrogen dynamics in humus and soil beneath sitka spruce (*Picea sitchensis* (Bong) Carr) planted in pure stands and in mixture with scots pine (*Pinus sylvestris* L). Plant Soil 144:77–84

Wilson DS, Puettmann KJ (2007) Density management and biodiversity in young Douglas-fir forests: challenges of managing across scales. For Ecol Manag 246:123–134

Wimmer R, Johansson M (2014) Effects of reaction wood on the performance of wood and wood-based products. In: Gardiner B, Barnett J, Saranpää S, Gril J (eds) The biology of reaction wood. Springer series in wood science. Springer, Berlin, pp 225–248

Zhang Y, Chen HYH (2015) Individual size inequality links forest diversity and above-ground biomass. J Ecol 103:1245–1252

Zhao J, Maguire DA, Mainwaring D et al (2014) Thinning mixed-species stands of Douglas-fir and western hemlock in the presence of Swiss needle cast: guidelines based on relative basal area growth of individual trees. For Sci 60:191–199

Ziegenhagen B, Gneuss S, Rathmacher G et al (2008) A fast and simple genetic survey reveals the spread of poplar hybrids at a natural Elbe river site. Conserv Genet 9:373–379

Zobel B (1984) The changing quality of the world wood supply. Wood Sci Technol 18:1–17

# Forest Management Planning in Mixed-Species Forests

# 10

Hans Pretzsch and Thomas Knoke

**Abstract**

In this chapter, we outline how forest management planning (FMP) develops in order to meet the requirements or even support the transition from far-from-nature monocultures to close-to-nature polycultures. We first introduce the various concepts and main steps of FMP in practice. The second section deals with the principle of participation, a key aspect of multiple-purpose FMP, which includes examples of how models may support the change towards complex mixed-species stands. The fourth section sketches how FMP has been enhanced to address mixed-species forests as well as problems that are yet to be addressed. In the last three sections, we demonstrate an example of FMP in a close-to-nature forest estate, including the assessment of biodiversity and landscape aesthetics based on grid-based inventory data. A second example demonstrates the components of a size-class model, which is applicable in mixed, uneven-aged forest. The chapter presents the further development of FMP which will be necessary in view of the close-to-nature management paradigm.

## Contents

10.1 The Practice of Forest Management Planning .......... 505
10.1.1 Prescriptive, Optimisation-Based Approaches .......... 507
10.1.2 Planning of Sustainable Yield Using a Descriptive Approach .......... 508
10.1.3 Planning of Sustainable Production Using a Descriptive Approach .......... 510

H. Pretzsch (✉)
Chair for Forest Growth and Yield Science, Faculty of Forest Science and Resource Management, Technical University of Munich, Hans-Carl-von-Carlowitz-Platz 2, 85354 Freising, Germany
e-mail: Hans.Pretzsch@lrz.tum.de

T. Knoke
Faculty of Forest Science and Resource Management, Institute of Forest Management, Technical University of Munich, Hans-Carl-von-Carlowitz-Platz 2, 85354 Freising, Germany
e-mail: knoke@tum.de





H. Pretzsch et al. (eds.), *Mixed-Species Forests*, DOI 10.1007/978-3-662-54553-9_10

10.2 Concept for Participatory Planning in Mixed-Species Forest Stands: The Role of Models and Inventory Data ........ 511
10.2.1 Models for Decision Support ........ 513
10.3 Extension of Forest Management Planning to Uneven-Aged Forests and Problems to Be Addressed ........ 516
10.4 Examples of Planning in Mixed or Uneven-Aged Forest Estates ........ 521
10.4.1 Scenario-Based Forest Management Planning in Complex Forests Owned by the Bavarian State ........ 521
10.4.2 Deriving the Annual Harvest in an Uneven-Aged Mixed Forest Using a Size-Class Approach ........ 524
10.5 Assessment of Biodiversity ........ 527
10.5.1 Composition, Structure, and Functioning at Different Scales as Components of Biodiversity ........ 530
10.5.2 Forest Inventories as Database ........ 531
10.5.3 Example of an Integrated View on Biodiversity Based on Forest Inventory Data ........ 532
10.6 Visualisation of Forest Management Planning Scenarios ........ 534
10.6.1 Visualisation Tools TREEVIEW and L-VIS ........ 535
Take-Home Messages ........ 538
References ........ 539

Since the beginning of sustainable forestry some 300 years ago (von Carlowitz 1713), the demand for forest functions and services has grown continuously. Restricted to mainly hunting, beekeeping, livestock grazing in forests, and wood harvesting in the seventeenth century, it has since widened to require ecological, economic, and social ecosystem services, such as, inter alia, preservation of carbon storage; provision of habitat; regulation of climate, soil, and water quality; supply of timber and jobs; and maintenance of an aesthetically attractive landscape for recreation. We follow the Millennium Ecosystem Assessment (2005) when using the term ecosystem services, meaning the comprehensive benefits humans obtain from ecosystems, for which biodiversity is the central precondition. However, we are aware that other more differentiated terms may be more appropriate, for example, when accounting is the aim (Boyd and Banzhaf 2007), a distinction would be made between ecosystem functions, processes, services, and benefits in place of using only the broad term 'ecosystem services'.

In rather densely populated regions like Central Europe, these services should be achieved by managing for multiple objectives on the same forest area rather than segregating between national parks, where management for timber and non-timber products is fully excluded, and far-from-nature plantations, which are managed for commercial purposes only. If management is subject to multiple objectives, mixed-species forests may be of particular importance. Evidence is growing that mixed-species forest stands can supply many ecological, economic, and sociocultural services in a similar or even superior way to far-from-nature monocultures (Hector and Bagchi 2007; Gamfeldt et al. 2013). Tree species richness may trigger diversity of habitats and other forest plant and animal species (Noss 1990) as well as improve humus conditions and soil fertility (Augusto et al. 2002; Heinsdorf 1999), resilience to disturbances (Griess and Knoke 2011; Neuner et al. 2015), and stand productivity

(Piotto 2007; Pretzsch et al. 2010; Pretzsch and Rötzer, 2016). Since the middle of the twentieth century, mixed-species stands have become the new paradigm of forest management (Puettmann et al. 2009).

In this chapter, we outline how forest management planning (FMP) develops in order to meet the requirements or even support the transition from far-from-nature monocultures to close-to-nature polycultures. We first introduce the various concepts and main steps of FMP in practice. The second section deals with the principle of participation, a key aspect of multiple-purpose FMP, which includes examples of how models may support the change towards complex mixed-species stands. The fourth section sketches how FMP has been enhanced to address mixed-species forests as well as problems that are yet to be addressed. In the last three sections, we demonstrate an example of FMP in a close-to-nature forest estate, including the assessment of biodiversity and landscape aesthetics based on grid-based inventory data. A second example demonstrates the components of a size-class model, which is applicable in mixed, uneven-aged forest. Rather than referring to the general basis of FMP—which is already well covered in existing textbooks (Bettinger et al. 2009; Knoke et al. 2012; Oesten and Roeder 2012)—the concluding take-home messages, like the chapter as a whole, summarise the further development of FMP which will be necessary in view of the close-to-nature management paradigm.

## 10.1 The Practice of Forest Management Planning

FMP may be perceived as a tool for forest owners and stakeholders that informs important forest management decisions at the estate level relating to aspects such as harvest levels and future forest structures to sustain multiple forest services. It is an instrument delivering a plan for a forest that formulates, analyses, evaluates, and improves management decisions at the forest (company) level and thus considers multiple silvicultural management units (forest stands) simultaneously (Bettinger et al. 2009; Knoke et al. 2012). FMP methodology was developed early in Central Europe, as described extensively by Fernow (1911), and usually includes stand-level planning as one inductive planning component. An 'inductive' planning method generates a management plan for a whole forest estate from a bottom-up perspective. It concludes with an aggregation of all the results from planning for each single stand to form an estate-level plan, assuming the whole consists of the sum of its parts. In contrast, a 'deductive' planning method derives the forest management plan from a top-down perspective, usually based on a planning model for the forest estate.

Our practical examples refer to the Central European region, whereas more recent examples are also adopted from operations research-based approaches that have been applied in Scandinavia and northern America. The modern approaches still use some basic principles invented by forest scientists, such as Hartig (1795), Cotta (1804), Hundeshagen (1826), or Judeich (1871), to model the development of a forest in space and time. These principles still play a fundamental role in modern forest practice.

Traditionally, decisions in forest management planning could be split into two types (cf Baader 1945). Long-term decisions to sustain the production of forest goods and services are of strategic importance. Their consequences can extend over 100 years or more. These decisions are summarised under forest 'production planning' and are addressed in more detail in Chap. 11 in relation to the example of choosing tree species and stand types (mixed or pure) to be established. The planning of forest production in particular will benefit from the growing knowledge of the consequences of mixed-species forests, which have been addressed from the organ to the stand level in this textbook (Chaps. 1–9). The second type of decisions relates to 'yield planning' for a forest estate. This is essentially about the regulation of timber harvesting (how much, in which forest locations and in which periods) to produce the desired forest goods and services. Chapter 10 is mainly focussed on the planning of forest yield and methods to describe quantitatively forests' structures and their services at the company level, which often comprises planning periods between 10 and 20 years. Production planning is mentioned only occasionally.

In general, we can outline the process of forest management planning as follows. On the basis of sound information from forest surveys and predictive tools provided by forest growth and yield science, the process of forest management planning formulates, analyses, evaluates, and improves management decisions at the forest (company) level and thus considers multiple silvicultural management units (forest stands) simultaneously. The planning process always starts with an analysis of owners' and stakeholders' objectives (Fig. 10.1). This also entails considering potential objectives of future forest owners and stakeholders (intergenerational fairness). The next step consists of collecting information on the actual forest status using stand-by-stand or enterprise-level inventories. The data collected forms

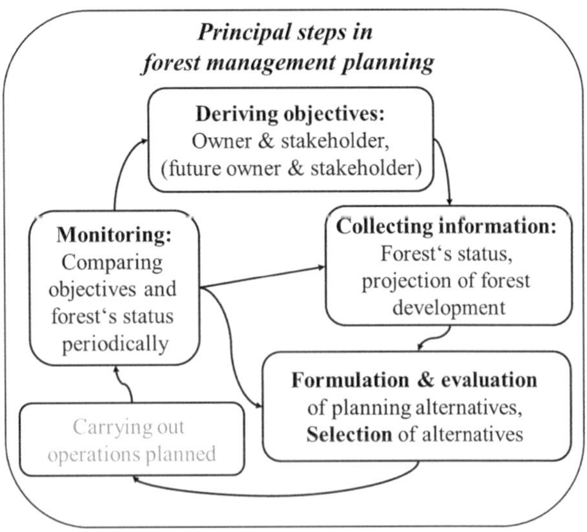


**Fig. 10.1** Steps in forest management planning (Knoke et al. 2012, adjusted)

starting points to project future forest development. For this purpose, traditional yield tables or more modern computer-based growth models are used. The core of forest management planning is then a proposal of enterprise-level management alternatives. This includes information about the maximum allowable harvest and its distribution across space and time and information about the optimum long-term tree species composition. Periodically (usually every 10 years), a new forest management plan is set up to compare the forest status that has been achieved following the implementation of the management activities with the original objectives.

This general procedure in forest management planning is commonly applied to pure or mixed forests. Owners' and stakeholders' objectives are usually incorporated into the planning process via a basic field inspection of the forest that is subject to planning. Here, all actors involved in the planning process take part and articulate their demands in relation to the management planning. Usually, silvicultural strategies at the level of typical stands are discussed during this basic forest inspection to offer a guideline for the forest management planner. The type of forest inventory is also decided on this occasion.

There are several approaches for supporting this planning process. Mendoza (2005) has classified these into two basic methods. One is prescriptive, algorithmic (optimisation-based), and highly structured. The other is descriptive, soft, and qualitative. Although common in literature, prescriptive, optimisation-based approaches are applied relatively seldom in multiple-purpose forest management in Europe, possibly because these approaches are quite data demanding. Except in the case of monospecific plantations, there is a general reluctance in forest practice to accept optimisation-based models because optimisation approaches do not adequately address numerous uncertainties (Hasenauer 2006). The second, 'softer' and more descriptive, approach is most popular in forest practice. In the following, we provide first international examples of the prescriptive approach. Subsequently, we present examples of descriptive, 'soft' approaches.

### 10.1.1 Prescriptive, Optimisation-Based Approaches

Among the more modern approaches to forest management, planning systems prevail that are based on forest growth simulators. These are equipped with capabilities to optimise the planning with regard to various objectives. They can be split into scenario- and optimisation-based examples. Scenario-based approaches predefine possible management strategies and simulate the associated forest development, whereas optimisation-based approaches define objective functions and constraints and set out to find the optimal forest treatment in light of the objectives and constraints. Härtl et al. (2013) mention scenario-based examples, such as the Landscape Management System, LMS, developed for applications in the USA (McCarter et al. 1998). The LMS is a computer program, which generates scenarios at a forest landscape level by integrating forest inventories, spatial information, growth models, visualisation, summarisation, and analysis of trade-offs between scenarios. Another scenario-based example is a decision support system for forest management planning in Southern Austria that is described by Lexer et al. (2005). Their core concept is a characterisation of preformulated management alternatives, which are stored in a database. In order to support FMP, various silvicultural stand treatments have been simulated for representative stands by means of an individual tree growth model. An evaluation of the alternatives by the decision maker is supported by the opportunity to assess the alternatives against the background of his/her own goals and preferences. The evaluation is facilitated by an additive utility model, which uses an analytic hierarchy process. A limitation of scenario-based methods is that the optimal scenario may not be included in the set of scenarios considered. Optimisation-based approaches consist of approximate (heuristic) and exact methods. Approximate methods generate many possible solutions and select the best one in light of a predefined objective. However, it is not possible to be certain that the best option has actually been selected. Thus, heuristic methods may achieve only approximate global optima. Examples of approximate methods include the application of simulated annealing, a heuristic method to achieve optimisation. This method has been applied to find an optimum timber harvest and its spatial distribution (Chen and Gadow 2002) for a Chinese case study. Heureka is a prominent example of an exact optimisation-based method. It is a FMP package from Sweden, where FMP program packages were developed at an early stage (cf Jonsson et al. 1993). Heureka allows for stand-level, estate-level, and regional-level planning. It addresses economic and other objectives and is coupled with an option for basic multi-criteria decision analyses (Lämås and Eriksson 2003; Wikström et al. 2011). The system is only applicable for Sweden because it is firmly interlinked with Swedish yield data.

There are many other examples, as outlined by Härtl et al. (2013) for FMP and by Uhde et al. (2015) with a special focus on multiple criteria methods, for addressing problems related to FMP. The implications of mixing tree species, however, are only addressed if they are taken into account in the biophysical data used to set up the scenarios. This is generally not the case, meaning that it is difficult to evaluate how modern forest management tools apply to mixed-species forests. Some rare instances relating to example effects of mixed-species forests will be described in Sects. 10.2 and 10.3.

### 10.1.2 Planning of Sustainable Yield Using a Descriptive Approach

No unique FMP practice is applied universally worldwide. There is even variation at the European level with different European states applying a diversity of FMP practices (European Commission 2013). Below, we describe an inductive example based on the forest stand level, which represents typical FMP practice in various German forest estates. This stand-level focus is also prominent in many other European countries (European Commission 2013). A stand-by-stand inventory is often the most pragmatic means of obtaining data for yield planning at the level of the silvicultural management unit. Relascope sampling (Bitterlich 1984) is a standard technique to acquire quantitative data on the tree species' basal areas per hectare, which enables the planner to compute the corresponding volumes via yield tables, if age and yield class (site index) are known. In mixed forests, the areas occupied by a certain species are usually treated as if they were pure forests. A mixed forest is, thus, virtually deconstructed into its species, and no interactions—neither those controlling growth nor those responsible for a stand's resistance—are considered. In Europe, growth projections are usually carried out with yield tables, even though it is known that these often underestimate recent growth (Pretzsch et al. 2014).

In larger forest estates, an enterprise-level inventory may form an alternative to the stand-by-stand inventory approach. The size of the forest estate for which this inventory type is meaningful depends on the inventory design, particularly on sampling density. If inventory information is required for forest strata, it is advisable to have at least 30–50 plots per stratum, depending on the variability of the target variable. For enterprise-level inventories, a systematic grid is used to select sampling units, where information about trees is recorded. Sample plot density usually ranges between 1 and 0.25 sampling plots per hectare, and the trees are mostly selected using concentric circles, where smaller trees are recorded in smaller circles and bigger trees in larger circles. This form of inventory is cost-effective and allows for computing the inventory's standard error (precision) at the enterprise level. It can also be applied in mixed, near-natural forests and—if designed as a continuous forest inventory—will facilitate realistic estimates of the (past) timber increment. However, enterprise-level inventory has the disadvantage that it does not deliver information at the stand level. Given that stands become less important under a more near-natural forest management, this disadvantage is likely to progressively disappear, and enterprise-level inventories could become standard in near-natural forest management.

Terrestrial inventories can be supported by remote sensing (RS) methods. Comprehensive guidelines exist to support the practical use of RS approaches (White et al. 2013). RS may be used for pre-stratification of the forest area to enhance sampling precision. It is an excellent tool to support mapping of forest areas and can be used to estimate classical forest information, such as height, tree numbers, and volumes. For example, Wallner et al. (2015) have shown how standing timber volumes can be derived from RS data based on regression curves. In addition, Elatawneh et al. (2014) offered an RS-based opportunity to update forest maps after severe calamities. However, the exact identification of tree species is still a major challenge (Elatawneh 2015). Moreover, RS methods are often widespread in countries with extensive but rather homogenous forest types. Special applications to support the management of mixed forests are rare.

The formulation of planning alternatives usually consists of a combination of stand-level (inductive method) and enterprise-level (deductive method) planning. However, the age-class planning system still prevails as the main deductive method in practical forest management planning in many European countries (European Commission 2013). It is based on the assumption that a so-called 'fully regulated' forest is the aim of any forest management planning to guarantee perfectly continuous timber supply in perpetuity. The age-class distribution in a 'fully regulated' forest is only controlled by the rotation period. Such a forest consists of age classes which are all endowed with an identical area, derived by the quotient of enterprise area and rotation, multiplied by the number of years represented in one age class. Enterprise area divided by rotation would also result in the annual allowable area for final harvests either carried out as real clear-cuts or distributed over the total area of a mature stand or parts of it as small gaps to create regeneration. Allowable harvest area may be modified according to stand-level planning (inductive method) and other indicators. This area multiplied by the average standing volume on a representative hectare of mature stands at the midpoint of the planning period is then the allowable timber harvest from the final crop stands.

The calculus described also feeds back to the methodology of stand-level planning in final crop stands, where harvest areas—as opposed to volumes—are planned in order to calculate harvest amounts by multiplying the stand's volume (inventory at the midpoint of the planning period) with the harvest area planned. The harvests in younger stands through commercial and pre-commercial thinnings are estimated by an experimental marking of those trees which should be removed. Experimental markings may be carried out for larger strata to obtain the silviculturally desirable harvest in immature stands. The procedure of forest management planning is usually repeated every 10 years, sometimes only every 20 years.

For uneven-aged, mixed-forest types, alternatives have been developed, such as the size-class method according to Mantel (1959). Here, the model of a fully regulated forest consisting of age classes is replaced by the model of a fully regulated distribution of tree numbers to size classes. However, no theoretically sound model has yet been developed to derive this fully regulated size-class distribution, which could serve to compute the mean allowable harvest. The size-class method has until now been seldom applied in practical forest management planning, with Fuchs (1996) being one of the rare examples.

### 10.1.3 Planning of Sustainable Production Using a Descriptive Approach

Another important part of practical forest management planning is the development of a long-term tree species composition for a forest enterprise. This is usually based on a site map, where homogeneous site units are delineated. For these site units, various proposals for adequate stand types are made, often with a clear preference for stand types very close to the potential natural forest cover. From these proposals, considering the actual forest composition and possible climate change-related risks for tree species (e.g. Kölling et al. 2009), the planner will derive the 'optimal' long-term composition of tree species for the forest enterprise. However, there is no special consideration given to tree species interactions in mixed forests, and economic models are not integrated (though see Chap. 11 for an economic approach).

The practical procedure described for forest management planning is based mainly on biophysical considerations. In contrast to Speidel's (1972) philosophy—which stressed the economic character of forest management planning—actual plans consider economic aspects and optimisation only rarely. Moreover, practical forest management planning does not include appropriate adaptations for enterprises consisting mainly of mixed forests.

## 10.2 Concept for Participatory Planning in Mixed-Species Forest Stands: The Role of Models and Inventory Data

There are several contrasting FMP perspectives, some of which have already been mentioned. For example, we have compared the prescriptive, algorithmic (optimisation-based), and highly structured approach with the descriptive, soft, and qualitative approach. We can also consider the more prescriptive approaches as being top-down, starting with the idea of a whole forest-level optimisation model, thus following a deductive perspective, which is often based on highly aggregated data. This approach aims at generating information on how a forest should be composed in the long run to deliver the required goods and services in a sustainable way. However, a bottom-up approach starting at the forest stand level is also informative, where the actual composition and the actual potentials of a forest are addressed more accurately. This perspective follows an inductive approach, which may be upscaled to the estate level, as mentioned above, by simple aggregation of the results of stand-level planning. The inductive planning perspective forms a whole picture about the forest estate from planning for its single compartments (forest stands) and is often part of the descriptive and qualitative approach. In this section, we demonstrate a stand-level planning process to convert a pure forest into a mixed-species forest involving the participation of multiple interested groups (stakeholders).

We mentioned above that the demand for multiple forest ecosystem services is growing and that this demand may be well satisfied by mixed-species forests. The participation of forest owners (who may be considered as the forest shareholders) and other interested groups (the forest stakeholders) is thus of particular importance when planning is carried out for pure forest stands which are to be converted into mixed-species forest stands. By elucidating the multivariable, long-term consequences of planning decisions, scenario analysis facilitated by growth models can support both decision making of experts and participation of owners and stakeholders. We illustrate this using the simplified concept for planning and decision making in Fig. 10.2. Starting with a particular actual state of a forest (forest stand, management block, whole enterprise), forest management planning involves the development of a target state for the system and the transformation of the actual into the target state. The development of a target state—in our example a mixed-forest stand of Norway spruce and European beech—results from negotiations among the people concerned, namely, the forest owners and stakeholders. In Fig. 10.2 (right), the negotiation process is symbolised by the

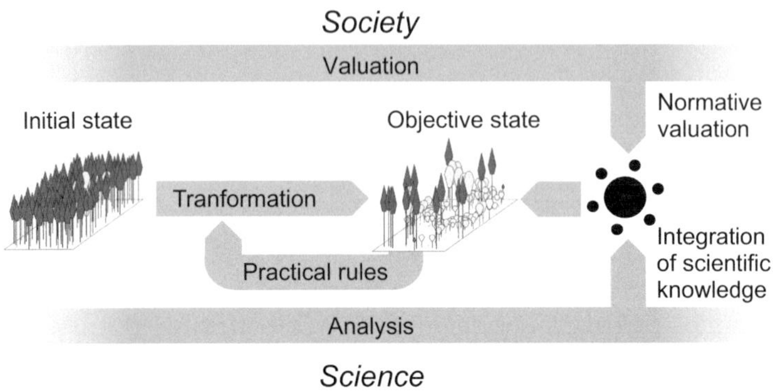


**Fig. 10.2** Concept for stand-based forest ecosystem management and the role of knowledge and models: given an initial state (forest stand, landscape unit), the aim of forest management is to transform a system into a target state. The normative values of society and scientific knowledge contribute to the development and achievement of the objectives

round table. Especially in the densely populated area of Central Europe, where the level of public interest in forests is high, both forest owners and stakeholders have a say when developing the planning objective. Excluding stakeholders from the planning process may ostensibly simplify planning but may trigger protest, countermovement, and departure from the objective when putting the plan into action. Many recent examples of isolated planning in the ivory tower of state forest administrations, forest companies, or municipal authorities provide evidence of planning and implementation failure and stress the importance of participatory forest management planning (Buchy and Hoverman 2000; Kangas 1994; Nordström et al. 2010; Pykäläinen et al. 1999).

Negotiations and development of the objectives tend to be dominated by normative societal values rather than by scientific knowledge. Vague arguments such as 'European beech forests are good because they are attractive and natural' and 'Norway spruce forests are bad because they are unecological and artificial' often play a much more decisive role in this negotiation process than arguments based on scientific knowledge. However, forest science should bring as much system knowledge as possible into the negotiation and decision making process. If the target state is defined clearly and formulated quantitatively, practical rules can be developed as guidelines for realising the transformation (feedback loop in the middle of Fig. 10.2).

The concept presented in Fig. 10.2 reveals at least three promising gateways for introducing scientific knowledge into FMP. The first is the supply of *target knowledge* for the development of objectives, for instance, deciding which species mixture is considered suitable for optimising the expected forest services as well as identifying economic expectations and demand for stand stability, for example, in the face of storms when forest is located close to residential buildings. The second gateway is the supply of *transformation knowledge* that follows after establishing further development objectives, for example, which practice to use to transform a pure stand into a mixed stand, which stand thinning regime to use so that a maximum number of trees with a prescribed threshold diameter can be harvested, or which stand treatments should be implemented to maximise stability against windthrow. The third way to instil valuable system knowledge in the planning process involves the application of *management models and simulators* to illustrate the long-term consequences of planning decisions (see Chap. 8) which are often difficult to foresee for experts and even more so for stakeholders, particularly those who are relatively unfamiliar with forestry.

### 10.2.1 Models for Decision Support

The strength of models in forest research and management is that they display the consequences of applying management options 'in quick motion'. In contrast to many other branches of natural sciences, scenario analyses and models are extremely important in forest research because the lifespan of trees is greater than that of other organisms such as herbaceous plants (Burkhart and Tomé 2012). In forests, we cannot commence experiments each time we have a new management idea or question and then await the results.

To support FMP, forest growth research derives tree growth relationships from already established experimental plots or inventories and aggregates the results in forest growth models. These models enable forest stand growth for the desired silvicultural regime to be modelled in time lapse. The yield-related, financial, and ecological consequences of new management regimes or disturbances can be predicted without establishing new experiments.

When used as a research tool, growth models can replace experiments to a certain degree. Within defined limits, models can predict forest development under changed site conditions, variations in silvicultural concepts, or changes in the intensity and frequency of disturbances. This is not possible in practical experiments. Ecophysiological process models, on the other hand, facilitate the determination of a range of options or a framework of forest management strategies for a given site. They assist the definition of boundaries within which forest management practices do not endanger the stability of the system (Fig. 10.3). The transition from management models such as yield tables, diameter distribution models, or position-dependent individual tree models to models such as succession models or ecophysiological process models requires a more complex database for model generation and parameterisation. The development of models over time—commencing with experience tables and yield tables as a prototype of descriptive stand growth models and progressing to mechanistic ecophysiological process models—simultaneously describes the fundamental change in understanding of forest systems, as well as changes in the information supply and demand in forest management. This should not be regarded as a chronological sequence in which poorer models are succeeded by better models each time. Rather the best model is the one that most effectively fulfils its purpose. Thus, yield tables may be the best approach for forest management in countries where available forest growth data are

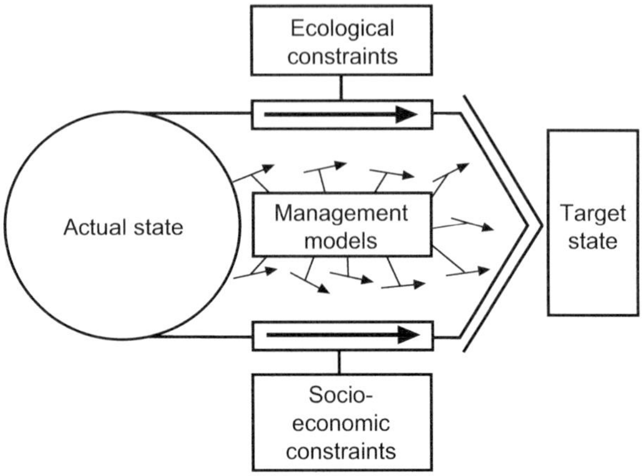


**Fig. 10.3** Management models support decisions within a given range of possibilities (*framed arrows*) by predicting the long-term consequences of treatment variants (*mobile arrows*). The different treatment options can be analysed with mechanistic model approaches at stand and estate level

incomplete or where volume production is given priority. For the sustainable management of pure and mixed-forest stands with detailed structural information, site-sensitive individual tree models may be more appropriate in some circumstances.

The most important applications of stand-level models in forestry are in strategic silvicultural planning for the generation of biophysical data to feed management models. Ecosystem analyses are increasingly being carried out with ecophysiological process-based models. Stand models pave the way for simulation calculations for the entire enterprises, landscapes, or regions (Lemm 1991; von Gadow 2005; Pretzsch and Kahn 1998). Here, the long-term effects of control measures such as thinning, fertilisation, and tree species mixture on forestry may also be predicted.

The main benefit of models lies in the scenario calculations, which display how a management unit (stand, management block, whole enterprise) with a given initial state will develop with respect to system variables ($I_1 \ldots I_n$) if treatment A, B, C, or D were applied (Figs. 10.4 and 10.5). In this example, we distinguish just four scenarios (A) no management, i.e. self-thinning, (B) moderate thinning, (C) target diameter harvest at the end of the rotation period, and (D) the classical clear-cut system. How do these alternative treatments affect a given vector of indicator variables? For each scenario considered, advanced models deliver information about the achievement of objectives, such as carbon stock, stability, growth and yield, biodiversity, protective value, and potential forest recreation uses (see Sects. 10.5 and 10.6). The scenario calculations can be repeated for different treatment options. Given a certain weighting of the different indicator variables, the total value of each option can be assessed, compared with other scenarios, and ranked (Hanewinkel 2001). This ranking based on a comprehensive set of indicators requires the application of multi-criteria decision analysis methods, which will be briefly discussed in later sections of this chapter.

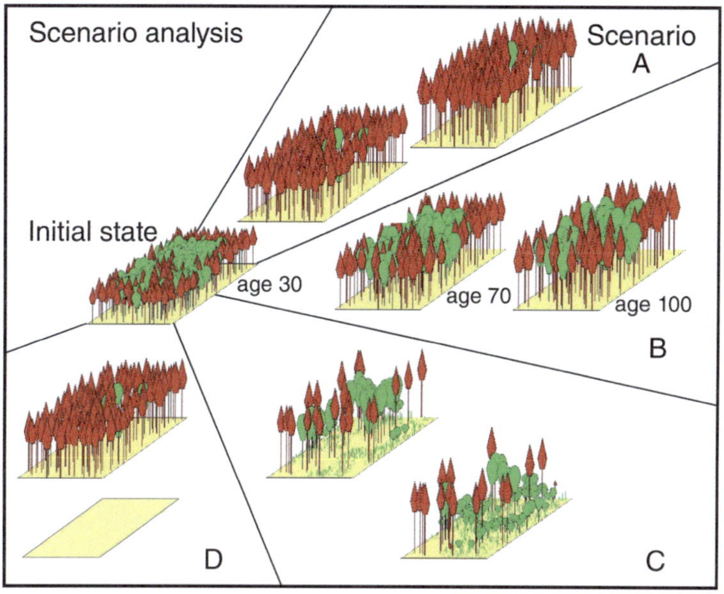


**Fig. 10.4** Scenario analysis with forest stand models: given an initial state of an ecosystem, models display the long-term consequences of the different management options A, B, C, and D for achieving different objectives and target states

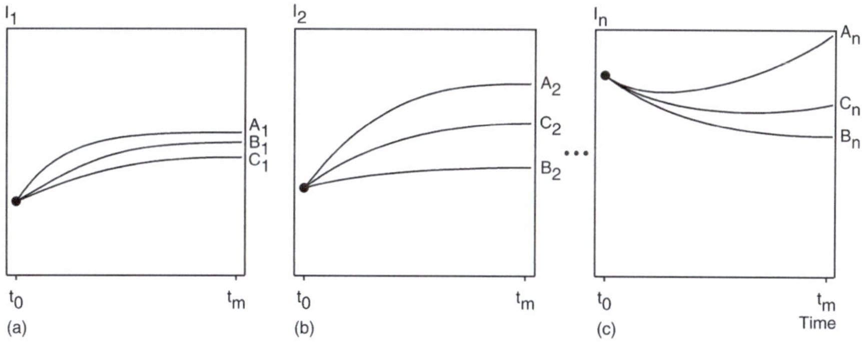


**Fig. 10.5** Application of the indicators $I_1, \ldots, I_n$ (e.g. standing volume, annual cut, biodiversity) for strategic forest management planning. Trajectories $A_i$, $B_i$, and $C_i$ show the simulated dynamics of $I_1, \ldots, I_n$ in the time span $t_0, \ldots, t_m$ as a result of alternative management regimes (e.g. various regimes of thinning, tending, stand establishment, final cut)

## 10.3 Extension of Forest Management Planning to Uneven-Aged Forests and Problems to Be Addressed

Following the above description of 'bottom-up' (stand-level) planning examples, we now address a 'top-down' (enterprise-level) perspective. Based on the age-class system and the original idea of area control invented by Cotta (1804), many quantitative models for harvest scheduling (yield planning) and other forest management problems have been developed worldwide that are based on operations research methods, some of which were mentioned in Sect. 10.1. However, such models have only rarely been adopted in practical forest management planning, and the majority of these are deterministic in nature (Bettinger et al. 2013). Moreover, they insufficiently cover the complexities involved in mixed-species forest growth modelling, the implications of mixing for stand resistance, and the management of mixed-species forests.

In an attempt to overcome these limitations at least in part, Härtl et al. (2013) designed the YAFO model. This estate-level planning model is a typical prescriptive, algorithmic (optimisation-based), and consistently modularised model, which incorporates survival curves for various stand types (mixed and pure) and comprehensively addresses fluctuation in net revenue flows caused by stand failure and timber price volatility. The classic forest planning problem of allocating silvicultural activities to space (forest stands) and time (planning periods) already identified by Cotta (1804) and Judeich (1871) is considered by YAFO to maximise an objective function. The traditional decision variables which indicate the extent to which activities are allocated are area or volume.

In the overwhelming majority of applications, the optimal allocation is achieved through linear programming (Bettinger and Chung 2004). According to such a programming perspective, the single stands forming a forest enterprise and the management activities considered appropriate for them would be perceived as components of a finance portfolio. The objective function can then be maximised for all scheduled management activities simultaneously. Based on precalculated financial coefficients—possibly derived from stand-based growth simulations—linear programming would find an allocation of area or volume to management activities in all considered forest stands distributed over the whole planning horizon under consideration in order to maximise the net present value. However, this approach would not allow for the integration of multiple uncertainties.

In contrast to the widely used linear programming approach, Härtl et al. (2013) use a non-linear programming routine and consider uncertainties relating to timber price fluctuations and natural hazards. Unlike other optimisation programmes and decision support systems, the approach by Härtl et al. (2013) integrates specific survival curves for mixed forests (adopted from Griess et al. 2012) which are able to capture the increase in resistance of a coniferous stand that is achieved through the admixture of broadleaves. Moreover, the model formulation supports various diversification effects to hedge against risks. To achieve risk diversification, the objective function was formulated following the maximin decision rule, which favours very cautious decision making. This formulation follows the precautionary principle to secure sustainable management. In fact, quasi-worst-case scenarios were considered by Härtl et al. (2013) to maximise the outcome under adverse uncertainty. Practically, a predefined quantile at the left end of the distribution of possible economic outcomes is maximised to obtain a result which is most likely to be achieved or even exceeded in reality. This favours a diverse portfolio of forest stands (especially mixed forests) of different ages and also management schedules which distribute harvesting amounts more or less continuously over time to cancel out price risks. Stands are thus not necessarily harvested as a whole by means of clear-cutting, but rather regeneration takes place for smaller parts of forests stands felled in successive operations, which are distributed over more than only one period.

YAFO is an interesting extension of operations research approaches to forest company level planning under multiple uncertainties. Recently, it has been used to analyse realistic timber supply scenarios for Bavaria under various timber price developments, where some price development scenarios were coupled with the oil price development (Härtl and Knoke 2014). Under increasing timber prices, this Bavarian level approach predicted decreasing timber supply in the near future and increasing supply in the more distant future. A further interesting application of the YAFO model that was published by Hahn et al. (2014) investigated the effects of continuous and constant harvesting under multiple risks. This publication showed that continuous and constant harvesting forms an effective hedge against timber price fluctuation. YAFO may be run with various sources of biophysical data. It has recently been applied by Härtl et al. (2016) for an Austrian case study (biophysical data generated with the Austrian PICUS model, Lexer 2001) and a Slovakian case study (biophysical data generated with the Slovakian SIBYLA model, Fabrika 2007). The biophysical data was generated for various climate scenarios, which allowed analysis of the effects of climatic change on the optimal yield planning in various forest enterprises. Interestingly, the YAFO modelling approach revealed that alternative management options, which resemble a close-to-nature, mixed-species forest management, are optimal under climatic change. For example, an ingrowth of more (tree species) diverse stand types was preferred by the optimisation approach over the other ingrowth options, which comprised stand types consisting of fewer tree species. This may be explained by the higher stand resistance in the stand types consisting of more tree species. However, even if the YAFO approach is able to cover some interactions in mixed even-aged forests, it can hardly be applied to already uneven-aged forest enterprises because age and the age-class planning system are still very dominant in the YAFO model (consider Box 10.1 for reasons why age is not a good variable for forest management planning in uneven-aged forests).

**Box 10.1 Age as the Central Variable for Controlling Forest Management Planning Is Not Practical for FMP in Uneven-Aged Mixed Forests**

In age-class forests the model of a fully regulated forest has been successfully applied. Age of forest stands is the main structural variable for controlling the management. For example, the rotation period—the average time until forest stands are considered to be mature for final harvesting—is the single variable determining the indented structure of age classes. It is also of primary importance for the allowable amount of harvestable timber. However, the tree age is no longer important in uneven-aged forests. From a practical perspective, information on tree ages is not achievable at an affordable cost because coring would be necessary. It must also be considered that age is not relevant for planning decisions because trees of comparable sizes could either be very old or still young. A small tree, which has been suppressed for decades, may still develop like a young tree when enough light and other resources become available.

In uneven-aged forests, the tree size (mostly in terms of tree dbh) is a better variable for controlling management. An optimal tree species and size structure (distribution of tree numbers to size classes) may be derived and compared with the actual distribution to derive information on the optimal treatment of the forest.

A practicable alternative to age-class planning systems is a size-class approach, which uses dbh classes instead of age classes. This planning system is well prepared to predict forest development and allowable harvests in uneven-aged, mixed forests. As mentioned earlier, such size-class models were already proposed by Mantel (1949) in his FMP textbook. In the scientific literature, size-class models have been proposed, for example, by Buongiorno and Michie (1980) and Buongiorno et al. (1994).

In size-class models, the forest enterprise is represented by a stem frequency distribution over dbh classes, which may cover the whole area of the forest enterprise or smaller units, such as forest subdivisions of an area from 5 to 35 ha. Forest inventory must therefore deliver structural information about tree dbh, which can be derived from sample plots. If relascope sampling is applied, it is important to measure dbh of the included trees to obtain their representation factor in order to compute the number of trees per hectare (resulting from angle count factor divided by the tree's actual basal area). Moreover, information regarding the expected diameter increment per tree species and dbh class for various stand densities is needed to compute transition probabilities that indicate how likely it is that trees would move from one dbh class into the other. Ingrowth to fill up the smallest dbh class must be estimated and also tree mortality per species and dbh class. Knoke (1998) demonstrated how such a size-class model can be used to estimate the recommendable timber harvest for every tree species and size class in an uneven-aged forest (see our example in Sect. 10.4.2). To be applicable in forest practice, a target size-class distribution is a necessary precondition to apply this approach, which may be derived from economic considerations. However, deriving a target size-class distribution is still problematic even though various proposals exist (Hanewinkel 1999).

Rößiger et al. (2016) applied this approach in a more complex form, while considering the variable diameter increment of trees of the same species and size class. He formulated his management plan as an economic optimisation problem to maximise the net present value from timber harvesting over a long time period of 100 years for a Slovenian beech-fir-spruce forest. Based on transition probabilities from one size class to the next higher size class, on mortality rates and on the ingrowth into the smallest size class, the natural dynamic was simulated. The non-linear model approach computed whether the decision to leave a tree of a specific size class and growing characteristic in the forest and not to harvest it increased the total net present value. If this was not the case, harvesting of the tree was simulated. Decision variables were thus the number of trees per species and size class to be harvested and the number of trees to be left within the forest. Interestingly, the economic optimum was never to clear-cut the forest. The great variation in the optimal felling diameter of trees, varying between tree species but also within the species, depending on the vigour of the trees, led to a significant variation in the scheduled felling times. This approach delivers realistic results and can be adapted by means of constraints on noneconomic requirements of the forest practice. The size-class planning approach would provide an advantageous option for forest management planning in mixed and uneven-aged forests. However, as a limitation of the above-described approach, we should keep in mind that all uncertainties have been excluded.

Size-class models (also called size-structured models) have often been applied in Scandinavian or US American studies to optimise forest management (e.g. Tahvonen et al. 2010; Hyytiäinen and Haight 2010). However, these sophisticated models often consider only one tree species, for example, Norway spruce (Tahvonen et al. 2010), and thus implicitly exclude the consequences of mixed-species forests. Or the initial stand condition is formed by a mixed forest (Hyytiäinen and Haight 2010), but tree species interactions are only rarely accounted for (Haight and Monserud 1990a, b).

Another important pathway to improving FMP for mixed species is the integration of multiple services, as mixed-species forests are often expected to provide higher levels of ecosystem services (Gamfeldt et al. 2013). The more sophisticated approaches to FMP often consider only one objective function and integrate other services, such as protection against gravitational hazards, via constraints (e.g. Härtl et al. 2016). However, multiple objectives are quite common in forest management and numerical approaches for considering more than one objective would be attractive. Although there are many techniques available to consider multiple objectives (e.g. Uhde et al. 2015), their application is largely limited to discrete methods. Discrete methods help to evaluate mutually exclusive alternatives such as various management scenarios, for example, by means of additive utility approaches as a prominent method of multi-criteria decision analysis (e.g. Lexer et al. 2005). However, even if discrete methods deliver valuable information, they may not be able to find the true optimal solution if this solution is not included in the scenarios or alternatives considered (e.g. Seppelt et al. 2013).

In contrast, continuous optimisation methods are not limited to comparing mutually exclusive options. These methods are often more or less closely related to goal programming (e.g. Romero 2001). This type of optimisation means that the set of alternatives is unlimited, which allows, for example, for land resources to be allocated in various proportions to several land-cover alternatives, similar to the programming methods reported in Chap. 11. For instance, Estrella et al. (2014) tested continuous methods to optimise afforestation planning. A substantial gap in multi-objective optimisation is, however, the current lack of methods to actively integrate uncertainty into the optimisation process.

Box 10.2 contains a model formulation, which allows an optimisation of multiple indicators for ecosystem services by means of allocating land area to various possible future forest stand types, considering mixed-species and pure stand types. It has been developed and applied for land-use planning on abandoned tropical lands by Knoke et al. (2016). This model uses a portfolio approach, which aims at minimising the non-achievement of (maximum) indicator values at the estate level. It would be challenging, but also quite interesting, to apply this approach in order to allocate forest area to be regenerated into new future stand types. This would be a completely new approach to multiple objective production planning at the estate level.

**Box 10.2 Suggestion for a Continuous Multiple-Objective Method for Optimising Various Indicators for Ecosystem Services Under Uncertainty that Is Suitable for Allocating Area to a Number Of Possible (Mixed-Species or Pure) Future Forest Stand Types (Concept Published by Knoke et al. 2016; Method for Consideration of Uncertainties Adopted from Knoke et al. 2015)**

$$
\begin{aligned}
&\max\ O = \sum_{i \in I, u_i \in U_i} p_{iu} \\
&\text{subject to} \\
&D_{u,\min} \geq 100 - p_{iu} \quad \forall i \in I \quad \forall u_i \in U_i \\
&0 \leq D_{u,\min} \leq 100 \\
&u_{li} = \pm f_U \cdot \mathrm{SEM}_{li} \\
&\text{if}'' \text{more is better}'' : \\
&p_{iu} = \frac{R_{iu} - \min(R_{liu})}{\delta_{\max,\min}} \cdot 100 \\
&\text{if}'' \text{less is better}'' : \\
&p_{iu} = \frac{\max(R_{liu}) - R_{iu}}{\delta_{\max,\min}} \cdot 100 \\
&R_{liu} = R_{li} + u_{li} \\
&R_{iu} = \sum_{l \in L} R_{liu} \cdot a_l \\
&\sum_{l \in L} a_l = 1 \\
&a_l \geq 0
\end{aligned}
$$

(continued)

**Box 10.2** (continued)

| | |
|---|---|
| $O$ | Objective function |
| $p_{iu}$ | Normalised indicator at estate level in percent, scaled between minimum and maximum |
| $i, I$ | Indicator for ecosystem services, $i$, as a member of the set of indicators considered, $I$ |
| $u_{li}, U_i$ | Indicator deviation from the recorded indicator value and set of all indicator deviations (uncertainty scenarios) |
| $D_{u,\min}$ | Minimised tolerated distance of a normalised indicator, $p_{iu}$, from the maximum achievement level of 100% |
| $f_U$ | Factor to determine the level of the uncertainty deviation considered. $f_U = 0, 0.5, 0.75, \ldots, 2.5, 2.75, 3.0$ |
| $SEM_{li}$ | Standard error of the mean for recorded indicator, $i$, and stand type, $l$ |
| $R_{iu}$ | Forest estate-level indicator value |
| $R_{li}$ | Originally recorded biogeochemical, economic, or social preference indicator value for a stand type, $l$ |
| $R_{liu}$ | $R_{li} \pm$ the considered deviation for given indicator deviations, $u_{li}$ |
| $\min(R_{liu})$ | Minimum indicator value for given indicator deviations, $u_{li}$ |
| $\max(R_{liu})$ | Maximum indicator value for given indicator deviations, $u_{li}$ |
| $\delta_{\max.\min}$ | Range of indicator values between stand type alternatives under given indicator deviations, $u_i$, $\max(R_{liu})$-$\min(R_{liu})$ |
| $a_l$ | Area proportion allocated to a forest stand type alternative, $l$, with $L$ being the set of stand types considered |

## 10.4 Examples of Planning in Mixed or Uneven-Aged Forest Estates

Here, we present an example application of the spatially explicit growth simulator SILVA that is routinely used for FMP in southern Germany (see Pretzsch et al. 2002; Pretzsch 2009). For analogous approaches see, among others, Nagel (1999), Ledermann and Sterba (2006), and Sterba and Ledermann (2006). In addition, a top-down planning approach based on a size class model is demonstrated.

### 10.4.1 Scenario-Based Forest Management Planning in Complex Forests Owned by the Bavarian State

This section demonstrates how to apply simulations at forest enterprise level based on data from continuous forest inventory plots. This is a scenario-based approach coupling the inventory data with a growth model to derive possible management

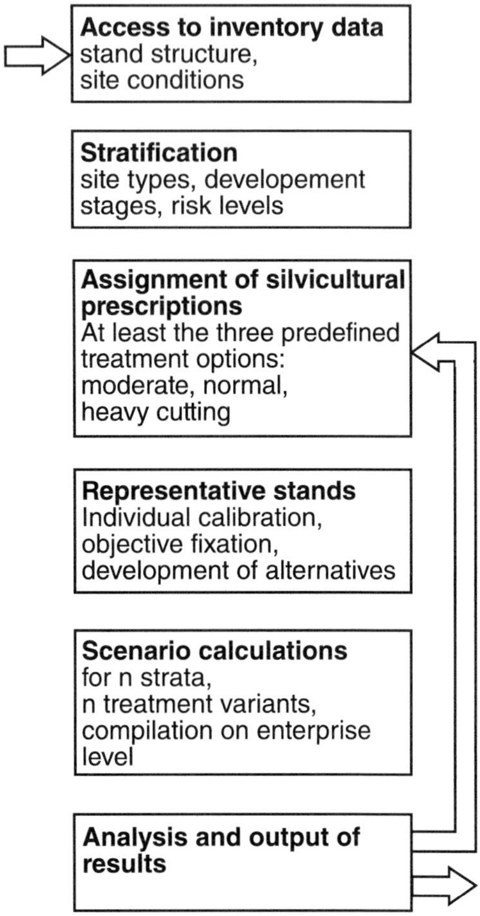


**Fig. 10.6** Concept for the application of growth models for recursive strategic planning at the enterprise level. Simulation helps in understanding the long-term effects of alternative silvicultural treatments on the development of the whole enterprise. Simulation models enable recursive testing of different combinations of treatment options and their effect on enterprise level. In this way, the option that best approximates the overall planning objective can be selected and stipulated in the management plan

scenarios for mixed and complex forests. It involves the following six steps (Fig. 10.6).

#### 10.4.1.1 Access to the Inventory Database

The inventory database contains the essential information about forest structure at the time of the last inventory along with important site characteristics. For the majority of Central European forest enterprises, such data are available in the form of temporary or continuous (sometimes already repeated) grid sample inventories, long-term growth and yield experiments, or at least indicator plots and detailed maps of the site conditions. Such precise information, acquired at a high cost, has until now rarely been exploited in FMP. It could well support FMP in mixed-species forests.

#### 10.4.1.2 Stratification

All sampling plots are allocated to certain strata in a cross-classification process; these strata may refer to site types, stand development stages, or a combination of both. This procedure represents a compromise between a coarse and therefore statistically ineffective stratification and an overly detailed stratification where the number of strata is equal to the number of stands. Stratification serves to reduce the complexity arising from a diversity of sites, tree species mixtures, development stages, etc. to a number of manageable planning units. The criteria 'tree species composition', 'development stage', or 'site conditions' may, for example, be used in stratification. Under relatively homogeneous site conditions in southern Bavaria, we can use tree species composition and stand development stage to stratify the forest area. The strata formed by tree species composition (i.e. stand type) may, for example, be (1) Norway spruce, (2) Norway spruce/hardwood species, and (3) hardwood species.

The stand development stages may be: (1) Juvenile stage (stands with mean diameter at breast height below 7 cm); (2) Growth stage 1 (stands with mean diameter at breast height above 7 cm, with increasing volume increment, up to 35 years old); (3) Growth stage 2 (stands with mean diameter at breast height above 7 cm, with increasing volume increment, over 35 years old); (4) Mature stage (stands with mean diameter at breast height above 7 cm, abating volume increment, no natural regeneration with development potential); (5) Regeneration stage (stands with adequate natural regeneration); (6) Selection stage (stands with at least three layers, with adequate natural regeneration).

Through this procedure, $3 \times 6 = 18$ strata are distinguished, and all sample plots are assigned to one of these 18 strata, which form the basic units for prediction.

#### 10.4.1.3 Assignment of Silvicultural Prescriptions

A specific silvicultural treatment scenario (e.g. selective or schematic thinning, diameter limit or crop tree thinning, thinning from below) is assigned to each stratum. To analyse the long-term effects of different silvicultural prescriptions, standard applications include the assignment of three silvicultural prescriptions to each stratum: a standard variant, a more conservative and low harvest treatment, and a more exploitation-like management.

The growth simulator SILVA 2.2 provides a whole set of predefined treatment programmes, including standard programmes that are widely distributed in forestry practice as well as more conservative or progressive forest management concepts. These treatment programmes are stored in a database. They define, for every stage of a stand/stratum, from the seedling stage up to the regeneration stage, the kind (e.g. thinning from above or below), severity (e.g. light, moderate, or heavy thinning), and intensity (e.g. repetition of thinning in 3-, 5-, or 10-year intervals) of treatment.

#### 10.4.1.4 Representative Stands

For dominant strata or even for all strata, representative stands can be chosen as examples of inspection, discussion, and calibration and as the basis for the development and allocation of silvicultural prescriptions. The representative stands can be visited by the forest owners and stakeholders involved in the planning process to demonstrate the silvicultural procedures planned and to substantiate negotiations of treatment alternatives. They further serve as a reference and training aid for the assigned silvicultural prescriptions.

#### 10.4.1.5 Scenario Calculations

To facilitate the simulation, a representative stand of each stratum is generated on the basis of the inventory points for each respective stratum. The growth prediction over several decades is then carried out strata-wise and for selected treatment alternatives. This makes it possible to analyse the consequences of stand or strata-related treatments for the long-term development of the enterprise as a whole. To prevent undesirable developments at the enterprise level, the treatment regimes applied to a stratum may be modified if required. Consequently, a feedback loop between the allocation of silvicultural prescriptions and the analysis of how they affect whole enterprise development enables what Speidel (1972) called a recursive procedure of forest planning.

#### 10.4.1.6 Analysis and Compilation of the Results

Important results for supporting the planning practice include site-related yield tables with gross felling budgets and maps of the biophysical, economic, and ecological variables for the selected treatment regimes. The results extend from the initial forest state to the state of the forest at the end of the forecasting period and describe the development of indicators on single plots and strata and within the enterprise as a whole.

The advantage of enterprise simulation runs performed by growth models lies in improved planning flexibility: undesired dynamics at the enterprise level can be identified and corrected through changes to silvicultural planning at the stand or enterprise level. Without the help of a simulator, the recursive procedure for which Speidel (1972, p. 162) strived—'corrections are repeated as long as a convenient adjustment between singular and total planning is attained'—has proven to be very elaborate and barely feasible within the planning of sustainable timber production, even in even-aged forests. In uneven-aged forests with a wider range of management models and sustainable development criteria, a recursive procedure is simply impossible without the support of simulation models.

### 10.4.2 Deriving the Annual Harvest in an Uneven-Aged Mixed Forest Using a Size-Class Approach

Here, we use an example adopted from Knoke et al. (2012, pp. 35–43) to illustrate how size-class-based forest management planning works in principle. The example is the Kreuzberg municipal forest, an uneven-aged forest consisting of spruce, fir, beech, and occasionally some maple trees. We need a system of three equations to form a simple prototype for a size-class model, which may be traced back to Prodan

(1949a, b), to derive the annual allowable harvest for an uneven-aged forest. These equations characterise the ingrowth in a size class (Eq. 10.1), the remaining trees (Eq. 10.2), and the harvest (Eq. 10.3). To keep the example simple, we do not distinguish between the mentioned tree species here. In the practice of FMP, the obtained numbers of trees scheduled for harvesting would be distributed across the existing tree species according to their actual proportions.

$$\text{Ingrowth } i_x = n_{x-1}\frac{id_{x-1}}{w} \tag{10.1}$$

$$\text{Remaining } r_x = n_x - n_x\frac{id_x}{w} \tag{10.2}$$

$$\text{Harvest } u_x = i_x + r_x - t_x \tag{10.3}$$

$i_x$ Number of trees growing into the next bigger size class, $x$
$n_{x-1}$ Number of trees in the next smaller size class, $x - 1$
$id_{x-1}$ Diameter increment of the trees from the next smaller size class, $x - 1$
$w$ Width of the size classes (in our case 4 cm)
$id_{x-1}$÷w Probability of ingrowth (transition probability) into a size class, $x$
$r_x$ Number of trees remaining in a size class, $x$
$n_x$ Number of actual trees present in a size class, $x$
$id_x \div w$ Probability of leaving a size class, $x$
$h_x$ Number of stems to be harvested in size class, $x$
$t_x$ Target stem number in size class, $x$, to be achieved in future

Table 10.1 presents an example of such a consideration for the case that the target stem number is already achieved. Data was obtained from Knoke (1998). The target stem numbers are crucial for this model. Knoke (1998) proposed a statistical approach to estimate a target distribution, which we used for this consideration. Further, we assumed an average diameter increment of 0.4 cm per year, leading to a probability for ingrowth and for leaving a size class of 0.1. It was also assumed that the ingrowth into the smallest size class would always fill up the stem numbers to completely compensate the number of trees leaving. Note that the sum of trees that remain in a size class and of those that grow into a size class (predicted actual stem number) is now compared with the target number of trees per size class. If the difference between the predicted actual stem number and the target stem number is positive, this figure represents the number of trees to be harvested. Finally, this number has to be multiplied by the trees' volume to estimate the number of cubic metres to be harvested.

The size-class approach works well if the actual and the target distribution of tree sizes to size classes is already similar. If they are rather different, the approach would lead to quite discontinuous harvest levels. We carried out the same consideration for an actual distribution of tree sizes to size classes and obtained a harvest

**Table 10.1** Computation of the allowable harvest per year for a selection forest, where the target stem number has already been achieved

| | Stem numbers (per ha) | | | | | | Timber volume (cubic metres outside bark) | |
|---|---|---|---|---|---|---|---|---|
| Size class (cm) | Actual ($n_x$) | Target ($t_x$) | Ingrowth ($i_x$) | Remaining ($r_x$) | Predicted actual $i_x + r_x$ | Harvest ($h_x$) | Per stem | Harvest |
| 16 | 59.58 | 59.58 | 5.96 | 53.64 | 59.58 | 0.00 | 0.10 | 0.00 |
| 20 | 48.60 | 48.60 | 5.96 | 43.74 | 49.70 | 1.10 | 0.18 | 0.20 |
| 24 | 39.70 | 39.70 | 4.86 | 35.73 | 40.59 | 0.89 | 0.30 | 0.26 |
| 28 | 32.40 | 32.40 | 3.97 | 29.16 | 33.13 | 0.73 | 0.46 | 0.33 |
| 32 | 26.40 | 26.40 | 3.24 | 23.76 | 27.00 | 0.60 | 0.65 | 0.39 |
| 36 | 21.60 | 21.60 | 2.64 | 19.44 | 22.08 | 0.48 | 0.90 | 0.43 |
| 40 | 17.60 | 17.60 | 2.16 | 15.84 | 18.00 | 0.40 | 1.21 | 0.48 |
| 44 | 14.40 | 14.40 | 1.76 | 12.96 | 14.72 | 0.32 | 1.55 | 0.49 |
| 48 | 11.80 | 11.80 | 1.44 | 10.62 | 12.06 | 0.26 | 1.95 | 0.51 |
| 52 | 9.60 | 9.60 | 1.18 | 8.64 | 9.82 | 0.22 | 2.41 | 0.53 |
| 56 | 7.80 | 7.80 | 0.96 | 7.02 | 7.98 | 0.18 | 2.83 | 0.51 |
| 60 | 6.40 | 6.40 | 0.78 | 5.76 | 6.54 | 0.14 | 3.38 | 0.47 |
| 64 | 5.20 | 5.20 | 0.64 | 4.68 | 5.32 | 0.12 | 4.00 | 0.48 |
| 68 | 4.30 | 4.30 | 0.52 | 3.87 | 4.39 | 0.09 | 4.64 | 0.42 |
| 72 | 3.50 | 3.50 | 0.43 | 3.15 | 3.58 | 0.08 | 5.34 | 0.43 |
| 76 | 2.80 | 2.80 | 0.35 | 2.52 | 2.87 | 0.07 | 6.07 | 0.43 |
| 80 | 2.30 | 2.30 | 0.28 | 2.07 | 2.35 | 0.05 | 6.83 | 0.34 |
| 84 | 0.23 | | 0.23 | | 0.23 | 0.23 | 7.70 | 1.77 |
| **Timber harvest in** $m^3\ ha^{-1}\ year^{-1}$: | | | | | | | | **8.48** |

Size classes are 4 cm wide and an annual diameter increment of 0.4 cm was assumed level of more than 19 cubic metres per ha for the first year (Table 10.2) but only about 4 cubic metres per ha for the second (Fig. 10.7).

Negative numbers in Table 10.2 indicate that the predicted actual tree number is lower than the target tree number. Thus, no harvest is possible in those size classes. Seventeen years would be needed for the target stem number in each size class to be achieved.

The average harvest level for 10 years (the normal planning period) amounts to 7.4 cubic metres/ha/year. This could be an indicator for planning an appropriate annual harvest level. The simplified procedure following the size-class model is actually helpful for forest management planning in uneven-aged forests. As Rößiger (2014) has shown, this approach can be further developed into an economic optimisation approach. However, research still needs to be carried out to cover all the interactions between tree species and their influence on growth, ingrowth, competition, resistance, mortality, and finally also the economic results.

Modelling of the ingrowth is still a big challenge. Many studies assume that there is a simple relation between the forest basal area and the ingrowth, with a higher ingrowth when basal area declines. These relations often obtain a maximum with zero basal area, which means that no seed trees are available (see Kant 1999 as one example). This assumption may lead to quite unrealistic prescriptions. Thus, enhancement of the biophysical modelling of ingrowth dynamics would be quite essential to improve the model results (cf Rößiger et al. 2016).

## 10.5 Assessment of Biodiversity

Caring for biodiversity may preserve endangered species for ethical reasons, and mixed-species forests have been considered beneficial for this aim (e.g. Knoke et al. 2008). Maintaining biodiversity may affect the functional, compositional, and structural features of ecosystems in a way that many desired forest functions and services—such as stability, resilience, productivity, or aesthetics—are improved. Both ethical and pragmatic aspects of biodiversity are considered increasingly crucial for sustainable forest ecosystem management (MCPFE 1993; Haber 1982; Hector and Bagchi 2007; Sterba and Ledermann 2006). For monitoring, planning, and management in multifunctional forests, biodiversity is becoming a criterion of equal rank with standing timber stock or annual cut, at least in the public forests. The biodiversity aspect is not the only extension of the classical sustainability paradigm. Carbon storage (Rötzer et al. 2010), nutrient cycling (Blanco et al. 2005), non-wood forest products (Pukkala 2013), wildlife management (Reimoser et al. 2009; Vospernik and Reimoser 2008), water protection (Wolfslehner et al. 2005), erosion control (Brauner et al. 2005), and recreation functions (Pretzsch and Seifert 1999) are also gaining increasing attention (Gamfeldt et al. 2013).

**Table 10.2** Computation of the harvest in one specific year for a selection forest, where the actual and target stem numbers are different

| Size class (cms) | Stem numbers (per ha) | | | | | | Timber volume (cubic metres outside bark) | | Stem numbers (per ha) |
|---|---|---|---|---|---|---|---|---|---|
| | Actual ($n_x$) | Target ($t_x$) | Ingrowth ($i_x$) | Remaining ($r_x$) | Predicted actual $i_x + r_x$ | Harvest ($h_x$) | Per stem | Harvest | Resulting numbers |
| 16 | 66.2 | 59.58 | 5.96 | 59.58 | 65.54 | 5.96 | 0.10 | 0.60 | 59.58 |
| 20 | 44.0 | 48.60 | 6.62 | 39.60 | 46.22 | −2.38 | 0.18 | 0.00 | 46.22 |
| 24 | 34.0 | 39.70 | 4.40 | 30.60 | 35.00 | −4.70 | 0.30 | 0.00 | 35.00 |
| 28 | 27.8 | 32.40 | 3.40 | 25.02 | 28.42 | −3.98 | 0.46 | 0.00 | 28.42 |
| 32 | 22.5 | 26.40 | 2.78 | 20.25 | 23.03 | −3.37 | 0.65 | 0.00 | 23.03 |
| 36 | 18.6 | 21.60 | 2.25 | 16.74 | 18.99 | −2.61 | 0.90 | 0.00 | 18.99 |
| 40 | 16.7 | 17.60 | 1.86 | 15.03 | 16.89 | −0.71 | 1.21 | 0.00 | 16.89 |
| 44 | 14.3 | 14.40 | 1.67 | 12.87 | 14.54 | 0.14 | 1.55 | 0.22 | 14.40 |
| 48 | 13.3 | 11.80 | 1.43 | 11.97 | 13.40 | 1.60 | 1.95 | 3.12 | 11.80 |
| 52 | 10.8 | 9.60 | 1.33 | 9.72 | 11.05 | 1.45 | 2.41 | 3.49 | 9.60 |
| 56 | 9.2 | 7.80 | 1.08 | 8.28 | 9.36 | 1.56 | 2.83 | 4.41 | 7.80 |
| 60 | 7.2 | 6.40 | 0.92 | 6.48 | 7.40 | 1.00 | 3.38 | 3.38 | 6.40 |
| 64 | 5.7 | 5.20 | 0.72 | 5.13 | 5.85 | 0.65 | 4.00 | 2.60 | 5.20 |
| 68 | 3.9 | 4.30 | 0.57 | 3.51 | 4.08 | −0.22 | 4.64 | 0.00 | 4.08 |
| 72 | 2.5 | 3.50 | 0.39 | 2.25 | 2.64 | −0.86 | 5.34 | 0.00 | 2.64 |
| 76 | 1.4 | 2.80 | 0.25 | 1.26 | 1.51 | −1.29 | 6.07 | 0.00 | 1.51 |
| 80 | 1.8 | 2.30 | 0.14 | 1.62 | 1.76 | −0.54 | 6.83 | 0.00 | 1.76 |
| 84 | 0.0 | 0.00 | 0.18 | 0.00 | 0.18 | 0.18 | 7.70 | 1.39 | 0.18 |
| **Timber harvest in $m^3\ ha^{-1}$ in a specific year:** | | | | | | | | **19.21** | |

Size classes are 4 cm wide and an annual diameter increment of 0.4 cm was assumed

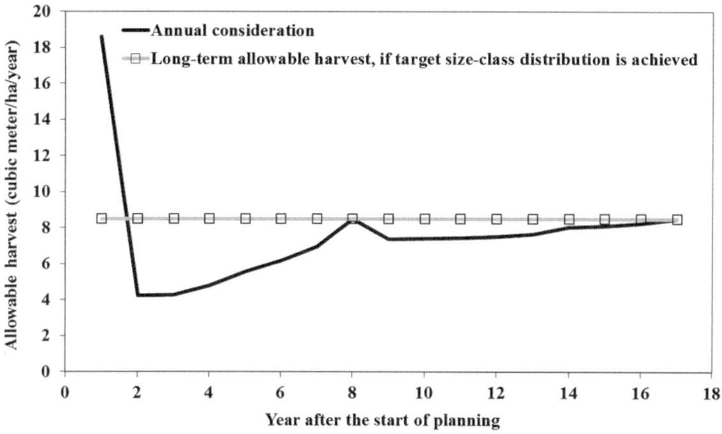


**Fig. 10.7** Harvest levels per year derived for the transition phase in order to approach the target stem numbers (adopted from Knoke et al. 2012)

Classical sustainable forest management has been focussed on forest growth and yield in all its essential steps, from inventory and planning to continuous operation and control. The ongoing integration of biodiversity and other aspects into planning certainly requires additional and new information—e.g. regarding occurrence of species or their habitats—which is currently relatively limited.

This section shows how to exploit available classical forest inventory data in order to support the management and conservation of biodiversity. Because of their size, firm position, and longevity, trees are the founder species in ecosystems (Whitham et al. 2006) and determine the habitat conditions, functions, and services of ecosystems (Hector and Bagchi 2007). Therefore, even classical forest inventories for timber production—especially when based on fixed grid points, including spatially explicit tree data, and when repeated over time—include valuable information on rarely extracted aspects of functional, compositional, and structural diversity. For example, the annual growth rate not only provides information about the allowable cut but also correlates closely with leaf biomass turnover, which is essential for feeding all herbivores and decomposers. Tree size and structure not only determine standing stock and wood quality but also form the habitat of many animals such as birds, bats, and spiders.

Of course, the aspects extractable from forest inventory data are not sufficient for a comprehensive quantification of biodiversity. However, the following set of indices represent another step forward from a rather constricted dendro-centric to a broader multiple-purpose forest planning and management.

### 10.5.1 Composition, Structure, and Functioning at Different Scales as Components of Biodiversity

Following Noss (1990), we refer to biodiversity as the setup of an ecosystem with respect to its composition, functioning, and structure, as sketched in Fig. 10.8. Compositional characteristics are the number of species, genotypes, communities, and landscape types. Functional aspects include, among others, biomass production, growth, competition, mortality, decomposition, and natural regeneration. Structural attributes are plant morphology, stand structure, gap patterns after disturbances, spatial distribution of stand development phases, and ecosystem fragmentation caused by roads, circuit lines, channels, or densely populated urban areas. All three components are relevant for a comprehensive assessment of biodiversity. For example, an ecosystem may be composed of a high number of occurring species but the competitive strength and growth of these may be so low that they are likely to vanish from the forest community. In another scenario, the present functioning of an ecosystem—e.g. a high biomass productivity—may on its own suggest vitality and fitness, but this may be based on nitrogen import which reduces mechanical tree and stand stability.

Furthermore, all three components should be considered at multiple spatial (individuals, stand, forest, landscape level) and temporal (vegetation period,

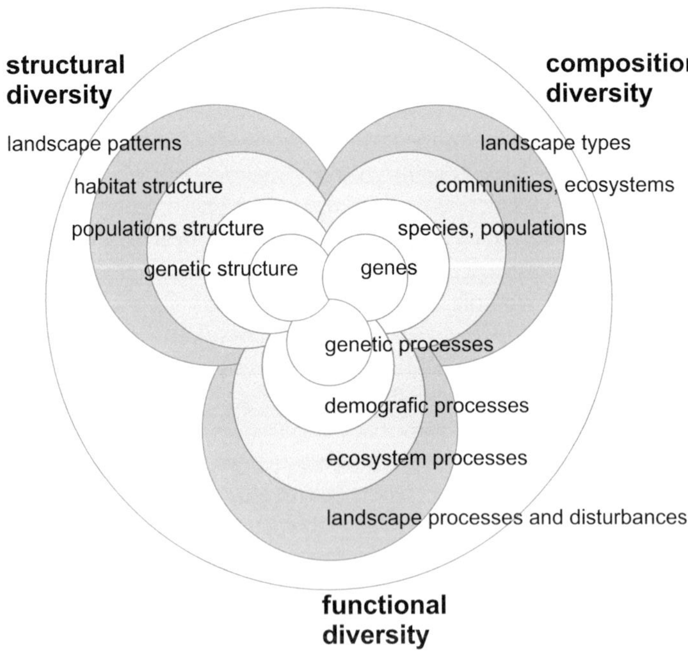


**Fig. 10.8** Composition, structure, and functioning at different scales as components of biodiversity (according to Noss 1990)

**Table 10.3** Indicating biodiversity with compositional, functional, and structural characteristics extracted from forest inventory data

| Composition | Structure | Functioning |
|---|---|---|
| Tree species richness | Tree morphology | Growth |
| Tree species diversity | Tree size distribution | Productivity |
| Species set in main layer | Height layering | Mortality |
| Species set in natural | Horizontal pattern | Disturbance |
| Regeneration | Stand mosaic at the forest level | Regeneration |

years, decades, inventory intervals) scales. The relevance of a given tree species number can differ depending on the spatial scale it refers to. For instance, interspecific interaction (competition, facilitation) between tree species may be intensive and cause overyielding in individual mixtures—less intensive in group or cluster mixture—and negligible when species are separated from one another in pure stands. The temporal scale of biodiversity becomes obvious through, for instance, the temporary occurrence of pioneer species after windthrow or other large-scale disturbances. That those species are transient and vanish after replanting or natural regeneration might be recorded by successive monitoring or inventories.

### 10.5.2 Forest Inventories as Database

The essential aspects of biodiversity—composition, functioning, and structure—can partly be recorded using classical or further developed forest inventories. Forest inventory data comprises as yet barely exploited information about all three mentioned components of biodiversity (Table 10.3). Classical forest inventories recorded primarily standing volume and volume growth in order to derive the sustainable annual cut. The information unit was the stand or district level, and by summarising all stand information, standing stock and growth of volume were available for each age phase, tree species, etc. of the respective forest area or forest enterprise.

Modern grid-based inventories (e.g. raster width 100–200 m) with successive measurements on continuous sample plots (e.g. plot area 250–500 $m^2$) provide additional data for extracting biodiversity information (see Fig. 10.9). This applies especially for advanced inventories (Tomppo et al. 2010) which consider tree species, tree coordinates, regeneration, disturbances, standing and lying dead wood beyond classical records of diameter, basal area, and tree height. As tree and forest stand characteristics determine many compositional, functional, and structural traits, they can be used as proxies for biodiversity because they can enrich habitat diversity (Tews et al. 2004) as well as plant and animal species richness (Ishii et al. 2004; Roth 1976).

When grid based, the inventory data provides information on trees and at the individual plot level (e.g. 500 $m^2$) but also at larger spatial scales (management block, enterprise, landscape). Through integrated evaluation, the records on the

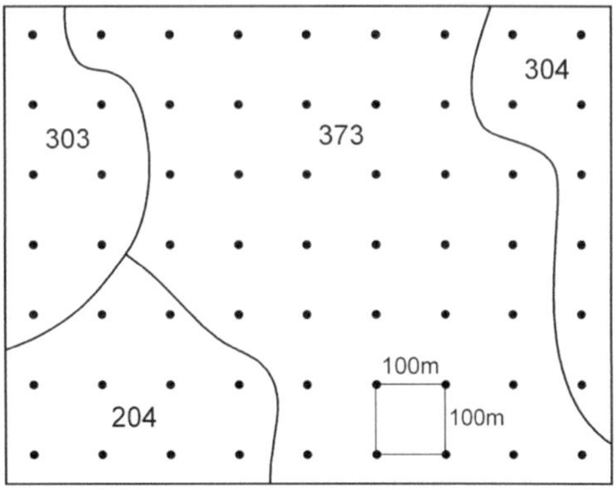


**Fig. 10.9** Grid-based permanent forest inventories as a basis for evaluation of the state and dynamics of forest biodiversity at various spatial and temporal scales. The raster width is 100 m; the three-digit numbers represent site classification units

plots can deliver a spatial pattern of biodiversity aspects. The greater the number of successive surveys (normally at intervals of 10–20 years), the better is the information on the temporal changes in biodiversity. In some areas in southern Germany, four successive surveys provide exciting insight into how close-to-nature forestry changed the species composition from rather patchily structured, conifer-dominated towards more homogeneous-structured, deciduous forests.

### 10.5.3 Example of an Integrated View on Biodiversity Based on Forest Inventory Data

The database for the following model example comprises successive, grid-based forest inventories at forest enterprise level (10,000–20,000 ha). Grid width is 100–200 m, so that each inventory plot represents 1–4 ha. On the 500 $m^2$-sized plots, the size and position of standing and removed trees are recorded on concentric circles. On an inner circle ($r = 2.82$ m, 25 $m^2$), trees of all sizes including regeneration are measured. On the medium ($r = 6.31$ m, 125 $m^2$) and outer ($r = 12.62$ m, 500 $m^2$) circle, trees with diameter $<12$ cm, 12–29.9 cm, and $\geq 30$ cm, respectively, are measured.

Based on this data, we derived indices for biodiversity at three spatial scales: for each single inventory plot (represented area 1–4 ha), for each plot including its four nearest plot neighbours (5–20 ha), and for each plot including its nearest 25 neighbours (26–100 ha). This evaluation addresses any variation in biodiversity compounds with increasing spatial scale. The temporal variation in biodiversity may be revealed through evaluation of indices based on successive inventories, e.g. five inventories in 10-year intervals.

We use species richness, maximum tree diameter, and stand productivity as indicators for composition, structure, and functioning, respectively. Table 10.4 shows the development of these three indicators from 1993 to 2010 for a large region in southern Germany as a whole (152,657 grid points) and a specific forest

**Table 10.4** Change in selected compositional (mean species number), structural (mean $d_{max}$), and functional (mean periodic annual volume increment) components of forest biodiversity in a forest enterprise from inventory spring 1993 to autumn 2010 compared with the characteristics in a large region in southern Germany

| | Level 1 | | Level 2 | | Level 3 | |
|---|---|---|---|---|---|---|
| Variable | 1993 | 2010 | 1993 | 2010 | 1993 | 2010 |
| Species richness (number) | | | | | | |
| Region | 2.3 | | 4.8 | | 9.1 | |
| Enterprise | 2.6 | 2.8 | 5.6 | 6.1 | 10.1 | 11.3 |
| Maximum diameter (cm) | | | | | | |
| Region | 44.6 | | 59.7 | | 71.8 | |
| Enterprise | 44.4 | 50.9 | 59.0 | 64.5 | 71.0 | 76.2 |
| Productivity, PAI ($m^3$ $ha^{-1}$ $year^{-1}$) | | | | | | |
| Region | 7.3 | | 8.2 | | 8.7 | |
| Enterprise | 6.5 | | 6.1 | | 5.9 | |

The changes are shown for the three spatial scales 3, 14, and 69 ha

enterprise (4531 grid points) at the three spatial scales of 3, 14, and 69 ha, respectively.

Among other compositional variables, the number of tree species and their occurrence at different spatial scales is indicative for biodiversity. Tree species richness often increases the variety of structures and habitats, triggers emergent properties and interactions between neighbouring species, decreases risks, and increases resilience to disturbances. Species richness can be relevant and achieved at different spatial and temporal scales; while plants or animals with a narrow zone of influence may need species richness at plot or stand level, larger mammals or birds and also humans may benefit from species richness at a wider scale.

The analysis reflects that, compared with the large region as a whole, this enterprise has a high species richness, which has even increased over the last 20 years. For the region as a whole as well as for the enterprise level, the finding that species richness increases considerably with an increase in the spatial scale shows the strong spatial heterogeneity. Whereas at the plot level the species richness amounts to just two to three species, the species assemblages vary strongly in space both within the state and the enterprise.

The maximum tree diameter is an important structural stand characteristic that indicates how far developed a stand is in terms of vertical expansion and system age. In addition, $d_{max}$ reflects an important habitat element for bats and birds and the potential of natural stand regeneration. The maximum tree diameters at the enterprise level are similar to the mean in the large region in southern Germany. However, at the enterprise level the silvicultural treatment has increased the maximum tree diameter over the last 20 years by about 5 cm.

As a key indicator for forest functioning in terms of growth, supply of biomass, and turnover, we use the mean and standard deviation of the periodic annual volume growth per hectare, PAI, in the respective inventory period. The productivity in terms of mean annual volume growth per unit area is correlated with many functional traits such as leaf area, leaf mass turnover, soil fertility, energy flow through the system, etc. The mean volume productivity and its coefficient of variation tell much about the absolute level of biomass synthesis and how evenly it is distributed over space. The productivity at the enterprise level was below the average in the large region as a whole; productivity increased at the regional level but decreased at the enterprise level.

Advanced management models use the spatially explicit inventory data as initial values to simulate and derive the sustainable annual cut. When models incorporate the compositional, functional, and structural indicators of biodiversity and model their development in a spatially explicit way, they have the capacity to also predict the effect of different management options on the biodiversity characteristics at different spatial scales (Sterba and Ledermann 2006).

In order to mitigate the ongoing biodiversity loss and extinction of species, aspects of biodiversity need to be integrated in forest planning and participatory decision making (Fig. 10.2). An important requisite for this integration is the quantification of the compositional, structural, and functional aspects of biodiversity. Compositional and structural aspects can be used as proxies for the living conditions of key species or functional groups of plants and animals. By quantifying those aspects of biodiversity, they become manageable for being integrated in inventory, modelling, optimisation, decision making, and forest management planning.

## 10.6 Visualisation of Forest Management Planning Scenarios

Visualisation is of particular importance for forest management planning in mixed-species stands where the aesthetic impression achieved through specific species assemblages or vertical stand structures may be essential for the recreation value (Stölb 2005). The easiest way to explain the long-term effect of planning options to nonspecialised stakeholders in the course of a participatory planning process (Fig. 10.2) is by means of visualisation (Pretzsch and Seifert 1999; Seifert 2006, 2008). Very helpful for a participatory planning process is the combination of stand simulation and visualisation, which illustrates the long-term effects of different silvicultural options. This paves the way for the integration of aesthetic aspects in long-term planning and decision making.

We consider four main criteria for an effective forest stand and landscape visualisation tool. Firstly, visualisation needs to cover the temporal scales suited to both the level of human perception and tree growth and forest stand development. This necessity arises from the fact that people's perceptions of landscapes are rather short term compared to forest regeneration cycles. Secondly, visualisation should be based on solid data (e.g. Sheppard and Harshaw 2001). Thirdly, visualisation must be realistic. To support the intuitive recognition pattern of the human brain, it is essential to display plants and landscapes as realistically as possible (Meitner et al. 2005). This implies, for instance, that visualisation of forests needs to be based on individual trees to account for structural differentiation which is an important element of forest recognition. Finally, visualisation must allow people to freely choose their perspective. A single static view is insufficient for gaining an adequate impression of a forest stand or landscape. Different viewpoints and perspectives are required for an assessment of aesthetic value for decision making. Being able to choose freely between the equally important bird's eye view and the human perspective from the forest floor is important for experiencing the properties and aesthetics of a landscape (Bell 2001).

### 10.6.1 Visualisation Tools TREEVIEW and L-VIS

As examples of visualisation tools, we present TREEVIEW, a software optimised for fast, spatially explicit interactive visualisation of forest stands, and L-VIS, the forest landscape visualisation software (Pretzsch et al. 2008; Seifert 2008).

TREEVIEW is designed for realistic and thematic forest visualisation at the stand level. It can directly display the outputs of spatially explicit statistical or process-based individual tree models for monocultures and mixed-species stands (Fig. 10.10). With TREEVIEW, it is possible to display interactively the simulation results of individual tree models, perform a virtual flight through the stands, and manually select future trees or trees for interactive thinning by mouse click (Fig. 10.11). TREEVIEW displays geometrically modelled trees with the same geometry as the virtual objects in the individual tree simulation models. To create a more realistic appearance, species-specific textures are applied to the tree models.

The landscape visualisation system L-VIS (Seifert 2006, 2008) was developed to create realistic views of forest landscapes up to an area of $5 \times 5$ km$^2$. It uses the

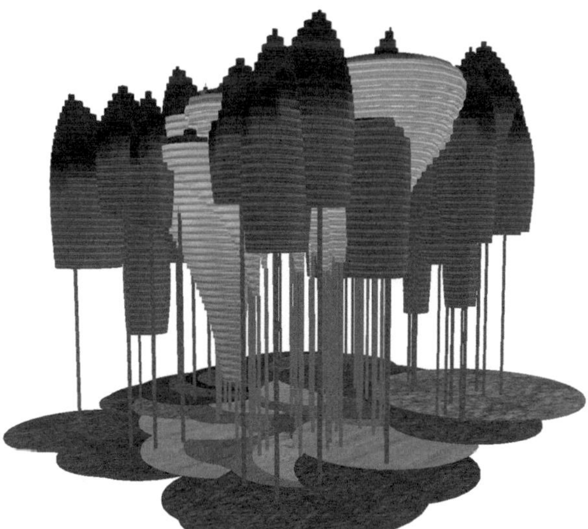

**Fig. 10.10** TREEVIEW can visualise the output of ecophysiological models (e.g. nutrient content in different organs, water content, and leaf area density). In this example, leaf area density in a mixed stand of Norway spruce and European beech is indicated by colour tone (the *darker* the *shade*, the *higher* the *leaf* area density)

**Fig. 10.11** TREEVIEW enables the interactive elimination of trees to promote selected future crop trees in monocultures and mixed-species stands

simulation results of individual tree simulators, such as SILVA 2.0, BWIN, or MOSES (Pretzsch et al. 2008), to visualise changes in a landscape, e.g. according to defined management or climate scenarios. It can be used in participatory planning and for public dissemination of scientific results on expected landscape changes.

The user may obtain a visual impression of the scenario displayed from an interactive walk through the forest and may also track virtual change in the forest over time as shown in Fig. 10.12. This example shows the importance of appropriate silvicultural treatment for maintaining planned mixing proportions during long-term stand development. Figure 10.12 shows a section of the 'Bürgerwald' in Traunstein. Initially, the stand comprises a 25-year-old mixed stand of Norway spruce and European beech. This stand, with an area of about 5 ha, is displayed firstly in its existing condition by the visualisation program L-VIS (Fig. 10.12, upper line). The growth simulator SILVA is used to simulate the stand development from 25 to 125 years in five-year time steps. Three management alternatives are compared in three corresponding simulation runs: (1) development with no silvicultural management, (2) moderate promotion of beech by thinning from above, and (3) strong promotion of beech by thinning from above. The results of these alternatives are visualised with L-VIS (Fig. 10.12, left, middle and right column, respectively). With no silvicultural intervention, a homogeneous pure Norway spruce stand evolves (left column). Beech cannot compete against Norway spruce and is lost almost completely by the age of 125 years due to self-thinning. With moderate promotion, European beeches account for 20% of the basal area at the end of the simulation period. With strong promotion, this figure is 50%.

Visualisation may evolve into an effective tool in participatory planning and decision making processes by assisting forest managers and stakeholders to make decisions about which forest management strategies to adopt.

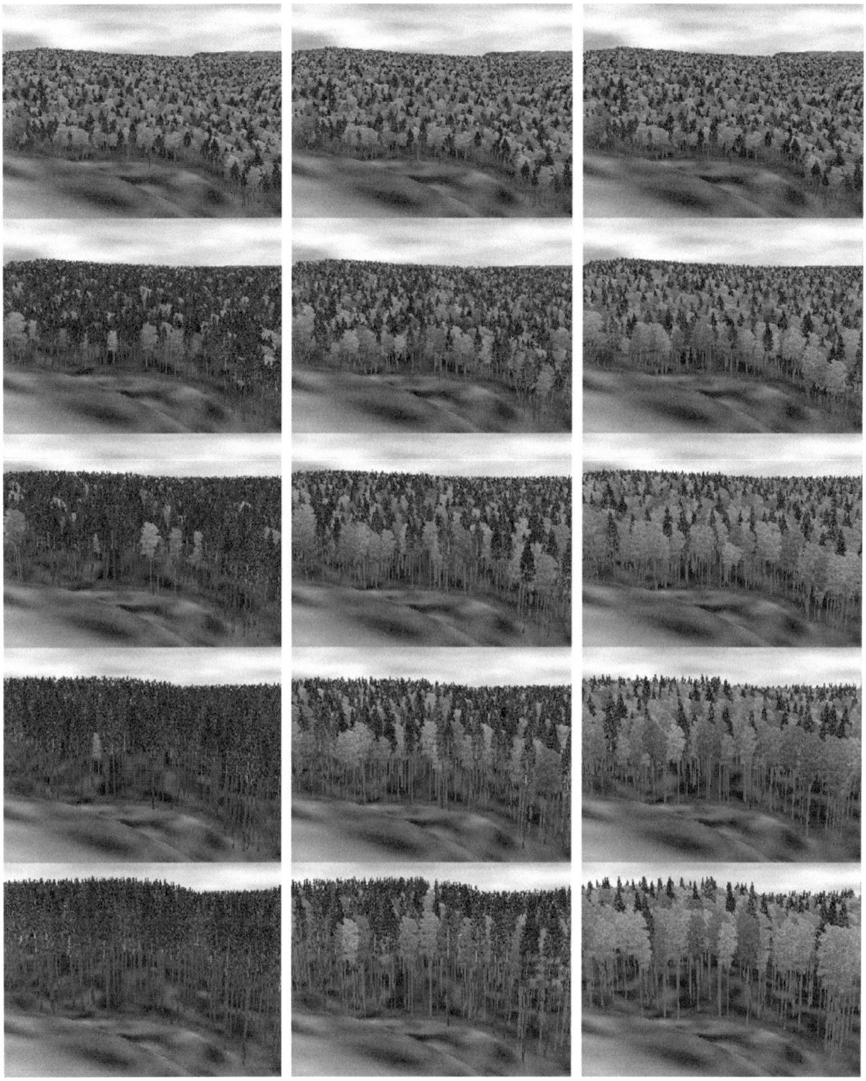

**Fig. 10.12** Visualisation at the landscape level by L-VIS (Seifert 2006, 2008). Development of Norway spruce-European beech mixed stands in the Traunstein forest estate, Germany, from age 25 to 125 (*from top to bottom*). Without management (*left column*), moderate promotion of beech through thinning from above (*middle column*) and strong promotion of European beech (*right column*)

**Take-Home Messages**

1. The practice of forest management planning is mainly based on traditional models, such as the age-class model. An adaptation to the requirements of mixed-species forests—for example, considering interactions between tree species—and to uneven-aged stand structures is still largely lacking.
2. Forest owners and stakeholders have a say when developing the planning objective. Neglecting stakeholders in the planning process may cause deviation from the objective when putting the plan into action.
3. Models allow the prediction of long-term consequences of alternative silvicultural treatments.
4. If different treatments or growth conditions are tested comparatively, this is referred to as a scenario analysis. Such scenario analyses deliver 'if-then' statements and thereby support the selection of the silvicultural treatment and the definition of annual cut.
5. As it illustrates the multivariable long-term consequences of planning decisions, scenario analysis using management models supports both decision making of experts and participation of stakeholders. Simulation models do not produce the 'perfect' treatment variant, but make it possible to choose among a set of considered variants, those that best match the desired objective.
6. Scientific models based on operations research methods have been developed for forest management planning mainly for age-class forests. In advanced planning models, an increased resistance in mixed forests and risk interactions are considered. Existing planning models for uneven-aged forests are based on the size-class approach.
7. Size-class modelling allows for an adequate prediction of the allowable harvests in uneven-aged, mixed forests. Still, these approaches have to be developed to appropriately capture the multiple interactions between trees and tree species in mixed, uneven-aged forests. A theory for deriving desirable target structures is currently unavailable.
8. Based on spatially explicit inventories and modelling at plot, stand, and enterprise level, information about compositional, functional, and structural aspects of biodiversity can be derived and integrated into forest management planning.
9. By visualising scenario outcomes at stand or landscape level, the long-term consequences of different species selection, thinning guidelines, or regeneration systems can be illustrated with respect to scenic beauty. Thus, non-experts in forest management at the round table are also able to gauge the effect on forest aesthetics and recreation value.
10. European forests integrate a multitude of different functions without clear spatial segregation. Forest ecosystem management in Europe must be sustainable, participatory, and transparent. Thus, ecological, economic, and social functions of forests should be considered together, trade-offs ought

to be analysed, and decisions made to achieve multipurpose objectives. The principle of integration requires greater knowledge, more negotiation, and compromise than segregation, which is equivalent to a spatial or temporal decoupling of different forest functions.

## References


Augusto L, Ranger J, Binkley D, Rothe A (2002) Impact of several common tree species of European temperate forests on soil fertility. Ann For Sci 59:233–253

Baader G (1945) Forsteinrichtung als nachhaltige Betriebsführung und Betriebsplanung. Sauerländer's, Frankfurt a.M.

Bell S (2001) Can a fresh look at the psychology of perception and the philosophy of aesthetics contribute to a better management of forest landscapes. In: Sheppard SRJ, Harshaw HW (eds) Forests and landscapes: linking ecology, sustainability and aesthetics. CABI Publishing, New York

Bettinger P, Chung W (2004) The key literature of, and trends in, forest-level management planning in North America, 1950–2001. Int For Rev 6:40–50

Bettinger P et al (2009) Forest management and planning. Academic press/Elsevier, San Diego/ Amsterdam, 331 p

Bettinger P et al (2013) Forest management planning technology issues posed by climate change. For Sci Technol 9:9–19

Bitterlich W (1984) The relascope idea. Commonwealth Agricultural Bureaux, Norwich

Blanco JA et al (2005) Sustainability of forest management practices: evaluation through a simulation model of nutrient cycling. For Ecol Manag 213:209–228

Boyd J, Banzhaf S (2007) What are ecosystem services? The need for standardized environmental accounting units. Ecol Econ 63:616–626

Brauner M et al (2005) Forest management decision support for evaluating forest protection effects against rockfall. For Ecol Manag 207:75–85

Buchy M, Hoverman S (2000) Understanding public participation in forest planning: a review. For Policy Econ 1:15–25

Buongiorno J, Michie BR (1980) A matrix model for uneven-aged forest management. For Sci 26:609–625

Buongiorno J et al (1994) Tree size diversity and economic returns in uneven-aged forest stands. For Sci 40:83–103

Burkhart HE, Tomé M (2012) Modeling forest trees and stands. Springer, Dordrecht, 457

Chen BW, Gadow K (2002) Timber harvest planning with spatial objectives, using the method of simulated annealing. Forstw Cbl 121:25–34

Cotta H (1804) Taxation der Waldungen. Johann Daniel Sander, Berlin

Elatawneh A (2015) Investigations into the potentials of hyperspectral and multi-seasonal/ multispectral satellites data for forest parameter determination. Dissertation thesis, Technische Universität München. https://mediatum.ub.tum.de/node?id=1246704. Accessed 17 Mar 2016

Elatawneh A, Wallner A, Manakos I, Schneider T, Knoke T (2014) Forest cover database updates using multi-seasonal rapideye data—storm event assessment in the Bavarian Forest National Park. Forests 5:1284–1303

Estrella R, Cattrysse D, van Orshoven J (2014) Comparison of three ideal point-based multi-criteria decision methods for afforestation planning. Forests 5:3222–3240

European Commission (2013) Forest management plans or equivalent instruments. //www.google.de/url?sa=t&rct=j&q=&esrc=s&source=web&cd=1&ved=0ahUKEwj87YC2vMfLAhWCDZoKHWyXC6kQFggiMAA&url=http%3A%2F%2Fec.europa.eu%2Fenvironment%2Fforests%2Fpdf%2Ffmp_table.pdf&usg=AFQjCNEbnD6azzh08zq3ktmHm6DAYnP9PA&cad=rja. Accessed 17 Mar 2016

Fabrika M (2007) Modelling of forest production at climate change by growth model SIBYLA. In: Střelcová K, Škvarenina J, Blaženec M (eds) Bioclimatology and natural hazards international scientific conference, Polana nad Detvou, Slovakia, 17–20 September, 2007. isbn 978-80-228-17-60-8
Fernow BE (1911) A brief history of forestry, Revised and Enlarged Edition. Forestry Quarterly: Cambridge University Press, Toronto, 511 p
Fuchs A (1996) Forsteinrichtung im Kreuzberger Plenterwald. Forstw Cbl 115:51–62
Gamfeldt L et al (2013) Higher levels of multiple ecosystem services are found in forests with more tree species. Nat Commun 4. doi:10.1038/ncomms2328
Griess VC, Knoke T (2011) Can native tree species plantations in Panama compete with Teak plantations? An economic estimation. New For 41:13–39
Griess VC et al (2012) Does mixing tree species enhance stand resistance against natural hazards? A case study for spruce. For Ecol Manag 267:284–296
Haber W (1982) Was erwarten Naturschutz und Landschaftspflege von der Waldwirtschaft? Schriftenreihe des Deutschen Rates für Landespflege H. 40:962–965
Hahn A et al (2014) Financially optimized management planning under risk aversion results in even-flow sustained timber yield. For Policy Econ 42:30–41
Haight RG, Monserud RA (1990a) Optimizing any-aged management of mixed-species stands. I. Performance of a coordinate-search process. Can J For Res 20:15–25
Haight RG, Monserud RA (1990b) Optimizing any-aged management of mixed-species stands: II. Effects of decision criteria. For Sci 36:125–144
Hanewinkel M (1999) Kritische Analyse von auf der Basis von Gleichgewichtsmodellen hergeleiteten Zielreferenzen für Plenterwälder im Wuchsgebiet Schwarzwald. Allg Forst-Jagdztg. 170:87–98
Hanewinkel M (2001) Neuausrichtung der Forsteinrichtung als strategisches management instrument. Allg Forst- Jagdztg 172:203–211
Hartig GL (1795) Anweisung zur Taxation der Forste oder zur Bestimmung des Holzertrags der Wälder. Georg-Ludwig-Hartig-Stiftung, Wiesbaden, p 1996
Härtl F, Knoke T (2014) The influence of the oil price on timber supply. For Policy Econ 39:32–42
Härtl F et al (2013) Risk-sensitive planning support for forest enterprises: the YAFO model. Comput Electron Agric 94:58–70
Härtl F et al (2016) Multifunctionality in European mountain forests – an optimization under changing climatic conditions. Can J For Res 46:163–171
Hasenauer H (ed) (2006) Sustainable forest management: growth models for Europe. Springer, Berlin
Hector A, Bagchi R (2007) Biodiversity and ecosystem multifunctionality. Nature 448:188–190
Heinsdorf D (1999) Das Revier Sauen – Ein Beispiel für erfolgreichen Waldumbau, Schriftenreihe des Ministeriums für Ernährung, Landwirtschaft und Forsten, Eberswalder Forstliche Schriftenreihe, Landesforstanstalt Eberswalde, Ministerium für Ernährung, Landwirtschaft und Forsten, Brandenburg, Band VI, 30 p
Hundeshagen JC (1826) Die Forstabschätzung auf neuen, wissenschaftlichen Grundlagen, nebst einer Chrakteristik und Vergleichung aller bisher bestandenen Forsttaxations-Methoden. H. Laupp, Tübingen, 428 p
Hyytiäinen K, Haight RG (2010) Evaluation of forest management systems under risk of wildfire. Eur J For Res 129:909–919
Ishii HT et al (2004) Exploring the relationships among canopy structure, stand productivity, and biodiversity of temperate forest ecosystems. For Sci 50:342–355
Jonsson B, et al. (1993) The forest management planning package. Theory and application. Studia Forestalia Suecica 189. 56 p. issn 0039-3150, isbn 91-576-4698-8
Judeich F (1871) Die Forsteinrichtung. Schönfelds Verlagsbuchhandlung, Dresden
Kangas J (1994) An approach to public participation in strategic forest management planning. For Ecol Manag 70:75–88

Kant S (1999) Sustainable management of uneven-aged private forests: a case study from Ontario. Canada Ecol Econ 30:131–146
Knoke T (1998) Analyse und Optimierung der Holzproduktion in einem Plenterwald – zur Forstbetriebsplanung in ungleichaltrigen Wäldern. Forstliche Forschungsberichte München Nr. 170
Knoke T et al (2008) Admixing broadleaved to coniferous tree species: a review on yield, ecological stability and economics. Eur J For Res 127:89–101
Knoke T et al (eds) (2012) Forstbetriebsplanung als Entscheidungshilfe. Ulmer, Stuttgart
Knoke T et al (2015) Optimizing agricultural land-use portfolios with scarce data – a non-stochastic model. Ecol Econ 120:250–259
Knoke T et al (2016) Compositional diversity of rehabilitated tropical lands supports multiple ecosystem services and buffers uncertainties. Nat Commun 7:11877
Kölling C et al (2009) Überlegungen zum Risiko des Fichtenanbaus in Deutschland vor dem Hintergrund des Klimawandels. Forstarchiv 80:42–54
Lämås T, Eriksson LO (2003) Analysis and planning systems for multiresource, sustainable forestry: the Heureka research programme at SLU. Can J For Res 33:500–508
Ledermann T, Sterba H (2006) Evaluating management regimes and their impact on commercial timber supply using individual-tree growth models and scenario analysis. In: Hasenauer H (ed) Sustainable forest management. Growth models for Europe. Springer, Berlin, pp 195–209
Lemm R (1991) Ein dynamisches Forstbetriebs-Simulationsmodell. Prognosen von betriebsspezifischen Waldentwicklungen, Waldschäden und deren monetäre Bewertung unter variablen Einflussgrößen. PhD thesis No 9369, Eidg TH Zürich, Dep Forsteinrichtung und Waldwachstum, 235 p
Lexer MJ (2001) Simulation der potentiellen natürlichen Vegetation für Österreichs Wälder. Forstliche Schriftreihe Universität für Bodenkultur, Band 16, Österreichische Gesellschaft für Waldökosystemforschung und experimentelle Baumforschung, Universität für Bodenkultur Wien, 166 p (in German)
Lexer MJ et al (2005) A decision support tool to improve forestry extension services for small private landowners in southern Austria: decision support systems for forest management decision support in multiple purpose forestry. Comput Electron Agric 49:81–102
Mantel W (1949) Forsteinrichtungslehre. Neumann-Neudamm, Hamburg
Mantel W (1959) Forsteinrichtungslehre. Neumann-Neudamm, Hamburg
McCarter JB et al (1998) Landscape management through integration of existing tools and emerging technologies. J For 96:17–23
MCPFE (1993) Resolution H1: general guidelines for the sustainable management of forests in Europe. In: Proceedings of 2nd ministerial conference on the protection of forests in Europe, Helsinki, Finland, p 5
Meitner MJ, Gandy R, Sheppard SRJ (2005) Reviewing the role of visualization in communicating and understanding forest complexity. In: Proceedings of 9th international conference on information visualisation (IV'05), pp 121–128
Mendoza GA (2005) Recent developments in decision analysis for forest management. In: Innes JL, Hickey GM, Hoen HF (eds) Forestry and environmental change, vol 5. CABI Publications/ Association with the International Union of Forestry Research Organizations, New York/ Wallingford, pp 239–256
Nagel J (1999) Konzeptionelle Überlegungen zum schrittweisen Aufbau eineswaldwachstumskundlichen Simulationssystems für Nordwestdeutschland.Schr Forstl Fak Univ Göttingen u Niedersächs Forstl Versuchsanst 128. JD Sauer-länder's Verlag, Frankfurt am Main
Neuner S, Albrecht A, Cullmann D, Engels F, Griess VC, Hahn WA, Knoke T (2015) Survival of Norway spruce remains higher in mixed stands under a dryer and warmer climate. Glob Chang Biol 21(2):935–946
Nordström EM et al (2010) Integrating multiple criteria decision analysis in participatory forest planning: experience from a case study in northern Sweden. For Policy Econ 12:562–574

Noss RF (1990) Indicators for monitoring biodiversity: a hierarchical approach. Conserv Biol 4:355–364

Oesten G, Roeder A (2012) Management von Forstbetrieben, Band I–III, Institut für Forstökonomie der Universität Freiburg, www.ife.uni-freiburg.de/lehre/lehrbuch-1 respec tively lehrbuch-2 und -3, Volume I–III: 383 p, 365 p, and 291 p

Piotto D (2007) A meta-analysis comparing tree growth in monocultures and mixed plantations. For Ecol Manag 255:781–786

Pretzsch H (2009) Forest dynamics, growth and yield. From measurement to model. Springer, Berlin, 664 p

Pretzsch H, Block J, Dieler J, Dong PH, Kohnle U, Nagel J, Spellmann H, Zingg A (2010) Comparison between the productivity of pure and mixed stands of Norway spruce and European beech along an ecological gradient. Ann For Sci 67:712

Pretzsch H, Kahn M (1998) Wuchsmodelle für die Unterstützung der Wirtschaftsplanung im Forstbetrieb, Anwendungsbeispiel: Variantenstudie Fichtenreinbestand versus Fichten/ Buchen-Mischbestand. Allg. Forst Z. Waldwirtsch. Umweltvorsorge 51:1414–1419

Pretzsch H, Rötzer T (2016) Indicating forest ecosystem and stand productivity: from deductive to inductive concepts. In: Larocque GR (ed) Ecological forest management handbook. CRC Press, Taylor & Francis Group, UK, pp 415–444

Pretzsch H, Seifert S (1999) Wissenschaftliche Visualisierung des Waldwachstums. Allg Forst Z Waldwirtsch Umweltvorsorge 54:960–962

Pretzsch H et al (2002) The single tree based stand simulator SILVA. Construction, application and evaluation. For Ecol Manag 162:3–21

Pretzsch H et al (2008) Models for forest ecosystem management: a European perspective. Ann Bot 101:1065–1087

Pretzsch H et al (2014) Forest stand growth dynamics in Central Europe have accelerated since 1870. Nat Commun 5(4967)

Prodan M (1949a) Normalisierung des Plenterwaldes? Schriftenreihe der Badischen Forstlichen Versuchsanstalt, Heft 7

Prodan M (1949b) Die theoretische Bestimmung des Gleichgewichtszustandes im Plenterwalde. Schweiz Z Forstwes 100:81–99

Puettmann KJ, Coates KD, Messier C (2009) A critique of silviculture: managing for complexity. Island Press, Washington

Pukkala T (ed) (2013) Multi-objective forest planning, vol 6. Springer, Berlin

Pykäläinen J et al (1999) Interactive decision analysis in participatory strategic forest planning: experiences from state owned boreal forests. J For Econ 5:341–364

Reimoser S et al (2009) Roe-deer habitat suitability and predisposition of forest to browsing damage in its dependence on forest growth-Model sensitivity in an alpine forest region. Ecol Model 220:2231–2243

Romero C (2001) Extended lexicographic goal programming: a unifying approach. Omega 29:63–71

Rößiger J (2014) Minderung forstlicher Risiken durch ökologische Stabilisierungseffekte zwischen Baumarten. Allg Forst Z Waldwirtsch Umweltvorsorge 69:26–27

Rößiger J et al (2016) Variability in growth of trees in uneven-aged stands displays the need for optimizing diversified harvest diameters. Eur J For Res 135:283–295

Roth RR (1976) Spatial heterogeneity and bird species diversity. Ecology:773–782

Rötzer T et al (2010) Productivity and carbon dynamics in managed Central European forests depending on site conditions and thinning regimes. Forestry 83:483–496

Seifert S (2006) Visualisierung von Waldlandschaften. Allgemeine Forstzeitschrift 61:1170–1171

Seifert S (2008) Modellierung und Visualisierung des Waldwachstums auf Landschaftsebene. PhD thesis, Univ Göttingen, Dep Ökoinformatik, Biometrie und Waldwachstum, Göttingen, 120 p

Seppelt R et al (2013) Identifying trade-offs between ecosystem services, land use, and biodiversity: a plea for combining scenario analysis and optimization on different spatial scales. Curr Opin Environ Sustain 5:458–463

Sheppard SRJ, Harshaw HW (2001) Forests and landscapes: linking ecology, sustainability and aesthetics, IUFRO Res Series. CABI Publishing, New York, 294 pp

Speidel G (1972) Planung im Forstbetrieb. Grundlagen und Methoden der Forsteinrichtung. Paul Parey, Hamburg

Sterba H, Ledermann T (2006) Inventory and modelling for forests in transition from even-aged to uneven-aged management. For Ecol Manag 224:278–285

Stölb W (2005) Waldästhetik: über Forstwirtschaft. Naturschutz und die Menschenseele, Kessel, 400 p

Tahvonen O et al (2010) Optimal management of uneven-aged Norway spruce stands. For Ecol Manag 260:106–115

Tews J et al (2004) Animal species diversity driven by habitat heterogeneity/diversity: the importance of keystone structures. J Biogeogr 31:79–92

Tomppo E et al (2010) National forest inventories – pathways for common reporting. Springer, Heidelberg, 607 pp

Uhde B et al (2015) Hybrid MCDA methods to integrate multiple ecosystem services in forest management planning: a critical review. Environ Manag 56:373–388

von Carlowitz HC (1713) Sylvicultura Oekonomica oder Haußwirthliche Nachricht und Naturmäßige Anweisung zur wilden Baum-Zucht. JF Braun, Leipzig

von Gadow K (2005) Forsteinrichtung. Analyse und Entwurf der Waldentwicklung. Univ-Verlag Göttingen, Göttingen, 342 p

Vospernik S, Reimoser S (2008) Modelling changes in roe deer habitat in response to forest management. For Ecol Manag 255:530–545

Wallner A et al (2015) Estimation of forest structural information using RapidEye satellite data. Forestry 88:96–107

White JC et al (2013) A best practices guide for generating forest inventory attributes from airborne laser scanning data using an area-based approach. Natural Resources Canada, Canadian Forest Service, Canadian Wood Fibre Centre: Information Report FI-X-010

Whitham TG et al (2006) A framework for community and ecosystem genetics: from genes to ecosystems. Nat Genet 7:510–523

Wikström P et al (2011) The Heureka forestry decision support system: an overview. Math Comput For Nat Resour Sci 3:87–94

Wolfslehner B et al (2005) Application of the analytic network process in multi-criteria analysis of sustainable forest management. For Ecol Manag 207:157–170

# Economics of Mixed Forests

# 11

Thomas Knoke

**Abstract**

This chapter introduces the economic framework of, but also challenges to, modern financial theory when applied in environmental and particularly forest science studies. It demonstrates the characteristics of financial models to capture the effects of mixing tree species. In addition, the chapter demonstrates how a portfolio-based approach for deriving the long-term forest structure as applied to the example of species composition and timber quality could form a new tool for forest management planning at the company level. Furthermore, it is shown how transitioning to the stand level through the consideration of interactions between tree species will alter the results compared to forest-level studies. The chapter concludes with advances in modelling the economics of near-natural forestry, where mixing of tree species is a key issue. This is done by means of optimising the distribution of timber harvests over time simultaneously with the composition of tree species.

## Contents

11.1 Motivation ........ 547
11.2 Financial Theory of Mixing Assets and Its Integration in Forest Economics ........ 548
11.2.1 Scientific Background ........ 553
11.2.2 Applications in Forestry and Related Areas ........ 554
11.3 Analogies and Differences Between Financial Stocks and Forests ........ 556
11.4 Modelling Financial Effects of Mixing ........ 557
11.4.1 Construction of Timberland Returns Based on a 'Financial Stock' Perspective ........ 558
11.4.2 Monte Carlo Simulation of Discounted Net Revenue Flows ........ 559

T. Knoke (✉)
Faculty of Forest Science and Resource Management, Institute of Forest Management, Technical University of Munich, Hans-Carl-von-Carlowitz-Platz 2, 85354 Freising, Germany
e-mail: knoke@tum.de



H. Pretzsch et al. (eds.), *Mixed-Species Forests*, DOI 10.1007/978-3-662-54553-9_11

11.4.3 An Alternative, Less Data-Demanding Portfolio Approach .................. 561
11.5 Deriving Long-Term Mixtures at the Forest Level (Excluding Interactions Between Species) .................. 563
11.6 The Effect of Tree Species Interactions from Stand-Level Mixtures .................. 565
11.7 Transitioning to Continuous Cover Forestry .................. 568
11.7.1 Background Information .................. 568
11.7.2 Does Improved Economic Modelling and Risk Aversion Lead to Mixed, Uneven-Aged Management? .................. 570
Take-Home Messages .................. 572
References .................. 572

Various motivations and objectives guide decisions in forest management. For example, tradition, culture, lifestyle, and similar factors are important for the decision making of forest managers and owners. However, the economic perspective is also important, because people's response to economic opportunities is often a central driver for their land-use activities (Lambin et al. 2001) and consequently also for forest management (Pearce et al. 2003). The choice to grow mixed or pure forests will have important economic but also ecological consequences. The economic consequences are, however, addressed only rarely in scientific studies (Knoke et al. 2008; Griess and Knoke 2013). The choice of tree species and whether to establish them in mixed or in pure forests addresses a classical problem of sustaining long-term forest production. From an economic perspective, this type of planning entails an investment perspective. It is typically different from the planning of forest yield, as addressed in Chap. 10. Yield regulation aims at planning timber harvesting over space and time, but does not address the long-term production of timber (Baader 1945). This chapter is mainly focussed on the planning of forest production, for which the choice of tree species and stand type (mixed or pure) is essential. Nevertheless, it will conclude with an attempt to amalgamate production and yield planning through a case study about a simultaneous optimisation of the tree species composition and the distribution of timber harvests over time. However, this will still be a stand-level consideration and has not been achieved at the level of a forest enterprise.

This chapter will start with an introduction to the economic framework of, but also challenges to, modern financial theory when applied in environmental and particularly forest science studies. Differences between forests and financial assets and necessary adjustments for the application of financial theory to forestry and land-use topics in general will be outlined (second step). The characteristics of modelling financial effects of mixing tree species will be presented in a third step referring to economic land-use models going back to von Thünen (1842). In a fourth step, the practical use of portfolio theory to plan the long-term forest structure as applied to the example of species composition and timber quality at the company level is introduced as a new tool for forest management planning. A key issue will be, in a fifth step, to show how transitioning to the stand level through the consideration of interactions between tree species will alter the results compared to forest-level studies. Finally, the advances in modelling the economics of near-natural forestry, where mixing of tree species is a key issue, will be reported.

Here, the distribution of timber harvests over time is simultaneously optimised. The methods demonstrated by means of steps one to five are all combined in the concluding part of this chapter. The hypothesis to be tested in this sixth step is formulated as follows:

> "The integration of risks and uncertainties into a biological-economic model will implicitly lead to mixed stands and diversified harvesting sequences, if the decision maker is a risk avoider."

## 11.1 Motivation

From an economic perspective, mixed forests can be perceived in several ways. (I) Mixed forests can be defined very broadly as a portfolio of tree species mixed at the forest estate level. From this perspective, forest estates could still consist of pure forest stands when various trees species are present at the forest estate level. These 'mixed' forests would be comparable with portfolios of financial stocks in various regards. A portfolio is a combination of financial or real assets, for which an expected economic return and an economic risk can be derived. For the estate-level case, no interactions between tree species will have an impact on the expected portfolio return. (II) Tree species may also be mixed at the stand level, forming real mixed-species forest stands. These stands would differ from portfolios of financial stocks because we may expect an impact of one species on the other regarding resistance, growth, or timber quality (Knoke and Seifert 2008). (III) A mixture of 'age classes' on small forest areas (uneven-aged forests) could also form a kind of a mixed forest. Here, various tree sizes are mixed leading to the opportunity for continuous harvesting. All three categories of 'mixed' forests are considered in this chapter. However, at present, it is only possible to speculate about the attractiveness of these mixed forests with regard to many economically relevant issues (see Table 11.1, which refers mainly to stand-level mixed forests).

From Table 11.1, we realise that mixed forests appear to promote mainly ecological aspects, whereas some economic aspects appear to play against the practice of creating mixed forests. Among the economic aspects in favour of mixed forests, the reduced economic risk may be a strong supportive argument. However, at the stand level, only more recent scientific studies have used quantitative data to investigate this aspect. Also, the complex interactions between various tree species in mixed stands have to date only been considered in very few studies. Given this background, it appears interesting to provide a state-of-the-art review about the economics of mixed forests, particularly given that in recent years, some methodological advances have been made which enable an improved quantification of the economic consequences of mixed forests. However, this review will also serve to make explicit the existing research gaps.

**Table 11.1** Statements and speculations about the consequences of growing mixed as opposed to pure forests (referring mainly to stand-level mixtures)

| | Consequence of mixed (as opposed to pure) stands | Reference | Advantage gained from mixed stand |
|---|---|---|---|
| Costs of establishment | Higher | Bristow et al. (2006) | No |
| Timber quality | Lower | Beimgraben (2002), Seifert (2004), Röhrig et al. (2006), Bleile (2006) | No |
| Resistance against hazards and pests | Higher | Schmid-Haas and Bachofen (1991), König (1995), Jactel et al. (2005), Mayer et al. (2005), Schütz et al. (2006), Griess and Knoke (2011), Griess et al. (2012), Neuner et al. (2015), Chap. 7 in this textbook | Yes, but mostly only for the susceptible species |
| Growth performance | Higher-indifferent-lower | Chaps 4 and 6 | Ambiguous, site and species dependent |
| Harvesting costs | Higher | Pausch (2005) | No |
| Costs of regeneration after a hazard | Lower | Hypothesis of forest practicioners | Yes |
| Average net revenues | Lower | Ripken and Spellmann (1980), Brandl (1989), Möhring (2004), Spellmann (2005) | No |
| Financial risk | Lower | Knoke et al. (2005, 2008), Knoke and Seifert (2008), Hildebrandt et al. (2010), Beinhofer and Knoke (2010), Griess and Knoke (2013) | Yes |
| Flexibility | Higher | Lohmander (1993), Jacobsen and Thorsen (2003) | Yes |
| Soil properties | Better | Rothe and Binkley (2001), Rothe (2005) | Yes |
| Tree nutrition | Improved | Richards et al. (2010) | Yes |
| Biodiversity | Higher | Franklin et al. (1999), Halpern et al. (1999) | Yes |

## 11.2 Financial Theory of Mixing Assets and Its Integration in Forest Economics

Forest-related diversification to buffer economic risks is possible in many ways (Table 11.2). For example, many forest owners also manage farmlands. This forestry-agriculture combination (portfolio) has many advantages (Knoke et al. 2012a, b). For example, working capacity may be utilised more efficiently in forest estates located in the temperate and boreal zones where work is demanded in summer (agriculture) and in winter (forestry). Machinery may be used on

**Table 11.2** Options to diversify forest-related risks (aspects included in this chapter in bold)

| Level of diversification | Type of risks | Diversification option |
|---|---|---|
| All assets of a shareholder or company | Typical forest risks | Forming a portfolio together with other asset classes, e.g. agriculture, financial stocks |
| Country/ regions | Political situation | Holding parts of the whole forest estate in different countries/regions |
| Ecology | Forest susceptibility to calamities, e.g. regional weather extremes, climatic risks | **Mixed species forest stands**, enhanced single tree stability, uneven-aged forest structure, spatial arrangement of forest stands |
| Time | Erosion of option values: Concentration of the timber harvest on only one point in time reduces options for future harvests | **Continuous and regular harvesting operations of moderate intensity** |
| Timber production | Fluctuating demand for various timber species, qualities and log sizes | **Establishment of a portfolio of tree species**, planting densities, tending, thinning, pruning, rotation |
| Products | Typical risks of timber production | Non-timber forest products, payments for ecosystem services |

agricultural lands and to skid timber logs. Moreover, the price fluctuations of agricultural and forestry products are hardly synchronous (Lönnstedt and Svennson 2000), which results in a buffering effect when both product categories are produced (see below for a more theoretical explanation). Other options for diversification are, inter alia, producing various timber qualities and log sizes (Beinhofer and Knoke 2010). This chapter focusses on diversification effects to be expected from mixing tree species (at the forest level and at the stand level) and from a more even distribution of timber harvests over time, as is usually the case in uneven-aged forest stands.

Before we start, it is important to outline the mechanisms behind diversification. Diversification may help to achieve a desired level of economic return associated with significantly reduced risk when compared to single assets. When considering the effects of diversification, it is important to note that the effectiveness of risk reduction depends heavily on the correlation of financial return between the assets mixed (tree species in our case). If we regard assets 1 and 2 in Fig. 11.1 as two different tree species, which can be combined in different proportions, we see that the correlation coefficient, $r_{1,2}$, has a great impact on the risk associated with the combined assets (portfolio). If the correlation is $r = +1$, this indicates two extremely similar species—a situation which in reality might not exist. However, Scots pine (*Pinus sylvestris*) and Austrian pine (*Pinus nigra*) could, for example, form mixed forests with highly positively correlated returns. Their timber would be delivered to the same markets, and their biophysical risks (e.g. fire) would be similar. Thus, if the timber price was low for species 1, it would also be low for

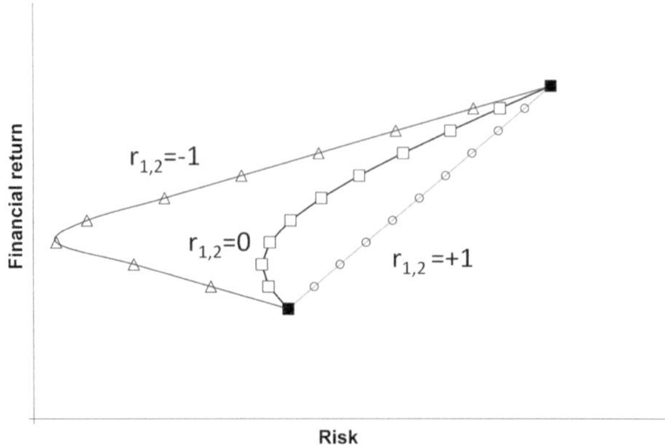


**Fig. 11.1** According to the theory of portfolio selection, correlation matters when mixing assets (tree species in our case); the *black squares* represent our assets, which are two different tree species in our example

species 2. If a fire attacked species 2, species 1 would also be vulnerable to this damage. We would expect very similar return fluctuations for both species as a consequence of their similarity. Thus, in this case, mixing would be not efficient because the risk of a mixed portfolio would only be a proportional combination of the risks of both species forming a straight line to connect species 1 with species 2.

A more realistic assumption would be a correlation of $r_{1,2} = 0$, as was, for example, derived by Knoke et al. (2005) for Norway spruce and European beech using Monte Carlo simulation. Here, we have independent return fluctuation because the spruce and beech timber markets are very different. Spruce is mainly used for construction timber, whereas beech is to a large part used for firewood or the beech stem wood is exported to foreign markets. The biophysical risks are also very different. For example, spruce is prone to bark beetle attacks, whereas beech usually is not. In addition, beech is much more wind firm than spruce. There is therefore a good chance that even if one of these species is affected by damage or low timber prices, the other one will still be economically successful. This situation would result in a risk compensatory effect (Fig. 11.1). If we, for example, mix 20% of species 2 and 80% of species 1, we can improve the return and reduce the risk at the same time. This mixture would yield the minimum risk. From this point onwards, further admixture of species 2 would increase financial return but also risk.

An extreme negative correlation, $r_{1,2} = -1$, is not found in forestry. Such a situation may be illustrated schematically. Assume a company produces umbrellas and swimming pools. This production programme could be regarded as almost free of risk because in a rainy summer, the company will sell the umbrellas, which people need as protection against the rain. And, vice versa, when the summer is hot and dry, there will be no demand for the umbrellas, but the company will sell swimming pools.

Evidence from several empirical analyses actually shows, on the one hand, that forestry is a suitable investment in which to diversify the portfolios of investors and, on the other hand, that diversification is also an effective instrument for stabilising forest returns. For example, Thorsen (2010) has shown very low correlation coefficients of around zero between forest and stock or bond returns. Importantly, he was able to use a very long period, starting in the year 1947. To stabilise forest returns when forestry is considered as an isolated investment, the correlation between forest stand types or between mixed tree species is important. This correlation depends, inter alia, on correlation between timber prices, on productivities, and on calamities. From an analysis of timber price data, Beinhofer (2009, p. 54) generally derived moderate to low correlations between prices for various tree species and timber grades (Table 11.3). Among all 66 tree species/log class combinations analysed, 44 showed correlations below 0.6. While correlations were found to be rather high among industrial wood of all four tree species and between spruce and pine saw logs, the beech saw log prices show only very low correlation to other timber classes. An exception is the very high correlation between beech saw logs and high-graded pine timber. However, only small timber volumes could be analysed for the high-graded timber classes. Similar to the data analysed by Beinhofer (2009), Jacobsen (personal communication) found a correlation between spruce and oak timber prices of 0.49 (which was 0.58 in Beinhofer's (2009) data). However, correlation between timber prices depends very much on the period analysed. For example, Knoke et al. (2005) found a correlation as strong as +0.85 between prices for spruce and beech saw logs, when analysing time series data from 1953 to 1979. But for the period from 1980 to 2001, the correlation obtained was −0.62. This shows how carefully data on correlation should be interpreted and used in optimisations.

Another important factor is the risk of damages presented by calamities. This aspect is often excluded from the economic analyses despite its very high importance. For example, Hyytiäinen and Penttinen (2008) found a high correlation between the return of various forest stands and concluded that only a small potential for silvicultural diversification exists. However, calamities, such as windthrow, snow breakage, fire, or insect attacks were not included in their study. When including these natural calamities, Knoke et al. (2005) always simulated very low correlations between the economic returns of spruce and beech.

Finally, the fluctuation in stand productivity may also have an impact on the decisions of forest owners or managers. However, possible effects have rarely been analysed. One example is the study by Dragicevic et al. (2016), who analysed the tree species mixtures for 90 French administrative departments. They found that the actual tree species composition differs from that resulting from optimisation of stand productivity and from that obtained from economic optimisation. Empirically chosen tree species compositions appear to be a compromise between productivity and economic considerations.

In summary, we can conclude that many studies point to a significant potential for silvicultural diversification as a means of stabilising economic returns from forestry. This diversification could, for example, entail mixing tree species at the forest or stand level but also uneven-aged stand structures, which have potential to improve the relation between economic return and risks (the latter is shown later).

**Table 11.3** Pearson correlation coefficients among timber prices for various tree species and log grades (Beinhofer 2009, p. 54)

| | Spruce saw logs | Spruce industrial wood | Pine high grade | Pine saw logs | Pine industrial wood | Beech high grade | Beech saw logs | Beech industrial wood | Oak high grade | Oak saw logs | Oak industrial wood |
|---|---|---|---|---|---|---|---|---|---|---|---|
| *Spruce high grade* | *0.32* | *0.59* | *0.58* | *0.37* | *0.3* | *0.17* | *0.33* | *0.72* | *0.63* | *0.48* | *0.69* |
| Spruce saw logs | | 0.49 | *0.19* | 0.87 | 0.73 | *0.01* | 0.15 | 0.46 | *0.49* | 0.58 | 0.38 |
| Spruce industrial wood | | | *0.25* | 0.38 | 0.74 | *−0.32* | −0.05 | 0.83 | *0.83* | 0.73 | 0.77 |
| *Pine high grade* | | | | *0.04* | *0.02* | *0.69* | *0.85* | *0.53* | *0.24* | *0.24* | *0.61* |
| Pine saw logs | | | | | 0.62 | *−0.06* | 0.02 | 0.39 | *0.47* | 0.53 | 0.26 |
| Pine industrial wood | | | | | | *−0.3* | −0.08 | 0.61 | *0.68* | 0.65 | 0.49 |
| *Beech high grade* | | | | | | | *0.92* | *0.11* | *−0.25* | *−0.31* | *0.21* |
| Beech saw logs | | | | | | | | 0.36 | *−0.01* | −0.12 | 0.43 |
| Beech industrial wood | | | | | | | | | *0.82* | 0.54 | 0.96 |
| *Oak high grade* | | | | | | | | | | *0.72* | *0.76* |
| Oak saw logs | | | | | | | | | | | 0.5 |

The analysed period is 1972–2004, and prices were recorded for Bavarian state forests. Data on high graded logs is presented in italics because the timber quantities analysed are only small. Tree species are *Picea abies* (spruce), *Pinus sylvestris* (pine), *Fagus sylvatica* (beech), *Quercus pertraea* or *Quercus robur* (oak)

### 11.2.1 Scientific Background

Financial theory has been dealing with risk and uncertainty for a long time. As a starting point, the seminal paper by Markowitz (1952) used mathematical statistics to explain and improve financial decision making. This study proposed an analysis of the expected financial return and the return's variance to describe (positive sense) and improve (normative sense) the behaviour of decision makers. It states that, by means of selecting a combination of financial assets (i.e. a portfolio) which maximises financial return without exceeding an acceptable risk (measured by the portfolio's return variance or standard deviation), it is possible to sufficiently satisfy the financial investor's desire to maximise expected utility (Markowitz 2010, 2014). While Markowitz's theory was initially limited to financial assets, Sharpe (1964) extended this theory to all kinds of assets, including real assets, such as buildings and also forests. For example, Sharpe analysed the relation between financial return and variance (or standard deviation) for single funds via the 'reward-to-variability ratio' (Sharpe 1966). Later, the aspect of the investor's flexibility was included in the evaluation of investments in capital assets (Dixit and Pindyck 1995). Critical perspectives have been offered by, inter alia, Fama and French (1993) on the appropriateness of the Sharpe model and by Mandelbrot and Hudson (2005) on the general assumption of normally distributed returns. We discuss the challenges and potential alterations to the theory of portfolio selection—at least in part—in later sections of this chapter.

Diversification effects form a central issue in Markowitz' theory and were recognised by financial sciences quite early—at least in theory. However, the theory of portfolio selection may also be realistically applied to land-use and environmental studies, where decision makers are actually known as risk avoiders (e.g. Pichón 1996). In fact, there are a considerable number of environmental studies applying the theory of portfolio selection (see Matthies et al. 2015 and Raes et al. 2016 for recent examples). In theory, diversification is a prominent means of hedging risks—risk-avoiding decision makers are usually identified by their tendency to diversify assets (Hirshleifer and Riley 2002). A precondition for positive effects through diversification of assets is that their returns are not too strongly positively correlated (e.g. Fig. 11.1). We have learnt that if returns are uncorrelated, a high return from asset A may be associated with a low return from asset B and vice versa. This leads to an averaging effect, while reducing the volatility of portfolio returns. A negative correlation would even enhance the effect of reducing return volatility. Only fully positively correlated returns between mixed assets would yield no diversification effects. However, at least in its classical understanding, diversification would always lead to a lower expected return. Decision makers thus pay a price for reducing their risks by sacrificing a part of the maximum possible expected return, which may be regarded as an insurance premium. One may also describe this issue from the perspective of expected return: A risk avoider would always demand a premium to accept higher risk. Only if she gets a more than proportionally increased return would she be willing to accept an additional unit of risk. This investment behaviour supports a positive correlation between return and risk, which is often observed. The higher the expected return, the higher the associated risk.

### 11.2.2 Applications in Forestry and Related Areas

The diversification effects described above have long been ignored in the economic evaluation of silvicultural options. While diversification between forestry as a whole and other assets (e.g. financial stocks or agriculture) was investigated relatively early (Mills and Hoover 1982; Thomson 1991; Wagner and Rideout 1991, 1992; Wippermann and Möhring 2001; Weber 2002; Hyytiäinen and Penttinen 2008), an application to silviculture and mixed forests has only been approached rather recently (see, e.g. Schou 2012). This is surprising given that, on the basis of intuition, Karl Gayer (1886) already identified diversification effects as prominent advantages of mixed forests. However, his ideas were not incorporated into forest economics for a long time. It was, for example, Deegen et al. (1997) who correctly argued that an application of the portfolio theoretic approach to stand-level mixed forests is impossible if the interactions between the species in mixed stands cannot be modelled adequately. While the financial theory gained ground in forest science, the portfolio theoretic approach has been combined in a sophisticated fashion with the classical Thünen approach (von Thünen 1842) in agriculture to obtain an optimum land allocation under risk (Macmillan 1992 and studies cited therein). The theoretical background of this combination of the theory of portfolio selection and the Thünen approach is described as an application of mathematical statistics, for the example of spatially non-explicit land allocation to land-use options in Box 11.1.

Other studies applying the theory of portfolio selection in agriculture address topics such as the Common Agricultural Policy in the European Union (Roche and McQuinn 2004; Havlik et al. 2008), linking landscape diversity and the resilience of agricultural returns (Abson et al. 2013), optimising the structural composition of farmlands (Kaplan 1985), and deriving conservation payments (Castro et al. 2013). Assessing afforestation on marginal land under irrigated farming systems (Djanibekov and Khamzina 2016), improving irrigation water management under uncertain conditions (Paydar and Qureshi 2012), improving flood management (Aerts et al. 2008) as well as water planning (Marioni et al. 2011), and analysing the economic attractiveness of organic bananas in Ecuador (Castro et al. 2015) are further example studies based on the theory of portfolio selection. For the field of forest science, Hildebrandt and Knoke (2011) provide a comprehensive overview of the consideration of financial uncertainties. Although now numerous environmental studies adopt the theory of portfolio selection, including in the scientific fields of fisheries (Edwards et al. 2004) and grassland sciences (Koellner and Schmitz 2006), we can conclude that a sound analysis of the economic effect of mixing tree species in forest management is still at its beginnings.

**Box 11.1 Mathematical Background of the Theory of Portfolio Selection, When Applied to Land-Use Based Allocation Problems (Adopted from Knoke et al. 2015)**

Classical portfolio optimisation derives an efficient frontier formed by portfolios which maximise the economic returns for various pre-defined levels of economic risk. The efficient frontier is based on the so-called "mean-variance" optimisation. The standard deviation of the economic return is classically adopted as a measure of risk. Using this theory for an optimisation of land allocation to various land-use options—which may also be various tree species—in a portfolio-theoretic Thünen framework (which is, however, not spatially explicit in our examples), one can formulate the optimisation problem as follows:

$$
\begin{aligned}
&\max\ E(Y_{\mathrm{L}}) = \sum_{i \in L} E(y_i) a_i \\
&\text{subject to} \\
&S_{\mathrm{L}} \leq S_{\mathrm{A}} \\
&S_{\mathrm{L}} = \sqrt{\sum_{i \in L} \sum_{j \in L} a_i a_j \mathrm{cov}_{i,j}} \\
&\sum_{i \in L} a_i = 1 \\
&\mathrm{cov}_{i,j} = k_{i,j} s_i s_j \\
&a_i \geq 0
\end{aligned}
$$

where,

| | |
|---|---|
| $E(Y_{\mathrm{L}})$ | Expected annual economic return of the entire land-use portfolio (monetary units ha$^{-1}$ year$^{-1}$) |
| $E(y_i)$ | Expected land rent of a single land-use option, $i$ (monetary units ha$^{-1}$ year$^{-1}$) |
| $a_i$ | Area proportion allocated to the land-use option $i$ |
| $L$ | Set of land-use (tree species) options considered |
| $S_L$ | Standard deviation of the uncertain land-use portfolio return |
| $S_A$ | Pre-defined, accepted standard deviation of the uncertain portfolio return |
| $cov_{i,j}$ | Covariance of uncertain returns of land-use options $i$ and $j$ |
| $k_{i,j}$ | Correlation coefficient of uncertain returns of land-use options $i$ and $j$ |
| $s_i$ | Standard deviation of the return of land-use option $i$ |

The frontier of efficient portfolios may be derived through a step-by-step process that starts by requiring $S_{\mathrm{A}}$ (acceptable level of risk) to be the risk minimum. Subsequently, increasingly higher risk levels will be allowed while still maximizing return for each risk level analysed, until finally, the maximum risk is achieved with a portfolio consisting of only the single land-use

(continued)

**Box 11.1** (continued)
option with the maximum risk (and usually also the maximum economic return. We see that the formulation used is stochastic because we consider the economic returns of the land-use options to be the expected values of random variables having a standard deviation and showing a specific covariance with the return of all other land-use options.

## 11.3 Analogies and Differences Between Financial Stocks and Forests

Similar to financial stocks, the economic result of forest stands is volatile because forest stands are exposed to many risks and uncertainties. For example, market prices and production costs are subject to significant fluctuation. Timber prices in particular are hard to predict and are extremely volatile. Moreover, storms, snow breakage, drought, or fire as well as insects and other calamities may cause severe damage to forest stands. As we have previously argued, diversifying structure and treatment of forest stands may help to reduce these risks. If the timber of two tree species is delivered to independent markets, for example, if one species is exported to foreign markets and the other is sold on domestic markets, the resulting timber prices may be independent. Also, in mixes of broadleaves and conifers, the two species groups are unlikely to be affected by the same biotic and abiotic risks, or at least not to the same extent. For example, while Norway spruce is very susceptible to bark beetles, most broadleaves are not. Finally, the diversification of harvesting times (Knoke et al. 2001), which is an option for small forest properties at the stand level or at the estate level (Hahn et al. 2014), may also serve as an effective hedge against volatile timber prices. Given all these characteristics, we may well expect significant positive diversification effects when mixing tree species, timber products, and silvicultural treatment. However, in forest economics, it has often been disregarded that mixed-species forests may actually show such diversification effects.

There are some important differences between mixed forests and portfolios of stocks (Knoke 2008). In contrast to stocks, forests are exposed not only to market price risks but also to biophysical risks, as previously mentioned. These risks are of great importance because the investment periods in forestry are usually much longer when compared to those of investments in financial stocks. This fact certainly limits the opportunities for adjusting the composition of the portfolio of tree species according to current market developments. However, when mixed at the stand level, some adjustments would be possible through thinning (see Lohmander 1993). It may be optimal to maintain this flexibility for such adjustments, by means of mixed forests, over long time periods (Jacobsen and Thorsen 2003).

Another issue is the lack of financial records about forests. There is only one timberland index reported in the USA (https://www.ncreif.org/data-products/timberland/), which has some major implicit weaknesses (Rinehart 2010). More importantly for the issue of mixed forests, the existing NCREIF timberland index does not allow statements about single silvicultural options, because it only reports average returns for a selection of American forest estates. The construction of a model-based timberland index (called 'forest performance index' by Weber 2002) to derive the internal rate of return earned from timberland is subject to important limitations, which we shall discuss later. Given these limitations, an analysis of diversification effects may alternatively be carried out on the basis of bioeconomic models, which are based on discounted net revenue flows. For example, Dieter et al. (2001) presented one of the first example applications for this type of analysis to optimise timber harvesting regimes based on the theory of portfolio selection.

Despite the limitations mentioned above, mixed forests are a very interesting case for economic analyses. Studies often assume that species are independent as is the case for species diversification at the estate level in which monocultures of species are planted in different locations to protect against uncertainty in economic returns that is induced by fluctuating market prices or biophysical damage (Weber 2002; Knoke et al. 2005, 2008; Knoke and Wurm 2006; Koellner and Schmitz 2006). According to the assumption of independence, the risk does not combine proportionally under species mixtures but rather the expected (economic) performance changes proportionally when the fractions of species are altered. In this case, diversification is (almost) analogous to that of financial assets (Koellner and Schmitz 2006). However, if species interact because of ecological interdependence, the direct analogy to financial assets vanishes (Deegen et al. 1997; Koellner and Schmitz 2006). Estimating the expected average performance of such a portfolio with interdependent components, in our case a mixed-forest stand, would require more than knowing the proportions of species in the portfolio. Moreover, it would seem particularly challenging in the case of long-lived forest stands. This chapter may contribute to addressing this scientific challenge.

## 11.4 Modelling Financial Effects of Mixing

Given the lack of stock market data, basically two approaches have been used to simulate or derive the expected economic return from forests as well as its fluctuation and covariances with other assets or between parts of the forest (stands or tree species). The first approach is based on a calculation of forest returns which resembles the logic for deriving return on financial stocks. The second approach is based on discounted net revenues, similar to that applied in the original approach by Markowitz (1952). Both approaches are described and discussed below.

### 11.4.1 Construction of Timberland Returns Based on a 'Financial Stock' Perspective

In financial sciences, the relative periodical change in stock value (as the stock's internal rate of return) is often a central indicator to quantify financial return (e.g. Fama and French 1993). This data is provided by the related stock market, which does not exist for forests or for the various tree species forming a forest. To derive a similar return indicator for US timberland, Hoover and Mills (1982) calculated the annual change in the market value of the standing timber stock and land, divided by the market value of timber and land at the beginning of the year. Weber (2002) computed a similar index for German tree species to derive an indicator for 'forest performance', and Hyytiäinen and Penttinen (2008) applied the same perspective in their analyses of forest assets in Finland.

The variability of the index is simulated by building the annual valuations (in some studies for each tree species or tree species group) on historical timber price time series, while disturbances from wind, fire, snow, or insects are usually excluded. However, obtaining the perspective of financial stock returns for forests poses several challenges. To derive an accurate return, the actual market price of the forest must be known for each considered period and this also for the various tree species (stand types) which form the forest. Using the calculated market value of the standing timber and the forest land (see Weber 2002) is only a meaningful proxy for the forest's actual market value if the forest consists solely of old, mature stands (Hyytiäinen and Penttinen 2008). However, this is rarely the case in actual forests, which usually also contain young stands where standing timber might even have zero or negative timber value even though the actual forest stand may achieve much higher market value when sold. Moreover, the suggested logic of using the calculated market value of the standing timber assumes that all timber could be felled and sold immediately. That timber and forest soil could be marketed separately is another implicit assumption. However, this all appears quite unrealistic.

An alternative would be to estimate the forest value of younger stands by means of their expected net present values (NPVs), as suggested by Hyytiäinen and Penttinen (2008). However, in this case, the interest rate used for discounting the annual net returns is predefined despite the fact that obtaining the (internal) interest rate of return is the aim of the whole calculation. In this case, the internal economic return obtained depends on the discount rate applicable in this case, which is undesirable. Using a predefined discount rate contradicts the logic of deriving an internal rate of return. In light of this, it appears problematic to use calculated forest values—which, as noted above, depend strongly on the discount rate—to compute the annual internal rate of return for forests. Moreover, further problematic assumptions have been made in order to consistently apply the internal rate of return perspective. For example, Weber (2002) assumes that loans may be obtained for quite different interest rates, whereby the interest rate for each loan corresponds to the internal rate of return of the tree species to be financed.

Finally, an important technical issue for the analysis of mixed forests is that, according to the 'financial stock' perspective, the portfolio weights of the mixed tree species must be based on their proportion of the (unknown) forest market value. These proportions may substantially deviate from the area proportions occupied by the tree species. This complication has significant consequences. First, the appropriate portfolio weights which are supposed to be the central outcome of optimisation cannot be implemented in forest management planning because they are only loosely related to forest area. This is problematic given that forest areas must be covered by stand types or tree species mixtures and not money. In addition, if the expected economic portfolio return and risk for the case of internal rates of return are based on area proportions of tree species as practised, for example, by Mills and Hoover (1982) or Weber (2002), the results obtained may be severely biased. In summary, when portfolio applications to land allocation are based on the 'financial stock' perspective, careful interpretation is advisable.

### 11.4.2 Monte Carlo Simulation of Discounted Net Revenue Flows

An alternative to deriving forest returns based on timberland indices is the Monte Carlo simulations of discounted net revenues under risk, which may be aggregated to NPV. This approach produces data on financial return (NPV in this case), risks, and return correlation. Based on empirical growth data—either from experimental areas or from growth models—the economic valuation may be carried out under fluctuating timber prices and under the risk of stand failure (Knoke et al. 2012b). To simulate volatile timber prices, bootstrapping may be applied as a well-tested method (Roessiger et al. 2011). Stand failure can be simulated based on drop-out probabilities (hazard rates) derived from available curves for survival probability. As a standard assumption, a binomial distribution can be used with the hazard rate as the expected drop-out probability.

The results of such Monte Carlo simulations are many possible scenarios for the economic outcome—often expressed as the sum of discounted cash flows, i.e. the NPV or the associated annuity (Fig. 11.2).

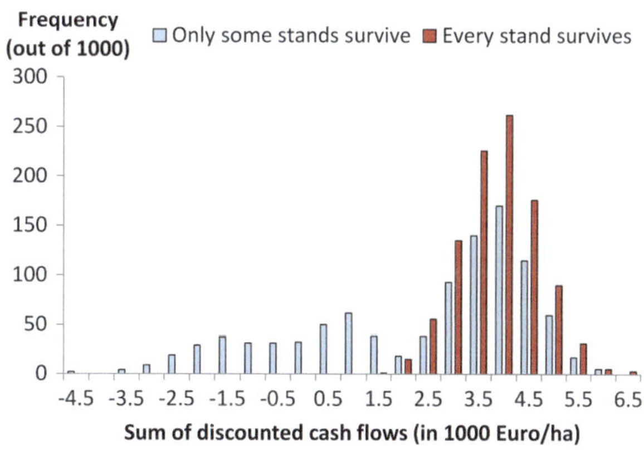


**Fig. 11.2** Exemplary output produced by 1000 Monte Carlo simulation runs for the case of a Norway spruce plantation in southern Bavaria (Tertiary Hills) showing high risk of stand failure; 100-year rotation period and interest rate of 2% are assumed for discounting purposes

From Fig. 11.2, it is also evident that the stand failure triggered by natural events may have a very important impact. It reduces the average economic performance and leads to an extremely inflated dispersion of the possible results. The loss by stand failure, for example, through wind, arises from broken timber, reduced timber prices in the case of storms after which large volumes of timber enter the market, and increased logging costs in the case of salvage harvesting. A reduction of 50% of the timber value achievable under 'normal' harvesting and market conditions may be assumed (e.g. Dieter 2001).

Finally, it is important to note that the optimisation considers area proportions—in the simplest case of tree species. This is appropriate only under the premise that discounted net revenues calculated on a per-hectare basis are used to simulate economic return. The forest or stand area is allocated to tree species in order to achieve a favourable correlation between the expected return and the associated risk to be accepted. All area proportions have to sum up to 1. In a more complicated approach (cf Roessiger et al. 2011), area proportions are also allocated to final harvesting operations (see below). This allows harvesting of the final crop trees through a sequence of harvest operations which are distributed over a period of time.

Considering area proportions rather than proportions of invested money differs from the original portfolio theory. In the meantime, this assumption has become standard in economic land-use modelling as is, for example, described in detail by Macmillan (1992) and also applied, inter alia, by Dieter et al. (2001), Moog and Knoke (2003), Knoke and Moog (2005), Halpern et al. (2011), Clasen et al. (2011), Knoke et al. (2009, 2011), Djanibekov and Khamzina (2016), Raes et al. (2016), and Dragicevic et al. (2016). Macmillan (1992) reformulated the portfolio theoretic approach to be used in a von Thünen economy, where the allocation of land is controlled by economic forces. For example, in the oldest approaches, the soil rent depending on the distance to the market is to be maximised through an appropriate land allocation (von Thünen 1842). Or declining marginal net revenues with expanding area allocated to different land-use options lead to an allocation problem in relation to maximising average net revenues of land-use portfolios. This approach was extended to 'Optimized Land-Use Diversification' considering market risks, where the 'reward-to-variability ratio' is maximised and with which the link to the financial portfolio theory was established (Knoke et al. 2011).

Combining the NPV method with portfolio optimisation, as is suggested here, has led to critical considerations. For example, the combination of a multiperiod, dynamic calculus (net present value) with a one-period, static model (portfolio approach) has met with concern (see Kruschwitz 2005, p. 389). However, Thomson (1991) has shown that one-period and multiperiod portfolio models yield similar results for his analyses of forest assets. Moreover, whether or not the critique holds true depends much on what 'dynamic' means. We can confirm that a multiperiod model is appropriate for describing dynamic economics in contrast to static economics (one period), but not every multiperiod model is truly dynamic. 'Dynamic' in a closer economic sense means that decisions in one period grow out of recent developments in a foregoing period. This considers the fact that decision makers must act now, knowing that one period later, more knowledge will be available. This is not taken into account in classical NPV applications. Depending on the new knowledge, revised decisions are possible for the next period (Samuelson 1969). Unless we consider the relationships between economic variables from period to period, we usually do not talk about a truly dynamic approach.

In this sense, the NPV approach—as is often applied to forest economic problems—is static if it disregards relationships between variables relating to different periods. A static nature is particularly obvious for the soil expectation value (SEV), which is frequently used in forestry as an economic calculus (e.g. Chang 1998; Chang and von Gadow 2010) and also in studies about mixed forests (Clasen et al. 2011; Griess and Knoke 2013). It considers the period of one rotation, over which returns accumulate as future values. The sum of all future values is then capitalised as an infinite, periodical rent, giving the SEV (also called willingness to pay for land). SEV is the same as the sum of an infinite series of appropriately discounted NPVs, each of which applies to the same rotation period. We can thus say that SEV is a one-period model given that the time period taken into account is the rotation period. As such, the economic variables do not change from one rotation to another. In contrast, a dynamic approach would consider possible adjustments after having made the initial investment, which could possibly increase the NPV of the investment. Castro et al. (2015) discuss the limitations and problems of such an approach extensively. In summary, we may conclude that there are no inconsistencies in standard combinations of NPV and the portfolio model, when applied to mixed-species forests.

### 11.4.3 An Alternative, Less Data-Demanding Portfolio Approach

It is important to keep in mind that the basic portfolio model is an application of long established statistical theory to asset allocation under financial risks, rather than a fundamentally new economic model. The behaviour of statistical deviations from a mean (or expected) value (i.e. the statistical error) is analysed, when various sources of error are combined to achieve the aggregated financial risk of portfolio investments. This explains why Markowitz (1990) noted in his Nobel lecture[1]: ‘When I defended my dissertation as a student in the Economics Department of the University of Chicago, Professor Milton Friedman argued that portfolio theory was not Economics, and that they could not award me a Ph.D. degree in Economics for a dissertation which was not in Economics. I assume that he was only half serious, since they did award me the degree without long debate. As to the merits of his arguments, at this point I am quite willing to concede: at the time I defended my dissertation, portfolio theory was not part of Economics. But now it is’.

[1] http://www.nobelprize.org/nobel_prizes/economic-sciences/laureates/1990/markowitz-lecture.pdf

Portfolio theory considers the economic return to be the expected value of a random portfolio return. It quantifies the associated risk by the variance or standard deviation of the return (Goldfarb and Iyengar 2003). The theory provides a stochastic model and the associated mathematics to combine all covariances of economic returns, which requires a normal distribution of economic returns. To solve problems associated with the lack of historical financial data on forest investments, techniques such as Monte Carlo simulation are being applied as described above. These methods are usually data demanding and may ultimately show that the distribution of returns is not normal (e.g. Knoke and Wurm 2006), mainly because of natural damage to the forest. As Mandelbrot and Hudson (2005) pointed out in their popular book, the requirement of normality is generally hard to achieve in real applications. The lack of stability of estimates of covariances is also problematic because only small perturbations of these input parameters may dramatically change the obtained solutions (e.g. Goldfarb and Iyengar 2003). Given these problems, alternative, non-stochastic representations of portfolio problems are under discussion, which are based on predefined uncertainty spaces. One recent example designed to solve problems of land allocation under uncertainty is briefly described in Box 11.2. Explicit information on return covariance is not required for this model type.

**Box 11.2 Mathematical Background of a Non-stochastic, Robust Formulation to Achieve Optimised Land Allocation Under Uncertainty, When Applied to Land-Use-Based Allocation Problems (Adopted from Knoke et al. 2015)**

The formulation of the non-stochastic optimisation problem is proposed as follows:

$$\max\ Y_L = \sum_{i \in L} y_i a_i$$

subject to

$$\sum_{i \in L} y_{si} \cdot a_i \geq \max(y_{si}) - \beta_U \cdot \delta_{\max,\min} \quad \forall s \in S$$

$$y_{si} = y_i \pm u_{si}$$

$$0 \leq \beta_U \leq 1$$

$$\sum_{i \in L} a_i = 1$$

$$a_i \geq 0$$

where,

| | |
|---|---|
| $y_i$ | Nominal economic return of land-use (tree species) option $i$ |
| $a_i$ | Area proportion allocated to the land-use (tree species) option $i$ |
| $y_{si}$ | Possible return of land-use option $i$ after subtracting or adding the deviation considered, $u_{si}$ |
| $s$ | Specific deviation scenario constructed by various combinations of deviations $u_{si}$, for each $y_i$ from a given set of land-use options, $L$, for example for tree species |
| $\max(y_{si})$ | Maximum return for scenario $s$ |

(continued)

**Box 11.2** (continued)

$\beta_U$ Control factor to adjust the degree to which the constraint has to be relaxed. Through calibration, a minimum of $\beta_U$ is sought under which all deviation scenarios meet their constraint. $\beta_U$ is kept constant for each level of uncertainty, $U$

$U$ Is introduced to indicate that $\beta$ depends on the size of the uncertainties considered

$\delta_{\max,\min}$ Range of economic returns within scenario $s$ $\max(y_{si})$-$\min(y_{si})$

$S$ Set of all deviation scenarios

$u_{si}$ Possible maximal deviation from the initial return parameter, $y_i$

This formulation contains no stochastic components. The objective function is considered certain, while uncertainty is represented through the uncertainty margins (referred to as deviation scenarios from here onwards) considered via the constraint data. Such an expression of the uncertainty problem has been called a purely "constraint-wise" construction of a robust optimisation problem (Ben-Tal et al. 2009, p. 11). Thus, this formulation implies the simultaneous consideration of one constraint for each deviation scenario. The scenarios for possible deviations have to be defined by the decision makers, see Knoke et al. (2015) for details.

This approach has led to greater diversity of land-use portfolios beyond classically derived land-use portfolios. However, the possible costs are very moderate (Knoke et al. 2015). These robust modelling techniques are promising, and their suitability to address forestry problems and in particular to evaluate mixed forests should be investigated further.

## 11.5 Deriving Long-Term Mixtures at the Forest Level (Excluding Interactions Between Species)

Having outlined various theoretical considerations, we now present explicit results from the economic analyses of mixed-species forests. In an initial approach to applying financial theory to forest management, interactions between tree species are ignored. When considering species mixtures at the forest (estate) level, e.g. among forest stands, disregarding stand-level mixtures in a first step may be justified. Stand-level mixtures will be addressed later (Sect. 11.6). Securing long-term forest production is a classical task in forest management planning. Here, the perspective of silvicultural investment is addressed rather than that of financing through optimised timber harvesting. For example, a decision needs to be made about which composition of tree species the whole forest company should have in the long run, for example, after one rotation period.

In German forest practice, this planning is mainly carried out on the basis of a site classification, where suitable tree species are proposed for specific site units.

However, financial considerations are not taken into account in this practice (Knoke et al. 2012b). In contrast, the international literature suggests that the internal rate of return of the tree species is often used as an indicator for selecting appropriate tree species (Cubbage et al. 2007). However, this calculus would lead to pure stands on a homogenous site unit because diversification and other mixing effects are regularly ignored in such calculations.

We currently lack an aid for investment planning to select tree species proportions while taking into consideration associated risks, including the important risk interactions that depend on the correlation of risks as described in Fig. 11.1. In an attempt to fill this gap, Neuner et al. (2013) showed how, based on Monte Carlo simulation of annualised net present values, the expected economic return may be maximised for a range of predefined risks—where risk is classically defined as the standard deviation of financial return.

The forest composition leading to maximum economic return for predefined levels of risk is derived through non-linear programming, with an objective function to maximise financial return subject to a maximum acceptable risk and being implemented via a constraint (see Box 11.1). Starting from the tree species composition associated with a minimum of risk, the risk constraint is relaxed step by step—and maximum economic return compositions are derived—until we have the often single tree species associated with the maximum risk and usually also the maximum return (Fig. 11.3).

From Fig. 11.3, we see that low-risk compositions of tree species are highly diversified and dominated by broadleaved tree species, oak and beech in our example. If the forest owner is willing to accept more risk, conifers enter the tree species portfolios, ending with a pure and profitable forest of pruned Douglas-fir. This option, however, bears the greatest financial risk because the investment required for the establishment of Douglas-fir is high and a successful establishment is uncertain. After planting, drought poses a great risk for young plants, and replanting is often necessary. Moreover, pruning entails acceptance of additional risky investment. The potential difference in future market price for pruned and unpruned timber is highly uncertain.

By including only 25% (area proportion) of broadleaves in the conifer tree species portfolio, the risk—given by the standard deviation of the expected annuities (Euro/ha/year)—may already be reduced to less than half of that associated with pure, pruned Douglas-fir. The financial return, however, would only drop by 12% when including 25% broadleaves (mainly beech and only a very small proportion of oak).

Figure 11.3 also shows that product diversification may be considered through the described financial planning approach. Depending on the degree of accepted risk, high-value timber products of oak as well as pruned logs of pine, spruce, and Douglas-fir are recommended as a means of diversifying market risks. It is well known that niche markets for valuable timber often develop independently from main markets for staple products (Mosandl and Knoke 2002). This provides an effective hedge against timber price uncertainty.

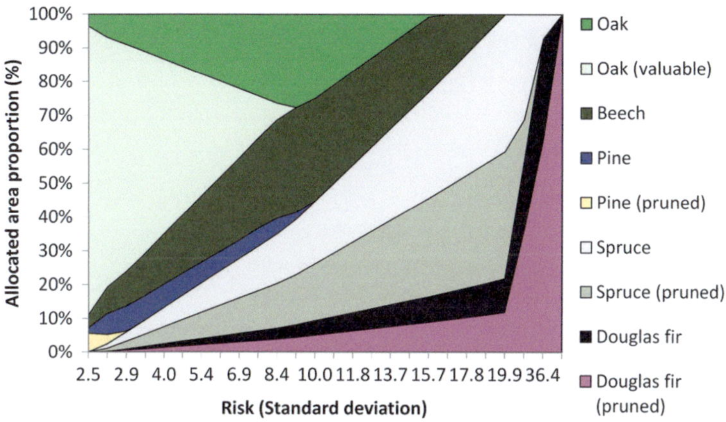


**Fig. 11.3** Forest-level mixtures promising the highest financial return for predefined levels of risk, where risk is derived as standard deviation of annuitized net present value (Data adopted from Neuner et al. 2013, with alterations)

In summary, we can derive benefits in the form of significant reduction of economic risk from the classical portfolio model, even if we mix blocks or stands with various pure tree species at the level of a forest company. However, silviculture usually only considers stand-level mixtures as mixed forest. While the results presented above should lead us to rethink this limited definition of a mixed forest, a more in-depth economic analysis of stand-level mixtures is also of high interest.

## 11.6 The Effect of Tree Species Interactions from Stand-Level Mixtures

In a conceptual study to prioritise information needs for the management of mixed forests, Knoke and Seifert (2008) pointed to the importance of stand resistance for the economic consequences of mixed forests. Based on their results, research efforts were undertaken to model the impact of broadleaf admixture on the ecological resistance of Norway spruce (Griess and Knoke 2011). Using exemplary data for Rhineland-Palatinate, Griess et al. (2012) were able to confirm a significant increase in survival probability of Norway spruce with increasing admixture of European beech (Fig. 11.4).

This study has been extended by Neuner et al. (2015) based on a much larger dataset and on the inclusion of climate variables. The recent study shows that the survival of spruce in pure forests is likely to decrease significantly under a predicted dryer and warmer climate, whereas the expected decrease is much less severe in mixed forests. This new evidence still needs to be combined with economic analyses, which so far have not taken climate effects into account.

Monte Carlo simulations to quantify economic return, covering the effect of increased resistance of spruce in mixed forests, were subsequently based on

**Fig. 11.4** The effect of European beech admixture on the survival probability of Norway spruce on medium sites, where trees achieve a maximum height of 40 m after 100 years (modified from Griess et al. 2012); curves were derived on the basis of forest damage surveys in Rhineland-Palatinate

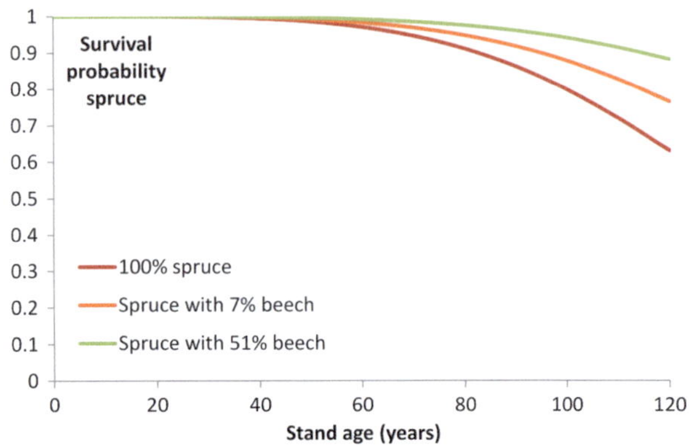


evidence of the survival of spruce in pure and mixed forests by Griess et al. (2012). The results show the tendency of a shift in frequencies from lower to higher NPV when comparing pure and mixed forests with 7% beech admixture (Fig. 11.5). On average, the admixture of 7% beech leads to a ~400 € per ha higher NPV than in a pure spruce forest (Table 11.4). Moreover, the standard deviation of NPV is also reduced significantly by means of the small 7% beech admixture.

However, when transitioning towards a mixed stand dominated by beech (51% versus 49% spruce), some potential for high economic return is lost (Fig. 11.5). This stand type achieves an average NPV ~1100 € per ha lower than that achieved in a pure spruce stand, but its NPV is ~1200 € per ha higher in comparison to pure beech. Moreover, the lowest simulated NPV (worst-case scenario) is +500 €, while for all other stand types, a zero or negative NPV represented the worst-case scenario. Thus, among the worst financial results covered by our simulations, the mixed forest dominated by European beech (51% plus 49% Norway spruce) still showed the relatively best performance. Preferring this stand type on the basis of risk aversion would be very conservative but could be appropriate for very cautious decision makers.

The standard deviation of the broadleaved stand with 49% spruce is with ±836 much lower than in any other stand type; compared to pure spruce (±1820), it is less than 50% (Table 11.4). These results indicate a very effective hedge against financial risks from growing one broadleaved and one coniferous tree species in relatively balanced proportions. Hildebrandt and Knoke (2009) demonstrated a version of robust portfolio optimisation, which also indicated acceptable financial results under a great variety of future conditions from a balanced, approximately 50:50 mixture. The quotient between average financial return and risk to be accepted for the stand type composed of 51% beech and 49% spruce is also favourable compared to the other stand types.

The economic effect of increased spruce resistance becomes clear when comparing the results between forest (estate)-level mixtures—i.e. mixtures in large blocks without higher resistance of spruce—and stand-level mixtures

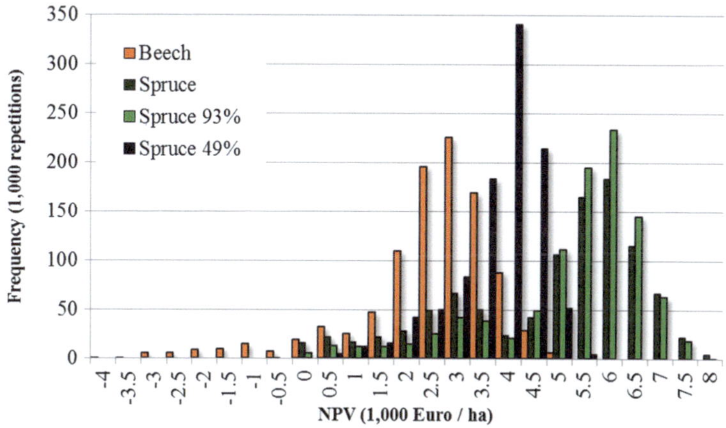


**Fig. 11.5** Frequency of simulated economic return and risk for four stand types (adopted from Griess and Knoke 2013)

**Table 11.4** Economic return of pure and mixed forest stands (adopted from Griess and Knoke 2013)

| | Average NPV [€/ha] | Median [€/ha] | Standard Deviation [€/ha] |
|---|---|---|---|
| Monocultures | | | |
| (1) Beech | 2506 | 2783 | ±1380 |
| (2) Spruce | 4850 | 5454 | ±1820 |
| Groupwise mixtures at stand level | | | |
| (3) Spruce 93% | 5255 | 5648 | ±1494 |
| (4) Spruce 49% | 3743 | 3912 | ±836 |
| Mixtures in blocks at forest (estate) level | | | |
| (1) Spruce 93% | 4686 | 5261 | ±1688 |
| (2) Spruce 49% | 3654 | 3968 | ±1096 |

(Table 11.4). A forest-level mixture of 93% spruce and 7% beech showed a ~570 € per ha lower performance combined with a standard deviation ~200 units higher than the corresponding stand-level mixture. We see a similar tendency for the 51% beech and 49% spruce mixture, where standard deviation is significantly reduced; economic return, however, increased only slightly when stand-level and forest-level mixtures were compared for the 51% beech and 49% spruce mixture. That means that, in general, stand-level mixing would increase economic return and reduce standard deviation at the same time. The effect on standard deviation is higher in mixture dominated by beech, whereas the impact on economic return is higher in spruce-dominated mixtures, thus demonstrating the considerably improved economic efficiency of stand-level mixtures.

## 11.7 Transitioning to Continuous Cover Forestry

### 11.7.1 Background Information

Continuous cover forestry usually aims at more or less uneven-aged stands that contain a mixture of tree species (Roessiger et al. 2011). As this method has long been confused with exploitative harvesting practices, until recently it has met with a great deal of resistance, for example, in Finland (Siiskonen 2007). The relatively low-volume productivity of selective logging (Assmann 1970),[2] the risk of exploitation due to overemphasis on resource extraction and ignoring long-term sustainability (Kuuluvainen et al. 2012), and the inferior economic performance that may occasionally occur (Nordström et al. 2013) are common arguments against continuous cover forestry (Knoke 2012). However, the number and quality of bioeconomic analyses on silvicultural alternatives have increased significantly. Some of the more advanced studies reveal—at least in part—a different picture of the economics of silvicultural alternatives which abstain from clear-cutting. But even the most sophisticated models applied to analyse the economics of continuous cover forestry usually ignore uncertainties and the effects of mixing tree species (cf Tahvonen et al. 2010).

The results obtained through stand-level studies depend strongly on the assumptions made, as is the case with any economic optimisation/evaluation. It is clear that transitioning to continuous cover forestry from already financially mature forest stands may not be advisable from an economic point of view. This was, for example, underlined by Andreassen and Øyen (2002) and also supported by Wikström (2000), who used heuristic methods to assess silvicultural treatments but experienced problems finding adequate growth models.

Strong assumptions have had impacts on the results of other studies. For example, Kant (1999) searched for an efficient economic equilibrium of ingrowth, standing timber stock, increment, and timber harvest but assumed that ingrowth was always costless. This costless ingrowth was simulated, even where intensive harvesting activities included all seed trees. This is a limitation common to many studies on the economics of continuous cover forest management (Sonnemann 2008). Finally, a more recent study by Nordström et al. (2013) also showed how strongly the results of scenario analyses may be influenced by the assumptions used, particularly by the selection of the stands and treatments to be compared.

In light of these issues, it would be useful to further develop the approaches used in the studies under discussion in order to increase the appropriateness of economic assessments of silvicultural management alternatives. Ideally, bioeconomic

[2] Assmann (1970, p. 469) concluded from his comparison of the increment in rotation forests (called high forests) and selection forests (as an important representative of continuous cover forests): 'The high forest stand, . . ., is not only superior in current increment over the last 16 years, and that by as much as 19%, but assuredly also in m.a.i. [mean annual increment] for the same harvesting age. Contrary to current widely prevailing views, the pronounced vertical layering of selection forests has an efficiency-reducing effect . . .'.

modelling approaches should allow for appropriate definition and modelling of the economic objectives of decision makers. Based on such an optimisation approach, suitable management strategies could then be derived. Taking this normative perspective, a predefinition of only a limited number of silvicultural treatments could be avoided. Instead, this approach would allow for an objective assessment of numerous management options. However, such advanced approaches would require models which consider only small components of a forest, either single trees or groups of trees (e.g. size-class models). The complexity of bioeconomic modelling will increase considerably under these newer approaches compared to more simple rotation forest considerations, which are still often applied for the evaluation of silvicultural alternatives (cf Cubbage et al. 2007). A more complex but also more biologically realistic forest economic approach has been advocated by, for example, Scandinavian studies (cf Tahvonen et al. 2010; Hyytiäinen and Haight 2010).

Mixing tree species, as addressed in the previous sections of this chapter, is an important attribute of continuous cover forestry (Pommerening and Murphy 2004). This topic is still of great interest, as a recent contribution on silvicultural alternatives to conventional even-aged forest management shows (Puettmann et al. 2015). However, studies analysing simultaneously the economic consequences of mixing tree species and timber harvesting for natural regeneration, which is spread over a long time period to achieve an uneven-aged structure in the future stand, are scarce. Roessiger et al. (2011) presented a new approach for bioeconomic optimisation to enable this, into which they integrated the effect of European beech admixture on the resistance of Norway spruce (Roessiger et al. 2013). This approach extends the diversification effects by mixing tree species by means of diversifying timber harvesting. Spreading the harvesting of timber over time instead of harvesting all at once (clear-cut) results in an effective compensation of market risks, called 'time diversification' and demonstrated by Knoke et al. (2001, 2005). 'Time diversification' yields an averaging effect through cancelling out extremely low and high timber prices.

The previously mentioned market risks may also be considered as an opportunity for the improvement of economic returns. By means of postponing the harvest until an appropriate timber price can be achieved, the economic return may be enhanced in an order of magnitude of 20 to 30%, as Knoke and Wurm (2006) have shown for mixed-species forests. However, it will often only be possible for the owners of small forest properties—who have very resistant forest stands that are not susceptible to damages—to make use of this option. Larger forest estates will have long-standing customer relations with the wood industry. For them, it would not be an advisable policy to strive for a volatile timber supply.

The mentioned studies by Roessiger et al. (2011, 2013) extend the Thünen approach, where proportions of land are allocated to land-use options according to economic criteria in order to model uneven-aged silviculture. While in the case of even-aged silviculture, land proportions are merely allocated to various tree species, the new bioeconomic approach not only allocates land to more than one tree species but also to various regeneration harvests carried out at different points in time on a given area of allocated land. That means that timber of one tree species is not necessarily cut completely in one operation (clear-cut). Rather, the stand's area allocated to one tree species is divisible, and the standing timber may be harvested in various cutting operations, which may be spread over time. By distributing the regeneration harvests over a long period of time, natural regeneration will be stimulated, and multiple age cohorts will be established. At the end of the total regeneration period, however, all the land allocated to regeneration harvests must be equal to the total land allocated to the relevant tree species (see Roessiger et al. 2011 for details). The outlined modelling is very close to forest management practice, where utilisation quotas indicate how much of a stand's area should be harvested in the coming and in the following planning periods (Knoke et al. 2012a, b).

The optimisation is finally carried out by non-linear programming to maximise the stand's annuitized net present value. This means that the annuity is formed on the basis of net present value and maximised, subject to predefined levels of acceptable risk, as was described earlier. We assume that a risk-averse decision maker would not be willing to accept the maximum risk. Rather, lower than maximum risk levels would be acceptable for risk avoiders. We analyse below whether the maximum return under lower than maximum acceptable risks is more likely to be achieved through mixed, uneven-aged or through pure, even-aged forest stands.

### 11.7.2 Does Improved Economic Modelling and Risk Aversion Lead to Mixed, Uneven-Aged Management?

Assumptions about a tendency towards risk aversion are commonly used in explaining human decision making (Pindyck and Rubinfeld 2009), particularly in the case of land use (see Knoke et al. 2011 and references therein). On this basis, we assume that maximum risk is unacceptable for risk-avoiding decision makers in forestry.

Given this background, we aimed to test the following hypothesis:

> "The integration of risks and uncertainties in a biological-economic model will implicitly lead to mixed stands and diversified harvesting sequences, if the decision maker is a risk avoider."

According to Roessiger et al. (2013), a pure spruce stand would yield a maximum average economic return of 222 € per ha per year, measured with its annuity, for a rotation period of 80 years (discount rate 2%). However, due to possible stand failure caused by natural disturbances such as storm or bark beetle attack as well as timber price volatility, the annuity's standard deviation, with a value of ±57, is also at a maximum for this pure spruce stand. To adopt the perspective of a risk-avoiding

**Table 11.5** Financial return and risk for variants of uneven-aged, mixed and even-aged, pure forests (adopted from Roessiger et al. 2013, with alterations)

| | Resistance of spruce improved | Resistance of spruce not improved | | |
|---|---|---|---|---|
| Management strategy | Mixed, uneven-aged | Mixed, uneven-aged | Pure, even-aged spruce | Pure, even-aged beech |
| Spruce proportion (%) | 80 | 70 | 100 | – |
| Beech proportion (%) | 20 | 30 | – | 100 |
| Average return (annuity) €/ha/year | 195 | 171 | 222 | 99 |
| Standard deviation return | 25 | 25 | 57 | 17 |
| Minimum return €/ha/year | 35 | 25 | −8 | −26 |
| Maximum return €/ha/year | 276 | 246 | 359 | 138 |
| Regeneration period (age when regeneration begins and ends) | 60–120 | 50–120 | 80 | 120 |

decision maker, the optimisation carried out by Roessiger et al. (2013) allowed for a maximum risk of ±25—whereby this threshold for acceptable risk served only as one possible example. Given this constraint, the authors subsequently maximised the average annuity by allocating stand area to tree species and to operations for regeneration harvesting, which could be spread across various points in time.

Interestingly, the resulting stand was no longer a pure stand managed by means of clear-cutting. Rather, a stand composed of 80% Norway spruce and 20% European beech maximised economic return without exceeding a standard deviation of ±25 (Table 11.5). Also, the regeneration harvest is not a clear-cut but a sequence of partial harvests spreading from age 60 to 120. A 60-year regeneration period ensures that the emerging natural regeneration will consist of diverse age cohorts, thus forming an uneven-aged stand. Compared to pure Norway spruce, the economic return is only reduced by 12%, while the standard deviation was decreased to only 44% compared to pure Norway spruce. In conclusion, we can say that the uneven-aged, mixed forest is more efficient under the chosen model assumptions, if we consider both financial risk and return.

An interesting methodological result can be seen in the modelling exercises, which did not consider the improved resistance of spruce when an admixture of beech is present (Table 11.5, column 2). This model approach would thus exclude the interactions between tree species. Assuming the absence of tree species interactions, the stand needs an increased beech proportion (from 20 to 30%) to achieve the low standard deviation of ±25. For the same standard deviation, the average financial return without considering tree species interaction was reduced by 24 € per ha per year. Ignoring tree species interactions would thus mean a significant underestimation of the performance of mixed, uneven-aged forests.

Considering all these economic effects, we should bear in mind that only the effect of beech admixture on the resistance of spruce was addressed here. The effect on growth when mixing tree species was not considered. Also, when modelling partial regeneration harvests, it was not taken into account that the remaining trees would benefit from removing some crop trees. Given these limitations, we can assume that an uneven-aged stand could possibly yield equal or even higher average economic return compared to the pure stand.

**Take-Home Messages**

1. First attempts to achieve a more realistic bioeconomic representation of mixed and even more complex, uneven-aged forests present an exciting and promising field of research. These approaches may even lead us to a more realistic and more positive evaluation of the economic effects of mixed and uneven-aged stands.
2. Many challenges to improving bioeconomic model approaches still exist. Of central importance is to transfer a revised portfolio approach to forest science studies, which would enable more robust results and would be less demanding in terms of the input data required.
3. The next steps to improve the biological realism of the bioeconomic model approaches will include integrating growth interactions (cf Biber et al. 2013) and the possible consequences of climatic change in the existing approaches.
4. Moreover, it will be important to identify challenges for forest production and yield planning, which will arise from the more complex structures of forests in the phase of transition to mixed, uneven-aged forests (cf Hanewinkel 2004).
5. Transitioning to the landscape level to consider many forest stands and maybe even non-forest land-use options simultaneously will pose another challenge.

## References


Abson DJ, Fraser EDG, Benton TG (2013) Landscape diversity and the resilience of agricultural returns: a portfolio analysis of land-use patterns and economic returns from lowland agriculture. Agric Food Secur 2:2

Aerts JCJH, Botzen W, van der Veen A, Krywkow J, Werners S (2008) Dealing with uncertainty in flood management through diversification. Ecol Soc 13:41

Andreassen K, Øyen BH (2002) Economic consequences of three silvicultural methods in uneven-aged mature coastal spruce forests of central Norway. Forestry 75:483–488

Assmann E (1970) The principles of forest yield study. Studies in the organic production, structure, increment and yield of forest stands. Oxford University Press, Oxford

Baader G (1945) Forsteinrichtung als nachhaltige Betriebsführung und Betriebsplanung. Sauerländer's, Frankfurt a.M.

Beimgraben T (2002) Auftreten von Wachstumsspannungen im Stammholz der Buche (Fagus sylvatica L.) und Möglichkeiten zu deren Verminderung. PhD-thesis at the Albert-Ludwigs-University, Freiburg im Breisgau

Beinhofer B (2009) Zur Anwendung der Portfoliotheorie in der Forstwissenschaft – Finanzielle Optimierungsansätze zur Bewertung von Diversifikationseffekten. PhD-thesis at the Technische Universität München

Beinhofer B, Knoke T (2010) Finanziell vorteilhafte Douglasienanteile im Baumartenportfolio. Forstarchiv 81:255–265

Ben-Tal A, El Ghaoui L, Nemirovski A (2009) Robust optimization. Princeton University Press, Princeton

Biber P, Knoke T, Pretzsch H (2013) Eine waldwachstumskundlich-ökonomische Simulationsstudie zu Effekten der Baumartenmischung Fichte-Buche: Konzept und erste ertragskundliche Ergebnisse. DVFFA, Sektion Ertragskunde: Beiträge zur Jahrestagung 2013

Bleile K (2006) Vorkommen und Analyse von Zugholz bei Buche (Fagus sylvatica L.) als Ursache von Spannungen im Rundholz und Verwerfungen des Schnittholzes. PhD-thesis at the Albert-Ludwigs-University, Freiburg im Breisgau

Brandl H (1989) Ergänzende Untersuchungen zur Ertragslage der Baumarten Fichte, Kiefer, Buche und Eiche in Baden-Württemberg. Allgemeine Forst- und Jagd-Zeitung 160:91–98

Bristow M, Nichols JD, Vanclay JK (2006) Improving productivity in mixed-species plantations. For Ecol Manag 233:193–194

Castro LM, Calvas B, Hildebrandt P, Knoke T (2013) Avoiding the loss of shade coffee plantations: how to derive conservation payments for risk-averse land-users. Agrofor Syst 87:331–347

Castro LM, Calvas B, Knoke T (2015) Ecuadorian banana farms should consider organic banana with low price risks in their land-use portfolios. PLoS One 10(3):e0120384. doi:10.1371/journal.pone.0120384

Chang SJ (1998) A generalized Faustmann model for the determination of the optimal harvest age. Can J For Res 48:652–659

Chang SJ, von Gadow K (2010) Application of the generalized Faustmann model to uneven-aged forest management. J For Econ 16:313–325

Clasen C, Griess VC, Knoke T (2011) Financial consequences of losing admixed tree species: a new approach to value increased financial risks by ungulate browsing. Forest Policy Econ 13 (6):503–511. doi:10.1016/j.forpol.2011.05.005

Cubbage F, Mac Donagh P, Sawinski Junior J, Rubilar R, Donoso P, Ferreira A, Hoeflich V, Morales Olmos V, Ferreira G, Balmelli G, Siry J, Noemi Baez M, Alvarez J (2007) Timber investment returns for selected plantations and native forests in South America and the Southern United States. New For 33:237–255

Deegen P, Hung BC, Mixdorf U (1997) Ökonomische Modellierung der Baumartenwahl bei Unsicherheit der zukünftigen Temperaturentwicklung. Forstarchiv 68:194–205

Dieter M (2001) Land expectation values for spruce and beech calculated with Monte Carlo modelling techniques. For Policy Econ 2:157–166

Dieter M, Moog M, Borchert H (2001) Considering serious hazards in forest management decision-making. In: Gadow KV (ed) Risk analysis in forest management. Kluwer, Dordrecht, pp 201–232. doi:10.1007/978-94-017-2905-5_8

Dixit AK, Pindyck RS (1995) The options approach to capital investment. Havard Bus Rev 73:105–115

Djanibekov U, Khamzina A (2016) Stochastic economic assessment of afforestation on marginal land in irrigated farming system. Environ Resour Econ 63:95–117

Dragicevic A, Lobianco A, Leblois A (2016) Forest planning and productivity-risk trade-off through the Markowitz mean-variance model. For Policy Econ 64:25–34

Edwards SF, Link JS, Rountree BP (2004) Portfolio management of wild fish stocks. Ecol Econ 49:317–329

Fama EF, French KR (1993) Common risk factors in the returns on stocks and bonds. J Financ Econ 33:3–56

Franklin JF, Norris LA, Berg DR, Smith GA (1999) The history of DEMO: an experiment in regeneration harvest of Northwestern forest ecosystems. Northwest Sci 73:3–11

Gayer K (1886) Der gemischte Wald. Parey, Berlin

Goldfarb D, Iyengar G (2003) Robust portfolio selection problems. Math Oper Res 28:1–38

Griess VC, Knoke T (2011) Growth performance, wind-throw, and insects: meta-analyses of parameters influencing performance of mixed-species stands in boreal and northern temperate biomes. Can J For Res 41:1141–1159

Griess VC, Knoke T (2013) Bioeconomic modelling of mixed Norway spruce – European beech stands: economic consequences of considering ecological effects. Eur J For Res 132:511–522

Griess VC, Acevedo R, Härtl F, Staupendahl K, Knoke T (2012) Does mixing tree species enhance stand resistance against natural hazards? A case study for spruce. For Ecol Manag 267:284–296

Hahn WA, Härtl F, Irland LC, Kohler C, Moshammer R, Knoke T (2014) Financially optimized management planning under risk aversion results in even-flow sustained timber yield. For Policy Econ 42:30–41

Halpern CB, Evans SA, Nelson CR, McKenzie D, Liguori DA (1999) Response of forest vegetation to varying levels and patterns of green-tree retention: an overview of a long-term experiment. Northwest Sci 73:27–44

Halpern BS, White C, Lester SE, Costello C, Gaines SD (2011) Using portfolio theory to assess tradeoffs between return from natural capital and social equity across space. Biol Conserv 144:1499–1507

Hanewinkel M (2004) Spatial patterns in mixed coniferous even-aged, uneven-aged and conversion stands. Eur J For Res 123:139–155

Havlik P, Enjolras G, Boisson J-M, Jacquet F, Lherm M, Veysset P (2008) Environmental good production in the optimum activities portfolio of a risk-averse farmer. Rev Agric Environ Stud 86:9–33

Hildebrandt P, Knoke T (2009) Optimizing the shares of native tree species in forest plantations with biased financial parameters. Ecol Econ 68:2825–2833

Hildebrandt P, Knoke T (2011) Investment decisions under uncertainty-a methodological review on forest science studies. For Policy Econ 13:1–15

Hildebrandt P, Kirchlechner P, Hahn A, Knoke T, Mujica HR (2010) Mixed species plantations in Southern Chile and the risk of timber price fluctuation. Eur J For Res 129:935–946

Hirshleifer J, Riley JG (2002) The analytics of uncertainty and information. University Press, Cambridge. 6th reprint, first published 1992

Hyytiäinen K, Haight RG (2010) Evaluation of forest management systems under risk of wildfire. Eur J For Res 129:909–919

Hyytiäinen K, Penttinen M (2008) Applying portfolio optimisation to the harvesting decisions of non-industrial private forest owners. For Policy Econ 10:151–160

Jacobsen JB, Thorsen BJ (2003) A Danish example of optimal thinning strategies in mixed-species forest under changing growth conditions caused by climate change. For Ecol Manag 180:375–388

Jactel H, Brockerhoff E, Duelli P (2005) A test of the biodiversity-stability theory: meta-analysis of tree species diversity effects on insect pest infestation, and re-examination of responsible factors. In: Scherer-Lorenzen M et al (eds) Forest diversity and function: temperate and boreal systems. Springer, Berlin, pp 235–262

Kant S (1999) Sustainable management of uneven-aged private forests: a case study from Ontario, Canada. Ecol Econ 30:131–146

Kaplan HM (1985) Farmland as a portfolio investment. J Portf Manag 11:73–78

Knoke T (2008) Mixed forests and finance – methodological approaches. Ecol Econ 65:590–601

Knoke T (2012) The economics of continuous cover forestry. In: Pukkala T, von Gadow K (eds) Continuous cover forestry, Managing forest ecosystems, vol 23, pp 167–193. doi:10.1007/978-94-007-2202-6_5

Knoke T, Moog M (2005) Timber harvesting versus forest reserves – producer prices for open-use areas in German beech forests (*Fagus sylvatica* L.) Ecol Econ 52:97–110

Knoke T, Seifert T (2008) Integrating selected ecological effects of mixed European beech-Norway spruce stands in bioeconomic modelling. Ecol Model 210:487–498

Knoke T, Wurm J (2006) Mixed forests and a flexible harvest strategy: a problem for conventional risk analysis? Eur J For Res 125:303–315

Knoke T, Moog M, Plusczyk N (2001) On the effect of volatile stumpage prices on the economic attractiveness of a silvicultural transformation strategy. For Policy Econ 2:229–240

Knoke T, Stimm B, Ammer C, Moog M (2005) Mixed forests reconsidered: a forest economics contribution to the discussion on natural diversity. For Ecol Manag 213:102–116

Knoke T, Ammer C, Stimm B, Mosandl R (2008) Admixing broadleaved to coniferous tree species: a review on yield, ecological stability and economics. Eur J For Res 127:89–101

Knoke T, Calvas B, Aguirre N, Román-Cuesta RM, Günter S, Stimm B, Weber M, Mosandl R (2009) Can tropical farmers reconcile subsistence demands with forest conservation? Front Ecol Environ 7:548–554

Knoke T, Steinbeis OE, Bösch M, Román-Cuesta RM, Burkhardt T (2011) Cost-effective compensation to avoid carbon emissions from forest loss: an approach to consider price–quantity effects and risk-aversion. Ecol Econ 70:1139–1153

Knoke T, Roman Cuesta RM, Weber M, Haber W (2012a) How can climate policy benefit from comprehensive land-use concepts? Front Ecol Environ 10:438–445

Knoke T, Schneider T, Hahn A, Grieß V, Rößiger J (2012b) Forstbetriebsplanung als Entscheidungshilfe. Ulmer, Stuttgart

Knoke T, Paul C, Härtl F, Castro LM, Calvas B, Hildebrandt P (2015) Optimizing agricultural land-use portfolios with scarce data – a non-stochastic model. Ecol Econ 120:250–259

Koellner T, Schmitz OJ (2006) Biodiversity, ecosystem function, and investment risk. Bioscience 56:977–985

König A (1995) Sturmgefährdung von Beständen im Altersklassenwald: Ein Erklärungs- und Prognosemodell. Sauerländer's, Frankfurt a.M.

Kruschwitz L (2005) Investitionsrechnung. 10, überarbeitete und erweiterte Auflage. Oldenbourg, Munich

Kuuluvainen T, Tahvonen O, Aakala T (2012) Even-aged and uneven-aged forest management in boreal fennoscandia: a review. Ambio 41:720–737

Lambin EF, Turner BL, Geist HJ, Agbola SD, Angelsen A, Bruce JW, Coomes OT, Dirzo R, Fischer G, Folke C, George PS, Homewood K, Imbernon J, Leemans R, Li X, Moran EF, Mortimore M, Ramakrishnan PS, Richards JF, Skanes H, Steffen W, Stone GD, Svedin U, Veldkamp TA, Vogel C, Xuet J (2001) The causes of land-use and land-cover change: moving beyond the myths. Glob Environ Chang 11:261–269

Lohmander P (1993) Economic two stage multi species management in a stochastic environment: the value of selective thinning options and stochastic growth parameters. Syst Anal Model Simul 11:287–302

Lönnstedt L, Svensson J (2000) Return and risk in timberland and other investment alternatives for NIPF owners. Scand J For Res 15:661–669

Macmillan WD (1992) Risk and agricultural land use: a reformulation of the portfolio-theoretic approach to the analysis of a von Thünen economy. Geogr Anal 24:142–158

Mandelbrot BB, Hudson RL (2005) Fraktale und Finanzen: Märkte zwischen Risiko, Rendite und Ruin. Piper, München

Marioni O, Adkins P, Hajkowicz S (2011) Water planning in a changing climate: joint applications of cost utility analysis and modern portfolio theory. Environ Model Softw 26:18–29

Markowitz H (1952) Portfolio selection. J Financ 7:77–91

Markowitz H (1990) Foundations of portfolio theory. http://www.nobelprize.org/nobel_prizes/economic-sciences/laureates/1990/markowitz-lecture.pdf

Markowitz H (2010) Portfolio theory: as I still see it. Annu Rev Financ Econ 2:1–23

Markowitz H (2014) Mean–variance approximations to expected utility. Eur J Oper Res 234:346–355

Matthies BD, Kalliokoski T, Ekholm T, Hoen HF, Valsta LT (2015) Risk, reward, and payments for ecosystem services: a portfolio approach to ecosystem services and forestland investment. Ecosyst Serv 16:1–12

Mayer P, Brang P, Dobbertin M, Hallenbarter D, Renaud J-P, Walthert L, Zimmermann S (2005) Forest storm damage is more frequent on acidic soils. Ann For Sci 62:303–311
Mills WL, Hoover WL (1982) Investment in forest land: aspects of risk and diversification. Land Econ 58:33–51
Möhring B (2004) Betriebswirtschaftliche Analyse des Waldumbaus. Forst Holz 59:523–530
Moog M, Knoke T (2003) Zur betriebswirtschaftlichen Bewertung von Einschränkungen der Waldbewirtschaftung. Forstwissenschaftliches Centralblatt vereinigt mit Tharandter forstliches Jahrbuch 122(1):59–77
Mosandl R, Knoke T (2002) Holzpreisschwankungen als Problem der Forstwirtschaft. Allg Forst Z Waldwirtsch Umweltvorsorge 57:118–119
Neuner S, Beinhofer B, Knoke T (2013) The optimal tree species composition for a private forest enterprise – applying the theory of portfolio selection. Scand J For Res 28:38–48
Neuner S, Albrecht A, Cullmann D, Engels F, Griess VC, Hahn A, Hanewinkel M, Härtl F, Kölling C, Staupendahl K, Knoke T (2015) Survival of Norway spruce remains higher in mixed stands under a dryer and warmer climate. Glob Chang Biol 21:935–946
Nordström E-M, Holmström H, Öhman K (2013) Evaluating continuous cover forestry based on the forest owner's objectives by combining scenario analysis and multiple criteria decision analysis. Silva Fennica 47:1046. doi:10.14214/sf.1046
Pausch R (2005) Ein System-Ansatz zur Darstellung des Zusammenhangs zwischen Waldstruktur, Arbeitsvolumen und Kosten in naturnahen Wäldern Bayerns. Forstliche Forschungsberichte München 199
Paydar Z, Qureshi ME (2012) Irrigation water management in uncertain conditions – application of modern portfolio theory. Agric Water Manag 115:47–54
Pearce D, Putz FE, Vanclay JK (2003) Sustainable forestry in the tropics: panacea or folly? For Ecol Manag 172:229–247
Pichón FJ (1996) The forest conversion process: a discussion of the sustainability of predominant land uses associated with frontier expansion in the Amazon. Agric Hum Values 13:32–51
Pindyck RS, Rubinfeld DL (2009) Mikroökonomie. Pearson Education, München
Pommerening A, Murphy ST (2004) A review of the history, definitions and methods of continuous cover forestry with special attention to afforestation and restocking. Forestry 77:27–44
Puettmann KJ, Wilson SMG, Baker SC, Donoso PJ, Drössler L, Amente G, Harvey BD, Knoke T, Lu Y, Nocentini S, Putz FE, Yoshida T, Bauhus J (2015) Silvicultural alternatives to conventional even-aged forest management – what limits global adoption? For Ecosyst 2(8)
Raes L, D'Haese M, Aguirre N, Knoke T (2016) A portfolio analysis of incentive programmes for conservation, restoration and timber plantations in Southern Ecuador. Land Use Policy 51:244–259
Richards AE, Forrester DI, Bauhus J, Scherer-Lorenzen M (2010) The influence of mixed tree plantations on the nutrition of individual species: a review. Tree Physiol 30:1192–1208
Rinehart J (2010) U.S. Timberland post-recession – is it the same asset? R&A Investment Forestry. www.investmentforestry.com
Ripken H, Spellmann H (1980) Modell-Berechnungen der Reinerträge der wichtigsten Baumarten sowie der gesamten Holzproduktion in den Niedersächsischen Landesforsten. Der Forst- und Holzwirt 35:153–165
Roche MJ, McQuinn K (2004) Riskier product portfolios under decoupled payment. Eur Rev Agric Econ 31:111–123
Roessiger J, Griess VC, Knoke T (2011) May risk aversion lead to near-natural forestry? A simulation study. Forestry 84:527–537
Roessiger J, Griess VC, Härtl F, Clasen C, Knoke T (2013) How economic performance of a stand increases due to decreased failure risk associated with the admixing of species. Ecol Model 255:58–69
Röhrig E, Bartsch N, von Lüpke B (2006) Waldbau auf ökologischer Grundlage, 7. Auflage edn. Ulmer, Stuttgart

Rothe A (2005) Tree species management and nitrate contamination of groundwater: a central Europe perspective. In: Binkley D, Menyailo O (eds) Tree species effects on soils: implications for global change. Kluwer, Dordrecht, pp 71–83
Rothe A, Binkley D (2001) Nutritional interactions in mixed species forests: a synthesis. Can J For Res 31:1855–1870
Samuelson PA (1969) Lifetime portfolio selection by dynamic stochastic programming. Rev Econ Stat 51:239–246
Schmid-Haas P, Bachofen H (1991) Die Sturmgefährdung von Einzelbäumen und Beständen. Schweizerische Zeitschrift für das Forstwesen 142:477–504
Schou E (2012) Transformation to near-natural forest management, climate change and uncertainty. Department of Food and Resource Economics, University of Copenhagen
Schütz J-P, Götz M, Schmid W, Mandallaz D (2006) Vulnerability of spruce (*Picea abies*) and beech (*Fagus sylvatica*) forest stands and consequences for silviculture. Eur J For Res 125:291–302
Seifert T (2004) Einfluss der waldbaulichen Behandlung auf die Holzqualität von Fichte und Buche in Rein- und Mischbeständen. Report of the project X 33 – part II of the Bavarian Forest Service
Sharpe WF (1964) Capital asset prices: a theory of market equilibrium under conditions of risk. J Financ 14:425–442
Sharpe WF (1966) Mutual fund performance. J Bus 39:119–138
Siiskonen H (2007) The conflict between traditional and scientific forest management in 20th century Finland. For Ecol Manag 249:125–133
Sonnemann D (2008) Das ideale Plentergleichgewicht – Leitbild oder Luxus? (The ideal equilibrium state in a selection forest – vision or luxury?) (Essay). Schweizerische Zeitschrift für das Forstwesen 159: 1–7
Spellmann H (2005) Produziert der Waldbau am Markt vorbei? Allg Forstzeitschrift/Der Wald 60:454–459
Tahvonen O, Pukkala T, Laiho O, Lähde E, Niinimäki S (2010) Optimal management of uneven-aged Norway spruce stands. For Ecol Manag 260:106–115
Thomson TA (1991) Efficient combinations of timber and financial market investments in single-period and multiperiod portfolios. For Sci 37:461–480
Thorsen BJ (2010) Risk, returns and possible speculative bubbles in the price of Danish forest land? In: Helles F, Nielsen PS (eds) Scandinavian forest economics no. 43. Proceedings of the biennial meeting of the scandinavian society of forest economics Gilleleje, Denmark, May 2010
von Thünen JH (1842) Der isolierte Staat in Beziehung auf Landwirtschaft und National-ö konomie, 2nd edn. Leopold, Rostock
Wagner JE, Rideout DB (1991) Evaluating forest management investments: the capital asset pricing model and the income growth model. For Sci 37:1591–1604
Wagner JE, Rideout DB (1992) The stability of the capital asset pricing model's parameters in analysing forest investments. Can J For Res 22:1639–1645
Weber M-W (2002) Portefeuille- und Optionspreis-Theorie und forstliche Entscheidungen, Schriften zur Forstökonomie Band, vol 23. Sauerländer's, Frankfurt a.M.
Wikström P (2000) A solution method for uneven-aged management applied to Norway spruce. For Sci 46:452–463
Wippermann C, Möhring B (2001) Exemplarische Anwendung der Portefeuilletheorie zur Analyse eines forstlichen Investments. Forst Holz 56:267–272

# Perspectives for Future Research on Mixed-Species Systems

# 12

Hans Pretzsch, Thomas Knoke, Carola Paul, Jürgen Bauhus, and David I. Forrester

**Abstract**

This chapter proposes how to supplement and integrate existing knowledge in order to proceed from rather fragmentary mosaic pieces of knowledge to a more substantial base for ecology and management of mixed-species stands. First we sketch how the empirical basis and understanding of mixed-species stand structure and dynamics may be systematically improved through concepts, answering the right questions, and experimental approaches. We then outline how better knowledge might be acquired from experiments and inventories in future. In addition we sketch the necessary next steps from the description and analysis of ecological functioning of mixed-species stands to systematically designing mixed-species stands. The subsequent sections address the limited body of scientific knowledge on the effects of mixing native and exotic tree species and the importance of interactions between trees and other crops for integrated land use. To better understand the economic effects of mixed and uneven-aged

H. Pretzsch (✉)
Chair for Forest Growth and Yield Science, Faculty of Forest Science and Resource Management, Technical University of Munich, Hans-Carl-von-Carlowitz-Platz 2, 85354 Freising, Germany
e-mail: Hans.Pretzsch@lrz.tum.de

T. Knoke • C. Paul
Faculty of Forest Science and Resource Management, Institute of Forest Management, Technical University of Munich, Hans-Carl-von-Carlowitz-Platz 2, 85354 Freising, Germany
e-mail: knoke@tum.de; carola.paul@tum.de

J. Bauhus
Chair of Silviculture, Faculty of Environment and Natural Resources, University of Freiburg, Tennenbacherstr. 4, 79106 Freiburg, Germany
e-mail: juergen.bauhus@waldbau.uni-freiburg.de

D.I. Forrester
Swiss Federal Institute for Forest, Snow and Landscape Research WSL, Zürcherstrasse 111, 8903 Birmensdorf, Switzerland
e-mail: david.forrester@wsl.ch

 
H. Pretzsch et al. (eds.), *Mixed-Species Forests*, DOI 10.1007/978-3-662-54553-9_12

stands, research is required into the economic efficacy of mixed-forest management, non-probabilistic approaches to economic uncertainty, and economies of scale, which have largely been neglected in economic analyses so far. The final section addresses how changes experienced by tree species under climate change can be predicted and modelled.

## Contents

12.1 Advanced Approaches to Studying the Underlying Mechanisms of Mixing Effects . 581
12.1.1 Disjunct Experimental Plots Established Along Gradients of Productivity and Environmental Conditions .... 581
12.1.2 Experiments for Detecting the Causes of Mixing Effects: Water Exclusion Experiments Covering Mixed and Pure Stand Conditions .... 583
12.1.3 Future Long-Term Experiments for Revealing Patterns and Processes in Mixed-Species Stands .... 584
12.2 Disentangling the Effects of Compositional and Structural Diversity in Mixed, Uneven-Aged Forests .... 586
12.3 How Does Stand Density Influence Productivity and Overyielding of Mixed-Species Stands? .... 589
12.4 From Analysing to Designing Mixed-Species Stands Based on Species Traits .... 592
12.4.1 Criteria, Knowledge Gaps, and Main Information Sources for the Design of Mixed-Species Stands .... 592
12.5 Advantages and Disadvantages of Mixtures with Native and Exotic Species .... 594
12.6 Mixing Trees with Other Crops in Integrated Land-Use Systems .... 596
12.7 Analysing the Economic Risk Premium for Some Mixed-Forest Alternatives .... 598
12.8 Testing New Approaches to Uncertainty .... 599
12.9 Considering Economies of Scale .... 600
12.10 Survival in Mixed and Pure Forest Stands Under Changing Climate .... 601
Take-Home Messages .... 601
References .... 602

Existing knowledge of ecology and management of mixed-species stands is growing rapidly but we are still far from having accepted general rules or laws such as those that are available for monospecific stands. This chapter aims at developing perspectives on how to supplement and integrate existing knowledge in order to proceed to a more substantial base for ecology and management of mixed-species stands.

First, we sketch how the empirical basis and understanding of mixed-species stand structure and dynamics may be systematically improved through concepts, answering the right questions, and experimental approaches. Section 12.1 deals with new experimental approaches for studying the underlying mechanisms of mixing effects. A study along a productivity gradient and a water retention experiment covering intra- and interspecific interactions serve as examples of how to reveal whether improved conditions of light, water or nutrients are behind overyielding effects in forest stands. Disentangling the effects of compositional and structural diversity in mixed, uneven-aged forests as addressed in Sect. 12.2 is relevant for both theory development and silvicultural management of mixed-species stands. In Sect. 12.3, we stress that knowledge of the effect of mixing on stand density and on over- and underyielding is largely missing and explain why such knowledge is essential for understanding, modelling, and silvicultural regulation of mixed-species stands. We also outline how new knowledge might be acquired from experiments and inventories. In Sect. 12.4, we sketch the necessary next step from the description and analysis of ecological functioning of mixed-species stands to systematically designing mixed-species stands, i.e. establishment, silvicultural management, and economic evaluation. Section 12.5 points to the limited body of scientific knowledge on the effects of mixing native and exotic tree species. Section 12.6 outlines the importance of interactions between trees and other crops for integrated land use. Section 12.7 addresses the analysis of the economic efficacy of mixed-forest management and the need for a sound biophysical update of the bioeconomic models applied so far. In Sect. 12.8 we sketch out a new, non-probabilistic approach to economic uncertainty, which has never before been applied to the analysis of mixed-forest management. Section 12.9 documents the importance of economies of scale, which have largely been neglected in economic analyses so far. In Sect. 12.10 we address how changes experienced by tree species under climate change can be predicted and modelled.

## 12.1 Advanced Approaches to Studying the Underlying Mechanisms of Mixing Effects

### 12.1.1 Disjunct Experimental Plots Established Along Gradients of Productivity and Environmental Conditions

Single experiments based on the experimental designs introduced in this book (see Chaps. 2 and 4) provide only limited evidence of mixing effects or their dependency on site conditions. Generalisations about any over- or underyielding of mixed versus pure stands require replicated experiments at different sites that cover the range of variation of the population of mixed-species stands to which the results are meant to be extrapolated (Forrester and Pretzsch 2015). In addition, these experiments need to be conducted over many decades. Here, we call experimental designs with replicates located at different sites along an ecological gradient 'disjunct experiments'. Whereas in the past, disjunct experiments mostly developed through gradual expansion of the regional and large-scale networks of experimental plots, in recent decades many trial series with replications over a wide range of site conditions have been established through national and international research cooperation (Fig. 12.1). These have often been existing forest stands or experiments that were then joined for the purpose of analysis. As such, they may not resemble a design with a priori selection of site conditions to capture the variation in the population of interest (Fig. 2.1).

One such model example of a disjunct experiment is the transect study in mixed stands of Scots pine and European beech (*Pinus sylvestris* and *Fagus sylvatica*) (del Río et al. 2017, Pretzsch et al. 2015). An important basis for this approach is a set of triplets of pure and mixed stands of these species located along an ecological gradient through Europe. Altogether 15 European countries are involved: Austria

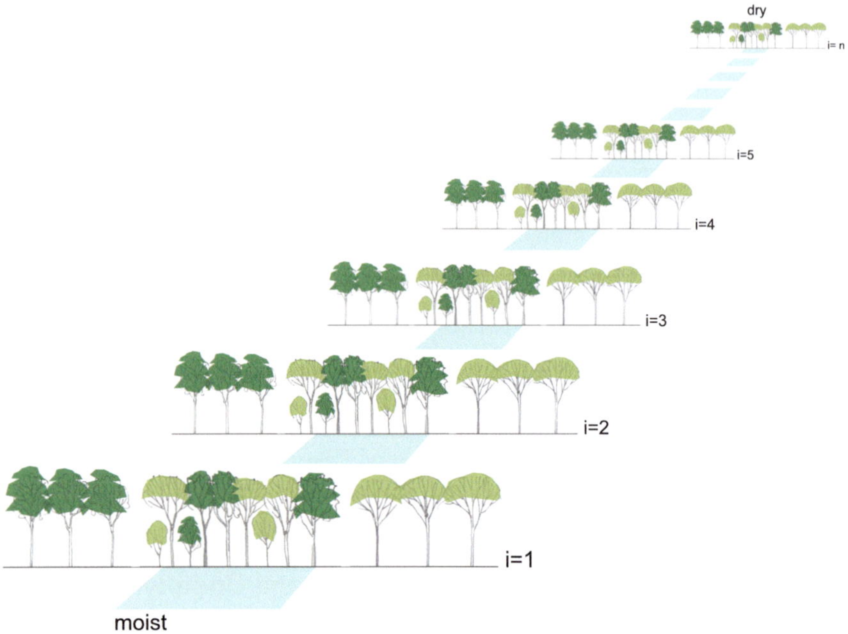


**Fig. 12.1** Design of a disjunct experiment comprising triplets (pure Scots pine, pure European beech, and mixed stands of Scots pine and European beech) located along a precipitation gradient from moist to dry sites across Europe. This design facilitates the analysis of the influence of environmental conditions on mixing effects (Pretzsch et al. 2015)

(1), Belgium (2), Bulgaria (1), Czech Republic (2), France (1), Germany (6), Italy (2), Lithuania (2), Poland (5), Serbia (1), Slovakia (1), Spain (2), Sweden (2), the Netherlands (1), and Ukraine (1) [number of plots are given in brackets].

As the rather Atlantic distribution range of European beech and the more continental range of Scots pine overlap, mixed stands containing these species are of considerable importance in Europe. Scots pine covers about $12 \times 10^6$ ha and European beech $49 \times 10^6$ ha. The potential area where these species can mix amounts to $32 \times 10^6$ ha, but the current area of mixed stands of Scots pine/European beech is only $1.7 \times 10^6$ ha (Brus et al. 2011).

The selected triplets of stands cover the overlapping geographical natural distribution of Scots pine and European beech ranging from the northern border in Lithuania to the southern range in Bulgaria and from the far southwest region in Spain to the eastern border in the Ukraine. The highest concentration of plots is in the Central European region comprising Germany, Belgium, The Netherlands, Czech Republic, and Poland. The mean annual temperature at these sites ranges from 6 to 10.5 °C, the total annual precipitation from 520 to 1175 mm per year, and the elevation from 20 to 1,250 m a.s.l.

Based on such designs, the dependency of mixing effects on environmental conditions can be analysed statistically. Although the site index, typically measured

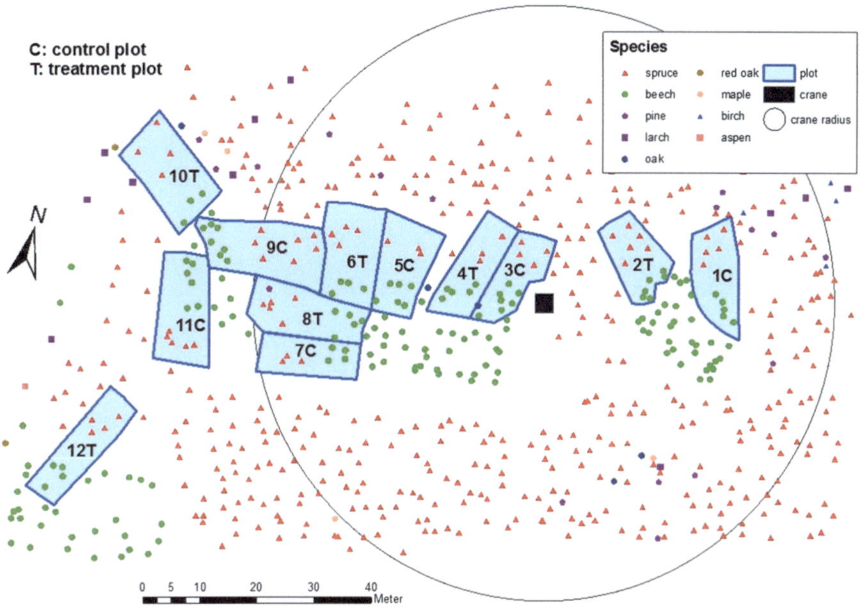


**Fig. 12.2** Map of the 12 subplots of the Kranzberg Forest water exclusion experiment. Such experiments pave the way for understanding the effect of climate trends and weather events on mature monospecific and mixed-forest stands (following Pretzsch et al. 2014b, 2016a)

as dominant tree height at a reference age, is a rather non-specific indicator of environmental conditions, the statistical relationship between overyielding and site index can provide a first indication of environmental influences on mixing effects. However, statistical relationships should not be confused with causal effects; many other factors such as provenance, silvicultural treatment or stress events may vary and be confounding factors. Additional records and measurements of the light, water, and mineral nutrients at such triplet sites will facilitate this shift in focus from statistical to causal relationships.

### 12.1.2 Experiments for Detecting the Causes of Mixing Effects: Water Exclusion Experiments Covering Mixed and Pure Stand Conditions

A model example of an experimental rather than a statistical approach to studying mixing effects is the water retention experiment in the Kranzberg Forest. On six plots with roofs and six control plots, the growth of *Picea abies* and *F. sylvatica* is analysed under intra- and interspecific conditions (Fig. 12.2). In order to avoid experimental artefacts, such as a modification of the within-stand bioclimate under the roofs, the roofs are kept open when there is no precipitation and only close

**Fig. 12.3** Water exclusion experiment to study how drought modifies the inter- and intraspecific interactions between Norway spruce and European beech (*Picea abies* and *Fagus sylvatica*) (Pretzsch et al. 2014b, Pretzsch et al. 2016a)

automatically during rainfall periods (Fig. 12.3). So, under ceteris paribus conditions, the water supply can be reduced through water retention.

This enables the analysis of the extent to which supply, capture, and water-use efficiency is modified through species mixing and whether a modified water regime in interspecific tree neighbourhoods contributes to the overyielding of mixed versus pure stands (Pretzsch et al. 2014b, Pretzsch et al. 2016a).

### 12.1.3 Future Long-Term Experiments for Revealing Patterns and Processes in Mixed-Species Stands

The need for information about mixed stand growth and yield and the lack of appropriate long-term experiments has resulted in the alternative use of forest inventory data (Condés et al. 2013; 2016; Liang et al. 2016; Toïgo et al. 2015). Furthermore, age series (Thurm and Pretzsch 2016) and triplets (del Río et al. 2017, Pretzsch et al. 2015) in mixed-species stands that have been established in an ad hoc manner may substitute for long-term experiments. Forest inventory data can provide information about the overyielding of mixed-species stands versus monocultures (Gritti et al. unpublished) and the statistical correlation between overyielding and species richness (Liang et al. 2016) or site condition (Toïgo et al. 2015). However, because of the restricted length of survey and plot size, the history and emergence of the present mixing pattern can not easily be traced back for inventory plots. The routine management of forests containing inventory plots means that plots with a maximum density or very low density, which would be important for analysing the density-growth relationship, are rare or missing. Even the most elaborate inventories can only shed light on statistical relationships. To reveal the mechanisms of species interactions, which are essential for the design of resource-use efficient mixed-species production systems, long-term experiments would be necessary. Other limitations of inventories, such as the dominance of species-poor plots and the limited number of plots with high tree species richness, were outlined in Chap. 2.

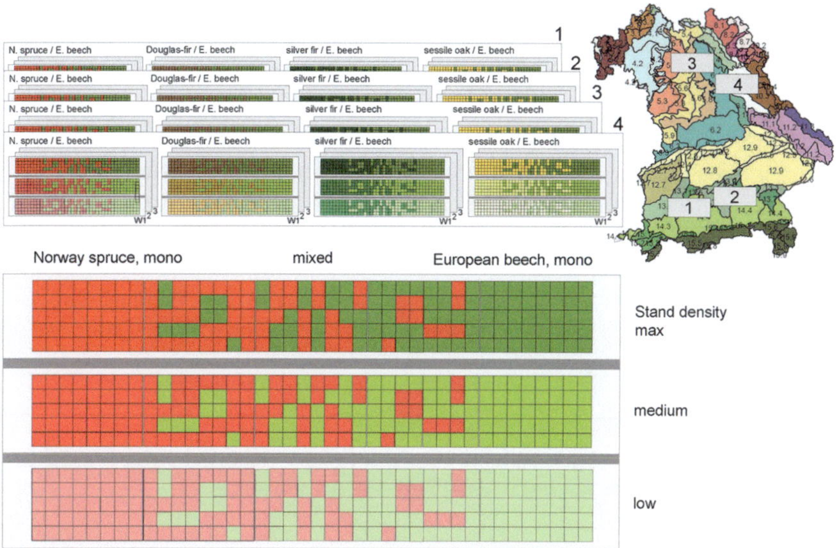


**Fig. 12.4** Concept of a new generation of disjunct long-term experimental plots exploring mixing of different tree species as modified by initial stand density, thinning, mixing proportion, mixing pattern, and site conditions. For that purpose, large-scale experiments are established in several regions differing in site conditions (see 1–4 on map *top left*). In each region, a large-scale, long-term experiment is established exploring the four most dominant tree species combinations (see Norway spruce/European beech, Douglas-fir/European beech, silver fir/European beech, and sessile oak/European beech, *top left*)

Existing long-term experiments require further development in order to provide better insight into, e.g. the causes of over- and underyielding and the dependency of these on initial stand density, mixing pattern, density regulation through thinning, and site conditions. To overcome the limitations of inventory data or other temporary plots, a new generation of long-term experiments is needed for mixed-species forest research and training; a considerable number of such experiments have been established in recent years in several parts of the world (e.g. Scherer-Lorenzen et al. 2005; Tobner et al. 2014; Uhl et al. 2014; Verheyen et al. 2016).

Another model example of a new generation of mixed-species experiments is the set of long-term experimental areas now established in Bavaria/Germany. This set of large-scale experiments covers several species combinations, stand densities, mixing proportions and patterns, and thinning regimes (Fig. 12.4). In order to reveal the dependency of mixing effects on the site conditions, the large-scale experiments need to be conducted at different sites along a gradient of water and nutrient supply (see Fig. 12.4, sites 1–4). At each of the four sites, at least four practically important species mixtures are being analysed. For each species mixture, the effects of initial density, thinning, mixing proportion, and mixing pattern are analysed. To reduce the space requirement for the analyses of the multiple factors, oblong subplots representing pure stand conditions of species 1 and 2 at the ends and a gradient of mixed stand conditions with different mixing proportions and pattern of both species in between are used (see Fig. 12.4). To generate different mixing proportions and mixing patterns, all plots are based on 6 × 6 m miniplots consisting of initially pure plantations with 1 × 2 m and 2 × 2 m initial spacing. The differences in stand density will be increased through future thinning.

For a sound statistical evaluation, the setup for each species combination (Fig. 12.4 unit shown below) is being replicated three times. Spatially explicit, long-term measurements of structure and growth as well as supply, capture and use efficiency of resources will allow the main essential questions to be addressed, at least for the most practically relevant two-species mixed stands. Future extension of this concept to mixtures with more than two species and multilayered structure will follow.

## 12.2 Disentangling the Effects of Compositional and Structural Diversity in Mixed, Uneven-Aged Forests

Many modern silvicultural guidelines are aimed at creating and maintaining heterogeneous stand structures (e.g. Utschig et al. 2011; O'Hara 2016; Schütz 1997; Zenner 2015). The assumption behind this is that structural heterogeneity is beneficial for a diversity of forest functions and services (Noss 1990; Letcher et al. 1998; McKelvey et al. 1993; Ammer and Schubert 1999; Müller et al. 2007, 2012). However, some studies investigating fast-growing, shade-intolerant tree species have found that productivity of monospecific stands decreases with structural heterogeneity (Luu et al. 2013; Soares et al. 2016). Promoting structural diversity in these cases may reduce productivity (Bourdier et al. 2016). In contrast, in mixed-species stands with shade-tolerant species, several studies found a higher productivity in more heterogeneous stand structures (Dănescu et al. 2016; Pretzsch and Schütze 2014; Pretzsch et al. 2016a, b; Liang et al. 2016). This raises the question of whether the relationship between structural diversity and productivity is related to stand composition (monospecific versus mixed) or to the presence of certain functional types (i.e. shade-tolerant species). The extent to which the overyielding in mixed versus pure stands is caused by structural heterogeneity and tree species diversity, or a combination of both is also of interest.

Figure 12.5 emphasises that tree species assemblage can have a direct effect (thin arrow) on stand productivity but also an indirect effect (bold arrow) on stand productivity via stand structure (Dănescu et al. 2016; Zhang and Chen 2015).

The figure shows the essential feedback between stand structure and stand productivity and that it is affected by tree species diversity in two different ways. Of course, the effect of stand structure is also caused indirectly by tree species mixing. However, the difference between a direct and indirect effect is important for the process of understanding, modelling, and managing mixed-species stands.

Several mechanisms related to light absorption may contribute to this increasing productivity with increasing structural complexity (Dănescu et al. 2016). On the one side, species and structurally diverse forests may simply support a higher leaf

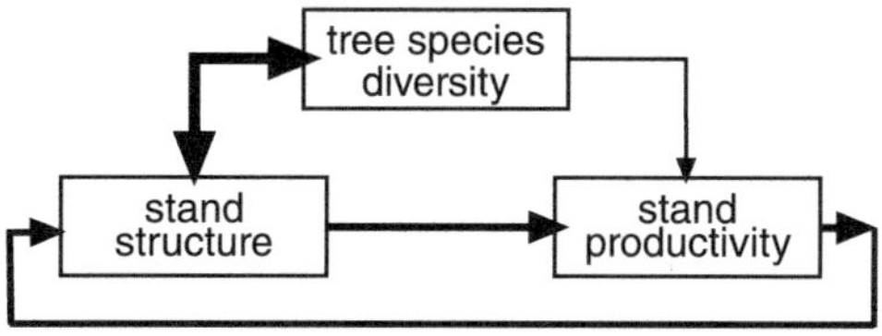


**Fig. 12.5** Tree species diversity can have a direct effect (*thin arrow*) on stand productivity but also a strong indirect effect (*bold arrow*) via stand structure. The conceptual graph shows the essential feedback between stand structure and stand productivity and that it can be affected by tree species diversity in two different ways

biomass and leaf area (e.g. Hardiman et al. 2011). On the other side, canopy complexity may increase the efficiency of photosynthetic tissue such that the same amount of foliage absorbs light more efficiently in heterogeneous than in homogeneous stand structures (Hardiman et al. 2011, 2013). The high light-use efficiency may be partially promoted through increased nutrient use efficiency caused by greater light availability per unit of leaf area, resulting from deeper light penetration to lower parts of the canopy (Hardiman et al. 2013).

Disentangling these effects of structural versus compositional diversity is important for theory development and practical forest management. If it is mainly the stand structure which causes the overyielding in mixed stands, structural homogenisation by certain types of thinning—analogous to crop tree treatment in monospecific stands—may be disadvantageous for stand productivity. If mainly structural heterogeneity is behind the overyielding, species identity is less relevant than the functional traits represented by the mixture. So, structural heterogeneity seems to have a positive influence not only on habitat diversity, stress resilience, protection, and aesthetics but also on productivity. Silvicultural guidelines promoting structural heterogeneity may foster a broad spectrum of forest functions and services including productivity. Thus, the productivity reduction (penalty) caused by structural diversity in pure stands would turn into an increase (benefit) in mixed-species stands (Larson et al. 2008).

Many studies found homogeneous stand structures under dry and therefore water-limited conditions as well as under moist and light-limited conditions (Zobel et al. 1976; Ferris et al. 2000; Ares and Fownes 1999; Wang et al. 2006; Webb 1968). On dry sites, the homogeneity may result from dominant trees being under permanent water stress which benefits the smaller neighbours. In the long term, the sizes of trees converge and the homogeneity results from symmetry of water competition. On moist sites, the tall trees can outcompete their smaller neighbours by pre-empting the light; so homogeneity results from mortality of the small individuals (Fig. 12.6).

On very rich sites where pure stands tend towards homogeneity (Fig. 12.7a, left), stand structuring through silvicultural interference may decrease stand productivity. For instance, though thinning from above may promote future crop trees and increase the structural heterogeneity by accelerating the size growth of tall trees, it usually decreases the stands' productivity (Assmann 1970, pp. 227–245; Pretzsch

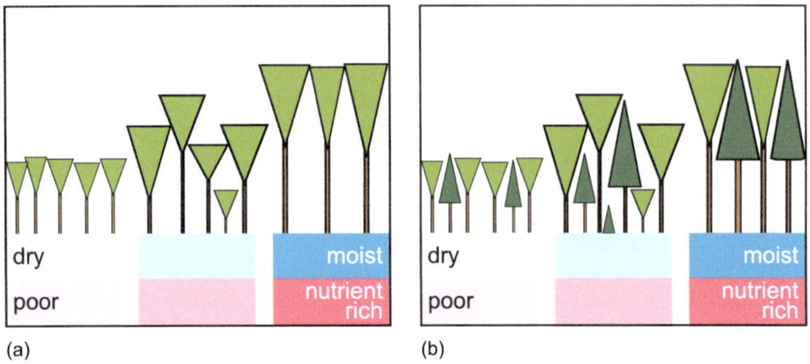


**Fig. 12.6** Hypotheses on the effect of site conditions on stand structure in (**a**) monospecific and (**b**) mixed-species stands. (**a**) Monospecific stands tend towards homogeneous stand structures on extremely low and high productive sites (**a**, *left* and *right*), whereas structural heterogeneity is promoted by alternating limitations in water, nutrient or light supply (**a**, *centre*). (**b**) On all sites, mixed-species stands are generally more heterogeneous, especially in species assemblages with strong niche complementarity

2005). On very poor sites, stands may also tend towards homogeneity (Fig. 12.7a, right). However, silvicultural interference may increase the productivity, especially if some of the less efficient smaller trees are eliminated in order to increase water availability for their tall neighbours. In mixed-species stands (Fig. 12.7b) on both rich and poor sites, silvicultural measures which cause more heterogeneous structures may increase productivity due to a spatially and temporally diversified resource exploitation (Chaps. 3 and 6). The strong effect of structural heterogeneity on productivity means that part of the potential of mixed stands lies in the crown plasticity and morphological adaptability of the species assemblage. The heterogeneity may need time to develop as trees only gradually occupy the space around them to the full extent. It also means that any silvicultural measures which homogenise stand density, vertical layering, and size distribution may cause a decrease in productivity. On poor sites, the below-ground structural heterogeneity may be more important in order to remedy the water and nutrient limitation. Here, above-ground heterogeneity may be less useful or may even be disadvantageous because of the increased water consumption of the ground vegetation. So, on fertile moist sites small trees can contribute considerably to stand growth, whereas on poor and dry sites, they can become competitors and their elimination may increase stand growth.

To further analyse and disentangle the effect of structural and compositional diversity, experimental plots in monospecific and mixed-species stands growing at the same site with similar density and standing volume but varying stand structure and species composition may be employed (Pretzsch et al. 2016a, b; Pretzsch and Schütze 2014). If such stands are available for different site conditions, the effects of stand structure and tree species mixing on productivity and its modification by site conditions can be analysed. In addition to experimental plots, forest inventory

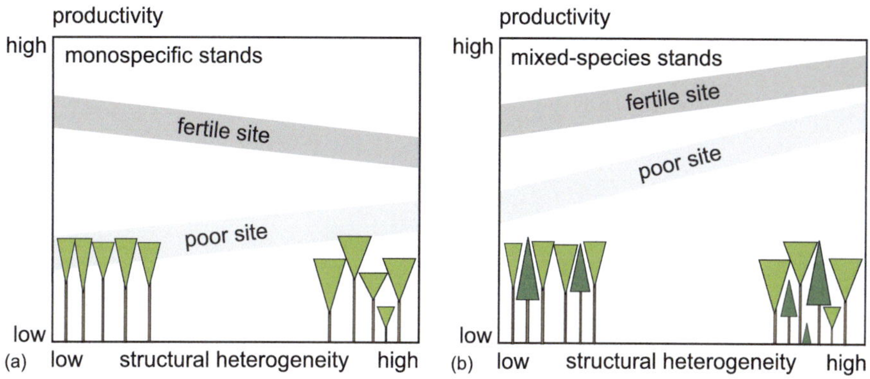


**Fig. 12.7** Hypotheses about the relationship between structural heterogeneity and productivity in (**a**) monocultures and (**b**) mixed-species stands. In monospecific stands, structural heterogeneity may decrease productivity on fertile sites and just slightly increase productivity on poor sites. For mixed-species stands (**b**) recent studies found an increase of productivity with increasing structural heterogeneity

data may facilitate analysis of this relationship as they often represent a broad range of stand structures and species mixtures growing under different site conditions.

## 12.3 How Does Stand Density Influence Productivity and Overyielding of Mixed-Species Stands?

Knowledge of the effect of mixing on stand density and over- and underyielding is essential for understanding, modelling, and silvicultural management of mixed-species stands.

When mixed stands are very sparsely stocked, trees of the different species may have limited interaction and are therefore unlikely to have a direct effect on one another. As such, these stands may not have higher productivity than comparable pure stands of the same density. However, if their stand density increases, crowns and roots interact and hence there is a greater probability for positive interactions above and below ground that may lead to overyielding as shown by Amoroso and Turnblom (2006) and Condés et al. (2013). In addition, species mixing may increase the maximum stand density, i.e. the carrying capacity, above the level of neighbouring monospecific stands (Chap. 4).

Without any positive or negative intra- or interspecific interactions between the trees, stand growth would increase proportionally with stand density. Stand growth would follow the course of a straight line to the origin when maximum density is reduced through thinning (see broken straight lines in Fig. 12.8a–c). In Chap. 4 we showed that removing trees may accelerate the growth of the remaining trees in such a way that the density-growth relationship follows a rather concave or even a unimodal curve (grey $\widehat{p}_{1,2}(\text{dens})$ curve in Fig. 12.8). This means that if stand density

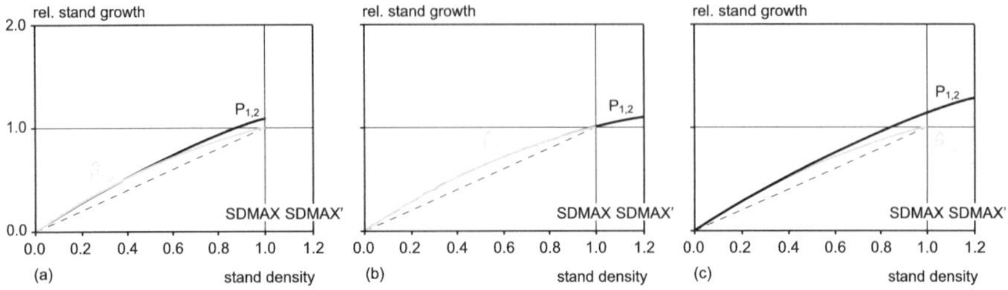


**Fig. 12.8** Hypotheses on the effect of species mixing on the growth-density relationship of forest stands. *The black solid curve* $(\widehat{p}_{1,2})$ represents the expected stand growth when the mixed stand achieves simply the weighted mean productivity of the mixed species $(\widehat{p}_{1,2}(\text{dens}) = p_1(\text{dens}) \times m_1 + p_1(\text{dens}) \times m_1)$. The *grey curve* shows the productivity of the mixed-species stand as a whole $(p_{1,2})$ for three basic modes of overyielding: (**a**) overyielding is zero when stocking is sparse and increases with increasing stand density $(p_{1,2} > \widehat{p}_{1,2})$. The maximum stand density is equal in pure and mixed stands (SDMAX = SDMAX′); (**b**) overyielding is zero over the whole range of pure stand density $(p_{1,2} = \widehat{p}_{1,2})$ but mixed stands can support a higher stand density (SDMAX′ > SDMAX); and (**c**) overyielding of mixed versus pure stands is based on both increased growth $(p_{1,2} > \widehat{p}_{1,2})$ and increased maximum density of mixed stands (SDMAX′ > SDMAX)

is halved, stand productivity may only be reduced by 25%, i.e. it may still be 75% of the productivity of fully stocked stands (Assmann 1970, pp. 227–235).

If two species with species-specific growth-density relations [$p_1$(dens), $p_2$(dens)] are mixed, and we simply assume that both have the same growing space requirement, then the weighted mean productivity $(\widehat{p}_{1,2}(\text{dens}) = p_1(\text{dens}) \times m_1 + p_1(\text{dens}) \times m_1)$ ($m_1$, $m_2$ stand for the mixing proportions) represents the expected mixed stand growth. This productivity $\widehat{p}_{1,2}(\text{dens})$ (grey curve in Fig. 12.8a–c) would result if the reaction in mixed stands is just the weighted mean of the pure stands' reaction.

However, mixing may alter this simple additive reaction pattern in the following way: Fig. 12.8 shows the basic hypotheses on the effect of species mixing on the growth-density relationship of forest stands. The hypothesis would be analogous for underyielding, which is also essential to know in order to avoid growth losses caused by species mixing. In the first case (Fig. 12.8a), overyielding increases with stand density, but the maximum stand density, SDMAX, is equal in pure and mixed stands. In the second case (Fig. 12.8b), there is no overyielding over the whole range of pure stand density, but it emerges at a higher maximum stand density in mixed stands. Finally, overyielding of mixed in comparison to pure stands is based on both increased stand growth *and* increased maximum density in mixed-species stands (Fig. 12.8c).

These hypotheses are based on knowledge of the respective relationships in pure stands which were thinned (density reduction) or fertilised (density increase). However, to further substantiate them, we need clarification of the maximum natural stand density in mixed stands, the natural relative density, and the implications of species-specific growing area requirements and mixing proportions.

Neglecting possible increases in productivity and maximum stand density with increasing tree species diversity and application of simple thinning guidelines developed for monospecific stands to mixed-species stands may result in severe growth losses and undesirable mixing proportions.

To reveal any general reaction pattern of stand growth, careful analysis and separation of the effects of relative and maximum stand density or species mixing proportions, as shown above, are essential. The same principles are relevant for the regulation of density and mixing proportions in long-term experiments as well as in regular forest stands. In experiments and even more so in practical forestry, density and mixing regulation has to follow rather simple rules, e.g. based on tree numbers, tree-to-tree distances, stand basal area, and simple indicators of tree vitality, quality, and stability, e.g. height-to-diameter ratios, relative crown length, or branch-free bole length.

If mean tree dimensions and species-specific growing area requirements differ between the associated species, this should be taken into account by equivalence factors when calculating mixing proportions (see Chap. 4). Species-specific space requirements need to be considered for quantifying stand density as well as for regulating and managing stand density in mixed-species stands (Dirnberger et al. 2016).

To better understand and manage mixed-species stands, we need knowledge of general relationships between growth and density. However, these may remain disguised as long as evaluation of existing experiments and design and regulation of new experiments apply absolute stand densities. Relative stand density measures that are standardised in reference to the maximum natural density may be more meaningful because these are more closely linked to resource availability and competition. The latter should be explored using fully stocked unmanaged plots on the respective sites (Dirnberger et al. 2016). Neglecting species-specific space requirements when regulating mixing proportions, e.g. based on tree numbers, will always cause spatial abundance and the spread of that species with the wider crown extension and higher space occupation. In Central Europe this is mostly European beech.

Neglecting the potential of higher stand densities in mixed stands and following, for example, the guideline for regulating stand density for monospecific stands may cause understocking, underuse of resources and significant growth losses, especially in advanced stand development phases in which the density of mixed stands often exceeds that of monospecific stands.

As the relative natural density is most suitable for analysing, modelling, and regulating stand density in terms of competition for resources, future mixed-species experiments should include unmanaged plots. They will reveal the often unknown maximum density in mixed compared with pure stands and serve as a reference (see Fig. 12.4).

**Table 12.1** Criteria, knowledge gaps, and main information source for improving the design of mixed-species stands

| Criteria for design | Knowledge gap | Main information source |
|---|---|---|
| Potential species combination matching the range of site conditions (environmental factors, resource supply) | Site-growth relationship, niche-extension or niche restriction by species combination, niche of introduced species | Inventory data |
| Species complementarity in terms of resource use | Relationship between species traits and complementarity, properties emerging in mixtures | Long-term experimental plots |
| Variation of the mixing effect within the range of site conditions | Dependency of mixing effects on site conditions, long-term effect of mixing on site properties | Inventory data, long-term experimental plots |
| Site-specific mixing effect of the species regarding productivity, stability, and quality | Relationship between species mixture and stand productivity, stability and wood quality | Long-term experimental plots, bio-group and individual tree analyses |
| Options for regulating the mixture through spatial or temporal separation of the species using silvicultural treatment | Relationship between spatial and temporal mixing structure and mixing effects | Combined mixing, spacing, and thinning trials established along a gradient of site conditions |

## 12.4 From Analysing to Designing Mixed-Species Stands Based on Species Traits

Our knowledge of the dependency of mixing effects on tree species identity, site conditions, stand establishment, thinning, and other silvicultural measures is still rather empirical. For systematic design of resource-efficient mixed-species stands at a given site, knowledge is still fragmentary. Forest management is in urgent need of rules and guidelines on how to design, establish, and treat mixed-species stands and of models to simulate their long-term performance. This requires both further empirical studies but also integration, systematisation, and generalisation based on the available knowledge of mixing reactions.

### 12.4.1 Criteria, Knowledge Gaps, and Main Information Sources for the Design of Mixed-Species Stands

The criteria compiled in Table 12.1 are most relevant for the design of mixed-species stands but far from sufficiently substantiated through existing knowledge or data. Columns 2 and 3 summarise general relationships which still need to be developed for improving the design of mixed-species stands and suitable data sources.

The selection of the potential species assemblages at a given site is usually based on relationships between site conditions and species occurrence derived for monocultures. However, mixing can significantly modify (extend or reduce) the niche of a species (Fig. 2.4d, e). This may extend the possibility of species combinations, e.g. some conifers may be cultivated in mixtures on sites where they would not succeed in monocultures. One species may benefit from frost reduction, nitrogen supply or protection against biotic or abiotic disturbances provided by the other species. When selecting the potential species for tree mixtures, their beneficial effect should be considered. Further studies should systematically analyse how mixing can modify the range of site conditions under which the species can grow. So far, the knowledge of the relationship between site conditions and species performance is mainly based on monospecific stands.

The species complementarity in resource use is a decisive criterion for species selection. Although many species traits (shade tolerance, rooting depth, early successional/late-successional type, wood density, early/late leaf flush, conifer/deciduous, high/low morphological plasticity, atmospheric $N_2$ fixation, low/strong swinging caused by wind) are well known, their relationship to complementarity requires further systematisation and generalisation. General relationships between traits of species or functional species groups and their complementarity would be helpful for species selection regarding beneficial multiplicative mixing effects. The development of such relationships for functional species groups has the advantage that the probable mixing effects for individual species which have not yet been intensively analysed can be predicted on the basis of their specific traits, assuming that they will behave analogously to species with similar traits. Species' complementarity and mixing effects ultimately depend on the species' traits (e.g. shade tolerance, rooting depth, morphological plasticity). General relationships between traits of species or functional species groups and their complementarity and mixing effects still need to be identified to facilitate decisions about useful or promising species combinations.

How selected species combinations and the respective mixing effects depend on the site conditions is a further relevant criterion. Many studies have shown that mixing effects can increase but also decrease with the soil fertility and moisture. The kind of reaction ultimately depends on the respective limiting factor and how a species combination can remedy the limitation through complementarity of the mixed species.

Analyses of mixing effects have to date focused on productivity but should be extended to also cover the stability and quality of stands and trees, respectively. So, in addition to the size growth of trees and the volume growth of stands, stability criteria such as ratio of slenderness or roughness of the canopy or quality measures such as crown width, height to the crown base or mean tree-ring width need to be considered.

Many species combinations require a spatial (e.g. establishment in groups, clusters) or temporal (non-simultaneous establishment of species) separation in order to reduce competition and keep all combined species in the stand as long as planned. We need clear silvicultural prescriptions as to which kinds of spatial and temporal separation are successful and which kinds of mixtures should be avoided on the basis that they need too much silvicultural input and care.

Species-specific mixing effects on productivity, stability, and wood quality and their dependency on site conditions, spatial and temporal design, and silvicultural treatment need to be integrated into stand simulators for better prognoses and scenario analyses of the long-term effects of various silvicultural options.

## 12.5 Advantages and Disadvantages of Mixtures with Native and Exotic Species

Globalisation does not only connect global economies, but it also builds (both deliberately and unintentionally) bridges between ecosystems. In many parts of the world, foresters have introduced exotic tree species because of their faster growth, higher volume production or lower susceptibility to pests or diseases compared to native tree species (Evans and Turnbull 2004). One quarter of the globally planted forests are established using introduced tree species (FAO 2010). In tropical regions, this share is much higher. For example, in South America, 97% of the planted forests comprise introduced species, with *Tectona grandis*, *Eucalyptus* spp. or *Pinus* spp. being the most commonly planted tree species (Evans and Turnbull 2004; FAO 2010). The main reason for this is the often lacking international markets for native timber products, a lack of seed sources and experience (Onyekwelu et al. 2011) and the fast initial growth of exotic timber species, which considerably reduces management costs, for example, for the labour-intensive weeding (Paul et al. 2015). Based on economies of scale, exotic timber plantations are mostly established in large-scale monocultures (see Sect. 12.9). The ecological value of such reforestation attempts is debated (see Brockerhoff et al. 2008; Bremer and Farley 2010).

While research on tree mixtures with native species has evolved in the tropics during recent decades (e.g. Piotto et al. 2004; Potvin and Dutilleul 2009; Bauhus et al. 2000), investigations of the effects of mixing native with exotic species are still missing in both tropical and temperate regions. Given the increasing importance of reforestation/afforestation for restoring ecosystem services, this could be an important field of future research. Mixing exotic with native tree species may create a compromise between ecological and economic interests. Such mixtures could either be established during planting or could result from a stand development using exotic nurse trees. Past research has demonstrated that exotic tree species can successfully serve as nurse trees on areas where native species are often not competitive, for example, by using pines to shade out pasture grass (Feyera et al. 2002; Weber et al. 2008; Ashton et al. 1998).

Investigating interactions between native and exotic species might offer a better understanding of over- and underyielding effects, which might be more pronounced

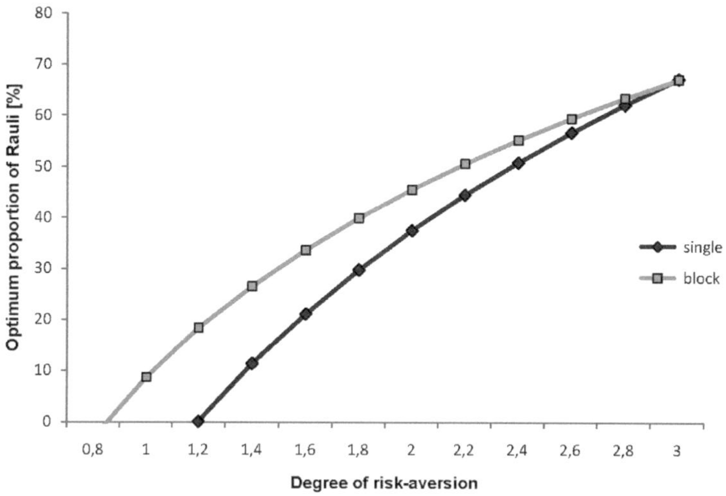


**Fig. 12.9** Optimal proportions of the native tree species Rauli (*Nothofagus alpina* (Poepp. and Endl.) Oerst.), when mixed with Douglas-fir (*Pseudotsuga menziesii*), depending on the degree of relative risk aversion and the type of mixture (single, species interact on tree level; block, species are separated into blocks, avoiding interactions at tree level); figure adopted from Hildebrandt (2011)

in native-exotic mixtures where evolutionary adaptive processes have been absent (Kawaletz et al. 2013).

One of the rare examples in which the establishment of native and exotic plantation forests has been investigated from a silvicultural and economic perspective is the study by Hildebrandt et al. (2010). This study underlined the importance of considering tree species interactions in mixed-forest stands. In their study, a single tree mixture (species mixed in rows) led to lower optimal proportions of the native tree species (Rauli, *Nothofagus alpina*) in Douglas-fir-Rauli stands for the same level of biotic or abiotic risk aversion (Fig. 12.9). However, this type of study is, so far, very rare.

On the other side of the coin, when introducing competitive exotic tree species, alien species may become invasive to natural and managed forests (Richardson and Rejmánek 2011). Such invasions of forest ecosystems can severely affect species composition. Examples of such unintended invasions are *Acacia* spp. from Australia to Europe (Lorenzo et al. 2010), South Africa and South America (Le Maitre et al. 2011), *Robinia pseudoacacia* and *Prunus serotina* from North America to Europe (Kawaletz et al. 2013) as well as *Acer platanoides* from Europe to North America (Martin 1999). Better understanding of competitive effects between exotic and native tree species can not only help to understand the potential threat resulting from further spreading of these species but can also support the development of management practices, which may allow for productive use of these species while reducing adverse effects on natural forest composition (Kawaletz et al. 2013).

In the face of climate change, introducing non-native species as well as natural shifts in species ranges might strongly affect forest ecosystems (Neuner et al. 2015; Hanewinkel et al. 2013). A better understanding of advantages and disadvantages of mixtures between native and non-native forest species will be crucial for future forest management.

## 12.6 Mixing Trees with Other Crops in Integrated Land-Use Systems

Agroforestry (AF) is a form of land use combining perennial, woody plants (mostly trees) with annual crops or cattle. Specific spatial or temporal combinations of woody and agricultural crop species (Lundgren and Raintree 1982) may be used to achieve particular land management goals. However, it is important that the combination leads to synergistic interactions and complementary use of soil and water resources between the mixed components (Rao et al. 1997). Closed nutrient cycles in these systems may maintain (or even increase) soil fertility and thus ensure sustainability of land use (Ong et al. 2015). AF systems have been proposed to reconcile food production and conservation of biodiversity (Perfecto and Vandermeer 2010).

Agroforestry is often associated with introducing single trees in agricultural landscapes (see the classification by Nair 1985). However, given the accelerating need for food and energy resources, integrating food or energy crops in timber production could also be considered (Table 12.2). Economic analyses show that admixing crops into forest plantations may increase their net present value substantially (Paul et al. 2015). Interplanting young trees with food or energy crops did not increase tree mortality; in fact, the height growth of some species was positively affected by admixing maize or shrubby crops such as pigeon peas to young trees (Paul and Weber 2016). This demonstrates the potential of tree-crop mixtures ("agrisilvicultural" systems) to provide economic and silvicultural advantages. While the effects of trees on crops have been intensively studied during recent decades (Ong et al. 2015), effects of crops on trees from a silvicultural perspective have largely been disregarded.

In addition to the intense mixing of trees and crops on the same parcels, other concepts propose mixing trees and crops on separate parcels (Knoke et al. 2012; Fig. 12.10). This practice reduces the sizes of agricultural parcels and with this the exposure of bare soils to wind and water. Using a land-use optimisation approach, Knoke et al. (2016) demonstrated that integrated mosaic landscapes might help to achieve a balanced provision of multiple ecosystem services with both ecological and socio-economic importance (Knoke et al. 2016). In conclusion, integrated land-use systems are of enormous importance and represent a future research field with high scientific potential.

**Table 12.2** Characteristics of agrisilvicultural practices for the purpose of timber production (taken from Paul 2014)

| | Agrisilvicultural practices for timber production | | |
|---|---|---|---|
| Characteristics | Traditional taungya system | Alley cropping with valuable timber species | Shade trees for woody crops |
| Schematic figure<br>= Tree<br>= annual crop<br>= perennial shade-tolerant woody crop |  |  |  |
| Synonyms | Sequential intercropping, plantation intercropping | Intercropping, hedgerow intercropping, plantation intercropping | Plantation intercropping |
| Primary production aim | Tropical hardwoods | Hardwoods (temperate and tropical regions), firewood, fodder, used as windbreak for improved crop production | Production of shaded coffee/cacao or other shaded crops, hardwoods |
| Associated crops | Annual | Annual/perennial | Perennial |
| Time frame of intercropping | Short-term association of trees and crops Until canopy closure (2–5 years) | Long-term association of trees and crops during whole rotation period of trees | Long-term association of trees and crops during whole rotation period of trees |
| Planting density of trees | Common density 3 × 3 m ~ 1111 trees/ha in tropical systems | Between tree rows: 3–50 m | ~150 trees/ha |
| Tree planting design | Regular | Regular | Regular to scattered |

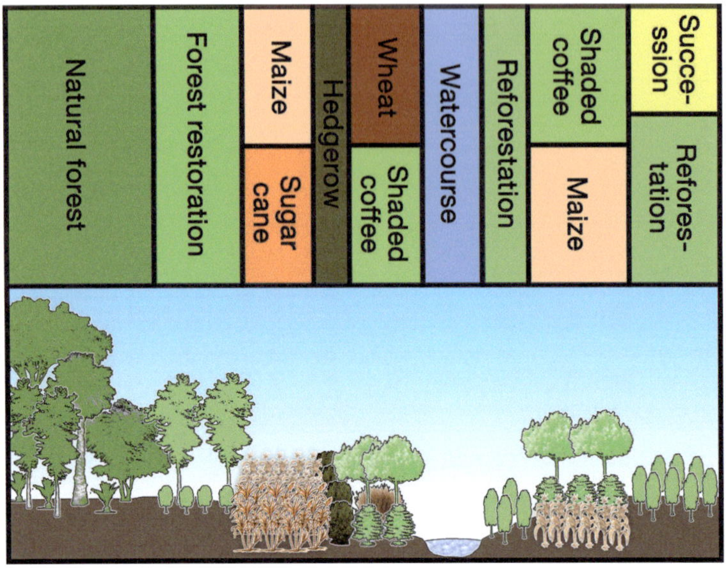


**Fig. 12.10** Example of an integrated land-use system, where agricultural crops and trees are mixed on separate parcels. This concept leads to reduced parcel sizes, which has positive ecological effects through minimising erosion caused by wind and water. With kind permission from Wiley, where the original version of this figure appeared in Frontiers in Ecology and the Environment (Knoke et al. 2012)

## 12.7 Analysing the Economic Risk Premium for Some Mixed-Forest Alternatives

A particularly challenging aspect of the bioeconomic modelling efforts to improve the economic analysis of mixed forests is appropriate modelling of the decision maker's economic objectives. Moving from the common assumption of a risk-neutral decision maker to a, certainly more realistic, risk-avoiding forest owner has already led to interesting results. Under the assumption of risk aversion, improved forest economic modelling may lead to mixed-forest and a continuous cover forest structure without clear-cutting, even if the growth modelling used has only stand-level resolution and, consequently, cannot consider the individual growth of single trees (e.g. Roessiger et al. 2011). However, this simple biophysical model approach applied in the mentioned study disregarded tree species interactions, for example, stabilisation effects in mixed forests (e.g. Griess et al. 2012) and thus required a very high proportion of broadleaves to achieve a low risk. This type of forest management might only be attractive for very cautious forest owners with only small properties because such a careful management strategy means a high "insurance" premium must be paid in comparison to management approaches which just maximise the average net present value. According to Roessiger et al. (2011), this

**Table 12.3** Results for simulated silvicultural management strategies that achieved the highest economic return in relation to the economic risk (adopted from Roessiger et al. 2013)

| Stand type | Mixed forest without clear-cuts | Spruce monoculture |
|---|---|---|
| Proportion spruce (%) | 80 | 100 |
| Proportion beech (%) | 20 | – |
| Average annuity (€/ha/J) | 195 | 222 |
| Standard deviation | ±25 | ±57 |
| Minimum annuity (€/ha/year) | 35 | −8 |
| Regeneration period (from age to years) | 60–120 | 80 (clear-cut) |

premium may be as high as 65 euro/ha/year and refers to the difference between the average annuity of a high-risk, pure *Picea abies* clear-cutting system (rotation 80 years) and the average annuity of a low-risk variant of continuous cover forestry. However, integrating improved models for tree survival, Roessiger et al. (2013) could show that the above-mentioned risk premium may be reduced to 27 euro/ha/year (Table 12.3). This result underlines the economic potential of integrating synergistic ecological interactions between tree species in mixed stands.

## 12.8 Testing New Approaches to Uncertainty

Sophisticated approaches to analysing more complex forest management systems (e.g. Tahvonen et al. 2010) usually assume that information on all biophysical information (including tree growth and possible tree mortality from stress and disturbance) and all financial- and market-related information (interest rates, timber prices, logging operations, logging costs) is perfectly complete. However, it is somewhat unrealistic to assume complete information, and thus, it is important to develop a methodology to achieve robust forest economic results from various silvicultural approaches, even if the input data were to change. Consequently, approaches to achieving and evaluating the robustness of the management recommendations derived from the improved models are indispensable and must be further developed.

A conventional approach to uncertainty which uses stochastic optimisation based on assumed probability distributions for the uncertain parameters was presented in Chap. 11, for example. As an alternative to this conventional perspective, the approach of robust optimisation, for example, as published by Ben-Tal and Nemirovski (2000), has become popular in many fields of research. Using the robustness approach, a less than optimal solution is accepted for the objective function if the solution remains almost optimal or at least acceptable when the input data changes (Bertsimas and Sim 2004). Here, optimisation is carried out over the complete predefined uncertainty space (see Fig. 12.11 for a simplified example). These approaches have appeal. However, they have rarely been used to address

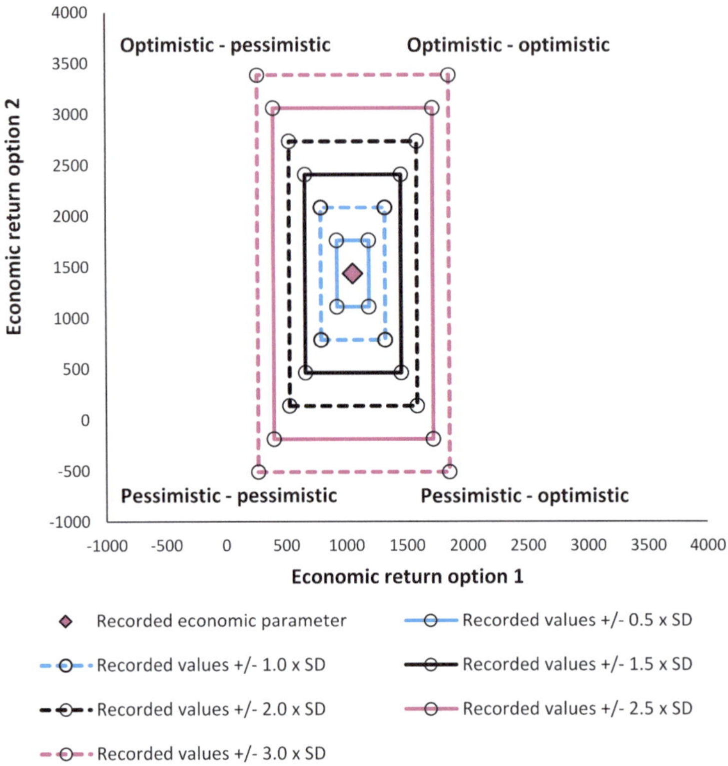


**Fig. 12.11** Schematic illustration of the construction of uncertainty spaces for economic input parameters (economic return, e.g. net present value), if only two options are combined in a portfolio analysis (modified after Knoke et al. 2015). Starting from the recorded/modelled values for economic return (*magenta diamond*), box uncertainty spaces are constructed by adding/subtracting possible deviations (proportional to the measured standard deviation, SD) from the recorded indicator values. To test the impact of the size of the uncertainty spaces, one may enlarge the size of the boxes, for example, from 0.5 to 3.0 times the SD. Addressing the parameter combinations at the corner points of the rectangular boxes ensures all uncertainty combinations are considered

long-term bioeconomic optimisation in forest science (Hildebrandt and Knoke 2011), even though they seem to be particularly appropriate in this context.

## 12.9 Considering Economies of Scale

Finally, an aspect which has rarely been addressed in the analysis of the economics of mixed forests is the consideration of ‘economies of scale’, which could, perhaps, explain why mixed-forest management is not practised very frequently around the world. Instead, industrialised and highly mechanised forms of rotation forest management, often with very short rotations, are found worldwide and produce a

**Table 12.4** Example of increases in the survival probability of Norway spruce on good soils caused by admixing various proportions of broadleaves (100% minus Spruce (*Picea abies*) ratio); adopted from Neuner et al. (2015)

| Survival probability up to age 120 | Favourable climate (wet and cold) | Unfavourable climate (dry and warm) |
|---|---|---|
| Spruce ratio 100% | 0.8865 | 0.7339 |
| Spruce ratio 81–99% | 0.9334 (+5%) | 0.8379 (+14%) |
| Spruce ratio <= 80% | 0.9624 (+9%) | 0.9063 (+23%) |

large part of the world's industrial roundwood (e.g. Cubbage et al. 2007). Thus, future formal optimisation approaches should address the question of how possible cost advantages obtained through 'economies of scale' could shift the economic attractiveness among silvicultural management alternatives.

## 12.10 Survival in Mixed and Pure Forest Stands Under Changing Climate

The next goal in the analysis of mixed forests could be to carry out in-depth analyses of tree survival in mixed and pure forest stands, for example, under the impact of climatic change. These analyses could serve to update the stand-level economic approaches mentioned above with new mortality models that are able to consider dryer and warmer climates as well as the effects of tree species mixture (Neuner et al. 2015; Table 12.4). Moreover, an update of available growth data, as, for example, reported in Chaps. 4–6 of this textbook, is also necessary, both because tree growth has changed substantially in recent decades (Pretzsch et al. 2014a, b) and because the impact of tree species mixture on growth performance is strong (Pretzsch et al. 2010).

**Take-Home Messages**

1. As an empirical basis for better understanding, modelling, and silvicultural treatments, we need long-term mixing experiments with various species combinations established on sites with different resource supply.
2. For theory development and practical forest management, the effects of compositional and structural diversity on stand growth need to be disentangled.
3. Better understanding of how stand density influences productivity and overyielding of mixed-species stands is essential for understanding, modelling, and developing silvicultural prescriptions for mixed-species stands.
4. The transition from analysing and describing to designing and regulating mixed-species stands based on species traits is overdue.

5. Mixing exotic and native species has rarely been analysed from an economic point of view.
6. Combining trees with crops in agroforestry systems could not only help to reduce land-use conflicts but might also have silvicultural advantages. Diversification of landscapes may also be conducted on separated parcels.
7. When adequately addressing tree species interactions and their impact on tree survival and growth, we can expect significant changes in the expected economic return.
8. New approaches to uncertainty will increase the robustness of the programming solutions for an optimal mixed-forest management.
9. The inclusion of economies of scale will improve the acceptance of economic models in forest management practice.
10. Mortality probabilities and models for forest stands are still mainly based on monocultures and need to be extended to mixed-species stands under normal conditions and different kinds of stress.

## References


Ammer U, Schubert H (1999) Arten-, Prozeß- und Ressourcenschutz vor dem Hintergrund faunistischer Untersuchungen im Kronenraum des Waldes. Forstw Cbl 118:70–87

Amoroso MM, Turnblom EC (2006) Comparing productivity of pure and mixed Douglas-fir and western hemlock plantations in the Pacific Northwest. Can J For Res 36(6):1484–1496

Ares A, Fownes JH (1999) Water supply regulates structure, productivity, and water use efficiency of Acacia koa forest in Hawaii. Oecologia 121(4):458–466

Ashton P, Gamage S, Gunatilleke I, Gunatilleke C (1998) Using Caribbean pine to establish a mixed plantation: testing effects of pine canopy removal on plantings of rain forest tree species. For Ecol Manag 106(2–3):211–222. doi:10.1016/S0378-1127(97)00314-9

Assmann E (1970) The principles of forest yield study. Pergamon Press, Oxford, NY, 506 p

Bauhus J, Khanna PK, Menden N (2000) Aboveground and belowground interactions in mixed plantations of *Eucalyptus globulus* and *Acacia mearnsii*. Can J For Res 30(12):1886–1894. doi:10.1139/x00-141

Ben-Tal A, Nemirovski A (2000) Robust solutions of linear programming problems contaminated with uncertain data. Math Program 88:411–424

Bertsimas D, Sim M (2004) The price of robustness. Oper Res 52:35–53

Bourdier T, Cordonnier T, Kunstler G, Piedallu C, Lagarrigues G, Courbaud B (2016) Tree size inequality reduces forest productivity: an analysis combining inventory data for ten European species and a light competition model. PLoS One 11(3):e0151852

Bremer L, Farley K (2010) Does plantation forestry restore biodiversity or create green deserts? A synthesis of the effects of land-use transitions on plant species richness. Biodivers Conserv 19 (14):3893–3915. doi:10.1007/s10531-010-9936-4

Brockerhoff EG, Jactel H, Parrotta JA, Quine CP, Sayer J (2008) Plantation forests and biodiversity: oxymoron or opportunity? Biodivers Conserv 17(5):925–951. doi:10.1007/s10531-008-9380-x

Brus DJ, Hengeveld GM, Walvoort DJJ, Goedhart PW, Heidema AH, Nabuurs GJ, Gunia K (2011) Statistical mapping of tree species over Europe. Eur J For Res 131(1):145–157

Condés S, del Rio M, Sterba H (2013) Mixing effect on volume growth of *Fagus sylvatica* and *Pinus sylvestris* is modulated by stand density. For Ecol Manag 292:86–95

Condés S, Vallet P, Bielak K, Bravo-Oviedo A, Coll L, Ducey M, Pach M, Pretzsch H, Sterba H, Vayreda J, del Rio M (2016) Climate influences on the maximum size density relationship in

Scots pine (*Pinus sylvestris* L.) and European beech (*Fagus sylvatica* L.) stands. For Ecol Manag 385:295–307. doi:10.1016/j.foreco.2016.10.059
Cubbage F, Mac Donagh P, Sawinski Junior J, Rubilar R, Donoso P, Ferreira A, Hoeflich V, Morales Olmos V, Ferreira G, Balmelli G, Siry J, Noemi Baez M, Alvarez J (2007) Timber investment returns for selected plantations and native forests in South America and the Southern United States. New For 33:237–255
Dănescu A, Albrecht AT, Bauhus J (2016) Structural diversity promotes productivity of mixed, uneven-aged forests in southwestern Germany. Oecologia 182:1–15
Dirnberger G, Sterba H, Condés S, Ammer C, Annighöfer P, Avdagic A, Bielak K, Brazaitis G, Coll L, Heym M, Hurt V, Kurylyak V, Motta R, Pach M, Ponette Q, Ruiz-Peinado R, Skrzyszewski J, Šrámek V, Streel G, Svoboda M, Zlatanov T, Pretzsch H (2016) Species proportions by area in mixtures of Scots pine (*Pinus sylvestris* L.) and European beech (*Fagus sylvatica* L.) Eur J For Res 136:1–13
Evans J, Turnbull JW (2004) Plantation forestry in the tropics: the role, silviculture, and use of planted forests for industrial, social, environmental and agroforestry purposes, 3rd edn. Oxford University Press, Oxford
FAO (2010) Global forest resources assessment 2010: main report. Food and Agriculture Organization of the United Nations, Rome
Ferris R, Peace AJ, Humphrey JW, Broome AC (2000) Relationships between vegetation, site type and stand structure in coniferous plantations in Britain. For Ecol Manag 136(1):35–51
Feyera S, Beck E, Lüttge U (2002) Exotic trees as nurse-trees for the regeneration of natural tropical forests. Trees 16(4):245–249. doi:10.1007/s00468-002-0161-y
Forrester DI, Pretzsch H (2015) Tamm review: on the strength of evidence when comparing ecosystem functions of mixtures with monocultures. For Ecol Manage 356:41–53
Griess VC, Acevedo R, Härtl F, Staupendahl K, Knoke T (2012) Does mixing tree species enhance stand resistance against natural hazards? A case study for spruce. For Ecol Manag 267:284–296
Gritti ES, Drössler L, Gary C, Mason B, Morin X, Pretzsch H, Jactel H (unpublished) Diversity-productivity: looking for a unified theory through meta-analysis
Hanewinkel M, Cullmann DA, Schelhaas M-J, Nabuurs G-J, Zimmermann NE (2013) Climate change may cause severe loss in the economic value of European forest land. Nat Clim Chang 3 (3):203–207. doi:10.1038/nclimate1687
Hardiman BS, Bohrer G, Gough CM, Vogel CS, Curtis PS (2011) The role of canopy structural complexity in wood net primary production of a maturing northern deciduous forest. Ecology 92:1818–1827
Hardiman BS, Gough CM, Halperin A, Hofmeister KL, Nave LE, Bohrer G, Curtis PS (2013) Maintaining high rates of carbon storage in old forests: a mechanism linking canopy structure to forest function. For Ecol Manag 298:111–119
Hildebrandt P (2011) Die Wahl von Baumartenmischungen als forstökonomische Entscheidung unter Unsicherheit. Bewertungsbeispiele aus der Waldwirtschaft in Chile. Freising, PhD thesis Center of Life and Food Sciences Weihenstephan, p 210
Hildebrandt P, Knoke T (2011) Investment decisions under uncertainty – a methodological review on forest science studies. For Policy Econ 13:1–15
Hildebrandt P, Kirchlechner P, Hahn A, Knoke T, Rodrigo Mujica H (2010) Mixed species plantations in Southern Chile and the risk of timber price fluctuation. Eur J For Res 129 (5):935–946. doi:10.1007/s10342-009-0284-4
Kawaletz H, Mölder I, Zerbe S, Annighöfer P, Terwei A, Ammer C (2013) Exotic tree seedlings are much more competitive than natives but show underyielding when growing together. J Plant Ecol 6(4):305–315. doi:10.1093/jpe/rts044
Knoke T, Schneider T, Hahn A, Grieß VC, Roessiger J (2012). Forstbetriebsplanung als Entscheidungshilfe. Ulmer, Stuttgart. isbn:9783800176113

Knoke T, Paul C, Härtl F, Castro LM, Calvas B, Hildebrandt P (2015) Optimizing agricultural land-use portfolios with scarce data – a non-stochastic model. Ecol Econ 120:250–259. doi:10.1016/j.ecolecon.2015.10.021

Knoke T, Paul C, Hildebrandt P, Calvas B, Castro LM, Härtl F, Döllerer M, Hamer U, Windhorst D, Wiersma YF, Curatola Fernandez GF, Obermeier WA, Adams J, Breuer L, Mosandl R, Beck E, Weber M, Stimm B, Haber W, Fürst C, Bendix J (2016) Compositional diversity of rehabilitated tropical lands supports multiple ecosystem services and buffers uncertainties. Nat Commun 7:11877. doi:10.1038/ncomms11877

Larson AJ, Lutz JA, Gersonde RF, Franklin JF, Hietpas FF (2008) Potential site productivity influences the rate of forest structural development. Ecol Appl 18(4):899–910

Le Maitre DC, Gaertner M, Marchante E, Ens E-J, Holmes PM, Pauchard A, O'Farrell PJ, Rogers AM, Blanchard R, Blignaut J, Richardson DM (2011) Impacts of invasive Australian acacias: implications for management and restoration. Divers Distrib 17(5):1015–1029. doi:10.1111/j.1472-4642.2011.00816.x

Letcher BH, Priddy JA, Walters JR, Crowder LB (1998) An individual-based, spatially-explicit simulation model of the population dynamics of the endangered red-cockaded woodpecker, *Picoides borealis*. Biol Conserv 86:1–14

Liang J, Crowther TW, Picard N, Wiser S, Zhou M, Alberti G, Schulze E-D, McGuire AD, Bozzato F, Pretzsch H, de-Miguel S, Paquette A, Hérault B, Scherer-Lorenzen M, Barrett CB, Glick HB, Hengeveld GM, Nabuurs GJ, Pfautsch S, Viana H, Vibrans AC, Ammer C, Schall P, Verbyla D, Tchebakova N, Fischer M, Watson JV, Chen HYH, Lei X, Schelhaas M-J, Lu H, Gianelle D, Parfenova EI, Salas C, Lee E, Lee B, Kim HS, Bruelheide H, Coomes DA, Piotto D, Sunderland T, Schmid B, Gourlet-Fleury S, Sonké B, Tavani R, Zhu J, Brandl S, Vayreda J, Kitahara F, Searle EB, Neldner VJ, Ngugi MR, Baraloto B, Frizzera L, Balazy R, Oleksyn J, Zawila-Niedzwiecki T, Bouriaud O, Bussotti F, Finér L, Jaroszewicz B, Jucker T, Valladares V, Jagodzinski AM, Peri PL, Gonmadje C, Marthy W, O'Brien T, Martin EH, Marshall AR, Rovero F, Bitariho R, Niklaus PA, Alvarez-Loayza P, Chamuya N, Valencia R, Mortier F, Wortel V, Engone-Obiang NL, Ferreira LV, Odeke DE, Vasquez RM, Lewis SL, Reich PB (2016) Positive biodiversity-productivity relationship predominant in global forests. Science 354. doi:10.1126/science.aaf8957

Lorenzo P, González L, Reigosa MJ (2010) The genus Acacia as invader: the characteristic case of Acacia dealbata Link in Europe. Ann For Sci 67(1):101. doi:10.1051/forest/2009082

Lundgren BO, Raintree JB (1982) What is agroforestry? Agrofor Syst 1:7–12

Luu TC, Binkley D, Stape JL (2013) Neighborhood uniformity increases growth of individual Eucalyptus trees. For Ecol Manag 289:90–97

Martin PH (1999) Norway maple (*Acer platanoides*) invasion of a natural forest stand: understory consequence and regeneration pattern. Biol Invasions 1(2/3):215–222. doi:10.1023/A:1010084421858

McKelvey K, Noon BR, Lamberson RH (1993) Conservation planning for species occupying fragmented landscapes: the case of the northern spotted owl. In: Kareiva PM, Kingsolver JG, Huey RB (eds) Biotic interactions and global change. Sinauer, Sunderland, MA, pp 424–450 (480 p)

Müller J, Hothorn T, Pretzsch H (2007) Long-term effects of logging intensity on structures, birds, saproxylic beetles and wood-inhabiting fungi in stands of European beech *Fagus sylvatica* L. For Ecol Manag 242(2):297–305

Müller J, Mehr M, Bässler C, Fenton MB, Hothorn T, Pretzsch H, Klemmt H-J, Brandl R (2012) Aggregative response in bats: prey abundance versus habitat. Oecologia 169(3):673–684

Nair PKR (1985) Classification of agroforestry systems. Agrofor Syst 3:97–128. doi:10.1007/BF00122638

Neuner S, Albrecht A, Cullmann D, Engels F, Griess VC, Hahn A, Hanewinkel M, Härtl F, Kölling C, Staupendahl K, Knoke T (2015) Survival of Norway spruce remains higher in mixed stands under a dryer and warmer climate. Glob Chang Biol 21:935–946

Noss RF (1990) Indicators for monitoring biodiversity: a hierarchical approach. Conserv Biol 4 (4):355–364
O'Hara KL (2016) What is close-to-nature silviculture in a changing world? Forestry 89(1):1–6
Ong CK, Black C, Wilson J (2015) Tree-crop interactions: agroforestry in a changing climate. CABI, Wallingford
Onyekwelu J, Stimm B, Evans J (2011) Review plantation forestry. In: Günter S, Weber M, Stimm B, Mosandl R (eds) Silviculture in the tropics, vol 8. Springer, Berlin, pp 399–454
Paul C (2014) Timber-based agrisilvicultural systems to facilitate reforestation in Panama – a silvicultural and economic evaluation. Dissertation, Technische Universität München
Paul C, Weber M (2016) Effects of planting food crops on survival and early growth of timber trees in eastern Panama. New For 47(1):53–72. doi:10.1007/s11056-015-9477-5
Paul C, Griess V, Havardi-Burger N, Weber M (2015) Timber-based agrisilviculture improves financial viability of hardwood plantations: a case study from Panama. Agrofor Syst 89 (2):217–235. doi:10.1007/s10457-014-9755-9
Perfecto I, Vandermeer J (2010) The agroecological matrix as alternative to the land-sparing/ agriculture intensification model. Proc Natl Acad Sci 107(13):5786–5791
Piotto D, Vıquez E, Montagnini F, Kanninen M (2004) Pure and mixed forest plantations with native species of the dry tropics of Costa Rica: a comparison of growth and productivity. For Ecol Manag 190(2–3):359–372. doi:10.1016/j.foreco.2003.11.005
Potvin C, Dutilleul P (2009) Neighborhood effects and size-asymmetric competition in a tree plantation varying in diversity. Ecology 90(2):321–327. doi:10.1890/08-0353.1
Pretzsch H (2005) Stand density and growth of Norway spruce (*Picea abies* (L.) Karst.) and European beech (*Fagus sylvatica* L.). Evidence from long-term experimental plots. Eur J For Res 124:193–205
Pretzsch H, Schütze G (2014) Size-structure dynamics of mixed versus pure forest stands. For Syst 23(3):560–572
Pretzsch H, Block J, Dieler J, Dong PH, Kohnle U, Nagel J, Spellmann H, Zingg A (2010) Comparison between the productivity of pure and mixed stands of Norway spruce and European beech along an ecological gradient. Ann For Sci 67:712
Pretzsch H, Biber P, Schütze G, Uhl E, Rötzer T (2014a) Forest stand growth dynamics in Central Europe have accelerated since 1870. Nat Commun 5:4967. doi:10.1039/ncomms5967
Pretzsch H, Rötzer T, Matyssek R, Grams TEE, Häberle KH, Pritsch K, Kerner R, Munch JC (2014b) Mixed Norway spruce (*Picea abies* [L.] Karst) and European beech (*Fagus sylvatica* [L.]) stands under drought: from reaction pattern to mechanism. Trees 28:1305–1321
Pretzsch H, del Río M, Ammer C, Avdagic A, Barbeito I, Bielak K et al (2015) Growth and yield of mixed versus pure stands of Scots pine (*Pinus sylvestris* L.) and European beech (*Fagus sylvatica* L.) analysed along a productivity gradient through Europe. Eur J For Res 134 (5):927–947
Pretzsch H, Bauerle T, Häberle KH, Matyssek R, Schütze G, Rötzer T (2016a) Tree diameter growth after root trenching in a mature mixed stand of Norway spruce (*Picea abies* [L.] Karst) and European beech (*Fagus sylvatica* [L.]). Trees 30:1–13. doi:10.1007/s00468-016-1406-5
Pretzsch H, del Rio M, Schütze G, Ammer C, Annighöfer P, Avdagic A, Barbeito I, Bielak K, Brazaitis G, Coll L, Drössler L, Fabrika M, Forrester DI, Kurylyak V, Löf M, Lombardi F, Matovic B, Mohren F, Motta R, den Ouden J, Pach M, Ponette Q, Skzyszewski J, Sramek V, Sterba H, Svoboda M, Verheyen K, Zlatanov T, Bravo-Oviedo A (2016b) Mixing of Scots pine (*Pinus sylvestris* L.) and European beech (*Fagus sylvatica* L.) enhances structural heterogeneity, and the effect increases with water availability. For Ecol Manag 373:149–166
Rao M, Nair P, Ong C (1997) Biophysical interactions in tropical agroforestry systems. Agrofor Syst 38(1):3–50. doi:10.1023/A:1005971525590
Richardson DM, Rejmánek M (2011) Trees and shrubs as invasive alien species – a global review. Divers Distrib 17(5):788–809. doi:10.1111/j.1472-4642.2011.00782.x

del Río M, Pretzsch H, Ruíz-Peinado R, Ampoorter E, Annighöfer P, Barbeito I, Fabrika M (2017) Species interactions increase the temporal stability of community productivity in *Pinus sylvestris–Fagus sylvatica* mixtures across Europe. J Ecol. doi:10.1111/1365-2745.12727

Roessiger J, Griess VC, Knoke T (2011) May risk aversion lead to near-natural forestry? A simulation study. Forestry 84:527–537

Roessiger J, Griess VC, Härtl F, Clasen C, Knoke T (2013) How economic performance of a stand increases due to decreased failure risk associated with the admixing of species. Ecol Model 255:58–69

Scherer-Lorenzen M, Körner C, Schulze E-D (2005) Forest diversity and function. In: Ecological studies, vol 176. Springer, Berlin, p 399

Schütz JP (1997) Sylviculture 2. La gestion des forets irrégulières et mélangées. Presses Polytechniques et Universitaires Romandes, Lausanne, 178 p

Soares AA, Leite HG, Souza AL, Silva SR, Lourenço HM, Forrester DI (2016) Increasing stand structural heterogeneity reduces productivity in Brazilian Eucalyptus monoclonal stands. For Ecol Manag 373:26–32

Tahvonen O, Pukkala T, Laiho O, Lähde E, Niinimäki S (2010) Optimal management of uneven-aged Norway spruce stands. For Ecol Manag 260:106–115

Thurm EA, Pretzsch H (2016) Improved productivity and modified tree morphology of mixed versus pure stands of European beech (*Fagus sylvatica*) and Douglas-fir (*Pseudotsuga menziesii*) with increasing precipitation and age. Ann For Sci 73:1–15

Tobner CM, Paquette A, Reich PB et al (2014) Advancing biodiversity–ecosystem functioning science using high-density tree-based experiments over functional diversity gradients. Oecologia 174:609–621

Toïgo M, Vallet P, Perot T, Bontemps JD, Piedallu C, Courbaud B (2015) Overyielding in mixed forests decreases with site productivity. J Ecol 103(2):502–512

Uhl E, Kölling Ch, Pretzsch H (2014) Überlegungen zur Weiterentwicklung des Ertragskundlichen Versuchswesens in Bayern. In: Bayerische Landesanstalt für Wald und Forstwirtschaft (Hrsg) Das forstliche Versuchswesen in Bayern, LWF Wissen, p 76, 100 S

Utschig H, Neufanger M, Zanker T (2011) Das 100-Baum-Konzept als Einstieg für Durchforstungsregeln in Mischbeständen. Allgemeine Forstzeitschrift für Waldwirtschaft und Umweltvorsorge AFZ-Der Wald AFZ 21:4–6

Verheyen K, Vanhellemont M, Auge H et al (2016) Contributions of a global network of tree diversity experiments to sustainable forest plantations. Ambio 45:29–41

Wang X, Fang J, Tang Z, Zhu B (2006) Climatic control of primary forest structure and DBH–height allometry in Northeast China. For Ecol Manag 234(1):264–274

Webb LJ (1968) Environmental relationships of the structural types of Australian rain forest vegetation. Ecology 49(2):296–311

Weber M, Günter S, Aguirre N, Stimm B, Mosandl R (2008) Reforestation of abandoned pastures: silvicultural means to accelerate forest recovery and biodiversity. In: Beck E, Bendix J, Kottke I, Makeschin F, Mosandl R, Caldwell MM, Heldmaier G, Jackson RB, Lange OL, Mooney HA, Schulze E-D, Sommer U (eds) Gradients in a tropical mountain ecosystem of Ecuador, vol 198. Springer, Berlin, pp 431–441

Zenner EK (2015) Differential growth response to increasing growing stock and structural complexity in even-and uneven-sized mixed *Picea abies* stands in southern Finland 1. Can J For Res 46(999):1–10

Zhang Y, Chen HYH (2015) Individual size inequality links forest diversity and above-ground biomass. J Ecol 103:1245–1252

Zobel DB, McKee A, Hawk GM, Dyrness CT (1976) Relationships of environment to composition, structure, and diversity of forest communities of the central western Cascades of Oregon. Ecol Monogr 46(2):135–156

# Glossary

| Term or concept | Definition/explanation | Source |
|---|---|---|
| Abiotic stress | Environmental conditions that reduce growth and yield below optimum levels | Cramer et al. (2011) |
| Absolute stand density | A quantitative measure of stocking expressed absolutely in terms of number of trees, basal area, or volume per unit area | Helms (1998) |
| Absorbed photosynthetically active radiation | The light intercepted by the canopy that has wavelengths between 400 and 700 nm; often with units of photons (mmol of photons $m^{-2}$ $s^{-1}$), or units of energy (MJ $m^{-2}$ $s^{-1}$) | McCree (1971) |
| Acclimation | A physiological or morphological reversible adjustment of an organism to changing environmental conditions | Merriam-Webster (2016), Helms (1998) |
| Acid rain | The deposition of a variety of acidic pollutants in either wet or dry forms | Helms (1998) |
| Adaptation | Permanent genetic adjustments in structure or functioning that fit individuals, populations, or species to their environment | Helms (1998) |
| Additive mixing effect | If the total growth of species in mixture is just the weighted mean growth of the two species when growing in monoculture, the difference between mixed-species stands results just from the species selection (see difference between multiplicative mixing effect and additive mixing effect) | Forrester and Pretzsch (2015) |
| Additive effect | See additive mixing effect | |
| Afforestation | Forests established artificially by afforestation on land which previously did not carry forest within living memory where there were no records or within 50 years if records existed. Clearly a man-made forest | Carle and Holmgren (2003) |

(continued)



H. Pretzsch et al. (eds.), *Mixed-Species Forests*, DOI 10.1007/978-3-662-54553-9

| Term or concept | Definition/explanation | Source |
|---|---|---|
| Age-class model | Works with a fully regulated forest. Forest area-based, and idealised model of a forest, where all age classes (often comprising 10 or 20 years) are endowed with an identical area. Allowable harvest is derived on the basis of this model | Bettinger et al. (2009), Knoke et al. (2012) |
| Agroforestry (system) | Intentional growing of trees and shrubs in combination with crops or forage on the same land | |
| Agroforestry plantation | Managed stands or assemblages of trees established in an agricultural matrix for subsistence or local sale and for their benefits on agricultural production; usually regular and wide spacing or row planting | CIFOR (2001) |
| Allelopathy | Allelopathy refers to the beneficial or harmful effects of one plant on another plant, both crop and weed species, by the release of chemicals from plant parts by leaching, root exudation, volatilisation, residue decomposition and other processes in both natural and agricultural systems | UF-IFAS (2016) |
| Allocation, allocation of carbon, allocation of biomass | The process of distributing carbon or biomass to different parts of a plant, e.g. roots, stem, or foliage | Helms (1998) |
| Allometric adjustment | Elimination of simple size effects caused by the allometric drift of plants before comparing their structure or functioning in dependence on environmental conditions | Pretzsch (2010) |
| Allometry | The relation between the size of an organism and the size of any of its parts | Ridley (2004) |
| APAR | See absorbed photosynthetically active radiation | |
| Artificial time series, chronosequence | Setup of sample plots of different stand ages at a given site in such a way that they represent the stand development over age | Pretzsch (2009) |
| Artificial regeneration | A forest crop raised artificially, either by sowing or planting, but without natural regeneration | |
| Biodiversity | The variability among living organisms from all sources including, inter alia, terrestrial, marine, and other aquatic ecosystems and the ecological complexes of which they are part; this includes diversity within species, between species, and of ecosystems | UN (1992) |

(continued)

| Term or concept | Definition/explanation | Source |
|---|---|---|
| Bioeconomic model | A bioeconomic model links models describing biological resource management decisions to models that describe current and alternative production possibilities in terms of economic inputs to achieve certain economic outputs and associated externalities | Janssen and van Ittersum (2007) |
| Biome | A regional ecosystem with a distinct assemblage of vegetation, animals, microbes, and physical environment often reflecting a certain climate and soil | Helms (1998) |
| Canopy stratification | The stratification of the canopy refers to the formation of vertical crown layers in forest stands | |
| Carbon mitigation | An intervention to decrease the sources of or improve the sinks for carbon-based greenhouse gases | |
| Carbon sequestration | The process of removing carbon from the atmosphere and depositing it in a reservoir | UNFCCC (2014) |
| Carbon sink | Any process, activity, or mechanism which removes a carbon-based greenhouse gas, an aerosol, or a carbon-based precursor of a greenhouse gas from the atmosphere. Forests and other vegetation are considered sinks because they remove carbon dioxide through photosynthesis | UNFCCC (2014) |
| Carbon stock | A component of ecosystems that has the capacity to store, accumulate, or release carbon. Bogs, soils, and forests are examples of carbon stocks or reservoirs | |
| Catchment (area) | An area that collects and drains rainwater | |
| Clear-cut (or clearfell) | A stand in which essentially all trees have been removed in one operation | Helms (1998) |
| Close-to-nature forest management | Close-to-nature forest management (CTNFM) is a production system that developed from concepts of multifunctional forest management. It aims to combine management goals related to productivity, diversity, and especially continuity and stability of forest conditions at small spatial scales by adhering to a set of principles | Bauhus et al. (2013) |

(continued)

| Term or concept | Definition/explanation | Source |
|---|---|---|
| Cohort | A group of trees developing after a single disturbance, commonly consisting of trees of similar age, although it can include a considerable range of tree ages of seeding or sprout origin and trees that predate the disturbance | Helms (1998) |
| Commensalism | A facilitative ecological interaction, in which one species benefits from the interaction, whereas the other is unaffected | Stachowicz (2001) |
| Community | In ecology: an assemblage of plants and animals living together and occupying a given area. | Helms (1998) |
| Compatibility of tree species | Tree species are compatible if they can coexist in the same stand or can be cultivated in mixture with a low level of management efforts to control competitive effects | |
| Competition | Ecological interaction in which one organism or species consumes a resource that would have been available to and could have potentially been consumed by the other, and hence the fitness of one is lowered by the presence of another | Begon et al. (1996) |
| Competition index | An index used as a proxy for the competition experienced by an individual tree within a stand. Can be calculated in many ways, such as the sum of the basal area of all trees within a given distance of the focal tree | |
| Competitive reduction | Occurs when the interspecific competition for a limiting resource in mixtures is less than the intraspecific competition for this resource in the monocultures. The species in mixture therefore utilise this resource more efficiently for growth | Kelty (1992), Vandermeer (1992) |
| Complementarity | The combined effects of competitive reduction (or niche differentiation) and facilitation | Loreau and Hector (2001) |
| Complexity, complex | Behaviour of a system or a model with many components which interact in manifold ways and follow just local, often non-linear, rules. Thus, understanding the functioning of each single part does not imply to understand the whole system | Forrester (1961), Johnson (2001), Perona and Miragliotta (2004) |

(continued)

| Term or concept | Definition/explanation | Source |
|---|---|---|
| Complicacy, complicate | Characteristic of a system or model where the components are knotted and entwined with each other. But to understand a complicated system, all there is to do is to subdivide it in all its components. Understanding all components will lead to understanding the system as a whole | Perona and Miragliotta (2004) |
| Conspecific | Belonging to the same species | |
| Constant final yield | The final maximum standing volume which depends on the site productivity and is achieved independently of the spacing and thinning at higher stand ages | Oliver and Larson (1996) |
| Continuous cover forestry | Type of management of forests which does not apply clear-cuts. Continuous cover forestry usually aims at more or less uneven-aged stands that contain a mixture of tree species | Pommerening and Murphy (2004) |
| Continuous optimisation | Uses continuous decision variables (e.g. in forest science: area, tree numbers, or timber volume) which are allocated to forest stands, strata, or size classes to model decision making. Continuous optimisation considers in principle an unlimited number of solution alternatives | Uhde et al. (2015) |
| Cross diagram | Diagram for visualising mixing effects on the productivity of two-species stands depending on the mixing proportions. The productivity of the monocultures of the two species is shown on the left and right ordinate, respectively. Productivity of mixed stands of different mixing proportions is shown in between | Harper (1977), Kelty (1992) |
| Crown architecture | The structure of a tree crown including all branches, twigs, and leaves. For example, the architecture may be defined in terms of branch angles and diameters, allometric relationships, shape, leaf area density, etc. | |
| Crown eccentricity | Crown eccentricity indicated by the distance between the gravity centre of the crown projection area and the position of the tree stem base | Pretzsch (2014) |

(continued)

| Term or concept | Definition/explanation | Source |
|---|---|---|
| Crown extension | Lateral crown extension of a tree often quantified as ratio of the crown diameter, cd, to the crown length, cl (cd/cl, degree of crown spread), as ratio of the crown diameter to the tree diameter, *d*, (cd/*d*, crown projection ratio), or as the ratio of the square of the crown diameter to the square of the tree diameter (cd2/d2, quotient of ground cover area) | Assmann (1970) |
| Crown jaggedness | The ratio $r_{min}/r_{max}$ between the shortest, $r_{min}$, and the longest, $r_{max}$, of the 4 or 8 measured crown radii of a tree can be used for indicating its crown jaggedness | Pretzsch (2014) |
| Crown plasticity, crown variability | The extent to which the crown structure and allometry of a genotype may vary due to the respective environmental conditions (e.g. resource supply, abiotic stress, mechanical abrasion) | Dieler and Pretzsch (2013) |
| Crown projection area | The cross-sectional area of a crown that is projected towards the ground, usually expressed in square meters (e.g. average crown radius$^2 \times \pi$) | |
| Crown projection area of a stand; relative sum of crown projection area, RSCPA | Sum of the crown projection areas of all trees in a stand; RSCPA is the ratio between the sum of the crown projection areas of a stand and the stand area | Pretzsch (2014) |
| Crown shyness | The crowns of neighbouring trees of similar height do not interdigitate but rather are generally separated by spaces called 'crown shyness' gaps (see also mechanical abrasion) | Putz et al. (1984) |
| Current annual increment CAI | Total increment of a tree or a stand (standing plus removal volume) in a specific one-year period. It results from the volume difference at the end, V2, and beginning, V1, of the period plus the removed volume, Vr, during the period (CAI = V2 − V1 + Vr) | Assmann (1970) |
| Deforestation | Forest conversion to another land use or the long-term reduction of tree canopy cover below the 10% threshold | Carle and Holmgren (2003) |
| Degree of crown disengagement | Ratio of the growing area of a tree to the crown projection area. Indicates the extent to which a tree is released from competing neighbours | Assmann (1970) |
| Discrete optimisation | Is based on predefined decision alternatives (e.g. management options) | Uhde et al. (2015) |

(continued)

| Term or concept | Definition/explanation | Source |
|---|---|---|
| Disturbance | Any relative discrete event in time that disrupts ecosystems, community, or population structure and changes resources, substrate availability, or the physical environment | Helms (1998) |
| Diversification | Spreading of investments, assets, or strategies in order to stabilise economic return. Effectiveness in reducing risk does much depend on the correlation of economic return between the components mixed | Elton et al. (2007) |
| Diversity | See 'biodiversity' | |
| Dominant height | The mean height of the larger trees within a stand. The larger trees are often the largest-diameter or tallest 50 or 100 trees $ha^{-1}$ | |
| Early successional species | Tree species that grow best in open areas within forests, shortly following a disturbance, or on abandoned agricultural land | |
| Ecological insurance | The idea that if some species are lost, others are available to fill their functional roles, due to functional redundancy (closely linked to biodiversity) | |
| Economic return | In financial sciences the relative periodical change in stock value (as the stock's internal rate of return) is often a central indicator to quantify economic return. A similar indicator for forestry may be approached by the annual change in the market value of the standing timber stock and land, divided by the market value of timber and land at the beginning of the year. However, this perspective has several limitations. Another indicator is the net present value of financial inflows and outflows | Fama and French (1993), Mills and Hoover (1982) |
| Economic risk | Is usually quantified by the standard deviation of the economic return. However, many more indicators do exists, as discussed in Hildebrandt and Knoke (2011) | Markowitz (1952), Hildebrandt and Knoke (2011) |
| Economical maturity | In the context of its use here, economical maturity is the period of time (or the end of that period) until the most economic benefit can be obtained from harvesting | |

(continued)

| Term or concept | Definition/explanation | Source |
|---|---|---|
| Ecosystem approach | A strategy for the integrated management of land, water, and living resources that promotes conservation and sustainable use in an equitable way. It is based on the application of appropriate scientific methodologies focused on levels of biological organisation which encompass the essential processes, functions, and interactions among organisms and their environment, while recognising that humans, with their cultural diversity, are an integral component of ecosystems | UNEP (2007) |
| Ecosystem functioning | Capacity of natural processes and components to provide goods and services that satisfy human needs, directly or indirectly | |
| Ecosystem goods and services | The comprehensive benefits humans derive from ecosystem functions | MEA (2005) |
| Efficiency of space exploitation, efficiency of stand area exploitation | Ratio of tree growth (annual or periodic mean growth) to the occupied stand area or crown projection area. Indicates how efficiently a tree makes use of the occupied area or space to grow in size | Pretzsch and Schütze (2009) |
| Efficiency of space occupation, efficiency of stand area occupation | Ratio of the stand area or crown projection area of a tree to its tree volume or tree biomass. Indicates how efficiently a tree uses its body mass for occupying growing space or stand area | Pretzsch and Schütze (2009) |
| Efficient frontier | Is formed by portfolios of assets which maximise economic return for given levels of economic risk. These portfolios are called 'Pareto optimal', because we cannot obtain higher economic return without accepting higher risk | Elton et al. (2007) |
| Emergence, emergent properties | Characteristic and behaviour of a system as a whole which can not be derived or predicted, even when all information about the system components and their interactions are known | Johnson (2001) |
| Empirical model | See Sect. 8.2.2 | |
| Endemic | Restricted to a particular area: used to describe a species or organism that is confined to a particular geographical region, for example, an island or river basin | |

(continued)

| Term or concept | Definition/explanation | Source |
|---|---|---|
| Evapotranspiration | The combined process of water evaporation from the Earth's surface and transpiration from vegetation | |
| Evenness | Evenness is a measure of the relative abundance of the different species making up the richness of an ecological community. For the same number of species, diversity increases with increasingly similar proportions of the participating species. Both species richness and evenness are considered in Shannon's diversity index | Shannon (1948) |
| Facilitation | Facilitation is a form of ecological interaction between different species that benefits at least one of the participants and causes no harm to the others. Facilitations can be mutualistic, when both species benefit from the interaction, or commensalistic, when only one species benefits and the other remains unaffected | Stachowicz (2001) |
| Fine root | Roots smaller than a certain, small diameter, which is often set at 2 mm for trees. See also 'root' | |
| fitness | The relative competitive ability of a genotype (survivorship × fecundity) expressed as the average number of surviving progeny of this genotype compared with the average number of surviving progeny of the competing genotypes | Helms (1998) |
| Flux | In ecology: the transfer of a substance from one pool or compartment to another (see also pool). Flux can also relate to the flow of energy into and out of systems or its compartment | |
| Forest certification | A number of market-based instruments that link trade in forest products to the sustainable management of the forest resource, by providing buyers with information on the management standards of the forests from which the timber came | FAO (2001) |

(continued)

| Term or concept | Definition/explanation | Source |
|---|---|---|
| Forest-level mixture (economic view) | Independence is assumed between mixed components (most often tree species) as it is the case for diversification at the estate level, in which, for example, monocultures of species are planted in different locations to protect against uncertainty in economic returns, induced by fluctuating market prices or bio-physical damage. Under the assumption of independence, the expected (economic) performance changes proportionally with the proportion of species. In this case, diversification is (almost) analogous to that of financial assets | Koellner and Schmitz (2006), Knoke and Seifert (2008) |
| Forest management planning (FMP) | FMP delivers a plan for a forest that formulates, analyses, evaluates, and improves management decisions at the forest (ownership) level and thus considers multiple silvicultural management units (forest stands) simultaneously | Bettinger et al. (2009), Knoke et al. (2012) |
| Foundation species | A species that has a strong role in structuring a community and therefore plays a key role for the resilience of the community. In forests, these are tree species that define and structure forest ecosystems through their influences on associated organisms and ecosystem processes | Dayton (1972), Ellison et al. (2005) |
| Functional diversity | Refers to the diversity of functions rather than species within a community or ecosystem. Different functions are often quantified as traits. The effects of functional diversity are expected to increase with the magnitude of the differences among functional traits within a community. There are several approaches to calculate functional diversity | See Chap. 2 |
| Fundamental niche | The subset of *n*-dimensional environmental space of all possible conditions in which a species can maintain itself in the absence of competition | McGill et al. (2006) |
| Gap dynamics, gap occupation | The change in space and time in the pattern, frequency, size, and successional processes of forest canopy gaps caused by the fall, death, or harvest of one or more canopy trees. Of special relevance is the occupation of the gap by the crowns and roots of remaining trees and upcoming natural regeneration | Helms (1998), Bugmann (2001) |

(continued)

| Term or concept | Definition/explanation | Source |
| --- | --- | --- |
| Generalist (species) | A species that uses a relatively large proportion, or in extreme cases all, of the available resource types | Calow (1998) |
| Gini coefficient | Measure for the degree of inequality of size, growth, or resource supply among the individuals of a population | Weiner and Thomas (1986) |
| Ground cover by crowns | Indicates the ground cover by crowns when looking vertically from above. It is generally analysed by dot count analysis of crown maps | Assmann (1970) |
| Growth dominance coefficient, GDC | Measure indicating the relationship between the share in the stem volume and the share in the stem volume growth that trees of different sizes have in a stand. It indicates how efficiently trees can use their superior size for advantageous growth | Binkley et al. (2006) |
| Growth resilience | The ability and extent of a plant after stress to achieve the same level of growth it had before the stress event | Lloret et al. (2011) |
| Heterogeneous | Exhibiting dissimilarity among members of a group | Helms (1998) |
| Heterospecific | Belonging to different species | |
| Holobiont | Functional entity consisting of a host organism (e.g. higher plant) and associated microorganisms (e.g. fungi), altogether representing a network of biotic interactions | Zilber-Rosenberg and Rosenberg (2008) |
| Hybrid model | See Sect. 8.2.2 | |
| Hydraulic lift or hydraulic redistribution | The process where roots take up water from moist soil and release it into drier soil | Neumann and Cardon (2012), Prieto et al. (2012) |
| Infiltration | Flow of water from the land surface into the subsurface | USGS (2016) |
| Interaction or ecological interaction | Ecological interactions are the effects that organisms/species in a community have on one another. These interactions can be neutral, negative (antagonism, competition), or positive (facilitation) | |
| Interception (of precipitation) | The process by which precipitation is caught and held by foliage, twigs, and branches of trees, shrubs, and other vegetation, and lost by evaporation. This portion of precipitation never reaches the surface of the ground. Interception equals the precipitation on the vegetation minus streamflow and through fall | |

(continued)

| Term or concept | Definition/explanation | Source |
|---|---|---|
| Interspecific | Occuring among members of different species | Helms (1998) |
| Intraspecific | Occuring among members of the same species | Helms (1998) |
| Invasive | Any species that is non-native to a particular ecosystem and whose introduction and spread causes, or is likely to cause, sociocultural, economic, or environmental harm or harm to human health | FAO (2015) |
| Inventory | A forest inventory involves the systematic collection of data to quantify many different aspects of the forest (e.g. species composition, growth, biomass, carbon stocks, etc.) | |
| Kleiber's rule | Assumption that a plant's metabolic rate, $A$, scales with 3/4 over plant mass, $m$, or volume, $v$ ($A \propto m^{3/4}$) | Kleiber (1947) |
| Land equivalent ratio, LER | Relative land requirements for intercrops versus monoculture. It is equivalent to the relative total yield. The ratio of the area needed under monoculture to a unit area of intercropping at the same management level to give an equal amount of yield (see also relative yield total, RYT) | Vandermeer (1992) |
| Late successional species | Tree species that grow best in forests that have undergone succession and developed without major disturbances for a long time; typically shade-tolerant species | Puettman et al. (2009) |
| Leaf area, leaf area index | The leaf area of a tree or stand is usually expressed as the one-sided leaf area (because only one side of the leaf can be facing the sun at any given time). The leaf area index is the one-sided leaf area per unit of ground surface area ($m^2\ m^{-2}$) | |
| Leaf area density | The leaf area of a crown per unit volume of that crown (usually $m^2\ m^{-3}$) | |
| Light compensation point | The light compensation point is the amount of light intensity on the light response curve where the rate of photosynthesis exactly matches the rate of respiration | Larcher (2003) |
| Light saturation point | The quantity of solar radiation beyond which further increases in solar radiation will not lead to an increase in the rate of photosynthesis | Larcher (2003) |

(continued)

| Term or concept | Definition/explanation | Source |
|---|---|---|
| Litterfall | The leaves, bark, twigs, small branches, and reproductive tissues that fall to the ground. May also refer to the roots that have died | |
| Lorenz curve | Cumulative tree volume or tree volume growth depicted as a function of cumulative tree number visualises the inequality of tree volume or tree volume growth in a forest stand | Kramer (1988) |
| Martonne index, M | An index that is widely used to quantify the drought condition or aridity in a given region. $M$ = annual precipitation (mm)/(mean annual temperature (°C) +10) | Martonne (1926) |
| Masting | A synchronised, reproductive pattern in which an entire population of trees reproduce at the same time | |
| Maximum, optimum, and critical stand density | Maximum density is the ceiling stand density in terms of, e.g. stand basal area, volume, or SDI, in a given phase in a fully stocked and unmanaged even-aged stand<br>Optimum density is the stand density that results in the highest increment in the respective phase of stand development<br>Critical stand density refers to the stand density reduced by thinning that still ensures 95% of the highest increment achieved under optimum density | Assmann (1970) |
| Mean annual increment, MAI | The total increment of a tree or a stand (standing plus removal volume) until a given age divided by that age | Assmann (1970) |
| Mechanical abrasion | Abrasion of buds, leaves, and twigs due to crown movement of neighbouring trees (trees knocking into one another) can create and maintain the 'crown shyness' gaps around each tree crown (see also crown shyness) | Putz et al. (1984) |
| Mixed forest | A forest stand of two or more prominent species | Helms (1998) |
| Mode of competition | The following modes of competition, which describe the relationship between plant size and growth of trees in a stand, are recognised: Complete symmetry means that all trees have the same growth even when different in size. Size symmetry means that growth is proportional to size. Size asymmetry means that tall trees grow over-proportionally to size (see also size symmetry vs. size asymmetry of competition) | Hara (1992, 1993) |

(continued)

| Term or concept | Definition/explanation | Source |
|---|---|---|
| Mode of tree mortality | Indicates whether in a population mortality eliminates mainly the small trees (beginning from the left side of the size distribution), tall trees (from the right side), or trees of all sizes to the same extent (trees of all sizes evenly) | Hara (1992, 1993) |
| Modern portfolio theory (MPT) | Statistical theory of combining the uncertainty of several variables applied to support the optimal choice of a collection of economic assets (normative use) but also used to describe investment behaviour (descriptive use). The collection of various financial and/or real assets may include not only, inter alia, stocks but also enterprises, buildings, breweries, or forests. The essential conclusion of MPT is that investments should not be analysed in isolation but always together with other investment opportunities | Elton et al. (2007) |
| Monoculture, monospecific (forest) | A forest stand composed of a single dominant (>90% of basal area or volume) tree species | Puettman et al. (2009), Schuck et al. (2002) |
| Morphology | Morphology is the study of organism form and structure | |
| Multifunctional | A coexistent provision of different products and benefits at the same area | McArdle (1953) |
| Multiple crown coverage | Phenomenon that considerable parts of the stand area can be covered more than once by multiple tree crown layers | Preuhsler (1981), Pretzsch (2009) |
| Multiplicative mixing effect | Interaction between different species which cause an outcome, e.g. in terms of growth, that differs from the weighted mean of the growth of both species when growing in monoculture (see difference between multiplicative mixing effect and additive mixing effect) | Forrester and Pretzsch (2015) |
| Mutualism | A facilitative ecological interaction, in which both/all species benefit from the interaction | Stachowicz (2001) |
| Native forest | Forests composed of indigenous tree species | |
| Natural disturbance | See disturbance | |
| Natural regeneration | Re-establishment of a forest stand by natural means, i.e. through seed fall or vegetative regeneration, roots suckers, or coppice shoots. It may be assisted by human intervention, e.g. by scarification of the soil or fencing to protect against wildlife or domestic animal grazing | |

(continued)

| Term or concept | Definition/explanation | Source |
|---|---|---|
| Naturalness | It is the degree to which ecosystems resemble natural, not anthropogenically modified ecosystems in terms of their structure, function, and composition | |
| Net competition, net facilitation | Combined occurrence of competition and facilitation between plants where negative effects on growth (net competition) or positive effects (net facilitation) have the upper hand | Maestre et al. (2009) |
| Niche | A niche is a specific segment of an n-dimensional environmental space that is occupied by a single species. The dimensions represent environmental conditions and resources that define the requirements of an individual or a population of a species to persist. See also Chap. 2 | Hutchinson (1957) |
| Niche complementarity | The ability of different species that use different resources or the same resources but at different times or different points in space to fill the available limited space to a greater extent than a single species | |
| Non-timber forest products | Products of biological origin, other than wood, derived from forests, other wooded land, and trees outside forests. NTFP may be gathered from the wild or produced in forest plantations, from agroforestry schemes, and from trees outside forests | FAO (2010) |
| Nurse trees, nurse crop | A cohort of fast-growing tree species, either naturally or artificially regenerated in large open areas created through clearfelling or other disturbances, which is suited to improve the microclimate to facilitate establishment and growth of slower growing (frost-sensitive) species and to improve soil conditions | |
| Ontogenetic drift | Changes in biomass allocation as a function of plant size (i.e., apparent plasticity). Such size-dependent changes can be separated from adjustment of ontogenetic trajectory in relation to environmental influences (true plasticity) | Evans (1972) |
| Ontogeny | The developmental history of an individual organism over its life cycle from the zygote to maturity | Helms (1998) |

(continued)

| Term or concept | Definition/explanation | Source |
|---|---|---|
| Overstory | That portion of the trees, in a forest of more than one story, forming the upper or uppermost canopy layer | Helms (1998) |
| Overyielding effect | When the production of biomass in species mixtures exceeds the productivity expected on the basis of the yields of the contributing species when grown in a monoculture. Transgressive overyielding can occur when the yield of the mixture is greater than the yield of monocultures of all species included in the mixture. See also 'underyielding' | Forrester and Pretzsch (2015) |
| PAR | See photosynthetically active radiation | |
| Paradigm | A philosophical and theoretical framework of a scientific school or discipline within which theories, laws, and generalisations and the experiments performed in support of them are formulated | Merriam Webster Dictionary (2016) |
| Partitioning | The process of distributing carbon, biomass, or growth among the plants of a population. Partitioning refers to the flux of a resource to a given pool (e.g. carbon to the aboveground biomass pool) as a fraction of the total flux and is expressed either as a percentage (%) or a proportion (0–1, no units) | Litton et al. (2007), Pretzsch and Biber (2010) |
| Paterson index, CVP | Index for quantifying the climatic conditions for forest growth. $CVP = T_v/T_a \times P \times G/12 \times E/100$ is based on the $T_v$ (mean temperature of the warmest month in °C), $T_a$(temperature amplitude calculated by the difference of the mean temperature of the warmest month minus mean temperature of the coldest month in °C), $P$ (mean annual precipitation in mm), $G$ (number of months out of 12 with mean temperature ≥3 °C), and $E$ (evapotranspiration intensity as a function of the latitude read out off a nomogram; see Paterson (1956, p. 74), where $E\%$ is given as a function of the geographical latitude in degrees) | Paterson (1956) |
| Periodic annual increment, PAI | The increment of a tree or stand (standing plus removed volume) observed over a specific time period divided by the length of this period | Helms (1998) |

(continued)

| Term or concept | Definition/explanation | Source |
| --- | --- | --- |
| Pest | Any organism that is out of place or causes stress or damage to a desired organism | NCCE (2013) |
| Phenology | Phenology is the study of periodic events in the life cycle of a plant (or animal) and the effects of abiotic and biotic factors on those, e.g. flowering and its relation to temperature and day length | |
| Photosynthetically active radiation | Light with wavelengths between 400 and 700 nm. Often quantified in using units of photons (mmol of photons $m^{-2}$ $s^{-1}$) or units of energy (MJ $m^{-2}$ $s^{-1}$) | McCree (1971) |
| Physiological response curve | A relationship giving fitness (or a component of fitness) as a function of one (occasionally several) environmental variables | McGill et al. (2006) |
| Pioneer | A plant which is capable of invading bare soils, e.g. newly exposed soil, and persisting there or colonising them until supplanted by successional species | Helms (1998) |
| Planning perspectives | An 'inductive' planning perspective generates a management plan for a whole forest estate from a bottom-up perspective. It concludes with an aggregation of all the results from planning for each single stand, to form an estate-level plan, assuming the whole consists of the sum of its parts. In contrast, a 'deductive' planning method derives the forest management plan from a top-down perspective, usually based on a planning model for the forest estate. The 'inductive' approach as often applied in forest practice is rather descriptive, soft, and qualitative. The 'deductive' approach is more prescriptive, algorithmic (optimisation-based), and highly structured | Baader (1945), Mendoza (2005) |
| Plantation | Forest stands established by planting and/or seeding in the process of afforestation or reforestation. They are either of introduced species (all planted stands) or intensively managed stands of indigenous species, which meet all the following criteria: one or two species at planting, even age class, regular spacing | Carle and Holmgren (2003) |

(continued)

| Term or concept | Definition/explanation | Source |
|---|---|---|
| Planted forests (contrast with 'plantations') | Forests predominantly composed of trees established through planting and/or after deliberate seeding of native or introduced species | Carle and Holmgren (2008) |
| Plenterforest or plenterwald | A particular form of single tree selection forest that is characterised by the presence of all size and age classes and a high degree of vertical heterogeneity at small spatial scale. In Europe, Plenterforests comprise mostly shade-tolerant species such as *Abies alba*, *Picea abies*, and *Fagus sylvatica* | |
| Polyculture | A forest stand with more than one tree species | |
| Polyphagous | Feeding on different types of food (e.g. a pest species that feeds on different host species) | |
| Pool | Within an ecosystem, a pool (often also called compartment) is the stored portion of a given resource, such as carbon, water, a nutrient, etc. In contrast, a flux is the flow of that resource into (input) or out of (output) the given pool. An example of a carbon pool is the carbon stored within the aboveground biomass | |
| Pre-emption of resources, e.g. light, water | A phenomenon that vectorial resources coming from above, such as light or precipitation, can be primarily exploited by those trees which have their crowns or their roots, respectively, above those of competing neighbours | Schwinning and Weiner (1998) |
| Primary forest | Naturally regenerated forest of native species, where there are no clearly visible indications of human activities and the ecological processes are not significantly disturbed | FAO (2010) |
| Process-based model | See Sect. 8.2.2 | |
| Production ecology equation | An equation describing tree or stand productivity as the product of resource availability, the proportion of the resource that is acquired, and the efficiency with which the resource is used | Monteith (1977) |
| Production forest | A forest where the extraction of forest products, usually wood and fibre, is the predominant management objective, including both wood and non-wood forest products | Carle and Holmgren (2003) |

(continued)

| Term or concept | Definition/explanation | Source |
|---|---|---|
| Production planning | Is concerned with the long-term forest production. Planning issues are of strategic nature, for example, long-term tree species composition of a forest estate, but also pruning of trees, and tending of forest stands | Baader (1945), Knoke et al. (2012) |
| Productivity | The rate at which biomass is produced per unit area by any class of organisms | Helms (1998) |
| Provisioning ecosystem services | Provisioning services are ecosystem services that describe the material or energy outputs from ecosystems. They include food, water, and other resources | TEEB (2016) |
| Pure stand | A forest stand existing of predominantly one tree species, which contributes at least to 90% of stand density | Helms (1998), Schuck et al. (2002) |
| Realised niche | The subset of n-dimensional environmental space where a species is present. It is usually assumed that the realised niche is a subset of (smaller than) the fundamental niche | McGill et al. (2006) |
| Redundancy or functional redundancy | The concept of functional redundancy is based on the observation that some species can be removed from an ecosystem without impairing the functioning of that system. Complete functional redundancy only occurs if, following the removal of one species, there is density compensation among the remaining species | Walker (1992) |
| Reforestation | Forests established artificially by reforestation on land that carried forest within the previous 50 years or within living memory and involved the replacement of a previous crop by a new and essentially different crop | Carle and Holmgren (2003) |
| Regulating ecosystem services | Regulating services are the services that ecosystems provide by acting as regulators, e.g. regulating the quality of air and soil or by providing flood and disease control | TEEB (2016) |
| Reineke's rule | Allometric decrease of tree number, $N$, per unit area with increasing mean diameter, $d$, in fully stocked, unmanaged monocultures following $N \propto d^{-1.605}$. It means that due to competition for resources in a stand tree number decreases by 1.605% when mean tree diameter increases by 1%. The concept has been subsequently broadened and described as the self-thinning rule | Reineke (1933) |

(continued)

| Term or concept | Definition/explanation | Source |
|---|---|---|
| Relative stand density | The ratio, proportion, or percent of absolute stand density to a reference level<br>The reference level might be derived from a yield table or a maximum stand density index (SDI) | Helms (1998) |
| Relative total yield, relative yield total, RYT | Yield of mixed-species stands per unit area divided by the yield of monocultures of the respective species per unit area. The RYT is identical with the land equivalent ratio, LER. The former is based on the yield, whereas the latter is based on the required area for mixed stands compared with monocultures of a species assemblage (see land equivalent ratio, LER) | Vandermeer (1992) |
| Removal trees, removal stand | Share eliminated from the tree population before the end of a rotation by mortality, silvicultural practices, or disturbances. The removal stand plus the remaining stand add up to the total yield | Johann (1993) |
| Resilience | The capacity of an ecosystem to return to the precondition state following a perturbation, including maintaining its essential characteristics taxonomic composition, structures, ecosystem functions, and process rates | Holling (1973) |
| Resistance | The capacity of the ecosystem to absorb disturbances and remain largely unchanged | Holling (1973) |
| Resource-use efficiency | The growth of a tree or stand per unit of resource used. Occasionally the resource use is the supply, but in most cases (including in this book), the resource use is defined as the amount of resource actually used by the plants | Binkley et al. (2004) |
| Root (coarse or fine) | Roots are often divided into coarse and fine roots. Definitions vary, but often fine roots are those with a diameter of <2 mm, medium roots have between 2 and 5 mm diameter, and coarse roots range from 5 to 50 mm. Roots above these diameters are regarded as structural roots, which are sometimes also classified as coarse roots. However, the classifications are not well defined and change with the purpose of the study | |
| Root-shoot relationship, root-shoot allometry, root-shoot ratio | Relationship between the growth and biomass of the below-ground and above-ground parts of a plant | Harper (1977), Pretzsch et al. (2012) |

(continued)

| Term or concept | Definition/explanation | Source |
|---|---|---|
| Rotation length | Period between regeneration establishment and final cutting of even-aged forests. In uneven-aged forests, the term production cycle reflects better the time needed to grow trees to economic maturity | Helms (1998) |
| Rubner's rule | Assumption that a plant's metabolic rate, $A$, scales with 2/3 over plant mass, $m$, or volume, $v$ ($A \propto m^{2/3}$), whereas respiration, $R$, increases linearly with mass or volume ($R \propto m$) | Rubner (1931) |
| Runoff | That part of the precipitation, snow melt, or irrigation water that appears in uncontrolled surface streams, rivers, drains, or sewers | USGS (2016) |
| Sampling effect | The effect, which is also called selection probability effect, describes a situation in which treatments which have the highest diversity have a greater probability of being dominated by the most productive species of the entire species pool. The possible artefact caused by this effect can be overcome by including monocultures of all species used in the species pool of biodiversity experiments | Wardle (1999) |
| Sapwood | The outermost part of the stem that conducts the water from the roots to the leaves (i.e. the xylem). The sapwood area is the cross-sectional area of the sapwood | |
| Scenario analyses | Deliver 'if-then' statements and thereby provide a decision support of among others the silvicultural treatment, the amount of the annual cut, and the procedure of natural regeneration methods (see Chap. 10) | |
| Self-thinning, alien-thinning | Self-thinning describes the density-dependent mortality in fully stocked monocultures; in mixed-species populations, alien-thinning means that the mortality in one species is caused by the density of an associated species. In mixed-species stands, density-dependent mortality may be driven by both processes | Harper (1977) |
| Self-thinning | A progressive decline in the density of a population of growing plants | Allaby (2010) |

(continued)

| Term or concept | Definition/explanation | Source |
|---|---|---|
| Self-tolerance | The ability of trees to grow and to survive under the stress of intraspecific competition. The relative rate of volume growth with respect to the decrease of the number of trees can be used as its measure | Zeide (1998) |
| Seminatural forest | A forest of mostly native tree species established through planting, seeding, or assisted natural regeneration. Typically a managed natural forest which, over time, has taken on a number of natural characteristics (such as layered canopy, enriched species diversity, random spacing, etc.) or planted forests which acquire more natural characteristics over time (e.g. abandoned plantation forests that diversify with age and natural regeneration of indigenous species) | Carle and Holmgren (2003) |
| Shade tolerance | The capacity of a given plant to tolerate low light levels | Valladares and Niinemets (2008) |
| Silviculture | Silviculture is the deliberate manipulation of forest structure, composition, and dynamics to provide ecosystem goods and services on a sustainable basis | Bauhus and Pyttel (2015) |
| Site (site index, site quality) | A site is a particular location in a forest that is characterised by a suite of environmental factors including soil properties, topography, rainfall, wind, temperature, etc. Sites are typically classified according to their potential for tree growth using a site index or yield class for a given tree species | Barnes et al. (1982) |
| Site preparation | Any treatment of a forest site to prepare the soil for the establishment of a new crop of trees by either planting or natural means | Puettman et al. (2009) |
| Size distribution, size frequency distribution | The structuring of a stand in terms of tree numbers of different tree sizes; can be quantified by, e.g. mean, range, and standard deviation of the distribution | Prodan (1965) |
| Size growth-size relationship | Relationship between annual or periodic annual increment of the trees in a stand and their size at the beginning of the respective period | Prodan (1965), Pretzsch and Biber (2010) |
| Size hierarchy | Size hierarchy is equivalent to size inequality. The degree of inequality can be visualised by the Lorenz curve or quantified by the Gini coefficient (see also Lorenz curve, Gini coefficient, growth dominance coefficient) | Weiner and Solbrig (1984) |

(continued)

| Term or concept | Definition/explanation | Source |
|---|---|---|
| Size symmetry vs. size asymmetry of competition | Contested resources between plants are divided in proportion to size vs. contested resources are divided over-proportionally with respect to size | Weiner et al. (1997), Schwinning and Weiner (1998) |
| Size-class model | Is also called size-structured model. In size-class models, the forest enterprise is represented by a stem frequency distribution over dbh classes, which may cover the whole area of the forest enterprise or smaller units, such as forest subdivisions of an area from 5 to 35 ha. Information of the expected diameter increment per tree species and dbh class for various stand densities is needed to compute transition probabilities indicating how likely trees move from one dbh class into the other. Ingrowth to fill up the smallest dbh class must be estimated and also tree mortality per species and dbh class. To be applicable in forest practice, a target size-class distribution is a necessary precondition to apply this approach, which may be derived from economic considerations | Roessiger et al. (2016) |
| Solitary trees, open grown trees | Trees that grow without competition by neighbouring trees, represent maximum crown extension, and are often used as reference for quantifying the effect of competition | Pretzsch (2014) |
| Space-/area-related running costs | Ratio of a tree's respiration to the occupied stand area or crown projection area. Indicates how much energy a tree spends just to keep its stand area or crown area occupied | Pretzsch and Schütze (2009) |
| Specialist (species) | A species that uses a relatively small proportion (in extreme cases only one) of the available resources types | Calow (1998) |
| Species diversity | The total number and distribution of species in an area: includes components of both species richness and the amount of evenness or dominance among the species present | |
| Species richness | The total number of species in a given area. Species richness is typically calculated for taxonomic groups, e.g. tree species richness | |
| Species selection effect | If the total growth of species in mixture is just the weighted mean growth of the component species when growing in monoculture, the difference between mixed-species stands results just from the species selection. (see also sampling effect and true mixing effect) | Forrester and Pretzsch (2015) |

(continued)

| Term or concept | Definition/explanation | Source |
|---|---|---|
| Specific leaf area | The one-sided area of a leaf per unit of its mass (often expressed as $m^2\ kg^{-1}$) | |
| Stability (here ecological stability of forest stands) | Ecological stability encompasses both resistance, the maintenance of a condition in the face of disturbance, and resilience, the capacity to return to a previous pre-disturbance state for example via regeneration. Stability is specific in relation to different types of disturbance, e.g. wind or fire, and may refer to stability of processes, populations, communities, or structures | |
| Stand area of a tree, growing area of a tree | Physical area available and utilised by a tree. Proxy for the trees share of the resources of the site. Stand area or growing area is not directly measurable but often obtained by an area allocation based on tree coordinates and crown projection areas and using mathematical methods (e.g. Thiessen and Voronoi polygons) | Assmann (1970), Helms (1998) |
| Stand density index, SDI | A widely used measure developed by Reineke (1933) that expresses the relative stand density in terms of the relationship of a number of trees to stand quadratic mean diameter. In order to eliminate the effect of stand development phase, the ratio is allometrically adjusted to an index diameter of 20 cm | Helms (1998) |
| Stand level mixture (economic view) | If mixed tree species interact because of ecological interdependence, the direct analogy between tree species portfolios and portfolios of financial assets vanishes. This will be the case when tree species are mixed more intimately at the stand level. In such a situation, important characteristics may change compared to a pure forest, including stand resistance, yield, and timber quality. Estimating the expected average performance of such a portfolio with interdependent components, in our case a mixed forest stand, requires more information than the simple proportion of species in the portfolio | Knoke and Seifert (2008) |
| Stand or forest stand | The basic unit of forest mapping and management; a contiguous area of forest characterised by a sufficiently uniform species composition, age, structure, and condition to be considered a homogeneous unit for management purposes | |

(continued)

| Term or concept | Definition/explanation | Source |
|---|---|---|
| State variables | In models the state variables are usually the ecosystem pools and these will be updated as the simulation progresses. Examples could be the biomass, nutrient, and water pools | |
| Stochastic | A term used to describe systems that are unpredictable due to the influence of a factor that occurs randomly | |
| Stress (in plants) | A state in which increasing demands made upon a plant lead to an initial destabilisation of functions, followed by normalisation and improved resistance. If the limits of tolerance are exceeded and the adaptive capacity is overtaxed, permanent damage or even death may result | Larcher (1987) |
| Subgressive underyielding, degressive underyielding | The extent to which a mixed stand falls below the productivity of the least productive monoculture of any of the component species at a given site | Río et al. (2015) |
| Susceptibility | The inability of a plant to resist the effect of damaging biotic or abiotic factors | |
| Sustainability | Characteristic by which a process or state can be maintained at a certain level indefinitely. In the environmental context, the term refers to the potential longevity of vital human ecological support systems and the various systems on which they depend in balance with the impacts of our unsustainable or sustainable design | |
| Sustainable forest management | The stewardship and use of forests and forest lands in a way and at a rate that maintains their biological diversity, productivity, regeneration capacity, vitality and their potential to fulfil, now and in the future, relevant ecological economic and social functions, at local, national, and global levels, and that does not cause damage on other ecosystems | FAO (2000) |
| Tending | The tending of forest stands comprises all silvicultural activities between the successful establishment of regeneration and the harvesting of trees. The most important tending practices include thinning and pruning | Bauhus and Pyttel (2015) |
| Thinning | Partial removal of trees in an immature stand to select for a specific species or to increase the growth rate and quality of the remaining trees | Puettman et al. (2009) |

(continued)

| Term or concept | Definition/explanation | Source |
|---|---|---|
| Thünen's land location theory | Spatial economic modelling approach to maximise society's well-being in a hypothetical environment. Land allocation depends on land rent, the economic surplus of each considered land-use alternative. Land rent is sensitive to transport costs, which grow with increasing distance to the market place. Thus, land use is allocated in concentric cycles around the city centre (market place) | Sasaki and Box (2003) |
| Total yield, gross yield | The sum of the intermediately removed volume (see removal trees) and the final standing volume | Helms (1998) |
| Trainer trees or trainer species | Trees that grow typically in the subcanopy with the purpose to shade the stems of faster-growing, valuable tree species. The purpose of trainer trees is typically to foster branch shedding and to suppress the development of epicormics branches in crop trees. In addition, they may be used to shade the ground to reduce development of an understorey which may pose an impediment to regeneration | |
| Trait | A well-defined, measurable property of organisms, usually measured at the individual level and used comparatively across species. A functional trait is one that strongly influences organismal performance | McGill et al. (2006) |
| Transgressive overyielding | The extent to which a mixed stand excels even the productivity of the most favourable growing species in monoculture at a given site | Pretzsch (2009) |
| Transpiration | The evaporation of water vapour from the surfaces of leaves through stomata | |
| Tree | A woody perennial with a single main stem or, in the case of coppice, with several stems, having a more or less definite crown | Schuck et al. (2002) |
| Tree allometry, crown allometry | Quantitative relations between key characteristics of trees, e.g. between stem diameter and crown diameter, often described by double logarithmic equations. (see also 'allometry') | Niklas (1994) |
| Trophic level | A position in the food chain assessed by the number of energy transfer steps to reach that level | Helms (1998) |

(continued)

| Term or concept | Definition/explanation | Source |
|---|---|---|
| True mixing effect | Interaction between different species which cause an outcome, e.g. in terms of growth, that differs from the weighted mean of the growth of both species when growing in monoculture (see also selection effect) | Forrester and Pretzsch (2015) |
| Underyielding | When the production of biomass in species mixtures falls below the productivity expected on the basis of the yields of the contributing species when grown in a monoculture | |
| Uneven-aged stand | A stand with trees of three or more distinct age classes, either intimately mixed or in small groups | Helms (1998) |
| Uniqueness | A measure of relative difference for a species from others | |
| Vapour flux | The flow rate of a gaseous substance | |
| Water quality | A term used to describe the chemical, physical, and biological characteristics of water, usually in respect to its suitability for a particular purpose | USGS (2016) |
| Woodland | 1. Forest area. 2. A plant community in which, in contrast to a typical forest, the trees are often small, characteristically short-boled relative to their crown depth and forming only an open canopy with the intervening area being occupied by lower vegetation, commonly grass | Helms (1998) |
| YAFO-optimisation model | An estate-level planning model is a typical prescriptive, algorithmic (optimisation based), and consistently modularised model, which incorporates survival curves for various stand types (mixed and pure) and comprehensively addresses fluctuation of net revenue flows caused by stand failure and timber price volatility. The classic forest planning problem of allocating silvicultural activities to space (forest stands) and time (planning periods) is considered by YAFO to maximise an objective function | Härtl et al. (2013) |
| Yield planning | Is concerned with the organisation of timber harvests, usually over the next 10 to 20 years. Yield planning prescribes in which forest stands (strata) and in which periods how much timber should be harvested | Baader (1945), Knoke et al. (2012) |

(continued)

| Term or concept | Definition/explanation | Source |
|---|---|---|
| Yield table | A table showing the expected timber yield until a certain age for even-aged stands, usually by site index classes, and typically including variables such as quadratic mean diameter (DBH), height, number of stems, basal area, and standing volume per unit area; yield tables may also include volume of thinnings, CAI, MAI, and other information | Helms (1998) |
| Yoda's rule | Allometric decrease of mean plant volume, *v*, or weight, with increasing plant number per unit area in fully stocked, unmanaged monocultures following $v \propto N^{-3/2}$. Also called the −3/2-power rule as mean volume decrease by −3/2% when plant number increases by 1%. Yoda's rule describes the phenomenon of self-thinning based on the variables mean *v* and *N*, whereas Reineke describes it based on *d* and *N* (see also Reineke's rule) | Yoda et al. (1963) |

## References


Allaby M (2010) A dictionary of ecology. Oxford University Press, Oxford

Assmann E (1970) The principles of forest yield study. Pergamon Press, Oxford, 506 p

Baader G (1945) Forsteinrichtung als nachhaltige Betriebsführung und Betriebsplanung. Sauerländer's, Frankfurt a.M.

Barnes BV, Pregitzer KS, Spies TA, Spooner VH (1982) Ecological forest site classification. J For 80:493–498

Bauhus J, Pyttel P (2015) Managed forests. In: Peh KS-H, Corlett RT, Bergeron Y (eds) Routledge handbook of forest ecology. Routledge, Oxon, pp 75–90

Bauhus J, Puettmann KJ, Kühne C (2013) Close-to-nature forest management in Europe: does it support complexity and adaptability of forest ecosystems? In: Messier C, Puettmann KJ, Coates KD (eds) Managing forests as complex adaptive systems: building resilience to the challenge of global change. Routledge, The Earthscan Forest Library, New York, pp 187–213

Begon M, Harper JL, Townsend CR (1996) Ecology: individuals, populations and communities. Blackwell Science, Oxford

Bettinger P, Boston K, Siry JP, Grebner DL (2009) Forest management and planning. Academic Press/Elsevier, Amsterdam, 331 p

Binkley D, Stape JL, Ryan MG (2004) Thinking about efficiency of resource use in forests. For Ecol Manag 193:5–16

Binkley D, Kashian DM, Boyden S et al (2006) Patterns of growth dominance in forests of the Rocky Mountains, USA. For Ecol Manag 236:193–201

Bugmann H (2001) A review of forest gap models. Clim Chang 51:259–305

Calow P (1998) The encyclopedia of ecology and environmental management. Blackwell Science, Oxford

Carle J, Holmgren P (2003) Definitions related to planted forests. Forest Resources Assessment Programme, Working paper 79, 24 p

Carle J, Holmgren P (2008) Wood from planted forests: a global outlook 2005–2030. For Prod J 58:6–18
CIFOR (2001) Typology of planted forests. CIFOR InfoBrief. Center for International Forestry Research (CIFOR), 4 p
Cramer GR, Urano K, Delrot S, Pezzotti M, Shinozaki K (2011) Effects of abiotic stress on plants: a systems biology perspective. BMC Plant Biol 11:163–176
Dayton PK (1972) Toward an understanding of community resilience and the potential effects of enrichments to the benthos at McMurdo Sound, Antarctica. In: Proceedings of the colloquium on conservation problems. Allen Press, Lawrence, Kansas, p 81–96
Dieler J, Pretzsch H (2013) Morphological plasticity of European beech (*Fagus sylvatica* L.) in pure and mixed-species stands. For Ecol Manag 295:97–108
Ellison AM, Bank MS, Clinton BD et al (2005) Loss of foundation species: consequences for the structure and dynamics of forested ecosystems. Front Ecol Env 3:479–486
Elton JE, Gruber MJ, Brown SJ, Goetzmann WN (2007) Modern portfolio theory and investment analysis, 7th edn. Wiley, New York
Evans GC (1972) The quantitative analysis of plant growth, vol 1. University of California Press, Berkley, CA, 734 p
Fama EF, French KR (1993) Common risk factors in the returns on stocks and bonds. J Financ Econ 33:3–56
FAO (2000) Asia-Pacific forestry commission: development of national-level criteria and indicators for the sustainable management of dry forests of Asia. Food and Agriculture Organization, Workshop Report, RAP Publication 2000/07
FAO (2001) Criteria and indicators for sustainable forest management: a compendium. Paper compiled by Froylán Castañeda, Christel Palmberg-Lerche and Petteri Vuorinen, May 2001. Forest management working papers, Working paper 5. Forest Resources Development Service, Forest Resources Division. FAO, Rome
FAO (2010) Global forest resources assessment 2010. http://www.fao.org/docrep/013/i1757e/i1757e.pdf
FAO (2015) Invasive species: impacts on forests and forestry. http://www.fao.org/forestry/aliens/en/. Accessed 11 Sept 2016
Forrester JW (1961) Industrial dynamics. MIT Press, Cambridge, MA
Forrester DI, Pretzsch H (2015) Tamm review: on the strength of evidence when comparing ecosystem functions of mixtures with monocultures. For Ecol Manag 356:41–53
Hara T (1992) Effects of the mode of competition on stationary size distribution in plant populations. Ann Bot 69:509–513
Hara T (1993) Mode of competition and size-structure dynamics in plant communities. Plant Species Biol 8(2-3):75–84
Harper JL (1977) Population biology of plants. Academic Press, New York, 892 p
Härtl F, Hahn A, Knoke T (2013) Risk-sensitive planning support for forest enterprises: the YAFO model. Comput Electron Agric 94:58–70
Helms JA (1998) The dictionary of forestry. The Society of American Foresters, Bethesda, MD
Hildebrandt P, Knoke T (2011) Investment decisions under uncertainty—a methodological review on forest science studies. Forest Policy Econ 13:1–15
Holling CS (1973) Resilience and stability of ecosystems. Ann Rev Ecol System 4:1–23
Hutchinson GE (1957) Concluding remarks. Cold Spring Harb Symp Quant Biol 22(2):415–427
Janssen S, van Ittersum MK (2007) Assessing farm innovations and responses to policies: a review of bio-economic farm models. Agric Syst 94:622–636
Johann K (1993) DESER-Norm 1993. Normen der Sektion Ertragskunde im Deutschen Verband Forstlicher Forschungsanstalten zur Aufbereitung von waldwachstumskundlichen Dauerversuchen. In: Proc Dt Verb Forstl Forschungsanst, Sek Ertragskd, in Unterreichenbach-Kapfenhardt, pp 96–104
Johnson S (2001) Emergence: the connected lives of ants, brains, cities and software. Simon and Schuster, New York, 288 p

Kelty MJ (1992) Comparative productivity of monocultures and mixed-species stands. In: Kelty MJ, Larson BC, Oliver CD (eds) The ecology and silviculture of mixed-species forests. Springer, Netherlands, pp 125–141

Kleiber M (1947) Body size and metabolic rate. Physiol Rev 27:511–541

Knoke T, Seifert T (2008) Integrating selected ecological effects of mixed European beech–Norway spruce stands in bioeconomic modelling. Ecol Model 210:487–498

Knoke T, Schneider T, Hahn A, Grieß V, Rößiger J (eds) (2012) Forstbetriebsplanung als Entscheidungshilfe. Ulmer, Stuttgart

Koellner T, Schmitz OJ (2006) Biodiversity, ecosystem function, and investment risk. BioScience 56:977–985

Kramer H (1988) Waldwachstumslehre. Paul Parey, Hamburg

Larcher W (1987) Stress bei Pflanzen. Naturwissenschaften 74:158–167

Larcher W (2003) Physiological plant ecology, 4th edn. Springer, Berlin

Litton CM, Raich JW, Ryan MG (2007) Carbon allocation in forest ecosystems. Glob Chang Biol 13:2089–2109

Lloret F, Keeling EG, Sala A (2011) Components of tree resilience: effects of successive low-growth episodes in old ponderosa pine forests. Oikos 120:1909–1920

Loreau M, Hector A (2001) Partitioning selection and complementarity in biodiversity experiments. Nature 412:72–76

Maestre FT, Callaway RM, Valladares F, Lortie CJ (2009) Refining the stress-gradient hypothesis for competition and facilitation in plant communities. J Ecol 97:199–205

Markowitz H (1952) Portfolio selection. J Financ 7:77–91

Martonne de E (1926) Une novelle fonction climatologique: L'indice d'aridité. La Météorologie 21:449–458

McArdle RF (1953) Multiple-use benefits. J For 51:323–325

McCree KJ (1971) The action spectrum, absorptance and quantum yield of photosynthesis in crop plants. Agric Meteorol 9:191–216

McGill BJ, Enquist BJ, Weiher E, Westoby M (2006) Rebuilding community ecology from functional traits. Trends Ecol Evol 21:178–185

MEA (Millennium Ecosystem Assessment) (2005) Ecosystems and human well-being: synthesis. World Resources Institute, Island Press, Washington, DC, 137 p

Merriam-Webster (2016) Merriam-Webster online dictionary. http://www.merriam-webster.com/dictionary. Accessed 10 Sept 2016

Mills WL, Hoover WL (1982) Investment in forest land: aspects of risk and diversification. Land Econ 58:33–51

Monteith JL (1977) Climate and the efficiency of crop production in Britain. Philos Trans R Soc Lond Ser B 281:277–294

NCCE (2013) North Carolina Cooperative Extension. www.ces.ncsu.edu/forestry/pdf/WON/won26.pdf. Accessed 17 May 2013

Neumann RB, Cardon ZG (2012) The magnitude of hydraulic redistribution by plant roots: a review and synthesis of empirical and modeling studies. New Phytol 194:337–352

Niklas KJ (1994) Plant allometry. University of Chicago Press, Chicago, IL

Oliver CD, Larson BC (1996) Forest stand dynamics. Wiley, New York

Paterson SS (1956) The forest area of the world and its potential productivity. The Royal University of Göteborg, Sweden, Department of geography, Lerums Boktryckeri, Sweden, p 216

Perona M, Miragliotta G (2004) Complexity management and supply chain performance assessment. A field study and a conceptual framework. Int J Prod Econ 90:103–115

Pommerening A, Murphy ST (2004) A review of the history, definitions and methods of continuous cover forestry with special attention to afforestation and restocking. Forestry 77:27–44

Pretzsch H (2009) Forest dynamics, growth, and yield. Springer, Berlin

Pretzsch H (2010) Re-evaluation of allometry: state-of-the-art and perspective regarding individuals and stands of woody plants. Prog Bot 71:339–369

Pretzsch H (2014) Canopy space filling and tree crown morphology in mixed-species stands compared with monocultures. For Ecol Manag 327:251–264
Pretzsch H, Biber P (2010) Size-symmetric versus size-asymmetric competition and growth partitioning among trees in forest stands along an ecological gradient in central Europe. Can J For Res 40:370–384
Pretzsch H, Schütze G (2009) Transgressive overyielding in mixed compared with pure stands of Norway spruce and European beech in Central Europe: evidence on stand level and explanation on individual tree level. Eur J For Res 128:183–204
Pretzsch H, Uhl E, Biber P, Schütze G, Coates D (2012) Change of allometry between coarse root and shoot of Lodgepole pine (*Pinus contorta* DOUGL. ex. LOUD.) along a stress gradient in the sub-boreal forest zone of British Columbia. Scand J For Res 27:532–544
Preuhsler T (1981) Ertragskundliche Merkmale oberbayerischer Bergmischwald-Verjüngungsbestände auf kalkalpinen Standorten im Forstamt Kreuth. Forstwiss Centralbl 100:313–345
Prieto I, Armas C, Pugnaire FI (2012) Water release through plant roots: new insights into its consequences at the plant and ecosystem level. New Phytol 193:830–841
Prodan M (1965) Holzmeßlehre. JD Sauerländer's Verlag, Frankfurt am Main, 644 p
Puettman KJ, Coates KD, Messier C (2009) A critique of silviculture: managing for complexity. Island Press, Washington, DC, 189 p
Putz FE, Parker GG, Archibald RM (1984) Mechanical abrasion and intercrown spacing. Am Midl Nat 112:24–28
Reineke LH (1933) Perfecting a stand-density index for even-aged forests. J Agric Res 46:627–638
Ridley M (2004) Evolution. Blackwell/Oxford University Press, Malden, MA/Oxford
Rio del M, Pretzsch H, Alberdi I et al (2015) Characterization of the structure, dynamics, and productivity of mixed-species stands: review and perspectives. Eur J For Res 135:23–49
Roessiger J, Ficko A, Clasen C, Griess VC, Knoke T (2016) Variability in growth of trees in uneven-aged stands displays the need for optimizing diversified harvest diameters. Eur J For Res 135:283–295
Rubner M (1931) Die Gesetze des Energieverbrauchs bei der Ernährung. In: Proc preuß Akad Wiss Physik-Math Kl 16/18, Berlin, Wien, 1902 p
Sasaki Y, Box P (2003) Agent-based verification of von Thünen's location theory. J Artif Soc Soc Sim 6. http://jasss.soc.surrey.ac.uk/6/2/9.html
Schuck A, Päivinen R, Hytönen T, Pajari B (2002) Compilation of forestry terms and definitions. European Forest Institute, Internal Report No. 6, 48 p
Schwinning S, Weiner S (1998) Mechanisms determining the degree of size asymmetry in competition among plants. Oecologia 113:447–455
Shannon CE (1948) A mathematical theory of communication. Bell Syst Tech J 27 (379–423):623–656
Stachowicz JJ (2001) Mutualism, facilitation, and the structure of ecological communities. BioScience 51:235–246
TEEB (2016) The economics of ecosystems and biodiversity. http://www.teebweb.org/resources/ecosystem-services/. Accessed 11 Sept 2016
UF-IFAS (2016) Allelopathy: how plants suppress other plants. University of Florida-Institute of Food and Agricultural Sciences. http://solutionsforyourlife.ufl.edu/. Accessed 11 Sept 2016
Uhde B, Andreas Hahn W, Griess VC, Knoke T (2015) Hybrid MCDA methods to integrate multiple ecosystem services in forest management planning: a critical review. Environ Manag 56:373–388
UN (1992) Convention on biological diversity. https://www.cbd.int/convention/
UNEP (2007) In-depth review of the application of the ecosystem approach. Convention on Biological Diversity. https://www.cbd.int/doc/meetings/sbstta/sbstta-12/official/sbstta-12-02-en.pdf

UNFCCC (2014) Glossary of climate change acronyms. United Nations Framework Convention on Climate Change. http://unfccc.int/. Accessed 11 Sept 2016
USGS (2016) Water science glossary of terms. United States Geological Survey. http://ga.water.usgs.gov/edu/dictionary.html#G. Accessed 11 Sept 2016
Valladares F, Niinemets Ü (2008) Shade tolerance, a key plant feature of complex nature and consequences. Annu Rev Ecol Evol Syst 39:237–257
Vandermeer J (1992) The ecology of intercropping. Cambridge University Press, New York
Walker BH (1992) Biodiversity and ecological redundancy. Conserv Biol 6:18–23
Wardle DA (1999) Is "Sampling Effect" a problem for experiments investigating biodiversity-ecosystem function relationships? Oikos 87:403–407
Weiner J, Solbrig OT (1984) The meaning and measurement of size hierarchies in plant populations. Oecologia 61:334–336
Weiner J, Thomas SC (1986) Size variability and competition in plant monocultures. Oikos:211–222
Weiner J, Wright DB, Castro S (1997) Symmetry of below-ground competition between Kochia scoparia individuals. Oikos:85–91
www.lfi.ch/glossar/glossar-en.php?frmWordList=336_Gloss. Accessed 17 May 2013
www.lfi.ch/glossar/glossar-en.php?frmWordList=418_Gloss. Accessed 17 May 2013
www.sswm.info/category/implementation-tools/reuse-and-recharge/hardware/recharge-and-disposal/short-rotation-plant#reference_book506. Accessed 17 May 2013
Yoda KT, Kira T, Ogawa H, Hozumi K (1963) Self-thinning in overcrowded pure stands under cultivated and natural conditions. J Biol Osaka City Univ 14:107–129
Zeide B (1998) Fractal analysis of foliage distribution in loblolly pine crowns. Can J For Res 28:106–114
Zilber-Rosenberg I, Rosenberg E (2008) Role of microorganisms in the evolution of animals and plants: the hologenome theory of evolution. FEMS Microbiol Rev 32:723–735

# Index

## A

*Abies alba,* 97, 98, 106, 145, 152, 168, 171, 187, 197, 198, 200, 294, 295, 464, 484, 486
  silver fir, 585
*Abies balsamea,* 366
Abiotic disturbances, 361, 374
Absorbed photosynthetically active radiation (APAR), 608
*Acacia mangium,* 83, 86, 97, 439
*Acacia mearnsii,* 83, 86, 90, 91, 153, 155, 156
Acclimation, 607
*Acer pseudoplatanus,* 58, 294
  sycamore maple, 295
Acid rain, 15
Adaptation, 607
Additive effect, 135
Agroforestry, 596
Alien thinning, 160–176, 627
  interspecific competition, 171
Alien-thinning slopes, 170
Allelopathy, 608
Allometric adjustment, 141–145, 321
  allometric exponents, 143
  direct mixing effects, 145
  elimination of initial size differences, 144
  indirect mixing effect, 144
  mean tree productivity, 144
  size difference, 145
  theory of allometry, 144
Allometric drift, 143
Allometric exponents
  allometry between stand area and tree diameter, 291
  $\alpha_{\mathrm{cpa,\ d}}$, 296
  $\alpha_{\mathrm{cpa/v,\ v}}$, 320
  $\alpha_{\mathrm{iv/cpa,\ v}}$, 320
  $\alpha_{\mathrm{iv/v,\ v}}$, 320
  $\alpha_N$, *d,* 162
  $\alpha_{N,\bar{d}}$, 170
  $\alpha_{N,\bar{v}}$, 144, 165
  $\alpha_{\bar{p},\bar{v}}$, 143, 144
  $\alpha_{\mathrm{sa/v,\ v}}$, 320
  $\alpha_{\mathrm{TY}}$,$h_q$, 178
  $\alpha_V$,$h_q$, 178, 180
  anabolistic term, 276
  catabolistic term, 276
  coarse root growth *vs.* stem diameter growth, 303
  crown projection area (cpa), 295, 299
  geometrical scaling, 139
  h–d allometry, 296
  ln(cpa)–ln*(d)* relationships can, 293, 294
  ln*(N)*–ln($d_q$) slope, 161
  scaling between hcb and d, 296
  stem diameter (d), 295
  stem volume *vs.* tree growing area, 278
  tree diameter (d), 299
Allometric relationship, 297
  general, 143
Allometric scaling, 143
Allometric theory, 296
Allometry, 608
  crown, 632
  equation, 275
  ratio, 626
  root-shoot, 626
  tree, 632
*Alnus glutinosa,* 188, 294
*Alnus rubra,* 97
Angiosperms, 291
Antagonistic interaction, 185
Artificial regeneration, 438

 
H. Pretzsch et al. (eds.), *Mixed-Species Forests*, DOI 10.1007/978-3-662-54553-9

Asymmetry of competition, 233
Average net revenues, 548

**B**
*Betula papyrifera,* 366
*Betula pendula,* 294
Biodiversity, 608
  α, 7
  β, 7
  composition, 8, 530
  different management options, 534
  at different spatial scales, 533
  forest inventory, 531
  functional, 6
  functioning, 8, 530
  γ, 7
  indicators, 532
  integrated in forest planning, 534
  maximum diameter, 533
  maximum tree diameter, 532
  species richness, 532
  stand productivity, 532
  structure, 8, 530
  temporal changes, 532
Biological diversity, 15
Biome, 609
Biophysical damage, 557
Biotic disturbance, 374
Boreal forests, 2, 5, 15, 34, 36, 64, 84, 147, 186, 275, 345, 350, 370, 391, 421, 456, 464, 470, 483, 548
Branchiness, 443
Bud break, 389

**C**
Calamities, 551
Canopy climate, 389
Canopy closure, 315
Canopy gap occupation, 313–316
Canopy ground cover, 249–250
Canopy space filling, 246–257, 259
  absorbed photosynthetically active radiation, 253
  light interception, 253
  light wells, 253
  mixed *vs.* monospecific stands, 248–255
  temporal change, 253–255
Canopy space overpacking, 252
Canopy stratification, 397
Carbon
  mitigation, 609
  sequestration, 609
  sink, 609
  stock, 609
*Carpinus betulus,* 294
Carrying capacity, 471, 589
*Castanopsis sclerophylla,* 422, 423
cd/*d* ratio, 248, 275
  mixed stands *vs.* monocultures, 248
Chemical stratification, 81
Chronosequence, 608
cl/*h* ratio, 248
Climate change, 19, 601
Close-to-nature forestry, 14
Close-to-nature polycultures, 505
Coarse roots, 302
Cohort, 610
Commensalism, 610
Community, 610
Companion species, 490
Competition, 45, 272, 273, 278, 280, 610
  asymmetry, 233–236
  asymmetry by species mixing, 233
  below-ground, 354
  index, 610
  intraspecific, 37
  iv-*v* relationship modification, 235
  net, 621
  reduction, 610
  size asymmetry, 629
  size symmetry, 629
  stand density, 280
  symmetry, 233–236
Competition indices, 397
Competition only, 280
Competition reduction, 273, 299
Competition release, 312
Competitive reduction, 45, 47, 49, 75, 123, 182
Complementarity, 81, 97, 100, 101, 104, 281, 324, 439, 592, 610
Complementarity hypothesis, 34, 35
Complexity, 11, 610
Complicacy, 611
Concept for planning and decision making
  actual state of a forest, 511
  guidelines, 512
  management models and simulators, 513
  negotiations, 511
  normative societal values, 512
  owners and stakeholders, 511
  participatory, 512
  target knowledge, 512
  target state, 511
  transformation, 511
  transformation knowledge, 512

Conservation, 435
Continuous cover forestry, 568–572
Costs of establishment, 548
Costs of regeneration, 548
cpa/sa ratio, 275
Cross diagram, 135, 140, 186, 188, 611
Crown
behaviour in mixed and pure stands, 309
degree of crown engagement, 306
development, 305
disengagement, 612
eccentricity, 308, 611
expansion, 292
expansion of individual trees in intra-and interspecific environments, 306–308
extension, 275, 612
fractal dimension, 309
inner crown structures, 309
inner structure, 310
jaggedness, 612
maps, 307
in mixed compared with monospecific stands, 310
morphology, 275
periphery, 310
plasticity, 306, 612
projection area, 612
radial increment, 304
recession, 307
restriction and regression, 312
rotundity, 308
shyness, 612
simulation, 305
space filling, 309
Crown allometry, 246–257, 295–299
in interspecific local environments, 296–303
Crown area growth efficiency, 319
Crown density, 255–256
Crown diameter, 292
Crown engagement
link between tree and stand level evaluation, 322–324
ratio, 249–250
Crown expansion
in inter-*vs.* intraspecific environments, 303–306
Crown extension, 264
lateral and vertical, 296
Crown map, 250
Crown measurement, successive, 303
Crown morphology, 298
in mixed *vs.* monospecific stands, 298, 303
size differences, 298
Crown plasticity, 252, 293, 296, 314
interspecific variation, 294
relative measure, 293
Crown projection area, 249–251, 263, 312–313, 316–317
mixed *vs.* monospecific stands, 251
Crown restriction, 299
Crown shape, 313
interspecific variation, 290–292
Crown size
intraspecific variation, 292–296
mixed compared with monospecific stands, 290–311
*Cunninghamia lanceolata,* 422, 423
Current annual increment (CAI), 612

**D**
Dauerwald concept, 14
Decision making, 535, 570
Degree of crown engagement, 312–313, 319
Degressive underyielding, 135
Density-growth relationship, 204, 584
monospecific and mixed stands, 193
Design of mixed-species stand, species selection, 593
Distribution extrapolation models, 387
Disturbance, 41, 341, 613
abiotic, 593
biotic, 593
definition, 340
duration, 341
large-scale, 11
magnitude, 341
natural, 620
recovery, 341
resilience, 341
resistance, 341
Diversification, 13, 548, 549, 551, 613
Diversity, 33, 613
compositional, 588
effects, 345
functional, 616
structural, 588
taxonomic, 6
$d_r/d_s$ ratio, 275
Drought, 47, 343–351, 373
basal area increment, 348
responses, 346–351
Drought response, effects of thinning, 344
Drought stress
recovery of tree growth, 346–350
resilience, 346–350
resistance, 346–350

Drought tolerance
mixed *vs.* monospecific stands, 347–348

E
Early experience tables, 387
Ecological disturbances, 339
Ecological insurance, 16, 17
Ecological stability, 18
Economic returns
covariances, 562
uneven-aged stand, 572
Economic risk, 547
Economies of scale, 600–601
Ecosystem
bottom-up control, 31
causal explanations, 32
functioning, 16, 33
goods and services, 614
hierarchical levels, 31
hierarchy, 51
levels of system hierarchy, 32
multiple hierarchical levels, 32
services, 520
spatial and temporal scales, 31
stability, 16
structures and processes, 31
top-down control, 31
Efficiency
of crown space occupation, 319
resource-use, 626
space exploitation, 614
stand area exploitation, 614
Efficiency parameters
elimination of the size effect, 319–320
Eichhorn rule (1902), 176–178
Elasticity of growth, 194
Elimination of size effects, 142–144
Emergent properties, 18, 614
Empirical model, 392
integration of mixing effects, 396–398
Endemic, 614
Environmental stress, 339
Epicormic branch, 441, 488
*Eucalyptus*
*E. globulus,* 83, 86, 90, 91, 153, 155, 156
*E. grandis,* 83, 86, 97, 103
*E. pilularis,* 103
*E. saligna,* 83, 97
*E. salinga,* 103
*E. urophylla x grandis,* 86, 97, 436, 439
Evapotranspiration, 615
Evenness, 615
Expansion factors, 201
Expected productivity, 136
Experiment
along an ecological gradient disjunct, 581
mixed-species, 581
unmanaged plots, 591
water exclusion, 583–584
water retention, 584
Experimental marking, 510
Exploratories, 63

F
Facilitation, 37, 47, 49, 75, 123, 182, 272, 273, 278, 279, 615
net, 621
net effect, 279
only, 280
species mixing, 281
stand density, 280
*Fagus sylvatica,* 42, 121, 123, 124, 134, 136, 145, 148, 151, 152, 158, 168–171, 181, 184–187, 193, 195, 197, 200, 219, 222, 226, 234, 243, 247, 252, 260, 294, 348, 349, 371, 410, 419, 441, 449, 452, 468, 474, 484, 584
European beech, 287, 293, 297, 299, 303, 309, 315, 415–417, 535, 537, 552, 566, 567, 571, 582, 585
*Falcataria moluccana,* 83, 97, 103
Feedback loop, 283
environmental conditions, 282
stand structure, 282
tree functioning, 282
Feedback on stand productivity, 587
Financial return, 559, 565
Financial risk, 548
Financial stock, 558–559
Financial theory, 546
Fine-root growth, 81
Fire, 373
Fire risk, 356–361
changes in fuel conditions, 358–359
changes in microclimate, 359
crown fires, 357
demographic mechanisms, 360
fire occurrence, 358
flammability, 358
ground fires, 357
mixed-species vegetation, 359
plant functional types, 357
post-fire regeneration, 358
resprouters, 358

seeders, 358
severity of fire, 358
species identity effects, 360
tree species mixtures, 357
Fitness, 36, 278, 615
Forest
certification, 15, 615
diverse, 5
functions, 18
growth simulator, 385
managed, 10
mixed, 5
normal, 12
secondary, 10
seminatural, 8
species-rich, 5
Forest inventory, 53, 509
Forest-level mixture, 567
Forest management, 262–263
Forest management planning (FMP), 505–511, 515, 616
age-class planning, 518
application of growth models, 522
collecting information, 506
composition of tree species, 510
deductive, 505
enterprise-level, 509
enterprise-level management alternatives, 507
estate-level planning, 516
examples, 521–527
forest owners and stakeholders, 506
forest status, 507
growth simulator SILVA, 521
inductive, 505
linear programming, 516
mixed even-aged forests, 517
mixed forests, 516
multi-criteria decision analyses, 508
objectives, 506
optimal treatment, 518
optimisation, 508
participatory planning, 511–515
planning process, 507
repeated every 10 years, 510
scenario-based, 521–524
size-class approach, 518
stand-level, 509
steps, 506
trade-offs, 508
uneven-aged forests, 516–521
visualisation, 534–537
YAFO model, 517
Forest owner, 598
Forest returns, 551
Forestry
continuous cover, 611
evidence-based, 28
Forest shareholders, 511
*Fraxinus excelsior,* 294, 468
ash, 299
Functional contrast, 3
Functional diversity, 34, 42–44
Functional groups, 3, 7
Functional trait, 43
Fungal pathogens, 365
Future climatic conditions, 446

**G**

Gaia hypothesis, 14
Gap closure, 314
Gap cutting, 314
Gap dynamics, 314, 316, 616
Gap models, 388
Gap occupation, velocity, 315
General yield level (Assmann (1961)), 178
General yield level (Gehrhardt (1923) and Assmann (1961)), 176–177
Generation of initial structures, 414
Genetic dilution, 446
Genetic improvement, 437
Gini coefficients, 239–242, 617
mixed-species stands *vs.* monocultures, 245
Global change, 341
Gross yield, 632
Ground cover by crowns, 252
Ground vegetation, 441
Growing area, 275, 290, 317, 591, 630
Growing space efficiency
in mixed *vs.* monospecific stands, 323
Growing space requirement, species-specific, 291
Growth and yield stand characteristics
even-aged mixed stands, 197–199
even-aged monocultures, 199–201
monocultures, 197–199
uneven-aged mixed-species stands, 199–201
Growth-density relationship
hypotheses, 590
species-specific, 590
Growth distribution between trees, 229–233
Growth dominance, 263
differentiation stage, 241
fragmentation and regeneration stage, 241
mature stage, 241
mixed *vs.* monospecific stands, 241, 246
open-grown young stage, 241

Growth dominance coefficient (GDC), 239–242, 617
Growth elasticity, 194
Growth hierarchy, 239–240
Growth partitioning, 233
Growth reaction patterns, 288–290
Growth resilience, 195–197
  mixed-species stands, 196
  monospecific stands, 196
Growth-structure elasticity, 245–246
  mixed *vs.* monospecific stands, 245
Gymnosperms, 291

**H**
Harvesting, 437, 444
Harvesting costs, 548
*h*/*d* ratio, 248, 275, 481
Heat, 343
Heathlands, 443
Helsinki process, 15
Herbivores, 343
Hierarchy of tree size, 239–246
Hierarchy theory, 31
Holism, 386
Holism paradigm, 31
Holobiont, 273
Holobiont-like systems, 273
Hologenome theory, 273
Hybridisation, 446, 450
Hybrid model, 394–395, 413–414
  application for scenario analysis, 414–420
  prediction algorithm, 414
Hybrid stand-level models, 389
Hydraulic lift, 349, 617
Hydraulic redistribution, 389, 617

**I**
Increment cores, 302
Indicator variables, 514
Individual tree models, 388
Initial stand structure, 415
Insect herbivores, 373
Insurance hypothesis, 39
Interactions in mixtures, 74
  accelerating rates of nutrient cycling, 79–80
  anisohydric/isohydric, 93–94
  canopy microclimate, 94
  carbon partitioning, 85
  changes in stand structure, 84
  changes in tree allometry, 85
  changes in tree physiology, 85
  contrasting sources, 93
  contrasting water requirements, 93
  density change, 107
  differences in phenology, 93
  framework for understanding, 95–98
  hydraulic redistribution, 93
  increased storage, 92
  interception losses, 92
  light absorption and use, 102
  light availability, 82–83
  light/nutrient availability and use, 94
  nutrient availability, 77
  nutrient mineralisation, 78
  nutrient uptake, 80–81
  nutrient-use efficiency, 81–82
  ontogeny, 87
  photosynthetic capacity, 87
  resource availability, 87
  simultaneously occurring, 104
  spatial and temporal dynamics, 95
  spatial and temporal variability, 106
  spatial dynamics, 100–101
  spatial effects, 101
  stand water availability, 90
  symbiotic N fixation, 78–79
  temporal dynamics, 102–103
  water availability, 88–92, 101
  water use, 89
  water-use efficiency, 88
Intercropping, 597
  sequential, 597
Interest rate, 558
Interspecific competition, 123
Invasive tree species, 446, 448, 618
Inventory data, 585
Investment planning, 564
ITTO process, 15

**K**
Kleiber's rule, 618
Knottiness, 248
Knowledge base about mixed-species stands, 28
  expert opinions, 29
  hierarchy of confidence, 29
  meta-analyses, 29
  replicated experiments, 29
Knowledge gap, 592

**L**
Land equivalent ratio (LER), 618
Landscape Management System, 507
Land-use options, 555
*Larix decidua,* 123, 152, 181, 294, 410, 452
  European larch, 299
*Larix europaea,* 58
*Larix laricina,* 366
Leaf area density, 618
Leaf area index, 618
Leaf pathogens, 367
Leaf turnover, 389
Light compensation point (LCP), 253, 618
Light flecks, 389
Light saturation point (LSP), 253, 618
Light-use efficiency, 82
Limiting factor, 439, 593
*Liquidambar formosana,* 422
Logging costs, 560
Log price, 551
Long-term experiment
  future, 584–586
  initial stand density, 585
  mixing pattern, 585
  mixing proportion, 585
  new generation, 585
  site conditions, 585
  thinning, 585
Lorenz curve, 239–240, 619

**M**
Main causes of overyielding, 324–330
  allometry of trees, 329–330
  atmospheric $N_2$ fixation, 327–328
  canopy space filling, 325
  deep-rooting, 327–328
  hydraulic lift, 326–327
  hydraulic redistribution, 326–327
  mineral nutrient exploitation, 328
  modification of growth partitioning, 329–330
  root space filling, 325
  temporal complementarity of niches, 328–329
  tree packing density, 325
  wood density, 330
Main crop species, 489
Management decisions, 506
Management models, 514
Marginal net revenue, 560
Market prices, 557
Market risks, 569
Markowitz's theory, 553
Martonne index, 619
Masting, 619
Maturity, 613
Maximum return, 571
Maximum stand density, 172, 190, 476, 589, 590
  common tree species mixtures in Central Europe, 168
  equivalence coefficients, 165
  meta-analysis, 167
  mixed *vs.* monospecific stands, 167
Maximum standing stock, 161
Mean annual increment (MAI), 619
Mean tree productivity, 171–176
Mechanical abrasion, 619
Mediterranean forests, 55, 64, 93, 147, 154, 186, 316, 345, 350, 360, 370, 421
Merchantable wood volume, 199
Meta-analysis, 186
Minimum return, 571
Mixed-species experiments, 55–62
  additive design, 60, 62
  functional diversity, 58
  gradients of species richness, 58
  initially substitutive, 61, 62
  non-random extinction scenarios, 56
  planted experiments, 56–59
  planting density, 58
  pot trials, 61
  substitutive design, 59
Mixed-species models
  bottom-up, 400
  carbon partitioning, 410
  complementarity effect, 422
  design silvicultural systems, 422
  functioning, 399–412
  initial stand parameters, 413
  light and phenology, 406
  maximum density and mortality, 408–410
  mineral nutrients, 407–408
  mixed-species stands and models, 399–412
  output variables, 416
  3-PG, 421
  schematic representation, 399
  SILVA, 413–414
  silvicultural prescriptions, 413
  simulation run, 414, 417
  site variables, 413
  spatial and temporal resolutions, 412–423
  stand structure, 401–405
  structure, 399–412
  temporal and spatial resolutions, 400–406

Mixed-species models (*cont.*)
  top down, 400
  water balance, 407
  within-stand environment, 399–412
Mixed-species stands
  artificial regeneration, 460
  creating and maintaining, 445–470
  criteria for design, 592
  design, 592
  native and non-native tree species, 445–451
  native tree species, 448
  natural regeneration, 460
  regeneration methods, 459–470
  sowing, 461
  spacing of planted seedlings, 461
Mixed-species stand simulation
  analysis and compilation of the results, 524
  assessment of biodiversity, 527–534
  assignment of silvicultural prescriptions, 523
  average harvest level, 527
  composition, 530–531
  functioning, 530–531
  ingrowth, 525
  inventory database, 522
  recursive strategic planning, 522
  representative stands, 523–524
  scenario calculations, 524
  silvicultural prescriptions, 524
  size-class approach, 524–527
  stand development stages, 523
  stems to be harvested, 525
  stratification, 523
  structure, 530–531
  transition probability, 525
  trees remaining, 525
Mixing effect(s), 272
  additive, 607
  alien thinning, 167–176
  alien-thinning line, 168–171
  alien-thinning slope, 170
  along a gradient from poor to fertile sites, 186
  annual growth, 371–372
  annual growth rhythms, 126
  annual volume growth, 126
  change in growing space efficiency, 172
  change in stand density, 172
  course of stand growth, 126
  course of total yield, 126
  drought, 188
  economic effect, 554
  facilitative, 154
  fast early growth, 156
  final yield, 126
  general yield level, 178–180
  levels of organisation, 272
  long-term survey yield, 157
  mixing proportion, 184, 287
  modified stand density, 180
  multiplicative, 620
  native with exotic species, 594
  within rotation period, 154–156
  self-thinning, 167–176
  separation of indirect from direct, 142
  short-term productivity, 157
  short-term surveys, 160
  site conditions, 183–187
  site fertility, 184
  site index, 177–181
  site quality, 184
  size effect, 142–145
  slope of self-thinning, 168–171
  spacing, 190–197
  spatial and temporal dynamics, 190
  spatial and temporal variation, 181–190
  spatial variability, 183–187
  species growth curves, rhythm and course of, 155
  species-specific, 185
  stand density, 167–176, 193, 287
  stand density index, 169
  during stand development, 157–160
  stand growth, 133–140
  standing volume, 178
  stand total volume yield, 178–180
  temporal variability, 187–190, 371–372
  thinning, 190–197
  total yield, 133–140, 160, 176–181
  total yield-based findings, 160
  tree diameter increment, 286
  true, 633
  water supply, 189
  whole-stand, 185
  yield level, 176–181
Mixing effects at the stand level, 201–203
  graphical abstract, 202
  increase in maximum stand density, 202
  increase in mean tree productivity, 202
  individual tree mixtures, 202
  maximum stand density, 202
Mixing portion, diminishing marginal utility principle, 201
Mixing proportion, 136, 137, 139
  growing space requirements, 138
  leaf area, 138

living tree biomass, 138
species-specific crown allometry, 138
specific wood density, 138
stand basal area, 138
stand density index, 140
standing stem volume, 138
tree number, 138
Mixing tree species, consequences, 569
Mixture
native-exotic, 595
tree-crop, 596
Mixture comparisons
community-level analyses, 77
levels of analyses, 76–77
neighbourhood-level analyses, 76
species-level, 77
stand-level analysis, 77
tree-level, 76
Model(s), 385
age-class, 608
bioeconomic, 557, 609
decision support, 513–515
diameter distribution models, 513
economic land-use, 546
ecophysiological process, 513
empirical, 614
hybrid, 617
management options, 513
mortality, 601
position-dependent individual tree, 513
process-based, 624
size-class, 518, 519, 629
succession, 513
YAFO-optimisation, 633
yield tables, 513
Modelling, bioeconomic, 598–599
Modelling mixed-species stands
indirect incorporation of mixing effects, 396
mixed-species forest growth by models, 396
multipliers affecting growth rates, 396
process-based representation, 396
weighted mean of monocultures, 396
Model output
assortment structures, 416
carbon and nutrient contents, 416
crown development, 416
financial yield, 417
growth and yield, 416
habitat and species diversity, 418
mean annual value increment, 417
species intermingling, 419
stem quality, 416
vertical species profile, 418
Mode of competition, 229–233, 619
completely symmetrical size-growth relationship, 232
convex size asymmetry, 232
mixed *vs.* monospecific stands, 232
partial size asymmetry, 232
partial size symmetry, 232
perfect size symmetry, 232
size symmetry, 232
Mode of mortality
mixed stand and the neighbouring monoculture, 236
mixed *vs.* monospecific stands, 237–239
modified by mixing, 239
removal and remaining stand, 238
Monte Carlo simulation, 559–561
Montreal process, 15
Morphological plasticity, 593
Multi-criteria decision analysis, 514, 519
Multifunctional forest management, 14
Multilayering, 264
Multiple crown coverage, 251
Multiple ecosystem functions, 435
Multiple-objective method, 520–521
Multiple services, 519
Multiplicative mixing effect, 135
Mutualism, 620
Mutualistic mixing interaction, 184
Mycorrhizal associations, 447
Mycorrhizal fungi, 50, 447


**N**


Natural density, 591
Nelder trials, 281
Net competition, 280, 281
Net effect of mixing, 288
Net facilitation, 280, 281
Net plant growth, 277
Net primary productivity, 391
Neutral interaction, 185
$N_2$ fixation, 154
N-fixing species, 438
Niche, 621
coexistence, 39
complementarity, 16, 47, 588, 621
differentiation, 39, 43, 47
ecological, 36–41
fundamental, 36–41, 616
markets, 564
occupation, 261

Niche (*cont.*)
realised, 37, 38, 625
similarities, 43
Non-isometric relationships, 298
*Nothofagus alpina,* 595
Nurse crops, 440–444
Nurse crop species, 489
Nurse trees, 621
exotic, 594
Nutrient cycling, 442

**O**
Ontogenetic drift, 275, 621
allometric relationship, 297–298
ratios, 297–298
Ontogeny, 621
Optimal density, 168
Optimal partitioning theory, 299
Optimisation, 551
continuous, 611
discrete, 612
land-use, 596
non-linear programming, 570
non-stochastic, 562
portfolio, 555
stochastic, 599
Optimized land-use diversification, 560
Overyielding, 17, 135, 153, 201, 264, 370
big data analysis, 149–150
correction factors, 153
current reviews and meta-analyses, 146–149
empirical evidence, 146–160
forest inventory data, 149
generic mixing responses, 175
growth conditions, 370
increases with the density of stocking and canopy, 263
size distribution, 258
size-structure dynamics, 258–259
stand and species level, 258
stands containing $N_2$-fixing species, 153–154
stand structure, 586, 587
structural heterogeneity, 587
temperate and boreal forests, 146
temperate forest species, 150–153
three-species mixture, 147
transgressive below-ground, 80
tree species mixtures in Central European forests, 152
water availability, 187

**P**
Partial damage
mixed *vs.* monospecific stands, 194–195
Participatory planning, 534
Partitioning, 622
Paterson index, 622
Pathogens, 343, 364–368, 373, 450–451
abundant antagonists, 368
antagonists, 364
effect of neighbourhood tree species, 365
exotic, 364
mixedspecies forests, 364
tree species diversity, 368
Periodic annual increment (PAI), 622
Pest insects, 361–364
effects of tree species diversity, 363
host-finding ability, 362
parasitoids, 362
positive effect of tree species richness, 361
predators, 362
reduced tree apparency, 362
resource concentration hypothesis, 362
Pests, 450–451
Phenology, 623
Photosynthetically active radiation, 82
Phylogenetic diversity, 44
Physiological amplitude, 446
Phytometer, 390
*Picea abies,* 42, 58, 97, 98, 106, 121, 123, 134, 136, 145, 148, 151, 152, 158, 168, 170, 171, 181, 187, 188, 193, 195, 197, 200, 226, 234, 243, 252, 294, 348, 349, 371, 409, 419, 464, 474, 484, 486, 584, 601
Norway spruce, 287, 295, 297, 309, 315, 415–417, 535, 537, 552, 559, 566, 567, 571, 585
*Picea mariana,* 366
*Pinus sylvestris,* 58, 106, 123, 124, 145, 152, 168, 169, 171, 181, 187, 189, 197, 226, 234, 243, 260, 294, 410, 464, 467, 552
Scots pine, 287, 295, 582
Pioneer, 623
Planning process, 507
descriptive, 507
optimisation-based, 507
scenario-based, 507
Plantation forests, 436
Plenterforest, 484, 624
Polyculture, 624
Polyphagous, 624
Portfolio, 553
analysis, 600

land-use, 555
returns, 553
selection, 550, 553
theory, 546, 561, 620
Post-disturbance recovery, 374
Potential growth, 284
facilitation and competition, 286–288
mixing effects, 285
Potential height growth, 391
Pre-commercial thinnings, 510
Pre-emption
light, 624
resources, 624
water, 624
Price risks, 517
Primary forest, 624
Process-based models, 388, 393–394, 422–423
confidence in performance, 420–421
integration of mixing effects, 398–399
Production, 435
ecology equation, 74–75, 624
planning, 506
Productivity, 625
Proxy variables, 391
Pruning, 437
*Pseudotsuga menziesii,* 97, 152, 168, 193, 197, 226, 234, 243, 449, 595
Douglas-fir, 303

**Q**

Quantifying mixing effects on stand productivity
basic variables, 126–127
degressive underyielding, 127
land equivalent ratio, 129
mixed *vs.* monospecific stand productivity, measures of RPA, 127
mixed *vs.* monospecific stand productivity, ratio RPR for, 129
cross diagrams, 128
mixing proportions, 128
overyielding, 127
reference, 129
RPA *vs.* RPR, 130
transgressive overyielding, 127
triplets, 130
underyielding, 127
*Quercus petraea,* 123, 145, 148, 152, 158, 168, 170, 171, 181, 184–187, 197, 222, 238, 247, 252, 294, 295, 348, 349, 371, 441, 458, 467, 552
sessile oak, 287, 299, 309
*Quercus pyrenaica,* 106
*Quercus robur,* 58, 294, 295, 467

**R**

Radiation, photosynthetically active, 623
Reductionism, 386
Reductionist paradigm, 30
Redundancy, functional, 625
Redundancy hypothesis, 34, 35
Reforestation, 594
Regeneration
competitive relationship, 465
diverse, 461–463
establishment, 462
gaps, 467
germination, 462
groups, 467
growth relationships, 465
harvesting, 571
height growth advantage, 464
height growth differences, 465
masting, 462
microsites, 463
mixed, 465
seedbed conditions, 463
spatial separation, 466–470
species-rich stands, 461
temporal separation, 463–466
Regeneration of mixtures
differentiation in height growth, 469
effect of browsing, 470
emulate natural disturbance, 470
gap-based silvicultural approaches, 469
ground or shrub vegetation, 469
spatial separation, 468
Reineke's rule, 163, 625
Relationship
biodiversity ecosystem functioning, 44
competitive, 17
facilitative, 17
mutualistic, 17
neutral, 17
parasitic, 17
structural heterogeneity and productivity, 589
Relative density, 166
Relative productivity, 98
Relative total yield, 626
Remote sensing, 509
Removal stand, 626
Replacement series, 46
Resilience, 350, 626
growth, 617
mixed-species communities, 373
Resistance, 446, 548, 571, 626
effects of mixing tree species, 373
tree species diversity, 373
Resource acquisition, 106
Resource availability, 106

Resource exploitation, 588
Resource-use efficiency, 106
Restoration planting, 444
Reward-to-variability ratio, 560
Risk, 12, 19, 547, 549, 555, 556, 564, 595, 613
  aversion, 570
  avoiders, 570
  biophysical, 550
  fire, 443
  frost, 442
  management, 340
  minimum, 550
Root channels, 389
Root grafting, 353
Rooting depth, 344, 593
Root pathogens, 365
Root-shoot allometry, 299–303
Root-shoot ratio, 300
Root-shoot relationship
  allometric relationship, 301–302
  mixed *vs.* monospecific stands, 301–302
  silvicultural treatment, 300
  site conditions, 300
  supply of nutrients and water, 300
Rotation length, 627
Rubner's rule, 627
Rule by Kleiber (1947), 276
Rule by Rubner (1931), 276, 627

S
Salvage harvesting, 560
Sampling effect, 34, 627
Sapwood, 627
Scale-overlapping analysis, 30
Scenario analysis, 515, 627
Scenario calculations, 514
Scorch of bark, 272
Segregation, 13
Selection effects, 345
Selection forest, 484
  declining exponential distribution, 485
  diameter distribution, 485
  equilibrium curve, 485
  process of recruitment, 485
  target-diameter harvesting, 486
  target equilibrium curves, 484
Self-thinning, 160–176, 299, 471, 627
  intraspecific competition, 171
Self-thinning line, 162, 166
  intercept, 162
  lateral crown extension, 170
  maximum standing stock, 162
  slopes, 162
Self-tolerance, 170, 628
Shade tolerance, 593
Shannon index, 5
Shelterwood removal, 466
Silvicultural diversification, 41
Silvicultural guidelines, 263, 476, 586
  applied to mixed-species stands, 263
Silvicultural management, 599
Silvicultural mixtures, 438
Silvicultural options, 433–490, 554
  heterogeneity, 434–437
  homogeneity, 434–437
Silvicultural prescriptions, 451–459, 594
  data from experiments, 453
  development of maximum heights, 452
  growth advantage, 452
  growth models, 453
  intensity of competition, 455
  maintaining rare european tree species, 456–459
  sacrificial species, 453
  spatial and temporal separation, 594
  spatial separation, 455
  species compatibility, 453
Silvicultural systems, 435, 436
  increasing complexity, 437
  mixed-species forests, 436
Silvicultural treatment, 592
Site fertility
  class, 390
  dominant height, 390
  forest floor vegetation as the indicator, 391
  mean height, 390
Site index, 628
Site quality, 628
Size-asymmetric competition, 264
Size asymmetry, 263
Size-class distribution, 519
Size effect, elimination, 141–145
Size frequency distribution, 628
Size-growth curve
  interindividual competition, effect of, 230–231
  schematic representation, 231
  site conditions, 231
  tree size, 230
Size-growth relationship, 231
Size hierarchy, 628
Size inequality, 263, 264
Size-structure dynamics
  additive effect, 218
  growth distributions, 217
  heterogeneous stands, 217
  homogeneous monospecific stands, 217
  mixed-species, characteristics of, 218

mixed stand *vs.* monoculture, 221
monocultures as a reference, 218–221
mortality, 217
multiplicative effect, 218
removal trees, 218
thinning, 217
tree size distribution, 217
Size structure, link between tree-level and stand-level, 261–262
Snow sliding, 272
Soil fertility, 389
Solitary growth, 280
Solitary tree, 629
*Sorbus aucuparia,* 294
*Sorbus domestica,* 458
Space occupation, 311–316
Spacing and thinning experiments, 192–194, 282
Spatial niche occupation, 215
Species, 632
early successional, 613
ecological, 613
exotic, 11
fast growing, 44
foundation, 616
generalist, 617
late successional, 618
light-demanding, 44
$N_2$-fixing, 203
nitrogen-fixing, 44, 48
non-nitrogen fixing, 44
ranges, 3
regional extinctions, 3
shade-tolerant, 44
shallow-and deeprooting, 203
specialist, 629
Species diversity, 2, 629
maintain, 9
Species identity effect, 345
Species interactions, 279, 389
drought years, 135
height growth curve, 119
interspecific interactions, 370
intra-annual course of size growth, 134–135
mean diameter, 122
mean height, 122
net effect, 279
ratio between $h_q/d_q$, 123
slowing down of height growth, 120
species-specific trajectories, level and rhythm, 119
temporal variation, 370
tree quality, 488
wood quality, 125
Species mixture
evergreen/deciduous, 186
evergreen/evergreen, 186
Species richness, 2, 4, 33, 50, 629
Species richness-productivity curve, 182
Species selection effect, 629
Species-specific packing density, 168
Species traits, 42–44
Specific leaf area, 630
Stability, ecological, 630
Stand area, 316–317, 630
Stand area growth efficiency, 319
Stand density, 94–95, 104–105, 160–176, 255–256
absolute, 607
critical, 191–192, 619
local, 287
maximum, 191–192, 619
optimum, 191–192, 619
overyielding, 589–592
productivity, 589–592
quantification, 162–166
regulating and managing, 591
relative, 626
standardisation, 162–166
stand density index, 162
Stand density index (SDI), 162, 630
quantify the density, 165
Reineke exponent, 165
Yoda exponent, 165
Stand density in terms of volume, 319
Stand density management diagrams (SDMD), 476
Stand density-productivity relationship, 196
Stand-level mixture, 567
Stand management diagrams, 168
Stand stability, 355
Stand structure, 260–261, 587
modified by water availability, 260
State variables, 631
Stem quality, 490
Stiffness, 248
Stochastic evolution models, 387
Stocking density, 263
Storm, 373
Storm breakage, 272
Storm damage, 351–356
critical tree heights, 351
following thinning, 351
root systems, 353

Storm damage (*cont.*)
stand density, 353
susceptibility of trees and stands, 352–355
susceptibility to damage, 353
vulnerability, 352
water-logged horizons, 353
Strategic silvicultural planning, 514
Stress, 631
abiotic, 607
events, 347
mixed-species forests, 342
organ losses, 350
Stress and disturbance
experimental design, 369
interactions, 368–369
Stress factors, 340
Stress-gradient hypothesis, 48, 182, 370
Stress reactions
monospecific and mixed stands, 349
species-specific, 349
Structural heterogeneity, 251, 587
site conditions, 588
Structure generator, 414
Study designs
comprehensiveness, 52
orthogonality, 52
representativeness, 52
Subtropical forests, 3, 29, 84, 133, 147, 186, 362, 421, 422
Sum of crown projection area (SCPA), 249–250
Surrogate variables, 391
Survival, 36
Survival probability, 566, 600–601
Susceptibility, 631
Sustainability, 631
Sustainable forest management, 631
criteria, 15
indicator, 15
Sustainable yield, 13
System simulation, 385

**T**

Target-diameter harvesting, 466
Temperate forests, 3, 4, 10, 50, 155, 157–160, 202, 216, 228–230, 248, 314, 325, 421
Temporal stabilisation, 370
Tending, 470–487, 631
Thinning, 55, 437, 444, 470–487, 631
carrying capacity, 477
conservation of biodiversity, 482
crop tree approaches, 478–483
desired species proportions, 480
habitat trees, 483
individual tree stability, 482
maximum stand density, 477
pre-commercial, 482
prescriptions, 473–474
prescriptions for mixed-species stands, 474–478
production goals, 481
purpose, 471–473
of quantitative information, 490
relative growth rates, 480
spatial and temporal mixing pattern, 480
spatial arrangement, 479
species-specific target diameters, 480
suboptimal stand densities, 477
uneven-aged stands, 483–487
value increment, 480
Thünen framework, 555
Thünen's land location theory, 632
*Tilia cordata,* 294
Timberland returns, 558–559
Timber prices, 550, 551
Timber price volatility, 570
Timber quality, 489, 548
Time diversification, 569
Total yield, 632
Trainer species, 440–442
Trainer tree, 488, 632
Trait, 632
Transect study, 581
Transgressive overyielding, 135, 156
Transition probabilities, 518
Transpiration, 345, 632
Treatment options, 514
Tree crown plasticity, 389
Tree growth, 278
generic patterns, 289
intra-*vs.* interspecific neighbourhoods, 288–290
modified by competition and facilitation., 289
stand productivity, 321
Tree growth partitioning
mixed *vs.* monospecific stands, 242–245
Tree ontogeny
basic relationships, 275
growth and yield curves, 276
principles, 273–278
solitarily growing tree, 273
stem volume, 275
stem volume increment, 275
Tree productivity, 278
Tree quality, 487–489
Tree-ring analysis, 302
Tree size distribution, 213–216
biodiversity, 215

crown surface roughness, 215
density, 225
effect of mixing, 228
growth partitioning, 215
kurtosis, 223–224
location, 225
mixed stands *vs.* monocultures, 228
mixed *vs.* monospecific stands, 223–229
niche separation, 215
shape, 225
size inequality, 215
skewness, 223–224
structural parameters, 215
tree height variation, 215
Tree species diversity, 19
productivity, 369–373
Tree species richness, 149
forest productivity, 150
Tree stand area, 291, 312–313
Tree–tree interaction, 282
feedback loop, 282
Triplet design, 135
Triplets, 224, 255, 323, 324, 581
Trophic level, 632
Tropical forests, 2, 3, 5, 7, 10, 11, 29, 50, 84, 93, 133, 147, 154, 421, 439, 444, 475, 488, 520, 594, 597
*Tsuga heterophylla,* 193

**U**
*Ulmus glabra,* 294
Uncertainty, 19, 517, 520, 547, 563, 599–600
Underyielding, 135, 137, 146, 370, 633
degressive, 631
subgressive, 631
Uneven-aged forests, 483
Unimodal growth curve, 277
genetic changes, 277
growth respiration, 277
hydraulic limitations, 277
maintenance respiration, 277
nutrient limitation, 277

**V**
Vapour flux, 633
Vertical layering, 588
Vertical stratification, 261
Vertical structuring, 261
Visualisation, 534
forest stands, 535
interactive thinning, 535
landscape, 535, 537
management alternatives, 536
SILVA, 536
virtual flight through, 535
Volatility, 553
Vulnerability to storms, 355–356
mixed-species stands, 355, 356
structurally diverse stands, 356
tree species, 355

**W**
Water stress, 94–95
species diversity, 351
species identity, 351
Water translocation, 389
Water use
efficiency, 345
mixing effects, 345
Willingness to pay for land, 561
Windbreak, 597
Wind speed, 389
Windthrow
high crown density, 354
low wood strength, 354
regeneration, 356
shallow root systems, 354
stem taper, 354
susceptibility, 353
swaying characteristics, 354
tree height, 354
Wood density, 201, 275
Wood properties, 487
Wood quality, 594
eccentricity of crowns, 489
knottiness, 489
in mixed *vs.* pure stands, 489
mixtures of species, 488
species identity, 489
structural complexity, 489
Wood strength, 248

**Y**
Yield
constant final, 611
planning, 506
tables, 290, 387, 509, 634
Yoda's rule, 163, 634

MIX
Papier aus verantwortungsvollen Quellen
Paper from responsible sources
FSC® C105338

If you have any concerns about our products,
you can contact us on
**ProductSafety@springernature.com**

In case Publisher is established outside the EU,
the EU authorized representative is:
**Springer Nature Customer Service Center GmbH**
**Europaplatz 3, 69115 Heidelberg, Germany**

Printed by Libri Plureos GmbH
in Hamburg, Germany